# SUSTAINABLE BIOCHAR FOR WATER AND WASTEWATER TREATMENT

अज्जनमुस्तोशीरैः शराजकोशातकामलकचूर्णैः।
कतकफलसमायुक्तैर्योगः कूपे प्रदातव्यः।। Vr.S.54.121 ।।

कलुषं कटुकं लवणं विरसं सलिलं यदि वाशुभगन्धि भवेत्।
तदनेन भवत्यमलं सुरसं सुसुगन्धि गणैरपरैश्च युतम।। Vr.S.54.122 ।।

# SUSTAINABLE BIOCHAR FOR WATER AND WASTEWATER TREATMENT

Edited by

**DINESH MOHAN**
School of Environmental Sciences, Jawaharlal Nehru University, New Delhi, India

**CHARLES U. PITTMAN, JR.**
Department of Chemistry, Mississippi State University, Mississippi State, MS, United States

**TODD E. MLSNA**
Department of Chemistry, Mississippi State University, Mississippi State, MS, United States

ELSEVIER

Elsevier
Radarweg 29, PO Box 211, 1000 AE Amsterdam, Netherlands
The Boulevard, Langford Lane, Kidlington, Oxford OX5 1GB, United Kingdom
50 Hampshire Street, 5th Floor, Cambridge, MA 02139, United States

Copyright © 2022 Elsevier Inc. All rights reserved.

No part of this publication may be reproduced or transmitted in any form or by any means, electronic or mechanical, including photocopying, recording, or any information storage and retrieval system, without permission in writing from the publisher. Details on how to seek permission, further information about the Publisher's permissions policies and our arrangements with organizations such as the Copyright Clearance Center and the Copyright Licensing Agency, can be found at our website: www.elsevier.com/permissions.

This book and the individual contributions contained in it are protected under copyright by the Publisher (other than as may be noted herein).

**Notices**

Knowledge and best practice in this field are constantly changing. As new research and experience broaden our understanding, changes in research methods, professional practices, or medical treatment may become necessary.

Practitioners and researchers must always rely on their own experience and knowledge in evaluating and using any information, methods, compounds, or experiments described herein. In using such information or methods they should be mindful of their own safety and the safety of others, including parties for whom they have a professional responsibility.

To the fullest extent of the law, neither the Publisher nor the authors, contributors, or editors, assume any liability for any injury and/or damage to persons or property as a matter of products liability, negligence or otherwise, or from any use or operation of any methods, products, instructions, or ideas contained in the material herein.

ISBN: 978-0-12-822225-6

For information on all Elsevier publications
visit our website at https://www.elsevier.com/books-and-journals

*Acquisitions Editor:* Louisa Munro
*Editorial Project Manager:* Mica Ella Ortega
*Production Project Manager:* Kumar Anbazhagan
*Cover Designer:* Miles Hitchen
*Cover Design Courtesy:* Chanaka Navarathna

Typeset by STRAIVE, India

  Working together to grow libraries in developing countries

www.elsevier.com • www.bookaid.org

# Contents

Contributors . . . . . . . . . . . . . . . . . . . . . . . . . . . . . . . . . . . . . . . xv
Preface . . . . . . . . . . . . . . . . . . . . . . . . . . . . . . . . . . . . . . . . . . . xxi
About the editors . . . . . . . . . . . . . . . . . . . . . . . . . . . . . . . . . . xxv

**Chapter 1  Comprehensive biomass characterization in preparation for conversion** . . . . . . . . . . . . . . . . . . . . 1
Sergio C. Capareda

1 Introduction . . . . . . . . . . . . . . . . . . . . . . . . . . . . . . . . . . . . . . 1
2 Biomass conversion processes . . . . . . . . . . . . . . . . . . . . . . . . 3
3 Biomass properties in preparation for biological conversion . . . . . . . 6
4 Biomass properties in preparation for thermochemical conversion . . . . . . . . . . . . . . . . . . . . . . . . . . . . . . . . . . . . . . 16
5 Conclusion . . . . . . . . . . . . . . . . . . . . . . . . . . . . . . . . . . . . . 32
Questions and problems . . . . . . . . . . . . . . . . . . . . . . . . . . . 33
References . . . . . . . . . . . . . . . . . . . . . . . . . . . . . . . . . . . . . 36

**Chapter 2  Biomass carbonization technologies** . . . . . . . . . . . . . . . . . 39
Manuel Raul Pelaez-Samaniego, Sohrab Haghighi Mood, Jesus Garcia-Nunez, Tsai Garcia-Perez, Vikram Yadama, and Manuel Garcia-Perez

1 Introduction . . . . . . . . . . . . . . . . . . . . . . . . . . . . . . . . . . . . 39
2 Char physicochemical properties . . . . . . . . . . . . . . . . . . . . . 43
3 Effect of operational variables on the yield and quality of charcoal . . . . . . . . . . . . . . . . . . . . . . . . . . . . . . . . . . . . 44
4 Carbonization reactions . . . . . . . . . . . . . . . . . . . . . . . . . . . 49
5 Equipment for carbonization processes . . . . . . . . . . . . . . . . 50
6 Conclusions . . . . . . . . . . . . . . . . . . . . . . . . . . . . . . . . . . . . 79
Questions and problems . . . . . . . . . . . . . . . . . . . . . . . . . . . 80

Acknowledgment . . . . . . . . . . . . . . . . . . . . . . . . . . . . . . . . . . . . . . . 80
References . . . . . . . . . . . . . . . . . . . . . . . . . . . . . . . . . . . . . . . . . . . . 80

**Chapter 3 Physicochemical characterization of biochar derived from biomass** . . . . . . . . . . . . . . . . . . . . . . . . . . . . . . . . . . . . . 93
**Sergio C. Capareda**
1 Introduction . . . . . . . . . . . . . . . . . . . . . . . . . . . . . . . . . . . . . . . . . 93
2 Thermal conversion processes for the production of biochar . . . . . 95
3 Physical properties of biochar . . . . . . . . . . . . . . . . . . . . . . . . . . . 98
4 Important thermal related properties . . . . . . . . . . . . . . . . . . . . . 103
5 Summary of standard methods for biochar analysis . . . . . . . . . . 118
6 Biochar conversion into activated carbon and performance analysis . . . . . . . . . . . . . . . . . . . . . . . . . . . . . . . . . . . . . . . . . . . 121
7 Conclusions . . . . . . . . . . . . . . . . . . . . . . . . . . . . . . . . . . . . . . . 130
Questions, concepts, definitions, and problems . . . . . . . . . . . . . . 131
References . . . . . . . . . . . . . . . . . . . . . . . . . . . . . . . . . . . . . . . . . 133

**Chapter 4 Biochar characterization for water and wastewater treatments** . . . . . . . . . . . . . . . . . . . . . . . . . . . . . . . . . . . . . . . 135
**Balwant Singh, Tao Wang, and Marta Camps-Arbestain**
1 Introduction . . . . . . . . . . . . . . . . . . . . . . . . . . . . . . . . . . . . . . . 135
2 Laboratory Analysis of biochar . . . . . . . . . . . . . . . . . . . . . . . . . 137
3 Summary . . . . . . . . . . . . . . . . . . . . . . . . . . . . . . . . . . . . . . . . . 146
Questions and problems . . . . . . . . . . . . . . . . . . . . . . . . . . . . . . . 148
References . . . . . . . . . . . . . . . . . . . . . . . . . . . . . . . . . . . . . . . . . 148

**Chapter 5 Biochar adsorption system designs** . . . . . . . . . . . . . . . . 153
**Dinesh Mohan, Abhishek Kumar Chaubey, Manvendra Patel, Chanaka Navarathna, Todd E. Mlsna, and Charles U. Pittman, Jr.**
1 Introduction . . . . . . . . . . . . . . . . . . . . . . . . . . . . . . . . . . . . . . . 153
2 Sorption isotherm models . . . . . . . . . . . . . . . . . . . . . . . . . . . . . 156
3 Sorption kinetic models . . . . . . . . . . . . . . . . . . . . . . . . . . . . . . 173
4 Sorption column design . . . . . . . . . . . . . . . . . . . . . . . . . . . . . . 178

|   |   |
|---|---|
| 5 Fixed-bed operation and design | 182 |
| 6 Conclusions | 193 |
| Questions and Problems | 194 |
| Acknowledgments | 199 |
| References | 199 |

## Chapter 6  Techno-economic analysis of biochar in wastewater treatment .......... 205
**Arna Ganguly and Mark Mba Wright**

|   |   |
|---|---|
| 1 Introduction | 205 |
| 2 Biochar markets and economics | 207 |
| 3 Biochar costs in wastewater facilities | 211 |
| 4 Biochar production | 217 |
| 5 Biochar market drivers | 224 |
| 6 Conclusions | 225 |
| Questions and problems | 226 |
| References | 227 |

## Chapter 7  Retention of oxyanions on biochar surface .......... 233
**Santanu Bakshi, Rivka Fidel, Chumki Banik, Deborah Aller, and Robert C. Brown**

|   |   |
|---|---|
| 1 Introduction | 233 |
| 2 Motivation for removing oxyanions of interest | 234 |
| 3 Chemical properties of biochar influencing oxyanion sorption | 238 |
| 4 Feedstock and pyrolysis temperatures influence on surface chemistry and physical properties of biochar | 241 |
| 5 Physical properties of biochar influencing oxyanion sorption | 248 |
| 6 Applications of biochar for removing oxyanions from water and wastewater | 250 |
| 7 Conclusions and future perspectives | 265 |
| Questions and problems | 266 |
| References | 267 |

## Chapter 8  Arsenic removal from household drinking water by biochar and biochar composites: A focus on scale-up ... 277
Jacinta Alchouron, Amalia L. Bursztyn Fuentes, Abigail Musser, Andrea S. Vega, Dinesh Mohan, Charles U. Pittman, Jr., Todd E. Mlsna, and Chanaka Navarathna

1 Introduction ... 277
2 Biochar-based adsorbent synthesis ... 281
3 Adsorbent performance testing for scaled-up drinking water arsenic treatment ... 291
4 Conclusions ... 310
Questions and problems ... 311
References ... 313

## Chapter 9  Application of biochar for the removal of actinides and lanthanides from aqueous solutions ... 321
Amalia L. Bursztyn Fuentes, Beatrice Arwenyo, Andie L.M. Nanney, Arissa Ramirez, Hailey Jamison, Beverly Venson, Dinesh Mohan, Todd E. Mlsna, and Chanaka Navarathna

1 Introduction ... 321
2 Biochar and engineered biochar ... 324
3 Sorption performance and dynamics ... 333
4 Scaling up: From batch to column configuration ... 349
5 Conclusions, gaps, and future perspectives ... 351
Questions and problems ... 352
References ... 353

## Chapter 10  Microplastic removal from water and wastewater by carbon-supported materials ... 361
Virpi Siipola, Henrik Romar, and Ulla Lassi

1 Introduction ... 361
2 Microplastics—Sources and effects in the aquatic environment ... 361
3 Analysis methods for microplastics ... 367

 4 Microplastic retention by biochars and activated carbons ....... 374
 5 Conclusions ................................. 386
 Questions and problems .......................... 387
 References ................................... 387

**Chapter 11 Sorptive removal of pharmaceuticals using sustainable biochars** ........................ 395
 Manvendra Patel, Abhishek Kumar Chaubey, Chanaka Navarathna, Todd E. Mlsna, Charles U. Pittman, Jr., and Dinesh Mohan
 1 Introduction ................................. 395
 2 Pharmaceuticals: Emerging contaminants ............... 396
 3 Biochar as an adsorbent ......................... 398
 4 Pharmaceutical sorption onto biochar ................. 406
 5 Conclusions and recommendations .................. 420
 Acknowledgments .............................. 421
 Questions and problems .......................... 421
 References ................................... 422

**Chapter 12 Dye removal using biochars** ................ 429
 Gordon McKay, Prakash Parthasarathy, Samra Sajjad, Junaid Saleem, and Mohammad Alherbawi
 1 Introduction ................................. 429
 2 Classification of dyestuffs ........................ 431
 3 Dye removal technologies ........................ 439
 4 Adsorption technology .......................... 442
 5 Dye removal using adsorption onto biochars ............. 443
 6 Summary ................................... 459
 Questions and problems .......................... 461
 Acknowledgment ............................... 461
 References ................................... 462
 Further reading ................................ 471

## Chapter 13  Biochar as a potential agent for the remediation of microbial contaminated water .................. 473
Jayani J. Wewalwela, Prasad Sanjeewa, and Sameera R. Gunatilake

1 Introduction ............................................... 473
2 Water sources and microbial contamination ................. 474
3 Mechanisms of microbial decontamination using biochar ..... 478
4 Recommendation and conclusion ............................ 484
Questions and problems ...................................... 484
References .................................................. 485

## Chapter 14  Biochar-based constructed wetland for contaminants removal from manure wastewater ................. 487
Valentine Nzengung and Stefanie Gugolz

1 Introduction ............................................... 487
2 Objectives ................................................. 492
3 Methods ................................................... 493
4 Batch sorption of nutrients in CAFO wastewater by biochar .. 493
5 Greenhouse study .......................................... 496
6 Greenhouse CTW results and discussion .................... 502
7 Conclusions ............................................... 519
Questions and problems ...................................... 521
References .................................................. 521

## Chapter 15  Biochar and biochar composites for oil sorption ...... 527
Chanaka Navarathna, Prashan M. Rodrigo, Vishmi S. Thrikawala, Arissa Ramirez, Todd E. Mlsna, Charles U. Pittman, Jr., and Dinesh Mohan

1 Introduction ............................................... 527
2 Uptake performance ........................................ 538
3 Uptake mechanisms ........................................ 550

|  |  |
|---|---|
| 4 Conclusions, challenges, and gaps | 550 |
| Questions and problems | 551 |
| References | 551 |

## Chapter 16  Biochar and biochar composites for poly- and perfluoroalkyl substances (PFAS) sorption . . . . . . . . . . 555

**Chanaka Navarathna, Michela Grace Keel, Prashan M. Rodrigo, Catalina Carrasco, Arissa Ramirez, Hailey Jamison, Dinesh Mohan, and Todd E. Mlsna**

|  |  |
|---|---|
| 1 Introduction | 555 |
| 2 Breakthrough column studies and pilot scale testing | 585 |
| 3 Sorption interactions | 589 |
| 4 Conclusions, gaps, and future studies | 591 |
| Questions and problems | 591 |
| References | 592 |

## Chapter 17  Simple modeling approaches to monitor and predict organic contaminant adsorption by biochar . . . . . . . . . 597

**Kyle K. Shimabuku and R. Scott Summers**

|  |  |
|---|---|
| 1 Introduction | 597 |
| 2 Relationships between OC adsorption capacity with surface area and H:C ratios | 599 |
| 3 Modeling the influence of OC concentration on adsorption by biochar | 602 |
| 4 Estimating biochar adsorber performance from batch adsorption data | 605 |
| 5 Using ultraviolet absorbance to monitor OC breakthrough in biochar column adsorbers | 606 |
| 6 Summary | 608 |
| Questions and problems | 609 |
| References | 609 |

## Chapter 18  Nanoscale zero-valent iron-decorated biochar for aqueous contaminant removal . . . . . . . . . . . . . . . . . . . . 611

Xuefeng Zhang, Tharindu Karunaratne, Chanaka Navarathna, Jilei Zhang, and Charles U. Pittman, Jr.

1 Introduction . . . . . . . . . . . . . . . . . . . . . . . . . . . . . . . . . . . . . 611
2 Synthesis of nZVI and biochar-supported nZVI . . . . . . . . . . . . . 612
3 Advantages of biochar as the nZVI support . . . . . . . . . . . . . . . 618
4 Characterization of nZVI@BC . . . . . . . . . . . . . . . . . . . . . . . . 619
5 nZVI@BC for water remediation . . . . . . . . . . . . . . . . . . . . . . 622
6 Conclusions and future perspectives . . . . . . . . . . . . . . . . . . . 631
Questions and problems . . . . . . . . . . . . . . . . . . . . . . . . . . . . 633
Acknowledgments . . . . . . . . . . . . . . . . . . . . . . . . . . . . . . . . 633
References . . . . . . . . . . . . . . . . . . . . . . . . . . . . . . . . . . . . . 633

## Chapter 19  An insight into the sorptive interactions between aqueous contaminants and biochar . . . . . . . . . . . . . . . 643

Chathuri Peiris, Janeshta C. Fernando, Y. Vindula Alwis, Namal Priyantha, and Sameera R. Gunatilake

1 Introduction . . . . . . . . . . . . . . . . . . . . . . . . . . . . . . . . . . . . 643
2 Structural changes that occur in biomass during biochar production . . . . . . . . . . . . . . . . . . . . . . . . . . . . . . . . . . . . . 644
3 Common functional moieties in biochar and adsorbates . . . . . . 646
4 Biochar-based sorptive interactions . . . . . . . . . . . . . . . . . . . . 648
5 Factors effecting the BC-adsorbate interactions . . . . . . . . . . . . 655
6 Prediction of contaminant-biochar interactions . . . . . . . . . . . . 660
7 Conclusions . . . . . . . . . . . . . . . . . . . . . . . . . . . . . . . . . . . . 662
Questions and problems . . . . . . . . . . . . . . . . . . . . . . . . . . . . 663
References . . . . . . . . . . . . . . . . . . . . . . . . . . . . . . . . . . . . . 663

## Chapter 20  Nanobiochar for aqueous contaminant removal . . . . . . . 667

Tej Pratap, Abhishek Kumar Chaubey, Manvendra Patel, Todd E. Mlsna, Charles U. Pittman, Jr., and Dinesh Mohan

1 Introduction . . . . . . . . . . . . . . . . . . . . . . . . . . . . . . . . . . . . 667
2 Nanobiochar preparation methods . . . . . . . . . . . . . . . . . . . . . 668

3 Comparative evaluation of pristine and nanobiochar properties .. 676
4 Nanobiochars for aqueous contaminant removal ............. 684
5 Conclusions and future recommendations .................. 694
Questions and problems ................................. 697
References ............................................ 698

**Chapter 21 Life cycle analysis of biochar use in water treatment plants ........................... 705**
**Md Mosleh Uddin and Mark Mba Wright**

1 Introduction ......................................... 705
2 Fundamentals of life cycle analysis (LCA) ................. 706
3 LCA of biochar production ............................. 714
4 Case study: Biochar production from corn stover .......... 719
5 LCA of biochar use in wastewater treatment .............. 728
6 Critical discussion .................................... 730
7 Conclusions ......................................... 732
Questions and problems ................................. 732
Appendix ............................................. 733
References ............................................ 733

**Chapter 22 Optimizing biochar adsorption relative to activated carbon in water treatment ................... 737**
**Matthew J. Bentley, Anthony M. Kennedy, and R. Scott Summers**

1 Introduction ......................................... 737
2 Non-adsorptive qualities ............................... 739
3 Adsorptive performance comparison in water and wastewater treatment ........................................... 745
4 Emerging biochar enhancement methods ................. 758
5 Summary ............................................ 766
Questions and problems ................................. 768
References ............................................ 769

**Chapter 23 Biochar-assisted advanced oxidation processes for wastewater treatment** .................. **775**
Rahul Kumar Dhaka, Karuna Jain, Charles U. Pittman, Jr., Todd E. Mlsna, Dinesh Mohan, Krishna Pal Singh, Pooja Rani, Sarita Dhaka, and Lukáš Trakal

1 Introduction ................................................. 775
2 Biochar-based catalysts ................................. 777
3 Biochar application in AOPs .......................... 786
4 Summary and future perspectives ................. 798
Questions and problems ................................. 800
References ..................................................... 801
Further reading .............................................. 806

Index ............................................................. 809

# Contributors

**Jacinta Alchouron** Facultad de Agronomía, Departamento de Recursos Naturales y Ambiente, Cátedra de Botánica General, Universidad de Buenos Aires, Buenos Aires, Argentina

**Mohammad Alherbawi** Division of Sustainable Development, College of Science and Engineering, Hamad Bin Khalifa University, Qatar Foundation, Doha, Qatar

**Deborah Aller** Agriculture Program, Cornell Cooperative Extension of Suffolk County, Riverhead, NY, United States

**Y. Vindula Alwis** College of Chemical Sciences, Institute of Chemistry Ceylon, Rajagiriya, Sri Lanka

**Beatrice Arwenyo** Department of Chemistry, Mississippi State University, Mississippi State, MS, United States; Department of Chemistry, Gulu University, Gulu, Uganda

**Santanu Bakshi** Bioeconomy Institute, Iowa State University, Ames, IA, United States

**Chumki Banik** Bioeconomy Institute; Department of Agricultural and Biosystems Engineering, Iowa State University, Ames, IA, United States

**Matthew J. Bentley** Environmental Engineering, University of Colorado Boulder, Boulder, CO, United States

**Robert C. Brown** Bioeconomy Institute, Iowa State University, Ames, IA, United States

**Amalia L. Bursztyn Fuentes** Technological Center of Mineral Resources and Ceramics (CETMIC - CIC/CONICET/UNLP), M.B. Gonnet, Argentina

**Marta Camps-Arbestain** New Zealand Biochar Research Centre, School of Agriculture and Environment, Massey University, Palmerston North, New Zealand

**Sergio C. Capareda** Texas A&M University, College Station, TX, United States

**Catalina Carrasco** Department of Chemistry, Mississippi State University, Mississippi State, MS, United States

**Abhishek Kumar Chaubey** School of Environmental Sciences, Jawaharlal Nehru University, New Delhi, India

**Rahul Kumar Dhaka** Department of Chemistry and Centre for Bio-Nanotechnology (COBS&H), Chaudhary Charan Singh Haryana Agricultural University, Hisar, Haryana, India

**Sarita Dhaka** Department of Chemistry, Sanatan Dharm (PG) College, Muzaffarnagar, Affiliated to Chaudhary Charan Singh University, Meerut, Uttar Pradesh, India

**Janeshta C. Fernando** College of Chemical Sciences, Institute of Chemistry Ceylon, Rajagiriya, Sri Lanka

**Rivka Fidel** Department of Environmental Science, University of Arizona, Tucson, AZ, United States

**Arna Ganguly** Department of Mechanical Engineering, Iowa State University, Ames, IA, United States

**Jesus Garcia-Nunez** CENIPALMA, Bogota, Colombia

**Manuel Garcia-Perez** Department of Biological Systems Engineering, Washington State University, Pullman, WA, United States

**Tsai Garcia-Perez** Department of Applied Chemistry and Systems of Production, Faculty of Chemical Sciences, Universidad de Cuenca, Cuenca, Ecuador

**Stefanie Gugolz** GEI Consultants, Inc., Atlanta, GA, United States

**Sameera R. Gunatilake** College of Chemical Sciences, Institute of Chemistry Ceylon, Rajagiriya, Sri Lanka

**Karuna Jain** Department of Chemistry and Centre for Bio-Nanotechnology (COBS&H), Chaudhary Charan Singh Haryana Agricultural University, Hisar, Haryana, India

**Hailey Jamison** Department of Chemistry; Department of Biochemistry, Molecular Biology, Entomology and Plant Pathology, Mississippi State University, Mississippi State, MS, United States

**Tharindu Karunaratne** Department of Sustainable Bioproducts, Mississippi State University, Starkville, MS, United States

**Michela Grace Keel** Department of Chemistry, Mississippi State University, Mississippi State, MS, United States

**Anthony M. Kennedy** Corona Environmental Consulting, LLC, Louisville, CO, United States

**Ulla Lassi** University of Oulu, Research Unit of Sustainable Chemistry, Oulu, Finland

**Gordon McKay** Division of Sustainable Development, College of Science and Engineering, Hamad Bin Khalifa University, Qatar Foundation, Doha, Qatar

**Todd E. Mlsna** Department of Chemistry, Mississippi State University, Mississippi State, MS, United States

**Dinesh Mohan** School of Environmental Sciences, Jawaharlal Nehru University, New Delhi, India

**Sohrab Haghighi Mood** Department of Biological Systems Engineering, Washington State University, Pullman, WA, United States

**Abigail Musser** Department of Civil and Environmental Engineering, Mississippi State University, Starkville, MS, United States

**Andie L.M. Nanney** Department of Chemistry, Mississippi State University, Mississippi State, MS, United States

**Chanaka Navarathna** Department of Chemistry, Mississippi State University, Mississippi State, MS, United States

**Valentine Nzengung** Department of Geology, College of Arts and Sciences, University of Georgia, Athens, GA, United States

**Prakash Parthasarathy** Division of Sustainable Development, College of Science and Engineering, Hamad Bin Khalifa University, Qatar Foundation, Doha, Qatar

**Manvendra Patel** School of Environmental Sciences, Jawaharlal Nehru University, New Delhi, India

**Chathuri Peiris** College of Chemical Sciences, Institute of Chemistry Ceylon, Rajagiriya, Sri Lanka

**Manuel Raul Pelaez-Samaniego** Department of Applied Chemistry and Systems of Production, Faculty of Chemical Sciences, Universidad de Cuenca, Cuenca, Ecuador; Composite Materials and Engineering Center, Washington State University, Pullman, WA, United States

**Charles U. Pittman, Jr.** Department of Chemistry, Mississippi State University, Mississippi State, MS, United States

**Tej Pratap** School of Environmental Sciences, Jawaharlal Nehru University, New Delhi, India

**Namal Priyantha** College of Chemical Sciences, Institute of Chemistry Ceylon, Rajagiriya; Department of Chemistry, University of Peradeniya, Peradeniya, Sri Lanka

**Arissa Ramirez** Department of Chemistry; Department of Biochemistry, Molecular Biology, Entomology and Plant Pathology, Mississippi State University, Mississippi State, MS, United States

**Pooja Rani** Department of Chemistry and Centre for Bio-Nanotechnology (COBS&H), Chaudhary Charan Singh Haryana Agricultural University, Hisar, Haryana, India

**Prashan M. Rodrigo** Department of Chemistry, Mississippi State University, Mississippi State, MS, United States

**Henrik Romar** University of Oulu, Research Unit of Sustainable Chemistry, Oulu, Finland

**Samra Sajjad** Centre for Advanced Materials, Qatar University, Doha, Qatar

**Junaid Saleem** Division of Sustainable Development, College of Science and Engineering, Hamad Bin Khalifa University, Qatar Foundation, Doha, Qatar

**Prasad Sanjeewa** College of Chemical Sciences, Institute of Chemistry Ceylon, Rajagiriya, Sri Lanka

**Kyle K. Shimabuku** Department of Civil Engineering, Gonzaga University, Spokane, WA, United States

**Virpi Siipola** VTT Technical Research Centre of Finland Ltd, Espoo, Finland

**Balwant Singh** School of Life and Environmental Sciences, Faculty of Science, The University of Sydney, Eveleigh, NSW, Australia

**Krishna Pal Singh** Biophysics Unit, College of Basic Sciences & Humanities, G.B. Pant University of Agriculture & Technology, Pantnagar, Uttarakhand; Mahatma Jyotiba Phule University, Bareilly, Uttar Pradesh, India

**R. Scott Summers** Department of Civil, Environmental, and Architectural Engineering; Environmental Engineering, University of Colorado Boulder, Boulder, CO, United States

**Vishmi S. Thrikawala** College of Chemical Sciences, Institute of Chemistry Ceylon, Rajagiriya, Sri Lanka

**Lukáš Trakal** Department of Environmental Geosciences, Faculty of Environmental Sciences Czech University of Life Sciences Prague, Praha, Suchdol, Czech Republic

**Md Mosleh Uddin** Department of Mechanical Engineering, Iowa State University, Ames, IA, United States

**Andrea S. Vega** Facultad de Agronomía, Departamento de Recursos Naturales y Ambiente, Cátedra de Botánica General, Universidad de Buenos Aires, Buenos Aires, Argentina

**Beverly Venson** Department of Chemistry; Department of Biological Sciences, Mississippi State University, Mississippi State, MS, United States

**Tao Wang** CAS Key Laboratory of Mountain Surface Processes and Ecological Regulation, Institute of Mountain Hazards and Environment, Chinese Academy of Sciences, Chengdu, China

**Jayani J. Wewalwela** Department of Agricultural Technology, Faculty of Technology, University of Colombo, Colombo, Sri Lanka

**Mark Mba Wright** Department of Mechanical Engineering and Bioeconomy Institute; Department of Mechanical Engineering, Iowa State University, Ames, IA, United States

**Vikram Yadama** Composite Materials and Engineering Center, Washington State University, Pullman, WA, United States

**Jilei Zhang** Department of Sustainable Bioproducts, Mississippi State University, Starkville, MS, United States

**Xuefeng Zhang** Department of Sustainable Bioproducts, Mississippi State University, Starkville, MS, United States

# Preface

This First Edition was motivated by the recent enormous growth in biochar research. From 2010 to the date this Preface was written, the Web of Science database shows 18,224 publications on biochar have appeared. This activity is accelerating. In 2010, 119 publications appeared. This grew to 3975 in 2020. By September 2021, 3588 biochar papers had already been counted for that year. The first use of biochar, as distinct from charcoal or activated carbon, for water remediation appeared in the *Journal of Colloid and Interface Science* in 2007 [JCIS, 310(1)2007, 57–73]. Fast pyrolysis by-product biochar obtained from biooil production in an auger-fed reactor was used to remove heavy metals from water. Following this publication, many reports on slow and fast pyrolysis biochar as a substitute for activated carbon in water purification have appeared. Previously, biochar had been used for soil enrichment and carbon sequestration.

What is driving this interest? Biochar's use for adsorbents to remove pollutants, soil amendment and enhancement of agriculture, and carbon sequestration are the major drivers. Activated carbon (AC) has long been the "go-to" standard carbonaceous adsorbent in commerce. AC has been derived mostly from coconut shells and some from coal. AC requires very high-temperature pyrolysis. Coconuts come largely from Sri Lanka and Indonesia, which means long-distance supply lines exist to many markets. Thus, AC's cost limits many applications. The growth of anthropogenic water pollution with rising living standards and human population growth has vaulted water security to a high global priority, raising the need for large quantities of readily available, low-cost carbon-based adsorbents. High surface area adsorbents can also be used to highly disperse additional adsorbent species on large surface areas. Hybrid adsorbents are needed to adsorb multiple sorbates simultaneously. Biochar can be made from municipal and agricultural wastes, energy crops, and virtually all available biomass. They are available worldwide. Their use on a large scale could diminish various waste streams.

During the period 2010–21, 6245 (34%) of the 18,224 biochar papers featured adsorption themes. A total of 9877 papers (54%) covered agriculture-related soil developments or carbon sequestration as an approach to combat global warming. Huge amounts of biochar are required for sequestration to play a serious role in reducing global warming. This use would be tied closely to agricultural soil improvement and land use patterns. Agricultural and carbon sequestration economics would be entwined and synergistic if large-scale biochar sequestration was adopted. The biochar supply for adsorbents would skyrocket. Feedstock sources would diversify. Biochar costs would drop, since they would be tied to factors besides adsorbent markets. *Sustainable Biochar for Water and Wastewater Treatment*, as the title implies, deals primarily with biochar as an adsorbent, but the reader should keep in mind the other "elephants in the room": soil enrichment (agriculture) and carbon sequestration (climate change). Biochar sequestration potential is in the billion-ton (Gt) scale, measured as $GtCO_2$/year, according to "Biochar, Climate Change and Soil: A Review to Guide Future Research," CSIRO Land and Water Science Report 05/09, February 2009. The United Nations Intergovernmental Panel on Climate Change (IPCC) 2021 report estimated sequestering carbon and improving soil could be implemented at scales of 3.7–6.6 $GtCO_2$/year using abandoned cropland, while improving depleted soils for future use and enhancing crop yields (particularly in the tropics by 25%). Soil benefits include raising soil carbon, decreasing $N_2O$ emissions, lowering nitrogen fertilizer use, raising soil water holding capacity, stimulating mycorrhizal fungi and earthworms, increasing nutrient retention, and enhancing root systems. In 2015, all UN member states adopted the 2030 Agenda for Sustainable Development, a 15-year plan to achieve its goals. Many of these goals are closely connected with the uses of biochar, particularly Goal 6: Clean Water and Sanitation, Goal 13: Climate Action, and Goal 2: Zero Hunger. Many others are also intertwined including Goal 3: Good Health and Well-being and Goal 7: Affordable and Clean Energy.

A common problem with using biochar for adsorbents is that every feedstock produces a different biochar, hence a different set of properties. Biomass must be characterized before conversion; the technology for conversion will differ and the chars produced need characterization. How do the different properties of these products meet the needs of the wastewater treatments at hand? These topics are discussed in this book, along with aspects of the adsorption system design and biochar optimization versus activated carbon. Water pollutants span an enormous contaminant range. All these cannot be considered here, but many are, including removal of dyes, pharmaceuticals, oil, actinides, lanthanides, organic compounds, microbes, microplastics, arsenic and other oxyanion species, and perfluoroalkyl substances. Nano-sized particulate biochar, zero-valent iron-decorated biochar, and biochar-assisted advanced oxidation processes are discussed.

Biochar's attractive sorptive interactions with sorbates are introduced. While commercial processes employing biochar adsorbents are not yet available for study, economic analysis and lifecycle considerations of biochar in water treatment are included. A biochar-based constructed wetland for manure removal represents a process partly based on adsorption. It is included for variety. This represents only a portion of the water pollution spectrum, but we hope it provides readers with a taste for the field.

A unique feature was added to each chapter. A set of questions with answers has been provided. These review various points made throughout each chapter, so they are not cumulative. Thus, chapters can be read in any order and the questions can still be used to review only that chapter. Newcomers to adsorption or carbonaceous adsorbents may find this feature helpful to reinforce readings. Spreadsheets, videos, and answers to the questions are also available at the companion site. This is password-protected. The companion site weblink is provided at the end of the preface section.

A book on biochar at this time constitutes a "shot in the dark." Thousands of biochar papers have been published, but the commercial carbon adsorption world currently belongs to activated carbon. No defined commercial sphere exists for biochar to provide side-by-side comparisons to activated carbon. Where this topic is heading is hard to define. Can we predict specific future applications and scales? It is easy to see the forces of anthropogenic effects acting on water supplies, land use, agricultural progress, food security, and climate change. They are all synergistically interrelated and bearing down on us. What will the outcomes be? How will biochar participate? Biochar use in adsorption and water purification might just be the tail being wagged by a dog, in this case a very large and rapidly growing agricultural and sequestration dog. Will biochar sequestration ever reach a fraction of that mentioned in the IPCC report? If so, biochar adsorbents will follow. It would be interesting to pull this book off the shelf in 2040 or 2050 and see then what had occurred.

We are indebted to the many contributing authors. Each chapter has been written by an expert actively working and contributing to biochar development. Useful suggestions have been made by many colleagues and experts (at Jawaharlal Nehru University, India, and Mississippi State University, United States) regarding the scope and style of this book to make it a teaching source and working reference. We are also thankful to the University Grant Commission (UGC), New Delhi, for providing financial assistance under the 21st Century Indo-US Research Initiative

2014 to Jawaharlal Nehru University, New Delhi, and Mississippi State University, United States, in the form of a project "Clean Energy and Water Initiatives" [UGC No. F.194-1/2014(IC)]. Both universities have established a long-term academic and research collaboration on clean energy and water.

We are also thankful to our family members for their support and understanding during the production of this book.

We welcome all suggestions and feedback for the improvement of this textbook.

**Dinesh Mohan**
**Charles U. Pittman, Jr.**
**Todd E. Mlsna**

URL: https://www.elsevier.com/books-and-journals/book-companion/9780128222256
Password: @dmohan@123

# About the editors

**Dinesh Mohan** is a professor in the School of Environmental Sciences at Jawaharlal Nehru University, New Delhi, India. He is an elected fellow of the Royal Society of Chemistry (FRSC), London, and of the National Academy of Agricultural Sciences (NAAS). He is also an adjunct professor at the Department of Chemistry, Mississippi State University, United States. He is a visiting professor at the Faculty of Environmental Sciences, Czech University of Life Sciences Prague, Czech Republic. He was an adjunct professor at the International Centre for Applied Climate Science University of Southern Queensland, Australia. He has earned his master's degree and PhD from the Indian Institute of Technology Roorkee (IITR). Professor Mohan also worked as a postdoctoral associate at Penn State University (1997) for 2 years and at Mississippi State University (2005) for more than 2 years. For the last more than 26 years, he has been involved in various research activities including water monitoring, assessment, modeling and remediation; sustainable treatment technologies for contaminants removal; climate change mitigation; and biomass conversion into biooil and biochar. Recently, he has developed a technology for a sustainable solution to the stubble burning problem in India. He has successfully completed more than 20 research projects and has published more than 150 research papers (total citations: >40,000 and h factor: 75) in the high impact factor journals.

Dr. Mohan has been a recipient of a number of academic and professional recognitions, including the 2007 Scopus Young Scientist Award (given by Elsevier), the Hiyoshi Environmental Award 2009 (given by Hiyoshi Corporation Japan), the USQ 2017 Research Giant (given by the University of Southern Queensland, Australia), and the Clarivate Analytics India Research Excellence Citation Awards 2019. He has been named "Outstanding Scientist" by CSIR, India. He has received global recognition as the Clarivate Highly Cited Researcher in 2014, 2015, 2016, 2017, 2018, 2019, 2020 and 2021. He has been named to the World's Most Influential Scientific Minds 2014 and 2015 published by Thomson Reuters. He is also listed in Two Categories (Environmental Science & Engineering as well as in Chemical Engineering) among Elsevier's list of most Highly Cited Researchers 2016 developed by ShanghaiRanking for Global Ranking of Academic Subjects.

**Charles U. Pittman Jr.** is an emeritus professor in the Department of Chemistry, Mississippi State University, United States. He is one of the world's premier synthetic materials chemists. His research is highly cited, having spanned the fields of organic, organometallic, polymer, inorganic chemistry, rocket propellant combustion, and composite materials utilizing techniques such as HPLC, GC, TLC, GPC, GCMS, NMR, IR, UV polarimetry, XPS, ISS, TEM/SEM, EXAFS, SANS, USANS, DMA, SAX, and photochemical methods. He has synthesized novel polymers, copolymers, and terpolymers and evaluated their radiation degradation behavior and sensitivity. He has published more than 900 research papers, chapters, and patents; has presented more than 420 invited lectures; and has been supported by NSF, ARO, AFOSR, ONR, PRF, EPA, DOE, NASA, Research Corporation, and numerous private sources.

**Todd E. Mlsna** is a full professor in the Department of Chemistry at Mississippi State University, United States. He graduated with a bachelor's degree from Albion College and a PhD from the University of Texas at Austin. From 1994 to 1998, he worked at the Naval Research Laboratory (NRL), Washington, DC, on the development of miniature analytical instrumentation. In 2003, Dr. Mlsna cofounded Seacoast Science where he served as president. He joined Mississippi State University in 2009. His work has been presented in many scientific conferences, and he has published 125 journal papers and book chapters. In 2016, he established Creekside Environmental Products to commercialize environmentally friendly adsorbents and soil amendments developed in his academic laboratory.

# Comprehensive biomass characterization in preparation for conversion

**Sergio C. Capareda**
*Texas A&M University, College Station, TX, United States*

### Learning objectives
Upon completion of this chapter, one should be able to:
a. Describe the various ways to convert biomass into valuable products,
b. Describe the various standard procedures performed on biomass samples in preparation for biological conversion,
c. Enumerate the equipment used for biomass characterization intended for biological conversion,
d. Describe the various standard procedures and facilities performed on biomass samples in preparation for thermochemical conversion, and
e. Enumerate the equipment used for biomass characterization intended for thermochemical conversion.

## 1 Introduction

Biomass materials come from plant and animal origin. The specific components found in biomass are rather complex. There are biomass resources high in cellulosic components, some high in sugar content, and there are biomass components high in oil contents. Hence, when one decides to convert biomass resources into valuable products, one must identify specific components of interest in the biomass materials. This chapter deals with various standard characterizations to prepare the biomass resources for conversion into valuable liquid fuels, high-value products, and materials.

The role of biomass for a sustainable low-carbon economy is depicted in Fig. 1.1. At present, the foremost important role of biomass is on climate change. Each country should reduce the use of

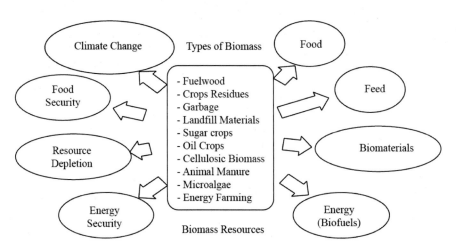

**Fig. 1.1** Role of biomass for a sustainable low-carbon economy.

fossil fuels to limit the carbon dioxide emissions and concentrations in the atmosphere. Climate change, as a result of rising greenhouse gas emissions, threatens the stability of the world's climate, economy, and population (Dagoumas and Barker, 2010). Evaluating biomass for its contribution to carbon dioxide emissions is to simply determine the amount of $CO_2$ displaced by the biomass per unit weight of biomass used. Biomass is likewise important for food security, resource depletion, and most importantly energy security as shown.

In this chapter, the focus is more on the conversion of biomass into energy in the form of biofuels and its co-products and additional output such as heat and power. Foremost biofuels produced from biomass in the world include bioethanol, biodiesel, and biogas. Other products of biomass also include heat and electricity production as well as the production of biochar to add to carbon in the soil. When biochar is added to agricultural soil, one is actually sequestering the carbon or what one may call the process a carbon capture strategy. Sometimes, one sees the acronym CCS and that stands for carbon capture and storage. Perhaps the only setback in the use of biomass is that it competes with our food and feed requirements.

When biomass is converted into liquid fuels such as biodiesel and bioethanol and gaseous fuels such as biogas, specific analysis must be made to determine the amounts of final biofuel products that may be potentially generated. For example, when one intends to produce bioethanol from biomass crops, the parameter of importance is the sugar contents. Sugar to bioethanol is the easiest pathway for making bioethanol biofuel with the aid of yeast from microorganisms. Bioethanol may likewise be produced from starchy materials, and hence, the amount of starch in the biomass must also be determined. This chapter is then designed to enumerate all biomass procedures for standard characterization including the equipment used for analysis. The procedures rely on standard analysis such as the American Standard for Testing Materials (ASTM). Hence, this chapter will begin describing important biomass properties for biological conversion such as those for making bioethanol.

## 2 Biomass conversion processes

There are numerous ways to convert biomass into useful products and co-products. While the most popular products in the market at present are liquid fuels such as bioethanol and biodiesel or gaseous fuels like biogas (a mixture of methane, $CH_4$, and carbon dioxide, $CO_2$), biomass may also be converted into highly valuable materials and specialized products. Fig. 1.2 shows the various pathways for biomass conversion as well as representative biomass for each pathway. In this figure, we have divided the conversion pathways into three sectors: (1) chemical conversion pathway, (2) biological conversion pathway, and (3) thermal conversion pathways. Specific biomass resources are directly suited to specific conversion pathways. Thus, the characterization of biomass is dependent upon what particular conversion pathway is chosen.

Commercial fossil-derived fuel may be replaced with biomass-based equivalents. Currently, diesel engines are blended with biodiesel derived from oil crops (approximately 5% in United States). Gasoline vehicles are blended with bioethanol (approximately 10% in United States). Biogas is purified, followed by the removal of carbon dioxide, and the resulting gas is equivalent to natural gas ($CH_4$).

The thermal conversion processes are more versatile than chemical or biological conversion. For example, bio-oil is a liquid product from biomass pyrolysis process and may be upgraded into crude oil, which may be fractionated into its gasoline, diesel, or aviation fuel components via the use of numerous catalysts for deoxygenation (removal of oxygen) and hydrogenation (addition of hydrogen). This is also a conversion pathway where most biomass wastes are suitable. The only major requirement is that the biomass material must be relatively dry of around 10% in moisture and the reduction of particle size.

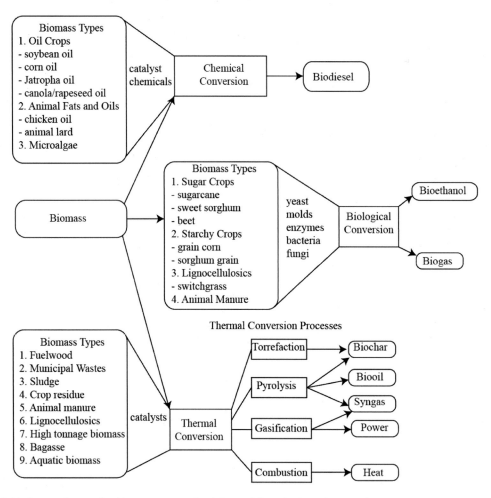

**Fig. 1.2** Various pathways for biomass conversion into useful products and co-products.

Thus, the characterization of biomass could vary depending upon what particular product is to be made. Thermal conversion (pyrolysis and gasification) may also generate electrical power via the combustion of synthesis gas in an internal combustion engine (ICE). Thus, biomass with high heating values (HHV) and low ash contents is suitable. The ash contents of biomass affect the operating conditions in many thermal conversion systems. The ash will melt at higher thermal conversion temperature leading to the formation of slag, which when deposited along the inner surfaces of conveying surfaces would lead to clogging or a technical term called fouling. Hence, biomass must be characterized for their propensity for slagging and fouling phenomena.

To generate power from biomass, the heating value and moisture of the biomass must be consistent. Moisture affects this heating value; hence, the requirement of determining the moisture content of the biomass materials is very important. The resulting product from thermal conversion (pyrolysis and gasification) is a solid product called biochar. This co-product will also be characterized or converted into valuable products such as activated carbon (Capareda, 1990). The characterization and analysis of biomass and biochar are very similar. Hence, complete biomass analysis described in this chapter may also be used to characterize biochar samples.

In biological conversion processes, it is the lignin that is a recalcitrant component for further biological conversion. These components are not readily converted by microorganisms into useful biofuels. Hence, this must be separated or removed or converted further into sugars via numerous [re-treatment options]. Fortunately, there are now high-value materials that may be made from lignin. Fig. 1.3 shows a printed circuit board (PCB) made out of pure lignin. This was a reported project from International Business Machines (IBM) Corporation and may revolutionize our use of biomass resources for biomaterials that will replace computer parts.

Another valuable material from biomass is furniture component parts. Fig. 1.4 is a photograph of lignin-based furniture part. This was a reported project from IKEA, a large home furniture distributor, and may also be made from lignin. Furniture made from

**Fig. 1.3** Printed circuit board made out of pure biomass lignin, a new high-value bio-based material.

**Fig. 1.4** Photograph of lignin-based furniture component part.

pure lignin was found to be more durable and stronger than regular wood material. In the future, one may be replacing all interior parts of a vehicle from lignin. At present, majority of interior component parts of vehicle are made of plastics from fossil fuels.

## 3 Biomass properties in preparation for biological conversion

The National Renewable Energy Laboratory (NREL) of the US Department of Energy has developed laboratory analytical procedures (or LAPs) for biomass compositional analysis. These procedures help scientists and analysts understand more about the chemical composition of raw biomass feedstocks and process intermediates for conversion to biofuels (Templeton David et al., 2016). These NREL LAPs are available online and are free to download. Table 4.2 (LAPs for Biomass Characterization) of the reference textbook by Capareda (2014) enumerated all the LAPs Numbers, the test procedure titles, and authors. This section will outline the basic procedures for the analysis and characterization, followed by simple examples for the calculations. The laboratory analysis is quite extensive and would require numerous equipment and facilities. Table 1.1 listed all these facilities including details of the minimum necessary requirements for specific equipment. Foremost of this equipment requirement is an analytical balance with an accuracy of at least 0.10 g. A spectrophotometer is required for scanning the functional groups found in the biomass. Whenever possible, an advanced Raman spectrophotometer should be available for rather complex biomass components. Convection drying ovens are required so as muffle furnaces that can convert the biomass quickly into ashes. The maximum temperature for the furnace must be around 1200°C. Water bath and autoclave are required

**Table 1.1 Facilities and equipment for compositional analysis of biomass.**

| No. | Facilities and equipment | Remarks |
|---|---|---|
| 1 | Analytical balance | Accuracy of 0.10 g |
| 2 | Spectrophotometer (diode array ultraviolet and visible range) | UV-VIS-NIR wavelength range |
| 3 | Vacuum solvent extraction system | Example is Buchi Rotavapor |
| 4 | Drying oven with temperature controller (e.g., at 105 ± 3°C) | Forced convection drying system |
| 5 | Muffle furnace (equipped with thermostat or ramping program) | Up to 1200°C |
| 6 | Water bath (with temperature controller) | Digital controller preferred |
| 7 | Autoclave (with temperature controller and mixer) | Pressure controller not required |
| 8 | Vacuum filtration system (Buchner funnel with glass microfiber) | 30 μm or less pore size |
| 9 | Desiccator with fresh desiccants | Multiple units needed |
| 10 | High-performance liquid chromatography (HPLC) system | Refractive index detector |
| 11 | Freeze-dryer system | For quick product freeze-drying |
| 12 | Standard laboratory knife mill (or intermediate Wiley mill, 1 mm) | Wiley Mill No. 4 with 2-mm pore |
| 13 | Sieve shaker (with horizontal and vertical axis motion) | Complete Tyler sieve set required |
| 14 | Riffler sampler with pans (sample splitter) | Rotary preferred |
| 15 | pH meter | Accurate to 0.01 pH unit |
| 16 | Vortex mixer | With speed controller |
| 17 | Centrifuge | 3000 rpm with 15 mL vial slots |
| 18 | Kjeldahl nitrogen digestion facility | Or elemental analyzer for N |
| 19 | Glucose analyzer | Any equivalent glucose analyzer |
| 20 | Incubator shaker | Maintaining temperature of 50 ± 1°C |
| 21 | Laminar fume hood | Or biosafety cabinet |
| 22 | Freezer | (Temperature down to −70°C) |
| 23 | Cell counting chamber slide | For yeast cell counts |
| 24 | Infrared ovens | For determining % solids |

with temperature controller, and for autoclave, a shaker is necessary. Vacuum filtration system is also needed with at least 30 μm or less pore size. The Buchner funnel with glass microfiber is appropriate. Desiccators with fresh desiccants are needed for cooling the sample prior to weighing and used after the ashing procedure. To measure the specific sugar contents of the biomass sample, a high-performance liquid chromatography (or HPLC) is required. The HPLC system must have a refractive index detector (RI detector) and the following columns: Shodex sugar SP0810 or Bio-Rad Aminex HPX-87P column or equivalent with ionic form H+/CO-3 deashing guard column and Bio-Rad Aminex HPX-87H column (or equivalent) equipped with an appropriate guard column.

A freeze-drying system is required for quick removal of water or moisture from the sample. Several size reduction equipment may be necessary such as several sizes of Wiley mill (e.g., Wiley Mill #4 with a 2-mm screen) or intermediate Wiley mill with 1-mm screen size. After milling, sieve shaker is needed to determine the various size fractions of the biomass. Hence, a complete set of Tyler sieves should be available. These Tyler sieves must also have a motor-driven shaker to separate biomass particles into various sizes. To divide the biomass samples properly, a riffler sampler with pans should be available.

Vortex mixer and centrifuge are needed to either mix samples or separate the biomass particles from the liquid media. When measuring the acidity of the liquid samples where some components are separated, a pH meter is required.

Proteins in biomass are measured by a special Kjeldahl nitrogen digestion facility, or sometimes, a simple elemental analyzer that measures the elements carbon (C), hydrogen (H), oxygen (O), nitrogen (N), and sulfur (S) will suffice. Glucose is the most important sugar found in biomass, and hence, a glucose analyzer may be a good addition to biomass characterization laboratory in addition to having an HPLC unit. Bioethanol production is normally done in temperature-controlled incubator shaker. Microbes are evaluated using cell counters. Other common facilities include infrared ovens and freezers. A freeze-drying facility is always beneficial for biological characterization methods. The other facilities are listed in Table 1.1 with explanation of needs and minimum requirements.

### 3.1 Biomass characterization in preparation for bioethanol production

To learn more of the complete biomass analysis in preparation for biological conversion, please refer to Fig. 4.1 of the textbook by Capareda (2014). This flowchart shows a problem-solving road map to establish compositional analysis of biomass. This road map will be described briefly in this section. The road map will guide the scientists with complete compositional analysis items and determine the effectiveness of future biofuel conversion.

In the compositional analysis, one will be looking for the three major components of the biomass, namely the cellulose, hemicellulose (or carbohydrates), and lignin. The latter is the most difficult and recalcitrant component for conversion and must be isolated or further converted into sugars whenever possible through pretreatments. Lignin is usually the cause for many inhibitions in many biological processes. For example, corn cobs have about 45% cellulose, 35% hemicellulose, and 15% lignin. Only the cellulose and hemicellulose components are converted into bioethanol via appropriate use of enzymes or yeasts catalysts.

The first step in the compositional analysis is the determination of % solids in biomass. If the biomass has high moisture, it must be carefully dried. Standards for drying include placing the sample in a forced convection oven with a temperature setting of around $105 \pm 3°C$ for at least 24h and noting the decline in weight after cooling in a desiccator. If the weight continues to go down, further drying is required until the final weight becomes constant. The moisture content (wet basis) of the material is then estimated using Eq. (1.1).

$$\text{Moisture content }(\%MC) = \frac{w_i - w_f}{w_i} \times 100\% \quad (1.1)$$

where
 $w_i$ = initial weight of the sample, g
 $w_f$ = final weight of the sample, g

Example 1.1 shows how moisture content of the material is calculated based on standard procedures.

---

**Example 1.1**: Rice straw was dried, and the following data were gathered. Initial weight of the sample = 5.00 g and final weight of the sample = 4.59 g. Determine the moisture content on wet basis.
**Solution:**
a. Use Eq. (1.1) for the calculations as follows:

$$\text{Moisture content }(\%MC) = \frac{5.00\,g - 4.59\,g}{5.00\,g} \times 100\% = 8.2\%$$

b. The moisture of this sample on wet basis is 8.2%.

---

In the road map, the first item to be determined is % dry solids in the biomass. The governing equation is given in Eq. (1.2).

$$\%\text{Total solids} = \frac{(W_{\text{dry pan + dry sample}} - W_{\text{dry pan}})}{W_{\text{sample as received}}} \times 100\% \quad (1.2)$$

where
 $W_{\text{dry pan+dry sample}}$ = weight of the pan and dry sample, g
 $W_{\text{dry pan}}$ = tare weight of the dry pan
 $W_{\text{sample "as received"}}$ = weight of the sample used for analysis, g

In the determination of percent solids in biomass, the material is first totally dried. The material is then totally milled to less than 2 mm. The material is then sieved through 20 and 80 mesh screens, and the fractions in the −20 to +80 range are used in subsequent tests. This "as received and dry basis" biomass is then used to estimate the % solids. Example 1.2 shows how this is calculated.

---

**Example 1.2**: Rice straw was dried and milled, and the following data were gathered. Initial weight of the "as received" sample = 5.00 g and final weight of the sample = 4.59 g. Determine the percent total solids of this sample.
**Solution:**
**a.** Use Eq. (1.2) for the calculations as follows:

$$\%\text{Total solids} = \frac{4.59}{5.00} \times 100\% = 91.8\%$$

**b.** Thus, the percent total solids of this sample are 91.8%.

---

The reader is then referred to the textbook by Capareda (2014) to follow and complete the compositional analysis of biomass such as the determination of extractives in biomass, percent protein, and the analysis of lignin and hemicellulose. Note that lignin has two types, the acid-insoluble lignin (AIL) and the acid-soluble lignin (ASL).

The carbohydrates in biomass are also of two types, the structural and nonstructural. Structural carbohydrates are bound in the matrix of the biomass, while the nonstructural carbohydrates can be removed using the simple extraction or washing steps alone. Nonstructural carbohydrates (NSC) in plant tissue are frequently quantified to make inferences about plant response to environmental conditions (Quentin et al., 2015). The determination of carbohydrates in biomass procedure is very similar to ASTM E-1758-01 (Standard Test Method for Determination of Carbohydrates in Biomass by High Performance Liquid Chromatography). The common sugar components of biomass high in sugar contents such as sweet sorghum and sugarcane include glucose, arabinose, mannose, galactose, and xylose.

When compositional analysis is complete for a certain biomass, the summary of components should be as close to 100% as possible. For example, corn stover was used for complete compositional analysis (Capareda, 2014) and the major components were found to be as follows: ash = 4.03%, extractives = 18.44%, lignin = 12.86%, protein = 5.8%, and structural carbohydrates = 57.8%. The extractives can be further separated into water-soluble units (16.23%) and ethanol-soluble fractions (2.21%). Likewise, the lignin component is differentiated between the acid-soluble fractions (12.04) and acid-insoluble fractions (0.82%). The sugars are broken down into the following components: glucose = 32.23%, xylose = 19.69%, galactose = 0.98%, arabinose = 2.29%, and mannose = 2.65%. Example 1.3 shows how the compositional analysis closure calculation is performed to show the effectiveness of the analytical procedure.

---

**Example 1.3**: Summarize the total compositional analysis of corn stover as described above and show the effectiveness of the total compositional analysis.
**Solution:**
1. The major components are added as follows:
Total compositional analysis = ash + extractives + lignin + protein + structural carbohydrates = 4.03% + 18.44% + 12.86% + 5.8% + 57.8% = 98.93%
2. The compositional analysis is quite effective with only 1.07% of components that are not known.

---

Once total sugars or carbohydrates are found, conversion to bioethanol would simply use yeast to convert those sugars into bioethanol. Please refer to other sources (e.g., Capareda, 2014) for specific sugar analysis using HPLC. The most popular yeast is *Saccharomyces cerevisiae* and commercially sold, for example, as Ethanol Red. Ethanol Red is a specially selected strain of *S. cerevisiae* that has been developed for the industrial bioethanol industry. This material has quite high ethanol tolerance, is fast acting, and displays higher alcohol yields even during high sugar content fermentation. Scientists should refer to the commercial distributors for effective dosage through either direct or indirect propagation (or technically called pitching).

Numerous bioethanol yields are reported in units of liters per hectare (or gallons per acre) and have been established with the highest reported yield that comes from Jerusalem artichoke at around 11,219 L/ha [1200 gal/acre] (Capareda, 2014). The lowest bioethanol yield comes from plums at around 204 L/ha [21.8 gal/acre]. Hence, when these values are established, one should be able to estimate the acreage needed for a certain yearly volume of production. Example 1.4 shows the simple way to estimate the acreage.

**Example 1.4**: Estimate the area required to establish a 3.7851 million liters/year [1 million gallon/year] bioethanol plant from sweet sorghum with a reported yield of 4674 L/ha [500 gal/acre].
**Solution:**
1. The calculation is straightforward as follows:

$$\text{Acreage (ha)} = \frac{3.785 \times 10^6 \text{L}}{\text{Year}} \times \frac{\text{ha}}{4674 \text{L}} = 810 \text{ ha}$$

$$\text{Acreage (acres)} = \frac{1 \times 10^6 \text{gal}}{\text{Year}} \times \frac{\text{acre}}{500 \text{ gal}} = 2000 \text{ acres}$$

2. Around 810 ha is required to achieve the targeted yearly output or around 2000 acres of land is required.

The governing equation for the conversion of sugar such as glucose ($C_6H_{12}O_6$) into bioethanol ($C_2H_6O$) or ethyl alcohol is shown in Eq. (1.3). In this equation, 2 mol of carbon dioxide are formed for every mole of glucose used. Hence, these $CO_2$ emissions must be handled properly and sequestered or captured to limit contributing to the global carbon. Example 1.5 shows the magnitude of contribution for every unit of glucose sugar used.

$$C_6H_{12}O_6 + \text{yeast} \rightarrow 2C_2H_6O + 2CO_2 + \text{Heat} \qquad (1.3)$$

**Example 1.5**: Determine the amount of $CO_2$ produced in metric tonnes per year for a 3.785 million L/year of ethanol production assuming all sugar comes from glucose. Assume the density of glucose is around 1.54 g/cm³, that of ethanol around 789.22 kg/m³, and that of $CO_2$ around 1.98 kg/m³.
**Solution:**
a. Use Eq. (1.3) to determine the ratio of $CO_2$ produced per given weight of bioethanol produced, which is as follows:

$$\frac{CO_2}{C_6H_6O} = \frac{((1 \times 12) + (2 \times 16))}{[(6 \times 12) + (6 \times 1) + (1 \times 16)]} = \frac{44}{94} = 0.468 \text{ kg} \frac{\text{kg } CO_2}{\text{kg } C_6H_6O}$$

b. The ratio of $CO_2$ production per given weight of bioethanol produced is 0.468 kg/kg.
c. If the densities of bioethanol and $CO_2$ are known, then the production rate is calculated as follows:

$$\frac{CO_2}{C_6H_6O} = 0.468\,kg\,\frac{kg\,CO_2}{kg\,C_6H_6O} \times \frac{789.2\,kg\,C_6H_6O}{m^3} \times \frac{1\,m^3}{1000\,L} \times \frac{3.785 \times 10^6\,L}{year}$$
$$\times \frac{tonne}{1000\,kg}$$
$$= 1{,}398\,\frac{tonnes\,CO_2}{year}$$

**d.** This amount of $CO_2$ must be sequestered or captured to reduce greenhouse gas (GHG) emissions.

## 3.2 Characterization of biomass in preparation for biodiesel production

Another common biofuel is biodiesel. Biomass characterization for biodiesel production simply includes the determination of the amount and quality of oil or triglyceride found in the biomass of a particular oil crop. In the United States, majority of biodiesel comes from soybean oil. Soybean oil has a seed yield of around 2245 kg/ha [2000 lbs./acre] and a biodiesel yield of around 449 L/ha [48 gal/acre] (Capareda, 2014). Canola or rapeseed oil is the most popular source of oil for biodiesel produced in Europe. Biomass characterization simply made use of various oil extraction facilities to evaluate oil (triglyceride) yield. The most common oil extraction method is by mechanical separation using either a mechanical press or an extruder. A better oil extruder should be the one with numerous sizes of apertures to accommodate various sizes of seeds. The crops with a reported highest oil yield in tonnes/ha/year are palm (5011 kg/ha) and coconut (2265 kg/ha) trees (Kemp, 2006). Example 1.6 shows the experiment to recover oil from peanut.

The vegetable oil or triglyceride may be analyzed further to determine the various fatty acids found in the oil. Hence, some common oils are technically named after the most abundant fatty acid found. For example, the dominant fatty acid in coconut oil is lauric acid with 12 carbons in its chemical formula and fully saturated with lauric acid composition of around 47%. On the other hand, palm oil's most dominant fatty acid is palmitic acid (perhaps where its name was derived or vice versa) with a fraction of around 45%, which is a 16-carbon chemical that is also saturated (Capareda, 2014). The technical name of palmitic acid is hexadecenoic acid and that for lauric acid is dodecanoic acid.

A gas chromatograph is the equipment used to measure fatty acids in oils.

---

**Example 1.6**: Determine the amount of oil extracted from peanut seeds. The initial amount of seed used was 150 g, and the amount of oil collected from the collection bottle was 67.5 g. The pressed meat weighed 78 g. Determine also the percent losses from this experiment.

**Solution:**

**a.** The percent oil yield is simply calculated from the ratio of the amount of oil collected to the initial amount of seed used, which is as follows:

$$\%\text{Oil yield }(\%) = \frac{67.5\,\text{g}}{150\,\text{g}} \times 100\% = 45\%$$

**b.** The percent meat is calculated similarly as shown below:

$$\%\text{Pressed meat }(\%) = \frac{78\,\text{g}}{150\,\text{g}} \times 100\% = 52\%$$

**c.** Thus, the losses are calculated as follows:

$$\%\text{Losses }(\%) = 100\% - (45\% + 52\%) = 3\%$$

---

Note that mechanical seed press will not recover all the oil. Solvent extraction is the process whereby close to 100% of the oil may be extracted. Unfortunately, this process is quite slow and the solvent (usually hexane) must be recovered since it is quite expensive. Biodiesel plants usually use mechanical seed presses for the purpose.

## 3.3 Characterization of biomass in preparation for biogas production

Biogas, a very popular biofuel, is usually produced from animal manure. The term biogas is simply a mixture of methane ($CH_4$) and carbon dioxide ($CO_2$). The methane-to-carbon dioxide ratio is around 65:35. The heating value of methane is around 24.2 MJ/m$^3$ [1000 Btu/ft$^3$], and hence, biogas has a heating value of around 24.2 MJ/m$^3$ [650 Btu/ft$^3$], which is 65% of the heating value of pure methane.

The potential of biomass for biogas production is evaluated using the parameter called biochemical methane potential (BMP) (Holliger et al., 2016). The biomass to be evaluated is usually composed of fresh manure or fresh manure mixed with biomass on a process called co-digestion. The ideal carbon-to-nitrogen ratio of a suitable substrate for biogas production is around 30:1. However, the carbon content of manure is around single digit such as 6.7 for poultry manure. Hence, the carbon-to-nitrogen ratio is enhanced with the addition of biomass with high carbon content such as wood chips with a carbon content of around 46.1 and a carbon-to-nitrogen ratio of 230:5 (Capareda, 2014). Even with a small amount of biomass used for co-digestion, the ideal C:N ratio is then easily achieved.

In BMP tests, all analyses must be carried out in triplicate, one must include a control, and an excellent inoculum must be sourced from an active biogas digester. The analysis simply measures the amount of biogas produced over time. However, the digester must have a pH between 7.0 and 8.5, a volatile fatty acid (VFA) concentration of less than 1.0 g acetic acid/L, a $NH_4^+$ concentration of less than 2.5 g N-$NH_4$/L, and an alkalinity of greater than 3 g $CaCO_3$/L (Holliger et al., 2016). An ideal substrate must generate close to a liter of methane gas per gram of volatile solids. Moody et al. (2011) reported a highest BMP of 810.7 ± 75.6 mL $CH_4$/g VS for food grease substrate but with a BMP efficiency of only 52%. Dairy manure has a BMP efficiency of around 85% but with a lower normalized methane production of only 264.3 ± 15.1 mL CH4/g VS. The reason for this is the chemical oxygen demand (COD) of the material affects the BMP efficiency following the use of Eq. (1.4) as shown. Example 1.7 shows the calculation of BMPs for certain substrates.

$$BMP_{eff} = \frac{V_{CH_4}}{M_{Sub\,COD}} \times \frac{1\,mg\,COD}{0.395\,mL\,CH_4} \times 100\% \quad (1.4)$$

---

**Example 1.7**: Determine the BMP efficiency of food grease and dairy manure as substrate. The following data are given: Dairy manure COD is 56,000 mg/L COD, while that for food grease is 165,200 mg/L COD. Normalized methane production potential for dairy manure is 264.3 mL methane/g VS, while that for food grease is 810.7 mL $CH_4$/g VS. The volatile solids (VS) for dairy manure are 71.2% and those for food grease are 41.5%.

**Solution:**

**a.** Use Eq. (1.4) for this calculation for dairy manure as follows:

$$BMP_{eff} = \frac{264.3\,mL\,CH_4}{g\,VS} \times \frac{L}{56,000\,mg\,COD} \times \frac{71.2\,g\,VS}{L}$$
$$\times \frac{1\,mg\,COD}{0.395\,mL\,CH_4} \times 100\% = 85\%$$

**b.** Use Eq. (1.4) for this calculation for food grease as follows:

$$BMP_{eff} = \frac{810.7\,mL\,CH_4}{g\,VS} \times \frac{L}{165,200\,mg\,COD} \times \frac{41.9\,g\,VS}{L}$$

$$\times \frac{1 \, \text{mg COD}}{0.395 \, \text{mL CH}_4} \times 100\% = 52\%$$

Another parameter for potential biogas production of biomass materials is the use of anaerobic toxicity assay (ATA). Most organic substrates added to the biogas digester or to animal manure to improve its methane production may inhibit biogas production. Hence, this evaluation procedure simply determines the effect of adding too much substrate that limits biogas production instead of improvement. The use of BMP alone is not a full proof measure for biogas production potential. For example, if the substrate releases small amounts of ammonia, it provides nutritious nitrogen to a digester for the microorganisms to improve its methane production power. However, too much ammonia disrupts methane production as well, and hence, ATA tests in conjunction with BMP test are the ideal couple of tests conducted to determine biochemical methane potential. We refer the reader to numerous ISO procedures to determine the inhibitory activity of numerous microorganisms in a digester. Examples are ISO 1364-1 (n.d.): Determination of inhibition of activity of anaerobic bacteria. Part 1: General Test. International Organization for Standardization, Geneva Switzerland, and ISO 1364-2 (n.d.): Determination of inhibition of activity of anaerobic bacteria. Part 2: Test at Low biomass concentration. International Organization for Standardization, Geneva Switzerland.

The ATA test is the most accurate method of identifying particular substrates that inhibit biogas production in a digester. In most instances, these compounds or materials come from external environment that the substrate is exposed to such as heavy metals, sulfates and sulfides, salty or acidic materials.

## 4 Biomass properties in preparation for thermochemical conversion

Thermal conversion of biomass such as pyrolysis, gasification and combustion require rather different parameters in preparation for conversion. Foremost of which is the heating value or energy content of biomass. Numerous biomass characterization procedures are similar for most thermal conversion processes enumerated. Hence, the following sections describe important characterization procedures true for all these processes.

## 4.1 Heating value analysis

Heat and power are the most common outputs or products of thermal conversion processes such as gasification and combustion. Hence, the heating value or calorific value of a biomass can easily be converted into its energy or power equivalent once the heating value is known. The only factor to consider is the conversion efficiency. Thus, the most common governing equation is the efficiency equation shown in Eq. (1.5). Example 1.8 uses this equation to estimate the power generated from a certain biomass if the heating value is known as well as overall conversion efficiency.

$$\text{Conversion efficiency}(\%) = \frac{\text{Output power}(MW)}{\text{Input power}(MW)} \times 100\% \quad (1.5)$$

---

**Example 1.8**: Determine the output power (units of MW) if wood chips with a heating value of 20 MJ/kg are converted into electrical power via combustion with an overall conversion efficiency of 50%. The material is being fed at a rate of 1000 kg/h.
**Solution:**
a. Eq. (1.5) is used directly and rearranged since two parameters are already given (i.e., input power and overall conversion efficiency).

$$\text{Output power}(MW) = \text{Input power}(MW) \times \text{Conversion efficiency (decimal)}$$

$$\text{Output power}(MW) = \frac{1000\,\text{kg}}{\text{h}} \times \frac{20\,\text{MJ}}{\text{kg}} \times \frac{1\,\text{h}}{3600\,\text{s}} \times \frac{MW}{MJ/s} \times 0.50 = 2.78\,MW$$

---

The heating value of biomass is best measured using a calorimeter or specifically a bomb calorimeter. The standard for the measurement of heating value of biomass is the ASTM E 711: Standard Test Method for Gross Calorific Value of Refuse Derived Fuel by Bomb Calorimeter. Note that the moisture content of the biomass must be known (use of Eq. 1.1) in order to convert the units into the high heating value (bone-dry sample or dry basis) and low heating value (or "as received") of biomass. The heating value of raw biomass with a given moisture is its heating value termed "as received." One other way of describing the energy content of biomass is its dry-ash-free basis. In this case, the ash content of biomass must also be known. The procedure for determining the ash content of biomass will be described in the next section. Eqs. (1.6) and (1.7) show the equation for determining the HHV and LHV of biomass materials. Example 1.9 shows the calculations to convert the heating value of biomass on "as received" basis, dry basis, and ash-free basis.

$$\text{HHV (dry basis)} = \frac{\text{HV (wet basis)}}{1 - \text{Moisture content (decimal)}} \quad (1.6)$$

$$\text{HHV (dry ash free basis)} = \frac{\text{HV (dry basis)}}{1 - \text{Ash content (decimal)}} \quad (1.7)$$

**Example 1.9**: The heating value of a certain biomass was reported as 18 MJ/kg "as received." The moisture content of this biomass was 10%. Convert this heating value number into its HHV equivalent. The ash content of this biomass was also found to be around 15%. Determine the heating value on a dry-ash-free basis as well.

**Solution:**

**a.** Use Eq. (1.6) to determine the HHV as shown:

$$\text{HHV (dry basis)} = \frac{\text{HV (wet basis)}}{1 - \text{Moisture content (decimal)}} = \frac{18}{1 - 0.10}$$

$$= 20 \, \frac{\text{MJ}}{\text{kg}}$$

**b.** Use Eq. (1.6) to convert the HHV of this biomass material into its dry-ash-free equivalent as shown:

$$\text{HHV (dry ash free basis)} = \frac{\text{HV (dry basis)}}{1 - \text{Ash content (decimal)}}$$

$$= \frac{20}{1 - 0.15} = 23.5 \, \frac{\text{MJ}}{\text{kg}}$$

## 4.2 Proximate analysis

The next parameter used to evaluate the properties of biomass in preparation for thermal conversion is the proximate analysis. This term comprised the measurement of the moisture content (MC), the volatile combustible matter (VCM), the fixed carbon (FC), and the ash. The measurement process is usually done in succession beginning from the determination of moisture content. Thermal conversion processes require the biomass to be relatively dry. One rule of thumb for pyrolysis and gasification processes is to use biomass with around 10% moisture. These thermal conversion processes require a little bit of moisture so that the hydrogen in water may be used for the water-gas shift reactions to generate combustible gases such as hydrogen, carbon monoxide, and methane. Each of these parameters has a specific ASTM standard procedure. Please refer to another textbook (e.g., Capareda, 2014) for more details of the ASTM number designations.

The moisture content (MC) equation has already been given in Eq. (1.1). The volatile combustible matter (VCM) equation is given in Eq. (1.8) and that of calculating the ash content is given in Eq. (1.9). The fixed carbon (FC) is calculated by difference assuming no losses. This is then given in Eq. (1.10). Note that in determining the VCM of a bone-dry biomass sample, a tube furnace is needed using a platinum crucible with a close-fitting cover and a contraption that could lower the crucible in the middle of the tube furnace that is set at around $950 \pm 20°C$ [ASTM E872-82 VCM Measurement for wood samples]. The sample is exposed at this temperature for around 7 min, cooled, and reweighed. This sample is then placed in a muffle furnace with a temperature setting of 600°C and with the cap removed and heat-treated for an hour. After treatment, it is then cooled and reweighed to calculate the ash content. The fixed carbon is determined by difference using Eq. (1.10). Example 1.10 shows how a complete proximate analysis is made.

$$\text{VCM}(\%) = \frac{v_i - v_f}{v_i} \times 100\% \qquad (1.8)$$

$$\text{Ash}(\%) = \frac{f_i - f_f}{v_i} \times 100\% \qquad (1.9)$$

$$\text{FC}(\%) = 100\% - (\%\text{VCM} + \%\text{Ash}) \qquad (1.10)$$

where

$v_i$ = initial weight of dry biomass before it is placed in a tube furnace, g
$v_f$ = final weight of the biomass after thermal treatment in tube furnace, g
$f_i$ = initial weight of the biomass before it is placed in a muffle furnace, g
$f_f$ = final weight of the biomass after thermal treatment in muffle furnace, g

**Example 1.10**: Fresh ground rice straw samples were used to estimate the complete proximate analysis. Show the proximate values on wet basis and dry basis. The following are the data from the experiment:
a. initial fresh sample weight = 4.9500 g
b. final sample weight after drying in convection oven = 4.5441 g
c. initial weight of sample placed in tube furnace = 4.5441 g
d. final weight of sample after treatment in tube furnace = 1.2028 g
e. initial weight of the sample placed in the muffle furnace = 1.2028 g
f. final weight of the sample after treatment in muffle furnace = 0.5198 g

**Solution:**
a. The moisture content is calculated as follows:

$$\text{Moisture content (\%MC)} = \frac{4.950\,\text{g} - 4.5441\,\text{g}}{4.5441\,\text{g}} \times 100\% = 8.2\%$$

b. The VCM is calculated as follows:

$$\text{VCM (\%)} = \frac{4.5441 - 1.2028}{4.5441} \times 100\% = 73.53\%$$

c. The % ash is calculated as follows:

$$\text{Ash (\%)} = \frac{1.2028 - 0.5198}{4.5441} \times 100\% = 15.03\%$$

d. The fixed carbon is calculated as follows:

$$\text{FC (\%)} = 100\% - (73.53\% + 15.03\%) = 11.44\%$$

e. To check the accuracy of values measured and calculated, we have, on dry basis as follows:
100% = 73.53% + 15.03% + 11.44% = 100%, verified correct.

f. To revise the calculation on wet basis, one simply needs to reduce the values of VCM, FC, and ash based on the moisture determined. In this case, the 8.2% MC shows that VCM, FC, and ash must comprise 91.8% of the wet sample. The following are the recalculations and checking of mass balance.

$$\text{MC (\%)} = 8.2\%$$

$$\text{VCM (\%)} = 73.53\% \times 0.918 = 67.5\%$$

$$\text{FC (\%)} = 11.44\% \times 0.918 = 10.50\%$$

$$\text{Ash (\%)} = 15.03\% \times 0.918 = 13.80\%$$

100% = 8.2% + 67.5% + 10.5% + 13.8% = 100%, verified correct.

g. Hence, one may report the proximate analysis either on dry basis or on wet basis as shown.

## 4.3 Ultimate analysis

Biomass sample may also be reported according to its ultimate or elemental analysis. The five primary elements in biomass are carbon (C), hydrogen (H), oxygen (O), nitrogen (N), and sulfur (S). All the other remaining components are lumped into its ash content. An ultimate analyzer is the instrument that determines the ultimate analysis of biomass. The importance of this process is to have an estimate of the "chemical formula" of the biomass such that combustion equation may be set up or that the amount of oxygen needed to thermally convert this biomass via gasification process may be estimated as well. Capareda (2014) listed all the ASTM standard procedures for this test.

In this analysis, approximately a gram of dried sample is placed in a receptacle for combusting the carbon as well as determining the percentages of the remaining component in similar fashion. The more expensive ultimate analyzer can handle higher weight samples, while the less expensive ones can only handle samples in microgram level. Some units can also handle liquid samples. Example 1.11 shows the process of determining the "chemical formula" of a biomass sample when the elemental analysis is known and also how to set up the complete combustion of this biomass to determine the air-to-fuel ratio either by weight or by volume.

---

**Example 1.11**: The results of ultimate analysis for a rice straw sample are given as follows: carbon=37.1%, hydrogen=5.7%, oxygen=41%, nitrogen=1%, and sulfur=0.2%. The ash content of this sample was found to be 15%. A 100 g dried rice straw sample was used to establish the chemical formula of this biomass. Determine this "chemical formula" and set up the stoichiometric combustion equation. Determine also the air-to-fuel ratio by weight and by volume.

**Solution:**

a. To determine the "chemical formula" of biomass sample given the ultimate analysis results, one simply needs to determine the mole of each element using their individual molecular weights. The molecular weights of these elements are as follows: carbon=12 g/mol, hydrogen=1 g/mol, oxygen=16 g/mole, nitrogen=14 g/mol, and sulfur=32 g/mol. Thus, the following is the number of moles for each element.

$$C = 37.1 \, g \times \frac{mol}{12 \, g} = 3.1$$

$$H = 5.7 \, g \times \frac{mol}{1 \, g} = 5.7$$

$$O = 41\,g \times \frac{mol}{16\,g} = 2.56$$

$$N = 1\,g \times \frac{mol}{14\,g} = 0.07$$

$$S = 0.2\,g \times \frac{mol}{32\,g} = 0.00625$$

b. Hence, the "chemical formula" for this biomass is shown below, where the moles are the subscripts for each element.

$$C_{3.1}H_{5.7}O_{2.56}N_{0.07}S_{0.00625}$$

c. When setting up the stoichiometric air-to-fuel ratio, we have the following equation:

$$C_{3.1}H_{5.7}O_{2.56}N_{0.07}S_{0.00625} + 1.82(O_2 + 3.76N_2)$$
$$\rightarrow 3.1CO_2 + 2.85H_2O + 13.7564N + 0.00625S$$

d. To calculate the air-to-fuel ratio by volume, the stoichiometric equation is actually the volumetric ratio, meaning about 1.82 L of air is needed for every liter of biomass combusted. The air-to-fuel ratio by weight is calculated as follows:

$$\frac{Air}{Fuel} = \frac{1.82 \times [(16 \times 2) + 3.76 \times (14 \times 2)]}{[(12 \times 3.1) + (1 \times 5.7) + (16 \times 2.56) + (0.07 \times 14) + (0.00625 \times 32)]}$$

$$= \frac{249.8496}{85.04} = 2.94 \frac{kg}{kg}$$

e. Hence, one will need around 3 kg of air for every kg of biomass combusted. The products are carbon dioxide, water, and heat.

Gasification processes require 30%–70% less air, and hence, when biomass is prepared for gasification process, less amount of air is required, thereby producing combustible gases called synthesis gas or syngas that comprised mostly of carbon monoxide (CO), hydrogen gas ($H_2$), methane gas ($CH_4$), and other low-molecular-weight hydrocarbons. The primary reason why most agricultural biomass cannot be combusted in open air with the use of stoichiometric amounts of air to generate heat or electrical power is the issue of slagging and fouling. Most agricultural biomass is high in ash content, and at higher combustion temperatures, these inorganic ashes will melt and form slag, a heavy rock-like material, which when cooled will adhere to conveying surfaces creating blockage. When these conduits are blocked because of the slag formation, the technical term is called fouling.

This will lead to dangerous processes in conveying systems and may lead to clogging and possible explosion. Hence, most biomass should be gasified instead of combusted for sustainable thermal conversion processes. The next section discusses the parameters for evaluating the potential of biomass for slagging or fouling issues.

## 4.4 Eutectic point analysis for biomass

The sustainability of any thermal conversion at elevated temperature is a strong function of the integrity or stability of the reactor. The reactor must be maintained at the proper temperature way below the melting point of the ash of the biomass. If the ash of the biomass melts, then slagging will start to form, and when these particles are conveyed and the temperature has decreased, it will most likely adhere to the surfaces of conveying systems and later on cause fouling. This phenomenon must be avoided for sustainable operation of thermal conversion systems.

One way to determine the proper operating temperature is to evaluate the melting point of the ash of the biomass and set criteria on operating temperature based on this analysis. Maglinao and Capareda (2010) have developed a process of evaluating the eutectic point of any biomass. The first step in the analysis is to generate large amounts of ash from the biomass in question and make pellets of appropriate size. These pellets are then exposed in the furnace at elevated temperatures common in combustion processes (usually between 1000°C and 2000°C). After the experiments, the pellets are tested for compressive strengths. The plot should look like a bell-shaped curve with the peak equivalent to the melting point of the ash of the biomass. The theory is that when the ash of the biomass melts, the material becomes brittle and the compressive strength would be lower. That peak should be an indication of the maximum temperature that the biomass should undergo to avoid slagging and fouling.

Coal power plants have also developed a procedure for this evaluation. Two parameters are used, namely the slagging ($R_s$) and fouling ($R_f$) factors given in Eqs. (1.11) and (1.12).

$$R_s = \left(\frac{\text{Base}}{\text{Acid}}\right) \times \%S \text{ (dry coal)} \qquad (1.11)$$

$$R_f = \left(\frac{\text{Base}}{\text{Acid}}\right) \times \%Na_2O \qquad (1.12)$$

In this biomass characterization, one determines the complete ash contents of biomass and the acidic and basic components of the ash measured and used in the numerator and denominator of these equations including percentage of sulfur of dry coal for slagging factor and percentage of sodium oxide ($Na_2O$) for the fouling factor as shown in the equations, respectively. Table 1.2 is used to determine the slagging and fouling types. The coal samples could exhibit low, medium, high, or severe slagging and fouling based on the resulting values for $R_s$ and $R_f$ (Winegartner, 1974). Example 1.12 shows the calculations for slagging and fouling factors for coal samples.

**Example 1.12:** Determine the slagging ($R_s$) and fouling ($R_f$) factors for coal samples designated as letter D in the following table. Describe the slagging and fouling types for this coal sample based on Table 1.2.

| % Component | A | B | C | D | D/E | E |
|---|---|---|---|---|---|---|
| Acids | (75.02%) | (82.43%) | (79.68%) | (76.15%) | (71.81%) | (68%) |
| $Al_2O_3$ | 25.4 | 20.4 | 30.8 | 26.6 | 22.8 | 20.9 |
| $SiO_2$ | 48.5 | 61 | 47.1 | 48.4 | 48 | 46.1 |
| $TiO_2$ | 1.12 | 1.03 | 1.78 | 1.15 | 1.01 | 1.0 |
| Bases | (18.01%) | (14.87%) | (12.33%) | (19.84%) | (21.55%) | (25.28%) |
| MgO | 2.24 | 1.61 | 1.65 | 2.85 | 2.5 | 1.11 |
| $P_2O_5$ | 0.52 | 0.30 | 1.44 | 0.61 | 0.56 | 0.37 |
| $K_2O$ | 2.13 | 1.96 | 0.66 | 2.39 | 0.79 | 0.88 |
| CaO | 2.13 | 3.07 | 6.73 | 4.32 | 4.97 | 4.87 |
| $Fe_2O_3$ | 11.1 | 7.6 | 3.15 | 9.11 | 11.1 | 16.6 |
| $Na_2O$ | 0.41 | 0.63 | 0.14 | 1.17 | 2.19 | 1.82 |
| $SO_3$ | 2.4 | 2.69 | 3.76 | 3.18 | 3.91 | 3.88 |
| %Sulfur | 0.81 | 0.77 | 0.47 | 0.80 | 1.35 | 2.45 |

**Solution:**
a. The slagging factor is calculated from Eq. (1.11) as follows:

$$R_s = \left(\frac{Base}{Acid}\right) \times \%S \text{ (dry coal)} = \left(\frac{19.84}{76.15}\right) \times 0.80 = 0.21$$

b. Based on Table 1.2, this coal sample has low slagging potential since $R_s$ value is less than 0.60.

c. The fouling factor is calculated from Eq. (1.12) as follows:

$$R_f = \left(\frac{Base}{Acid}\right) \times \%Na_2O = \left(\frac{19.84}{76.15}\right) \times 1.17 = 0.30$$

**Table 1.2** Slagging and fouling indices for coal samples undergoing combustion.

| Slagging and fouling types | $R_s$ | $R_f$ |
|---|---|---|
| Low | <0.60 | <0.20 |
| Medium | 0.6–2.0 | 0.2–0.5 |
| High | 2.0–2.6 | 0.5–1.0 |
| Severe | >2.6 | >1.0 |

 d. The fouling potential is medium since the $R_f$ value falls between 0.20 and 0.50.

Maglinao and Capareda (2010) recommended a different set of criteria for evaluating the eutectic point of the ash in agricultural biomass. During their extensive experiments, they discovered that the coal slagging and fouling potential are not appropriate for high ash biomass such as those found in agricultural crops. They have recommended three parameters as follows:

 a. Alkali index (AI)
 b. Base-to-acid ratio ($R_{b/a}$)
 c. Bed agglomeration index (BAI)

The alkali index was based on the "alkali" component in the ash and defined by Vamvuka and Zograf (2004) as the sum of the oxides $K_2O$ and $Na_2O$ and also referred to by Miles et al., 1995. It is given in Eqs. (1.13) and (1.14), and it expresses the quantity of alkali oxide in the biomass fuel per unit of fuel energy (kg alkali/GJ) for metric system and (lb alkali/mmBtu) for English system. Alkali index values lower than 0.17 kg/kJ [0.4 lb./mmBtu] should have low slagging risk and above 0.34 will have high risk of slagging.

$$AI = \frac{kg(K_2O + Na_2O)}{GJ} \quad (1.13)$$

$$AI = \frac{1 \times 10^6}{HV\left(\frac{Btu}{lb}\right)} \times \%ash \times \%alkali\ in\ ash = \frac{lb\ alkali}{mmBtu} \quad (1.14)$$

The base-to-acid ratio ($R_{b/a}$) is given in Eq. (1.15), and if this value is more than unity, slagging and fouling are certain to occur during thermal combustion of biomass. More specifically, low potential for $R_{a/b} < 0.5$; medium between 0.5 and 1; high between 1 and 1.75; and severe above 1.75. Note that this equation is simply ratios of base and acid components of the biomass ash.

$$R_{b/a} = \frac{\%(Fe_2O_3 + CaO + MgO + K_2O + Na_2O)}{\%(SiO_2 + TiO + Al_2O_3)} \quad (1.15)$$

The bed agglomeration index (BAI) on the other hand is given in Eq. (1.16), and agglomeration is said to occur when the BAI values are lower than 0.15.

$$BAI = \frac{\%Fe_2O_3}{\%(K_2O + Na_2O)} \quad (1.16)$$

Example 1.13 shows how to calculate these new slagging and fouling indices for agricultural biomass. This example also compares the slagging and fouling indices used for coal to show differences and validation of theory that slagging and fouling indices for coal are not appropriate for some agricultural biomass.

**Example 1.13**: Determine the alkali index (AI) in English units, base-to-acid ratio ($R_{b/a}$), and bed agglomeration index for the complete ash analysis of hardwood given in the table below partially taken from Lachman et al. (2021) with high heating value of biomass found from Lisy et al. (2020). Compare the potential slagging and fouling using $R_s$ and $R_f$ values. As additional exercise, do the calculations for the rest of biomass.

| % Component | Spruce | Softwood | Hardwood | Amaranth | Hay | Straw |
|---|---|---|---|---|---|---|
| Acids | (3.27%) | (10.28%) | (3.39%) | (2.87%) | (40.48%) | (50.28%) |
| $Al_2O_3$ | 0.51 | 1.56 | 0.55 | 0.22 | 3.91 | 4.97 |
| $SiO_2$ | 2.72 | 8.62 | 2.8 | 2.38 | 36.3 | 45 |
| $TiO_2$ | 0.04 | 0.10 | 0.04 | 0.27 | 0.27 | 0.31 |
| Bases | (56.46%) | (52.38%) | (57.01%) | (61.11%) | (39.01%) | (32.62%) |
| MgO | 10.9 | 3.03 | 4.34 | 6.74 | 4.28 | 3.20 |
| $P_2O_5$ | 1.99 | 5.55 | 5.47 | 3.7 | 5.72 | 5.07 |
| $K_2O$ | 12.5 | 8.71 | 9.81 | 14.4 | 20.9 | 18.8 |
| CaO | 32.1 | 39.3 | 42.2 | 39.2 | 11.6 | 8.1 |
| $Fe_2O_3$ | 0.71 | 0.94 | 0.49 | 0.48 | 1.58 | 1.94 |
| $Na_2O$ | 0.25 | 0.40 | 0.17 | 0.29 | 0.65 | 0.58 |
| $SO_3$ | 2.47 | 1.86 | 2.17 | 1.87 | 5.05 | 3.22 |
| %Sulfur | 0.99 | 0.74 | 0.87 | 0.75 | 2.02 | 1.29 |
| HV (MJ/kg) | 18.21 | 18.20 | 18.04 | 16.31 | 15.79 | 17.24 |
| HV (Btu/lb) | 7844 | 7841 | 7774 | 7027 | 6803 | 7427 |
| % Ash in biomass | 0.33 | 3.01 | 2.44 | 7.70 | 17.73 | 6.64 |
| % Alkali in biomass | 12.75 | 9.11 | 9.98 | 14.69 | 21.55 | 19.38 |

**Solution:**
a. The alkali index is calculated as follows:

$$\mathrm{AI}\left(\text{hardwood}\frac{\text{lb alkali}}{\text{mmBtu}}\right) = \frac{1\times 10^6}{7774\,\text{Btu/lb}} \times \frac{2.44\%}{100} \times \frac{9.98\%}{100}$$

$$= 0.31\frac{\text{lb alkali}}{\text{mmBtu}}$$

b. The AI index showed medium slagging and fouling potential because the value falls between 0.17 and 0.34.
c. The base-to-acid ratio is calculated as follows:

$$R_{b/a} = \frac{57.01}{3.39} = 16.82$$

d. The base-to-acid ratio ($R_{b/a}$) values do not expect any slagging and fouling issues.
e. The bed agglomeration index (BAI) is calculated as follows:

$$\mathrm{BAI} = \frac{\%\mathrm{Fe_2O_3}}{\%(\mathrm{K_2O + Na_2O})} = \frac{0.49}{9.81 + 0.17} = 0.049$$

f. The BAI expects no slagging and fouling issues.
g. The $R_s$ and $R_f$ values are as follows:

$$R_s = \left(\frac{\text{Base}}{\text{Acid}}\right) \times \%S\,(\text{dry coal}) = \left(\frac{57.01}{3.39}\right) \times 0.87 = 14.63$$

$$R_f = \left(\frac{\text{Base}}{\text{Acid}}\right) \times \%\mathrm{Na_2O} = \left(\frac{57.01}{3.39}\right) \times 0.17 = 2.86$$

h. Both of these values predicted severe slagging and fouling issues. In actual thermal conversion processes, these conditions may not occur for this type of biomass.

## 4.5 Other biomass characterization processes

There are numerous other properties of biomass that are important in the design of biomass conversion systems. Some of these parameters are used in the design of storage facilities for biomass as well as transport properties.

### 4.5.1 Bulk density

Bulk density is the unit weight of the biomass per unit of volume. The volume includes the void spaces occupied by the biomass. The standard method is ASTM E 873-82 (2006) Standard Test Method for Bulk Density of Densified Particulate Biomass

Fuels. A simple box measuring 305 × 305 × 305 mm [12 × 12 × 12 in.] is used with appropriate handles. A weighing scale is required with an accuracy of 100 g. The tare weight of the box must also be known with an accuracy of around 100 g. The biomass sample is then poured onto this box, and the whole box with biomass is dropped from a height of about 610 mm [2 ft]. The box with sample is dropped around 5 times to allow for biomass settling. Additional biomass samples are added, and excess samples are removed using a straight edge device. The box is then weighed to within 100 g. The bulk density is calculated using Eq. (1.17).

$$\text{Bulk density}\left(\frac{g}{cm^3}\right) = \frac{(\text{weight of box and sample}) - (\text{weight of box})}{(\text{volume of box})}$$

(1.17)

Biomass with high bulk density is preferred for long-range transport due to its lesser volume requirement but with heavier weight. The average bulk density of wood is in the range of 300–550 kg/m³ [18.7–34 lb./ft³]. For biochar, the reported range is 200–300 kg/m³ [12.5–18.7 lb./ft³] (Lata and Mande, 2009). Rice straws and rice hulls have a lower bulk density of about 70 kg/m³ [4.4 lb./ft³], while densified products (such as briquettes) may have bulk densities as high as 700–800 kg/m³ [43.6–50 lb./ft³] (Capareda, 2014).

### 4.5.2 Particle density

Particle density is the measure of the weight of the biomass particle per unit volume that the particle has occupied. This is the "true" particle density of the biomass. The void spaces in between the biomass particles are not used in the calculation. Thus, this measurement requires a specialized equipment called pycnometer where the volume of void space is measured by using an inert gas such as helium. The procedure is straightforward. A sample of ground and dried biomass is placed in a receptacle of known ad calibrated volume. The receptacle is then placed in an enclosed system where all the void spaces are removed by creating a total vacuum and then replaced with a measured amount of inert gas. The volume occupied by the biomass is then calculated by subtracting the overall volume of the receptacle minus the volume occupied by this inert gas. The degree of compactness of a biomass sample may be evaluated by the difference in values of the biomass bulk density and particle density. Hence, to improve the transport of biomass and save on fuel transport, biomass resources are usually densified through briquetting or simply by compaction, similar to baling of cotton to improve transport efficiencies of biomass over longer distances.

### 4.5.3 Particle size distribution (PSD)

Size reduction is a common biomass preprocessing. Thermal conversion processes usually require smaller average particle size, and hence, biomass must be ground and sieve out larger particles. Thermal biomass treatment processes are a strong function of particle size. The smaller the particle, the quicker the conversion. Hence, the particle size distribution is a way of determining the average particle size of the ground and dried biomass samples. The device used for measuring the PSD of the biomass sample is a set of standard sieves (e.g., Tyler sieves) with a motor-driven shaker and timer. Tyler sieves come in standard sizes from 37 μm [0.0015 in.] to 3360 μm [0.131 in.]. The shaking time is usually between 15 and 20 min.

A known weight of ground biomass samples is placed on top of Tyler sieve set and shaken for 15 min. The weight of the biomass retained in each of the sieve size is weight and the fraction for each size range measured. The mean particle size is given in Eq. (1.18).

$$\overline{d_p} = \frac{1}{\sum^{all\ i} \left(\frac{x}{d_p}\right)_i} \quad (1.18)$$

where
 $\overline{d_p}$ = mean particle size
 $x$ = weight fraction within a certain sieve size range
 $d_p$ = average particle size within a certain sieve size range.

Capareda (2014) shows an example calculation for determining the average particle size of some materials (Chapter 11, Section 11.2). For particles less than 100 μm, a different instrument is used such as the following:
a. Beckman Multisizer Coulter Counter, and
b. Malvern Mastersizer 3000

The above equipment differs in the characterization of particle size distribution (PSD) of fine biomass, but the results are quite comparable. For biomass samples of this small size, the common parameter that describes the size is the particle equivalent diameter (AED) in microns. The AED is calculated by knowing the equivalent spherical diameter (ESD), which is the value reported by the above-named instruments. The primary reason is that the above instruments assume the particle it is measuring is close to a sphere, which is usually not the case. The Malvern Mastersizer has a database of known particle shape but of known dust samples and not necessarily biomass fine particles. The equation to convert ESD to AED is given in Eq. (1.19).

$$\text{AED} = \text{ESD} \times \sqrt{\rho} \qquad (1.19)$$

where

$\rho$ = particle density, g/cm$^3$

The particle density is measured by a pycnometer described in Section 4.5.2.

### 4.5.4 Angle of repose

When biomass materials are naturally piled on a flat horizontal surface, an angle is formed due to the interaction among particles and forming like a hill. This angle is important in the design of biomass silos to optimize the angle of the top cone and ensure maximum containment in a silo. This particular angle is called the angle of repose. It is the steepest angle of descent relative to the horizontal plane in which the granular biomass is poured. It will create a constant angle depending upon the type of surface used as well as the moisture content of biomass, biomass density, particle shape and the coefficient of friction between the materials and the surface used.

The standard procedure in establishing the angle of repose of biomass sample is simply to pour biomass samples from a funnel on a particular flat horizontal surface and measure the angle formed by the pile. This parameter is also important in conveying the biomass on a belt conveyor to ensure that there will be no biomass spillage. In many fuel bins for biomass conversion into a reactor such as during thermal conversion, the angle of the bin must be carefully designed to prevent clogging or creation of pockets leading to feeding problems.

### 4.5.5 Angle of friction

The angle of friction is the minimum angle with which a pile of biomass will begin to slide when this particular material is raised from the horizontal. The application of this parameter is very important when biomass is being unloaded from, say a truck carrying large volume of biomass. For example, in large biomass-fueled power plants, the delivery of biomass is made by several high tonnage capacity trucks and unloading is an issue. Using conveyor belts from the floor of the trucks is not appropriate since it will take 15–20 min simply just to upload the biomass, say wood chips.

Numerous biomass power plants simply hook up the truck on a harness, open the back door, and lift the whole truck to unload the biomass in a shorter period of time. This lift angle must be greater than the angle of friction of the biomass to unload tons of biomass in a matter of seconds or a few minutes.

This parameter is also important in conveying systems for biomass to ensure that the biomass sample will not be clogging the delivery chute. The measurement of angle of friction is simply made by raising the biomass from the horizontal and measuring the minimum angle whereby all biomass samples slide from the surface. Hence, the surface used for the test must be the same or similar surface used in those conveying systems. Of course, some other parameters affecting the angle are similar to those enumerated in the angle of repose section.

### 4.5.6 Heat capacity and thermal conductivity of biomass

There is a general agreement on the proportional increase in biomass-specific heat capacity with temperature (Dupont et al., 2014). The unit used to describe the biomass-specific heat capacity is in J/kg K. The biomass heat capacity from numerous wheat and wood samples varies from a low of 1300 J/kg K to a high of 2000 J/kg K from temperature that ranges from 313 to 353 K (Dupont et al., 2014). For biochar samples treated at 773 K of from the same wheat and wood samples reported above, the heat capacity ranges from 1127 to 1193 J/kg K, showing narrow range than the original biomass it came from. At much elevated temperature of 1073 K, the heat capacities ranged between 1130 and 1330 J/kg K.

Data for heat capacities of biomass and biochar are usually used for thermal conversion modeling studies and projects.

In many other publications (Blokhin et al., 2011), heat capacities or enthalpy of biomass samples may also be reported in units of J/mol K. Microcrystalline cellulose samples have thermodynamic heat capacity that ranges from 30.7 to 377 J/mol K corresponding to the temperature range from 50 to 580 K. The importance of this parameter is in modeling the thermodynamic response of biomass materials and predict its crystallinity index and more importantly in pyrolysis or gasification processes. The authors reported that the adiabatic temperatures of pyrolysis or gasification processes and the energetic characteristics of the products are a strong function of the heat capacity values over the range of pyrolysis or gasification temperature used. The resulting product evaluated is the synthesis gas produced. Hence, scientists may be able to model those thermal conversion processes by having an excellent data on their heat capacities as temperature is increased.

Thermal conductivity is the ability of the material to conduct or heat transfer. Heat transfer of the biomass material with high thermal conductivity is higher than that with lower thermal conductivity. Consequently, materials with high thermal conductivity are used to store heat energy, while those with low thermal conductivity are used as insulating material (Capareda, 2014). Thus, some biomass materials may be a good choice as insulation than heat storage. Thermal conductivity has a unit of W/mK [Btu/h-ft]. Biomass thermal conductivity values are used to model biomass in storage systems to determine its ability to conduct heat through the enclosure as well as its ability to store energy. In some instances, this parameter is used to prevent auto-ignition of storage bins for dry biomass.

The thermal conductivity of some biomass ranged from 0.099 W/mK for softwood to as high as 0.419 W/mK for maple biomass (Mason et al., 2016). The procedure to determining the thermal conductivity of solid biomass is similar to that used in polymers and plastics. A heat source is usually applied to the extremity of one reference piece and a heat sink applied to the opposite end of its counterpart. The axial temperature gradient across the two reference pieces is measured and the heat flow in each determined. The heat flow through the test piece is taken to be the average of that in the two reference pieces. Given the dimensions of the test piece and the measured temperature differential across it, the thermal conductivity is derived (Mason et al., 2016).

One conclusion from the establishment of thermal conductivities of biomass and biochar samples is that torrefied biomass (torrefied pine and black pellet) has significantly higher thermal conductivity compared to the natural wood materials. Likewise, the data for thermal conductivity may also be used for modeling self-ignition of different biomass materials.

## 5 Conclusion

The sustainable utilization of biomass resources is a strong function of how well one measures the biomass properties that are important for the targeted product or co-products to be produced. Mankind has been growing and utilizing biomass for fuels, energy, and products for thousands of years (Davison, 2012), but we do not have complete understanding of its properties until the biofuels bloom in recent years. Improved characterization techniques lead to improved biomass conversion and vice versa.

Now, scientists have made significant improvements in the conversion of particular product through the management of these biomass resources whether they are native species or genetically modified. Numerous biofuel plants have been slightly modified through breeding leading to improved bioconversion efficiencies.

Biomass is a complex feedstock. One may look at its complex components such as cellulose, hemicellulose, and lignin including minor components such as proteins and ash for possible biological conversion into liquid products such as bioethanol. Improved characterization of the biomass resources will lead to better preparation for pretreatment processes and ultimately lead to improved conversion efficiencies. The reverse is true. Better products require ideal feedstock and may signal the modification of plant parts to reduce wastage.

Biomass characterization illustrates the chemistry and relative structure of the biomass components. However, one needs to make the characterization process faster, efficient, and reliable. Biofuel refineries also need advanced real-time imaging techniques instead of time-consuming standard processes. Hence, spectral response of biomass is an important area of research in the near future. High-throughput characterization techniques must be developed using nondestructive processes such as those using ultraviolet, visible, near, and far-infrared as well as newer techniques used in special equipment such as Raman spectroscopy. There are also newer techniques such as expanded neutron beam techniques and many more.

There are now numerous biomass techniques that scientists are using to better understand more diverse and detailed biomass properties important for conversion. Plant breeders are now equipped with genome of some biomass that allows tailoring of biomass for a particular pathway. This will lead to further development of unique and newer biomass characterization techniques.

Finally, there is still a major knowledge gap in biomass characterization and subsequent conversion since the conversion of lignocellulosic biomass has not yet met the target performance and cost requirements for large-scale production and marketability (Foston and Ragauskas, 2012).

## Questions and problems

**Problem 1.1** Rice straw from Texas was dried, and the following data were gathered. Initial weight of the sample = 6.00 g and final weight of the sample = 5.478 g. Determine the moisture content on wet basis.

**Problem 1.2** Rice straw from Texas was dried and milled, and the following data were gathered. Initial weight of the "as received" sample = 6.00 g and final weight of the sample = 5.478 g. Determine the percent total solids of this sample.

**Problem 1.3** Summarize the total compositional analysis of microalgae biomass with the following data and determine the effectiveness of the total compositional analysis: extractives = 48.09%, protein = 23.84%, ash = 3.90%, lignin = 4.64%, and structural carbohydrates = 7.17%.

**Problem 1.4** Estimate the area required to establish a 11.355 million liters/year [3 million gallon/year] bioethanol plant from corn with a reported yield of 2001 L/ha [214 gal/acre].

**Problem 1.5** Determine the amount of $CO_2$ produced in metric tonnes per year for a 11.355 L/year of ethanol production assuming all sugar comes from glucose. Assume the density of glucose is around 1.54 g/cm$^3$, that of ethanol around 789.22 kg/m$^3$, and that of $CO_2$ around 1.98 kg/m$^3$.

**Problem 1.6** Determine the amount of oil extracted from Jatropha seeds. The initial amount of seed used was 250 g, and the amount of oil collected from the collection bottle was 106.5 g. The pressed meat weighed 136.75 g. Determine also the percent losses from this experiment.

**Problem 1.7** Determine the BMP efficiency of poultry litter as substrate. The following data are given: Poultry litter COD is 32,520 mg/L COD. The normalized methane production potential for poultry litter is 244.8 mL methane/g VS. The volatile solids (VS) for poultry litter are 40.3%.

**Problem 1.8** Determine the output power (units of MW) if wood chips with a heating value of 22 MJ/kg are converted into electrical power via combustion with an overall conversion efficiency of 60%. The material is being fed at a rate of 1500 kg/h.

**Problem 1.9** The heating value of a certain biomass was reported as 20 MJ/kg "as received." The moisture content of this biomass was 12%. Convert this heating value number into its HHV equivalent. The ash content of this biomass was also found to be around 13%. Determine the heating value on a dry-ash-free basis as well.

**Problem 1.10** Fresh ground rice straw samples were used to estimate the complete proximate analysis. Show the proximate values on wet basis and dry basis. The following are the data from the experiment:

**a.** initial fresh sample weight = 6.000 g
**b.** final sample weight after drying in convection oven = 5.478 g
**c.** initial weight of sample placed in tube furnace = 5.478 g
**d.** final weight of sample after treatment in tube furnace = 1.782 g

e. initial weight of the sample placed in the muffle furnace = 1.782 g
f. final weight of the sample after treatment in muffle furnace = 1.4639 g

**Problem 1.11** The results of ultimate analysis for a rice straw sample from Texas are given as follows: carbon = 39.9%, hydrogen = 6%, oxygen = 24.7%, nitrogen = 1.3%, and sulfur = 0.2%. The ash content of this sample was found to be 17.9%. A 100 g dried rice straw sample was used to establish the chemical formula of this biomass. Determine this "chemical formula" and set up the stoichiometric combustion equation. Determine also the air-to-fuel ratio by weight and by volume.

**Problem 1.12** Determine the slagging ($R_s$) and fouling ($R_f$) factors for coal samples designated as letter E in the following table. Describe the slagging and fouling types for this coal sample based on Table 1.2.

| % Component | A | B | C | D | D/E | E |
|---|---|---|---|---|---|---|
| Acids | (75.02%) | (82.43%) | (79.68%) | (76.15%) | (71.81%) | (68%) |
| $Al_2O_3$ | 25.4 | 20.4 | 30.8 | 26.6 | 22.8 | 20.9 |
| $SiO_2$ | 48.5 | 61 | 47.1 | 48.4 | 48 | 46.1 |
| $TiO_2$ | 1.12 | 1.03 | 1.78 | 1.15 | 1.01 | 1.0 |
| Bases | (18.01%) | (14.87%) | (12.33%) | (19.84%) | (21.55%) | (25.28%) |
| MgO | 2.24 | 1.61 | 1.65 | 2.85 | 2.5 | 1.11 |
| $P_2O_5$ | 0.52 | 0.30 | 1.44 | 0.61 | 0.56 | 0.37 |
| $K_2O$ | 2.13 | 1.96 | 0.66 | 2.39 | 0.79 | 0.88 |
| CaO | 2.13 | 3.07 | 6.73 | 4.32 | 4.97 | 4.87 |
| $Fe_2O_3$ | 11.1 | 7.6 | 3.15 | 9.11 | 11.1 | 16.6 |
| $Na_2O$ | 0.41 | 0.63 | 0.14 | 1.17 | 2.19 | 1.82 |
| $SO_3$ | 2.4 | 2.69 | 3.76 | 3.18 | 3.91 | 3.88 |
| %Sulfur | 0.81 | 0.77 | 0.47 | 0.80 | 1.35 | 2.45 |

**Problem 1.13** Determine the alkali index (AI) in English units, base-to-acid ratio ($R_{b/a}$), and bed agglomeration index for the complete ash analysis of straw given in the table below partially taken from Lachman et al. (2021) with high heating value of biomass found from Lisy et al. (2020). Compare the potential slagging and fouling using $R_s$ and $R_f$ values. As additional exercise, do the calculations for the rest of biomass.

| % Component | Spruce | Softwood | Hardwood | Amaranth | Hay | Straw |
|---|---|---|---|---|---|---|
| Acids | (3.27%) | (10.28%) | (3.39%) | (2.87%) | (40.48%) | (50.28%) |
| $Al_2O_3$ | 0.51 | 1.56 | 0.55 | 0.22 | 3.91 | 4.97 |
| $SiO_2$ | 2.72 | 8.62 | 2.8 | 2.38 | 36.3 | 45 |
| $TiO_2$ | 0.04 | 0.10 | 0.04 | 0.27 | 0.27 | 0.31 |
| Bases | (56.46%) | (52.38%) | (57.01%) | (61.11%) | (39.01%) | (32.62%) |
| MgO | 10.9 | 3.03 | 4.34 | 6.74 | 4.28 | 3.20 |
| $P_2O_5$ | 1.99 | 5.55 | 5.47 | 3.7 | 5.72 | 5.07 |
| $K_2O$ | 12.5 | 8.71 | 9.81 | 14.4 | 20.9 | 18.8 |
| CaO | 32.1 | 39.3 | 42.2 | 39.2 | 11.6 | 8.1 |
| $Fe_2O_3$ | 0.71 | 0.94 | 0.49 | 0.48 | 1.58 | 1.94 |
| $Na_2O$ | 0.25 | 0.40 | 0.17 | 0.29 | 0.65 | 0.58 |
| $SO_3$ | 2.47 | 1.86 | 2.17 | 1.87 | 5.05 | 3.22 |
| %Sulfur | 0.99 | 0.74 | 0.87 | 0.75 | 2.02 | 1.29 |
| HV (MJ/kg) | 18.21 | 18.20 | 18.04 | 16.31 | 15.79 | 17.24 |
| HV (Btu/lb) | 7844 | 7841 | 7774 | 7027 | 6803 | 7427 |
| % Ash in biomass | 0.33 | 3.01 | 2.44 | 7.70 | 17.73 | 6.64 |
| % Alkali in biomass | 12.75 | 9.11 | 9.98 | 14.69 | 21.55 | 19.38 |

# References

Blokhin, A.V., Voitkevich, O.V., Kabo, G.J., Paulechka, Y.U., Shishonok, M.V., Kabo, A.G., Simirsky, V.V., 2011. Thermodynamic properties of plant biomass components. Heat capacity, combustion energy, and gasification equilibria of cellulose. J. Chem. Eng. Data 56, 3523–3531. ACS Publications Center, Washington, DC, USA.

Capareda, S.C., 1990. Studies on Activated Carbon Produced from Thermal Gasification of Biomass Wastes (PhD dissertation). Biological and Agricultural Engineering, Texas A&M University, College Station, TX. August 1990.

Capareda, S.C., 2014. Introduction to Biomass Energy Conversions. CRC Press Taylor and Francis Group, Boca Raton, FL, ISBN: 978-1-4665-1333-4 (Hardback).

Dagoumas, A.S., Barker, T.S., 2010. Pathways to a low carbon economy for the UK with the macro-econometric E3MG model. Energy Policy 38 (6), 3067–3077.

Davison, B.H., 2012. Overview: the increasing importance and capabilities of biomass characterization. Ind. Biotechnol. 8 (4), 1–2.

Dupont, C., Chiriac, R., Gauthier, G., Toche, F., 2014. Heat capacity measurements of various biomass types and pyrolysis residues. Fuel 115, 644–651.

Foston, M., Ragauskas, A.J., 2012. Biomass characterization: recent progress in understanding biomass recalcitrance. Ind. Biotechnol. 8 (4), 191–208.

Holliger, C., Alves, M., Andrade, D., Angelidaki, I., Astals, S., Baier, U., Bougrier, C., Buffière, P., Carballa, M., de Wilde, V., Ebertseder, F., Fernández, B., Ficara, E., Fotidis, I., Frigon, J.-C., de Laclos, H.F., Ghasimi, D.S.M., Hack, G., Hartel, M., Heerenklage, J., Horvath, I.S., Jenicek, P., Koch, K., Krautwald, J., Lizasoain, J., Liu, J., Mosberger, L., Nistor, M., Oechsner, H., Oliveira, J.V., Paterson, M., Pauss, A., Pommier, S., Porqueddu, I., Raposo, F., Ribeiro, T., Pfund, F.R., Strömberg, S., Torrijos, M., van Eekert, M., van Lier, J., Wedwitschka, H.,

Wierinck, I., 2016. Towards a standardization of biomethane potential tests. Water Sci. Technol. J. 71 (11), 2512–2522. Creative Commons Publishing, Mountain View, CA.

ISO 1364-1: Determination of Inhibition of Activity of Anaerobic Bacteria. Part 1: General Test. International Organization for Standardization, Geneva Switzerland, n.d.

ISO 1364-2: Determination of Inhibition of Activity of Anaerobic Bacteria. Part 2: Test at Low Biomass Concentration. International Organization for Standardization, Geneva Switzerland, n.d.

Kemp, W.H., 2006. Biodiesel Basics and Beyond: A Comprehensive Guide to Production and Use for the Home and Farm. Aztext Press, Ontario, Canada.

Lachman, J., Balas, M., Lisy, M., Lisa, H., Milcak, P., Elbl, P., 2021. An overview of slagging and fouling indicators and their applicability to biomass fuels. Fuel Process. Technol. 217, 1–10.

Lata, K., Mande, S.P., 2009. Bioenergy resources. In: Kishore, V.V.N. (Ed.), Renewable Energy Engineering and Technology: Principles and Practice. Earthscan Publishers, London, UK, pp. 593–672.

Lisy, M., Lisa, H., Jecha, D., Balas, M., Krizan, P., 2020. Characteristic properties of alternative biomass fuels. Energies 13 (6), 1448. https://doi.org/10.3390/en13061448.

Maglinao, A.L., Capareda, S.C., 2010. Predicting fouling and slagging behavior of dairy manure (DM) and cotton gin trash (CGT) during thermal conversion. Trans. ASABE 53 (3), 903–909.

Mason, P.E., Darvell, L.I., Jones, J.M., Williams, A., 2016. Comparative study of the thermal conductivity of solid biomass fuels. Energy Fuel 30, 2158–2163.

Miles, T.R., Miles Jr., T.R., Baxter, L.L., Bryers, R.W., Jenkins, B.M., Oden, L.L., 1995. Alkali Deposits Found in Biomass Power Plants: A Preliminary Investigation of their Extent and Nature. Summary report for the National Renewable Energy Laboratory (NREL) of the US Department of Energy (USDOE), Golden, Colorado, USA.

Moody, L.B., Burns, R.T., Bishop, G., Sell, S.T., Spacic, R., 2011. Using biochemical methane potential assay to aid in co-substrate selection for co-digestion. Appl. Eng. Agric. 27 (3), 433–439.

Quentin, A.G., Pinkard, E.A., Ryan, M.G., Tissue, D.T., Baggett, L.S., Adams, H.D., Maillard, P., Marchand, J., Landhäusser, S.M., Lacointe, A., Gibon, Y., Anderegg, W.R.L., Asao, S., Atkin, O.K., Bonhomme, M., Claye, C., Chow, P.S., Clément-Vidal, A., Davies, N.W., Dickman, L.T., Dumbur, R., Ellsworth, D.S., Falk, K., Galiano, L., Grünzweig, J.M., Hartmann, H., Hoch, G., Hood, S., Jones, J.-E., Koike, T., Kuhlmann, I., Lloret, F., Maestro, M., Mansfield, S.D., Martínez-Vilalta, J., Maucourt, M., McDowell, N.G., Moing, A., Muller, B., Nebauer, S.G., Niinemets, Ü., Palacio, S., Piper, F., Raveh, E., Richter, A., Rolland, G., Rosas, T., Joanis, B.S., Sala, A., Smith, R.A., Sterck, F., Stinziano, J.R., Tobias, M., Unda, F., Watanabe, M., Way, D.A., Weerasinghe, L.K., Wild, B., Wiley, E., Woodruff, D.R., 2015. Non-structural carbohydrates in woody plants compared among laboratories. Tree Physiol. 35, 1146–1165. https://doi.org/10.1093/treephys/tpv073.

Templeton David, W., Wolfrum, E.J., Yen, J.H., Sharpless, K.E., 2016. Compositional analysis of biomass reference materials: results from an interlaboratory study. Bioenergy Res. 9, 303–314. https://doi.org/10.1007/s12155-015-9675-1.

Vamvuka, D., Zografos, D., 2004. Predicting the behavior of ash from agricultural wastes during combustion. Fuel 83 (14–15), 2051–2057.

Winegartner, E.C. (Ed.), 1974. Coal Fouling and Slagging Parameters. Report prepared by the American Society of Mechanical Engineers (ASME) Research Committee on Corrosion and Deposits from Combustion Gases. Copyright by the ASME, Three Park Avenue, NY, 1974.

# Biomass carbonization technologies

Manuel Raul Pelaez-Samaniego[a,b], Sohrab Haghighi Mood[c], Jesus Garcia-Nunez[d], Tsai Garcia-Perez[a], Vikram Yadama[b], and Manuel Garcia-Perez[c]

[a]Department of Applied Chemistry and Systems of Production, Faculty of Chemical Sciences, Universidad de Cuenca, Cuenca, Ecuador. [b]Composite Materials and Engineering Center, Washington State University, Pullman, WA, United States. [c]Department of Biological Systems Engineering, Washington State University, Pullman, WA, United States. [d]CENIPALMA, Bogota, Colombia

## 1 Introduction

"Carbonization" is the appropriate name for thermochemical conversion technologies resulting in the production of high yields of carbonaceous products, although "pyrolysis" is still commonly used as an equivalent term. If the technology is intended to obtain "thermal depolymerization products" such as bio-oil or wood vinegar, the technology should be called "pyrolysis." Depending on operational conditions, thermochemical processes are classified into slow pyrolysis, fast pyrolysis, hydrothermal carbonization, gasification, torrefaction, and flash carbonization (Garcia-Nunez et al., 2017). To start a discussion on carbonization technologies, it is appropriate first to discuss the terms of "*char*," "charcoal," and "biochar." The term *charcoal* or "*char*" in its short form should be used to indicate a carbonaceous product from thermochemical processes, which is intended for cooking, heating, or industrial applications (e.g., smelting) (Lehmann and Joseph, 2015; Brown et al., 2015). *Biochar* is a carbonaceous material, intended for environmental management (i.e., air, water, and wastewater treatment) and/or soil amendment (Lehmann and Joseph, 2015; Brown et al., 2015). In this chapter, both terms *charcoal* and *biochar* are employed, depending on the context. Another term that is related to synthetic carbonaceous materials is *hydrochar*, also referred to as *hydrothermal carbon* (Titirici, 2012). This is a solid product of hydrothermal carbonization (HTC) or hydrothermal liquefaction (HTL) (Liu and Zhang, 2009; Han et al., 2016; Takaya et al., 2016).

The production of carbonaceous materials has been critical for human development. Wood carbonization to obtain charcoal has existed for as long as humankind history has been recorded. Charcoal was used for painting caves around 38,000 years ago (Antal et al., 2003; Antal and Grønli, 2003). The early Egyptian civilization employed charcoal for metal processing (Lucas, 1948). The Romans produced charcoal for iron melting around 2000 years ago (Ginzel, 1995; Historic England, 2018; Carrari et al., 2017). Most northern and central European countries reduced their dependency on charcoal by the 19th century due to the rapidly increasing and widespread use of coal, but the Mediterranean area was still producing charcoal, using ancient technologies during the Industrial Revolution and even during the 1950s, though it is still in practice today in some remote mountain areas (Carrari et al., 2017). Records also show that China used charcoal for medical purposes around 2000 years ago (Chen et al., 2019) and for iron melting by 300 BCE (Westra, 2020). In China, charcoal was a common fuel during the 12th century (Zhu et al., 2016). In Japan, charcoal production goes back to 12,000 BCE (Rikumo, 2016). Charcoal was the prime fuel source for the smelting of copper, tin, bronze, silver, and iron by 1000 BCE (Colwell, 2018). During the Edo period (by the beginning of the 17th century), charcoal was employed as a "natural purifying" agent (Rikumo, 2016), which is, probably, one of the few records on the early uses of this material for environmental purposes.

In the United States, charcoal was largely produced since the colonial times through part of the 20th century using European techniques that were brought by colonizers, especially for iron processing. "Charcoaling with mounds faded from our landscape around the turn of the 19th century" (Hvfarmscape, 2020). "Livingston's Ancram iron furnace, founded ca. 1749, burned charcoal; and smaller-scale operations likely existed before that. The traditional round charcoal mound was in use in England (where it was called a 'clamp') at least as early as the Middle Ages." (Hvfarmscape, 2020). Turf-covered stacks (the "clamps") were used for charcoal production in the United Kingdom since medieval times until the beginning of the Industrial Revolution (Hazel et al., 2017). In the wood distillation industry, residual small charcoal particles were used as a filtering medium (Klar and Rule, 1925). Rolando (1992) and Fitzpatrick (2020) are two very illustrative sources on the charcoal development industry in the United States.

Charcoal was also the alternative to oil-derived fuels for running the so-called gas producer vehicles during the World War II, due to the lack of oil-derived fuels, especially in Germany,

France, and Sweden (Ryan, 2009). The "Imbert gasifier" worked with charcoal (and sometimes directly with wood) as fuel. In Sweden, despite safety risks related to the operation of these gasification systems (due to the presence of CO), up to 25% of the vehicle fleet consisted of gas producer vehicles (Ryan, 2009).

Indigenous people in the Amazonian basin of Brazil used charcoal for improving soil quality more than a thousand years ago (Glaser et al., 2001). Even nowadays, the charcoal residues (fines) in some parts of the Amazonian region are intentionally added to soil to improve soil fertility and to increase agricultural yields (Coomes and Miltner, 2016). These examples show the importance of charcoal on the humankind progress and the perspectives for further use of this material. Table 2.1 shows some important milestones in the development of the pyrolysis technology, with emphasis on the production of charcoal.

Wood has been the preferred raw material for charcoal manufacture. Nevertheless, the use of other types of lignocellulosic materials such as agro-industrial and forest residues is gaining attention. In addition to agricultural residues (e.g., rice straw, corn stover, wheat straw) (Park et al., 2014), other raw materials for modern charcoal production include aquatic biomass species (Wang and Brown, 2013; Trinh et al., 2013), herbaceous crops (Gómez and Puigjaner, 2005), and waste biomass such as poultry manure (Zolfi-Bavariani et al., 2016), digested diary fiber (Ferraz et al., 2016; Pelaez-Samaniego et al., 2018), and food waste (Elkhalifa et al., 2019), to mention a few.

Despite the enormous and increasing attention of science on biochar for environmental management and soil amendment, information on quantities produced for these purposes is not abundant (FAO, 2017). The UNdata (2020) is the main source of information available on the global wood charcoal production, trade, and consumption. In 2018, the global wood charcoal production was approximately 53.2 million tons. Assuming an optimistic global efficiency of carbonization of 20%, the requirement of wood and wood residues for charcoal was larger than 266 million tons. Africa produces more than 64% of the world's wood charcoal (UNdata, 2020). Nigeria and Ethiopia, with 8.6% and 8.5% of the global production, respectively, are Africa's main charcoal producers (UNdata, 2020). According to the UNdata (2020), Brazil is the largest world wood charcoal producing country, with around 12% of the total. The main uses of charcoal are as raw material for value-added materials (e.g., activated carbon, silicon carbide, and catalyst), cooking fuel, and reducing agent. Compared to direct biomass combustion, charcoal used as cooking fuel results in the release of much lower quantities of

Table 2.1 Some historical milestones in the production and use of charcoal.

| Year(s) | Historical milestone |
| --- | --- |
| 35,000–26,000 BCE | Grotte Chauvet art with charcoal (Antal et al., 2003) |
| ~12,000 BCE | Japanese culture used charcoal as fuel |
| ~4000 BCE | Egyptians used charcoal for metallurgy (Lucas, 1948) |
| ~2000 years ago | Evidences of charcoal in China for medical purposes (Chen, 2019) |
| ~2000 years ago | The Romans produced charcoal for iron melting (Ginzel, 1995) |
| 500–1400s (Middle Ages) | Traditional round charcoal mounds ("clamp") used in England |
| ~1000 | Indigenous people in the Amazonian region used charcoal for soil amendment (Glaser et al., 2001) |
| ~1100 | Charcoal largely used in China as fuel |
| ~1500s | Charcoal production for metal processing in Brazil |
| 1556 | Detailed description of charcoal in Europe in metallurgy |
| ~1603 | Edo period in Japan. Use of charcoal as a natural purifying material |
| 1661 | R. Boyle describes wood distillation |
| 1600s–1800s | Charcoal used in the United States for iron processing |
| 1798 | Industrial wood distillation begins |
| 1850 | Wood distillation industry expands rapidly in Europe and North America |
| 1850 | Horizontal retorts were used mainly by Germany, England, and Austria, while France developed vertical retorts (Klar and Rule, 1925) |
| 1920–1960 | Char production via slow pyrolysis for metallurgy in Europe and the United States |
| 1930–1960s | First records on the use of portable metal kilns in the United States and Europe |
| 1930–1940s | Charcoal for running internal combustion engines (via charcoal gasification) |
| 1980s | Impetus on fast pyrolysis for liquid production |
| 1990s | Yield of fixed carbon is suggested in substitution of charcoal mass yield to measure the carbonization efficiency |
| 2000 | Oil prices and global warming create demand for biofuels and biochar |
| 2005 | Abundant studies on charcoal for carbon sequestration, soil amendment, and environmental remediation start to appear |

harmful smoke, which explains its large use as fuel. Charcoal's low sulfur content is the main reason for using it in the steel and iron industry. Most of the publications in last decade are related to biochar use as a tool to fight global warming.

Although there are excellent reviews on biomass pyrolysis technologies and the related economic and environmental aspects (e.g., Antal et al., 1990; Antal and Grønli, 2003; Mohan et al., 2006; Meyer et al., 2011; Jahirul et al., 2012; Patel et al.,

2016; Garcia-Nunez et al., 2017), there are only a limited number of publications (e.g., Emrich, 1985; FAO, 1987; Domac and Trossero, 2008; Panwar et al., 2019; Rodrigues and Junior, 2019) devoted to review equipment specifically developed for carbonization of lignocellulosic biomass. The objective of this work is to summarize old and new concepts and designs for carbonization reactors. Herein, we review reactions and operation parameters responsible for charcoal yield and properties and the evolution of the biomass carbonization technology.

## 2 Char physicochemical properties

Char has unique properties such as high surface area, fine porous structure, and surface functional groups (Tan et al., 2015), which determine its performance in different applications. Thus, knowledge on char physicochemical properties is critical to understand performance. Thermochemical processing conditions, feedstock type, and post-carbonization modification/activation are important factors affecting char properties (Shakoor et al., 2021). A broad range of feedstocks such as agricultural waste, woody biomass, municipal solid wastes, and food animal wastes with different properties have been used to produce char (Shakoor et al., 2021). The resulting materials are polyaromatic ring systems with different sizes containing a variety of functional groups (e.g., carbonyl, hydroxyl, and carboxyl). C, O, N, and H are the most important heteroatoms in char and can be quantified by elemental analysis (Haghighi Mood et al., 2021). The O and H content decreases and the C content increases as carbonization temperature increases (Zhao et al., 2017; Chaves Fernandes et al., 2020; Jafri et al., 2018). The removal of O results in a more hydrophobic char (Zhao et al., 2017). Raman spectroscopy and Fourier transform infrared (FTIR) spectroscopy are used to analyze chemical functionalities of char. For instance, the FTIR peaks between 3200 and 3500 $cm^{-1}$ can be used to follow how —OH stretching decreases as carbonization temperature increases. The dehydration of methylene group and disappearance of peaks corresponding to C=O and O—H functionalities can be observed at temperatures above 500°C (Kumar et al., 2020). Char structure has a crystalline and amorphous phase that can be evaluated using X-ray diffraction (XRD).

Char proximate analysis determined by thermogravimetric methods can be expressed in terms of moisture, volatiles, fixed carbon, and ashes. The elemental composition of ash is an important parameter influencing the char application. For example, the presence of Mg and Ca in char can contribute to adsorption of phosphate in aqueous phase (Haghighi Mood et al., 2020). The composition of ash can be detected using inductively coupled plasma (ICP) spectrometry. ICP can measure the concentration of K, Al, Ca, Cr, Ni, Cu, Fe, Zn, Pb, Ba, Cr, Mo, Mn, Ni, Mg, Mn, P, and Na in ash. Crystals of metal oxides (such as MgO and CaO) can be determined by XRD analysis. The surface area of char typically increases with increasing temperature and residence time due to volatilization of organic materials with increasing production temperature (Suliman et al., 2016). However, after certain temperature, the surface area decreases because of the porous structure collapse when too much matter is removed (Ayiania et al., 2019). Fu et al. (2019) studied the effect of pyrolysis temperature on morphology of char derived from food wastes. The results indicated that char produced at 200°C has a smooth surface structure and, as temperature increased to 300°C, the crack and pore structures began to develop. Similar results have been reported for torrefied softwood (Pelaez-Samaniego et al., 2014).

## 3 Effect of operational variables on the yield and quality of charcoal

The ideal method for char production should consider the possibility of a complete use of the by-products of carbonization as a way to reduce residues and improve the charcoal yields and economics of the carbonization process. Charcoal mass yield (calculated as $y_{char} = m_{char}/m_{biom}$, where $m_{char}$ is the mass of charcoal and $m_{biom}$ is the mass of the raw biomass, both in dry basis) has been the parameter normally used for evaluating the carbonization processes (Antal and Grønli, 2003; Garcia-Nunez et al., 2017). The following subsections present a brief discussion on the effect of pyrolysis operating conditions on charcoal yield and properties.

### 3.1 Effect of temperature

Temperature is one of the key parameters in thermochemical process affecting physicochemical properties of char. Generally, as temperature increases in thermochemical process (e.g., pyrolysis), the solid residue yield decreases due to degradation of lower and unstable parts of biomass macromolecules with the removal of oxygenated functional groups (hydroxyl, carboxylic, and carbonyl groups) (Haghighi Mood et al., 2020; Suliman et al., 2016).

This is attributed to thermal decomposition of lignocellulosic materials at this range of temperature and volatile release from converting biomass (Paris et al., 2005; Suliman et al., 2016; Chatterjee et al., 2020).

The relative char ash and fixed carbon content increases with increasing pyrolysis temperature, while volatile content in char decreases with increasing temperature (Tag et al., 2016; Enders et al., 2012) Likewise, O and H content decreases and C increases as pyrolysis temperature increases. Leng et al. (2021) explored the influence of pyrolysis temperature on char surface area and porosity. The surface area and porosity of char are highly affected by pyrolysis temperature. The pyrolysis of biomass at temperature below 400°C leads to slight change in surface area and total pore volume due to incomplete devolatilization of volatile compounds. Conversely, at temperatures between 400°C and 500°C, more volatiles are released from converting biomass and amorphous carbon converts to crystalline carbon through condensation reactions, which results in generating pores and larger surface are. As temperature raises, the surface area continues to increase. However, depending on the type of lignocellulosic raw material, increasing the pyrolysis temperature to above 700°C could cause surface area and pore volume decrease. This could be attributed to structural ordering and merging of pores, which lead to collapse of pore structure and decrease in surface area (Kumar et al., 2017; Ayiania et al., 2019). Pyrolysis temperature can impact aromaticity and polarity of char. As char production temperature increases, H/C ratios decrease, indicating that aromaticity of char increases (Xiao et al., 2016). Increasing pyrolysis temperature also decreases O/C ratio, showing decrease of char polarity (Hassan et al., 2020). The pH of the char is also influenced by pyrolysis temperature. High pH value of char produced at high pyrolysis temperatures is attributed to high concentration of non-pyrolyzed inorganic components in the feedstocks and the formation of basic surface oxides in char structure (Rehrah et al., 2014). Moreover, pyrolysis temperature has a direct impact on acidic and basic functional groups. Increasing pyrolysis temperature causes decrease of acidic functional groups (e.g., carboxylic groups). The reverse trend can be seen for basic functional groups (e.g., quinone). The number of basic functional groups also raises as the pyrolysis temperature increases (Shaaban et al., 2014).

## 3.2 Effect of ash content

The presence of ash in feedstock can affect the properties of char produced via thermochemical conversion processes. Biomass with high ash content generally has higher char yield compared to low ash content feedstock. Ash content in feedstocks can contribute to the formation of char since alkaline earth metals such as K, Ca, and Mg have a catalytic effect favoring cross-linking reactions leading to char formation (Yaman, 2004; Song and Guo, 2012). Surface area and total pore volume can be influenced by ash content (Ronsse et al., 2013). The presence of ash in char can cause pore blockage, resulting in lower surface area due to fusion of molten ash, which fills up pores in the char (Ronsse et al., 2013; Leng et al., 2021). For instance, the higher surface area of woody biomass compared to manure-based char could be explained by higher content of ash in manure (Haghighi Mood et al., 2021; Leng et al., 2021). The presence of ash in char structure can inversely affect the fixed carbon content. Fixed carbon in char derived from biomass with low ash content is higher than that in char produced from biomass with high ash content (e.g., char derived from manure) (Domingues et al., 2017). Na and K are also known gasification catalysts. Thus, they contribute to increase char gasification activity and result in the formation of additional pores and surface areas during char activation.

High ash content in charcoal could limit some uses. Pellets for domestic consumption, for instance, require ash content <1%, according to the Pellet Fuel Institute. Similar requirements may need to comply charcoal and/or charcoal pellets for other uses. The carbonization method and raw material not always guarantee low ash content. For instance, charcoal produced in Brazil can contain ash from 1.5% to almost 4%, as seen in Table 2.2, depending on the pyrolysis conditions.

**Table 2.2 Example of charcoal properties produced in Brazil using eucalyptus as raw material.**

| Carbonization temperature (°C) | Immediate analysis (dry basis) (%) | | | Reference |
| --- | --- | --- | --- | --- |
| | Fixed carbon | Volatiles | Ash content | |
| 450 | 75.06 | 21.03 | 3.91 | Froehlich and Moura (2014) |
| 550 | 86.53 | 10.12 | 3.33 | |
| 700 | 89.82 | 7.20 | 2.93 | |
| Not known | 78%–80% | – | – | Pinheiro (2016) |

## 3.3 Effect of acids

Acid treatment can affect the properties of char. Sulfuric acid, nitric acid, phosphoric acid, oxalic acid, hydrochloric acid, and citric acid have been used to improve porous structure and acidity of char surface (Rajapaksha et al., 2016). The acids catalyze the formation of water. The dehydration products are precursors of char production. Char obtained from pretreated biomass with phosphoric acid showed surface area and pore volume improvement and higher yields (Chu et al., 2018). Micropores and mesopores can be generated in modified chars through the interaction between phosphoric acid and carbon structure (Guo and Rockstraw, 2006; Zhao et al., 2017; Chu et al., 2018). Phosphoric acid pretreatment can also increase surface functional groups and aromatization (Taha et al., 2014). Generally, high content of carbon in biomass structure is lost in pyrolysis process because of thermal decomposition and volatilization (Zhao et al., 2014). However, pretreatment of biomass with phosphoric acid can improve carbon retention of char and reduce carbon loss during pyrolysis. In the pyrolysis processes, phosphoric acid can react with biomass to form stable structures (e.g., C—O—$PO_3$ or $(CO)_2PO_2$). Treatment of biomass with phosphoric acid before pyrolysis showed that this acid could reduce carbon loss during carbonization and decrease the microbial mineralization (Zhao et al., 2014). Acid pretreatment can modify biomass structure and introduce functional groups in the resulting char. Char obtained from pretreated eucalyptus sawdust with citric, tartaric, and acetic acids showed high surface carboxyl functional group (Sun et al., 2015).

## 3.4 Effect of heating rate

Pyrolysis heating rate is a key factor that affects char properties. Pyrolysis can be categorized as slow, fast, and flash pyrolysis depending mainly on heating rate. In slow pyrolysis, low heating rate (e.g., 0.1–0.8°C/s) and long residence time (minutes to days) are applied to maximize char production (Roy and Dias, 2017). In general, char yield in slow pyrolysis is higher than that in fast pyrolysis. In slow pyrolysis, decomposition of biomass occurs under low heating rate and long residence time, thus allowing organic compounds to be repolymerized leading to high solid yield (Kan et al., 2016). The surface area and total pore volume are influenced by heating rate. As heating rate increases, the char surface area and pore volume increase. Then, after a certain heating rate, the reverse trend occurs, and surface area and pore

**Table 2.3** Yield char produced at different heating rates.

| Feedstock | Heating rate (°C/min) | Process temperature (°C) | Yield (wt%) | Reference |
|---|---|---|---|---|
| Tea waste | 5 | 500 | 34.9 | Uzun et al. (2010) |
| Tea waste | 300 | 500 | 35.1 | |
| Tea waste | 500 | 500 | 34.9 | |
| Tea waste | 700 | 500 | 27 | |
| Rapeseed stems | 1 | 650 | 25.4 | Zhao et al. (2018) |
| Rapeseed stems | 5 | 650 | 27 | |
| Rapeseed stems | 10 | 650 | 26.8 | |
| Rapeseed stems | 15 | 650 | 26 | |
| Rapeseed stems | 20 | 650 | 25.1 | |

volume decrease at very high heating rate (Leng et al., 2021). Uzun et al. (2010) explored the effect of heating rate on char yield (Table 2.3), and the results indicated that as heating rate increased from 5 to 700 K/min, the char yield decreased from 34.3% to 27.1%. Zhao et al. (2018) studied the effect of pyrolysis heating rate on rapeseed stem-derived char properties (Table 2.3), and the results indicated that the char yield first increased with increasing heating rate (from 1°C/min to 5°C/min) and then decreased with increasing heating rate (from 5°C/min to 20°C/min). Moreover, the increasing trend of surface area can be observed as the heating rate increased from 10°C/min to 30°C/min, but the surface area decreased as the heating rate increased to 50°C/min.

### 3.5 Effect of pressure

Increasing pressure in pyrolysis process can improve formation of char and influence the morphology of char (Cetin et al., 2005). Qin et al. (2020) studied the effect of increasing pressure on properties of char derived from pine nutshell. Their results showed that char yield, oxygenated functional groups, and combustion characteristics can be enhanced by increasing the pyrolysis pressure. However, increasing pyrolysis pressure can have a negative effect on char surface area. Melligan et al. (2011) found that increasing pyrolysis pressure from 0.1 to 2.6 MPa (1 to 26 bar) resulted in decreasing surface area of char derived from *Miscanthus* × *giganteus* from 161.7 to 0.137 $m^2$/g.

## 4 Carbonization reactions

Biomass is composed of three main compounds: cellulose, hemicellulose, and lignin. Cellulose is a polysaccharide, which contains a linear D-glucose chain linked to β-(1,4)-glycosidic bonds to each other (Haghighi Mood et al., 2013). Hemicellulose is branched heterogeneous biopolymer consisting of pentoses (xylose and arabinose), hexoses (mannose), sugar acid (glucuronic acid), and organic acids (Saha et al., 2009). Lignin is a heterogeneous biopolymer, which is mainly composed of aromatic structures including p-coumaryl, coniferyl, and sinapyl alcohols (Wang et al., 2017). Cellulose, hemicellulose, and lignin behave differently under thermochemical conversion process (e.g., slow pyrolysis). As temperature starts to increase, thermal cleavage of cellulose, hemicellulose, and lignin bonds occurs (primary thermal decomposition). Secondary thermal decomposition reactions can occur to form volatiles. The species formed might undergo more reactions such as condensation and polymerization to form products such as char. Pyrolysis begins with biomass drying, and, as temperature raises, cellulose, hemicellulose, and lignin undergo thermal degradation. Hemicellulose and cellulose are decomposed in the temperature range of 220–315°C and 314–400°C, respectively. However, lignin decomposition occurs in a broader temperature range (160°C to 900°C) (Yang et al., 2007). Differential thermogravimetry (DTG) curves of lignin in air and nitrogen environment are almost similar at temperatures up to approximately 300°C, meaning that reactions typical of pyrolysis occur (releasing volatiles) and that lignin heated up to these temperatures, even under air conditions, does not undergo significant oxidation (Pelaez-Samaniego et al., 2013).

The main phases of conversion of cellulose in pyrolysis process include formation of the active cellulose or anhydrocellulose, depolymerization, and the char production process (Akhtar et al., 2018). Pyrolysis of cellulose starts with cleavage of the glycosidic groups and dehydration is the first degradation reaction (Zhang et al., 2011). The glycosidic bonds in cellulose structure tend to be cleaved as temperature increases (Wang et al., 2017). Anhydrocellulose is formed after dehydration reactions. Most of the cellulose weight loss occurs during dehydration reactions at temperature below 300°C, and dehydration reaction is considered as an important reaction for char formation (Collard and Blin, 2014). As temperature increases above 300°C, anhydro-oligosaccharides and anhydro-saccharides can be formed due to depolymerization of cellulose (Wang et al., 2012). The structure of char is altered significantly from 270°C to 400°C. Generally, cellulose charring process occurs at temperature between 380°C and 800°C (Collard and Blin, 2014). The initial stage of hemicellulose pyrolysis is dehydration and depolymerization of structure and formation of oligosaccharides, followed by cleavage of the glycosidic bonds in the xylan chain, resulting in production of depolymerized compounds after rearrangement reactions (Shen et al., 2010; Ghodake et al., 2021). The intermediate products are subjected to dehydration, aromatization, decarboxylation, decarboxylation, condensation, and finally, intramolecular rearrangement. The hemicellulose charring process occurs in the temperature range of 320–800°C (Collard and Blin, 2014; Shen et al., 2010; Zhou et al., 2017; Ghodake et al., 2021). The pyrolysis of lignin is more complicated compared to cellulose and hemicellulose due to complex structure of lignin. The yield of char derived from lignin is higher than that from cellulose and hemicellulose (Azadi et al., 2013). Lignin pyrolysis begins with cleavage of weaker bonds such as hydrogen bindings. As temperature increases, cleavage of stronger bonds such as β-O-4 linkages occurs (Liu et al., 2015). Increasing temperature to 300°C leads to unstable C—C bonds in alkyl chain. Most of the bonds between monomers are cleaved at temperature higher than 450°C (Sharma et al., 2004; Nakamura et al., 2008). Cleavage of β-O-4 linkage leads to the formation of free radicals. The reaction of these free radicals causes polyaromatic char to be formed through repolymerization of the radicals at temperatures higher than 350°C (Liu et al., 2015).

## 5 Equipment for carbonization processes

Charcoal production methods have not evolved substantially in the last thousand years. One of the most common methods for producing charcoal is by employing pit kilns, an ancient technology that is still used in several countries worldwide (Antal and Grønli, 2003; Pelaez-Samaniego et al., 2008; Camou-Guerrero et al., 2014; Rodrigues and Junior, 2019). Pit kilns are characterized by a low efficiency and the release of pollutant by-products to the environment (Antal et al., 2003; Pelaez-Samaniego et al., 2008). The use of slightly improved kilns, such as the "hot-tail" kiln that is largely employed in Brazil, has helped to increase the yields of carbonization, but these are still below expected yields. Despite the development of other carbonization kilns in the last 150 years and the large experience on their design and use, high investments required for carbonization equipment have hindered the adoption of more efficient systems, especially in developing countries (FAO, 2017). New research and demonstration plants are encouraging the carbonization industry to look for improved equipment as a way to increase efficiency and reduce the negative footprint of wood carbonization using old methods. Still, some companies are reproducing old concepts (i.e., some developed almost 200 years ago) to design new pyrolysis reactors (Garcia-Nunez et al., 2017).

Equipment for charcoal production can be divided into four groups: kilns, retorts, converters, and fast pyrolysis reactors. Important features of each technology are presented in Table 2.4. Since the beginning of the 20th century, it has been identified that charcoal greatly depends on the carbonization technology (Klar and Rule, 1925). Thus, both types of carbonization equipment and operation parameters should be considered when deciding for the carbonization process. The following subsections present a discussion on the equipment historically and currently used for carbonization.

**Table 2.4 Main characteristics of pyrolysis reactors.**

| Type | Common designs | Material used | Typical operation | Main product |
|---|---|---|---|---|
| Kiln | Earth kiln, Brazilian half-orange, Missouri kiln, Argentine | Trunks, cordwood, logs | Batch, capacity: up to 80 t/batch, carbonization time: up to 3 weeks | Charcoal |
| Retorts | Iron drum retorts[a], Lurgi, Lambiotte, Reichert, Wagon, Carbo Twin | Trunks, cordwood, logs | Continuous/semi-continuous, capacity up to 6000 t/year, carbonization time: up to 48 h | Charcoal |
| Converters | Herreshoff, Rotary drums, Chroren paddle design, Auger, Pyrovac moving bed | Pellets, Briquettes, fine particles | Continuous, capacity: up to 90,000 t/year | Charcoal + oil + heat |
| Fast pyrolysis | Cone reactor, ablative reactor, bubbling fluidized bed, conical spouted bed, fluid circulating bed | Small particles (less than 1 mm) | Continuous, capacity: up to 70,000 t/year | Primarily liquid oil. Charcoal is a by-product |

[a]Iron drum retorts were small equipment displaced by oven retorts at the beginning of the 20th century (Brown, 1917).

## 5.1 The "classical" methods for charcoal production

### 5.1.1 Kilns for wood carbonization

Earth kilns

There are excellent sources reporting details on earth kiln construction and operation worldwide (e.g., Emrich, 1985; FAO, 1987; Schure et al., 2019). Thus, only a synthesis of these kilns is presented in this section. Earth kilns include both mound and pit kilns that are used exclusively for producing charcoal. Mound kilns have been employed for centuries (see Section 1) and are still used in developing countries (Pelaez-Samaniego et al., 2008; Nogueira et al., 2009; Santos et al., 2017; Nabukalu and Gieré, 2019; Schure et al., 2019; Rodrigues and Junior, 2019). The main characteristic of earth kilns is the use of a soil coverage that prevents contact between the wood logs that are being carbonized and the oxygen from the environment to avoid combustion. During the carbonization process, the non-condensable gases are released to the atmosphere and the condensates are released to the soil (Emrich, 1985; Garcia-Nunez et al., 2017; Schure et al., 2019). Earth kilns are very simple to build, but their operation needs expertise, especially for controlling the required amount of air to circulate inside the kiln. One of the reasons for earth kiln use is the low capital investment. However, the difficulty of controlling the release of harmful environmental emissions is a deficiency of these systems. Fig. 2.1 shows schematics of both mound and pit kilns.

Mound earth kilns can be classified into vertical mound kilns, horizontal mound kilns, and improved mound kiln designs, such as the Casamance mound kiln. An important advantage of mound kilns is that these can be built in the same place the biomass is

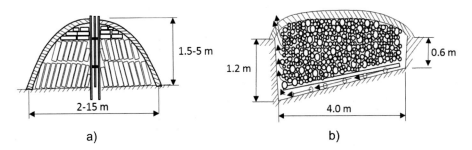

**Fig. 2.1** Schematic representation of (A) mound and (B) pit kilns (dimensions are provided for reference only). Source: Garcia-Nunez, J.A., Pelaez-Samaniego, M.R., Garcia-Perez, M.E., Fonts, I., Abrego, J., Westerhof, R.J.M., Garcia-Perez, M., 2017. Historical developments of pyrolysis reactors: a review. Energy Fuel, 31(6), 5751–5775. https://doi.org/10.1021/acs.energyfuels.7b00641.

available but they are labor intense. These kilns are not suitable for working with small wood particles or small particle size agricultural residues. Additionally, their operation is very sensible to weather conditions (Garcia-Nunez et al., 2017). The efficiency of earth kilns in some sub-Saharan countries can be as low as 11% (Schure et al., 2019), although the Casamance kiln's efficiency is higher than that of the other types of mound kilns. This kiln is equipped with a chimney made of oil drums that allow controlling the airflow in a better way. The yield is increased by partially recirculating the hot gases. The efficiency of mound kilns is affected by location, type of wood used, and operators' skills. Thus, the yields of these kilns vary from one region to another, as reported elsewhere (Schenkel et al., 1998; Kammen and Lew, 2005; Menemencioglu, 2013).

In Europe, the melting process used charcoal that was produced in part using earth kilns till the 19th century, when charcoal was substituted by coal (Tintner et al., 2020). The shapes of the piled logs to prepare earth kilns used in Europe at that time varied from one region to another. Elliptical shapes were common in the Mediterranean region, circular shapes in Belgium and Germany, and variable shapes (circular, oval, square, or irregular) in Norway (Carrari et al., 2017). Rectangular earth kiln shapes were also common in some European countries in the 18th century. However, the terms "square" and "rectangular" refer only to the shape of the piled wood logs prior to the carbonization process (i.e., not to the masonry "rectangular" kilns described later in Section "Brick, concrete, and metal kilns"). Such shapes were obtained with the help of small diameter wood poles placed vertically. Rectangular earth kilns were used in Sweden and Austria (Tintner et al., 2020). These old carbonization techniques have been the subject of research to keep track of the processes and equipment used in the past. Recently, Tintner et al. (2020) built a rectangular earth kiln with a volume of $103\,m^3$, using a mix of two wood species, and were able to obtain a charcoal yield of $29 \pm 1.9\%$.

Brick, concrete, and metal kilns

**Brick and concrete kilns** The use of bricks and concrete for kiln construction in substitution of earth kiln was an important evolution in the wood carbonization industry. Several variants of bricks and concrete kilns have been in use worldwide. These kilns are or have been built using two typical shapes: (a) round kilns (e.g., beehive kilns) and (b) hangar kilns, with rectangular or square shapes. Square kilns are not manufactured nowadays, but in the past, "square kilns were located at the Athens Research Center in

Georgia and on the Dukes Forest in Upper Michigan" (Forest Products Laboratory, 1961). In the United States, the so-called beehive kilns were built using bricks to substitute earth kilns and, thus, to increase carbonization productivity during the 19th century and the beginning of the 20th century. It appears that these are the first records on the use of masonry for manufacturing carbonization kilns. According to Bates (1922), brick beehive kilns could hold up to 90 cords of wood (i.e., up to 122 t of oven-dry wood) and the whole carbonizing process took from 15 to 25 days (Bates, 1922; Brown, 1917) (1 cord≈1360 kg of oven-dry wood).

Although there are not evidences that all beehive kilns were operated the same way, in these systems, "conversion to charcoal was carried out by burning wood or gas in a furnace underneath the pile and also by drawing air through to give a partial burning of the wood itself." In some variants, fans were used for removing vapors from the kilns and for helping circulation through a pipe condenser to recover acetate of lime and alcohol. However, charcoal was the targeted product (Bates, 1922). The use of masonry allowed long kilns' life span (up to 10 years) even under severe weather conditions. Their operation required skilled workers and high investment and operation costs. The release of pollutant gases is still not possible to avoid in these kilns. Beehive masonry kilns were still in use during the 1960s in the United States (Forest Products Laboratory, 1961). Capacities varied widely, and there was not a uniform design. The larger kilns were up to approximately 7.7 m high, with circular base with up to 9.6 m diameter and wall thickness of approximately 0.3 m. The dome-shaped ceiling had an opening for loading and firing. A number of air-inlet ports around and near the base of the kiln were employed to regulate air intake for partial combustion. Outlet ports located somewhat higher in the wall were used to allow smoke to escape after it has circulated throughout the charge (Forest Products Laboratory, 1961).

Two types of round shape masonry kilns still used in South America include the so-called Brazilian beehive kiln (Pelaez-Samaniego et al., 2008; Rodrigues and Junior, 2019) and the half-orange Argentine kiln (Argentinaforestal, 2020) (see Fig. 2.2). The Brazilian beehive-type kiln is also used in other places, such as Cambodia and Thailand (with small variations) (Kajina et al., 2019). Both the Brazilian and Argentine kilns are built with mud bricks, which allow using locally available clay/sand bricks and mud mortar. The life span of these systems can be up to 8 years. The operation of these kilns is relatively simple and can be placed in batteries of several kilns in the same place. Variants

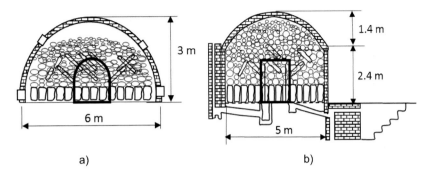

**Fig. 2.2** Schematic shape masonry kilns: (A) Argentinian kiln and (B) Brazilian kiln (dimensions are provided for reference only). Source: Garcia-Nunez, J.A., Pelaez-Samaniego, M.R., Garcia-Perez, M.E., Fonts, I., Abrego, J., Westerhof, R.J.M., Garcia-Perez, M., 2017. Historical developments of pyrolysis reactors: a review. Energy Fuel, 31(6), 5751–5775. https://doi.org/10.1021/acs.energyfuels.7b00641.

of the Brazilian masonry kiln are the surface kiln, the slope kiln, and the so-called hot-tail kiln (Rodrigues and Junior, 2019). The hot-tail kiln is largely used in Brazil for producing charcoal for the steel industry (Santos et al., 2017; Pelaez-Samaniego et al., 2008; Rodrigues and Junior, 2019). Both the beehive and the slope kiln have chimneys for gases removal. However, the hot-tail kiln does not possess chimneys. Holes along the kiln and its base are used to control air inlet and gases and vapor outlet. The carbonization process is controlled visually by observing the color and quantity of the smoke that is released through the holes (Rodrigues and Junior, 2019). In Brazil, the efficiency of all these brick kilns variants is approximately similar (de Vilela et al., 2014).

In Brazil, the development of the charcoal industry and the evolution of carbonization kilns have a long history that is linked to the iron and steel industry and started back in the 15th century. The steel industry in the 19th century boosted the charcoal production at larger scales (Nogueira et al., 2009). However, the industry evolved substantially during the 20th century due to the necessity of increasing steel production. Until recently, Minas Gerais, the state in which the iron industry began, still produced around 80% of the charcoal in Brazil, mainly for the iron and steel industry. More than 50% of the wood used for carbonization is obtained from planted eucalyptus forests and the rest from native forests (CGEE-Centro de Gestão e Estudos Estratégicos, 2015). In the country, there is a growing concern about charcoal supply reduction (Nogueira et al., 2009) and the negative effects of charcoal production (CGEE-Centro de Gestão e Estudos Estratégicos, 2015; PNUD, 2020). Therefore, the Brazilian carbonization industry has received a lot of attention from the academia, as seen from the rich bibliography on the topic (e.g., Santos et al., 2017; Costa, 2012; Coelho, 2013; Queiroz, 2014) that has focused on technical, economic, social, historical, and environmental analyses of wood carbonization. The government has also frequently conducted studies to promote a better and more efficient carbonization industry (CGEE-Centro de Gestão e Estudos Estratégicos, 2015; PNUD, 2020; INEE, 2021).

Currently, there is a large diversity of kilns built with bricks operating worldwide. Some designs of brick kilns are used in places such as the Philippines for producing charcoal from coconut residues (Charcoalkiln, 2013a) and in India for carbonizing bamboo (BTG, 2013). The designs of brick kilns can vary from a country to another. Brick kilns that are developed in Brazil, for example, by the Company MAGGI Ltda (Fornoparacarvao, 2021), have cylindrical or prismatic shapes and combine bricks (used for the walls) with metallic parts (used for doors, gases ducts, and covers). The yields of these kilns are up to 35%–40%, and the carbonization takes up to 3 days. Cooling the kilns prior to charcoal removal can take up to three more days (depending on the size of the kiln). Variants of the MAGGI kilns include burning systems to reduce vapors/gases released to the atmosphere and the recovery of liquid by-products.

The devclopment of *rectangular kilns* has roots in works in the United States during the 1940s. Single-stack rectangular masonry-block kilns were reported in 1946, and the first records on the operation of rectangular kilns at industrial scale appeared by the 1950s (Jarvis, 1960) (Forest Products Laboratory, 1961). Although variants of these kilns existed, typically, air input was carried out through ports along each sidewall at groundline and a single smokestack at the rear. The original design also featured a small stove or heating chamber that helped to improve the flow of air throughout the kiln (Forest Products Laboratory, 1961). Currently, two countries that use rectangular kilns for charcoal production at industrial scale are the United States and Brazil.

In the United States, the most known rectangular kiln is the Missouri-type kiln (also referred to as concrete kiln or batch-type charcoal kiln), which was developed in the early 1950s in Missouri (Garcia-Nunez et al., 2017). By 1959, there were 408 of these kilns operating in the State of Missouri (Jarvis, 1960). Jarvis (1960) presents a detailed and useful description of the early Missouri kilns design. These kilns (after improvements) are still responsible for a large fraction of the charcoal produced in the United States. The Missouri kiln has volumes up to $350\,m^3$ (typically between 150 and $200\,m^3$), thus allowing to use mechanized loading and unloading systems. The carbonization cycle consists of 4 days of loading/

unloading, 6 days of carbonization, and up to 20 days of cooling. Several improvements have been made to the original design. For instance, using thermocouples within the kiln contributes to the identification of cold ports and control of airflow. Additionally, the environmental impact of these kilns has been reduced through the use of afterburners (Garcia-Nunez et al., 2017). The Missouri kilns have proved that can perform adequately, from an environmental point of view, and have been approved for operation even under current strict federal environmental regulations (Environmental Protection Agency, 2020).

In Brazil, masonry rectangular kilns have been in use for a relatively short time. The design of these kilns is based on the Missouri kiln (Kajina et al., 2019; Rodrigues and Junior, 2019). These kilns were introduced in the country by the Acesita Company in 1990 (Pinheiro, 2016). Other large companies that produce both charcoal and pig iron (referred to as "integrated" companies) started the use of these kilns intending to increase the capacity of their carbonization kilns and to help mechanization of the loading and unloading operations (CGEE-Centro de Gestão e Estudos Estratégicos, 2015). In the last decade, integrated companies in states such as Minas Gerais, Mato Grosso do Sul, and Maranhao have been substituting masonry circular kilns by rectangular kilns to increase charcoal production capacity. In these kilns, the loading process takes up to 1 day, the carbonization process requires up to 5–6 days, and the cooling process takes up to 12–14 days (Pinheiro, 2016). Around 20% of the initial wood is burnt in the furnace to dry the materials and start the carbonization process. Some of these kilns possess water sprinkler systems that are used for faster charcoal cooling down, and the cooling process is reduced to 6–7 days. However, the resulting charcoal becomes more brittle when these cooling systems are used (Pinheiro, 2016).

The volume of rectangular kilns commonly used by integrated companies varies from $150\,m^3$, which is the case of, for example, the FR190 kiln (dimensions up to $13\,m$ long $\times\,4\,m$ wide) to up to $450\,m^3$ in the RAC700 kiln (dimensions up to $25\,m \times 8\,m$), able to hold up to 180 t of dry wood (CGEE-Centro de Gestão e Estudos Estratégicos, 2015). However, new developments are allowing to build rectangular kilns as large as 800 to $2000\,m^3$ (Kajina et al., 2019). Such large dimensions allow using trucks and tractors for loading raw materials and unloading the product, and, thus, the labor requirement is reduced (Pinheiro, 2016). Several conducts are employed for air input, and one or more chimneys are used for gases removal. A reported problem during the operation of these kilns is the difficulty of controlling the temperature to keep it uniform inside the kiln, which results in charcoal with non-uniform properties (Pinheiro, 2016). The carbonization cycle varies from 13 to 18 days in kilns that possess automatic temperature controlling systems. The yields are in the range of 32% to 35% (mass%), and the fixed carbon yield can reach up to 80% (commonly up to 78%). Additionally, up to 84 kg of tar per ton of charcoal can be recovered (Pinheiro, 2016). The charcoal production can vary from 750 t/year (for kiln FR190) to 2000 t/year (in kiln model RAC700). Other rectangular masonry kilns include the V&M and the Acesita Kilns (Santos et al., 2017). The V&M kiln has charcoal yields of up to 33%, capacities up to 42 t of dry wood (INEE, 2021), and the operation cycle can take up to 11 days (Santos et al., 2017). It appears that most of these systems possess systems for recovering by-products from the wood pyrolysis (tar and pyroligneous acid) (Pinheiro, 2016). In addition, some versions of these kilns present gas burners to reduce the negative impacts of carbonization (CGEE-Centro de Gestão e Estudos Estratégicos, 2015; Pinheiro, 2016). Thus, some of these systems could fit in the category of retorts, which are described later (Section 5.1.2). Fig. 2.3 shows examples of

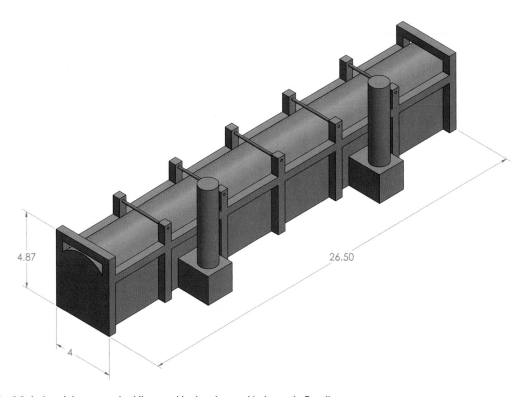

**Fig. 2.3** Industrial rectangular kilns used in the charcoal industry in Brazil. Source: Figueiró, C.G., De Oliveira Carneiro, A. C., Santos, G.R., Carneiro, A.P.S., Fialho, L. De F., Magalhães, M.A., da Silva, C.M.S., de Castro, V.R., 2019. Caracterização do carvão vegetal produzido em fornos retangulares industriais Revista Brasileira de Ciências Agrárias 14(3). https://doi.org/10.5039/agraria. v14i3a5659.

industrial rectangular kilns used in Brazil. An interesting source to learn about the construction and operation of brick rectangular kilns in Brazil is Barcellos (2017).

**Metal kilns** Metal kilns are relatively small kilns built from steel sheets with cylindrical or tapered shape. Metal kilns are portable and can be hauled from one place to another, which makes them of interest when, due to raw materials availability, the kiln needs to be moved accordingly. Two well-known metallic kilns are the so-called Tropical Products Institute (TPI) kiln and the UK tapered metal kiln. The widespread charcoal production using cylindrical transportable metal kilns originated in Europe during the 1930s and transferred to developing countries in the 1960s. A variant of the original European metal kiln is the Mark V kiln that was developed in Uganda (Saravanakumar and Haridasan, 2013). It also appears that a small number of a larger variant of the European metal kilns were still operating in Poland until recently (Reuters, 2016). Portable metal kilns with shapes other than cylindrical or tapered (e.g., square shape) have only sporadically been built and tested (Saravanakumar and Haridasan, 2013). An excellent overview on the most known metal kilns can be found in FAO (1987); thus, only a few general points on these kilns are presented in this section.

The TPI kiln is relatively easy to manufacture. It has a diameter of around 2.2 m, a height of around 2.3 m, and sheet thickness of 3 mm. The kiln consists of two interlocking cylindrical sections and a conical cover. Commonly, vapor release is through four equally spaced ports on the cover that can be closed off with plugs as required. Eight inlet/outlet channels, arranged radially around the base, serve to support the kiln. During charring, four smoke stacks are fitted onto alternate air channels (FAO, 1987). Although part of the wood is consumed during the charring process, the efficiency of these kilns (approximately 28%) is higher than that of mound and concrete kilns. The carbonization process can take up to 1 week. A possible deficiency of these kilns is that, due to the use of steel sheets, they are prone to corrosion, especially in high relative humidity environments (which are common in tropical countries). However, the modular construction makes them of interest due to the easiness for dismounting and transporting as required. Fig. 2.4 shows a picture of a TPI kiln that has been used to carbonize oil palm residues in Colombia. The TPI kiln has been of interest in other countries, as seen in works in, for instance, El Salvador (Guillen, 2011).

In the United States, the first records on the use of metallic portable kilns appeared by the 1940s. The Connecticut Agricultural

**Fig. 2.4** Picture of a research TPI kiln used for carbonization of oil palm residues in Colombia.

Experiment Station (Olson and Hicock, 1941) developed a portable "chimney kiln" of prismatic (rectangular) shape, which was intended to improve deficiencies of a previous chimney kiln with circular shape that possessed one air inlet and one gas outlet. Typical dimensions of the rectangular chimney kiln were, approximately, $2.5 \times 1.6 \times 1.6$ m (length × width × height). Variants of the circular chimney kilns were as follows: (1) the Black Rock Forest Kiln (Beglinger, 1956), which had a tapered shape and shallow lid. Dimensions were approximately 2.2 m diameter at the base and 1.3 m at the top. The kiln was equipped with 4 draft ports and 4 smoke flues alternately spaced through the bottom section at ground line, and (2) the New Hampshire-type kiln that first operated in 1938 (Baldwin, 1958; Forest Products Laboratory, 1961). One of the reported problems of metallic small kilns was the excessive heat loss by radiation that, apparently, resulted in lower yields and higher kilns cost per volume compared to masonry-block kilns (Forest Products Laboratory, 1961). Currently, to the best of the authors' knowledge, these kilns are not in use in the United States.

In addition to those previously described, metal kilns for carbonization include simple kilns built from 200 L (55 US gallon) drums that are in operation in some places (FAO, 1987; Charcoalkiln, 2013b). These kilns can operate with a diversity of raw materials, but the resulting char can contain high volatile content, especially when wood logs are used, due to the difficulty of achieving complete carbonization (FAO, 1987). One advantage of these kilns is the easy operation, which makes them possible to operate up to 10 drums at the same time by one worker only. However, these kilns require logs less than 30 cm long to achieve satisfactory results. Therefore, logs preparation is labor intensive. Up to 30 kg of charcoal per batch can be obtained using these kilns, but the yield will depend on the type of raw material employed (FAO, 1987).

Finally, the designs of metal kilns in some latitudes such as South Africa appear to focus on large metal kilns that are built using pre-formed sheets to obtain a half-cylinder shape kiln. This half-cylinder metallic kiln (referred to as "drum char retort") lays horizontally on the soil, and one or more small chimneys on the top are used to remove gases. It appears that these kilns have two doors and their dimensions are around 3 m high, 5–6 m wide, and 8 m long (or larger). An interesting characteristic of these kilns is that, in some variants, they can be assembled in the charcoal production site, using parts previously manufactured in mechanic shops. Moreover, it appears that metallic underground storage tanks (with cylindrical shape) are also employed to build variants of the described metallic kilns (Vuthisa, 2018).

### 5.1.2 Retorts to produce charcoal

The main difference between kilns and retorts is the way pyrolysis gases are managed. In kilns, the gases and vapors are released to the atmosphere but retorts are designed to condense the vapors and make use of the energy content of gases (Garcia-Nunez et al., 2017). This section puts together literature on retorts that have been developed and used in the past and retorts that are used nowadays in different countries. A key characteristic of retorts is that they work with wood logs as raw material.

Prior to 1922, the equipment used in the wood distillation industry included both kilns and retorts (Bates, 1922). Kilns were the cheapest option for charcoal making. However, the charcoal yields were lower in comparison with retorts. In retorts, even as a by-product, charcoal yield was the highest among the pyrolysis products, reaching up to 36% yield from dry wood (Brown, 1917; Bates, 1922; Klar and Rule, 1925). In the old wood distillation plants, charcoal was used as a fuel, mostly for the iron and steel industry (Brown, 1917; Bates, 1922). The other products of wood distillation were "wood gas" (~24% yield), a raw liquor mixture (around 14%–15% yield), and water of decomposition. Harwood species were preferred (Bates, 1922). In addition to wood, residues from the agro-industry, such as olive kernels and kernel of fruits (e.g., peaches), and residues from the forest industry (sawdust) have been used since prior to the beginning of the 20th century. For example, the yield of charcoal from olive kernels was around 35%, with ash content up to 9% (Dumesny and Noyer, 1908; Klar and Rule, 1925). Charcoal from olive kernels was recognized for its high quality and easy burning (Dumesny and Noyer, 1908). The waste or residual small charcoal particles (in the old carbonization industry referred to as "breeze") were used for briquette production (Dumesny and Noyer, 1908; Klar and Rule, 1925). The spirits industry used the breeze as a filtering medium, and sometimes, large charcoal particles were intentionally ground to obtain smaller particles for this purpose (Klar and Rule, 1925). The fines resulting from breaking up stick charcoal were also used for charcoal briquettes. These briquettes were burnt as fuel, for instance, for heating railroad carriages. Thus, filter charcoal and briquettes were produced side by side (Klar and Rule, 1925).

The use of retorts for wood distillation at industrial scale began with the work of Philip Lebon in 1798 (Dumesny and Noyer, 1908). Violette (referred by Dumesny and Noyer, 1908) identified that "slow" distillation produced approximately double the amount of charcoal compared to "quick" distillation. This finding could have impacted the way the carbonization equipment was developed and operated. During the 1800s, the early wood distillation industry in both the United States and Europe (also referred to as carbonizing industry) produced charcoal either as the main product or as a by-product (Klar and Rule, 1925). A remarkable development of the wood distillation equipment happened throughout the second half of the 19th century and the beginning of the 20th century. The evolution from "wasteful charcoal pits" to brick kilns helped to increase charcoal yields but not to recover condensable gases. An important further step was the development of iron retorts (Brown, 1917; Bates, 1922). Retort is a suitable solution to overcome the environmental issues of classical mound kilns (Sparrevik et al., 2015).

Early retorts were small cylindrical vessels manufactured initially with cast iron and later with steel sheets. Dimensions were approximately 1.25 m diameter and 2.9 m long, capable of holding around 0.85 t of dry wood. A single door was employed for loading and unloading. Clay was used to seal the entrance of oxygen after the heating process started. These systems were externally heated, and the carbonizing (distillation) process took from 22 to 24 h. The charcoal or other by-products from the process were used as fuel (Brown, 1917). Quoting to Veitch (1907), "the charcoal left in the retort when distillation is complete constitutes from 20 to

35 percent of the original weight of the wood, the quantity depending on the kind of wood and the manner of heating the charge." The development of other retorts variants in the United States was reported at the beginning of the 1960s. One of these variants included a rotary and a slightly inclined horizontal tube for feed agitation and travel. The carbonization heat was introduced into the tube from an outside source (Forest Products Laboratory, 1961).

The Adam retort (Adam, 2009) is well-known and relatively simple to build retort that has been developed in Africa and Asia. This retort is built with double walls of bricks or stabilized earth blocks and has prismatic shape (see Fig. 2.5). This system works in two phases. In the first phase, the system works as a traditional kiln built with bricks. A second fire chamber that burns waste wood is used to dry the wood used for carbonization. In the second phase, the gases produced are burned in a hot fire chamber. Thus, harmful emissions are reduced by up to 75% compared to traditional kilns and the heat generated in the second chamber is recycled to accelerate the carbonization process. The charring process is faster than in brick kilns (around 12h), and the efficiency of these retorts varies between 32% and 40%. Recently, a smaller mobile Adam retort has been developed and used successfully (Charcoalkiln, 2015).

**Fig. 2.5** Schematic of an Adam retort (Adam, 2009).

The literature is rich on describing old and new retorts that have been developed for wood carbonization, especially in Europe. Retorts commonly described in the literature include the wagon retort, Lambiotte French SIFIC, the Lurgi Process (Australia), the VMR (Norway), the Carbon Twin Retort, and the O.E.T. Calusco (former Carbolisi). Retorts can work in batch schemes or continuously. Literature on retorts is becoming abundant (Klar and Rule, 1925; Tomas et al., 2009; Garcia-Nunez et al., 2017; Rodrigues and Junior, 2019), and only a synthesis is presented in this section. Table 2.5 shows details on some retorts. Fig. 2.6 shows schematics of selected retorts, and Fig. 2.7 presents retorts' pictures. Illustrative presentations on the design and operation of retorts presented in Fig. 2.7 can be seen in Rousset (2006) and Ronsse et al. (2013).

New commercial designs and systems are available, such as Agritherm, Black Carbon, and Biogreen Energy (Ronsse, 2013). The main advantages of these systems include the high charcoal yield and high charcoal quality. In these systems, the process is carried out under slow pyrolysis conditions and atmospheric pressure. As expected, the by-products from the vapors can be recovered. The main disadvantages are related to the high capital costs, attrition problems, the need of external sources of energy, and the fact that most of these systems are not portable and require a concentrated supply of raw materials (Garcia-Nunez et al., 2017).

In Brazil, two interesting types of retorts that have been developed are the FCR (Rima container furnaces) and the Bricarbras Furnaces (Fornos Bricarbras), which are described in CGEE-Centro de Gestão e Estudos Estratégicos (2015). The first Rima container furnaces (developed back in 2001) were cylindrical retorts that had capacities of $5\,m^3$, productivity of up to $5\,kg/h$ of charcoal, and yields between 25% and 28%. Improved versions of this retort allowed to increase the yields to 35% and productivity up to $700\,kg/h$ due to increased retorts' volume (up to $400\,m^3$). The Bricarbras retorts are metallic cylinders that work in sets of, e.g., eight retorts, which are conveyed using overhead bridge cranes to make the process continuous. Pre-drying of the raw material is conducted using heat released by the combustion of the carbonization gases. During the carbonization process, combustion chambers hold the retorts. In these chambers, the heat for the carbonization process is obtained from the combustion of wood. Loading of the wood for carbonization is carried out manually, and the unloading is mechanized. The yields of these systems are around 33%. Illustrations on this retort system can be seen in CGEE-Centro de Gestão e Estudos Estratégicos (2015).

**Table 2.5 Characteristics and operational details of some retorts operating with logs**

| | Lurgi | Lambiotte French SIFIC | The wagon retort | Carbo Twin Retort | VMR | ACESITA | O.E.T. Calusco (former Carbolisi) |
|---|---|---|---|---|---|---|---|
| Production/processing capacity per unit reaction volume | 10 t/year/m³ | 16 t/year/m³ | | 70 t/year/m³ | | Production: 45 t/day of wood | 6000 t/year |
| Carbonization time | n/a | n/a | 25–35 h | 8 h (carbonization), 24–48 h (cooling) | 9–11 h (depending on moisture content) | 12 h | 23–24 h |
| Heating method | Contact with heat gases | | External heat and volatile combustion, an oil burner (or LPG) is used to provide heat for the initial start-up | | | | |
| Dimensions | Height: 27 m Diameter: 3 m Woodfeed size: 150 × 150 × 250 mm | Height: 16.3–18 m $D$ = 3–4.3 m Volume: 600 m³ | Trolleys: 12 m³ Length: 8–16 m Diameter: 2.5 m Tunnel capacity: 35–60 m³ Length: 45 m | Volume per Vessel: 5 m³ Six vessels are needed to keep the system running | Volume per retort: 4.5 m³, 1 VMR oven = 2 retorts | | |

*Continued*

Table 2.5 Characteristics and operational details of some retorts operating with logs—cont'd

| | Lurgi | Lambiotte French SIFIC | The wagon retort | Carbo Twin Retort | VMR | ACESITA | O.E.T. Calusco (former Carbolisi) |
|---|---|---|---|---|---|---|---|
| Reactor position | Vertical | Vertical | Horizontal | Horizontal | | | Continuous process with horizontal design |
| Loading and discharge methods | Mechanical | Mechanical | Use of wagons | | | Mechanical | Wagons are transported through a tunnel |
| Process control | Direct measurement of temperature | | | | | | |
| Mode of operation | Continuous | | | Semi-continuous | | Continuous | |
| Pretreatment needed | Pre-dried | | | | | | |
| Charcoal yields reported | 30–35 wt% | | 30–33 wt% | 33 wt% | | 40% | |

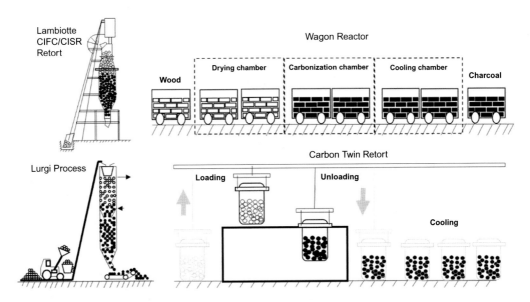

**Fig. 2.6** Schematics of some types of retorts. Adapted from Garcia-Nunez, J.A., Pelaez-Samaniego, M.R., Garcia-Perez, M.E., Fonts, I., Abrego, J., Westerhof, R.J.M., Garcia-Perez, M., 2017. Historical developments of pyrolysis reactors: a review. Energy Fuel, 31(6), 5751–5775. https://doi.org/10.1021/acs.energyfuels.7b00641.

## 5.2 Charcoal as by-product

Some technologies have been developed for producing pyrolysis bio-oil as the target product; i.e., in these systems, charcoal is not the main product but a by-product. In this section, we briefly discuss some of these systems, including auger pyrolysis reactors, drum reactors, and fast and flash pyrolysis reactors.

### 5.2.1 Auger pyrolysis and rotary drum reactors

The carbonization reactors described in previous sections are used mainly for wood logs processing, but are not appropriate for small biomass particles and chips. Auger reactors and rotary drums are two options to deal with small particle lignocellulosic biomass such as olive stones, coconut shell, oil palm fiber, oil palm shell, bark, twiglets, chips, and pellets (Garcia-Nunez et al., 2017). Both types of reactors are suitable not only for char production, but also to produce bio-oil and gases. These reactors are a type of pyrolysis equipment designed for continuous operation, which results from the use of a constant feeding system. The heating methods for these reactors can be external (using external electrical resistances) or internal (through the use of a heat carrier such as hot sand) (Boateng et al., 2015).

**Fig. 2.7** Pictures of some types of retorts: (A) French CML retort, (B) Lurgi, (C) Lambiotte, (D) Van Marion Retort (VMR), (E) O.E.T. Calusco (former Carbolisi), (F) Acesita Retort (Brasil) (Rousset, 2006; Ronsse, 2013).

Auger pyrolysis reactor

Fig. 2.8 shows a schematic of an auger pyrolysis reactor. In these systems, the biomass is fed through a hopper or a feeder screw. Then, a rotating screw conveys the biomass horizontally inside the body of the reactor where the pyrolysis occurs. At the end of the reactor, char is collected, and vapors and gases are extracted and led to a condenser. The capacity of these pyrolysis reactors varies widely from pilot scale to systems able to process up to 150 t/day of biomass (Meier et al., 2013). The heating method is through direct contact with hot gases or through indirect heating. One important advantage of this reactor is that it can be stationary or portable. The process control is carried out through direct measurement of the temperature. If the biomass is not small enough, it is necessary a pretreatment operation to convert it into chips (Garcia-Nunez et al., 2017).

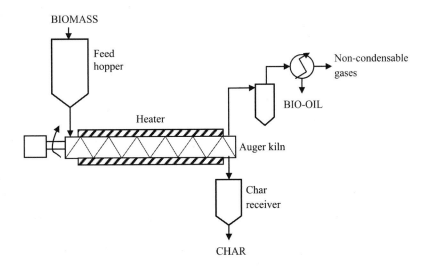

**Fig. 2.8** Auger/screw pyrolysis reactor. Source: Dhyani, V., Bhaskar, T. 2018. A comprehensive review on the pyrolysis of lignocellulosic biomass. Renew. Energy, 129, 695–716.

The auger reactor is suitable for small- and medium-scale processes. In large screw reactors, heat transfer limitations (due to size) could be an aspect that needs to be considered very carefully (Brassard et al., 2017). Henrich et al. (2016) reported a big scale reactor named the twin screw Lurgi-Ruhrgas mixer reactor using sand as a heat carrier for heating coal for producing town gas or olefins. Another screw pyrolysis reactor that has proven adequate for processing lignocellulosic biomass is the Haloclean pyrolysis reactor. This reactor was originally designed for treating electronic wastes and uses iron spheres as a heat carrier (Hornung et al., 2005, 2009). Other examples of externally heated auger reactor are the systems developed by Advanced Biorefinery, Inc., Pyreg GmbH, and the Kansai kiln (Boateng et al., 2015).

The yield of char, bio-oil, and gases from auger pyrolysis reactors depends on the operational conditions such as particle size, temperature, and ash content. Garcia-Nunez et al. (2017) summarized pyrolysis product yields for different kinds of biomass without and with heat carriers. Without heat carrier, the temperature ranged from 400°C to 500°C, obtained char from 12 to 30 wt%, bio-oil from 32 to 60 wt%, and gas from 12 to 40 wt%. Using heat carriers, the temperature reported for the operation of pyrolysis reactors with different capacities was 500°C. Char production ranged from 15 to 28 wt%, bio-oil yield ranged from 51 up to 69 wt%, and gas from 16 up to 24 wt%. It seems that with higher temperatures and using heat carriers, the production of bio-oil tends to be higher.

Rotary drum reactors

Rotary drums are designed to treat small biomass particles and chips. This reactor consists of an internal concentric steel tube and a cylindrical internally insulated mantle that makes up the rotary part. The solid and gaseous products are discharged by two fixed parts at the end of the rotary drum. The rotation speed and the angle of the drum control the residence time of the biomass into the reactor. The capacity of these reactors can be up to 288 t/day. The heating method is by direct contact of the biomass particles with hot gases or indirect heating, and can be portable or stationary (Garcia-Nunez et al., 2017). In this kind of reactor, the char yield varies from 19 to 38 wt% and bio-oil yields vary from 37 to 62 wt% (Garcia-Nunez et al., 2017). Rotary drum reactors have been in use at commercial scale in Europe and Japan for treating tires, sewage sludge, municipal solid waste, and plastics (Lemieux et al., 1987; Malkow, 2004).

### 5.2.2 Flash (high-pressure) pyrolysis reactors

Charcoal (weight) yield is not always a good indicator of the efficiency of the carbonization process. Thus, the yield of fixed carbon is recommended instead. The theoretical yield of char from biomass has been estimated to range from 44% to 55%, depending on the type of biomass (Mok et al., 1992). Research on how to increase the yields of charcoal and reach the theoretical one has extensively been conducted by the team work led by Prof. Antal, at the University of Hawaii (Mok and Antal, 1983; Mok et al., 1992; Antal et al., 1990; Wang et al., 2013; Boateng et al., 2015). Charcoal yields can be augmented by using high-pressure flash pyrolysis reactors that work with steam or an inert gas (Mok and Antal, 1983). At laboratory scale, the sample holder (reactor) is subjected to high pressure (e.g., 0.6–1 MPa) prior to the carbonization process (Mok and Antal, 1983; Antal et al., 2003). Then, a flash fire is ignited into a packed bed of biomass. The authors showed that this approach allows to reach fixed carbon yields of up to 100% of the theoretical limit in relatively short times (20 to 30 min) (Antal et al., 2003; Boateng et al., 2015). Thus, charcoal yields can be as high as 40%.

Also, at laboratory level, it was demonstrated that small particle size of biomass, combined with high pressure, results in the highest charcoal yields. Charcoal and fixed carbon yields of sawdust processed at high pressures, considering the effect of particle size as well, have been reported elsewhere (Wang et al., 2013). The effect of pressure on biomass carbonization yields was poorly studied prior to the 1980s (Mok and Antal, 1983). Even today, pressure is not always considered an important parameter to consider in review papers and research (see, for instance, Zhang et al. (2019)). However, the use of high-pressure reactors is a promising path to increase the yields of charcoal, as confirmed by recent studies. Pecha et al. (2017) showed that increasing pressure increases char yield, likely due to the conversion of lignin clusters with 2–5 aromatic rings into char.

### 5.2.3 Fast pyrolysis reactors

Fast pyrolysis reactors are designed mainly for bio-oil production. Char is obtained as a by-product, and often, it is combusted together with the pyrolysis gases, as a way of producing heat for the reactor. Some of the most important types of fast pyrolysis reactors are rotating cone, ablative, conical spouted bed, bubbling fluidized bed, and circulating bed. There are several excellent reviews on fast pyrolysis reactors (Czernik and Bridgwater, 2004; Bridgwater, 2012; Venderbosch and Prins, 2010; Meier and Faix, 1999). Therefore, we encourage the reader to consider these sources for further reading about these technologies.

### 5.2.4 Solar pyrolysis reactors

Solar energy is a promising alternative to energy generated from fossil fuel. Solar energy is the most renewable and abundant energy source that can help reduce the greenhouse gas emission (Sobek and Werle, 2019). Thermochemical conversion technologies generally need high amount of heat generated from fossil fuels or electricity. It is estimated that 32% of $CO_2$ emission can be diminished through incorporating solar energy (Joardder et al., 2014). In solar pyrolysis reactor, feedstock is heated either directly or indirectly. In directly irradiated solar reactors, biomass in a transparent reactor is subjected to concentrated solar radiation. In indirect reactors, the solar energy is absorbed and then transferred to biomass through conduction and convection with heat transferring fluid. This type of reactor is made from metals with high thermal conductivity (Weldekidan et al., 2018). Table 2.6 shows performance parameters of solar pyrolysis reactors.

**Table 2.6** Summary of solar-assisted biomass pyrolysis

| Ref. | Heat source | Power (kW) | Range of temp. (°C) | Heating rate (°C/s) | Sample studied | Yield summary |
|---|---|---|---|---|---|---|
| (Zeng et al., 2015a, 2015b, 2015c, 2015d) | Solar simulator | 1.5 | 600–2000 | 50 | Pellet wood | Gases (CO, CH4) 15.3%–37.1% and liquids 70.7%–51.6% |
| | Solar dish | 1.5 | 600–2000 | 5–450 | Beech wood | 28% liquid bio-oil, 10% char, and 62% gas (H2, CH4, CO, CO2, and C2H6) |
| | Solar dish | 1.5 | 800–2000 | 50–450 | Beech wood | Gases (H2, CO, CO2, CH4, and C2H6) heating value increased about five times (3527–14,589 kJ/kg) |
| | Solar simulator | 1.5 | 600–2000 | 5–450 | Wood | Char yield decreased with temperature and heating rate in the temperature range |
| (Authier et al., 2009) | Xenon lamp | | 907–487 | | Oak wood | Gas and char |
| (Morales et al., 2014) | Parabolic trough | | 465 | | Orange peel | Liquid (77.64 wt%), a non-condensable gas (1.43 wt%), and char 20.93 wt% |

## 5.3 Other types of charcoal for environmental remediation and other uses

Some segments of the agro-industry use their own generated residual biomass for the production of steam and heat through boilers and burners. Biomass burning often produces large amounts of ashes that could contain relatively high organic matter content. An example is the sugarcane industry in tropical countries that releases large amounts of sugarcane bagasse ash (SCBA) that is used as a soil fertilizer (Sales and Lima, 2010) or disposed of in landfills (Somna et al., 2012; Xu et al., 2018) (Patel, 2020). SCBA can serve to remove phenols (Mall et al., 2006) or dyes

Table 2.6 Summary of solar-assisted biomass pyrolysis—cont'd

| Ref. | Heat source | Power (kW) | Range of temp. (°C) | Heating rate (°C/s) | Sample studied | Yield summary |
|---|---|---|---|---|---|---|
| (Li et al., 2016) | Sun simulator | | 800–2000 | 50 | Sawdust, peach pit, grape stalk and grape marc | Gas (63 wt%), tar and char (37 wt%) |
| (Joardder et al., 2014) | | | 500 | 5 | Date seed | Liquid (50 wt%), solid char and gas (50 wt%) |
| (Ramos and Pérez-Márquez, 2014) | Parabolic trough | | >270 | 0.5 | Wood | Charcoal (39%) |
| (Zeaiter et al., 2015) | Fresnel lens | | 850 | | Scrap tires | Oil and gas |
| (Hazel et al., 2017) | Linear mirror | | 500 | | Agricultural wastes | Charcoal |
| (Soria et al., 2017) | Solar dish | 1.5–2 | 600–2000 | 10 and 50 | Beech wood | Char, tar, and gases |
| (Arribas et al., 2017) | 7 kW xenon short arc lamp | 5800 kW/m$^2$ | | | Algae, wheat straw, and sludge | 63–90 vol% syngas |
| (Weldekidan et al., 2020) | Parabolic dish | 1.5 | 800–1600 | 10–500 | Chicken-litter waste and rice husk | Gas yields from 10 to 39 wt%<br>Bio-oil yields from 48 to 41 wt%<br>Char yields from 42 to 18 wt% |
| (Rahman, 2020) | Parabolic dish | 1.4 | 250–600 | 0.33–1.66 | Algal biomass (*E. compressa*) | Bio-oil yield of 38.12 wt%<br>The maximum char yield of 65 wt% |
| (Xie et al., 2019) | Solar simulator | 4 | 450–850 | 10 | Cotton stalk | Char yields from 19.50 to 38.50 wt% |

from wastewater using the ash as received (Mall et al., 2006; Siqueira et al., 2020) or after an activation process (Valix et al., 2004). Sometimes, prior to some specific uses of SCBA, the inorganic part is separated from the organic one. The inorganic fraction can be used as an additive for the concrete and the ceramic industries (Teixeira et al., 2008; Sales and Lima, 2010; Frías et al., 2011; Xu et al., 2018), or as an alkali-activated binder (Castaldelli et al., 2013). The organic fraction, conversely, has been explored for wastewater treatment, after an activation process (Kaushik et al., 2017). A recent review (Patel, 2020) summarizes some uses of SCBA. The residues from the combustion of other types of biomass require further study. The oil palm industry in some tropical countries, for instance, also generates large amounts of ashes, which are normally added to the soils as fertilizers.

Another thermochemical operation to convert biomass into other useful products (e.g., syngas) is biomass gasification. As in the case of combustion, the thermochemical conversion in gasification occurs at high temperatures and one of the by-products is gasification char. This section presents concepts, both under development and commercially available, that integrate biomass gasification systems with the production of energy and charcoal, either as a by-product or as an intermediate product. Charcoal obtained in these systems can be used for environmental remediation and/or energy production. This section shows works on using gasification chars and suggests possible integration schemes of biomass gasification plants with the use of the resulting chars.

### 5.3.1 By-product gasification chars from commercial gasification plants

Gasification is a well-known technology used to convert lignocellulosic biomass into syngas, which is further used for combined heat and power (CHP) production. Commercial biomass gasification plants for CHP are available in several places across Europe (Hansen et al., 2015; Marchelli et al., 2019). The main products of biomass gasification are synthetic gases, charcoal, and tar. The charcoal from biomass gasification has sometimes been referred to as gasification char, char from biomass gasification, or gasifier char (Pelaez-Samaniego et al., 2020). The mass yield of gasification char in large gasification reactors can be as high as 10% (Ahmad et al., 2014; Vakalis et al., 2016). Gasification chars are considered residues with low value, and sometimes, these materials are even disposed of as waste (Benedetti et al., 2018; Marchelli et al., 2019). Thus, new options for adding value to gasification chars have been sought recently (Hansen et al., 2015; You et al., 2017; Pelaez-Samaniego et al., 2020).

Gasification charcoal can be used (1) directly as fuel, (2) for soil amendment and plant growth, (3) for carbon sequestration, (4) as adsorbent for removing heavy metals and organic pollutants in

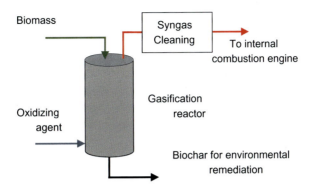

**Fig. 2.9** Schematic showing a direct use of gasification char for environmental remediation.

the environment or other chemicals such as acetaminophen and caffeine, after an activation process, (5) as an additive for anaerobic digestion (AD) to enhance methane production, (6) for tar removal or biodiesel, and (7) for electrochemical applications (Pelaez-Samaniego et al., 2020). In addition to the mentioned uses of gasification char, this material can be employed for biogas cleaning, specifically for hydrogen sulfide removal from biogas (Marchelli et al., 2019; Pelaez-Samaniego et al., 2020). Fig. 2.9 shows a schematic on the concept of using gasification chars for environmental remediation.

A key property of gasification chars is their high surface area, the presence of minerals and metals (e.g., Ca, K, and Fe) in the chars' ash, and the presence of oxygen-containing functional groups (You et al., 2017; Marchelli et al., 2019; Pelaez-Samaniego et al., 2020). Surface areas higher than 500 $m^2/g$ have been reported (Pelaez-Samaniego et al., 2020) for gasification chars. Thus, these materials offer potential for environmental applications directly (i.e., without the need of further activation processes). Minerals and metals on char's ash can form catalytic centers on the surface of the chars to adsorb $H_2S$ (Pelaez-Samaniego et al., 2020).

### 5.3.2 Intermediate gasification chars from CHP systems

Hrbek (2016) mentioned that, at the end of 2015, there were more than 60 gasification plants across the 10 countries participating in the IEA Bioenergy Task 33 (thermal gasification of biomass). In Europe, it appears that the main type of gasification reactor is the downdraft fixed-bed gasifier (Baratieri and Patuzzi, 2019). A major problem for syngas production in one-stage gasifiers is the presence of tars. Tars are a chief technical hurdle for the commercial implementation of biomass gasification. Thus, the operation of gasification equipment requires the adoption of expensive syngas cleaning processes prior to the syngas use (Hernandez et al., 2016). The development of two-stage gasifiers can solve the problem of producing synthesis gas with high tar contents (Pelaez-Samaniego et al., 2008). The processes developed by Choren Company of Germany and by the Technical University of Denmark (TUD, in cooperation with the Danish Fluid Bed Technology) are examples of this type of system. Two-stage gasifiers are formed by a pyrolysis unit, followed by a) a high-temperature step at more than 1200°C to convert the pyrolysis vapors into soot (Choren process), or b) a low-temperature unit using a circulating fluidized-bed gasifier (LT-CFB) (TUD process). In the former case, the soot and the charcoal are further converted into syngas in the presence of an oxidation agent through the use of a second gasifier. In the latter case, the char is separated from the pyrolysis gases and further gasified in a char gasification unit. The resulting gases are sent back to the pyrolysis unit where they are mixed with new biomass for pyrolysis (at around 650°C) and the pyrolysis gas is employed for energy production in engines (Hansen et al., 2015). Fig. 2.10 shows a schematic of the operation of the LT-CFB concept.

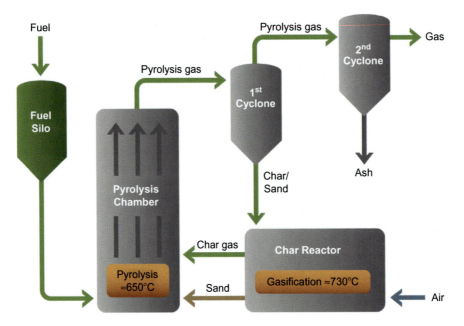

**Fig. 2.10** Schematic of the low-temperature circulating fluidized-bed gasifier (LT-CFB) (Hansen et al., 2015).

Gasification plants using a LT-CFB gasifier are employed to gasify wood chips and materials with high contents of low melting ash compounds (e.g., straw, manure, or sewage sludge) (Hansen et al., 2015). The LT-CFB technology has been proved in continuous operation, as a 6-MW demonstration plant, and a 2 MW commercial CHP plant that operates using the two-stage process. The 2 MW plant is expected to produce ~64 t/year of char residues, while a planned 60 MW full-scale commercial plant is expected to generate approximately 10,000 t/year of carbon-rich residues (Hansen et al., 2015). The potential scale-up of gasification plants needs to include options to add value to these large amounts of char. Therefore, this charcoal could be removed prior to the second (gasification) step. However, the system appears feasible for removing part of the char that is produced in the pyrolysis step.

### 5.3.3 Chars from hydrothermal carbonization

One of the limitations of biomass pyrolysis for char production when using biomass with high moisture content is the requirement of a preceding (often expensive) drying operation step. Thus, using high moisture content materials for pyrolysis is non-economically viable. For these materials, one alternative to pyrolysis is hydrothermal carbonization (HTC). HTC is a thermochemical process to convert biomass into chars (commonly called hydrochars) and other useful by-products using water under elevated pressures and temperatures at the lower region of liquefaction process. HTC has also been referred to as "wet pyrolysis" (Libra et al., 2011). Feedstocks for HTC have included woody landscape/forestry residues, agricultural residues, food processing residues, animal manures, or domestic wastewater sludges (Román et al., 2018). Fig. 2.11 shows a schematic of the HTC process.

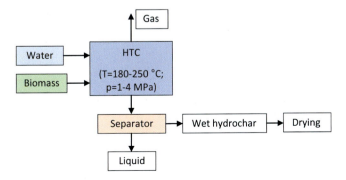

**Fig. 2.11** Schematic of a typical HTC process (Missaoui et al., 2017).

HTC is conducted at lower temperatures than in pyrolysis (e.g., around 180°C to 250°C), but at relatively higher pressures (Missaoui et al., 2017; Zhang et al., 2019). The main parameters for the process are reaction temperature, pressure, residence time, and water-to-biomass ratio. Yields of hydrochars are in the range from 40 to 70 wt% (Zhang et al., 2019). Hydrochars are different from chars in terms of chemical and physical properties. For instance, pH of hydrochars is approximately half the pH of char (Takaya et al., 2016; Wiedner et al., 2013). Hydrochars have a less stable structure (dominated by alkyl moieties) than chars (dominated by aromatics). "Hydrochars have lower proportions of aromatic compounds than chars (less stable) but are rich in functional groups (higher cation exchange capacity) than chars" (Wiedner et al., 2013).

HTC is a commercially available technology, especially in Europe, as seen in Table 2.7. Currently, several commercial plants are operating, especially in Europe. C Green, for instance, is installing a commercial full-scale HTC plant, using its OxyPower HTC. Ingelia also has HTC plants operating currently in Spain and with intended plants to be installed in the United Kingdom and in Italy (Nannoni, 2019). An HTC plant for processing sewage sludge is

**Table 2.7** Some companies on HTC at commercial level.

|  | Links |
|---|---|
| Grenol GmbH | http://www.grenol.org/ |
| Ingelia | https://ingelia.com/?lang=en |
| Terra Nova | https://terranova-energy.com/en/projects/ |
| SunCoal Industries GmbH | https://www.suncoal.com/ |
| HT Cycle | https://ipi.ag/en/htc-plant_14 |
| C Green (OxyPower HTC Technology) | https://www.c-green.se/ |
| Artec Biotechnologie GmbH | http://www.biomastec.com/partners/partner-profiles/enterprises/artec-biotechnologie-gmbh.html |
| AVA-$CO_2$ Schweiz AG | NA |
| CS Carbon Solutions Deutschland GmbH | NA |
| SmartCarbon AG | NA |
| KWT Rosenkranz GmbH | NA |

NA—not available.

being operated in China by TerraNova Energy GmbH with 14,000 t/year in the first project phase (up to 40,000 t/year in the following phase) (Ghosh, 2017) (TerraNova, 2015). Although most of the char produced at commercial-scale HTC plants is currently used as fuel, this material offers potential for further uses that involve environmental remediation and soil quality improvement (Han et al., 2016; Takaya et al., 2016). Takaya et al. (2016) compared the behavior of char and hydrochar for phosphate and ammonium adsorption from wastewater and showed that hydrochar performs the same as char.

## 6 Conclusions

Charcoal is the oldest material synthesized by the humankind. The techniques for producing charcoal and its uses, as presented at the beginning of this chapter, have substantially evolved during the centuries. The oldest methods for charcoal production, still practiced in some regions, are based on the use of earth to cover wood logs and limit the contact between wood logs and the oxygen from the air. The evolution of these rudimentary techniques started only in the last millennia, where improved earth kilns were used. Retorts, already in use in some countries at industrial scale, are an invention of the 19th century, where the old distillation industry realized that these systems performed adequately for recovering pyrolysis by-products. Unfortunately, the large expertise of the old wood distillation school was lost due to the production of methanol from coal and petroleum. This explains why the attempts on producing new equipment for wood pyrolysis had required enormous resources and time. Current retorts are the result of extensive research and hard work of researchers and companies that are looking for efficient carbonization methods. New research and demonstration plants are encouraging the carbonization industry to look for improved equipment as a way to increase efficiency and reduce the negative footprint of wood pyrolysis using old methods. The biorefinery concepts have been seen as potential pathways for making a better use of pyrolysis by-products. However, the liquid fraction (bio-oil) needs to be value added for better performance of these concepts. It is expected that metal kilns and retorts will continue to evolve and expand in the future due to the requirements of higher carbonization efficiencies and environmental regulations. It is of special interest the use of retorts and reactors with the capacity to process forest and agriculture wastes.

## Questions and problems

(1) Describe some historical milestones in the development of charcoal production and their impact on human development.
(2) What was the role of the "wood distillation industry" in the development of wood carbonization?
(3) Which country is currently producing large quantities of charcoal for its metallurgical industry? Please indicate the volume of the largest carbonization reactor used.
(4) Why is charcoal preferred over other carbonaceous materials for iron and steel processing?
(5) What happens to charcoal C and O content as the carbonization temperature increases?
(6) Discuss other consequence(s) of increasing carbonization temperature in charcoal properties.
(7) What is the role of ash on char yield and properties?
(8) Mention some impacts of using acids to pretreat biomass prior to pyrolysis.
(9) Can high ash content be considered a property that negatively impacts charcoal quality for specific uses?
(10) List some of the elements that you need to consider selecting a carbonization unit.

## Acknowledgment

T.Garcia-Perez and M.R.Pelaez-Samaniego acknowledge the support of the Department of Research of the Universidad de Cuenca (DIUC) for conducting part of this work, through project DIUC2020_014_UPS_PELAEZ_RAUL.

## References

Adam, J.C., 2009. Improved and more environmentally friendly charcoal production system using a low-cost retort-kiln (eco-charcoal). Renew. Energy 34, 1923–1925. https://doi.org/10.1016/j.renene.2008.12.009.

Ahmad, M., Rajapaksha, A.U., Lim, J.E., Zhang, M., Bolan, N., Mohan, D., Vithanage, M., Lee, S.S., Ok, Y.S., 2014. Biochar as a sorbent for contaminant management in soil and water: a review. Chemosphere. https://doi.org/10.1016/j.chemosphere.2013.10.071.

Akhtar, A., Krepl, V., Ivanova, T., 2018. A combined overview of combustion, pyrolysis, and gasification of biomass. Energy Fuel 32 (7), 7294–7318.

Antal, M.J., Grønli, M., 2003. The art, science, and technology of charcoal production. Ind. Eng. Chem. Res. 42 (8), 1619–1640. https://doi.org/10.1021/ie0207919.

Antal, M.J., Mochidzuki, K., Paredes, L.S., 2003. Flash carbonization of biomass. Ind. Eng. Chem. Res. 42 (16), 3690–3699. https://doi.org/10.1021/ie0301839.

Antal, M.J., Mok, W.S.L., Varhegyi, G., Szekely, T., 1990. Review of methods for improving the yield of charcoal from biomass. Energy Fuel 4 (3), 221–225. https://doi.org/10.1021/ef00021a001.

Argentinaforestal, 2020. Una mirada a la cadena de valor de la producción de carbón de madera en la Argentina. Retrieved April 2nd, 2021, from: https://www.argentinaforestal.com/2020/11/19/una-mirada-a-la-cadena-de-valor-de-la-produccion-de-carbon-de-madera-en-la-argentina/.

Arribas, L., Arconada, N., González-Fernández, C., Löhrl, C., González-Aguilar, J., Kaltschmitt, M., Romero, M., 2017. Solar-driven pyrolysis and gasification of low-grade carbonaceous materials. Int. J. Hydrog. Energy 42 (19), 13598–13606.

Authier, O., Ferrer, M., Mauviel, G., Khalfi, A.-E., Lédé, J., 2009. Wood fast pyrolysis: comparison of Lagrangian and Eulerian modeling approaches with experimental measurements. Ind. Eng. Chem. Res. 48 (10), 4796–4809.

Ayiania, M., Carbajal-Gamarra, F.M., Garcia-Perez, T., Frear, C., Suliman, W., Garcia-Perez, M., 2019. Production and characterization of $H_2S$ and $PO_4 3-$ carbonaceous adsorbents from anaerobic digested fibers. Biomass Bioenergy 120, 339–349.

Azadi, P., Inderwildi, O.R., Farnood, R., King, D.A., 2013. Liquid fuels, hydrogen and chemicals from lignin: a critical review. Renew. Sust. Energ. Rev. 21, 506–523.

Baldwin, H.I., 1958. The New Hampshire Charcoal Kiln. New Hampshire Forestry and Recreation Commission, Concord, NH.

Baratieri, M., Patuzzi, F., 2019. Case Study of South Tyrol (Italy): results of environmental and performance monitoring of gasification plants. In: EU-Japan Small-Scale Biomass Gasification Seminar 16 January 2019—Higashihiroshima City and Art Center Kurara, 14th Conference on Biomass Science.

Barcellos, D.C., 2017. Forno De Carvão: Como construir e adaptar para uma produção ecológica. Retrieved August 11, 2020, from: https://meunegocioflorestal.com/forno-de-carvao/.

Bates, J.S., 1922. Distillation of Hardwoods in Canada; Forestry Branch Bulletin no. 74. Canada Department of the Interior, Ottawa, ON.

Beglinger, E., 1956. Charcoal Production, Report No. 1666-11. Forest Products Laboratory, Madison, WI. Retrieved April 3rd, 2021, from https://ir.library.oregonstate.edu/downloads/r781wm27t.

Benedetti, V., Patuzzi, F., Baratieri, M., 2018. Characterization of char from biomass gasification and its similarities with activated carbon in adsorption applications. Appl. Energy 227 (August 2017), 92–99. https://doi.org/10.1016/j.apenergy.2017.08.076.

Boateng, A.A., Garcia-Perez, M., Mašek, O., Brown, R., del Campo, B., 2015. Biochar production technology, Ch. 3. In: Lehmann, J., Joseph, S. (Eds.), Biochar for Environmental Management, Science, Technology and Implementation, second ed. Routledge, NY.

Brassard, P., Godbout, S., Raghavan, V., 2017. Pyrolysis in auger reactors for biochar and bio-oil production: a review. Biosyst. Eng. 161, 80–92.

Bridgwater, A.V., 2012. Review of fast pyrolysis of biomass and product upgrading. Biomass Bioenergy 38, 68–94.

Brown, N.C., 1917. The Hardwood Distillation Industry in New York, The New York State College of Forestry at. Syracuse University. XVII(1).

Brown, R., del Campo, B., Boateng, A.A., Garcia-Perez, M., Mašek, O., 2015. Fundamentals of biochar production, Ch. 3. In: Lehmann, J., Joseph, S. (Eds.), Biochar for Environmental Management, Science, Technology and Implementation, second ed. Routledge, NY.

BTG, 2013. Charcoal production from alternative feedstocks. Retrieved April 9th, 2021, from: https://energypedia.info/images/2/20/Charcoal_Production_from_Alternative_Feedstocks_-_NL_Agency_2013.pdf.

Camou-Guerrero, A., Ghilardi, A., Mwampamba, T., Serrano, M., Avila, T.O., Vega, E., Oyama, K., Masera, O., 2014. Analysis of charcoal production in the basin of Lake Cuitzeo, state of Michoacan, Mexico: implications for a sustainable production. Investigación ambiental 6 (2).

Carrari, E., Ampoorter, E., Bottalico, F., Chirici, G., Coppi, A., Travaglini, D., Verheyen, K., Selvi, F., 2017. The old charcoal kiln sites in Central Italian forest landscapes. Quat. Int. 458 (2017), 214e223. https://doi.org/10.1016/j.quaint.2016.10.027.

Castaldelli, V.N., Akasaki, J.L., Melges, J.L.P., Tashima, M.M., Soriano, L., Borrachero, M.V., Monzó, J., Payá, J., 2013. Use of slag/sugar cane bagasse ash (SCBA) blends in the production of alkali-activated materials. Materials 6, 3108–3127. https://doi.org/10.3390/ma6083108.

Cetin, E., Gupta, R., Moghtaderi, B., 2005. Effect of pyrolysis pressure and heating rate on radiata pine char structure and apparent gasification reactivity. Fuel 84 (10), 1328–1334.

CGEE-Centro de Gestão e Estudos Estratégicos, 2015. Modernização da produção de carvão vegetal no Brasil: subsídios para revisão do Plano Siderurgia—Brasília. Centro de Gestão e Estudos Estratégicos. https://www.cgee.org.br/documents/10195/734063/Carvao_Vegetal_WEB_02102015_10225.PDF/a3cd6c7c-5b5b-450a-955b-2770e7d25f5c?version=1.3.

Charcoalkiln, 2013a. Making Charcoal from Coconut shells Using Brick Kiln. Retrieved March 20th, 2021, from: http://charcoalkiln.com/making-coals-coconut-shells-brick-kiln/.

Charcoalkiln, 2013b. Charcoal Production in 200 Liter Drum Kilns. Retrieved April 9th, 2021, from: http://charcoalkiln.com/charcoal-production-200-liter-drum-kilns/.

Charcoalkiln, 2015. Mobile Adam Retort Charcoal Kiln—Mobile Charcoal Kiln. Retrieved April, 9th, 2021, from: http://charcoalkiln.com/making-coals-coconut-shells-brick-kiln/.

Chatterjee, R., Sajjadi, B., Chen, W.-Y., Mattern, D.L., Hammer, N., Raman, V., Dorris, A., 2020. Effect of pyrolysis temperature on PhysicoChemical properties and acoustic-based amination of biochar for efficient CO2 adsorption. Front. Energy Res. 8 (85).

Chaves Fernandes, B.C., Ferreira Mendes, K., Dias Júnior, A.F., da Silva Caldeira, V.P., da Silva Teófilo, T.M., et al., 2020. Impact of pyrolysis temperature on the properties of eucalyptus wood-derived biochar. Materials (Basel, Switzerland) 13 (24), 5841.

Chen, Z., Ye, S.-Y., Li, Z.-Y., 2019. A review on charred traditional Chinese herbs: carbonization to yield a haemostatic effect. Pharm. Biol. 57 (1), 496–506. https://doi.org/10.1080/13880209.2019.1645700.

Chu, G., Zhao, J., Huang, Y., Zhou, D., Liu, Y., Wu, M., Peng, H., Zhao, Q., Pan, B., Steinberg, C.E.W., 2018. Phosphoric acid pretreatment enhances the specific surface areas of biochars by generation of micropores. Environ. Pollut. 240, 1–9.

Coelho, M.P., 2013. P. Desenvolvimento de Metodologia para o Dimensionamento de Câmaras de Combustão para Gases Oriundos do Processo de Carbonização da Madeira. Ph.D. Thesis, Federal University of Viçosa, Viçosa, Brasil.

Collard, F.-X., Blin, J., 2014. A review on pyrolysis of biomass constituents: mechanisms and composition of the products obtained from the conversion of cellulose, hemicelluloses and lignin. Renew. Sust. Energ. Rev. 38, 594–608.

Colwell, A., 2018. Japanese Charcoal Pit Kilns in the Gulf Islands: History, Archaeology, Anthropology by Stephen Nemtin. Retrieved June 17th, 2020, from: https://galianoclub.org/2018/04/japanese-charcoal-pit-kilns-in-the-gulf-islands-history-archaeology-anthropology-by-stephen-nemtin/.

Coomes, O.T., Miltner, B.C., 2016. Indigenous charcoal and biochar production: potential for soil improvement under shifting cultivation systems. Land Degrad. Dev. 00, 1–11. https://doi.org/10.1002/ldr.2500.

Costa, J.M.F.N., 2012. Temperatura Final de Carbonização e Queima dos Gases na Redução de Metano, Como Base à Geração de Créditos de Carbono. Master's Thesis, Federal University of Viçosa, Viçosa, Brasil.

Czernik, S., Bridgwater, A.V., 2004. Overview of applications of biomass fast pyrolysis oil. Energy Fuel 18 (2), 590–598.

Domac, J., Trossero, M., 2008. Industrial Charcoal Production. http://www.fao.org/tempref/GI/Reserved/FTP_FaoSeur/New%20REU/projects/tcp_cro_3101_en.pdf.

Domingues, R.R., Trugilho, P.F., Silva, C.A., Melo, I.C.N.A.D., Melo, L.C.A., Magriotis, Z.M., Sánchez-Monedero, M.A., 2017. Properties of biochar derived from wood and high-nutrient biomasses with the aim of agronomic and environmental benefits. PLoS One 12 (5), e0176884.

Dumesny, P., Noyer, J., 1908. Wood Products, Distillates and Extracts. Scott, Greenwood and Son, London.

Elkhalifa, S., Al-Ansari, T., Mackey, H.R., McKay, G., 2019. Food waste to biochars through pyrolysis: a review. Resour. Conserv. Recycl. 144, 310–320. https://doi.org/10.1016/j.resconrec.2019.01.024.

Emrich, W., 1985. In: Reidel, D. (Ed.), Handbook of Charcoal Making: The Traditional and Industrial Methods. Springer, Dordrecht, The Netherlands.

Enders, A., Hanley, K., Whitman, T., Joseph, S., Lehmann, J., 2012. Characterization of biochars to evaluate recalcitrance and agronomic performance. Bioresour. Technol. 114, 644–653.

England, H., 2018. Roman and Medieval Pottery and Tile Production: Introductions to Heritage Assets. Historic England, Swindon. Retrieved Mach 11th, 2021, from https://historicengland.org.uk/images-books/publications/iha-roman-medieval-pottery-tile-production/heag228-roman-medieval-pottery-tile-production/.

Environmental Protection Agency, 2020. Air Plan Approval; Missouri; Restriction of Emissions From Batch-Type Charcoal Kilns. Retrieved April 7th, 2021, from: https://www.govinfo.gov/content/pkg/FR-2020-02-05/pdf/2020-01300.pdf.

FAO, 1987. Simple Technologies for Charcoal Making, Rome. http://www.fao.org/3/X5328E/x5328e00.htm#Contents.

FAO, 2017. In: van Dam, J. (Ed.), The Charcoal Transition: Greening the Charcoal Value Chain to Mitigate Climate Change and Improve Local Livelihoods. Food and Agriculture Organization of the United Nations, Rome.

Ferraz, G.P., Frear, C., Pelaez-Samaniego, M.R., Englund, K., Garcia-Perez, M., 2016. Hot water extraction of anaerobic digested dairy fiber for wood plastic composite manufacturing. Bioresources 11 (4), 8139–8154.

Fitzpatrick, M., 2020. The Black Arte. Retrieved march 11th, 2021, from https://eaiainfo.org/2020/07/23/the-black-arte/.

Forest Products Laboratory, 1961. Charcoal Production, Marketing, and Use, Report No. 2213, Madison, WI. https://www.fpl.fs.fed.us/documnts/fplr/fplr2213.pdf.

Fornoparacarvao, 2021. Maggi Ecofornos. Retrieved April 1st, 2021, from: https://fornoparacarvao.com.br/fornos/.

Frías, M., Villar, E., Savastano, H., 2011. Brazilian sugar cane bagasse ashes from the cogeneration industry as active pozzolans for cement manufacture. Cem. Concr. Compos. 33 (4), 490–495. https://doi.org/10.1016/j.cemconcomp.2011.02.003.

Froehlich, P.L., Moura, A.B.D., 2014. Carvão Vegetal: Propriedades Físico-Químicas E Principais Aplicações. Tecnol. Tend. 13 (9), 1–19.

Fu, M.-M., Mo, C.-H., Li, H., Zhang, Y.-N., Huang, W.-X., Wong, M.H., 2019. Comparison of physicochemical properties of biochars and hydrochars produced from food wastes. J. Clean. Prod. 236, 117637.

Garcia-Nunez, J.A., Pelaez-Samaniego, M.R., Garcia-Perez, M.E., Fonts, I., Abrego, J., Westerhof, R.J.M., Garcia-Perez, M., 2017. Historical developments of pyrolysis reactors: a review. Energy Fuel 31 (6), 5751–5775. https://doi.org/10.1021/acs.energyfuels.7b00641.

Ghodake, G.S., Shinde, S.K., Kadam, A.A., Saratale, R.G., Saratale, G.D., Kumar, M., Palem, R.R., Al-Shwaiman, H.A., Elgorban, A.M., Syed, A., Kim, D.-Y., 2021. Review on biomass feedstocks, pyrolysis mechanism and physicochemical properties of biochar: state-of-the-art framework to speed up vision of circular bioeconomy. J. Clean. Prod. 297, 126645.

Ghosh, S.K., 2017. Waste valorisation and recycling. In: 7th IconSWM-ISWMAW. vol. 2. Springer Nature Singapore Pte Ltd. https://link.springer.com/book/10.1007/978-981-13-2784-1.

Ginzel, E.A., 1995. Steel in Ancient Greece and Rome. Retrieved March 11th, 2021, from http://dtrinkle.matse.illinois.edu/MatSE584/articles/steel_greece_rome/steel_in_ancient_greece_an.html.

Glaser, B., Haumaier, L., Guggenberger, G., Zech, W., 2001. The 'Terra Preta' phenomenon: a model for sustainable agriculture in the humid tropics. Naturwissenschaften 88, 37–41.

Gómez, C.J., Puigjaner, L., 2005. Slow pyrolysis of woody residues and an herbaceous biomass crop: a kinetic study. Ind. Eng. Chem. Res. 44 (17), 6650–6660. https://doi.org/10.1021/ie050474c.

Guillen, R.I., 2011. Adaptacion De Horno Metalico Para Fabricacion De carbon vegetal Aprovechando Raleos En Bosques Energéticos. In: Informe Final Del Proyecto ES 3.14. Retrieved November 12th, 2020, from https://www.sica.int/busqueda/busqueda_archivo.aspx?Archivo=info_86587_1_30052014.pdf.

Guo, Y., Rockstraw, D.A., 2006. Physical and chemical properties of carbons synthesized from xylan, cellulose, and Kraft lignin by $H_3PO_4$ activation. Carbon 44 (8), 1464–1475.

Haghighi Mood, S., Ayiania, M., Cao, H., Marin-Flores, O., Milan, Y.J., Garcia-Perez, M., 2021. Nitrogen and magnesium co-doped biochar for phosphate adsorption. In: Biomass Conversion and Biorefinery.

Haghighi Mood, S.H., Ayiania, M., Jefferson-Milan, Y., Garcia-Perez, M., 2020. Nitrogen doped char from anaerobically digested fiber for phosphate removal in aqueous solutions. Chemosphere 240, 124889.

Haghighi Mood, S., Hossein Golfeshan, A., Tabatabaei, M., Salehi Jouzani, G., Najafi, G.H., Gholami, M., Ardjmand, M., 2013. Lignocellulosic biomass to bioethanol, a comprehensive review with a focus on pretreatment. Renew. Sust. Energ. Rev. 27, 77–93.

Han, L., Ro, K.S., Sun, K., Sun, H., Wang, Z., Libra, J.A., Xing, B., 2016. New evidence for high sorption capacity of hydrochar for hydrophobic organic pollutants. Environ. Sci. Technol. 50 (24), 13274–13282. https://doi.org/10.1021/acs.est.6b02401.

Hansen, V., Müller-Stöver, D., Ahrenfeldt, J., Holm, J.K., Henriksen, U.B., Hauggaard-Nielsen, H., 2015. Gasification biochar as a valuable by-product for carbon sequestration and soil amendment. Biomass Bioenergy 72 (1), 300–308. https://doi.org/10.1016/j.biombioe.2014.10.013.

Hassan, M., Liu, Y., Naidu, R., Parikh, S.J., Du, J., Qi, F., Willett, I.R., 2020. Influences of feedstock sources and pyrolysis temperature on the properties of biochar and functionality as adsorbents: a meta-analysis. Sci. Total Environ. 744, 140714.

Hazel, Z., Crosby, V., Oakey, M., Marshall, P., 2017. Archaeological investigation and charcoal analysis of charcoal burningplatforms, Barbon, Cumbria, UK. Quat. Int. 458, 178–199. https://doi.org/10.1016/j.quaint.2017.05.025.

Henrich, E., Dahmen, N., Weirich, F., Reimert, R., Kornmayer, C., 2016. Fast pyrolysis of lignocellulosics in a twin screw mixer reactor. Fuel Process. Technol. 143, 151–161. https://doi.org/10.1016/j.fuproc.2015.11.003.

Hernandez, J.J., Lapuerta, M., Monedero, E., 2016. Characterisation of residual char from biomass gasification: effect of the gasifier operating conditions. J. Clean. Prod. 138, 83–93. https://doi.org/10.1016/j.jclepro.2016.05.120.

Hornung, A., Bockhorn, H., Appenzeller, K., Roggero, C.M., Tumiatti, W., 2005. Plant for the Thermal Treatment of Material and Operation Process Thereof. US Patent Number: 6,901,868.

Hornung, U., Schneider, D., Hornung, A., Tumiatti, V., Seifert, H., 2009. Sequential pyrolysis and catalytic low temperature reforming of wheat straw. J. Anal. Appl. Pyrolysis 85, 145–150.

Hrbek, J., 2016. Status Report on Thermal Biomass Gasification in Countries Participating in IEA Bioenergy task 33 (issue April). http://www.ieatask33.org/download.php?file=files/file/2016/Status report-corr_.pdf.

Hvfarmscape, 2020. Charcoal Pits. Retrieved December 20th, 2020, from: https://hvfarmscape.org/charcoal.

INEE, 2021. Projeto Carvão Verde—Fazenda São Domingos. Retrieved April 2nd, 2021, from: http://www.inee.org.br/biomassa_carvao.asp?Cat=biomassa.

Jafri, N., Wong, W.Y., Doshi, V., Yoon, L.W., Cheah, K.H., 2018. A review on production and characterization of biochars for application in direct carbon fuel cells. Process. Saf. Environ. Prot. 118, 152–166.

Jahirul, M.I., Rasul, M.G., Chowdhury, A.A., Ashwath, N., 2012. Biofuels production through biomass pyrolysis: a technological review. Energies 5, 4952–5001. https://doi.org/10.3390/en5124952.

Jarvis, J.P., 1960. The wood charcoal industry in the state of Missouri. Univ. Missouri Bull. 61 (21). Available at: https://mospace.umsystem.edu/xmlui/bitstream/handle/10355/66814/Jarvis1960-optimized.pdf?sequence=1&isAllowed=y.

Joardder, M.U.H., Halder, P.K., Rahim, A., Paul, N., 2014. Solar assisted fast pyrolysis: a novel approach of renewable energy production. J. Eng. 2014, 252848.

Kajina, W., Junpen, A., Garivait, S., Kamnoet, O., Keeratiisariyakul, P., Rousset, P., 2019. Charcoal production processes: an overview. J. Sustain. Energy Environ. 10, 19–25.

Kammen, D.M., Lew, D.J., 2005. Review of Technologies for the Production and use of Charcoal. Berkeley, CA, Renewable & Appropriate Energy Laboratory.

Kan, T., Strezov, V., Evans, T.J., 2016. Lignocellulosic biomass pyrolysis: a review of product properties and effects of pyrolysis parameters. Renew. Sust. Energ. Rev. 57, 1126–1140.

Kaushik, A., Basu, S., Singh, K., et al., 2017. Activated carbon from sugarcane bagasse ash for melanoidins recovery. J. Environ. Manage. 15, 29–34. https://doi.org/10.1016/j.jenvman.2017.05.060.

Klar, M., Rule, A., 1925. The Technology of Wood Distillation. Chapman & Hall, London.

Kumar, U., Maroufi, S., Rajarao, R., Mayyas, M., Mansuri, I., Joshi, R.K., Sahajwalla, V., 2017. Cleaner production of iron by using waste macadamia biomass as a carbon resource. J. Clean. Prod. 158, 218–224.

Kumar, A., Saini, K., Bhaskar, T., 2020. Hydochar and biochar: production, physicochemical properties and techno-economic analysis. Bioresour. Technol. 310, 123442.

Lehmann, Joseph, 2015. In: Lehmann, J., Joseph, S. (Eds.), Biochar for Environmental Management, Science, Technology and Implementation, Second ed. Routledge, NY.

Lemieux, R., Roy, C., de Caumia, B., Blanchette, D., 1987. Preliminary engineering data for scale up of a biomass vacuum pyrolysis reactor. Prepr. Pap. - Am. Chem. Soc., Div. Fuel Chem. 32 (2), 12–20.

Leng, L., Xiong, Q., Yang, L., Li, H., Zhou, Y., Zhang, W., Jiang, S., Li, H., Huang, H., 2021. An overview on engineering the surface area and porosity of biochar. Sci. Total Environ. 763, 144204.

Li, R., Zeng, K., Soria, J., Mazza, G., Gauthier, D., Rodriguez, R., Flamant, G., 2016. Product distribution from solar pyrolysis of agricultural and forestry biomass residues. Renew. Energy 89, 27–35.

Libra, J.A., Ro, K.S., Kammann, C., Funke, A., Berge, N.D., Neubauer, Y., Titirici, M.-M., Führer, C., Bens, O., Kern, J., Emmerich, K.H., 2011. Hydrothermal carbonization of biomass residuals: a comparative review of the chemistry, processes and applications of wet and dry pyrolysis. Biofuels 2 (1), 71–106. https://doi.org/10.4155/bfs.10.81.

Liu, W.-J., Jiang, H., Yu, H.-Q., 2015. Thermochemical conversion of lignin to functional materials: a review and future directions. Green Chem. 17 (11), 4888–4907.

Liu, Z., Zhang, F.S., 2009. Removal of lead from water using biochars prepared from hydrothermal liquefaction of biomass. J. Hazard. Mater. 167 (1–3), 933–939. https://doi.org/10.1016/j.jhazmat.2009.01.085.

Lucas, A., 1948. Ancient Egyptian Materials and Industries, third ed. Edward Arnold Publishers, London. Revised.

Malkow, T., 2004. Novel and innovative pyrolysis and gasification technologies for energy efficient and environmentally sound MSW disposal. Waste Manag. 24, 53–79.

Mall, I.D., Srivastava, V.C., Agarwal, N.K., 2006. Removal of orange-G and methyl violet dyes by adsorption onto bagasse fly ash—kinetic study and equilibrium isotherm analyses. Dyes Pigm. https://doi.org/10.1016/j.dyepig.2005.03.013.

Marchelli, F., Cordioli, E., Patuzzi, F., Sisani, E., Barelli, L., Baratieri, M., Arato, E., Bosio, B., 2019. Experimental study on $H_2S$ adsorption on gasification char under different operative conditions. Biomass Bioenergy 126, 106–116. https://doi.org/10.1016/j.biombioe.2019.05.003.

Meier, D., Faix, O., 1999. State of the art of applied fast pyrolysis of lignicellulosic materials-a review. Bioresour. Technol. 68, 71–77.

Meier, D., van de Beld, B., Bridgwater, A.V., Elliott, D.C., Oasmaa, A., Preto, F., 2013. State-of-the-art of fast pyrolysis in IEA bioenergy member countries. Renew. Sust. Energ. Rev. 20, 619–641. https://doi.org/10.1016/j.rser.2012.11.061.

Melligan, F., Auccaise, R., Novotny, E.H., Leahy, J.J., Hayes, M.H.B., Kwapinski, W., 2011. Pressurised pyrolysis of Miscanthus using a fixed bed reactor. Bioresour. Technol. 102 (3), 3466–3470.

Menemencioglu, K., 2013. Traditional wood charcoal productionlabour in Turkish forestry (Cankiri sample). J. Food Agric. Environ. 11 (2), 1136–1142.

Meyer, S., Glaser, B., Quicker, P., 2011. Technical, economical, and climate-related aspects of biochar production technologies: a literature review. Environ. Sci. Technol. 45 (2011), 9473–9483. https://doi.org/10.1021/es201792c.

Missaoui, A., Bostyn, S., Belandria, V., Cagnon, B., Sarh, B., Gökalp, I., 2017. Hydrothermal carbonization of dried olive pomace: energy potential and process performances. J. Anal. Appl. Pyrolysis 128 (April), 281–290. https://doi.org/10.1016/j.jaap.2017.09.022.

Mohan, D., Pittman Jr., C.U., Steele, P.H., 2006. Pyrolysis of wood/biomass: a critical review. Energy Fuel 20 (3), 848–889. https://doi.org/10.1021/ef0502397.

Mok, W.S.L., Antal, M.J., 1983. Effects of pressure on biomass pyrolysis. I. Cellulose pyrolysis products. Thermochim. Acta 68 (2–3), 155–164. https://doi.org/10.1016/0040-6031(83)80221-4.

Mok, W.S.L., Antal, M.J., Szabo, P., Varhegyi, G., Zelei, B., 1992. Formation of charcoal from biomass in a sealed reactor. Ind. Eng. Chem. Res. 31 (4), 1162–1166. https://doi.org/10.1021/ie00004a027.

Morales, S., Miranda, R., Bustos, D., Cazares, T., Tran, H., 2014. Solar biomass pyrolysis for the production of bio-fuels and chemical commodities. J. Anal. Appl. Pyrolysis 109, 65–78.

Nabukalu, C., Gieré, R., 2019. Charcoal as an energy resource: global trade, production and socioeconomic practices observed in Uganda. Resources 8, 183. https://doi.org/10.3390/resources8040183.

Nakamura, T., Kawamoto, H., Saka, S., 2008. Pyrolysis behavior of Japanese cedar wood lignin studied with various model dimers. J. Anal. Appl. Pyrolysis 81 (2), 173–182.

Nannoni, 2019. New HTC plant in Italy will produce biofertilizer and biocoal from sewage sludge. In: BE Sustainable Magazine. Retrieved June 14th, 2020, from http://www.besustainablemagazine.com/cms2/ingelia-new-htc-plant-will-be-installed-in-tuscany-italy-8051/.

Nogueira, L.A.H., Coelho, S.T., de Oliveira, A.U., 2009. Sustainable charcoal production in Brazil. In: Criteria and Indicators for Sustainable Woodfuels. FAO, Rome. Recuperado de http://www.fao.org/docrep/012/i1321e/i1321e04.pdf.

Olson, A.R., Hicock, H.W., 1941. A Portable Charcoal Kiln. Using the chimney principle. Bulletin 448. Connecticut Agricultural Experiment Station, New Heaven. https://portal.ct.gov/-/media/CAES/DOCUMENTS/Publications/Bulletins/B448pdf.pdf?la=en.

Panwar, N.L., Pawar, A., Salvi, B.L., 2019. Comprehensive review on production and utilization of biochar. SN Appl. Sci. 1, 168. https://doi.org/10.1007/s42452-019-0172-6.

Paris, O., Zollfrank, C., Zickler, G.A., 2005. Decomposition and carbonisation of wood biopolymers—a microstructural study of softwood pyrolysis. Carbon 43 (1), 53–66.

Park, J., Lee, Y., Ryu, C., Park, Y.-K., et al., 2014. Slow pyrolysis of rice straw: analysis of products properties, carbon and energy yields. Bioresour. Technol. 155, 63–70.

Patel, H., 2020. Environmental valorisation of bagasse fly ash: a review. RSC Adv. 10, 31611. https://doi.org/10.1039/d0ra06422j.

Patel, M., Zhang, X., Kumar, A., 2016. Techno-economic and life cycle assessment on lignocellulosic biomass thermochemical conversion technologies: a review. Renew. Sust. Energ. Rev. 53, 1486–1499. https://doi.org/10.1016/j.rser.2015.09.070.

Pecha, M.B., Terrell, E., Montoya, J.I., et al., 2017. Effect of pressure on pyrolysis of milled wood lignin and acid-washed hybrid poplar wood. Ind. Eng. Chem. Res. 56 (32), 9079–9089.

Pelaez-Samaniego, M.R., Garcia-Perez, M., Cortez, L.B., Rosillo-Calle, F., Mesa, J., 2008. Improvements of Brazilian carbonization industry as part of the creation of a global biomass economy. Renew. Sust. Energ. Rev. 12, 1063–1086. https://doi.org/10.1016/j.rser.2006.10.018.

Pelaez-Samaniego, M.R., Perez, J.F., Ayiania, M., Garcia-Perez, T., 2020. Chars from wood gasification for removing $H_2S$ from biogas. Biomass Bioenergy 142. https://doi.org/10.1016/j.biombioe.2020.105754, 105754.

Pelaez-Samaniego, M.R., Smith, M., Zhao, Q.Z., Garcia-Perez, T., Frear, C., Garcia-Perez, M., 2018. Charcoal from anaerobically digested dairy fiber for removal of hydrogen sulfide within biogas. Waste Manag. 76, 374–382. https://doi.org/10.1016/j.wasman.2018.03.011.

Pelaez-Samaniego, M.R., Yadama, V., Garcia-Perez, M., Lowell, E., McDonald, M., 2014. Effect of temperature during wood torrefaction on the formation of lignin liquid intermediates. J. Anal. Appl. Pyrolysis 109, 222–233. https://doi.org/10.1016/j.jaap.2014.06.008.

Pelaez-Samaniego, M.R., Yadama, V., Lowell, E., Espinoza-Herrera, R., 2013. A review of wood thermal pretreatments to improve wood composite properties. Wood Sci. Technol. 47, 1285–1319. https://doi.org/10.1007/s00226-013-0574-3.

Pinheiro, P.C., 2016. La Producción Del Carbón Vegetal. In: Energia—Investigaciones en América del Sur. Editorial de la Universidad Nacional del Sur, Ediuns, Bahia Blanca.

PNUD, 2020. Brasil. https://medium.com/@pnud_brasil/siderurgia-sustent%C3%A1vel-produ%C3%A7%C3%A3o-de-carv%C3%A3o-vegetal-4c022c0e7888.

Qin, L., Wu, Y., Hou, Z., Jiang, E., 2020. Influence of biomass components, temperature and pressure on the pyrolysis behavior and biochar properties of pine nut shells. Bioresour. Technol. 313, 123682.

Queiroz, L.A., 2014. Desempenho e Avaliação de uma Fornalha Metálica para Combustão dos Gases da Carbonização da Madeira. Master's Thesis, Federal University of Viçosa Viçosa Brasil.

Rahman, M.A., 2020. Valorizing of weeds algae through the solar assisted pyrolysis: effects of dependable parameters on yields and characterization of products. Renew. Energy 147, 937–946.

Rajapaksha, A.U., Chen, S.S., Tsang, D.C.W., Zhang, M., Vithanage, M., Mandal, S., Gao, B., Bolan, N.S., Ok, Y.S., 2016. Engineered/designer biochar for contaminant removal/immobilization from soil and water: potential and implication of biochar modification. Chemosphere 148, 276–291.

Ramos, G., Pérez-Márquez, D., 2014. Design of semi-static solar concentrator for charcoal production. Energy Procedia 57, 2167–2175.

Rehrah, D., Reddy, M.R., Novak, J.M., Bansode, R.R., Schimmel, K.A., Yu, J., Watts, D.W., Ahmedna, M., 2014. Production and characterization of biochars from agricultural by-products for use in soil quality enhancement. J. Anal. Appl. Pyrolysis 108, 301–309.

Reuters, 2016. Charcoal Burners Hope to Keep the Fire Alight. August 20th, 2021, from: https://widerimage.reuters.com/story/charcoal-burners-hope-to-keep-the-fire-alight.

Rikumo, 2016. A Kiln, a Branch and the Sea: The Making of Binchotan Charcoal. Retrieved May 5th, 2020, from https://journal.rikumo.com/journal/making-of-binchotan-charcoal.

Rodrigues, T., Junior, A.B., 2019. Charcoal: a discussion on carbonization kilns. J. Anal. Appl. Pyrolysis 143 (2019). https://doi.org/10.1016/j.jaap.2019.104670, 104670.

Rolando, V.R., 1992. Historical overview of charcoal making. In: 200 Years of Soot and Sweat (Ch. 5). Vermont Archaeological Society, Burlington.

Román, S., Libra, J., Berge, N., Sabio, E., Ro, K., Li, L., Ledesma, B., Alvarez, A., Bae, S., 2018. Hydrothermal carbonization: modeling, final properties design and applications: a review. Energies 11 (1), 1–28. https://doi.org/10.3390/en11010216.

Ronsse, F., 2013. Commercial biochar production and its certification. In: Interreg Conference. retrieved April 5th, 2021, from: http://archive.northsearegion.eu/files/repository/20140811101209_02_FrederikRonsse-Commercialbiocharproductionanditscertification.pdf.

Ronsse, F., van Hecke, S., Dickinson, D., Prins, W., 2013. Production and characterization of slow pyrolysis biochar: influence of feedstock type and pyrolysis conditions. GCB Bioenergy 5 (2), 104–115.

Rousset, P., 2006. Estado da arte das tecnologias de carbonização desenvolvidas na Europa. In: Seminário: Prática, Logística, Gerenciamento e Estratégias para o Sucesso da Conversão da Matéria Lenhosa em Carvão Vegetal para Uso na Metalurgia e Indústria. Belo Horizonte, nov 27–28. http://agritrop.cirad.fr/544698/1/document_544698.pdf.

Roy, P., Dias, G., 2017. Prospects for pyrolysis technologies in the bioenergy sector: a review. Renew. Sust. Energ. Rev. 77, 59–69.

Ryan, J.F., 2009. Wartime Woodburners. Gas Producer Vehicles in World War II. An overview. Schiffer Publishing Ltd., Atglen, PA.

Saha, B.C., Jordan, D.B., Bothast, R.J., 2009. Enzymes, industrial (overview). In: Schaechter, M. (Ed.), Encyclopedia of Microbiology, third ed. Academic Press, Oxford, pp. 281–294.

Sales, A., Lima, S.A., 2010. Use of Brazilian sugarcane bagasse ash in concrete as sand replacement. Waste Manag. https://doi.org/10.1016/j.wasman.2010.01.026.

Santos, S.F.O.M., Piekarski, C.M., Ugaya, C.M.L., Donato, D.B., Junior, A.B., Francisco, A.C., Carvalho, A.M.M.L., 2017. Life cycle analysis of charcoal production in Masonry Kilns with and without carbonization process generated gas combustion. Sustainability 9, 1558. https://doi.org/10.3390/su9091558.

Saravanakumar, A., Haridasan, T.M., 2013. A novel performance study of kiln using long stick wood pyrolytic conversion for charcoal production. Energ. Educ. Sci. Technol. 31 (2), 711–722.

Schenkel, Y., Bertauxa, P., Vanwijnbserghea, S., Carre, J., 1998. An evaluation of the mound kiln carbonization technique. Biomass Bioenergy 14 (5–6), 505–516. https://doi.org/10.1016/S0961-9534(97)10033-2.

Schure, J., Pinta, F., Cerutti, P.O., Kasereka-Muvatsi, L., 2019. Efficiency of charcoal production in sub-Saharan Africa: solutions beyond the kiln. Bois For. Trop. 340, 57–70. https://doi.org/10.19182/bft2019.340.a31691.

Shaaban, A., Se, S.-M., Dimin, M.F., Juoi, J.M., Mohd Husin, M.H., Mitan, N.M.M., 2014. Influence of heating temperature and holding time on biochars derived from rubber wood sawdust via slow pyrolysis. J. Anal. Appl. Pyrolysis 107, 31–39.

Shakoor, M.B., Ye, Z.-L., Chen, S., 2021. Engineered biochars for recovering phosphate and ammonium from wastewater: a review. Sci. Total Environ. 779, 146240.

Sharma, R.K., Wooten, J.B., Baliga, V.L., Lin, X., Geoffrey Chan, W., Hajaligol, M.R., 2004. Characterization of chars from pyrolysis of lignin. Fuel 83 (11), 1469–1482.

Shen, D.K., Gu, S., Bridgwater, A.V., 2010. Study on the pyrolytic behaviour of xylan-based hemicellulose using TG-FTIR and Py-GC-FTIR. J. Anal. Appl. Pyrolysis 87 (2), 199–206.

Siqueira, T.C.A., da Silva, I.Z., Rubio, A.J., et al., 2020. Sugarcane bagasse as an efficient biosorbent for methylene blue removal: kinetics, isotherms and thermodynamics. Int. J. Environ. Res. Public Health 17 (2). https://doi.org/10.3390/ijerph17020526.

Sobek, S., Werle, S., 2019. Solar pyrolysis of waste biomass: part 1 reactor design. Renew. Energy 143, 1939–1948.

Somna, R., Jaturapitakkul, C., Rattanachu, P., Chalee, W., 2012. Effect of ground bagasse ash on mechanical and durability properties of recycled aggregate concrete. Mater. Des. 36, 597–603. https://doi.org/10.1016/j.matdes.2011.11.065.

Song, W., Guo, M., 2012. Quality variations of poultry litter biochar generated at different pyrolysis temperatures. J. Anal. Appl. Pyrolysis 94, 138–145.

Soria, J., Zeng, K., Asensio, D., Gauthier, D., Flamant, G., Mazza, G., 2017. Comprehensive CFD modelling of solar fast pyrolysis of beech wood pellets. Fuel Process. Technol. 158, 226–237.

Sparrevik, M., Adam, C., Martinsen, V., Jubaedah, Cornelissen, G., 2015. Emissions of gases and particles from charcoal/biochar production in rural areas using medium-sized traditional and improved "retort" kilns. Biomass Bioenergy 72, 65–73.

Suliman, W., Harsh, J.B., Abu-Lail, N.I., Fortuna, A.-M., Dallmeyer, I., Garcia-Perez, M., 2016. Influence of feedstock source and pyrolysis temperature on biochar bulk and surface properties. Biomass Bioenergy 84, 37–48.

Sun, L., Chen, D., Wan, S., Yu, Z., 2015. Performance, kinetics, and equilibrium of methylene blue adsorption on biochar derived from eucalyptus saw dust modified with citric, tartaric, and acetic acids. Bioresour. Technol. 198, 300–308.

Tag, A.T., Duman, G., Ucar, S., Yanik, J., 2016. Effects of feedstock type and pyrolysis temperature on potential applications of biochar. J. Anal. Appl. Pyrolysis 120, 200–206.

Taha, S.M., Amer, M.E., Elmarsafy, A.E., Elkady, M.Y., 2014. Adsorption of 15 different pesticides on untreated and phosphoric acid treated biochar and charcoal from water. J. Environ. Chem. Eng. 2 (4), 2013–2025.

Takaya, C.A., Fletcher, L.A., Singh, S., Anyikude, K.U., Ross, A.B., 2016. Phosphate and ammonium sorption capacity of biochar and hydrochar from different wastes. Chemosphere 145, 518–527. https://doi.org/10.1016/j.chemosphere.2015.11.052.

Tan, X., Liu, Y., Zeng, G., Wang, X., Hu, X., Gu, Y., Yang, Z., 2015. Application of biochar for the removal of pollutants from aqueous solutions. Chemosphere 125, 70–85.

Teixeira, S.R., De Souza, A.E., De Almeida Santos, G.T., Peña, A.F.V., Miguel, Á.G., 2008. Sugarcane bagasse ash as a potential quartz replacement in red ceramic. J. Am. Ceram. Soc. https://doi.org/10.1111/j.1551-2916.2007.02212.x.

TerraNova, 2015. Project Kaiserslautern—Sludge Drying. Retrieved June 14th, 2020, from: https://terranova-energy.com/en/service/projects/.

Tintner, J., Preimesberger, C., Pfeifer, C., Theiner, J., Ottner, F., Wriessnig, K., Puchberger, M., Smidt, E., 2020. Pyrolysis profile of a rectangular kiln—natural scientific investigation of a traditional charcoal production process. J. Anal. Appl. Pyrolysis 146. https://doi.org/10.1016/j.jaap.2019.104757, 104757.

Titirici, M.A., 2012. Hydrothermal carbons: synthesis, characterization, and applications. In: Novel Carbon Adsorbents., https://doi.org/10.1016/B978-0-08-097744-7.00012-0.

Tomas, S., Planiniæ, M., Buciæ-Kojiæ, A., Biliæ, M., Veliæ, D., Èaèiæ, M., 2009. Technological solution for the sustainability of the destructive distillation of wood in classic horizontal retorts. Chem. Biochem. Eng. Q. 23 (4), 555–561.

Trinh, T.N., Jensen, P.A., Dam-Johansen, K., Knudsen, O.K., Sørensen, H.R., Hvilsted, S., 2013. Comparison of lignin, macroalgae, wood, and straw fast pyrolysis. Energy Fuel 27, 1399–1409. https://doi.org/10.1021/ef301927y.

UNdata, 2020. FAOSTAT. Wood charcoal. Retrieved April 19th, 2020, from: http://data.un.org/Data.aspx?q=charcoal&d=FAO&f=itemCode%3a1630.

Uzun, B.B., Apaydin-Varol, E., Ateş, F., Özbay, N., Pütün, A.E., 2010. Synthetic fuel production from tea waste: characterisation of bio-oil and bio-char. Fuel 89 (1), 176–184.

Vakalis, S., Sotiropoulos, A., Moustakas, K., Malamis, D., Baratieri, M., 2016. Utilisation of biomass gasification by-products for onsite energy production. Waste Manag. Res. https://doi.org/10.1177/0734242X16643178.

Valix, M., Cheung, W.H., McKay, G., 2004. Preparation of activated carbon using low temperature carbonisation and physical activation of high ash raw bagasse for acid dye adsorption. Chemosphere. https://doi.org/10.1016/j.chemosphere.2004.04.004.

Veitch, F.P., 1907. Chemical Methods for Utilizing Wood. U.S. Department of Agriculture, Bureau of Chemistry Circular No. 36.

Venderbosch, R.H., Prins, W., 2010. Fast pyrolysis technology development. Biofuels Bioprod. Biorefin. 4, 178–208.

de Vilela, A.O., Lora, E.S., Quintero, Q.R., Vicintin, R.A., Suza, T.P.S., 2014. A new technology for the combined production of charcoal and electricity through cogeneration. Biomass Bioenergy 69, 222–240. https://doi.org/10.1016/j.biombioe.2014.06.019.

Vuthisa, 2018. Charcoal Production Techniques In South Africa. Retrieved April 8th, 2021, from: https://vuthisa.com/2018/10/24/charcoal-production-techniques-in-south-africa/.

Wang, K., Brown, R.C., 2013. Catalytic pyrolysis of microalgae for production of aromatics and ammonia. Green Chem. 15, 675–681.

Wang, S., Dai, G., Yang, H., Luo, Z., 2017. Lignocellulosic biomass pyrolysis mechanism: a state-of-the-art review. Prog. Energy Combust. Sci. 62, 33–86.

Wang, S., Guo, X., Liang, T., Zhou, Y., Luo, Z., 2012. Mechanism research on cellulose pyrolysis by Py-GC/MS and subsequent density functional theory studies. Bioresour. Technol. 104, 722–728.

Wang, L., Skreiberg, Ø., Gronli, M., Specht, G.P., Antal, M.J., 2013. Is elevated pressure required to achieve a high fixed-carbon yield of charcoal from biomass? Part 2: the importance of particle size. Energy Fuel 27 (4), 2146–2156. https://doi.org/10.1021/ef400041h.

Weldekidan, H., Strezov, V., Li, R., Kan, T., Town, G., Kumar, R., He, J., Flamant, G., 2020. Distribution of solar pyrolysis products and product gas composition produced from agricultural residues and animal wastes at different operating parameters. Renew. Energy 151, 1102–1109.

Weldekidan, H., Strezov, V., Town, G., 2018. Review of solar energy for biofuel extraction. Renew. Sust. Energ. Rev. 88, 184–192.

Westra, A., 2020. Ancient Chinese Inventions that Will Surprise You. Vol. 10. Retrieved December 20th, from https://www.thecollector.com/ancient-chinese-inventions/.

Wiedner, K., Rumpel, C., Steiner, C., Pozzi, A., Maas, R., Glaser, B., 2013. Chemical evaluation of chars produced by thermochemical conversion (gasification, pyrolysis and hydrothermal carbonization) of agro-industrial biomass on a commercial scale. Biomass Bioenergy 59, 264–278. https://doi.org/10.1016/j.biombioe.2013.08.026.

Xiao, X., Chen, Z., Chen, B., 2016. H/C atomic ratio as a smart linkage between pyrolytic temperatures, aromatic clusters and sorption properties of biochars derived from diverse precursor materials. Sci. Rep. 6, 22644.

Xie, Y., Zeng, K., Flamant, G., Yang, H., Liu, N., He, X., Yang, X., Nzihou, A., Chen, H., 2019. Solar pyrolysis of cotton stalk in molten salt for bio-fuel production. Energy 179, 1124–1132.

Xu, Q., Ji, T., Gao, S.J., Yang, Z., Wu, N., 2018. Characteristics and applications of sugar cane bagasse ash waste in cementitious materials. Materials. https://doi.org/10.3390/ma12010039.

Yaman, S., 2004. Pyrolysis of biomass to produce fuels and chemical feedstocks. Energy Convers. Manag. 45 (5), 651–671.

Yang, H., Yan, R., Chen, H., Lee, D.H., Zheng, C., 2007. Characteristics of hemicellulose, cellulose and lignin pyrolysis. Fuel 86 (12), 1781–1788.

You, S., Ok, Y.S., Chen, S.S., Tsang, D.C.W., Kwon, E.E., Lee, J., Wang, C.H., 2017. A critical review on sustainable biochar system through gasification: energy and environmental applications. Bioresour. Technol. 246 (June), 242–253. https://doi.org/10.1016/j.biortech.2017.06.177.

Zeaiter, J., Ahmad, M.N., Rooney, D., Samneh, B., Shammas, E., 2015. Design of an automated solar concentrator for the pyrolysis of scrap rubber. Energy Convers. Manag. 101, 118–125.

Zeng, K., Flamant, G., Gauthier, D., Guillot, E., 2015a. Solar pyrolysis of wood in a lab-scale solar reactor: influence of temperature and sweep gas flow rate on products distribution. Energy Procedia 69, 1849–1858.

Zeng, K., Gauthier, D., Li, R., Flamant, G., 2015b. Solar pyrolysis of beech wood: effects of pyrolysis parameters on the product distribution and gas product composition. Energy 93, 1648–1657.

Zeng, K., Gauthier, D., Lu, J., Flamant, G., 2015c. Parametric study and process optimization for solar pyrolysis of beech wood. Energy Convers. Manag. 106, 987–998.

Zeng, K., Minh, D.P., Gauthier, D., Weiss-Hortala, E., Nzihou, A., Flamant, G., 2015d. The effect of temperature and heating rate on char properties obtained from solar pyrolysis of beech wood. Bioresour. Technol. 182, 114–119.

Zhang, M., Geng, Z., Yu, Y., 2011. Density functional theory (DFT) study on the dehydration of cellulose. Energy Fuel 25 (6), 2664–2670.

Zhang, Z., Zhu, Z., Shen, B., Liu, L., 2019. Insights into biochar and hydrochar production and applications: a review. Energy 171, 581–598. https://doi.org/10.1016/j.energy.2019.01.035.

Zhao, L., Cao, X., Zheng, W., Kan, Y., 2014. Phosphorus-assisted biomass thermal conversion: reducing carbon loss and improving biochar stability. PLoS One 9 (12), e115373.

Zhao, B., O'Connor, D., Zhang, J., Peng, T., Shen, Z., Tsang, D.C.W., Hou, D., 2018. Effect of pyrolysis temperature, heating rate, and residence time on rapeseed stem derived biochar. J. Clean. Prod. 174, 977–987.

Zhao, L., Zheng, W., Mašek, O., Chen, X., Gu, B., Sharma, B.K., Cao, X., 2017. Roles of phosphoric acid in biochar formation: synchronously improving carbon retention and sorption capacity. J. Environ. Qual. 46 (2), 393–401.

Zhou, X., Li, W., Mabon, R., Broadbelt, L.J., 2017. A critical review on hemicellulose pyrolysis. Energ. Technol. 5 (1), 52–79.

Zhu, R., Zhang, B., Liu, F., Cai, C., Wang, Z., 2016. A Social History of Middle-Period China. Cambridge University Press, The Song, Liao, Western Xia and Jin Dynasties, p. 2016.

Zolfi-Bavariani, M., Ronaghi, A., Ghasemi-Fasaei, R., Yasrebi, J., 2016. Influence of poultry manure-derived biochars on nutrients bioavailability and chemical properties of a calcareous soil. Arch. Agron. Soil Sci. 62 (11), 1578–1591. https://doi.org/10.1080/03650340.2016.1151976.

# Physicochemical characterization of biochar derived from biomass

### Sergio C. Capareda
*Texas A&M University, College Station, TX, United States*

#### Learning objectives
Upon completion of this chapter, one should be able to:
a. Describe the various methods of characterizing the initial physical and thermal properties of biomass important for biochar production
b. Describe the various procedures for estimating important biochar properties for various commercial uses
c. List the various equipment used for biochar production from biomass resources
d. Enumerate and describe the different standard methods for characterizing biochar
e. Describe other biochar properties to improve its commercial value and for other potential applications for contaminants adsorption.

## 1 Introduction

This chapter deals with important properties that are commonly used for evaluating the quality of biochar produced from biomass. Procedures are similar for estimating biomass properties. These characterization properties play a significant role in the improvement of the economic value of biochar. Biochar may be used for energy-related purposes such as biochar pellets for space heating in pelleted space heaters, cofiring with coal, cogasification with other biomass, as well as simple cooking and heating needs. Biomass resources are so diverse in their physical, chemical, and biological properties that there is not a single property that defines the resulting biochar. The International Biochar Initiative (IBI) listed numerous standards for commercial biochar as cited by Lehmann and Joseph (2009). For example, biochar may have certain storage class, fertilizer class, liming class, and particle size class. Other related properties include its potential for soil remediation and as a fertilizer use, carbon sequestration, and carbon credits economy. The sustainability of biochar use relies on certainty of its product properties. Hence, users and sellers must adhere to some form of standards to define the characteristics of a particular biochar. No two biochar materials are exactly alike. There are differences in the raw materials used, the type of preparation done prior to conversion, conversion temperature used, as well as rate of conversion. Numerous biochar characterization parameters will be discussed beginning with the physical properties and equipment needed for characterization. The physical and thermal properties discussed in this chapter including other important biochar properties are shown in Fig. 3.1. Note that the source of biochar must first be known and characterized. Hence, we refer the reader to previous publications where thorough biomass characterization is described (Capareda, 2014).

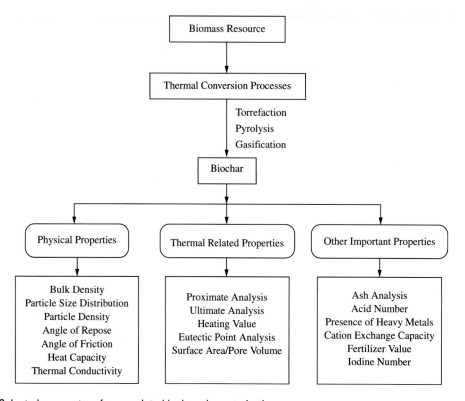

**Fig. 3.1** Selected parameters for complete biochar characterization.

The important physical properties of biochar include particle and bulk density, particle size, and particle size distribution. Other properties that are important in biochar handling and storage include angle of repose, friction, heat capacity, and heat conductivities. The major thermally related properties of biochar include proximate analysis, ultimate analysis, and heating values. One other important thermally related biochar property is the eutectic point or the fusion temperature of ash in the biochar. This parameter is important in slagging and fouling problems in biochar thermal conversion.

Biochar is an important material for water and wastewater treatments. The primary reason is its large surface area and pore volume, rich in organic carbon content and mineral components and abundance in functional groups with the ability for adsorption and absorption (Wang et al., 2020). However, its raw form is still inferior to commercial activated carbon adsorbents. There are now massive research publications on the application of biochar for water and wastewater treatment processes. However, numerous raw biochar materials require modification to improve its performance compared with commercial adsorption alternatives coming from high grade coal. One other issue concerning biochar is the presence of some toxic materials including heavy metals and must be removed prior to commercial use. This is where the importance of biochar standards becomes very useful.

There are numerous sources of biochar for commercial use. Biomass comes in many forms. The most common source of biochar comes from forest materials followed by agricultural wastes after harvest. Hence, the quality of biochar would depend on its initial origin. Commercial activated carbons usually come from high grade coal. The major source of activated carbon from terrestrial biomass comes from coconut shells. Coconut shells are hard and durable resulting in excellent biochar properties suitable for adsorption and absorption processes. Agricultural wastes such as those coming from rice and corn harvests will yield biomass with low bulk density and hence when converted into biochar will result in powdered form and may have inferior value to granular activated carbon coming from denser biomass. The type of biomass conversion processes also affects the quality of biochar output. Careful selection of initial biomass resource is key to a highly valuable biochar product.

## 2 Thermal conversion processes for the production of biochar

Biochar materials are primarily produced from thermal conversion of organic materials. To be most effective, the complete absence of oxidant is necessary. These thermal conversion

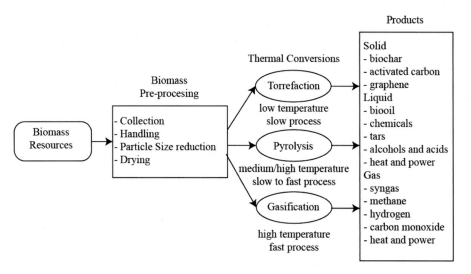

**Fig. 3.2** Various thermal conversion processes for biomass conversion into biochar.

processes are usually categorized as torrefaction, pyrolysis, and gasification (Fig. 3.2). Torrefaction process is conducted at low temperature and usually at slow rate while pyrolysis is at higher temperature and at higher rates of conversion and sometimes at elevated pressures. Both of these processes are done in complete absence of oxidant. Gasification on the other hand uses partial oxidant for the conversion of biomass into biochar. Hence, the latter process is more energy efficient, since part of the biomass is used for sustainable biochar production. Torrefaction and pyrolysis require external energy for the process to proceed. Fig. 3.2 shows the schematic of the various thermal conversion processes suitable for biochar production showing temperature ranges as well as rates of conversion including products produced.

When biomass materials are exposed to heat, irreversible chemical changes occur and biomass is transformed into either its solid, liquid, or gaseous components. The quality and amount of these products are related to the temperature ranges used as well as how fast the reaction occurred including variety of reactors.

The primary advantage of torrefaction as a conversion process for making biochar is its higher yield, lower energy required due to low temperature used in the process, and unwanted components such as moisture is removed. Little amounts of condensable liquid results from this process. Pyrolysis has lower yield that torrefaction due to higher temperature used but the throughput is increased substantially. Pyrolysis will result in liquid products called biooil, a complex mixture of organic acids and alcohols with 100 other compounds that require purification and upgrade to generate valuable fuels. Be mindful that both torrefaction and pyrolysis require external heat from other sources to accomplish the conversion process.

Gasification on the other hand is a thermal conversion process using partial oxidant. Biomass is partially converted to generate combustible synthesis gas (or also called syngas or producer gas); primary carbon monoxide (CO) and hydrogen ($H_2$) and are produced and used to further heat up the reactor for continuous sustainable conversion process. Only the initial heating is required in the reactor to begin the conversion process. Thus, gasification process is endothermic and hence more sustainable. However, yields are lower due to the partial use of biomass while gaseous products are produced reducing mass yields of biochar. Rule of thumb one could you to make an estimate of biochar yield by gasification is at least 20% (Capareda, 1990). The syngas must be used onsite and may be used for electrical power and heat. Tar is also produced during gasification and must be removed from the gas streams when the syngas is used in internal combustion engine (ICE) to generate mechanical or electrical power. Tar is a serious problem in gasification for power using ICE and must be addressed seriously to maintain high-quality biochar.

Reactor types vary for all these three conversion processes. All have batch type of reactors as well as advanced fluidized bed systems. However, in fluidized bed systems, inert gas must be used to fluidize the reactor bed. Slow torrefaction and pyrolysis processes may use an auger-based reactor to move the biomass while being heated to accomplish the conversion process. Numerous companies have used auger-based reactor for large-scale commercial production of biochar (e.g., Dynamotive Energy Systems Corporation, Richmond, Canada). However, the area footprint for this type of reactors is quite large with rather low throughput. Fluidized bed systems are highly ideal for large output in commercial facilities. For fluidized bed torrefaction and pyrolysis processes, nitrogen is the common inert gas used and usually recovered from air by removing the carbon dioxide and other minute unwanted gases to generate very high purity nitrogen fluidizing medium.

Combustion process, a thermal conversion process using complete stoichiometric amounts of oxidant, is never used for making biochar. The primary product of this process is heat, carbon dioxide ($CO_2$), moisture ($H_2O$), and not so valuable ash. When the product of biomass thermal conversion is grayish in color, the process is usually combustion. However, when the solid product is black, biochar is usually the end product. Technically, biochar is a material with at least 1.5% carbon in it. Hence, the resulting color of the final product is generally dark or black.

Note that when biochar is produced at elevated temperature, care must be made to ensure it does not come in contact with air or oxidant; otherwise, it will spontaneously turn into ash and would lose its valuable properties, especially its high carbon content.

## 3 Physical properties of biochar

There are numerous biochar physical properties that must be measured to report its quality. Usually, the important properties are associated with biochar storage, conveyance, and transport. One would like to have biochar with very high bulk or particle density and with uniform particle size range. The facilities to use as well as standard procedures for characterization are outlined in this section.

### 3.1 Particle density

Particle density is the measure of the biomass weight per unit volume ($kg/m^3$ or $lb./ft^3$). The most common device used for measurement of particle density is the pycnometer. One particular brand comes from Micromeritics (Model Accupyc 1330, Norcross, GA). The basic principle for particle density measurements is very easy to understand by noting the unit used. The numerator is the mass and the denominator is the volume occupied by the physical material. A biochar sample is simply weighed accurately and placed in a container of known and calibrated volume. This container is then placed in an enclosed system where all void spaces are filled with an inert gas (normally helium) and the volume occupied by the biomass is calculated. Thus, this parameter is the measure of the biomass particle weight per unit of volume which the biomass particle has occupied. Sometimes this is also referred to as the true particle density of the material. The measurement procedure is very fast and can be easily replicated with great accuracy.

### 3.2 Bulk or apparent density

The bulk or apparent density parameter has the same units as particle density ($kg/m^3$) but the volume unit includes the void spaces occupied by the biochar. The standard method for measuring bulk density of biomass may also be used to determine the bulk density of biochar. One such standard is the ASTM E 873-82 (2006) entitled

"Standard Test Method for Bulk Density of Densified Particulate Biomass Fuels." The bulk density of biomass or biochar is easily measured using a simple box measuring around 305 × 305 × 305 mm (12 × 12 × 12 in) with appropriate handle. The volume of this box should be within 16.39 cm$^3$ (1 in$^3$) for improved accuracy. A scale is needed with the capacity to weigh the box sample to within 100 g. The quantity of the sample must be large enough to be representative but not less than 45.45 kg (100 lbs). The tare weight of the box is first known (to within 100 g, 0.22 lb) and a certain amount of biochar is poured in this container from a height of about 610 mm (2 ft). The box is dropped about 5 times from a height of about 150 mm (6 in) to a nonresilient surface to allow settling. After dropping, one should replenish additional biochar samples and excess samples are removed from the top edge. The box is then weighed to within 100 g (0.22 lb). The bulk density is calculated using Eq. (3.1).

$$\text{Bulk density} \left(\frac{\text{g}}{\text{cm}^3}\right) = \frac{(\text{weight of box and sample}) - (\text{weight of box})}{(\text{volume of box})}$$

(3.1)

The volume occupied by the biochar is the denominator of the bulk density equation. The numerator is the weight of the biochar. The bulk density value varies with how well the biochar is packed in the container. Biochar with high bulk density is suitable for transport since this will require less storage volume and more weight.

The bulk density of biochar made from different type of woods processed in different types of traditional kilns ranged from 300 to 430 kg/m$^3$ [18.70–26.8 lbs/ft$^3$] (Downie et al., 2009). Some other research publications have different reported biochar bulk density range of 200–300 kg/m$^3$ [12.5–18.7 lbs/ft$^3$] (Lata and Mande, 2009). Densified products (or briquettes) may have bulk densities as high as 700–800 kg/m$^3$ [43.6–50 lbs/ft$^3$] (Capareda, 2014).

Bulk densities are also given in literature for activated carbons used for gas adsorption. They range from 400 to 500 kg/m$^3$ [24.94–31.17 lbs/ft$^3$]; while activated carbons used form decolorization, the range is around 250–750 kg/m$^3$ [15.59–46.76 lbs./ft$^3$] (Rodriguez-Reinoso, 1977).

## 3.3 Particle size and particle size distribution (PSD)

Biochar materials, after being thermally converted, will need to be processed for its particle size distribution. Some unwanted particle sizes must be separated to improve its marketability. Thus, freshly pyrolyzed biochar must be separated into its large, medium, or small finer particles for appropriate packaging and storage. Biochar materials are usually already in ground form. However, they are usually characterized for physical dimensions. Storing bulky materials is always costly.

The particle size or particle size distribution (PSD) of the biochar material is measured using various standard sieves. Powdered biochar materials are normally separated. The usual sieve size for ground biochar is around less than 100 μm in size or the lowest screen size in Tyler sieve set. The particle size distribution of biochar is first separated using sieve shaker.

Table 3.1 shows the list of standard sieve screen sizes from 37 to 3360 μm. Ground samples of biochar are placed on different sets of sieve combinations from higher pore space on top to the smaller screen size at the bottom. A known amount of biochar is placed on the top sieve and then subjected to several minutes of shaking (normally about 15–20 min). A mechanical sieve shaker is more appropriate for the purpose.

The biochar collected at each size range is weighed. The percentage of each size range is then calculated based on the original weight of the biochar used for the test. For particle sizes less than 100 μm, a different set of equipment is used: the Beckman Coulter Counter Multisizer 3 (Beckman Coulter, Brea, CA) and/or the Malvern Mastersizer 3000 (Malvern Instruments, Worcestershire, UK).

In the Beckman Coulter Counter unit, the biochar particles are dispersed using a sonicator in a solvent to break apart each component particle. Then, each particle is passed through an aperture where the measured particle size is plotted. The Malvern Mastersizer is operated similarly and has both dry and wet options for measuring the PSD of the material. In both cases, the particle equivalent spherical diameter (ESD, in microns) is first reported by the instrument. The equipment would usually assume spherical shape of the material. However, biochar samples are not spherical in nature. The particle aerodynamic equivalent diameter (AED, in microns) of the material is then calculated by multiplying the ESD (which is reported by the instruments) with the square root of the particle density (g/cm$^3$), as shown in Eq. (3.2).

$$AED = ESD\sqrt{\rho} \qquad (3.2)$$

The particle densities are those values measured using the AccuPyc 1330 Pycnometer. Particle sizes of this range are normally reported in AED, since biochar particles are not spherical in nature. However, most instruments approximate the particle sizes as a sphere. From this analysis, the particle size distribution of the sample may be plotted with the abscissa as the aerodynamic diameter (AED) and the ordinate as the cumulative percent

**Table 3.1** Table of standard sieve screen sizes.

| Tyler Mesh number | US standard ASTME 11-61 | British standard BSS 410 1969 | Aperture Inch | Aperture Microns ($\mu$) |
|---|---|---|---|---|
| 6 | 6 | 5 | 0.131 | 3360 |
| 7 | 7 | 6 | 0.111 | 2830 |
| 8 | 8 | 7 | 0.0937 | 2380 |
| 9 | 10 | 8 | 0.0787 | 2000 |
| 10 | 12 | 10 | 0.0661 | 1680 |
| 12 | 14 | 12 | 0.0555 | 1410 |
| 14 | 16 | 14 | 0.0469 | 1190 |
| 16 | 18 | 16 | 0.0394 | 1000 |
| 20 | 20 | 18 | 0.0331 | 840 |
| 24 | 25 | 22 | 0.0280 | 707 |
| 28 | 30 | 25 | 0.0232 | 595 |
| 32 | 35 | 30 | 0.0197 | 500 |
| 35 | 40 | 36 | 0.0165 | 420 |
| 42 | 45 | 44 | 0.0138 | 354 |
| 48 | 50 | 52 | 0.0117 | 297 |
| 60 | 60 | 60 | 0.0098 | 250 |
| 80 | 80 | 85 | 0.0070 | 177 |
| 100 | 100 | 100 | 0.0059 | 149 |
| 150 | 140 | 150 | 0.0041 | 105 |
| 200 | 200 | 200 | 0.0029 | 74 |
| 250 | 230 | 240 | 0.0025 | 63 |
| 400 | 400 | 400 | 0.0015 | 37 |

volume of the sample. Another way of plotting the particle size distribution (PSD) is the abscissa as the particle diameter in microns (from 0 to 100) and the ordinate as the percent mass of the sample. The ordinate may also be the cumulative mass of the sample to determine percentages of particles less than or greater than a particular size range.

## 3.4 Angle of repose

The angle of repose is the steepest angle of descent or the slope relative to the horizontal plane when a biochar material is poured on a horizontal surface of known surface finish or material. For example, when granular biochar materials are poured from a funnel and dropped on a horizontal surface, a mound or a conical pile is formed. The angle or slope of the mound is called the angle of repose. This angle is technically defined as the internal angle between the surface of the pile and the horizontal surface. The factors affecting this angle are the density of the material, the surface area and shape of the particles, as well as the coefficient of friction of the material. This parameter is important in the design of equipment for biochar material storage, particularly silos and hoppers for biochar storage and conveying systems. If a belt conveyor is used to transport the biochar material, the size or width of this belt is dependent upon the angle of repose for transporting a given volume of mass of the material without overflowing along the belt.

## 3.5 Angle of friction

The angle of friction, on the other hand, is that angle with which a pile of biochar will slide when raised from the horizontal. This term is also synonymous with the angle of friction of the biochar material on a given surface. One observes that if the surface is smooth, the biochar slides at a lower angle than when the surface is rather rough. This parameter is also important in the design of conveying systems and hoppers for biochar fuel bins. Proper use of angle of friction prevents bridging from biochar feed bins or conveying systems.

## 3.6 Heat capacity and thermal conductivity of biochar

The heat capacity of a material is defined as the ratio of the amount of heat energy transferred to the material to the resulting increase in temperature of this material. In SI system of units, the heat capacity is given in units of Joules/Kelvin. There is currently limited research and few publications about the heat capacity of biochar materials. Information is lacking because heat capacity is an extensive property, that is, its physical characteristics vary with the scale of the system. In some applications, specific heat capacity is more practical to use, since this is the heat energy per unit of mass at a given temperature. This parameter is now being used in the modeling of some biochar processes and utilization.

The thermal conductivity, on the other hand, is the ability of the material to conduct or transfer heat. Heat transfer of the material with high thermal conductivity is higher than that with lower thermal conductivity. Consequently, materials with high thermal conductivity are used to store heat energy (as heat sink) while those with low conductivity are used as insulation. Most biochar materials are more suitable for insulation than as heat sinks. The units of thermal conductivity are W/m-K [Btu/hr-ft]. Likewise, thermal conductivity is used for modeling biochar in storage systems to measure its ability to conduct heat through the enclosure as well as its ability to store heat energy. Recent studies show that wood-based biochar contributes to the thermal conductivity properties of soil. The reported general trend is that the average thermal conductivity and heat capacity of the soil-biochar mixture were increased by 27%–33% and 48%–68%, respectively. This substantial increase was attributed to the fact that the smaller particles partially fill in the spaces between larger particles and thus lesser content of air which has much lower thermal conductivity and heat capacity than the organic matter added. The value of the thermal conductivity of biochar mixture was reported at 0.132 W/mK and its heat capacity was reported at 0.913 MJ/$m^3$K. The authors also published thermal diffusivity of biochar samples with a value of 0.145 $mm^2$/s (Usowich et al., 2016).

## 4 Important thermal related properties

There are standard test methods for biochar characterization developed by the American Society for Testing Materials (ASTM). They were originally designed for biomass but are now extensive used to establish biochar properties as well. There are also methods developed by the US Department of Energy (USDOE) National Renewable Energy Laboratory (NREL). The ASTM standards originated from the characterization of wood, coal, or coke. The three common sets of standards commonly used for biochar characterization as it relates to its thermal conversion properties are the proximate, ultimate, and heating value analysis. For further thermal conversion of biochar, analysis of its melting point of the ash, or called the eutectic point of its ash, becomes important. The test methods used originally were for the evaluation of coal. The test methods described here may be used to establish the rank of biochar as fuel and to show the ratio of combustibles to incombustibles and sometimes to provide the basis for marketing the biochar for thermal conversion purposes.

### 4.1 Proximate analysis

Proximate analyses include the measurement of moisture content (MC), volatile combustible matter (VCM), fixed carbon (FC), and ash. Moisture is important for biochar transport to

**Table 3.2** List of ASTM standards used for biomass proximate analysis for wood fuel, coal, and coke.

| ASTM designation | Description |
| --- | --- |
| E 871-82 (Reapproved 2006) | MC for particulate wood fuels |
| E 872-82 (Reapproved 2006) | VCM for particulate wood fuels |
| D 1103-84 (Reapproved 2007) | Ash analysis of wood |
| D 3172-07a | Proximate analysis of coal and coke |
| D 3173-03 (Reapproved 2008) | MC for coal and coke |
| D 3174-04 | Ash in coal and coke |
| D 3175-07 | VCM for coal and coke |

biorefineries and would also impact the type of conversion process used. The moisture content could be as high as 50% for some biochar samples exposed in open atmosphere. Drying biochar would require some amounts of energy. Almost always, the most economical method for drying biochar is through the sun.

Table 3.2 lists the various ASTM standards for measurement of proximate analysis. Both the coal industry and the wood industry have their own standards, and the standards used for wood fuels are those that are closely related to the biomass. However, the differences in the procedures are quite minor. They are also used for biochar characterization without modifications.

### 4.1.1 Moisture content (MC)

To measure the biochar moisture, the sample (as received) is normally ground to reduce the particle size if the particle size is quite large. Replicated amounts are placed in a dry container in a forced convection oven with a temperature setting of 103°C (ASTM E871-82 [reapproved 2006] for wood). This is normally done overnight, or for at least 16 h, until the weight of the biochar (with the container) becomes constant. The equation to calculate the moisture content for both wet and dry basis is presented here. A similar standard is used for coal or coke samples (ASTM D 3173-03 [Reapproved 2008]). The difference is that the temperature used for coal and coke is between 104 and 110°C. Eq. (3.3) is used to measure the moisture content of biochar.

$$\text{Moisture content } (\%MC) = \frac{m_i - m_f}{m_i} \times 100\% \qquad (3.3)$$

The $m_i$ is the initial weight of the biochar sample before placing in the drying oven and $m_f$ is the final constant weight after drying. This moisture content, if the biochar is exposed to atmosphere over long periods of time, is referred to as the "as-received" basis. Moisture is very important in the transportation of biochar, as well as during thermal conversion. If the moisture content of biochar is quite high, additional energy is required to remove the water and this will definitely affect the conversion efficiency.

The equipment for the measurement of biochar moisture content is a simple gravimetric forced convection dryer. The unit must have the ability to accurately control and maintain the temperature within the range of 104°C–110°C. The door of the unit should contain a hole of approximately 3.2 mm (1/8 in.) in diameter near the bottom to permit the free flow of air through the oven space.

### 4.1.2 Volatile combustible matter (VCM)

To measure the volatile combustible matter (VCM), the bone-dry biochar sample is placed in a platinum crucible with a close-fitting cover and placed in a tube furnace. The temperature setting is usually around 950±20°C (ASTM E872-82 [reapproved 2006] for wood). The material is exposed for about 7 min, cooled and reweighed. The loss in weight is from the VCM of the biochar. Likewise, a similar procedure is used for coal or coke samples (ASTM D 3175-07). The temperature used is the same (950±20°C) including the time of exposure (7 min). Eq. (3.4) may be used to measure the volatile combustible matter (VCM) of the biomass.

$$\text{Volatile combustible matter } (\%\text{VCM}) = \frac{v_i - v_f}{v_i} \times 100\% \quad (3.4)$$

In Eq. (3.4), $v_i$ is the initial weight of the biochar sample before it is placed in the tube furnace, and $v_f$ is the final constant weight after cooling and removal of volatiles. Volatiles are important in thermal conversion, as they lead to more combustible gases during the conversion. The ASTM standard calls for the use of specially made platinum crucibles for the test. However, these crucibles are quite expensive. VCM determinations by some laboratories use alternate nickel-chromium alloy crucibles which are less expensive. However, if they are used, they must be calibrated with a platinum crucible to determine if relative bias exists. If a bias is identified, it will be used to correct the results when using the alternate crucibles.

### 4.1.3 Fixed carbon (FC) and ash

After the measurement of the VCM, the biochar material is placed in a muffle furnace for about an hour, with the cap removed and the temperature set to 600°C. After an hour, the crucible is placed in a desiccator until the material has cooled. Then, it is reweighed. One should ensure that there are no carbons left in the material after it is taken out of the furnace. Otherwise, one may add another 30 min until the color of the material becomes grayish white. The material left on the crucible is the amount of ash present in the biochar. The difference between the ash content and the VCM is the fixed carbon (FC). The ASTM standard for wood is designated as ASTM D 1102-84 (Reapproved 2007). The equivalent ash analysis for coal and coke is designated as ASTM D 3174-04. The main difference between these standards is the temperature used for ash analysis. Coal samples use temperatures between 700 and 750°C. For coke samples, the recommended temperature is 950°C. The total time for the ash analysis of coal and coke is about 4 h. It is very important to place the material in a desiccator to avoid absorption of moisture. In addition, the material must be cooled appropriately before weighing, in order to remove the effects of temperature on the material being weighed. Example 3.1 shows how proximate analysis is calculated and reported. The method should be replicated at least three times in order to calculate the mean. Eq. (3.5) may be used to measure the ash content of biomass.

$$\text{Ash}(\%\text{ash}) = \frac{f_i - f_f}{v_i} \times 100\% \quad (3.5)$$

In Eq. (3.5), $f_i$ is the initial weight of the biochar sample before placing in the muffle furnace and $f_f$ is the final constant weight after cooling and removal of the fixed carbon components. The parameter $v_i$ is the one used in Eq. (3.4).

The fixed carbon content is calculated by taking the difference between 100% and the sum of the VCM and ash percentages of the bone-dried sample, as shown in Eq. (3.6). The ash component is important in thermal conversion since high ash content biomass will tend to form slag and may cause fouling problems in many thermal conversion conveying parts. The ash contents of the biochar material are usually higher than the ash contents of the biomass itself. The reason is that ash components are not thermally converted and would keep its original weight. This will result in higher percentage of ash since some carbons have been converted into other forms.

$$\%\text{aAsh} = 100 - \%\text{VCM} - \%\text{FC} \quad (3.6)$$

The fixed carbon is also important for conversion. Carbon is the prime element useful for making liquid biofuels and if combined with hydrogen will generate hydrocarbon fuels.

---

**Example 3.1.** Proximate Analysis. Energy sorghum biochar samples after coming off a pressurized batch pyrolyzer processed at 400°C were used to estimate the proximate analysis. The following are the data from the experiment: (a) an initial sample weighing 5.00 g was used and placed in the drying oven set at 103°C overnight until the final weight becomes 4.85 g. About 1.05 g of bone-dry samples was placed in the tube furnace set at 950°C for 7 min and the resulting weight of the product was 0.8085 g. The same sample was then placed in a muffle furnace set at 600°C for 4 h with a resulting final weight of 0.4305 g. Calculate the proximate analysis for this sample (Santos and Capareda, 2015).

**Solution:**

a. The moisture content is calculated from the following equation:

$$\text{Moisture content }(\%MC) = \frac{m_i - m_f}{m_i} \times 100\% = \frac{5.00 - 4.85}{5.00} \times 100\% = 3\%$$

b. The volatile combustible matter is calculated from the following equation:

$$\%VCM = \frac{v_i - v_f}{v_i} \times 100\% = \frac{1.05 - 0.8085}{1.05} \times 100\% = 23\%$$

c. The % ash is calculated from the following equation:

$$\text{Ash }(\%ash) = \frac{f_i - f_f}{v_i} \times 100\% = \frac{0.8085 - 0.420}{1.05} \times 100\% = 37\%$$

d. Finally, fixed carbon is calculated by difference as shown.

$$\%FC = 100 - \%VCM - \%Ash$$

$$\%FC = 100 - 23 - 37 = 40\%$$

e. The material has significant amounts of carbon

---

## 4.2 Ultimate analysis

The ultimate analysis consists of measuring major elemental components of the biochar such as its carbon (C), hydrogen (H), oxygen (O), nitrogen (N), sulfur (S), and ash. The ash should have been established during the proximate analysis. The individual elements in the ultimate analysis have also its individual ASTM standards as listed in Table 3.3. Carbon and hydrogen are

### Table 3.3 List of ASTM standards used for biomass characterization.

| Designation | Title |
| --- | --- |
| ASTM E 777-08 | Standard Test Method for Carbon and Hydrogen in the Analysis Sample of Refuse-Derived Fuel |
| ASTM E 775-87 (2008) e1 | Standard Test Methods for Total Sulfur in the Analysis Sample of Refuse-Derived Fuel |
| ASTM E 778-08 | Standard Test Methods for Nitrogen in the Analysis Sample of Refuse-Derived Fuel |

integrally analyzed in accordance with ASTM E 777; the measurement of sulfur content follows ASTM E 775, and the measurement of nitrogen content follows ASTM E 778. The oxygen content of biomass is a calculated value and is the difference between 100% and the sum of C, H, N, S, and ash. One type of an ultimate analyzer is the Elemental Analyzer (Exeter Analytical, North Chelmsford, MA). Other companies manufacture ultimate analyzers, such as the TruSpec Micro Series CHNS/O Analyzer (Leco Corporation, St. Joseph, MI). The more expensive units are those that can handle biomass sample at the gram level rather than the microgram level. The microlevel ultimate analyzers must have a dedicated microbalance that must be accurate to the microgram level or lower. This will provide accurate readings. Example 3.2 shows how to calculate the atomic C/H and O/C ratios important to thermal conversion systems.

**Example 3.2.** Ultimate Analysis. Determine the atomic carbon-to-hydrogen ratio for sapwood biochar pyrolyzed at 500°C. Compare this with hardwood biochar. Calculate also the atomic O/C ratios. The ultimate analysis data for both biochar samples are given as follows: Softwood biochar: C (85.8%), H (2.4%), O (10.85%), N (0.35%), and S (0.35%); hardwood biochar: C (88.88%), H (2.6%), O (7.75%), N (0.5%), and S (0.40%) (Yang et al., 2015).

**Solution:**

a. The atomic C/H ratios are calculated by first determining the number of moles of carbon and hydrogen from data in table (assuming say, 100 g sample) as follows:

Softwood Biochar Carbon (mol) = 85.8/12 g/mol = 7.15

Softwood Biochar Hydrogen (mol) = 2.4/1 g/mol = 2.4

Thus, the atomic C/H ratio is calculated simply as C/H = 7.15/2.4 = 2.98

The number of moles of each element is in the same ratio as the number of atoms.

b. The atomic O/C ratios are calculated by first determining the number of moles of oxygen and carbon from data in table as follows:

Softwood Biochar Oxygen (mol) = 10.85/16 g/mol = 0.68

Softwood Biochar Carbon (mol) = 85.8/12 g/mol = 5.36

Thus, the atomic O/C ratio is calculated simply as O/C = 0.68/5.36 = 0.13

c. The atomic C/H ratio for hardwood is calculated the same way as follows:

Hardwood Biochar Carbon (mol) = 88.88/12 g/mol = 7.41

Hardwood Biochar Hydrogen (mol) = 2.6/1 g/mol = 2.6

Thus, the atomic C/H ratio is calculated simply as C/H = 7.41/2.6 = 2.85

d. The atomic O/C ratios are calculated as follows:

Hardwood Biochar Oxygen (mol) = 7.75/16 g/mol = 0.48

Hardwood Biochar Carbon (mol) = 88.88/12 g/mol = 7.41

Thus, the atomic O/C ratio is calculated simply as O/C = 0.48/7.41 = 0.065

## 4.3 Heating value analysis

One of the most important characteristics of biochar is its heating value or calorific value. The heating value measures the energy content of the sample and is measured simply by an equipment called a calorimeter. This test procedure combusts the sample in pure oxygen environment and the energy released during the combustion process is measured to indicate the heating value. The heating value of biochar is highly dependent upon the presence of moisture in the sample. Moisture has no heating value. Consequently, one should be very careful in reporting heating value numbers. When biomass heating value is reported and noted "as received," the moisture content must be provided. Otherwise, the heating value must be reported on either a higher or lower heating value basis as affected by moisture. In some disciplines, as in coal combustion, the heating value of feedstock is reported in either the dry basis (excludes moisture) or the dry ash-free basis (excludes moisture and ash). Example 3.3 shows the distinction between these two parameters. Eq. (3.7) shows the calculation of high heating value (HHV) as a function of moisture.

$$\text{HHV (dry basis)} = \frac{\text{HHV (wet basis)}}{1 - \text{moisture content (fraction)}} \quad (3.7)$$

The heating value calculation on an ash-free basis is given by Eq. (3.8).

$$\text{HHV (dry ash-free basis)} = \frac{\text{HHV (dry basis)}}{1 - \text{ash content (fraction)}} \quad (3.8)$$

The primary reason for using the high heating value of feedstock on a dry ash-free basis is because most power plants, such as coal power plants, are fed on a heating value basis and not on a mass basis, to ensure that the power output is always constant.

---

**Example 3.3.** Heating Value Calculations. The heating value of a certain biochar was measured by a bomb calorimeter and reported as 17.8 MJ/kg "as received." The moisture content of this biochar was 12%. Convert this heating value numbers into its HHV equivalent. The ash content of the biochar on a dry basis was found to be 13%, and determine the heating value on a dry ash-free basis as well.

**Solution:**

a. Use Eq. (3.7) for this calculation with the moisture content transformed into decimal equivalent.

$$\text{HHV (dry basis)} = \frac{\text{HHV (wet basis)}}{1 - \text{moisture content (fraction)}} = \frac{17.8}{1 - 0.12} = 20.2 \frac{\text{MJ}}{\text{kg}}$$

b. Use Eq. (3.8) for this calculation

$$\text{HHV (dry ash-free basis)} = \frac{\text{HHV (dry basis)}}{1 - \text{ash content (fraction)}} = \frac{20.2}{1 - 0.13} = 23.2 \frac{\text{MJ}}{\text{kg}}$$

---

The standard for the measurement of heating value is designated as ASTM E 711 (Standard Test Method for Gross Calorific Value of Refuse-Derived Fuel by Bomb Calorimeter). In this unit, pure oxygen is used to fill the stainless steel bomb canister for combustion of the sample. Approximately 1 g of relatively dry biochar sample is needed for the analysis. Note that the moisture content of the biochar sample to be tested must be known in order to convert the units into its high heating value (bone-dry sample) or lower heating values (as-received) basis.

If a calorimeter is not available, there are various ways of estimating the heating value of biochar that are based on other compositional parameters, as reported in the literature.

### 4.3.1 Heat of combustion

When biomass is converted thermally in the presence of excess amounts of air (i.e., combustion), the total energy released in the form of heat is termed as heat of combustion (or heating value/calorific value). The heating value of the biomass is reported in units of Btu/lb. or kJ/kg. Gasoline used as fuel for running internal combustion engines has a reported heating value of about 47 MJ/kg (20,250 Btu/lb) and diesel has a heating value of about 45 MJ/kg (19,400 Btu/lb). Biomass materials, on the other hand. May have heating values ranging from 15 to 30 MJ/kg (6500–13,000 Btu/lb). In the absence of devices to measure heating values of any biomass waste or residue, several methods are used. Primary information is needed to estimate heating values. There are two categories of information. One category is based on ultimate analysis, that is, the percentage of carbon, hydrogen, oxygen, nitrogen, and sulfur or CHONS. The second category is based on proximate analysis data, that is, moisture content (MC), volatile combustible matter (VCM), fixed carbon (FC), and ash.

### 4.3.2 Stoichiometry of reactions

In the analysis of thermal conversion processes, one must be able to write a complete reaction of carbon, hydrogen, and other combustible elements in the fuel that are in the presence of oxygen. For example, all carbon present is assumed to be burned to carbon dioxide, and all hydrogen is converted into water. No oxygen is present in the products of combustion. The complete combustion of a certain biochar is as follows: $CH_{1.13}O_{0.47}N_{0.01}$ (switchgrass biochar) is shown in Eq. (3.9).

$$CH_{1.13}O_{0.47}N_{0.011}S_{0.0009} + 1.05(O_2 + 3.76N_2) \rightarrow CO_2 + 0.565H_2O + 3.94N_2 + 0.0009S + \text{Heat} \quad (3.9)$$

The heat released, if measured, is termed as the heat of combustion of the fuel, the basic premise in calorimetry measurements. To balance the equation, begin by balancing the carbon, followed by hydrogen, and balancing the oxygen last. Note that if air is used as the oxidant, it is convenient to assume air is composed of 21% $O_2$ and 79% $N_2$; thus, the stoichiometric equation is presented in Eq. (3.9).

Example 3.4 shows how the chemical formula of this biochar is established. The elemental analysis by weight is first found and assuming say 100 g sample, the percentage of each element multiplied by 100 g is simply the weight of each compound and dividing by their molecular weight will result in estimating the moles of each element. The chemical formula is then normalized to the elemental carbon to come up with the estimated chemical formula of the biochar.

---

**Example 3.4.** Determine the chemical formula of switchgrass biochar pyrolyzed at 400°C for 3 h (Sadaka et al., 2014) and resulted in the following characteristics by weight: C=56.0%; H=5.3%; O=35.1%; N=0.76%; and S=0.12% and ash of 2.72%.

**Solution:**
a. Assuming 100 g sample, there will be around 56 g of carbon, 5.3 g of H, 35.1 g of O, 0.76 g of N, and 0.12 g of S. Hence, the molar ratios are as follows:

$$C = \frac{56 \text{ g}}{12 \text{ g/mol}} = 4.67.$$

$$H = \frac{5.3 \text{ g}}{1 \text{ g/mol}} = 5.3$$

$$O = \frac{35.1 \text{ g}}{16 \text{ g/mol}} = 2.19$$

$$N = \frac{0.76 \text{ g}}{14 \text{ g/mol}} = 0.05$$

$$S = \frac{0.12 \text{ g}}{32 \text{ g/mol}} = 0.004$$

b. The original chemical formula is as follows:

$$C_{4.67}H_{5.3}O_{2.19}N_{0.05}S_{0.004}$$

c. The normalized form is calculated by dividing the mole ratios with that of carbon and as follows.

$$CH_{1.13}O_{0.47}N_{0.011}S_{0.0009}$$

Setting up the chemical formula of this material is important when designing gasification system for the conversion of biochar, say into electrical power or syngas. Example 3.5 shows how to estimate the air-to-fuel ratio for the complete combustion of biochar. This stoichiometric air-to-fuel ratio will be used to determine the limits on the amounts of air that will be used when this material is gasified. In gasification process, only about 30%–70% of stoichiometric air is used. The biochar production via gasification may be increased by proper operation of the unit and operating at a rather high fuel-to-air ratio. Capareda (1990) conducted experiments using cotton gin trash as feedstock to optimize the production of biochar. By using higher fuel-to-air ratios, biochar production may be increased to around close to 40% yield instead of a conservative 20% yield. Hence, the possible ranges of biochar production via fluidized bed gasification ranges between 20% and 40%. Care should be done when operating the gasifier at higher fuel-to-air ratios since conversion temperatures may be too low to be sustainable. Some biomass may not be able to achieve the highest yield due to this limitation.

**Example 3.5.** For the biochar sample given in Example 3.4, determine the stoichiometric air-to-fuel ratio, by volume and by weight, if this material is combusted using 100% stoichiometric air.

**Solution:**

a. The chemical formula of this material is shown in Eq. (3.9); hence, the air-to-fuel ratio by volume is simply 1.259:1. Hence for every liter of switchgrass biochar, one would need at least 1.259 L of air to undergo complete combustion.

b. The air-to-fuel ratio (AFR) by weight is calculated as follows:

$$\begin{aligned} AFR &= \frac{1.259(O_2 + 3.76N_2)}{CH_{1.13}O_{0.47}N_{0.011}S_{0.0009}} \\ &= \frac{1.05[(32 + 3.76 \times 28)]}{[(12 \times 1) + (1 \times 1.13) + (16 \times 0.47) + (14 \times 0.011) + (32 \times 0.0009)]} \\ &= \frac{144.14}{20.83} = 6.92 \frac{kg}{kg} \end{aligned}$$

c. To combust this biochar, one would need about 7 kg of air per kg of the biochar material.

d. When used in gasification processes, one would only need between 2.1 kg and 4.84 kg of air to produce combustible synthesis gas.

## 4.4 Eutectic point analysis

The fusion (or melting) temperature of the ash integrally contained in the biomass or biochar sample is also an important parameter used for its suitability for thermal conversion. At the high temperature of the thermal conversion process, the ash components integral in the biochar will melt and upon cooling will form heavy slag; these particles will stick through the walls of the conveying surfaces and create a blockage. If they are used in fluidized bed gasification system with sand as bed material, this will agglomerate the sand and form slag as well. If the biochar has a very high content of ash, the problem becomes severe, causing a complete blockage of the conveying conduits, which leads to fouling. Fig. 3.3 shows the slag generated from gasification of high as biomass. This figure shows the formation of slag when high ash biomass such as dairy manure is gasified using fluidized bed systems.

The two most common indices for evaluating the potential of a biochar to agglomerate are the slagging ($R_s$) and fouling ($R_f$) factors. These factors are primarily used for coal samples but are sometimes used for biochar materials as well. Mathematically, these factors are defined in Eqs. (3.10) and (3.11). The percentage of sulfur and $Na_2O$ are based on dry coal samples.

$$R_s = \left(\frac{\text{Base}}{\text{Acid}}\right) \times \%S \qquad (3.10)$$

$$R_f = \left(\frac{\text{Base}}{\text{Acid}}\right) \times \%Na_2O \qquad (3.11)$$

**Fig. 3.3** Photo of slag generated from gasification of biomass or biochar sample with low eutectic points.

**Table 3.4** Slagging and fouling indices for coal samples (Winegartner, 1974).

| Slagging/fouling type | $R_s$ | $R_f$ |
|---|---|---|
| Low | <0.60 | <0.2 |
| Medium | 0.6–2.0 | 0.2–0.5 |
| High | 2.0–2.6 | 0.5–1.0 |
| Severe | >2.6 | >1.0 |

The total acid in coal ash is the sum of the percentages of $SiO_2$, $TiO_2$, and $Al_2O_3$, while the total base in coal ash is the sum of the percentages of $Fe_2O_3$, CaO, MgO, $Na_2O$, and $K_2O$ in the ash. The percentage of sulfur in coal is also measured independently so that it can be used on the slagging factor, or it could be based on the $SO_3$ components. Table 3.4 shows the range of values for $R_s$ and $R_f$ with expected four types of slagging and fouling conditions. Severe slagging is expected if the calculated slagging factor is greater than 2.6 while severe fouling is expected if the fouling factor is greater than 1.0. Unfortunately, these factors are not suitable for biomass, biochar, or lignite samples (Maglinao and Capareda, 2010). The following are other indices that are also used to evaluate the slagging and fouling potential of biomass samples:

a. Base-to-acid ratio ($R_{b/a}$)
b. Agglomeration index (BAI)
c. Alkali index (AI)

A more detailed calculation of these indices is shown in the following example. The slagging and fouling indices for coal samples are not appropriate for biomass or biochar samples that are high in inorganic components. The sample analysis starts by producing ash samples from the biomass or biochar in question. Then, a complete compositional analysis is made to determine the basic and acid components and applied to Eqs. (3.12)–(3.14). Example 3.6 shows how slagging and fouling factors are calculated. Data for ash composition of agricultural wastes such as cotton gin trash and dairy manure are shown in Table 3.5. These values may be used to make a preliminary estimate of slagging and fouling indices.

The alkali index (AI) expresses the quantity of alkali oxide in the biochar per unit of biochar energy (kg alkali/GJ). It is computed using Eq. (3.12) for metric system and 11.13 for English

**Table 3.5 Ash composition of cotton gin trash (CGT) and dairy manure ashes.**

| Inorganic ash | CGT ash (%) | Dairy manure ash (%) |
|---|---|---|
| $SiO_2$ | 21.7 | 32.46 |
| $K_2O$ | 24.62 | 5.28 |
| $CaO$ | 23.30 | 27.41 |
| $MgO$ | 5.69 | 10.90 |
| $Na_2O$ | 0.76 | 1.82 |
| $P_2O_5$ | 2.25 | 4.98 |
| $Fe_2O_3$ | 1.11 | 1.84 |
| $Al_2O_3$ | 3.46 | 3.12 |
| $SO_3$ | 7.40 | 6.12 |
| $TiO_2$ | 0.25 | 0.22 |
| $MnO$ | 0.06 | 10.90 |
| Total | 90.60 | 94.29 |
| % Ash in material | 11.86 | 18.62 |
| % Alkali in ash | 25.38 | 7.10 |

system. Note the use of % ash in the material is included in the calculation.

$$AI = \frac{kg(K_2O + Na_2O)}{GJ} \quad (3.12)$$

$$\frac{1 \times 10^6}{Btu/lb} \times \%ash \times \%alkali = \frac{lb\,alkali}{mmBtu} \quad (3.13)$$

When AI values are within the range 0.17–0.34 kg/GJ (Vamvuka and Zografos, 2004), fouling or slagging may or may not happen, but it is certain to occur when the values are above this range.

The base-to-acid ratio ($R_{b/a}$) in the ash is obtained as shown in Eq. (3.14).

$$R_{b/a} = \frac{\%(Fe_2O_3 + CaO + MgO + K_2O + Na_2O)}{\%(SiO_2 + TiO_2 + Al_2O_3)} \quad (3.14)$$

The label for each compound makes reference to its weight fractions in the ash. As $R_{b/s}$ increases or above 1, the fouling tendencies of the biochar fuel ash increase.

The bed agglomeration index (BAI) relates ash composition to agglomeration in fluidized bed reactors (Skrifvars et al., 1998). It is calculated using Eq. (3.15)

$$\text{BAI} = \frac{\%(\text{Fe}_2\text{O}_3)}{\%(\text{K}_2\text{O} + \text{Na}_2\text{O})} \qquad (3.15)$$

The bed agglomeration occurs when BAI values are lower than 0.15 (Vamvuka and Zografos, 2004).

---

**Example 3.6.** Slagging and Fouling Factor for Coal. Determine the slagging factor ($R_s$) and fouling factor ($R_f$) for CGT ash data as shown in Table 3.5. Discuss the slagging and fouling potential based on these indices.

**Solution:**
a. The acid components include the following: $SiO_2$ (21.70%), $Al_2O_3$ (3.46%), and $TiO_2$ (0.25%) and amounting to 25.41%.

b. The basic components include the following: $Fe_2O_3$ (1.11%), CaO (23.30%), MgO (5.69%), $Na_2O$ (0.76%), and $K_2O$ (24.62%) and this amounted to 55.48%.

c. Thus, the slagging factor is calculated from the equation shown. Note that the % sulfur in $SO_3$ is 40% based on their molecular weights (i.e., 1 mol sulfur = 32 g/mol and 3 mol oxygen = 16 × 3 = 48 g/mol to get 32/(32+48) × 100% = 40% sulfur in $SO_3$) and 40% of 7.40% × 0.40 = 2.96%.

$$R_s = \left(\frac{\text{Base}}{\text{Acid}}\right) \times \%S = \left(\frac{55.48\%}{25.41\%}\right) \times 2.96\% = 0.065$$

d. The fouling factor is calculated in the same way.

$$R_f = \left(\frac{\text{Base}}{\text{Acid}}\right) \times \%\text{Na}_2\text{O} = \left(\frac{55.48\%}{25.41\%}\right) \times 0.76\% = 0.016$$

e. Based on Table 3.4, this material has low potential for slagging and fouling.

f. It will be shown in next example that this material will have more tendency for slagging and fouling when using the other indices suited for biomass wastes which would be its behavior during actual thermal conversion.

---

**Example 3.7.** Slagging and Fouling Factor for CGT Ash. Determine the alkali index, base-to-acid ratio and agglomeration index for CGT ash data as shown in Table 3.5. Assume the heating value of the material is 16.67 MJ/kg [7182 Btu/lb]. Discuss the slagging and fouling potential based on these indices.

**Solution:**
a. The alkali index is calculated as follows and shows high slagging risk instead ($>0.34$ kg/GJ).

$$\text{AI}\left(\text{CGT}\frac{\text{lb alkali}}{\text{mmBtu}}\right) = \frac{1\times 10^6}{7{,}182\,\frac{\text{Btu}}{\text{lb}}} \times \frac{11.86\%}{100} \times \frac{25.38\%}{100} = 4.19\frac{\text{lb alkali}}{\text{mmBtu}}$$

$$\text{AI}\left(\text{CGT}\frac{\text{kg alkali}}{\text{GJ}}\right) = 4.19\frac{\text{lb alkali}}{\text{mmBtu}} \times \frac{\text{kg}}{2.2\,\text{lbs}} \times \frac{1\,\text{mm Btu}}{1\times 10^6\,\text{Btu}} \times \frac{\text{Btu}}{1055\,\text{J}}$$
$$\times \frac{1\times 10^9\,\text{J}}{\text{GJ}}$$
$$= 1.8\frac{\text{kg alkali}}{\text{GJ}}$$

b. The base-to-acid ratio is calculated as follows showing potential slagging and fouling incidence instead of low from using $R_s$ and $R_f$ indices. The value is more than 1.

$$R_{b/a} = \frac{\%(\text{Fe}_2\text{O}_3 + \text{CaO} + \text{MgO} + \text{K}_2\text{O} + \text{Na}_2\text{O})}{\%(\text{SiO}_2 + \text{TiO}_2 + \text{Al}_2\text{O}_3)}$$
$$= \frac{\%(1.11 + 23.30 + 5.69 + 24.62 + 0.76)}{\%(21.70 + 0.25 + 3.46)} = \frac{55.48}{25.41} = 2.18$$

c. The agglomeration index is calculated as follows showing bed agglomeration may occur as the value is less than 0.15.

$$\text{BAI} = \frac{\%(\text{Fe}_2\text{O}_3)}{\%(\text{K}_2\text{O} + \text{Na}_2\text{O})} = \frac{\%(1.11)}{\%(24.62 + 0.76)} = 0.04$$

# 5 Summary of standard methods for biochar analysis

There are numerous methods for charactering biomass, not all of which are discussed in detail in this chapter. However, Table 3.6 summarizes some of the additional methods that may be used for other applications. Most of the standard methods were developed by the America Society for Testing Materials (ASTM). However, some methods were developed by other organizations such as the American Society for Biological and Agricultural Engineering (ASABE), formerly the American Society of Agricultural Engineering (ASAE).

The fusibility of biomass ash is an important parameter. Table 3.7 lists some biomass types and the report of temperatures for their initial deformation, softening, formation of hemispherical

**Table 3.6 Other standard methods for biomass analysis.**

| Property/designation | | Title |
|---|---|---|
| Heating value | ASTM D 2015 | Gross calorific value of solid fuel by the adiabatic bomb calorimeter |
| Particle size distribution | ASTM E 828 | Designing the size of RDF-3 from its sieve analysis |
| | ASAE S 319 | Expressing fineness of feed materials by sieving |
| Specific gravity | ASTM D 2395 | Specific gravity of wood and wood-based materials |
| Proximate analysis | ASTM D 3172 | Proximate analysis of coal and coke |
| Moisture content | | Moisture content of wood |
| VCM | ASTM D 2016 | Volatile matter in the analysis sample of coke and coal |
| Ash | | Ash in the analysis sample of coal and coke |
| | ASTM D 3175 | |
| | ASTM D 3174 | |
| Ultimate analysis | ASTM D 3176 | Ultimate analysis of coal and coke |
| Thermal properties | ASTM C 351 | Mean specific heat of thermal insulation |
| | ASTM C 687 | Thermal resistance of low-density fibrous loose fill-type building insulation |
| Mechanical properties | ASTM D 2555 | Establishing clear wood strength values |

conditions, and transition to a fluid (Jenkins, 1989). These fusibility values are important for thermal conversion such as gasification, where the temperature of the reaction is (normally) slightly above 1000 °C. Thus, there are biomass materials listed in Table 3.7 whose initial deformation temperature is below 1000°C. One should be aware of slagging and fouling issues when converting these biomass or biochar materials under extremely high reaction temperatures. Problems may not happen at the conversion reactor but may occur when these materials are cooled downstream from the reactor.

In using the production of biochar for thermal conversion, such as those for power generation, Example 3.8 shows the conservative estimate of biochar production rate for a 1 MW wood to power production using fluidized bed gasification.

Table 3.7 Fusibility of some biomass ash (Jenkins, 1993).

| Biomass type | Temperature (°C) | | | |
|---|---|---|---|---|
| | Initial deformation | Softening | Hemispherical | Fluid |
| Alfalfa seed straw | 700 | | 1550 | |
| Almond shell | 790 | | 1440 | |
| Barley straw | 925 | | 1100 | |
| Bean straw | 900 | | 1150 | |
| Corn cobs | 900 | | 1020 | |
| Corn stalks | 820 | | 1091 | |
| Cotton gin trash | 1010 | | 1380 | |
| Grape prunings (oxidizing) | 1313 | 1368 | 1374 | 1424 |
| Grape prunings (reducing) | 1310 | 1360 | 1371 | 1382 |
| Madrone wood (reducing) | 1271 | 1330 | 1352 | 1404 |
| Manzanita wood | 1268 | 1299 | 1310 | 1321 |
| Olive pits | 1080 | | 1400 | |
| RDF pellets | 850 | | 1480 | |
| RDF (oxidizing) | 890 | 1092 | 1130 | 1193 |
| RDF (reducing) | 1065 | 1063 | 1131 | 1182 |
| Rice hull | 1024 | | 1097 | |
| Rice straw | 1439 | | >1650 | |
| Rice straw (oxidizing) | 1060 | 1421 | 1250 | >1500 |
| Rice straw (reducing) | 1027 | 1410 | | >1500 |
| Safflower straw | 770 | | 1430 | |
| Tan oak wood (oxidizing) | 1390 | 1440 | 1449 | 1457 |
| Tan oak wood (reducing) | 1377 | 1438 | 1446 | 1454 |
| Walnut shell | 820 | | 1225 | |

**Example 3.8.** Biochar Production via Gasification. As a rule of thumb, about 1.2 tonnes/h (1.32 dry tons per hour or 2.64 green tons per hour) of wood fuel can generate 1 MW of electrical power. This will result in the production of 20% biochar. Freshly harvested wood would normally have 40%–50% moisture when harvested in the summer. Assume an energy content of wood fuel as 20 MJ/kg [8616 Btu/lb] and an overall electrical power conversion efficiency of 15%. How much biochar is produced per hour, per day, and per year if the unit is operating 300 days a year [7200 h]. If this biochar is subsequently used to generate additional electrical power, how much additional power is generated? Assume the heating value of biochar is 22 MJ/kg.

**Solution:**

a. The theoretical power required from wood each hour is calculated as follows:

$$\text{Power (MW)} = \frac{1.2\,\text{tonne}}{\text{hr}} \times \frac{20\,\text{MJ}}{\text{kg}} \times \frac{1000\,\text{kg}}{\text{tonne}} \times \frac{1 \times 10^6\,\text{J}}{\text{MJ}} \times \frac{1\,\text{h}}{3600\,\text{s}} \times \frac{\text{W}}{\text{J/s}}$$
$$\times \frac{\text{MW}}{1 \times 10^6\,\text{W}}$$
$$= 6.67\,\text{MW}$$

b. Then, the energy output may be calculated from the efficiency equation as follows:

$$\text{Overall efficiency (\%)} = \frac{\text{Outpu power (MW)}}{\text{Input power (MW)}} \times 100\%$$

$$\text{Output power (MW)} = \text{Input power (MW)} \times \text{overall efficiency}$$
$$= 6.67\,\text{MW} \times 0.15 = 1\,\text{MW}$$

c. The biochar production per hour is calculated as follows:

$$\text{Biochar}\left(\frac{\text{tonnes}}{\text{h}}\right) = \frac{1.2\,\text{tonne}}{\text{h}} \times 0.20 = 0.24\,\frac{\text{tonnes}}{\text{h}}$$

d. The yearly biochar production from this facility is calculated as follows:

$$\text{Biochar}\left(\frac{\text{tonnes}}{\text{yr}}\right) = 0.24\,\frac{\text{tonnes}}{\text{h}} \times 7200\,\frac{\text{h}}{\text{yr}} = 1728\,\text{tonnes}$$

e. Hence, for every year of operation of a 1 MW biomass power plant using gasification, over 1728 tonnes of biochar is coproduced.

f. If this biochar is then consequently used to generate electrical power, via gasification as well, an additional 0.2 MW of power is produced as shown in the subsequent calculations.

$$\text{Biochar power (MW)} = \frac{0.24\,\text{tonnes}}{\text{h}} \times \frac{1000\,\text{kg}}{1\,\text{tonne}} \times \frac{22\,\text{MJ}}{\text{kg}} \times \frac{1\,\text{h}}{3600\,\text{s}} \times \frac{\text{MW}}{\text{MJ/s}}$$
$$\times 0.15$$
$$= 0.22\,\text{MW}\,[220\,\text{kW}]$$

g. An additional 220 kW is the potential produced when the biochar is concurrently used to generate electrical power.

# 6 Biochar conversion into activated carbon and performance analysis

Activated carbon is a highly valuable material with an average price of around $5600/tonne (Alhashimi and Aktas, 2017). Activated carbon is a billion-dollar industry with Calgon Carbon

Corporation (Pittsburgh, PA) named as the world's leader in commercial sale of activated carbon-related materials. Some research reports predicted that the activated carbon market size will be worth 14.07 billion by 2027 (Emergen Research, 2020). Raw biochar from pyrolysis or gasification process does not have enough surface area or pore volume to be categorized as activated carbon. Some means must be performed to improve its adsorptive or absorptive properties. There are physical and chemical methods that may be used to improve the adsorptive properties of biochar.

## 6.1 Physical activation processes

The simplest physical activation process is by steam reaction or by the use of other gases such as super critical carbon dioxide. Physical activation by steam uses superheated steam and not simply saturated steam. The latter will only adhere moisture to the surface and not generate or open up more pore surface areas. The activation process is performed under high temperature and exposure time. The resulting biochar should increase its surface area as well as pore volume. The theory behind this process is the removal of noncarbon materials, thereby increasing the surface area. During gasification or pyrolysis processes, the formation of tar causes blockage of some pores, thereby lowering its adsorptive properties. Thus, some means must be made to remove these blockages. One sophisticated device to measure the improvement in surface area and pore volume is a BET (Brunauer–Emmett–Teller) analyzer. The BET theory tries to explain the physical adsorption process on solid surfaces of the biochar. This serves as the basis for the technique and uses liquid nitrogen for the measurement of the specific surface area. There are also indirect ways to measure these adsorptive properties such as the use of iodine numbers or iodine adsorption values. This is a chemical method and relies also on color as indication of adsorptions.

## 6.2 Chemical activation process

Chemical activation process relies on the use of specific chemicals to improve the surface area and pore volume. The most popular is the use of zinc chloride ($ZnCl_2$) (Hernandez-Maglinao and Capareda, 2019). Others include the use of phosphoric acids and potassium hydroxide. The mechanisms to increase the surface area and pore volume by chemical means are not very clear. Apparently, these chemical compounds remove moisture in the sample, solubilize the tars formed, and evolve some volatile compounds present. This process enhances the performance of the adsorption process. After the activation process, the material undergoes other chemical adsorption performance index detection such as the following: methylene blue adsorption value, phenol adsorption value, carbon tetrachloride adsorption value, caramel adsorption value, quinine sulfate adsorption value, among many others.

Other additional physical and chemical performance index testing includes pH value, ash, moisture, ignition point, uncarbonized components, sulfide, chloride, cyanide, sulfate, acid soluble, alcohol soluble, and other metals including heavy metal contents. There are research methods using microbial detection, fluorescent substance detection, denitrification rate detection, and volatile component detection.

Perhaps the two most important activated carbon analysis is the BET analysis as well as adsorption isotherm development.

## 6.3 Adsorption isotherms for activated carbon

The performance characteristic curves of activated carbon can be developed by performing adsorption isotherm studies. Adsorption isotherms are developed by measuring the amount of target contaminants removed from water or wastewater after treatment with increasing amounts of activated carbon. The target contaminants may be chemical oxygen demand (COD), heavy metals, or specific chemical compound contaminant that is highly adsorptive the activated carbon surface. The data can be fitted into either the Langmuir, Freundlich, or BET isotherms (Capareda, 1990). The Langmuir isotherm is presented in Eq. (3.16).

$$\frac{X}{M} = \frac{b \times Q \times C}{1 + b \times C} \quad (3.16)$$

where
  $X$ = mass of adsorbate, mg;
  $M$ = mass of dry adsorbent, g;
  $b$ = constant (units of reciprocal of concentration);
  $Q$ = maximum capacity (assumed to be monolayer coverage of adsorbent surface), mg/g, and
  $C$ = solution concentration at equilibrium
  The linearized form of the equation is shown in Eq. (3.17).

$$\frac{M}{X} = \frac{1}{Q} + \frac{1}{b \times Q \times C} \qquad (3.17)$$

When the reciprocal of the adsorption capacity ($M/X$) is plotted against the reciprocal of the concentration ($1/C$), the result is a straight line with the slope equal to $1/bQ$ and the intercept $1/Q$.

The Freundlich equation is shown in Eq. (3.18).

$$\frac{X}{M} = k \times C^{(1/n)} \qquad (3.18)$$

where $k$ and $n$ are constant and $X$, $M$, and $C$ are defined as before.
The linearized form is given in Eq. (3.19).

$$\log \frac{X}{M} = \log k + \frac{1}{n} \log C \qquad (3.19)$$

The constants $k$ and $n$ define both the nature of the carbon and the adsorbate. A high value of $k$ and $n$ indicates high adsorption throughout the concentration range studied. A low value of $n$ (steep slope) indicates high adsorption at strong solute concentrations and poor adsorption at dilute concentrations.

While the generalized regression curve for activated carbon is the most meaningful technique to characterize the activated carbon material and predict its properties in a particular situation, it is generally not used by the public.

Traditional quality control parameters are used by the carbon industry in the manufacture of activated carbons. These parameters are usually all of the information available to a potential customer setting up a process control system. These parameters include iodine numbers or BET surface area and pore volume, bulk and particle density, ash content, moisture content, and particle size distribution (PSD). Table 3.8 shows the results of experiments to reduce the chemical oxygen demand of wastewater after treatment with increasing amounts of commercial activated carbon versus literature reported activated carbon, commercial Fisher activated carbon and cotton-gin-trash-based activated carbon. This table will be used to demonstrate how the adsorption isotherm curves are developed for a particular activated carbon. Commercial activated carbon and biomass-based activated carbon are described by their iodine numbers or biomass iodine numbers (BINs). Biomass iodine number is defined in Eq. (3.20).

$$\text{BIN} = \frac{\text{Iodine number}}{1 - \text{ash content}} \qquad (3.20)$$

This BIN may be compared with the iodine number of commercially available activated carbon. Usually, commercial activated

Table 3.8 Chemical oxygen demand (COD) of wastewater after treatment with increasing amounts of activated carbon.

| Carbon dose (g/100 mL) | Concentration remaining in solution (mg/L COD) | | |
|---|---|---|---|
| | Research report (Hutchins, 1981) | Fisher activated carbon (iodine number = 1200) (*interpolated data) | Cotton gin trash activated carbon (BIN = 700) |
| Control | 500 | 483 | 650 |
| 0.05 | 475 | 439 | 647 |
| 0.10 | 450 | 394 | 412 |
| 0.20 | 420 | 350 | 216 |
| 0.50 | 310 | 305 | 216 |
| 1.00 | 200 | 204 | 196 |
| 2.0 | 100 | 112 | 98 |
| 5.0 | 35 | 20 | 59 |
| 10.0 | 12 | 6 | 40 |
| 20.0 | 4 | 4 | 20 |
| **Freundlich parameter** | | | |
| $k$ (mg/L) | 1.042 | 1.455 | 0.0272 |
| $n$ (L/mg) | 1.60 | 1.729 | 0.659 |
| Correlation | 0.9985 | 0.9872 | 0.9677 |
| **Langmuir parameter** | | | |
| $Q$ (mg/L) | 31.13 | 53.13 | 75.25 |
| $b$ (L/mg) | 0.0207 | 0.0128 | 0.0023 |
| Correlation | 0.9853 | 0.9827 | 0.8449 |

From Capareda, S.C., 1990. Studies on Activated Carbon Produced From Thermal Gasification of Biomass Wastes. PhD Dissertation, Biological and Agricultural Engineering, Texas A&M University, College Station, TX.

carbons have very negligible amounts of ash. Thus, making this ash content correction will not affect its iodine number. However, biomass-based activated carbons have significant amounts of ash. These ash contents will not affect its adsorptive properties. However, the biochar loading will simply be increased due to its ash content to consider the true weight of carbon present in the material. For example, a biomass-based activated carbon with a 50% ash will only have around 50% carbon content. If its iodine number is 500, its biomass iodine number is 1000. The BIN of

1000 makes its performance equal to that of a commercial activated carbon with an iodine number of 1000. In actual usage, twice the amount of biomass-based activated carbon must be used for every amount of commercial activated carbon to achieve equal performance.

By incorporating the ash content of biomass-based activated carbon, one should be able to compare these two types of adsorptive materials and have meaningful loading rate calculations during adsorption process. Examples 3.9 and 3.10 show how adsorption isotherms are developed.

**Example 3.9.** Use the data in Table 3.8 to verify the reported Langmuir isotherm following the procedures outlined in this section. About 100 mL of contaminated water was treated with increasing amounts of activated carbon and the resulting COD adsorb was measured. Use the published data (1st column) by Hutchins (1981). Report the value of $Q$ (mg/L) and $b$ (L/mg) and explain the significance of the resulting values. For a specific case where it is desired to lower the COD of wastewater from 500 to 200 mg/L, what is the required loading rate in units of grams of carbon per liter of wastewater?

**Solutions:**

a. For establishing the Langmuir parameters ($Q$ and $b$), the linearized form of the equation is used (Eq. (3.17)). Hence, the values of $M/X$ and $1/C$ are shown in the table below. Note that since only 100 mL of sample was used, this units must be converted into g/L of sample to be consistent with concentration units of mg/L.

| Equilibrium concentration | Mass of adsorbate | Mass of dry adsorbent | Y<br>Ratio (M/X) | X<br>1/solutions concentration |
|---|---|---|---|---|
| C (mg/L) | X (mg/L) | M (g/L) | M/X | 1/C |
| 475 | (500 − 475) = 25 | 0.5 | 0.02000 | 0.00211 |
| 450 | (500 − 450) = 50 | 1 | 0.02000 | 0.00222 |
| 420 | (500 − 420) = 80 | 2 | 0.02500 | 0.00238 |
| 310 | (500 − 310) = 190 | 5 | 0.02632 | 0.00323 |
| 200 | (500 − 200) = 300 | 10 | 0.03333 | 0.00500 |
| 100 | (500 − 100) = 400 | 20 | 0.05000 | 0.01000 |

| 35 | (500 − 35) = 465 | 50 | 0.10753 | 0.02857 |
| 12 | (500 − 488) = 488 | 100 | 0.20492 | 0.08333 |
| 4 | (500 − 4) = 496 | 200 | 0.40323 | 0.25000 |

b. The next step is the regression of columns $Y$ and $X$ using suitable spreadsheet software to perform this task such as MS Excel. The results of regression analysis showed the values of intercept, $X$ variable (or slope), and multiple $R$ and $R^2$ as follows:

Intercept = 0.0321
Slope = 1.5541
Multiple $R$ = 0.9853
$R^2$ = 0.9708

c. To calculate the maximum capacity $Q$ in units of mg/g, we have the following calculations:

$$Q\left(\frac{mg}{g}\right) = \frac{1}{\text{intercept}} = \frac{1}{0.0321} = 31.15 \frac{mg}{g}$$

d. To calculate the constant b with units the reciprocal of concentration (L/mg), we have the following calculations:

$$b\left(\frac{L}{mg}\right) = \frac{1}{Q\left(\frac{mg}{L}\right) \times \text{Slope}} = \frac{1}{31.15 \frac{mg}{g} \times 1.5541} = 0.0207 \frac{L}{mg}$$

e. The resulting values of $Q$ and $b$ are consistent with published values in Table 3.8 including the value for multiple $R$.

f. For a specific case where it is desired to lower the COD of wastewater from 500 to 200 mg/L, the required loading rate in units of grams of carbon per liter of wastewater is calculated as follows:

$$\text{Loading rate}\left(\frac{g\,\text{carbon}}{L\,\text{wastewater}}\right) = X \times \left[\frac{1+bC}{1+bQC}\right]$$

$$M\left(\frac{g\,\text{carbon}}{L\,\text{wastewater}}\right) = (500-200) \times \left[\frac{1+0.0207 \times 200}{1+0.0207 \times 31.15 \times 200}\right]$$

$$= 11.9 \frac{g\,\text{carbon}}{L\,\text{wastewater}}$$

g. It will require around 12 g activated carbon per liter of wastewater to reduce the COD concentration from 500 to 200 mg/L.

This example shows consistent resulting adsorption isotherm values as reported in the literature. The following example shows similar calculation using the Freundlich isotherm and results are similarly compared. The results showed that the Langmuir isotherm is more conservative compared with Freundlich adsorption isotherm.

**Example 3.10.** Use the data in Table 3.8 to verify the reported Freundlich isotherm following the procedures outlined in this section. About 100 mL of contaminated water was treated with increasing amounts of activated carbon and the resulting COD adsorb was measured. Use the published data (1st column) by Hutchins (1981). Report the value of $k$ (mg/L) and $n$ and explain the significance of the resulting values. For a specific case where it is desired to lower the COD of wastewater from 500 to 200 mg/L, what is the required loading rate in units of grams of carbon per liter of wastewater?

**Solutions:**

a. For establishing the Freundlich parameters ($k$ and $n$), the linearized form of the equation is used (Eq. 3.19). Hence, the values of $M/X$ and $1/C$ are shown in the table below. Note that since only 100 mL of sample was used, the units must be converted into g/L of sample to be consistent with concentration units of mg/L.

| Equilibrium concentration (mg/L) | Mass of dry adsorbent (g/L) | Mass of adsorbate (mg) | Y | X |
|---|---|---|---|---|
| $C$ (mg/L) | $M$ (g/L) | $X$ (mg) | Log $(X/M)$ | Log $C$ |
| 475 | 0.5 | (500 − 475) = 25 | 1.69897 | 2.65321 |
| 450 | 1 | (500 − 450) = 50 | 1.69897 | 2.62325 |
| 420 | 2 | (500 − 420) = 80 | 1.60206 | 2.49136 |
| 310 | 5 | (500 − 310) = 190 | 1.57978 | 2.30103 |
| 200 | 10 | (500 − 200) = 300 | 1.47712 | 2.00000 |
| 100 | 20 | (500 − 100) = 400 | 1.30103 | 1.54407 |
| 35 | 50 | (500 − 35) = 465 | 0.96848 | 1.07918 |
| 12 | 100 | (500 − 488) = 488 | 0.68842 | 0.60206 |
| 4 | 200 | (500 − 4) = 496 | 0.39445 | 2.65321 |

b. The next step is the regression of columns $Y$ and $X$ using suitable spreadsheet software to perform this task such as MS Excel. The results of regression analysis showed the values of intercept, $X$ variable (or slope), and multiple $R$ and $R^2$ as follows:

Intercept = 0.017766
Slope = 0.62598
Multiple $R$ = 0.9985
$R^2$ = 0.9970

c. To calculate the parameter $k$ in units of mg/L, we have the following calculations:

$$k\left(\frac{mg}{L}\right) = 10^{intercept} = 10^{0.017766} = 1.042\frac{mg}{L}$$

d. To calculate the constant $n$, we have the following calculations:

$$n = \frac{1}{Slope} = \frac{1}{0.62598} = 1.5975$$

e. The resulting values of $k$ and $n$ are consistent with published values in Table 3.8 including the value for multiple $R$.
f. For a specific case where it is desired to lower the COD of wastewater from 500 to 200 mg/L, the required loading rate in units of grams of carbon per liter of wastewater is calculated as follows:

$$\text{Loading rate}\left(\frac{g\,carbon}{L\,wastewater}\right) = \frac{X}{kC^{(1/n)}}$$

$$M\left(\frac{g\,carbon}{L\,wastewater}\right) = \frac{(500-200)}{1.042 \times 200^{(1/1.5975)}} = 10.5\frac{g\,carbon}{L\,wastewater}$$

g. The calculation shows very similar values with the Langmuir isotherm (11.9 g/L) showing slightly conservative loading rates.

Performing adsorption isotherm studies is useful for estimating activated carbon loading. If the reader would complete the analyses for the data on other columns, the result would show that the Fisher activated carbon would behave similarly while the biomass-based activated carbon shows lower loading rates of about 1/3 or about 30% less amount per L of wastewater. The reason is that biomass-based activated carbons are better at removing COD levels at higher concentration levels. It is not good at the lower concentration levels for those processes requiring COD removal at very low concentration levels (Capareda, 1990).

## 7 Conclusions

This chapter has described some parameters for biochar characterization that are important for thermal conversion. The physical parameters such as bulk density, particle density, angle of repose, and angle of friction, including related thermal properties, are important in biochar transport and storage including conveying systems. The properties of biochar that are important for thermal conversion are heating value, proximate analysis, and ultimate analysis. Proximate analysis includes the determination of moisture content, volatile combustible matter, fixed carbon, and ash. Ultimate analysis includes elemental analysis of carbon, hydrogen, oxygen, nitrogen, sulfur, and ash. Perhaps the most important parameters for thermal conversion are those associated with the tendency of ash in biochar to melt or fuse at high temperature and upon cooling start to agglomerate and form slag. The slag may block conveying surfaces, leading to fouling of those conveying surfaces and ultimately clogging the pathways.

Heating value is another very important biochar parameter. It describes the energy content of the biochar in units of energy per unit of weight (MJ/kg or Btu/lb). This parameter is most important in determining the process efficiency when converting or using the biochar for energy purposes. Biochar will usually have higher energy content than the biomass feedstock itself. Care must be made in reporting the energy content of the biochar and its products. One should be familiar with reporting energy units on a dry basis (higher heating value), on an "as-received" basis (with a specification of moisture content), or with the "dry ash-free" basis. One application of biochar is cofiring with coal power plants. In this application, the slagging and fouling properties must be evaluated. Coal power plants use different slagging and fouling parameters compared with agricultural-based biochar.

The greatest value of biochar is when it is used for highly specialized processes such as contaminants removal. This chapter shows how adsorption isotherms are developed or established for a particular application such as chemical oxygen demand (COD) removal. Examples were shown to compare the performance of commercial activated carbon versus biomass-based activated carbon. One result showed that biomass-based activated carbons are excellent materials for removing higher contaminant levels such as COD in wastewater. More upgrading is required if biomass-based activated carbon is to be used for very low contaminant levels removal.

## Questions, concepts, definitions, and problems

### Biochar use for heat, energy, and power

1. Enumerate all possible uses of biochar for energy purposes. Describe the important biochar properties that will be used in preparation for the utilization of biochar for heat, energy, and electrical power production.

### Biochar use for nonenergy applications

2. Enumerate all possible uses of biochar for nonenergy purposes. Describe the important biochar properties that will be used in preparation for the utilization of biochar for these applications.

### Proximate analysis calculations

3. Sorghum biochar samples after coming off a pressurized batch pyrolyzer processed at 500°C were used to estimate the proximate analysis. The following are the data from the experiment: (a) an initial sample weighing 10.00 g was used and placed in the drying oven set at 104°C overnight until the final weight becomes 9.90 g. About 2.05 g of bone-dry samples was placed in the tube furnace set at 950°C for 7 min and the resulting weight of the product was 1.749 g. The same sample was then placed in a muffle furnace set at 600°C for 4 h with a resulting final weight of 0.9375 g. Calculate the proximate analysis for this sample (Santos et al., 2015).

### Ultimate analysis and Van Krevelen plot

4. Determine the atomic carbon-to-hydrogen ratio for carbonized switchgrass biochar pyrolyzed at 400°C for 3 h. Compare this with a sample pyrolyzed at 300°C for 3 h. Calculate also the atomic O/C ratios. The ultimate analysis data for both biochar samples are given as follows: switchgrass biochar at 400°C for 3 h: C (59.6%), H (4.7%), O (31.30%), N (1.00%), and S (0.14%); switchgrass biochar at 300°C for 3 h: C (48.8%), H (5.9%), O (41.6%), N (0.63%), and S (0.15%) (Sadaka et al., 2014). These values may be used to generate a van Krevelen plot to compare its migration with coal.

## Heating value calculations

5. The heating value of switchgrass biochar was measured by a bomb calorimeter and reported as 15.0 MJ/kg "as received." The moisture content of this biochar was 9.1%. Convert this heating value numbers into its HHV equivalent. The ash content of the biochar on a dry basis was found to be 11.3%, and determine the heating value on a dry ash-free basis as well.

## Biochar chemical formula

6. Determine the chemical formula of switchgrass biochar pyrolyzed at 300°C for 3 h (Sadaka et al., 2014) and resulted in the following characteristics by weight: C=48.8%; H=5.9%; O=41.6%; N=0.63%; and S=0.15% and ash of 2.92%.

## Stoichiometric air-to-fuel ratio (AFR) calculations

7. For the biochar sample given in Problem 6, determine the stoichiometric air-to-fuel ratio, by volume and by weight, if this material is combusted using 100% stoichiometric air.

## Slagging and fouling factor for coal

8. Determine the slagging ($R_s$) and fouling factor ($R_f$) for dairy manure ash data as shown in Table 3.5. Discuss the slagging and fouling potential based on these indices.

## Slagging and fouling factor for dairy manure ash

9. Determine the alkali index, base-to-acid ratio and agglomeration index for dairy manure ash data as shown in Table 3.5. Assume the heating value of the material is 15.93 MJ/kg [6863 Btu/lb]. Discuss the slagging and fouling potential based on these indices. The % ash in biomass was reported to be 18.62% and the alkali in ash is around 7.10%.

## Biochar production via gasification and subsequent biochar use

10. Determine the biochar production for a 5 MW fluidized bed gasification facility in units of tonnes/h. Assume biochar production rate is 25%. The feedstock is Douglas fir with 50% moisture when harvested. The energy content of this wood fuel was 21 MJ/kg [8616 Btu/lb] (LePori and Soltes, 1985)

and an overall electrical power conversion efficiency of 20%. How much biochar is produced per hour, per day, and per year if the unit is operating 350 days a year [8400 h]. If this biochar is subsequently used to generate additional electrical power, how much additional power is generated? Assume the heating value of biochar is 22.5 MJ/kg.

# References

Alhashimi, H.A., Aktas, C.B., 2017. Life cycle environmental and economic performance of biochar compared with activated carbon: a meta-analysis. Resour. Conserv. Recycl. 118, 13–26.

Capareda, S.C., 1990. Studies on Activated Carbon Produced From Thermal Gasification of Biomass Wastes. PhD Dissertation, Biological and Agricultural Engineering. Texas A&M University, College Station, TX.

Capareda, S.C., 2014. Introduction to Biomass Energy Conversions. CRC Press, Taylor and Francis Group, Boca Raton, FL, ISBN: 978-1-4665-1333-4 (Hardback).

Downie, A., Munroe, P., Crosky, A., 2009. Characteristics of biochar – physical and structural properties. In: Lehmann, J., Joseph, S. (Eds.), Biochar for Environmental Management Science and Technology. Earthscan Publishing, Sterling, VA, ISBN: 978-1-84407-658-1, pp. 13–29.

Emergen Research, 2020. Activated carbon market size worth USD 14.07 by 2027 growing at a CAGR of 9.6%. Intrado Globe Newswire. December 02, 2020. Available at: https://www.globenewswire.com/news-release/2020/12/02/2138482/0/en/Activated-Carbon-Market-Size-Worth-USD-14-07-Billion-by-2027-Growing-at-a-CAGR-of-9-6-Emergen-Research.html. (Accessed 1 March 2021).

Hernandez-Maglinao, J.R., Capareda, S.C., 2019. Improving the surface areas and pore volumes of bio-char produced from pyrolysis of cotton gin trash via steam activation process. Int. J. Sci. Eng. Sci. 3 (6), 15–18. Available at: file:///G:/My%20Drive/Publications/2019/73-IJSES-V3N4-Joan-AC-2019.pdf. Accessed on June 20, 2019.

Hutchins, R.A., 1981. Activated carbon and development of design parameters. In: Perrich, J.R. (Ed.), Activated Carbon Adsorption for Wastewater Treatment. CRC Press, Boca Raton, FL.

Jenkins, B.M., 1989. Physical properties of biomass. In: Kitani, O., Hall, C.W. (Eds.), Biomass Handbook. Gordon and Breach Science Publishers, New York, NY.

Lata, K., Mande, S.P., 2009. Bioenergy resources. In: Kishore, V.V.N. (Ed.), Renewable Energy Engineering and Technology: Principles and Practice. Earthscan Publishers, London, UK.

Lehmann, J., Joseph, S., 2009. Biochar for Environmental Management Science and Technology. Earthscan Publishing, Sterling, VA, ISBN: 978-1-84407-658-1.

LePori, W.A., Soltes, E.J., 1985. Thermochemical conversion for energy and fuel. In: Hiler, E.A., Stout, B.A. (Eds.), Biomass Energy: A Monograph. Texas Engineering Experiment Station, The Texas A&M University Press, College Station, TX.

Maglinao, A.L., Capareda, S.C., 2010. Predicting fouling and slagging behavior of dairy manure (DM) and cotton gin trash (CGT) during thermal conversion. Trans. ASABE 53 (3), 903–909. ASABE, St. Joseph, MI.

Rodriguez-Reinoso, F., 1977. Introduction to Carbon Technologies. Universidad de Alicante, University of Alicante Publications, Alicante, Spain.

Sadaka, S., Sharara, M.A., Ashworth, A., Keyser, P., Allen, F., Wright, A., 2014. Characterization of biochar from switchgrass carbonization. Energies 7, 548–567.

https://doi.org/10.3390/en7020548. Available at https://www.mdpi.com/1996-1073/7/2/548/htm. Accessed February 26, 2021.

Santos, B., Capareda, S.C., 2015. Energy sorghum pyrolysis using pressurized batch reactor. Biomass Convers. Biorefin. J. Available at http://link.springer.com/article/10.1007/s13399-015-0191-5.

Skrifvars, B.J., Ohman, M., Nordin, A., Hupa, M., 1998. Predicting bed agglomeration tendencies for biomass fuels fired in FBC boilers: a comparison of three different prediction methods. Energy Fuel 13 (2), 359–363.

Usowich, B., Lipiec, J., Lukowski, M., Marczewski, W., Usowich, J., 2016. The effect of biochar application on thermal properties and albedo of loess soil under grassland and fallow. Soil Tillage Res. 164, 45–51. Elsevier Science Direct, UK.

Vamvuka, D., Zografos, D., 2004. Predicting the behavior of ash from agricultural wastes during combustion. Fuel 83 (14–15), 2051–2057. Elsevier Publishing, UK.

Wang, X., Guo, Z., Zhang, J., 2020. Recent advances in biochar application for water and wastewater treatment: a review. PeerJ 8. https://doi.org/10.7717/peerj.9164, e9164.

Winegartner, E.C. (Ed.), 1974. Coal Fouling and Slagging Parameters. Report Prepared by the American Society of Mechanical Engineers (ASME) Research Committee on Corrosion and Deposits from Combustion Gases. Copyright 1974 by the American Society of Mechanical Engineers (ASME), Three Park Avenue, NY.

Yang, Z., Kumar, A., Hunke, R.L., Buser, M., Capareda, S., 2015. Pyrolysis of eastern redcedar: distribution and characteristics of fast and slow pyrolysis products. Fuel 166 (15 February 2016). https://doi.org/10.1016/j.fuel.2015.10.101.

# Biochar characterization for water and wastewater treatments

**Balwant Singh[a], Tao Wang[b], and Marta Camps-Arbestain[c]**

[a]School of Life and Environmental Sciences, Faculty of Science, The University of Sydney, Eveleigh, NSW, Australia. [b]CAS Key Laboratory of Mountain Surface Processes and Ecological Regulation, Institute of Mountain Hazards and Environment, Chinese Academy of Sciences, Chengdu, China. [c]New Zealand Biochar Research Centre, School of Agriculture and Environment, Massey University, Palmerston North, New Zealand

## 1 Introduction

Soil biochar application has been promoted for over a decade because of the long-term stability of biochar-carbon and other benefits to soil properties and crop productivity (Glaser et al., 2002; Biederman and Harpole, 2013; Creamer and Gao, 2016; Ye et al., 2019). More recently, several environmental applications of biochar have been reported. Such applications include adsorption of heavy metals, adsorption of organic pollutants, remediation of contaminated soils, rehabilitation of degraded soils, and improvement of soil physical properties (Kookana et al., 2011; Ahmad et al., 2014; Oliveira et al., 2017; Patra et al., 2021). Similarly, biochar and its activated derivatives have been found to be very efficient in removing a variety of contaminants from water and wastewater, including heavy metals, organic contaminants, and pathogenic organisms (Mohan et al., 2014; Mohanty et al., 2014; Reddy et al., 2014; Kaetzl et al., 2018, 2020).

The benefits and efficiency of biochar for environmental application, including water and/or wastewater, are strongly dependent on biochar properties, such as carbon content, specific surface area, porosity, cation- and anion-exchange capacity, and degree of fused aromaticity. The evaluation of key chemical and physical properties of biochar is thus crucial for the identification of most suitable biochar for a particular application. In fact, for biochar application to soil, there is a protocol and a classification system, which are based on selected biochar properties (IBI, 2012; Camps-Arbestain et al., 2015). It is well recognized that biochar properties are dependent on the type of the feedstock, i.e., initial organic waste used for biochar production and pyrolysis conditions (Singh et al., 2010; Weber and Quicker, 2018; Hassan et al., 2020). In addition to pyrolysis, other techniques, such as hydrothermal carbonization, gasification, and torrefaction (Meyer et al., 2011), have been used to convert organic wastes into charred material. However, the materials produced using these techniques do not always meet the biochar definition as specified in the guidelines for the European Biochar Certificate (EBC, 2012) and the International Biochar Initiative (IBI, 2012).

Additional treatments or modifications are done sometimes to biochar for increasing its efficiency for a particular application. Generally, gas or steam treatment is employed to increase the porosity of biochar; such treatments are particularly suitable for contaminants whose sorption is dependent on the availability of pore space (e.g. Kong et al., 2019; Krerkkaiwan and Fukuda, 2019; Kwak et al., 2019). Acids and oxidizing agents enhance oxygen-containing surface functional groups of biochar for greater sorption of contaminants in cationic forms (e.g. Mia et al., 2017a, b). The alkaline modification results in high aromaticity of the biochar, promoting the $\pi-\pi$ EDA interaction, which are involved in the sorption for some organic contaminants, such as dyes and antibiotics (Wang et al., 2020). Impregnations or coating of biochar with metal (such as Fe and Al) oxides is also done for increased sorption of anions such as $PO_4^{3-}$ (Wang et al., 2020). In general, biochars produced at high temperature of pyrolysis are effective in retaining nonpolar organic contaminants, where specific surface area, microporosity, and hydrophobicity are important. Biochars produced at low temperature have abundant surface functional groups and thus contribute to the retention of inorganic contaminants and also that of polar organic compounds (Ahmad et al., 2014). Moreover, in contrast to activated carbon, biochar contains an ash fraction that may also interact with soil contaminants, in particular metal and metalloid ions (Xu et al., 2017).

In this chapter, we identify key properties for biochar's use for water and wastewater treatments. The recommended procedures to determine the identified biochar properties are described. Additionally, data for relevant chemical and physical properties of selected biochars made from different feedstocks are presented.

## 2 Laboratory Analysis of biochar

**Sampling and storage:** It is vital to use a representative biochar sample for the laboratory analysis. Readers are referred to other resources that describe the theory of sampling and provide a comprehensive framework for obtaining a representative sample for the laboratory analysis (Gy, 2004; Petersen et al., 2005). Since biochar is a highly sorptive material, it can adsorb water, gases, and other compounds, which can modify biochar surfaces and other properties. Thus, care must be taken during transport, storage, and sample preparation of biochar for analysis. Hilber et al. (2017) outlined the procedure for obtaining a representative biochar sample for laboratory analysis. The exact methodology depends on various factors, particularly the type of biomass, the total quantity of biochar, and the number of pyrolysis batches. Once a representative biochar sample is obtained for laboratory analysis, it should be ground to <2 mm fraction and dried at 40°C. The dried sample is stored in an airtight container, to avoid exposure to any gases or liquid contaminants, at room temperature.

**Chemical properties:** The chemical properties of biochar that are most relevant to biochar's application in water and wastewater treatments includes pH, electrical conductivity (EC), ion exchange capacity, total carbon (and organic carbon ($C_{org}$)), $H/C_{org}$ atomic ratio (as a proxy of degree of fused aromaticity), inorganic carbon and ash concentration. Chemical properties for a set of biochars are provided in Table 4.1.

**pH and electrical conductivity:** pH is an important property of biochar which can have a significant direct or indirect influence on biochar's characteristics. The biochar pH can affect the adsorption–desorption of organic and inorganic contaminants on/from biochars (Fidel et al., 2018; Liu et al., 2018; Park et al., 2019). Additionally, biochar pH can influence precipitation-dissolution reactions of various inorganic contaminants. The EC of biochar should be considered when this is to be applied to soils under arid and semi-arid conditions (Kong, 2021).

Both pH and EC can be measured in the extract of 1:10 biochar: water ratio (Singh et al., 2017). We recommend shaking 5.0 g air-dry biochar sample (<2 mm) with 50 mL deionized water in a 100 mL centrifuge tube for 1 h at 25°C. Let the suspension to stand for about 20 min and measure EC (first) and pH using calibrated EC and pH meters.

**Total carbon:** The suitability of biochars as a tool to eliminate either inorganic or organic pollutants is indirectly related to its $C_{org}$ content, given that their properties as adsorbents are to a

**Table 4.1** Mean ($n = 3$ or more) value of pH (1:10 in water extract), electrical conductivity (EC, 1:10 in water extract), total carbon, lime equivalence, ash concentration, and cation exchange capacity of biochars produced from difference feedstocks pyrolyzed at different temperatures.

| Biochar—Feedstock and pyrolysis temperature (°C) | $pH_{H_2O}$ (1:10) | EC (1:10) (dS m$^{-1}$) | Total C (%) | Lime equivalence (%-CaCO$_3$-eq) | Ash (%) | CEC[a] (mmol$_c$ kg$^{-1}$) |
|---|---|---|---|---|---|---|
| Wheat straw 550°C | 10.38 | 3.41 | 67.6 | 5.7 | 20.54 | 98.5 |
| Wheat straw 700°C | 10.62 | 5.07 | 67.9 | 6.5 | 22.11 | 93.1 |
| Switch grass 400°C | 7.04 | 0.29 | 73.6 | 1.9 | 3.35 | 137.9 |
| Switch grass 550°C | 9.73 | 0.23 | 81.9 | 3.0 | 4.97 | 131.0 |
| Pine chips 400°C | 6.68 | 0.23 | 73.1 | 3.9 | 3.68 | 125.4 |
| Pine chips 550°C | 7.23 | 0.11 | 81.0 | 5.0 | 5.45 | 102.9 |
| Eucalyptus wood 450°C | 7.10 | 1.12 | 68.6 | 2.6 | 5.11 | 127.7 |
| Eucalyptus wood 550°C | 9.35 | 0.93 | 74.0 | 6.3 | 5.77 | 121.7 |
| Poultry litter 550°C | 9.58 | 3.99 | 41.4 | 11.8 | 44.71 | 114.0 |
| Digestate 700°C | 10.72 | 3.90 | 59.1 | 10.8 | 32.68 | 91.6 |
| Greenwaste 550°C | 7.81 | 0.10 | 79.7 | 1.8 | 8.64 | 102.7 |
| Rice husk 550°C | 10.79 | 1.00 | 46.5 | 1.5 | 48.18 | 129.3 |
| Rice husk 700°C | 10.75 | 0.92 | 47.0 | 1.9 | 48.26 | 128.3 |
| Miscanthus straw 550°C | 10.47 | 1.40 | 75.5 | 3.8 | 12.13 | 101.0 |
| Miscanthus straw 700°C | 10.35 | 3.69 | 78.5 | 5.6 | 11.50 | 96.2 |
| Mixed softwood 550°C | 7.39 | 0.16 | 83.7 | 1.5 | 1.48 | 63.2 |
| Mixed softwood 700°C | 8.87 | 0.34 | 89.1 | 2.3 | 1.67 | 86.5 |
| Tomato waste 550°C | 10.25 | 29.93 | 30.7 | 20.5 | 51.21 | 124.7 |
| Durian shell 400°C | 10.27 | 4.94 | 65.5 | 9.3 | 11.70 | 134.3 |

[a]CEC (cation exchange capacity) values are for washed biochars on dry weight basis.
From Singh, B., Camps-Arbestain, M., Lehmann, J., 2017. Biochar: A Guide to Analytical Methods. CRC Press, New York.

large extent related to their porosity and presence/absence of surface functional groups on the charcoal structure. Moreover, the larger their $C_{org}$ content along with a low H/$C_{org}$ atomic ratio (as explained in the following section), the smaller their degradability. In this chapter, we describe the procedures for the measurement of total and inorganic carbon and propose the calculation of $C_{org}$ concentration in biochar by the difference of two values.

Total carbon (and total H) in biochar can be measured by conventional elemental analyzers although there is no standard protocol for it (Bird et al., 2017). Due to the generally high carbon content of biochar, only a small quantity of biochar is needed as compare to other organic materials, including various feedstocks of biochar. Given the small sample size for the analysis, it is vital to take a representative and homogeneous biochar sample for this analysis. The homogenization of sample can be achieved by finely ground and well-mixed a relatively large amount of sample. In addition, as biochar can be quite hygroscopic, biochar sample should be dried at 105°C and maintained dry during the elemental analysis (Bird et al., 2017).

**Inorganic carbon:** This carbon form can be found in the ash fraction of biochar, mainly as calcite and/or dolomite, although generally in small amounts (Wang et al., 2014). The presence of these carbonates is important not only for the liming and pH-buffering properties of biochars (Yuan et al., 2011) but also because of the fertilizer value associated with the addition of calcium and/or magnesium therein (Camps-Arbestain et al., 2015). The presence of carbonates in biochar can affect the adsorption–desorption behavior of organic and inorganic contaminants directly or indirectly. The direct role of carbonates results from the sorption and precipitation of contaminants on calcite (and dolomite) surfaces (Calugaru et al., 2016; Guo et al., 2020), and the indirect effect results from their alkalizing properties, which favor the sorption of cationic contaminants onto biochar surfaces (Guo et al., 2020).

The carbonate concentration of biochar can be determined using a titrimetric method as described by Wang et al. (2014) and Calvelo Pereira et al. (2017). The method involves the determination of carbonate content of biochar after reacting the biochar with a dilute HCl acid and trapping evolved $CO_2$ in an alkali (NaOH) solution. In general, a small amount (1–2 g) of dry and finely ground biochar is used. Adequate reaction time (>24 h) is needed to ensure a complete dissolution of carbonates and capture of the resulting $CO_2$ gas.

**$H/C_{org}$ ratio:** As the proportion of condensed aromatic structures in biochar increases, so does (i) the longevity of the biochar, and (ii) the possibility for $\pi - \pi$ EDA type of interactions, which are involved in the sorption for some organic contaminants. During the condensation of aromatic structures, hydrogen atoms are displaced, thus the $H/C_{org}$ atomic ratio can be used as a proxy for the abundance of condensed aromatic C in biochar. For a carbonized material to fulfill the biochar definition, the $H/C_{org}$ atomic ratio must be <0.7 (IBI, 2012; EBC, 2012).

**Ash concentration:** The inorganic constituents of the ash fraction are metal carbonates, silicates, phosphates, sulfates, chlorides, and oxy-hydroxides (Singh et al., 2010; Vassilev et al., 2013a). This fraction of biochar where available nutrients are concentrated is generally soluble. Also, the liming properties of biochars are associated with some salts present in the ash fraction (Vassilev et al., 2013b). Precipitation reactions of some inorganic pollutants (e.g., Cd, Cu, Zn, Pb) can be influenced by the presence of these salts (Guo et al., 2020). The ash fraction becomes smaller over time as the salts dissolve, with solution pH and the presence of chelating substances having a strong influence on their solubilities.

Ash is operationally defined as the inorganic residue that remains after combustion (ASTM, 2012; Enders and Lehmann, 2017). It is determined after the decomposition of organic components by heating biochar in an oxidative atmosphere at a high temperature (e.g. 750°C) for 6h. Automated thermogravimetric analysis has been used to measure the ash content of biochar (Calvelo Pereira et al., 2011). However, the thermogravimetric analyzer might not be always readily accessible. Therefore, manual methods can be adopted to determine the ash content of biochar following the standard protocol developed for coal by American Society for Testing and Materials (ASTM) (ASTM, 2012). Briefly, a small quantity of (~1g) finely ground biochar dried at 105°C is weighed into a crucible and placed in a muffle furnace, heating from room temperature to 750°C at a rate of 5° C min$^{-1}$, and holding at 750°C for 6h. After cooling to 105°C, the crucible is transferred to a desiccator and cooled to room temperature. The weight of ash is determined by the difference in weights of the crucible plus ash and the empty crucible (Enders and Lehmann, 2017). Ash content is reported as the proportion of the 105°C dried mass of initial biochar (Enders and Lehmann, 2017).

**Ion exchange capacity:** Ion exchange capacity of biochar includes two aspects: cation exchange capacity (CEC) and anion exchange capacity (AEC). The CEC of biochar results from its negative charges; while AEC arises from the positively charged surface of biochar. In ubiquitous biochar pH range, CEC is more important than AEC. CEC of biochar plays an important role in the retention of heavy metals and other cationic nutrients (Graber et al., 2017).

The CEC of biochar can be measured according to a two-step procedure involving saturating the exchange sites with a cation (such as $NH_4^+$) and then exchanging the retained cation with another cation (such as $K^+$) (Graber et al., 2017). As many biochars contain soluble salts in the ash, it is recommended the soluble salts should be removed by repeated washing with deionized water before the cation exchange step. For CEC measurement, biochar sample is firstly repeatedly extracted (preferably 3 times) with 1M ammonium acetate ($NH_4OAc$) at pH 7, and excess $NH_4OAc$ is washed with 60% ethanol. Subsequently, biochar is mixed thoroughly with 2M KCl (three extractions if using batch extraction procedure) to replace adsorbed $NH_4^+$. The extracted $NH_4^+$ is then analyzed using a colorimetric or any other appropriate method. CEC (expressed as cmol$_c$ kg$^{-1}$ or mmol$_c$ kg$^{-1}$) of biochar is calculated from $NH_4^+$ concentration. Note that CEC is pH dependent. As NH$_4$OAc buffers the reaction at pH 7, it may overestimate the CEC of a biochar sample with a pH < 7.

**Physical properties:** Biochar has an extensive porosity and specific surface area, which make it readily able to bond with other substances. Biochar pores vary in size from nanometer to micrometer. Pores are classified based on IUPAC conventions (IUPAC, 2014) as: micropores (<2 nm), mesopores (2–50 nm), and macropores (>50 nm). Macropores account for most of the total porosity of biochar (often >95%) (Brewer et al., 2014; Gao and Masiello, 2017). These are generated during pyrolysis until temperature reaches *ca.* 350°C. As temperature of pyrolysis increases, total porosity remains relatively constant, but mesopores and nanopores are generated while O-containing functional groups are lost. Therefore, as temperature of pyrolysis increases, specific surface area (SSA) increases but CEC decreases. Biochars with high ash content have a small SSA due to the blockage of pores by the ash. Additionally, ash particles are not porous like the carbonaceous matrix in the biochar. The mechanisms through which nonionic organic compounds are retained in biochars are through adsorption and/or partition processes, with the former being dictated by porosity, SSA, and aromaticity. Adsorption reactions are more prominent in biochars produced at higher pyrolysis temperatures (Chen et al., 2008), whereas partition reactions are dominant in biochars produced at lower pyrolysis temperatures (Chen et al., 2008). The contribution of adsorption and partition reactions was governed by the relative proportion of condensed aromatic and noncondensed aliphatic C fractions, and surface and bulk properties of biochars. Ash and noncarbonized organic C contents are more important in partition reactions (Guo et al., 2020). Physical properties for a set of biochars are provided in Table 4.2.

**Total porosity and pore size distribution:** There is no specific technique able to precisely measure the pore volume across the entire pore range, due to the challenges of measuring pore sizes that can change over 5 orders of magnitude (Brewer et al., 2014). Methods used include gas adsorption (described below as used for specific surface area measurements), mercury porosimetry, scanning electron microscopy, X-ray computed tomography, and pycnometry (Gao and Masiello, 2017). Here, we propose the use of a technique based on density measurements (pycnometry) using two displacement fluids: helium gas to measure the skeletal density ($\rho_s$) and a granular solid fluid to measure the envelop density ($\rho_e$) (Brewer et al., 2014; Gao and Masiello, 2017). Combining these two measurements, biochar porosity ($\varphi$) can be calculated:

Table 4.2 BET surface area, mean pore size, micropore volume, and porosity of reference biochars. Surface area, pore size, and micropore volume were measured by $CO_2$ adsorption method, whereas porosity was measured by pycnometry.

| Biochar—Feedstock and pyrolysis temperature (°C) | Surface area (m²/g) | Mean pore size (Å) | Micropore volume (cm³/g) | Porosity (%) |
|---|---|---|---|---|
| Wheat straw 550°C | 94.7 | 2.45 | 65.5 | 63.7 |
| Wheat straw 700°C | 117.3 | 2.48 | 75.0 | 63.0 |
| Switch grass 400°C | 100.2 | 1.59 | 73.1 | 82.3 |
| Switch grass 550°C | 113.7 | 1.97 | 94.2 | 83.7 |
| Pine chips 400°C | 72.0 | 1.85 | 67.6 | 68.3 |
| Pine chips 550°C | 139.8 | 1.96 | 98.0 | 71.5 |
| Eucalyptus wood 450°C | 161.0 | 1.54 | 118.6 | 65.0 |
| Eucalyptus wood 550°C | 115.8 | 2.34 | 76.4 | na[a] |
| Poultry litter 550°C | 69.0 | 2.45 | 44.0 | 57.8 |
| Digestate 700°C | 328.6 | 1.05 | 186.6 | na |
| Greenwaste 550°C | 187.8 | 2.14 | 129.1 | 75.4 |
| Rice husk 550°C | 74.4 | 2.14 | 52.0 | 78.4 |
| Rice husk 700°C | 82.4 | 2.34 | 55.4 | 80.8 |
| Miscanthus straw 550°C | 104.2 | 2.20 | 70.0 | 59.7 |
| Miscanthus straw 700°C | 137.7 | 2.57 | 88.6 | 62.0 |
| Mixed softwood 550°C | 56.8 | 2.95 | 76.5 | 58.6 |
| Mixed softwood 700°C | 170.0 | 2.05 | 117.9 | 56.9 |
| Tomato waste 550°C | 5.4 | 5.85 | 5.1 | na |
| Durian shell 400°C | 63.8 | 3.42 | 36.6 | 79.9 |

[a] na = Samples not available.
From Singh, B., Camps-Arbestain, M., Lehmann, J., 2017. Biochar: A Guide to Analytical Methods. CRC Press, New York.

$$\varphi = 1 - \left(\frac{\rho_e}{\rho_s}\right)$$

Biochar samples are sieved and particles >2 mm are chosen for the measurements of both skeletal and envelop densities. It is important that biochar particles are free of volatiles and moisture as these can affect the results. Prior to analysis, samples are oven-dried at 105°C and then placed in a desiccator. Skeletal density is measured using a helium pycnometer. Envelop density is measured using an Envelop density analyzer.

The pycnometry-based technique quantifies porosity across the entire pore size range (nm–100 µm), whereas the gas adsorption procedure determines only micropores (<2 nm). Since different porosity measurement techniques may measure different pore size, the application of particular technique depends on the purpose or application of such analysis.

**Specific surface area (SSA)** of biochar is commonly measured by $N_2$ or $CO_2$ adsorption isotherms at 77 K using a Brunauer–Emmett–Teller (BET) surface area analyzer. However, it is recommended to use $CO_2$ adsorption because of its greater similarity with the C surface and greater penetration than $N_2$ (Donne, 2017). In fact, $CO_2$ adsorption measures biochar pores that are <0.7 nm, whereas $N_2$ adsorption measures pores that are <2 nm (Illingworth et al., 2013). In either case, only micropores (<2 nm) (when $N_2$ is used) or a fraction of the micropores (when $CO_2$ is used) are measured with this technique. Prior to the measurement, samples should be degassed overnight (18 h) at 300°C under vacuum to remove surface water and volatile organic species and kept under vacuum thereafter. For the samples described in this chapter (Table 4.2), the adsorption of $CO_2$ gas was carried out at 0°C using a Micromeritics Gas Adsorption and Porosity System. For the samples described here, micropore volume was found to be linearly correlated with the specific surface area (SSA) of biochars using the $CO_2$ adsorption procedure—Micropore volume $(cm^3/g) = 14.9 + 0.56 \times SSA$ $(m^2/g)$.

**Spectroscopic analysis for functional groups characterization:** The reactivity of biochar toward contaminants is largely determined by the functional groups on biochar surfaces and the nature of contaminants. Chargeable functional groups, such as carboxyl, phenolic, alcoholic, carbonyl, and other oxygen-containing moieties, which increasingly dissociate at increasing pH values and thereby creating negative surface charges which contribute to the affinity of biochar to metal and metalloid cations and positively charged organic contaminants (Bolan et al., 2014). Sorption of nonpolar organic contaminants by charred material can be described by a dual-mode model: partition and adsorption (Chen et al., 2008). The partition mode occurs mainly for materials thermally treated at low temperatures (<250°C). These contain amorphous carbon phase (Keiluweit et al., 2010) and likely consist of small (poly)aromatic and aliphatic functional groups (Inyang and Dickenson, 2015). The adsorption mode works for biochars made at higher heating temperature and becomes increasingly important as pyrolysis temperature increases (Chen et al., 2008; Keiluweit et al., 2010).

Functional groups of biochar can be investigated using spectroscopic techniques, among which the most frequently used techniques include Fourier-transform infrared (FTIR) spectroscopy (Singh et al., 2016), solid-state $^{13}C$ nuclear magnetic resonance (NMR) spectroscopy (Brewer et al., 2009), and X-ray photoelectron spectroscopy (XPS) (Singh et al., 2014). Biochar samples subjected to spectroscopic analyses are usually in a dried and powdered form. For NMR analysis, samples might be demineralized using a diluted hydrofluoric acid solution to remove paramagnetic ions prior to the analysis (Wang et al., 2013; Smernik, 2017).

FTIR analysis of biochar mainly uses two different sample presentation methods: diffuse reflectance (DR-FTIR) and attenuated total reflectance (ATR-FTIR) (Johnston, 2017). For DR-FTIR analysis, biochar samples are prepared by mixing with dry potassium bromide (KBr) in an agate mortar. For ATR-FTIR analysis, samples are run as fine powders or KBr-diluted powders pressed on to the surface of a diamond or germanium internal reflection elements (IRE) crystals. FTIR spectra can be collected with an ordinary FTIR spectrometer (e.g. Nicolet and Bruker) equipped with a medium-band liquid nitrogen-cooled MCT detector and a KBr beam splitter (Johnston, 2017). Depending on the IR method and the IRE crystal type if the ATR method is adopted, spectrum can be collected in the wavenumber range of $4000-400\,cm^{-1}$ (DR-FTIR) and $4000-600\,cm^{-1}$ (germanium IRE). Generally, a resolution of $4\,cm^{-1}$ is adequate, and the number of scans can be determined according to the quality of the spectrum. In most cases, the spectrum is automatically outputted in a Fourier-transformed form. The bands can be assigned according to their peak position or range (Table 4.3.).

X-ray photoelectron spectroscopy is an element specific technique and capable of obtaining quantitative information of functional groups within a depth of <10 nm from biochar surface (Smith, 2017). Gently crushed or finely ground biochar samples are mounted on a specialized XPS sample holder using clean metal screws and clamps. In some cases, samples are too fine or too fragile to be fixed on the sample holder; as a result, they can be mounted using a piece of double-sided adhesive tape. The samples and the sample holder are then placed in the vacuum chamber of the XPS instrument and analyzed under an ultra-high vacuum ($<5 \times 10^{-7}$ Pa). To obtain representative results, multiple and large-area analyses are required for large biochar particles and less homogenized samples (Smith, 2017). For finely ground and well-homogenized samples, a few small-area analyses are adequate to acquire a representative data. However, one should note that results for the well-ground samples tend to represent the bulk composition rather than that of biochar surfaces. Relative abundance of different chemical states (i.e. species, Table 4.4.) of the element of interest can be calculated by fitting the spectral curve using a Gaussian–Lorentzian function after the subtraction of baseline. Examples of the curve-fitting procedure and data interpretation can be referred to Smith (2017).

**Table 4.3** FTIR band assignments of biochar and associated inorganic phases.

| Band range (cm$^{-1}$) | Assigned band(s) |
|---|---|
| 3670–3630 | $\nu$(OH) from non-hydrogen bonded O—H groups |
| 3600–3200 | $\nu$(OH) from sorbed water and hydrogen-bonded biochar O—H groups |
| 3080–3020 | Aromatic $\nu$(CH) |
| 2990–2950 | Asymmetric aliphatic $\nu$(CH) from terminal -CH$_3$ groups |
| 2950–2920 | Asymmetric aliphatic $\nu$(CH) from -CH$_2$ groups |
| 2890–2870 | Symmetric aliphatic $\nu$(CH) from terminal -CH$_3$ groups |
| 2870–2840 | Symmetric aliphatic $\nu$(CH) from terminal -CH$_3$ groups |
| 1740–1650 | $\nu$(C=O) from carboxylic acids, amides, esters and ketones |
| 1650–1610 | H-O-H bending band of water (v2 mode) |
| 1610–1580 | $\nu$(C=C) |
| 1590–1520 | $\nu$(COO–) carboxylate anions, Amide-II vibrations |
| 1510–1485 | Aromatic skeletal vibrations |
| 1480–1440 | CH$_2$ deformation (scissor vib) |
| 1450–1400 | Carbonate (v3; asymmetric stretch) |
| 1390–1310 | Phenolic O—H bend, -C(CH$_3$) C—H deformation |
| 1280–1200 | Carboxylic acid C-OH stretch, O—H deformation, carboxyl, ester/amide region |
| 1160–1020 | $\nu$(C—O) polysaccharide, carbohydrate region |
| 1140–1000 | $\nu$(Si—O) from clay minerals associated with biochar |
| 940–820 | $\nu$(M-O-H) O—H bending bands from clay minerals associated with biochar |
| 900–700 | O-H bending |
| 800–780 | Quartz 'doublet' |

Adopted from Johnston, C.T., 2017. Biochar analysis by Fourier-transform infra-red spectroscopy. In: Singh, B., Camps-Arbestain, M. & Lehmann, J. (Eds) Biochar: A Guide to Analytical Methods. CRC Press, New York, pp. 199–213.

Solid-state $^{13}$C NMR spectroscopy is a powerful technique to analyze the functional groups of biochar based on variation in the local chemical environment of C atom. Due to the heterogenous nature of biochar C atoms, however, a $^{13}$C NMR spectrum of biochar often exhibits a series of broad peaks, each of which corresponds to a class of C atoms in a similar chemical environment rather than in a specific molecular structure. These peaks in $^{13}$C NMR spectrum are generally assigned to "alkyl C," "O-alkyl C," "aryl C," "O-aryl C," and "carbonyl C" among others as listed in Table 4.5 (Baldock and Smernik, 2002). So far, no standard protocols are available for $^{13}$C NMR analysis of biochar. Nevertheless, tens to hundreds of milligrams of finely ground and well-homogenized biochar sample is required for a successful solid-state $^{13}$C NMR analysis (Smernik, 2017). In order to obtain a spectrum of adequate quality, thousands of individual scans are often carried out for biochar analysis.

Table 4.4 Chemical shifts of different types of organic C in biochar.

| Chemical shift limits (ppm) | Proposed dominant type of C |
| --- | --- |
| 290–265 | Carbonyl and amide spinning side band |
| 265–245 | O-Aryl spinning side band |
| 245–215 | Aryl spinning side band |
| 215–190 | Ketone C |
| 190–165 | Carbonyl and amide C |
| 165–145 | O-Aryl C (phenolic and furan) |
| 145–110 | Aryl and unsaturated C |
| 110–90 | Di-O-Alkyl C |
| 90–65 | O-Alkyl C |
| 65–45 | Methoxyl and N-alkyl C |
| 45–0 | Alkyl C |

Adopted from Baldock, J.A., Smernik, R.J., 2002. Chemical composition and bioavailability of thermally altered Pinus resinosa (Red pine) wood. Org. Geochem. 33, 1093–1109.

Two different NMR techniques—cross-polarization (CP) and direct polarization (DP)—have been used for biochar characterization (Freitas et al., 1999; McBeath et al., 2014). For biochars produced at pyrolysis temperature below 600°C, the NMR spectra obtained by both techniques are similar (McBeath et al., 2014). However, the CP technique shows lower observability than the DP technique, in particular for biochar samples produced at 350°C and a higher temperature (McBeath et al., 2011). Therefore, for biochars produced at a pyrolysis temperature higher than 350°C, the DP technique is preferentially used. Moreover, the DP technique allows for a quantitative analysis of C composition in biochar. The quantification is accomplished by integrating the NMR spectrum according to the chemical shift limits shown in Table 4.5. Examples of biochar analysis by solid-state $^{13}$C NMR can be found in McBeath et al. (2011) and Smernik (2017).

## 3 Summary

We have outlined procedures for some chemical and physical properties of biochars. These properties may be useful for the use of biochar for water and wastewater treatments. For a more comprehensive details about analytical procedures of biochar, we refer readers to the 'Biochar—A Guide to Analytical Methods" book (Singh et al., 2017). From the chemical and physical

Table 4.5 Binding energies for chemical states of C1s, O1s, and N1s in biochar as investigated by XPS (adopted from Smith, 2017).

| Component and line | Binding energy (eV) |
|---|---|
| **C1s** | |
| Carbide | 282–284 |
| C-C | 284.4–284.7 |
| C-C | 285.0–285.2 |
| C-O | 286.3–286.6 |
| C=O | 287.8–288.0 |
| COO | 289.1–289.3 |
| Carbonate | 290.3–290.6 |
| π-π* | 291.3–291.7 |
| **O1s** | |
| Metal oxides | 529–532 |
| Metal hydroxides | 531–532 |
| O in C=O | 531.3–532.3 |
| O in C-O | 533.3–533.9 |
| **N1s** | |
| Nitride | 396.5–399 |
| Nitrile | 397.5–400.2 |
| C-N | 399–401 |
| Ammonium/quaternary N ion | 400.7–403 |
| Nitroso | 401.5–402 |
| Nitrite | 404–405 |
| Nitrate | 407–408 |

properties of biochars given in Tables 4.1 and 4.2, it is clear that biochar produced from woody biomass have lower electrical conductivity liming equivalence and ash content, and higher total C than the animal manure and crop residues biochars. Additionally, woody biomass biochars have greater porosity and specific surface area than the other biochars. The specific surface of biochars made from the same feedstocks tends to increase as temperature of pyrolysis increases, with the volume of micropores following the same trend, as opposed to their pore size values. From these analyses, woody biomass biochars in general are better for water treatments than the other biochars. Additionally, the biochars produced at high pyrolysis temperatures are suitable for the removal of nonpolar organic contaminants, and those produced at low pyrolysis temperatures are good for sorbing inorganic contaminants and polar organic contaminants from contaminated waters.

## Questions and problems

1. What factors should be considered for the preparation of biochar samples for laboratory characterization?
2. Using the data given in Table 4.1, evaluate relationships between ash content and pH, EC, total C, and lime equivalence of biochars. Briefly explain the reasons for these relationships.
3. How does inorganic carbon influence the removal of contaminants from contaminated waters?
4. The $H/C_{org}$ atomic ratio has been used as a proxy for the C aromaticity of biochar; why Corg and not the total C is used as the denominator in the calculation?
5. Some biochars, such as those made from sewage sludge, often contain metal oxyhydroxides; For such biochars, is the $H/C_{org}$ atomic ratio still a reliable proxy for C aromaticity?
6. When measuring the CEC of biochar, the sample is repeatedly extracted with 1 M ammonium acetate ($NH_4OAc$) at pH 7. What kind of problem we can encounter if we want to measure the CEC of a biochar sample which has a pH of 6?
7. Using the same type of feedstock, we produce biochar at different pyrolysis temperatures. We often find the specific surface area increases but CEC decreases as pyrolysis temperature increases. Why?
8. Which of the two commonly used NMR techniques, cross-polarization (CP) and direct polarization (DP), is preferentially used if we want to have a better observability of C and quantify the contributions of different C forms in biochars listed in Tables 4.1 and 4.2?
9. List some important chemical properties of biochar that may influence the application of the material in water and wastewater treatments. By comparing the properties of biochars listed in Tables 4.1 and 4.2, explain why woody biomass biochars usually perform better than poultry litter and tomato waste biochars for wastewater treatment?

## References

Ahmad, M., Rajapaksha, A.U., Lim, J.E., Zhang, M., Bolan, N., Mohan, D., Vithanage, M., Lee, S.S., Ok, Y.S., 2014. Biochar as a sorbent for contaminant management in soil and water: a review. Chemosphere 99, 19–33.

ASTM, 2012. Standard D3174-12: Standard test method for ash in the analysis sample of coal and coke from coal. West Conshohocken, PA.

Baldock, J.A., Smernik, R.J., 2002. Chemical composition and bioavailability of thermally altered *Pinus resinosa* (Red pine) wood. Org. Geochem. 33, 1093–1109.

Biederman, L.A., Harpole, W.S., 2013. Biochar and its effects on plant productivity and nutrient cycling: a meta-analysis. Glob. Change Biol. Bioenergy. 5, 202–214.

Bird, M., Keitel, C., Meredith, W., 2017. Analysis of biochars for C, H, N, O and S by elemental analyser. In: Singh, B., Camps-Arbestain, M., Lehmann, J. (Eds.), Biochar: A Guide to Analytical Methods. CRC Press, New York, pp. 39–50.

Bolan, N., Kunhikrishnan, A., Thangarajan, R., Kumpiene, J., Park, J., Makino, T., et al., 2014. Remediation of heavy metal(loid)s contaminated soils—to mobilize or to immobilize? J. Hazard. Mater. 266, 141–166.

Brewer, C.E., Schmidt-Rohr, K., Satrio, J.A., Brown, R.C., 2009. Characterization of biochar from fast pyrolysis and gasification systems. Environ. Prog. Sustain. Energy 28, 386–396.

Brewer, C.E., Chuang, V.J., Masiello, C.A., Gonnermann, H., Gao, X., Dugan, B., Driver, L.E., Panzacchi, P., Zygourakis, K., Davies, C.A., 2014. New approaches to measuring biochar density and porosity. Biomass Bioenergy 66, 176–185. https://doi.org/10.1016/j.biombioe.2014.03.059.

Calugaru, I.L., Neculita, C.M., Genty, T., Bussière, B., Potvin, R., 2016. Performance of thermally activated dolomite for the treatment of Ni and Zn in contaminated neutral drainage. J. Hazard. Mater. 310, 48–55.

Calvelo Pereira, R., Kaal, J., Camps-Arbestain, M., Pardo Lorenzo, R., Aitkenhead, W., Hedley, M., et al., 2011. Contribution to characterisation of biochar to estimate the labile fraction of carbon. Org. Geochem. 42, 1331–1342.

Calvelo Pereira, R., Camps-Arbestain, M., Wang, T., Enders, A., 2017. Inorganic carbon. In: Singh, B., Camps-Arbestain, M., Lehmann, J. (Eds.), Biochar: A Guide to Analytical Methods. CRC Press, New York, pp. 51–63.

Camps-Arbestain, M., Amonette, J.E., Singh, B., Wang, T., Schmidt, H.-P., 2015. A biochar classification system and associated test methods. In: Lehmann, J., Joseph, S. (Eds.), Biochar for Environmental Management: Science, Technology and Implementation. Routledge & CRC Press, New York, pp. 165–194.

Chen, B., Zhou, D., Zhu, L., 2008. Transitional adsorption and partition of nonpolar and polar aromatic contaminants by biochars of pine needles with different pyrolytic temperatures. Environ. Sci. Technol. 42, 5137–5143.

Creamer, A.E., Gao, B., 2016. Carbon-based adsorbents for post combustion $CO_2$ capture: a critical review. Environ. Sci. Technol. 50, 7276–7289.

Donne, S., 2017. Specific surface area and porosity measurements. In: Singh, B., Camps-Arbestain, M., Lehmann, J. (Eds.), Biochar: A Guide to Analytical Methods. CRC Press, New York, pp. 297–299.

EBC, 2012. Guidelines for a Sustainable Production of Biochar. European Biochar Certificate. European Biochar Foundation. http://www.european-biochar.org.

Enders, A., Lehmann, J., 2017. Proximate analyses for characterising biochars. In: Singh, B., Camps-Arbestain, M., Lehmann, J. (Eds.), Biochar: A Guide to Analytical Methods. CRC Press, New York, pp. 9–22.

Fidel, R.B., Laird, D.A., Spokas, K.A., 2018. Sorption of ammonium and nitrate to biochars is electrostatic and pH-dependent. Sci. Rep. 8, 1–10.

Freitas, J.C.C., Bonagamba, T.J., Emmerich, F.G., 1999. 13C high-resolution solid-state NMR study of peat carbonization. Energy Fuel 13, 53–59.

Gao, X., Masiello, C.A., 2017. 12 analysis of biochar porosity by pycnometry. In: Singh, B., Camps-Arbestain, M., Lehmann, J. (Eds.), Biochar: A Guide to Analytical Methods. CRC Press, New York, pp. 132–140.

Glaser, B., Lehmann, J., Zech, W., 2002. Ameliorating physical and chemical properties of highly weathered soils in the tropics with charcoal—a review. Biol. Fertil. Soils 35, 219–230.

Graber, E.R., Singh, B., Hanley, K., Lehmann, J., 2017. Determination of cation exchange capacity in biochar. In: Singh, B., Camps-Arbestain, M., Lehmann, J. (Eds.), Biochar: A Guide to Analytical Methods. CRC Press, New York, pp. 74–84.

Guo, M., Song, W., Tian, J., 2020. Biochar-facilitated soil remediation: mechanisms and efficacy variations. Front. Environ. Sci. 8, 183.

Gy, P., 2004. Sampling of discrete materials—a new introduction to the theory of sampling—I. Qualitative approach. Chemom. Intel. Lab. Syst. 74, 7–24.

Hassan, M., Liu, Y., Naidu, R., Parikh, S.J., Du, J., Qi, F., Willett, I.R., 2020. Influences of feedstock sources and pyrolysis temperature on the properties of biochar and functionality as adsorbents: a meta-analysis. Sci. Total Environ. 744, 140714.

Hilber, I., Schmidt, H.-P., Bucheli, T.D., 2017. Sampling, storage and preparation of biochar for laboratory analysis. In: Singh, B., Camps-Arbestain, M., Lehmann, J. (Eds.), Biochar: A Guide to Analytical Methods. CRC Press, New York, pp. 1–8.

IBI, 2012. Standardized Product Definition and Product Testing Guidelines for Biochar that Is Used in Soil. International Biochar Initiative. http://www.biochar-international.org.

Illingworth, J., Williams, P.T., Rand, B., 2013. Characterisation of biochar porosity from pyrolysis of biomass flax fibre. J. Energy Inst. 86, 63–70. https://doi.org/10.1179/1743967112Z.00000000046.

Inyang, M., Dickenson, E., 2015. The potential role of biochar in the removal of organic and microbial contaminants from potable and reuse water: a review. Chemosphere 134, 232–240.

IUPAC, 2014. Compendium of Chemical Terminology: Gold Book. International Union of Pure and Applied Chemistry. http://goldbook.iupac.org/PDF/goldbook.p.

Johnston, C.T., 2017. Biochar analysis by Fourier-transform infra-red spectroscopy. In: Singh, B., Camps-Arbestain, M., Lehmann, J. (Eds.), Biochar: A Guide to Analytical Methods. CRC Press, New York, pp. 199–213.

Kaetzl, K., Lübken, M., Gehring, T., Wichern, M., 2018. Efficient low-cost anaerobic treatment of wastewater using biochar and woodchip filters. Water 10, 818–835.

Kaetzl, K., Lübken, M., Nettmann, E., Krimmler, S., Wichern, M., 2020. Slow sand filtration of raw wastewater using biochar as an alternative filtration media. Sci. Rep. 10, 1229–1240.

Keiluweit, M., Nico, P.S., Johnson, M.G., Kleber, M., 2010. Dynamic molecular structure of plant biomass-derived black carbon (biochar). Environ. Sci. Technol. 44, 1247–1253.

Kong, C., 2021. Designing Technosols to Reduce Salinity and Water Stress of Crops Growing under Arid Conditions. PhD Dissertation, Massey University.

Kong, S.H., Lam, S.S., Yek, P.N.Y., Liew, R.K., Ma, N.L., Osman, M.S., Wong, C.C., 2019. Self-purging microwave pyrolysis: an innovative approach to convert oil palm shell into carbon-rich biochar for methylene blue adsorption. J. Chem. Technol. Biotechnol. 94, 1397–1405. https://doi.org/10.1002/jctb.5884.

Kookana, R.S., Sarmah, A.K., Van Zwieten, L., Krull, E., Singh, B., 2011. Biochar application to soil: agronomic and environmental benefits and unintended consequences. Adv. Agron. 112, 103–143.

Krerkkaiwan, S., Fukuda, S., 2019. Catalytic effect of rice straw-derived chars on the decomposition of naphthalene: the influence of steam activation and solvent treatment during char preparation. Asia Pac. J. Chem. Eng. 14, 15. https://doi.org/10.1002/apj.2303.

Kwak, J.H., Islam, M.S., Wang, S., Messele, S.A., Naeth, M.A., El-Din, M.G., Chang, S.-X., 2019. Biochar properties and lead(II) adsorption capacity depend on feedstock type, pyrolysis temperature, and steam activation. Chemosphere 231, 393–404.

Liu, Y., Lonappan, L., Brar, S.K., Yang, S., 2018. Impact of biochar amendment in agricultural soils on the sorption, desorption, and degradation of pesticides: a review. Sci. Total Environ. 645, 60–70.

McBeath, A.V., Smernik, R.J., Schneider, M.P.W., Schmidt, M.W.I., Plant, E.L., 2011. Determination of the aromaticity and the degree of aromatic condensation of a thermosequence of wood charcoal using NMR. Org. Geochem. 42, 1194–1202.

McBeath, A.V., Smernik, R.J., Krull, E.S., Lehmann, J., 2014. The influence of feedstock and production temperature on biochar carbon chemistry: a solid-state $^{13}$C NMR study. Biomass Bioenergy 60, 121–129.

Meyer, S., Glaser, B., Quicker, P., 2011. Technical, economical, and climate-related aspects of biochar production technologies: a literature review. Environ. Sci. Technol. 45, 9473–9483.

Mia, S., Dijkstra, F.A., Singh, B., 2017a. Aging induced changes in biochar's functionality and adsorption behavior for phosphate and ammonium. Environ. Sci. Technol. 51, 8359–8367.

Mia, S., Dijkstra, F.A., Singh, B., 2017b. Long-term aging of biochar: a molecular understanding with agricultural and environmental implications. In: Sparks, D.-L. (Ed.), Advances in Agronomy. vol. 141. Academic Press, Burlington, ISBN: 978-0-12-812423-9, pp. 1–51.

Mohan, D., Sarswat, A., Ok, Y.S., Pittman Jr., C.U., 2014. Organic and inorganic contaminants removal from water with biochar, a renewable, low cost and sustainable adsorbent e a critical review. Bioresour. Technol. 160. 191e202191-202.

Mohanty, S.K., Cantrell, K.B., Nelson, K.L., Boehm, A.B., 2014. Efficacy of biochar to remove *Escherichia coli* from stormwater under steady and intermittent flow. Water Res. 61, 288–296.

Oliveira, F.R., Patel, A.K., Jaisi, D.P., Adhikari, S., Lu, H., Khanal, S.K., 2017. Environmental application of biochar: current status and perspectives. Bioresour. Technol. 246, 110–122.

Park, J.-H., Wang, J.J., Kim, S.-H., Kang, S.-W., Jeong, C.Y., Jeon, J.-R., et al., 2019. Cadmium adsorption characteristics of biochars derived using various pine tree residues and pyrolysis temperatures. J. Colloid Interface Sci. 553, 298–307.

Patra, B.R., Mukherjee, A., Nanda, S., Dalai, A.K., 2021. Biochar production, activation and adsorptive applications: a review. Environ. Chem. Lett. https://doi.org/10.1007/s10311-020-01165-9.

Petersen, L., Minkkinen, P., Esbensen, K.H., 2005. Representative sampling for reliable data analysis: theory of sampling. Chemom. Intel. Lab. Syst. 77, 261–277.

Reddy, K.R., Xie, T., Dastgheibi, S., 2014. Evaluation of biochar as a potential filter media for the removal of mixed contaminants from urban storm water runoff. J. Environ. Eng. 140, 04014043.

Singh, B., Singh, B.P., Cowie, A.L., 2010. Characterisation and evaluation of biochars for their application as a soil amendment. Aust. J. Soil. Res. 48, 516. https://doi.org/10.1071/SR10058.

Singh, B., Fang, Y., Cowie, B.C.C., Thomsen, L., 2014. NEXAFS and XPS characterisation of carbon functional groups of fresh and aged biochars. Org. Geochem. 77, 1–10.

Singh, B., Fang, Y., Johnston, C.T., 2016. A Fourier-transform infrared study of biochar aging in soils. Soil Sci. Soc. Am. J. 80, 613–622.

Singh, B., Camps-Arbestain, M., Lehmann, J., 2017. Biochar: A Guide to Analytical Methods. CRC Press, New York.

Smernik, R.J., 2017. Analysis of biochars by 13C nuclear magnetic resonance spectroscopy. In: Singh, B., Camps-Arbestain, M., Lehmann, J. (Eds.), Biochar: A Guide to Analytical Methods. CRC Press, New York, pp. 151–161.

Smith, G.C., 2017. X-ray photoelectron spectroscopy analysis of biochar. In: Singh, B., Camps-Arbestain, M., Lehmann, J. (Eds.), Biochar: A Guide to Analytical Methods. CRC Press, New York, pp. 227–244.

Vassilev, S.V., Baxter, D., Andersen, L.K., Vassileva, C.G., 2013a. An overview of the composition and application of biomass ash. Part 1. Phase-mineral and chemical composition and classification. Fuel 105, 40–76. https://doi.org/10.1016/j.fuel.2012.09.041.

Vassilev, S.V., Baxter, D., Andersen, L.K., Vassileva, C.G., 2013b. An overview of the composition and application of biomass ash. Part 2. Potential utilisation, technological and ecological advantages and challenges. Fuel 105, 19–39. https://doi.org/10.1016/j.fuel.2012.10.001.

Wang, T., Camps-Arbestain, M., Hedley, M., 2013. Predicting C aromaticity of biochars based on their elemental composition. Org. Geochem. 62, 1–6.

Wang, T., Camps-Arbestain, M., Hedley, M., Singh, B.P., Calvelo-Pereira, R., Wang, C., 2014. Determination of carbonate-C in biochars. Soil Res. 52, 495–504.

Wang, X., Guo, Z., Hu, Z., Zhang, J., 2020. Recent advances in biochar application for water and wastewater treatment: a review. Peer J. 8. https://doi.org/10.7717/peerj.9164, e9164.

Weber, K., Quicker, P., 2018. Properties of biochar. Fuel 217, 240–261.

Xu, X., Zhao, Y., Sima, J., Zhao, L., Mašek, O., Cao, X., 2017. Indispensable role of biochar-inherent mineral constituents in its environmental applications: a review. Bioresour. Technol. 241, 887–899.

Ye, L., Camps-Arbestain, M., Shen, Q., Lehmann, J., Singh, B., Sabir, M., 2019. Biochar effects on crop yields with and without fertilizer: a meta-analysis of field studies using separate controls. Soil Use Manage. 36, 2–18. https://doi.org/10.1111/sum.12546.

Yuan, J.H., Xu, R.K., Qian, W., Wang, R.H., 2011. Comparison of the ameliorating effects on an acidic Ultisol between four crop straws and their biochars. J. Soils Sediments 11, 741–750. https://doi.org/10.1007/s11368-011-0365-0.

# 5

# Biochar adsorption system designs

Dinesh Mohan[a], Abhishek Kumar Chaubey[a], Manvendra Patel[a], Chanaka Navarathna[b], Todd E. Mlsna[b], and Charles U. Pittman, Jr.[b]

[a]School of Environmental Sciences, Jawaharlal Nehru University, New Delhi, India. [b]Department of Chemistry, Mississippi State University, Mississippi State, MS, United States

## 1 Introduction

Adsorption is a unit operation. Industrial wastewaters contain many organic and inorganic contaminants that are hard to remove using conventional secondary treatments. These pollutants are found in very small concentrations that make remediation by many other common techniques difficult. Adsorption is usually applied as a polishing step to tertiary wastewater prior to final discharge. Water adsorption decontamination is common because of its low cost, easy implementation, and excellent removal efficiencies (Choudhary et al., 2020; Kumar et al., 2019; Patel et al., 2019, 2021; Vimal et al., 2019). Adsorption is applied for the remediation of aqueous contaminants including dyes, inorganics, soluble organics, phenols, pesticides, pharmaceuticals, and many emerging contaminants. Adsorption is applied to a large variety of waters and wastewaters. The most important factors affecting sorption include solution pH, sorbent particle size, sorbent surface area, contact or residence time, solute solubility, solute affinity for the sorbent (biochar), and degree of sorbate ionization. Adsorption systems are operated in batch or semicontinuous and continuous flow modes. Batch sorption is performed in batch reactors at both lab and industrial scale. Batch sorption consists of contacting finely divided adsorbent with the contaminated water for a given time period in a mixing vessel.

Continuous flow studies are performed in dynamic sorption reactors with a fixed bed of adsorbent. Dynamic mode sorption experiments permit the estimation of bed life, adsorbent usage, and operational costs for commercial-scale water treatment (Merle et al., 2020; Ye et al., 2019). Different types of continuous stirred tank process configuration include single-stage downflow (Fig. 5.1A), single-stage upflow (Fig. 5.1B), multistage downflow (Fig. 5.1C), and multistage upflow (Fig. 5.1D) can be designed for conducting batch sorption experiments. Single-stage continuous flow tank is a continuously flowing system where adsorbent density is higher than the liquid. So liquid is stirred and slowly rises through the stirred sorbent. Multistage upflow can be a closed system under upflow pressure. Multistage upflow stirred columns (Fig. 5.1E) are designed so that the stirred liquid rises through dividers to each stage. So, sorbent particles can remain in stage. Some are designed so each stage can have fresh sorbent added and spent sorbent removed. The spacers' holes can be too small to allow sorbent to get through in some cases (depending on density, upflow rates, etc.).

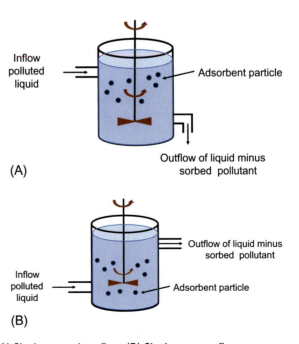

**Fig. 5.1** (A) Single-stage downflow. (B) Single-stage upflow.

*(Continued)*

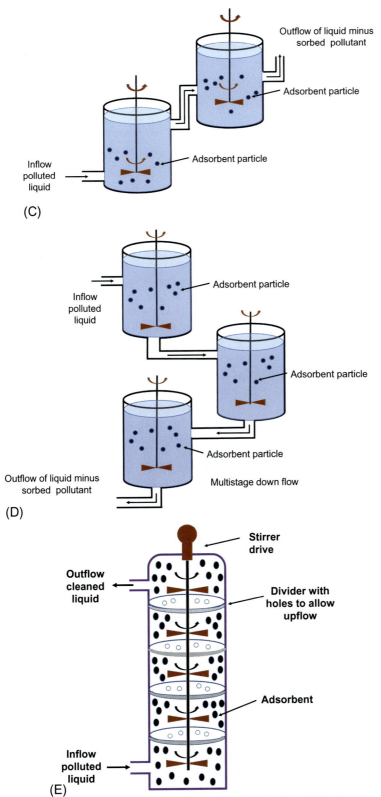

**Fig. 5.1,** Cont'd (C) Multistage upflow. (D) Multistage downflow. (E) Multistage upflow stirred columns.

## 2 Sorption isotherm models

The relation between an adsorbent and its equilibrium at a constant temperature is called the sorption isotherm. This is expressed in Eq. (5.1).

$$q = f(C_e) \qquad (5.1)$$

Here $q$ = mass of the sorbate sorbed/adsorbent mass (i.e., the concentration of the adsorbate at equilibrium on the solid adsorbent) and $C_e$ = equilibrium concentration of adsorbable species in solution.

The adsorption isotherm curve provides important information related to adsorbate mobility or retention phenomenon from an aqueous phase to a solid phase upon adsorbent (Foo and Hameed, 2010; Limousin et al., 2007). Laboratory batch tests are employed for rapid adsorbent screening. Equilibrium isotherm tests directly correlate with the full-scale plant performance. Laboratory tests help to determine the adsorption capacity to better design full-scale plant performance. The adsorption capacity helps to determine the capital cost because it determines the adsorbent dose which is one factor in scaling plant design. Capacity determines the adsorber vessel volume.

The relationship between an adsorbate concentration in solution ($C$) and the amount of adsorbate adsorbed per unit mass is determined for collecting equilibrium isotherm data using the well-known bottle point experiment. A discontinuous batch adsorption is shown in Fig. 5.2. A definite adsorbent dose ($m$) is

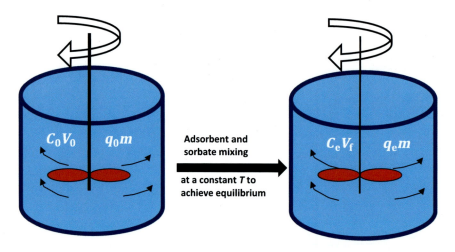

**Fig. 5.2** Schematic of a standard batch sorption operation.

placed in stirred vessel to generate a slurry with an initial feed solution (volume $V_0$ with an initial sorbate concentration, $C_0$). This suspension is thoroughly mixed at constant temperature until equilibrium is reached. After achieving equilibrium, the concentration of sorbate in solution drops to $C_e$ while solute concentration on sorbent is increased to $q_e$.

To determine the amount of sorbate bound to the sorbent when equilibrium ($q_e$) is reached, a mass balance is performed according to Eq. (5.2) (Geankoplis, 1998).

$$C_0 V_0 + q_0 m = C_e V_f + q_e m \qquad (5.2)$$

For a pristine sorbent, the adsorbate amount at the beginning is equal to 0, leading to Eq. (5.3):

$$C_0 V_0 = C_e V_f + q_e m \qquad (5.3)$$

or

$$q_e = \frac{C_0 V_0 - C_e V_f}{m} \qquad (5.4)$$

The aliquot taken for adsorbate quantification is negligible, thus $V_0 = V_F = V$. Therefore, the adsorbate adsorbed on the sorbent can be calculated using Eq. (5.5).

$$q_e = \frac{(C_0 - C_e)V}{m} \qquad (5.5)$$

These laboratory batch sorption experiments are required to determine the adsorbent quality by investigating the adsorption isotherm (Fig. 5.3.) at many operational parameters including

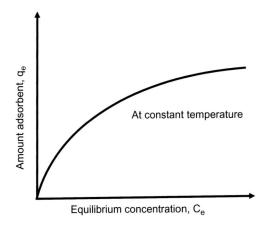

**Fig. 5.3** Generalized adsorption isotherm.

solution pH, temperature, adsorbent/adsorbate concentrations, sorbent particle size, and equilibrium time.

It is not advisable to compare sorbents based on examining just their percent of sorbate removal. This is a very crude method. It must not be used (Mohan and Pittman, 2006). Adsorption isotherms (Fig. 5.3) are more accurate than percentage removal. Four different adsorption experiments are shown in Fig. 5.4 (Mohan and Pittman, 2006). Final removal results depend on the selected equilibrium concentration used for calculating percent adsorption. If the percent removal (at initial adsorbate concentration = 250 mg/L and adsorbent dose = 2.5 g/L) was calculated at point no. 1, the percent removal capacity increases in the order $A < C < B < D$, while at point no. 2 the order is $A < B < C < D$ (Mohan and Pittman, 2006). Therefore, sorbents must not be compared based on percent contaminant adsorption. Adsorption calculations must be obtained and analyzed in terms of mg/g (using Eq. 5.1) and various sorption models (preferably Langmuir monolayer capacity) should be used for comparative evaluation of sorbents (Mohan and Pittman, 2006).

The equilibrium data are fitted to a sorption isotherm model. All isotherms are unique, varying for different sorbents and sorbates but they can exhibit common isotherms shapes (Brunauer et al., 1940). Six different isotherms including type 1, type II, type III, type IV, type V, and type VI exist (Limousin et al., 2007; Mohan and Singh, 2005; Sing et al., 1985) (Fig. 5.5).

A type I isotherm is obtained from adsorbents having only micropores (Sing et al., 1985). The type II isotherm is characteristic

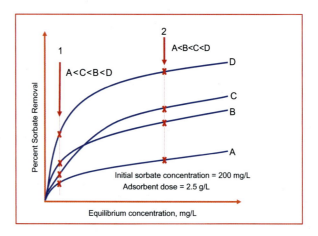

**Fig. 5.4** Comparative evaluation of adsorbents. Modified from Mohan, D., Pittman, C.U., 2006. Activated carbons and low cost adsorbents for remediation of tri- and hexavalent chromium from water. J. Hazard. Mater. 137(2), 762–811. Copy right.

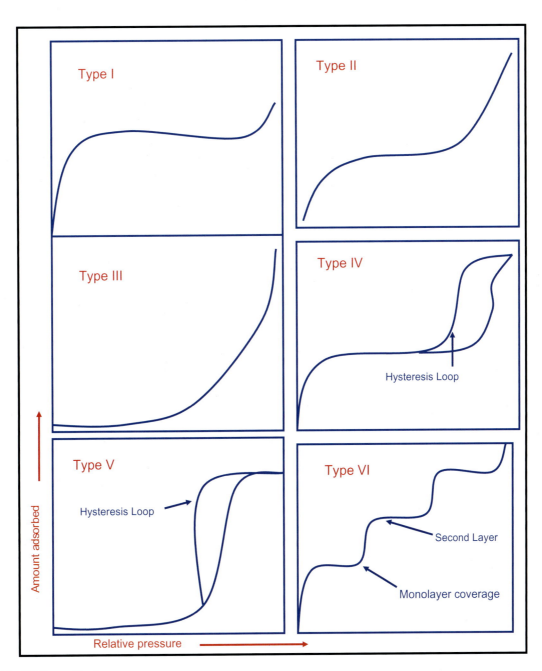

**Fig. 5.5** Type of isotherms.

of gas physical adsorption on nonporous solids. After initial monolayer coverage, multilayer adsorption follows at higher pressure (Sing et al., 1985). A type III isotherm is obtained from sorbents with both micropores and mesopores. Type IV isotherms are obtained from both nonporous and mesoporous solids. This class of isotherm occurs when sorbents condense gas inside narrow capillary pores. This phenomenon is called capillary condensation (Sing et al., 1985). Type V isotherms originate with mesoporous and microporous adsorbents. Type V isotherms are linked with type III isotherms. They are uncommon (Sing et al., 1985). A type VI isotherm is obtained from uniform nonporous surfaces where stepwise multilayer adsorption successively occurs. The step sharpness varies with both the system and its temperature. The step height is a measure of the monolayer capacity for each sorbed layer. The number of steps represents individual layers. The layer capacities (step heights) remain nearly constant for two or three sorbed layers.

A number of mathematical models were developed to explain various sorption equilibrium isotherms (Foo and Hameed, 2010; Limousin et al., 2007; Majd et al., 2021; Xu et al., 2013). Modeling all the details is not always required. A less sophisticated model may be sufficient. For example, if the aim is to simply fit a sorption isotherm for an adsorber design, the detailed adsorbate phase behavior is not required. Only coverage as a concentration function is needed. It is also possible to fit the sorption equilibrium data to different models. The Langmuir adsorption isotherm will be sufficient for small coverage range while a Freundlich, Sips, or Toth or any other model can be used for wider concentration ranges.

The common isotherm models are broadly classified into one-, two-, and three-parameter equations.

## 2.1 One-parameter isotherm

### 2.1.1 Henry's law

Henry's law is a one-parameter isotherm (Eq. 5.6). It can only be employed for sorption at low sorbate concentrations on a uniform surface. The linear relationship exists between the sorbate's concentration in the solvent and the adsorbed phase at equilibrium.

$$q_e = K_H C_e \qquad (5.6)$$

The proportionality constant ($K_H$) is the sorption equilibrium constant, known as Henry's constant. It is given in concentration terms (L/g). Henry's law is the thermodynamically required limiting case for isotherms for increasingly low concentrations ($c \to 0$). Furthermore, this $K_H$ is also used in geosorption and is referred to as the distribution coefficient, $K_d$.

## 2.2 Two-parameter isotherms
### 2.2.1 Freundlich isotherm model

This is an empirical model (Bemmelen, 1888; Freundlich, 1906). Freundlich described the adsorption isotherm relationship to be reversible, nonideal, and not limited to a monolayer. The Freundlich model is applied to nonuniform adsorption over heterogeneous surfaces exhibiting multilayer adsorption (Freundlich, 1906). The Freundlich isotherm has consistently been applied to heterogeneous adsorption systems.

The nonlinear form of the Freundlich isotherm model is shown in Eq. (5.7)

$$q_e = K_F C_e^{1/n} \quad \text{(Nonlinear form)} \tag{5.7}$$

The log of the nonlinear equation gives the linear Freundlich Eq. (5.8)

$$\log q_e = \log K_F + \frac{1}{n} \log C_e \quad \text{(Linear form)} \tag{5.8}$$

where $q_e$ is the sorption capacity (mg/g), $K_F$ is a constant indicative of the adsorbent's relative sorption capacity (mg/g), $C_e$ is the equilibrium sorbate concentration (mg/L), and $1/n$ represents an adsorption intensity constant that spans the range between 0 and 1. Adsorption becomes more heterogeneous as the slope approaches zero (Freundlich, 1906; Foo and Hameed, 2010). A plot of "log $q_e$" versus "log $C_e$" gave the values of the constant "$1/n$" (slope) and "log $K_F$" (intercept), respectively.

A simple mass balance equation (Eq. 5.9) derived from Freundlich model can be used to determine the theoretical breakthrough time (Faust and Aly, 1987; Ram et al., 1990).

$$\text{Breakthrough time} = \left(\frac{K_F C_0^{\frac{1}{n}}}{C_0}\right) \frac{W}{Q} \tag{5.9}$$

where $C_0$ is the influent concentration, $W$ gives the adsorbent weight, and $Q$ represents the volumetric flow rate.

### 2.2.2 Langmuir isotherm model

Irving Langmuir developed an isotherm model that describes gas-solid-phase adsorption on porous carbon materials (Langmuir, 1916). Since then, the model has been modified and is successfully used to find adsorption capacity of liquid-solid-phase systems on various sorbents (Foo and Hameed, 2010). This model assumes that adsorption takes place on adsorbent surface homogenously with no adsorbate transmigration (Foo and Hameed, 2010). Moreover, monolayer adsorption of adsorbate occurring on the sites of the adsorbents is finite, identical, equivalent, and localized with no lateral exchange or stearic hindrance between adsorbed molecules and even on the adjacent sites of adsorbent (Foo and Hameed, 2010).

The nonlinear and linear Langmuir equations (Langmuir, 1916) are shown in Eqs. (5.10), (5.11)

$$q_e = \frac{Q^0 b C_e}{1 + b C_e} \quad \text{(nonlinear form)} \tag{5.10}$$

$$\frac{1}{q_e} = \frac{1}{b C_e} + \frac{1}{Q^0 b} \quad \text{(Linear form)} \tag{5.11}$$

Here, $q_e$ is the sorbate quantity sorbed per unit weight of sorbent (mg/g), $Q^0$ represents the capacity achieved by monolayer sorption (mg/g), $b$ is a constant linked to the net sorption enthalpy ($b \propto e^{-\Delta H/RT}$), and $C_e$ is solute equilibrium concentration (mg/L). A linear plot of "$1/q_e$" versus "$1/C_e$" provided the constants "$b$" and "$Q^0$" from the slope and intercept, respectively.

The Langmuir constants can be utilized to develop a dimensionless constant called the separation factor ($R_L$) (Weber and Chakravorti, 1974). The separation factor is represented by Eq. (5.12). The $R_L$ gives types of equilibrium isotherm (Fig. 5.6)

$$R_L = \frac{1}{1 + b C_o} \tag{5.12}$$

### 2.2.3 Temkin isotherm model

The indirect effects of adsorbate-adsorbent interactions on adsorption isotherms were examined by Temkin and Pyzhev (1940). This led to the Temkin isotherm which assumes the sorption heat of all the molecules in a layer drops linearly except in very low and large adsorbate concentrations (Temkin and Pyzhev, 1940). The model is derived using a uniform binding energy distribution. The Temkin model was developed for gas adsorption onto platinum electrodes. This model is well suited

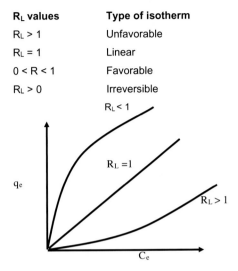

**Fig. 5.6** Types of equilibrium isotherms.

for gas phase equilibrium prediction (Kim et al., 2004). This model is not suitable for liquid phase adsorption isotherms (Kim et al., 2004).

The nonlinear and linear forms of Temkin isotherm model are given by Eqs. (5.13), (5.14).

$$q_e = \frac{RT}{b_{Te}} \ln(a_{Te} C_e) \text{(nonlinear form)} \quad (5.13)$$

$$q_e = B_{Te} \ln K_{Te} + B_{Te} \ln C_e \text{(linear form)} \quad (5.14)$$

Here, $b_{Te}$ is a constant related to the heat of sorption, $a_{Te}$ is the Temkin isotherm constant, $C_e$ is the sorbate equilibrium concentration (mg/L), while $R$ is the gas constant (8.314 J/mol•K). $T$ is the absolute temperature (K), and $B_{Te}$ and $K_{Te}$ are Temkin constants. The Temkin isotherms are plotted between "$q_e$" and "ln $C_e$". The slope and intercepts give the values of "$B_{Te}$" and "ln $K_{Te}$", respectively.

### 2.2.4 Hill isotherm model

The Hill equation describes the binding of adsorbate species on a homogeneous surface (Hill, 1910). The Hill isotherm assumes adsorption to be a cooperative phenomenon, where the adsorbate-adsorbent binding on one site may affect other binding sites of the adsorbent (Foo and Hameed, 2010). The Hill isotherm is given by Eq. (5.15). Here, $Q_{SH}$ is the maximum sorbate uptake at saturation (mg/g) and $K_D$ is the Hill constant. The $n_H$ is the cooperativity coefficient of binding interaction obtained from the slope of a Hill plot. A slope of >1 implies positive cooperative binding between sorbate and sorbent, while a slope <1 indicates negative cooperative binding.

$$q_e = \frac{Q_{s_H} C_e^{n_H}}{K_D + C_e^{n_H}} \qquad (5.15)$$

### 2.2.5 Dubinin-Radushkevich isotherm model

The Dubinin-Radushkevich isotherm was developed and applied to subcritical vapor adsorption onto micropore solids through a pore filling mechanism (Dubinin and Radushkevich, 1947). This model utilizes a Gaussian energy distribution on heterogeneous surfaces to characterize adsorption (Dabrowski, 2001; Günay et al., 2007). The model evaluates the sorption mean free energy and helps in distinguishing between physical or chemical adsorption. The Dubinin-Radushkevich isotherm is temperature-dependent. This model is expressed as Eq. (5.16). The parameters $Q^0$, $K_{ad}$, and $\varepsilon$ represent the maximum sorption capacity (mg/g), the Dubinin-Radushkevich isotherm constant (mol²/KJ²) associated with sorption energy, and sorption potential (kJ/mol) constant, respectively.

$$q_e = Q^0 \exp\left(-k_{ad}\varepsilon^2\right) \qquad (5.16)$$

where the parameter $\varepsilon$ is calculated through Eq. (5.17) (Hu and Zhang, 2019).

$$\varepsilon = RT \ln\left(\frac{C_s}{C_e}\right) \qquad (5.17)$$

$Q^0$ (mg/g) constitutes the D-R theoretical adsorption capacity. $C_e$, $C_s$, $T$, and $R$ are the equilibrium sorbate concentration, the solubility, the absolute temperature (K), and the gas constant (8.314 J/mol K), respectively.

### 2.2.6 Halsey model

The Halsey model (Halsey, 1948) is used for multilayer adsorption on heteroporous adsorbent and is expressed in Eq. (5.18),

$$q_e = \exp\left(\frac{\ln k_H - \ln C_e}{n}\right) \qquad (5.18)$$

where $K_H$ represents the Hasley isotherm constant and $n$ is Hasley's isotherm exponent.

## 2.3 Three-parameter isotherm models

### 2.3.1 *Sips (Langmuir-Freundlich) isotherm model*

Upon combining both Langmuir and Freundlich isotherms, the Sips isotherm was developed (Sips, 1948). It overcomes the limitation associated with the Freundlich isotherm model as the adsorbate concentration rises. This Sips model predicts sorption on heterogeneous systems similar to the Freundlich isotherm (Foo and Hameed, 2010). At higher adsorbate concentrations, the Sips model predicts monolayer adsorption capacity, a characteristic of the Langmuir isotherm model. When adsorbate concentrations are lowered, the Sips model is reduced to the Freundlich isotherm. The Sips model expresses the surface saturation at very high solute concentrations. The Sips model is represented by Eq. (5.19). Here, $q_e$ is the quantity of sorbate sorbed per unit of sorbent weight (mg/g). $C_e$ is the sorbate's equilibrium concentration (mg/L) in the solution. $K_s$, $a_s$, and $\beta s$ are the Sips isotherm constants.

$$q_e = \frac{K_s C_e^{\beta s}}{1 + a_s C_e^{\beta s}} \tag{5.19}$$

### 2.3.2 *Toth isotherm model*

The Toth isotherm model was developed to further improve the data fittings versus the Langmuir isotherm model. Thus, it satisfies both high- and low-end concentration boundaries and is applicable to heterogeneous adsorption (Toth, 1971). The Toth isotherm represents adsorption using a quasi-Gaussian asymmetrical energy distribution (Toth, 1971). This assumes that a majority of adsorption sites display energies lower than the mean value. The Toth isotherm is expressed as Eq. (5.20), where parameter n is adsorption intensity. $K_T$ (mg/g) and $a_T$ (L/mg) are the Toth isotherm constants.

$$q_e = \frac{K_T C_e}{(a_T + C_e)^{1/n}} \tag{5.20}$$

### 2.3.3 *Redlich-Peterson isotherm model*

The versatile Redlich-Peterson isotherm is employed for both heterogeneous and homogenous systems (Redlich and Peterson, 1959). This three-parameter model includes both Freundlich and Langmuir isotherms (Eq. 5.21). This equation's (Eq. 5.21) numerator expresses a linear concentration dependence, while the denominator characterizes how adsorption varies with concentration as an exponential function. $K_R$ (L/g), $a_R$ (L/mg) and $\beta$ are the Redlich-Peterson isotherm constants. $C_e$ is the equilibrium sorbate concentration (mg/L) in solution and $q_e$ represents the amount of sorbate sorbed per unit weight of sorbent (mg/g). The exponent parameter $\beta$ lies between 0 and 1. When $\beta = 0$, Eq. (5.21) reduced to Henry's law (Eq. 5.22), while when $\beta = 1$, Eq. (5.21) assumes the form of the Langmuir isotherm model (Eq. 5.23).

$$q_e = \frac{K_R C_e}{1 + a_R C_e^\beta} \tag{5.21}$$

When $\beta = 0$, Eq. (5.21) becomes Henry's equation.

$$q_e = \frac{K_R C_e}{1 + a_R} \tag{5.22}$$

When $\beta = 1$, Eq. (5.21) reduces to Langmuir isotherm equation.

$$q_e = \frac{K_R C_e}{1 + a_R C_e} \tag{5.23}$$

### 2.3.4 Radke-Prausnitz isotherm model

The Radke-Prausnitz isotherm works well with high chi-square values and high root mean square errors (Radke and Prausnitz, 1972). It is applicable over a broad concentration range (Radke and Prausnitz, 1972). The Radke-Prausnitz isotherm model is given by Eq. (5.24). Here, $q_e$ is amount of sorbate adsorbed per unit weight (mg/g) of adsorbent and $C_e$ is equilibrium sorbate concentration (mg/L) in solution. The $a_{RP}$, $r_R$ and $\beta_R$ terms are the Radke-Prausnitz sorption isotherm constants.

$$q_e = \frac{a_{RP} r_R C_e^{\beta_R}}{a_{RP} + r_R C_e^{\beta_R - 1}} \tag{5.24}$$

### 2.3.5 Koble-Corrigan isotherm model

The Koble-Corrigan isotherm is another Langmuir-Freundlich type isotherm model (Koble and Corrigan, 1952). The model is given by Eq. (5.25) (Koble and Corrigan, 1952). This model incorporates three parameters: $a$, $b$, and $\beta$.

$$q_e = \frac{a C_e^\beta}{1 + b C_e^\beta} \quad \text{(nonlinear form)} \tag{5.25}$$

The Koble-Corrigan equation (5.25) reduces to the Langmuir expression when the exponent $\beta$ becomes 1. If $bC_e^\beta$ drops to significantly below 1, the amount of sorbate that is sorbed becomes very low. At this point, the Koble-Corrigan equation changes into the Freundlich equation. If $bC_e^\beta$ is much higher than 1, sorption becomes extremely high and the Koble-Corrigan isotherm becomes the Freundlich isotherm. At low concentrations, Koble-Corrigan isotherm becomes linear (Aşçi et al., 2007).

## 2.4 Multilayer physisorption isotherms

### 2.4.1 Brunauer-Emmett-Teller isotherm (BET)

Brunauer-Emmett-Teller (BET) isotherm is one of the most commonly applied models for solid-gas equilibrium (Brunauer et al., 1938). The BET isotherm was initially developed for multilayer sorption systems. The BET model applied for solid-liquid interface sorption is given by Eq. (5.26).

$$q_e = \frac{Q^0 C_{BET} C_e}{(C_s - C_e)[1 + (C_{BET} - 1)(C_e/C_s)]} \tag{5.26}$$

$C_s$ and $C_e$ are monolayer adsorbate concentrations expressed in mg/L, obtained at saturation and at equilibrium, respectively, $C_{BET}$ is the BET adsorption isotherm (L/mg), $q_e$ and $Q^0$ are equilibrium adsorption capacity and theoretical isotherm saturation capacity, respectively, expressed in mg/g. Since, $C_{BET}$ and $C_{BET}(C_e/C_s)$ are larger than 1, Eq. (5.26) reduces to Eq. (5.27).

$$q_e = \frac{Q^0}{1 - (C_e/C_s)} \tag{5.27}$$

### 2.4.2 Frenkel-Halsey-Hill isotherm (FHH) model

The Frenkel-Halsey-Hill model (known as FHH theory) constitutes a multilayer model derived from potential theory (Frenkel, 1946; Hill, 1952). It is given in Eq. (5.28).

$$\ln\left(\frac{C_e}{C_s}\right) = -\frac{\alpha}{RT}\left(\frac{Q^0}{q_e d}\right)^r \tag{5.28}$$

The $\alpha$, $d$, and $r$ terms are the Frenkel-Halsey-Hill isotherm constant, the interlayer spacing sign, and inverse power of distance from the sorbent, respectively.

### 2.4.3 MacMillan-Teller isotherm (MET) model

MacMillan-Teller isotherm (MET) interprets the inclusion of surface tension effects in BET isotherm (Mcmillan and Teller, 1951). This model mathematical is expressed in Eq. (5.29).

$$q_e = q_s \left( \frac{k}{\ln(C_s/C_e)} \right)^{1/3} \tag{5.29}$$

Here, $k$ is the isotherm constant. Eq. (5.29) takes the form of Eq. (5.30) when $C_s/C_e$ approaches unity. $q_s$ is theoretical saturation capacity of the isotherm.

$$q_e = q_s \left( \frac{kC_e}{C_s - C_e} \right)^{1/3} \tag{5.30}$$

## 2.5 Multicomponent sorption equilibrium models

Multicomponent sorption systems are complicated due to the sorbate-sorbate competition and various sorbate-surface interactions. Multicomponent sorption occurs at active sorption sites where the emergence of solid-liquid phase equilibrium occurs exhibiting distinct differing capacities of each single sorbate present in the sorbate mixture. Multisolute systems generated distinctly less attention than single sorbates. Multicomponent adsorption predictions are challenging. Many models have been developed and implemented to determine batch sorption parameters required for fixed-bed adsorber design (Sontheimer et al., 1988). However, most models are founded either on unrealistic assumptions or on empirical equations without clear definitions (Alkhamis and Wurster, 2002; Mohan and Singh, 2005; Wurster et al., 2000). Some of the more widely used multicomponent sorption equilibrium models are discussed to understand the multi-sorbate (organic and inorganic contaminants) sorption on sorbents including biochars.

### 2.5.1 Butler and Ockrent model or extended Langmuir model

Butler and Ockrent developed a model in 1930 (Butler and Ockrent, 1930) for competitive sorption of two solutes (Sontheimer et al., 1988). This model is applicable when each single solute followed the Langmuir isotherm model. The solute capable of lowering the surface tension more will be adsorbed to a relatively greater extent than the other. In case of substantial difference, the more actively adsorbed solute will be adsorbed in larger quantity. If two solutes are present, the Langmuir sorption model is explained by Eqs. (5.31), (5.32).

$$q_1 = \frac{Q_1^0 b_1 C_{e1}}{1 + b_1 C_{e1} + b_2 C_{e2}} \tag{5.31}$$

$$q_2 = \frac{Q_2^0 b_2 C_{e2}}{1 + b_1 C_{e1} + b_2 C_{e2}} \quad (5.32)$$

In Eqs. (5.31) and (5.32), $Q_1^0$, $b_1$ and $Q_2^0$, $b_2$ are the Langmuir constants obtained from single-solute systems with sorbate no. 1 and sorbate no. 2, respectively. $C_{e1}$ and $C_{e2}$ are the sorbate equilibrium concentrations of 1 and 2 present, respectively.

This model holds well when the adsorption surface area available for sorbates 1 and 2 are similar. $Q_1^0$ and $Q_2^0$ describe the difference in sizes for the monolayer surfaces covered by the two sorbates. Furthermore, the adsorption free energy for this model is considered to be surface coverage independent.

For simultaneous adsorption of $N$ different sorbates from solution, the Langmuir multicomponent model is expressed in Eq. (5.33).

$$q_i = \frac{Q_i^0 b_i C_{ei}}{1 + \sum_{j=1}^{N} b_j C_{ej}} \quad (5.33)$$

The Butler and Ockrent model (Eq. 5.33) depends if single-solute adsorption is described by Langmuir model. Despite many shortcomings, this model has been deployed to describe many multicomponent sorptions on activated carbons (Hsieh et al., 1977).

### 2.5.2 Jain and Snoeyink model

Jain and Snoeyink (1973) further modified the Butler and Ockrent model (Butler and Ockrent, 1930) for the adsorption where one component of the mixture is noncompetitive (Sontheimer et al., 1988). Jain and Snoeyink advanced a binary systems model that accounts for this noncompetitive adsorption. They postulated that noncompetitive sorption takes place when $Q_1^0 \neq Q_2^0$ and where the number of available sorption sites that are not undergoing this competition is equal to $(Q_1^0 - Q_2^0)$ where $Q_1^0 > Q_2^0$. The Jain and Snoeyink model is described by Eqs. (5.34), (5.35).

$$q_1 = \frac{(Q_1^0 - Q_1^0) b_1 C_{e1}}{1 + b_1 C_{e1}} + \frac{Q_2^0 b_1 C_{e1}}{1 + b_1 C_{e1} + b_2 C_{e2}} \quad (5.34)$$

$$q_2 = \frac{Q_2^0 b_2 C_{e2}}{1 + b_1 C_{e1} + b_2 C_{e2}} \quad (5.35)$$

Here, $q_1$ and $q_2$ are the quantities of sorbate 1 and 2, respectively, that are adsorbed per unit of adsorbent weight when equilibrium is achieved. $C_{e1}$ and $C_{e2}$, respectively, are the sorbate concentrations in solution. $Q_1^0$ and $Q_2^0$ are the maximum quantities of sorbates 1 and 2, respectively, that can be adsorbed. $Q_1^0$ and $Q_2^0$ are found in experiments on the single-solute systems. The constants $b_1$ and $b_2$ are related to the adsorption energies of sorbates 1 and 2, when adsorbed from their individual solutions.

This model successfully explains the competitive sorption between the molecules characterized by different single-sorbate sorption capacities. The Jain and Snoeyink model has been invoked to describe multicomponent sorption. Huang and Steffens (1976) predicted the competitive sorption of organics onto activated carbons using this approach. Mutual equilibrium sorption suppression occurred due to competition between acetic acid and butyric acid. This caused the obtained sorption data to be better fit to values predicted using the Jain and Snoeyink model than those predicted by the Langmuir equation. However, the amount of suppression determined from experiments was greater than the predicted suppression for acetic acid, while it was smaller for butyric acid.

### 2.5.3 Mathews and Weber model

The Mathew and Weber empirical model (Mathews and Weber, 1975) is a modified and extended form of Redlich and Peterson model (Redlich and Peterson, 1959). It is a single-solute adsorption isotherm model with three parameters as given in Eq. (5.36) for an N solute mixture.

$$q_i = \frac{K_i C_{ei}}{1 + \sum_{j=1}^{N} a_j C_{ej}^{bj}} \tag{5.36}$$

The constants $K_j$, $a_j$, and $C_j$ are typically derived experimentally from single-solute isotherm data. To further understand the multicomponent data, an interaction parameter $\eta_i$ has been introduced which is calculated from multicomponent sorption data. The modified form of Eq. (5.36) becomes Eq. (5.37) with the incorporation of interaction parameter $\eta_i$.

$$q_i = \frac{K_j \left[\frac{C_{ei}}{\eta_i}\right]}{1 + \sum_{j=1}^{N} a_j \left[\frac{C_j}{\eta_i}\right]^{b_j}} \tag{5.37}$$

This interaction term $\eta_i$ is a constant. The value of $\eta_i$ differs for varying equilibrium compositions. Hence, this limits the model's success in physically explaining multicomponent system behavior.

### 2.5.4 Fritz and Schlunder multicomponent model

The empirical Fritz and Schlunder model (Fritz et al., 1981; Fritz and Schlunder, 1974) is for multicomponent systems like the Mathews and Weber model (Sontheimer et al., 1988). This is used for sorbates only when their single-solute isotherms conform to the the Freundlich isotherm and is expressed by Eqs. (5.38)–(5.41). These equations have 10 adjustable parameters.

$$q_1 = \frac{K_1 C_{e1}^{n_1+\beta_{11}}}{C_{e1}^{\beta_{11}} + \alpha_{\dot{1}2} C_{e2}^{\beta_{21}}} \tag{5.38}$$

$$q_2 = \frac{K_2 C_{e2}^{n_2+\beta_{22}}}{C_{e2}^{\beta_{22}} + \alpha_{\dot{2}1} C_{e1}^{\beta_{21}}} \tag{5.39}$$

$$\alpha_{\dot{1},2} = \frac{\alpha_{1,2}}{\alpha_{1,1}} \tag{5.40}$$

$$\alpha_{\dot{2},1} = \frac{\alpha_{2,1}}{\alpha_{2,2}} \tag{5.41}$$

In this model, $q_1$ and $C_{e1}$ stand for the sorbed and equilibrium concentrations of sorbate 1, respectively, while $q_2$ and $C_{e2}$ are the corresponding solid- and liquid-phase concentrations of sorbate 2. The $K_1$, $n_1$ and $K_2$, $n_2$ terms present the Freundlich constants from single sorbate 1 and sorbate 2 systems.

### 2.5.5 Dastgheib and Rockstraw model

Dastgheib and Rockstraw (2002) proposed a multicomponent model for use with binary systems as expressed by Eqs. (5.42), (5.43).

$$q_1 = \left[ \frac{K_1 C_{e1}^{n_1}}{K_1 C_{e1}^{n_1} + \alpha_{12} K_2 C_{e2}^{n_2} + b_{12} C_{e2}^{n_{12}}} \right] K_1 C_{e1}^{n_1} \tag{5.42}$$

$$q_2 = \left[ \frac{K_2 C_{e2}^{n_2}}{K_2 C_{e2}^{n_2} + \alpha_{21} K_1 C_{e1}^{n_1} + b_{21} C_{e1}^{n_{21}}} \right] K_2 C_{e2}^{n_2} \tag{5.43}$$

Here, $q_1$ and $C_{e1}$ are the sorbate 1 concentration in solid and liquid phases, respectively. $q_2$ and $C_{e2}$ are the corresponding sorbate 2 concentrations. $K_1$, $n_1$, $K_2$, $n_2$ are the Freundlich isotherm constants obtained from single-sorbate systems, while $\alpha_{12}$, $\alpha_{21}$, $b_{12}$, $b_{21}$, $n_{12}$ and $n_{21}$ are the interaction constants that are determined from a least-squares binary data analysis.

The term in parentheses (on the right) represents the total competition and interaction factor with a value $\leq 1$. When $C_2$ goes to zero, this interaction factor goes to 1. The terms $\alpha_{12} K_2$ and $\alpha_{21} K_1$ can be reduced to single terms, and then are taken as one constant. This model can be given by Eq. (5.44) when expressed for $i$th component systems.

$$q_i = \left[\frac{(K_i C_{ei}^{n_i})^2}{K_2 C_{e2}^{n_2} + \alpha_{21} K_1 C_{e1}^{n_1} + b_{21} C_{e1}^{n_{21}}}\right] K_2 C_{e2}^{n_2} \qquad (5.44)$$

Here $q_i$ and $C_1$ represent the sorbate 1 concentration in the sorbent and liquid phases, respectively; $C_j$ signifies the concentration of other sorbates in liquid phase, while $K_i$ and $n_i$, $K_j$ and $n_j$ are the single-component Freundlich constants. The binary interaction constants $\alpha_{ij}$, $b_{ij}$, and $n_{ij}$ are determined by a least-squares analysis of the multicomponent data where $\alpha_{ii} = b_n = 0$ and N indicates the sorbate numbers. $_{ij}$ are determined by a least squares analysis of the multicomponent data where $\alpha_{ii} = b_{ii} = 0$ and N indicates the sorbate numbers.

### 2.5.6 Sheindorf et al. model

The Sheindorf (Sheindorf et al., 1981) model is used with solutes, where all single sorbates conform to the Freundlich isotherm relationship. This model assumes that each component in the multicomponent system interacts with sorption sites where the sorption energies are exponentially disturbed. Furthermore, this distribution function is identical to a single-sorbate system. The Sheindorf model is given in Eqs. (5.45), (5.46).

$$q_1 = K_1 C_{e1} (C_{e1} + \eta_{12} C_{e2})^{n_1 - 1} \qquad (5.45)$$

$$q_2 = K_2 C_{e2} (C_{e2} + \eta_{21} C_{e1})^{n_2 - 1} \qquad (5.46)$$

Here, $q_1$ and $C_1$ are the sorbate 1 concentrations in the sorbent and liquid phases, respectively, while $q_2$ and $C_2$ represent the corresponding sorbate 2 concentrations. $K_1$, $n_1$, $K_2$, $n_2$ are the Freundlich constants obtained in single-sorbate systems, while $\eta_{12}$ and $\eta_{21}$ are the interaction constants.

### 2.5.7 Ideal adsorbed solution theory (IAST)

The ideal adsorbed solution theory is built upon sorption thermodynamics. The IAST model was first applied to determine multisolute sorption of gaseous mixtures (Myers and Prausnitz, 1965). This model was further modified by Radke and Prausnitz (1972) and widely used to calculate multicomponent adsorption parameters from liquids to solids. IAST model can be used for multicomponent analysis using many different single-solute isotherms. Furthermore, IAST model is more accurate than the Langmuir multicomponent isotherm model (Weber Jr. and Smith, 1987). The IAST model predicts multicomponent behavior from single-solute adsorption isotherm experiments which are conducted first (Crittenden et al., 1985; Sontheimer et al., 1988).

The IAST model presumes the thermodynamic equivalence of the spreading pressure $\pi$ (it is a difference in interfacial tension between the pure sorbate-solid interface and the bulk solute-solid interface at constant temperature) of each sorbate at equilibrium.

To calculate the spreading pressure of each sorbate, data from single-sorbate isotherms are needed. The model is expressed in Eq. (5.47)

$$\pi_i(C_i) = \frac{RT}{A} \int_0^{C_i^*} \frac{q_i}{C_i} dC_i \tag{5.47}$$

$C_i^*$ is the solute concentration of species $i$ obtained from the single-sorbate system giving the same spreading pressure as that of the mixture under study. $C_i$ is the solid-phase concentration of sorbate $i$ in the mixture. $T$ is the absolute temperature (K), $A$ is the adsorbent-specific surface area and $R$ is the ideal gas constant. When the single-sorbate data have been described by a suitable model, the spreading pressure can then be calculated using modified Eq. (5.48)

$$\pi_i(q_i) = \frac{RT}{A} \int_0^{-q_i^*} \frac{d\log C_i}{d\log q_i} dq_i \tag{5.48}$$

$q_i^*$ is the solid concentration of sorbate $i$ in the single-sorbate system giving a spreading pressure equal to that of the mixture. $q_i$ is concentration of sorbate $i$ on the solid adsorbent.

Additional equations essential for an IAST solution are expressed in Eqs. (5.49)–(5.53).

$$\pi_i = \pi \tag{5.49}$$

$$C_i = z_i C_i^* \tag{5.50}$$

$$\sum_{i=1}^{N} z_i = 1 \tag{5.51}$$

$$\frac{1}{q_T} = \sum_{i=1}^{N} \frac{z_i}{q_i^*} \tag{5.52}$$

$$q_i = z_i q_T \tag{5.53}$$

Here, $z_i$ is the sorbate's mole fraction $i$ adsorbed on the adsorbent, $q_T$ is the total adsorbed concentration that has been removed from the mixture and $N$ is the number of distinct sorbate species present in the system.

## 3 Sorption kinetic models

High sorption capacities and fast kinetics are key characteristics for sorbents to have in order to develop an excellent sorption system (Tan and Hameed, 2017). The sorption process is carried out in batch or column modes. To interpret material design and system selection, it is necessary to find the sorption rate limiting step.

The sorption rate of a molecule onto a surface can be expressed identically to any kinetic process. The rate-limiting step is determined by fitting sorption kinetic data into different kinetic equations. These parameters help in providing information useful for the design of fixed-bed reactors (Patel et al., 2021; Vimal et al., 2019). Kinetic models have been well reviewed (Revellame et al., 2020; Simonin, 2016; Tan and Hameed, 2017; Ho and McKay, 1999; Wang and Guo, 2020).

## 3.1 Pseudo-first order model

The pseudo-first order (PFO) rate equation was first expressed by Lagergren (1898) for oxalic and malonic acids sorption on charcoal. The Lagergren model is more widely used to determine the solute uptake kinetics (Liu and Shen, 2008; Mohan et al., 2006; Revellame et al., 2020; Simonin, 2016; Yuh-Shan, 2004) (Eqs. 5.54–5.56).

$$q_t = q_e(1 - e^{-k_1 t}) \quad \text{(Nonlinear form)} \tag{5.54}$$

or

$$\ln \frac{q_e - q_t}{q_e} = -k_1 t \tag{5.55}$$

or

$$\frac{dq_t}{dt} = k_1(q_e - q_t) \tag{5.56}$$

Here, $k_1$ is the first-order rate constant (min$^{-1}$), $q_e$ is the sorbate sorbed per gram of adsorbent and $q_t$ is the quantity of sorbate sorbed at time "$t$."

The nonlinear Eq. (5.56) rearranged and integrated with the boundary conditions ($q_t = 0$ at $t = 0$ and $q_t = q_t$ at $t = t$), forms the linear rate Eq. (5.57).

$$\log(q_e - q_t) = \log(q_e) - \frac{k_1}{2.303} t \quad \text{(linear form)} \tag{5.57}$$

An X–Y curve drawn between $\log(q_e - q_t)$ and t gives a straight line with $k_1$ as the slope and $\log(q_e)$ as the intercept.

## 3.2 Pseudo-second-order model

In pseudo-second-order (PSO) kinetics, both the sorbate concentration and the sorbent dosage control the rate (Ho and McKay, 1999). The reactions that follow pseudo-second-order kinetics depend greatly on the number of surface-active sites on sorbent surface (Ho and McKay, 1999; Mohan et al., 2007; Navarathna et al., 2019; Revellame et al., 2020). The mathematical expression for this kinetic model is given in Eq. (5.58).

$$\frac{dq_t}{d_t} = k_2(q_e - q_t)^2 \quad \text{(Nonlinear form)} \quad (5.58)$$

Here, $k_2$ (g mg$^{-1}$ min$^{-1}$) is the pseudo-second-order rate constant, $q_e$ (mg/g) is the adsorbate quantity adsorbed at reaching equilibrium, and $q_t$ (mg/g) is the sorbate quantity sorbed at time $t$.

Integrating the nonlinear equation with the boundary conditions $q_t = 0$ when $t = 0$ and $q_t = q_t$ at $t = t$, gives the linear Eq. (5.59).

$$\frac{t}{q_t} = \frac{1}{k_2 q_e^2} + \frac{t}{q_e} \quad \text{(Linear form)} \quad (5.59)$$

An X-Y plot between $(t/q_t)$ and $(t)$ provides a straight line. Comparing the experimental equation with that of a straight line gives $\frac{1}{k_2 q_e^2}$ as the slope and $1/q_e$ as the intercept. The second-order rate constants ($k_2$) and $q_e$ values can easily be determined from the slope and intercept.

Pseudo-first-order (PFO) and pseudo-second-order (PSO) models were reviewed (Simonin, 2016; Wang and Guo, 2020).

## 3.3 Revised pseudo-second-order model

A revised pseudo-second-order (rPSO) model was derived by Bullen and coworkers (Bullen et al., 2021). The rPSO model removes experimental conditions and focuses on initial sorbate concentration ($C_0$) and sorbent concentration ($C_s$). The rPSO model is expressed in Eq. (5.60).

$$\frac{dq_t}{dt} = k'C_t\left(1 - \frac{q_t}{q_e}\right)^2 \quad (5.60)$$

Here, $k'$ is the revised rate constant, $q_e$ (mg/g) is the quantity of sorbate adsorbed upon reaching equilibrium, and $q_t$ (mg/g) is the sorbate quantity sorbed at time $t$.

Wherever necessary, the rPSO model rate constant, $k'$, can be determined from linearized PSO kinetics using the expression (5.61).

$$k' = \frac{k_2 q_e^2}{C_0} \qquad (5.61)$$

### 3.4 Mixed-order (MO) model

A mixed 1, 2-order rate equation was deduced by Marczewski (2010) and discussed by Guo and Wang (2019) and Wang and Guo (2020). The MO model is expressed by Eq. (5.62).

$$\frac{dq_t}{dt} = k_1'(q_e - q_t) + k_2'(q_e - q_t)^2 \qquad (5.62)$$

The PFO and PSO rates of this MO model can be computed using Eqs. (5.63), (5.64)

$$\text{PFO rate} = k_1'(q_e - q_t) \qquad (5.63)$$

$$\text{PSO rate} = k_2'(q_e - q_t)^2 \qquad (5.64)$$

Here, $k_1'$ and $k_2'$ are rate constants, $q_e$ (mg/g) is the quantity of sorbate sorbed when equilibrium is reached. $q_t$ (mg/g) is the sorbate quantity sorbed at time $t$.

### 3.5 Elovich equation

The Elovich equation was proposed to investigate the existence of chemical interactions between the adsorbate and adsorbent (Roginsky and Zeldovich, 1934). The Elovich equation is applicable toward the end of sorption process when the process becomes slow. The concentration of sorbate has decreased and fewer active adsorption sites remain. This model makes the assumption that the sorbent surface is highly heterogeneous and favors chemisorption (Roginsky and Zeldovich, 1934). The model is suitable for use when the kinetics are far from having reached equilibrium. An equilibrium desorption no longer occurs due to low surface coverage (Roginsky and Zeldovich, 1934; Tan and Hameed, 2017). The Elovich is expressed as Eq. (5.65).

$$q_t = \frac{1}{\beta} \ln(1 + \alpha\beta t) \qquad (5.65)$$

The adsorption capacity at time $t$ (h) is $q_t$ (mg/g). At equilibrium, $\alpha$ (mg/g min) represents the initial sorbate uptake rate, $\beta$ (g/mg) is a desorption constant related to both the extent of surface coverage and chemisorption activation energy.

## 3.6 Intraparticle diffusion equation

Weber Jr. and Morris (1963) proposed using this equation for modeling adsorption kinetics characterized by intraparticle diffusion. It is known as intraparticle diffusion model (IPD). It is one of the most often used kinetic models to investigate the metal ion transport from solution into adsorbent pores (Tan and Hameed, 2017). The sorption rate-limiting step is assumed to occur by the adsorbate mass transfer via intraparticle diffusion. The model is represented by Eq. (5.66).

$$q_t = k_d \sqrt{t} + C \quad (5.66)$$

In Eq. (5.66), $q_t$ (mg/g) stands for the quantity of sorbate removed from solution at time $t$ (h), and $k_d$ (mg/g h$^{0.5}$) is the IPD rate constant. $C$ is the initial quantity of adsorbate already sorbed (mg/g). The $q_t$ versus $\sqrt{t}$ plot is linear. When adsorption occurs exclusively by IPD (e.g., $C=0$), that plot passes through the origin. A better fit and multilinearity can also be observed when these lines do not pass-through origin. A multilinear plot indicates that multiple mechanisms control the sorption process. The IPD model can be explained in three steps (Fig. 5.7). The first sorption step occurs on the external surface, followed intraparticle diffusion of adsorbate into pores of adsorbent. Diffusion occurs and afterwards sorption occurs until equilibrium is attained by the system.

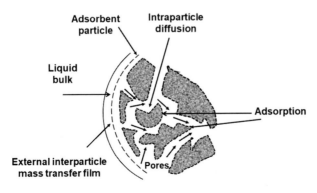

**Fig. 5.7** Multiple mechanisms during sorbate adsorption on an adsorbent.

## 3.7 Bangham equation

The Bangham model assumes that intraparticle diffusion is the only rate-controlling step (Aharoni et al., 1979) as represented in Eq. (5.67). This model is commonly applied to estimate the role of pore diffusion as a rate-controlling mechanism (Aharoni et al., 1979). A linear plot of this equation versus experimental data implicates pore diffusion as the main sorption-controlling mechanism.

$$\log\left(\log\frac{C_o}{C_o - q \cdot m}\right) = \log\left(\frac{k_0 m}{2.303 V}\right) + \alpha \log t \quad (5.67)$$

In Eq. (5.67), $k_o$ and $\alpha$ are the Bangham constants, $C_o$ is the initial sorbate concentration, $m$ is the sorbent quantity (g), $t$ is time (min), and $V$ is the solution volume.

## 3.8 Linear film diffusion equation

The linear film diffusion model is applied in adsorptive systems where adsorptive accumulation on the adsorbent surface constitutes the rate at which sorbate passes through the liquid film boundary layer before contacting the adsorbent's surface (Le Roux et al., 1991; Tan and Hameed, 2017). A system dominated by such film diffusion is expressed as Eq. (5.68).

$$\frac{C_e}{C_o} = \exp(-k_f t) \quad (5.68)$$

$C_o$ and $C_e$ stand for initial and equilibrium sorbate concentrations, $k_f$ is the film diffusion coefficient (min$^{-1}$), and $t$ is time in minutes.

## 4 Sorption column design

A typical water decontamination system does not work in the batch mode. Usually, it operates in continuous mode where contaminated water is passed through a granular adsorbent (biochar) column or fixed-bed or adsorber. The terms column, fixed-bed, and adsorber are used interchangeably throughout the book. Batch equilibrium sorption data are generally not applicable to dynamic treatment systems where contact time is not sufficient to reach equilibrium. The sorption capacity and other parameters determined from a batch system are only helpful in providing the preliminary information on the system effectiveness. Hence, there is a need to perform initial laboratory flow tests using mini columns (Low and Lee, 1991).

Subsequent laboratory experiments are usually carried out employing pilot-scale columns to determine the parameters necessary for full-scale design and operation. Most commonly used continuous sorption systems for water and wastewater treatment include (a) fixed-bed or expanded-bed sorption, (b) moving-bed sorption, and (c) fluidized-bed sorption.

Fixed-bed adsorbers are conducted in either downflow or upflow directions (Fig. 5.8). Downflow fixed-bed adsorbers (Fig. 5.8A) are more commonly applied in treating water and wastewater (Gupta et al., 1997). Fixed-bed systems can be operated as a single column or dual columns or as multiple columns (Mohan and Chander, 2006). These multiple columns may be operated either in parallel or in series (Figs. 5.9 and 5.10).

## 4.1 Fixed-bed or expanded bed adsorber (down flow or up flow)

Fixed-bed adsorption exhibits more complexity than a simple batch process. It is a time- and distance-dependent process. The fixed-bed column adsorption process is preferred because of its simplicity and efficient performance (Worch, 2012). In a fixed-bed adsorber, the adsorbent remains in place during this

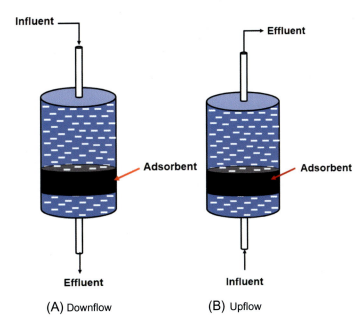

**Fig. 5.8** Schematic diagram for (A) downflow and (B) upflow fixed-bed adsorbers.

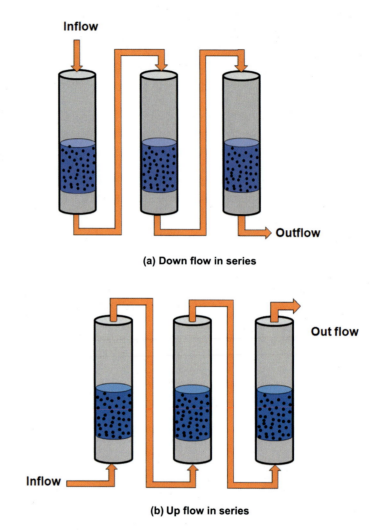

**Fig. 5.9** Schematic representation of (A) downflow and (B) upflow fixed-bed multiple columns in series.

operation. During the fixed-bed sorption, all sorbent particles in the bed adsorb the sorbate from the influent until the column reaches to equilibrium. The equilibrium state is achieved slowly and takes place on vertical sections down the columns (Worch, 2012). In a fixed-bed adsorption column, solution passes through the bed of adsorbent particles at a fixed flow rate (Sarswat and Mohan, 2016). The solute present in the feed water is adsorbed while it passes through the adsorbent (biochar) bed. A typical fixed-bed adsorber is displayed in Fig. 5.11 (Sarswat and Mohan, 2016).

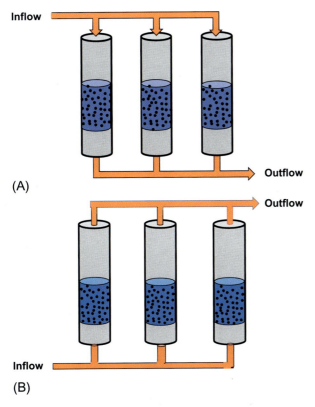

**Fig. 5.10** Schematic representation of (A) downflow and (B) upflow fixed-bed multiple columns in parallel.

## 4.2 Moving-bed or fluidized-bed adsorber

In a moving or fluidized-bed adsorber, influent enters the bottom of the column and moves upward through the adsorbent (activated carbon or biochar beds) while the spent adsorbent (activated carbon or biochar beds) is removed from the bottom and equal amount of virgin adsorbent (activated carbon or biochar beds) is added from the top of the adsorber (Fig. 5.12). The major shortcoming of fluidized-bed adsorbers is that it does not use the full capacity of its sorbent (Faust and Aly, 1987; Low and Lee, 1991). The volume of adsorbent (or its surface area per unit particle size) will be lower because, for a given bed volume, less adsorbent will be present than in a packed fixed bed. But

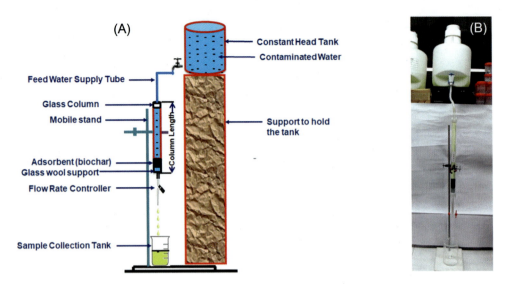

**Fig. 5.11** (A) Schematic diagram and (B) image of the fixed-bed downflow column set-up.

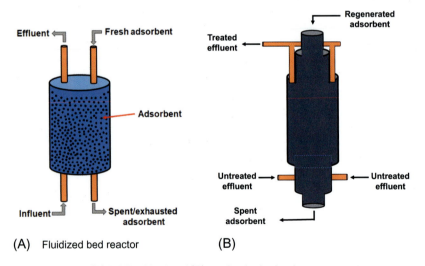

**Fig. 5.12** Schematic diagram of (A) fluidized bed and (B) moving bed adsorbers.

the capacity at equilibrium will be the same. It is likely that the mixing of particles in upper and lower parts will require different breakthrough removals but packed beds have flow rate limits that can also cause wide changes.

## 5 Fixed-bed operation and design

The breakthrough curve concept is most commonly applied when using a fixed-bed adsorber in a sorbate-sorbent system.

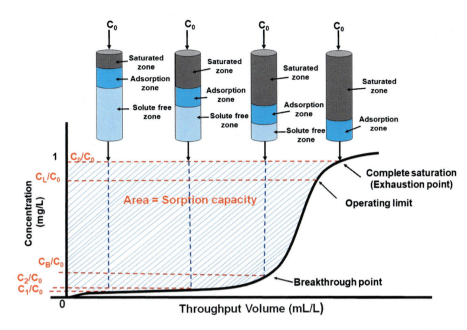

**Fig. 5.13** Breakthrough curve showing the movement of saturated zone, sorption zone, and the sorbate-free zone (where $C_0$, $C_e$, $C_B$, and $C_L$ are the influent, effluent, breakthrough, and operating limit concentrations, respectively).

This information is later used when designing the final technical fixed-bed reactor. An effluent concentration versus time plot gives an S-shaped curve referred to as *breakthrough curve* (Weber, 1972). A typical fixed-bed downflow system and adsorption zone movement is shown in Fig. 5.13. The adsorption zone is defined as the distance between the exhausted sorbent (e.g. biochar) at the top of the zone and the fresh sorbent (e.g. biochar) at the bottom of the zone. Sorbent (e.g. biochar) at the top of the adsorption zone is always in contact with the highly contaminated ($C_0$) contaminated feed water. As the solution passes through the sorbent (e.g. biochar) layers, solute that was not adsorbed by the top layers is adsorbed by the lower sorbent layers, and finally, solution is almost free from solute present in the solution when it comes out of the column ($C_f = 0$).

The biochar bed layer that encounters with the highest solute concentration solution is called the primary adsorption zone (PAZ) (Fig. 5.13). As the feed water flow continues down through the column, the upper biochar layers of the bed become saturated with the sorbate. Thus, PAZ advances downward into the column. Here, the solution with dissolved sorbate now encounters the sorbent (e.g. biochar), which is not spent and below that contains fresh sorbent (e.g. biochar). This sorbate-laden solution has a solute concentration, which is now highest for these layers. Less and less sorbate tends to get adsorbed as the PAZ moves downward along with the concentration front movement ($C_0$). A greater sorbent (e.g. biochar) column depth than the PAZ is required for effective sorbate removal. Otherwise, from the first instance of feed water addition, there will be sharp increase in the effluent solute concentration after its first discharge (Weber, 1972).

The concentration front movement rate as well as the wavelike primary adsorption zone movement is slower than the linear feed water velocity. With an increase in time or feed water volume, the effluent to the influent solution concentration ($C_f/C_0$) ratio increases (Fig. 5.13). At the point where equilibrium has been established between sorbate and sorbent (e.g. biochar), negligible or no sorbate sorption from feed water takes place. This point is considered to be the *breakpoint*. The curve $C_f/C_0$ versus time or feed water volume is known as *breakthrough curve*. Most fixed-bed sorption studies follow S-shaped curves. However, curve steepness and breakpoint vary with change in (a) sorbent (e.g. biochar) type used for preparing the fixed-bed, (b) sorbate concentration, (c) sorbent column depth and width, (d) flow rate used, and (e) sorbent (e.g. biochar) particle size. Usually, breakpoint time increases with the (a) low sorbate concentration, (b) greater depth and width of the sorbent bed, (c) a slow solution flow rate, and (d) a smaller sorbent (e.g. biochar) particle size which results in an increase in the surface area available for the sorption.

In continuous flow experiments, determining the sorbent bed's exhaustion rate is essential along with establishing the bed's lifetime before regeneration is necessary.

Fixed-bed, fluidized bed, and moving media reactors are the most used processes.

Many models are available for fixed-bed sorption column design. The most commonly used models include mass transfer zone (MTZ) (Weber, 1972), Thomas model (Thomas, 1944), empty bed residence time model (EBRT) (Lee et al., 2000; McKay and Bino, 1985, 1990), the Adams-Bohart model (Bohart and Adams, 1920), and bed-depth-service-time (BDST) (Lee et al., 2000).

## 5.1 Designing a fixed-bed adsorber applying a mass-transfer model

The simple mass transfer method described by Weber (1992) was employed to study fixed-bed adsorber design. This model was used to determine necessary parameters for large fixed-bed adsorber design (Choudhary et al., 2020; Gupta et al., 1997; Patel et al., 2021; Sarswat and Mohan, 2016). An ideal breakthrough curve has almost no sorbate during the beginning of

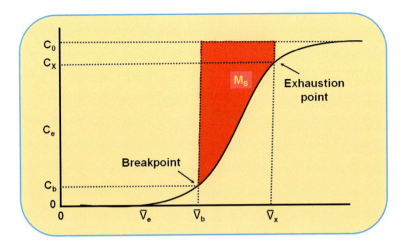

**Fig. 5.14** Ideal breakthrough curve showing the design parameters.

operations when the effluent concentration ($C_f$) is passed through a fixed bed of sorbent (e.g. biochar). However, under practical conditions, $C_f$ is not always zero due to the possibility of a small amount of leakage which may occur even at the start (Fig. 5.14). Thus, the ideal breakthrough curve is expressed in terms of the sorbate's mass concentration in the effluent, $C_f$, and the total mass quantity of sorbate-free water, $\overline{V_e}$, passing per unit of sorbent cross-sectional area. This curve depends on (a) the total mass of effluent per unit of sorbent (e.g. biochar) surface area at breakpoint, $\overline{V_b}$, and (b) the nature of the curve between, $\overline{V_b}$ and $\overline{V_x}$, where $\overline{V_x}$ is the total effluent mass per unit sorbent (e.g. biochar) area when the sorbent gets close to saturation. The effluent sorbate concentrations are $C_b$ and $C_x$ at $\overline{V_b}$ and $\overline{V_x}$, respectively. The constant zone length ($\delta$) is explained as part of the sorbent bed when the sorbate concentration is decreased from $C_x$ to $C_b$ (Weber, 1972).

The total time $t_x$ taken by the primary adsorption zone (PAZ) formation followed by the downward motion and out of the bed is calculated using Eq. (5.69).

$$t_x = \frac{\overline{V_x}}{F_m} \quad (5.69)$$

$F_m$ is the mass flow rate (mass per unit cross-sectional area of the fixed-bed).

The time, $t_\delta$, required for the zone moving down to its own length in the column is calculated using Eq. (5.70).

$$t_\delta = \frac{\overline{V_x} - \overline{V_b}}{F_m} \quad (5.70)$$

The ratio of sorbent bed depth ($D$) to time is calculated using Eq. (5.71).

$$\frac{\delta}{D} = \frac{t_\delta}{t_x - t_b} \quad (5.71)$$

$t_b$ is the time required for initial PAZ formation.

The fractional capacity ($f$) is found using Eq. (5.72).

$$f = 1 - \frac{t_b}{t_\delta} \quad (5.72)$$

The primary adsorption zone length ($\delta$) is calculated using Eq. (5.73).

$$\delta = D \left[ \frac{t_\delta}{t_b + t_\delta(f-1)} \right] \quad (5.73)$$

$\rho_p$ is the sorbent's apparent density, $q_\infty$ is the sorbate sorbed per unit sorbent weight, and $D$ is the sorbent bed depth. The percent saturation is computed using Eq. (5.74)

$$\text{Percent saturation} = \frac{D - \delta(f)}{D} \times 100 \quad (5.74)$$

The total adsorbent capacity is calculated by determining the area between the influent and effluent to the breakthrough point divided by the sorbent weight obtained in the fixed-bed constructed column. Similarly, the total column capacity is estimated by determining the total area to the point where effluent plot joins the effluent, divided by the sorbent quantity. The total column capacity is always useful and should be compared with batch sorption capacity for large-scale fixed-bed designs (Faust and Aly, 1987).

## 5.2 Empty bed contact time

Empty bed contact time (EBCT) (Faust and Aly, 1987; Lee et al., 2000; McKay and Bino, 1985, 1990) is explained as the total time during which the influent is in contact with the (sorbent (e.g. biochar) bed in the column. EBCT is determined using different design parameters given in Eqs. (5.75)–(5.78).

$$\text{Empty bed contact time} = \frac{\text{Bed volume}}{\text{Hydraulic surface loading} \times \text{bed surface area}} \quad (5.75)$$

or

$$\text{Empty bed contact time} = \frac{\text{Bed volume}}{\text{Flow rate}} \quad (5.76)$$

or

$$\text{Empty bed contact time} = \frac{\text{Bed depth}}{\text{Liner velocity}} \quad (5.77)$$

Bed volume (in Eqs. 5.75 and 5.76) is determined using Eq. (5.78)

$$\text{Bed volume} = \frac{\text{Weight } of \text{ sorbent (kg)}}{\text{Sorbent bulk density (kg/m}^3)} \quad (5.78)$$

The sorbent usage rate can be calculated using Eq. (5.79).

$$\text{Sorbent usage rate (kg/L)} = \frac{\text{Weight of sorbent in column (kg)}}{\text{Breakthrough volume (L)}} \times 1000 \quad (5.79)$$

## 5.3 Hutchins bed-depth-service-time (BDST) model

This model is used to determine the column efficiency when attaining the desired breakthrough curve under constant operational conditions. The time duration until the sorbate adsorption onto the adsorbent has reached before regeneration is defined as the service time (Hutchins, 1975; Lee et al., 2000, 2015). The BDST model was extended to predict breakthrough and exhaust time when the columns were operated at different flow rates and varied influent concentrations.

The influent-sorbate concentration, $C_0$, is the feed to the column, where it is desired to lower the sorbate concentration in the effluent to a value not more than $C_b$. At the start of the operation, the sorbent is fresh, the effluent concentration is actually less than the permitted concentration, $C_b$, but as the operation progresses and the sorbent reaches saturation, the effluent concentration approaches $C_b$. The condition, as discussed earlier, is known as the break point.

The initial work on bed-depth-service-time (BDST) model proposed a relation between the bed's depth and service time, $t$. This was expressed by Eq. (5.80) (Bohart and Adams, 1920) based on the process concentration and sorption parameters (Faust and Aly, 1987).

$$\ln\left(\frac{C_0}{C_b} - 1\right) = \ln\left(e^{\frac{K_a N_0 D}{V}} - 1\right) - K_a C_0 t \quad (5.80)$$

$C_0$ is the feed's influent-sorbate concentration delivered to the column; $C_b$ is the effluent-sorbate concentration that, when reached, the column will be regenerated; $V$ is linear flow rate (m/h); $D$ is the sorbent bed depth (meter); $K_a$ is the rate constant (m³/kg/h); and $N_0$ is the sorption capacity. At time t when the effluent-sorbate concentration reaches $C_b$, the influent sorbate would be stopped or shifted to a fresh sorbent column.

Hutchins (1975) proposed a linear relationship between bed depth and service time and expressed in Eq. (5.81) (rearrangement of Eq. 5.80)

$$t = \frac{N_0 D}{C_0 V} - \frac{1}{K_a C_0} \ln\left(\frac{C_0}{C_b} - 1\right) \quad (5.81)$$

The critical bed depth, $D_0$, is the theoretical sorbent depth that is large enough to stop the sorbate concentration from becoming greater than the value of $C_b$ at zero time ($t = 0$). This is calculated using Eq. (5.80) by substituting $t=0$ and solving for D. The exponential term $\frac{K_a N_0 D}{V}$ is usually $>1$; the unity term in the brackets on the right-hand term of Eq. (5.80) may be ignored.

$$D_0 = \frac{V}{K_a N_0} \ln\left(\frac{C_0}{C_b} - 1\right) \quad (5.82)$$

Eq. (5.82) facilitates the service time ($t$) determination of an adsorbent bed of depth D. Service time and bed depth are associated with the process parameters, solute flow rate, and sorption capacity. To obtain the time,

Eq. (5.82) can be revised to Eq. (5.83)

$$t = aX + b \quad (5.83)$$

where $a = \frac{N_0}{C_0 V}$ and $b = \frac{1}{K_a C_0} \ln\left(\frac{C_0}{C_b} - 1\right)$.

Thus, sorption capacity, $N_0$, and rate constant, $K_a$, are then determined by getting the slope and intercept (see Fig. 5.15) of the plot "$t$" versus "depth".

Thus, the BDST curve shows how the sorption wave front moves through the sorbent bed. The slope a of the line is the time required for the wave front to progress through 1 m of sorbent. The reciprocal of the slope then is the rate at which a sorbent bed of 1 m is utilized. This value is multiplied by the adsorbent's apparent bulk density to generate an approximation of the rate at which adsorbent must be regenerated. The intercept at the abscissa is the critical bed depth, $D_0$. This critical bed depth is the minimum bed length (depth) for containing a sufficient volume of effluent at time zero, where applying the test operating conditions. The intercept "b" of the ordinate is an estimation of the sorption rate. It is defined as the necessary time for the sorption wave front to pass through the critical bed depth.

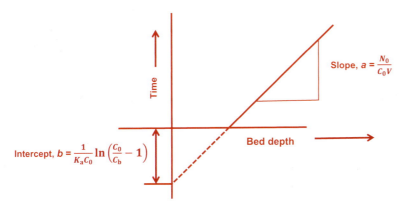

**Fig. 5.15** Theoretical BDST plot.

**Modification of BDST equation for different velocities and breakthrough concentrations**

1. The BDST equation may be revised to handle variations in the feed sorbate concentration. The relationships shown in Eq. (5.84), (5.85) display the needed modifications.

$$\text{New slope} = \text{old slope}\left(\frac{C_{0,\text{old}}}{C_{0,\text{new}}}\right) \quad (5.84)$$

$$\text{New Intercept}, b = \text{old } b \left(\frac{C_{0,\text{old}}}{C_{0,\text{new}}}\right) \frac{\ln\left[\left(\frac{C_{0,\text{new}}}{C_b}\right) - 1\right]}{\ln\left[\left(\frac{C_{0,\text{old}}}{C_b}\right) - 1\right]} \quad (5.85)$$

2. The BDST equation determined using column test results at one linear flow rate can also be changed and used to determine equations for predicting other rates. This is achieved by multiplying original slope, a, by the ratio of the original and new rates (Eq. 5.86)

$$\text{New Slope} = \text{old slope}\left(\frac{\text{original volumetric flow rate}}{\text{New volumetric flow rate}}\right) \quad (5.86)$$

The new intercept will remain unchanged. Thus, lab tests can be reliably scaled up to other flow rates without running pilot tests in larger columns.

In summary, BDST model has been used extensively by various researchers for predicting the breakthrough curve and designing fixed-beds. The BDST model has been, by far, the most successful way to analyze column test data (Aksu and Ferda, 2004; Faust and Aly, 1987; Maliyekkal et al., 2006; McKay, 1995; Sarin et al., 2006; Walker and Weatherley, 1997).

## 5.4 Thomas model

Thomas model (Thomas, 1944) was developed using the Langmuir isotherm model and pseudo-second-order kinetic equation. The model assumes the absence of axial adsorbate dispersion during transit through and adsorption on the column and that the Langmuir isotherm holds for this process. The Thomas model helps define both the sorption rate constant and maximum sorbate concentration adsorbed on the sorbent, where the internal and external diffusion resistance is very small.. The model is expressed in Eq. (5.87).

$$\frac{C_t}{C_0} = \frac{1}{1 + \exp\left[\dfrac{K_{Th}}{Q(q_{max} \cdot m - C_0 V T)}\right]} \quad (5.87)$$

The liner form of Thomas model is defined in Eq. (5.88)

$$\ln\left[\left(\frac{C_0}{C_t}\right) - 1\right] = \left(\frac{K_{Th} \cdot q_{max} \cdot m}{Q}\right) - K_{Th} C_0 t \quad (5.88)$$

where $C_0$ (mg/L) is the initial concentration of the influent, $C_t$ (mg/L) is the effluent concentration at time $t$, $t$ (min) is the sampling time, $K_{Th}$ (mL/min mg) is Thomas rate constant, $q_{max}$ (mg/g) is maximum sorption capacity, $m$ (g) is the quantity of the sorbent in the column, and $V_T$ represents the cumulative throughput volume of treated solution and $Q$ (mL/min) stands for the influent flow rate. The Thomas parameters $q_{max}$ and $K_{TH}$ are determined from the slope ($K_{TH} q_{max}$) and intercept $\left(\frac{K_{Th} \cdot q_{max} \cdot m}{Q}\right)$, respectively, obtained from the linear plot of $\ln\left[\left(\frac{C_0}{C_t}\right) - 1\right]$ versus "$t$" (Bohart and Adams, 1920; Soetaredjo et al., 1994).

## 5.5 Adams-Bohart model

The Adams-Bohart model (Bohart and Adams, 1920) was constructed to express the relationship of $C_t/C_0$ versus time for chlorine sorption on charcoal in a fixed-bed column. Sorption rate was proportional to the sorbate concentration and the residual sorbent capacity. The Bohart-Adams model explains the initial breakthrough curve ($C_t/C_0 \sim 0.5$). The Adams-Bohart model (Bohart and Adams, 1920; Chu, 2010) is expressed in Eq. (5.89)

$$\ln\left(\frac{C_t}{C_0} - 1\right) = \frac{k_{BA} N_0 H}{U} - k_{BA} C_0 t_b \quad (5.89)$$

where $C_o$ (mg/L) is the concentration of the initial influent, and the concentration of effluent at time $t$ is $C_t$ (mg/L). $N_0$ is the sorbent sorption capacity per unit bed volume, $t_b$ (min) is the breakthrough time, $K_{BA}$ is the Bohart-Adams rate constant, $H$ is the sorbent bed-depth, $U$ is the linear velocity $\left[\frac{\text{flow rate}}{\text{column sectional area}}\right]$.

The Bohart-Adams model did not fit to asymmetric breakthrough curves (Chu, 2010). The Thomas model and Bohart-Adams give similar results for highly favorable sorption isotherms. Thus, the Bohart-Adams model is considered to be the limiting form of the Thomas model (Chu, 2010, 2020).

## 5.6 Wolborska model

Wolborska model (Wolborska, 1989) was developed using the mass-transfer equation in the low concentration range of breakthrough curves (Eq. 5.90). This model predicts the break-through time to find out the required operating parameters for fixed-bed adsorbers.

$$\ln \frac{C_t}{C_0} = \frac{\beta_a C_0}{q} t - \frac{\beta_a H}{U} \tag{5.90}$$

where $C_o$ (mg/L) is the initial influent concentration, $C_t$ (mg/L) is the effluent concentration at time $t$, $N_0$ is the sorbent sorption capacity per unit bed volume, $t$ (min) is the breakthrough time, $\beta_a$ is the external mass transfer kinetic coefficient, $H$ is sorbent bed-depth, $U$ is the linear velocity. $\beta_a$ and $q$ can be determined from the liner plot obtained between $\ln \frac{C_t}{C_0}$ and "$t$." The Adams-Bohart and Wolborska models show the same linear dependence (Trgo et al., 2011). Thus, same plots are used to calculate the constants of both the model (Trgo et al., 2011).

## 5.7 Yoon-Nelson model

The Yoon and Nelson single-component sorption model (Yoon and Nelson, 1984) is simpler than other fixed bed models. This model assumed that decrease in the sorption rate for each sorbate is proportional to its sorption and the sorbent breakthrough.

If $A$ is sorbed sorbate fraction and $P$ is the sorbate fraction in the effluent, the sorption rate is given by Eq. (5.91)

$$-\frac{dA}{dt} = k_{YN}(t - \tau) \tag{5.91}$$

Substituting $P = 1 - A$ and integrating ($A = 0.5$ at a given time $\tau$) Eq. (5.70) gives Yoon and Nelson nonlinear sorption isotherm Eq. (5.92).

$$\ln\left(\frac{C_t}{C_0 - C_t}\right) = k_{YN}(t - \tau) \tag{5.92}$$

Here, $C_o$ (mg/L) is the initial influent concentration, $C_t$ (mg/L) represents its effluent concentration at a given time, $k_{YN}$ is the rate constant (min$^{-1}$), $\tau$ is the theoretical time needed for 50% sorbate breakthrough (min), and $t$ is the breakthrough time (min). The Yoon and Nelson liner sorption isotherm model is given by Eq. (5.93).

$$t = \tau + \frac{1}{K_{YN}} \ln\left(\frac{C_t}{C_0 - C_t}\right) \tag{5.93}$$

## 5.8 Clark's model

The Clark model (Clark, 1987) was developed by combining Freundlich sorption isotherm and mass-transfer concept (Eqs. 5.94–5.98) (Sahel and Ferrandon-Dusart, 1993).

$$C = \left(\frac{C_i^{n-1}}{1 + \left[\left(\frac{C_i^{n-1}}{C_b^{n-1}} - 1\right)e^{rt_b}\right]e^{-rt}}\right)^{\frac{1}{n-1}} \tag{5.94}$$

or

$$C = \left(\frac{C_i^{n-1}}{1 + Ae^{-rt}}\right)^{\frac{1}{n-1}} \tag{5.95}$$

where

$$A = \left(\frac{C_i^{n-1}}{C_b^{n-1}} - 1\right)e^{rt_b} \tag{5.96}$$

and

$$r = \left(\frac{K_T}{G_s}\right)V \tag{5.97}$$

The linear Clark model is expressed in Eq. (5.98).

$$\ln A - rt_b = \ln\left(\frac{C_i^{n-1} - C_b^{n-1}}{C_b^{n-1}}\right) \tag{5.98}$$

$C_i$ (mg/L) is a constant influent concentration on the sorbent bed; $C_b$ (mg/L) is breakthrough concentration; $t_b$ (min) is the breakthrough time; $n$ is the Freundlich constant; $K_T$ is the mass transfer coefficient (min$^{-1}$); $G_S$ is the sorbate flow (m$^3$/min/m$^2$); and $V$ is the sorption zone velocity. $A$ and $r$ are the Clark constants and determined using the slope and intercept of a plot between $\ln\left(\frac{C_i^{n-1}-C_b^{n-1}}{C_b^{n-1}}\right)$ versus time, respectively.

# 6 Conclusions

The chapter is written generally for all adsorbents. However, this is a book on BIOCHAR. Contaminant adsorption on biochar is relatively a new practice. A large number of publications appeared on biochar applications in water and wastewater purification. No consistency exists in biochar sorption data presentation and modeling. Despite the large number of publications on biochar adsorbents used for the remediation of inorganic and organic contaminants, limited papers exist on complete batch and fixed-bed column studies. The majority of publications are restricted to only laboratory batch experiments. Regeneration and multicomponent sorption studies are also lacking. It would be difficult to simulate or predict dynamic performance using batch equilibrium sorption isotherms.

BC particle sizes and size distributions, bulk density issues, column packing to get good solid packing fractions without raising back pressures or negatively influencing flow rates need to be carefully studied. Some of these properties are discussed in the first two chapters on biochar characterization. Of course, the feed source and the pyrolysis conditions followed by any size reduction (grinding, sieving) or size expansion (pelletization) must be studied. How will it likely compare with what is done currently with AC or GAC. This is key practical information that people using biochar (BC) and academic chemists, who want to compare their work with AC and GAC, need to know. Since BC sources and pyrolysis processes differ, this research must be undertaken. All those doing BC studies should be asked to develop this type of information in their future papers.

The shape of various biochar surface porosities from large to small to very small is needed. How will that affect the flow rates needed as a function of batch equilibrium times and column-packing density. The best approach would be to again compare to a standard AC or GAC (which will likely have different column packing densities, surfaces, pore size distributions, and particle sizes). Comparisons of different biochars are difficult due to the inconsistencies in the sorption isotherms which are used to explain the sorption and desorption processes between sorbents and sorbates. These isotherms are required to determine the adsorbent sorption capacity. Both basic and advanced equilibrium, and kinetic and fixed-bed sorption models are introduced in this chapter. Advance models provided in this chapter can be used to show how biochar will need to be compared eventually to AC or GAC. This comparison would be really valuable. This is just never seen in the literature. The increasing computer ability makes feasible sophisticated computational predictions to choose the right biochar (feed, process, and properties) may help get the best results. Thus, major efforts are needed to obtain a deeper understanding of biochar adsorption and to provide practical, useful, and cost-effective treatment technologies at many locations and various scales worldwide.

## Questions and Problems

1. Name the common two-parameter and three-parameter isotherm models.
2. Name the common kinetic equations.
3. What is the Freundlich isotherm? What does its slope represent?
4. What are the assumptions of the Langmuir isotherm model?
5. What does the separation factor ($R_L$) infer about the adsorption process?
6. Fill in the blanks.
   The Sips isotherm is derived from the.................. and .................. isotherm equations.
7. Write the Toth isotherm equation and its assumptions.
8. What are the advantages of dynamic sorption over batch sorption?
9. The maximum phenol concentration of 5 mg/L was determined in a wastewater [$K = 20$ $(mg/g)(L/mg)^{1/n}$ and $1/n = 0.5$]. Calculate the biochar dose needed to reduce the phenol concentration from 50 to 5 mg/L.
10. What is a breakthrough curve?
11. What are the four common column models and their equations? Explain the advantages and disadvantages of these four models. Where are they employed relative to each other?

12. Identify break point and exhaustion point on the following breakthrough curve.

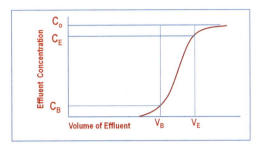

13. Identify A, B, C, D, isotherms obtained in batch sorption studies.

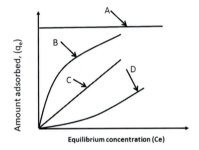

14. Plot a calibration curve using data (obtained from UV-visible spectrophotometer) provided in the table below. Calculate the concentration of an unknown solution at an absorbance of 0.75.

| No. | Absorbance | Solution concentration (mg/L) |
|---|---|---|
| 1. | 0.00 | 0.0 |
| 2. | 0.100 | 1.0 |
| 3. | 0.201 | 2.0 |
| 4. | 0.299 | 3.0 |
| 5. | 0.407 | 4.0 |
| 6. | 0.500 | 5.0 |

15. Calculate the percentage removal and adsorption capacity for a copper sorption system, if the initial and final concentrations are 10.23 and 1.98 mg/L, respectively. The adsorbent dose and copper solution volume used in sorption study were 1 g/L and 100 mL, respectively.

**16.** Construct a linear Freundlich isotherm for the data given below.

| S. no. | Final concentration $(C_e)$ (mg/L) | Sorption capacity $(q_e)$ (mg/g) |
|---|---|---|
| 1. | 0.02 | 0.72 |
| 2. | 0.03 | 1.81 |
| 3. | 0.73 | 6.23 |
| 4. | 2.11 | 7.59 |
| 5. | 3.54 | 8.71 |
| 6. | 5.049 | 9.53 |
| 7. | 9.549 | 11.70 |
| 8. | 13.849 | 13.37 |
| 9. | 22.69 | 15.36 |

**17.** Calculate the Freundlich parameters for the sorption data provided in Question 16.

**18.** Plot a linear Langmuir isotherm for the data given in the following table.

| S. no. | Final concentration $(C_e)$ (mg/L) | Sorption capacity $(q_e)$ (mg/g) |
|---|---|---|
| 1. | 0.02 | 0.72 |
| 2. | 0.03 | 1.81 |
| 3. | 0.73 | 6.23 |
| 4. | 2.11 | 7.58 |
| 5. | 3.54 | 8.71 |
| 6. | 5.04 | 9.53 |
| 7. | 9.54 | 11.70 |
| 8. | 13.84 | 13.37 |
| 9. | 22.69 | 15.36 |

**19.** Calculate the Langmuir constant and monolayer sorption capacity for the data provided in Question 18.

**20.** Adsorption of acetaminophen on biochar follows the Langmuir isotherm at 25°C. The biochar showed an adsorption of 80% when acetaminophen concentration is 10 mg/L, adsorbent dose is 2 g/L and the volume used for sorption experiment is 50 mL. Calculate the experimental adsorption capacity of the biochar at the provided conditions.

21. Plot the pseudo-first and pseudo-second order curves for the ciprofloxacin sorption on biochar using the data given in the following table.

    | S. no. | Time ($t$) (h) | $q_e$ (mg/g) |
    | --- | --- | --- |
    | 1. | 0.084 | 1.24 |
    | 2. | 0.167 | 1.31 |
    | 3. | 0.250 | 1.28 |
    | 4. | 0.500 | 1.37 |
    | 5. | 1.000 | 1.43 |
    | 6. | 2.000 | 1.62 |
    | 7. | 4.000 | 1.68 |
    | 8. | 6.000 | 1.88 |
    | 9. | 8.000 | 2.10 |
    | 10. | 12.00 | 2.16 |
    | 11. | 24.00 | 2.40 |

22. Calculate the pseudo-second order constants for the ciprofloxacin sorption on biochar using the data given in Question 21.
23. Calculate the separation factor ($R_L$) value using the sorption data provided in Question 18. The maximum initial concentration is 50 mg/L. Explain the nature of sorption process on the basis of its $R_L$ value.
24. Why are fixed-bed adsorption studies necessary?
25. What factors affect the mass-transfer zone in a breakthrough curve?
26. List the common mathematical models commonly use to predict the breakthrough capacities. Give the essential features of each model.
27. Why are the adsorption capacities typically low in breakthrough systems vs. batch?
28. What is meant by column channelling? What are the consequences of column channelling on adsorption?
29. What is the key difference between column breakthrough capacity and exhaustion capacity?
30. The following table displays the effluent concentrations for a very small fixed-bed column packed with biochar (0.5 g) for arsenate adsorption. The column flow rate is 5 mL/min, and the influent arsenate concentration ($C_0$) is 1 mg/L. The number of data points is reduced for simplicity.

| Time (min) | Arsenate concentration (mg/L) |
|---|---|
| 30 | 0.01 |
| 60 | 0.06 |
| 90 | 0.45 |
| 120 | 0.52 |
| 150 | 0.60 |
| 180 | 0.65 |
| 210 | 0.78 |
| 240 | 0.86 |
| 270 | 0.97 |
| 300 | 0.99 |

31. In perfluoroalkyl substance remediation from drinking waters, EPA has recommended achieving 70 ng/L of PFAS in the effluent. Suppose the effluent concentration from a biochar packed column is 100 ng/L. Suggest two different approaches to reduce the concentration below 70 ng/L.

32. Paraquet is adsorbed in a small biochar packed bed (dia. = 2 cm, length = 20 cm, and biochar quantity = 10 g). The inlet paraquet concentration ($C_0$) and flow rate are 100 mg/L and 5 L/h, respectively. Data measuring the outlet Paraquet concentration over time from the bed are shown in the breakthrough curve given below.

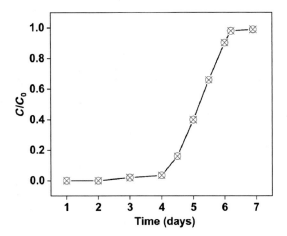

a. Determine the break-point time, the fraction of total capacity used up to the break point, and the length of

the unused bed. Also determine the saturation biochar loading capacity.
b. If the break-point time required for a new column is 7 days, what is the new column length required?

## Acknowledgments

The authors are also thankful to University Grant Commission (UGC), New Delhi for providing the financial assistance under 21st Century Indo-US Research Initiative 2014 to Jawaharlal Nehru University, New Delhi and Mississippi State University, USA in the project "Clean Energy and Water Initiatives" [UGC No. F.194-1/2014(IC)].

## References

Aharoni, C., Sideman, S., Hoffer, E., 1979. Adsorption of phosphate ions by collodion-coated alumina. J. Chem. Technol. Biotechnol. 29 (7), 404–412.

Aksu, Z., Ferda, G.F., 2004. Biosorption of phenol by immobilized activated sludge in a continuous packed bed: prediction of breakthrough curves. Process Biochem. 39 (5), 599–613.

Alkhamis, K.A., Wurster, D.E., 2002. Prediction of adsorption from multicomponent solutions by activated carbon using single-solute parameters. Part II—proposed equation. AAPS Pharm. Sci. Tech. 3 (3), E23.

Aşçi, Y., Nurbaş, M., Açikel, Y.S., 2007. Sorption of Cd(II) onto kaolin as a soil component and desorption of Cd(II) from kaolin using rhamnolipid biosurfactant. J. Hazard. Mater. B139, 50–56.

Bemmelen, J.M.V., 1888. Die adsorptionverbindungen und das adsorptionvermögen der ackererde. In: Die Landwirtschatlichen Versuchs-Stationen. vol. 35, pp. 69–136.

Bohart, G., Adams, E., 1920. Some aspects of the behavior of charcoal with respect to chlorine. J. Am. Chem. Soc. 42 (3), 523–544.

Brunauer, S., Emmett, P.H., Teller, E., 1938. Adsorption of gases in multimolecular layers. J. Am. Chem. Soc. 60, 309–319.

Brunauer, S., Deming, L.S., Deming, W.E., Teller, E., 1940. On a theory of the van der waals adsorption of gases. J. Am. Chem. Soc. 62 (7), 1723–1732.

Bullen, J.C., Saleesongsom, S., Gallagher, K., Weiss, D.J., 2021. A revised pseudo-second-order kinetic model for adsorption, sensitive to changes in adsorbate and adsorbent concentrations. Langmuir 37, 3189–3201.

Butler, J.A.V., Ockrent, C., 1930. Adsorption from solutions containing two solutes. Nature 125, 853–854.

Choudhary, V., Patel, M., Pittman Jr., C.U., Mohan, D., 2020. Batch and continuous fixed-bed lead removal using Himalayan pine needle biochar: isotherm and kinetic studies. ACS Omega 5 (27), 16366–16378.

Chu, K.H., 2010. Fixed bed sorption: setting the record straight on the Bohart–Adams and Thomas models. J. Hazard. Mater. 177, 1006–1012.

Chu, K.H., 2020. Breakthrough curve analysis by simplistic models of fixed bed adsorption: in defense of the century-old Bohart-Adams model. Chem. Eng. J. 380, 122513.

Clark, R.M., 1987. Evaluating the cost and performance of field-scale granular activated carbon systems. Environ. Sci. Technol. 21 (6), 573–580.

Crittenden, J.C., Luft, P., Hand, D.W., Oravitz, J.L., Loper, S.W., Ari, M., 1985. Prediction of multicomponent adsorption equilibria using ideal adsorbed solution theory. Environ. Sci. Technol. 19, 1030–1043.

Dabrowski, A., 2001. Adsorption from theory to practice. Adv. Colloid Interf. Sci. 93, 135–224.

Dastgheib, S.A., Rockstraw, D.A., 2002. A systematic study and proposed model of the adsorption of binary metal ion solutes in aqueous solution onto activated carbon produced from pecan shells. Carbon 40, 1853–1861.

Dubinin, M.M., Radushkevich, L.V., 1947. The equation of the characteristic curve of activated charcoal. Proc. Acad. Sci. USSR Phys. Chem. 55, 235–241.

Faust, S.D., Aly, O.M., 1987. Adsorption Process for Water Treatment. Butterworths.

Foo, K.Y., Hameed, B.H., 2010. Insights into the modeling of adsorption. Chem. Eng. J. 156 (1), 2–10.

Frenkel, J., 1946. Kinetic Theory of Liquids. Clarendon Press, Oxford.

Freundlich, H.M.F., 1906. Over the adsorption in solution. J. Phys. Chem. 57, 385–471.

Fritz, W., Schlunder, E.U., 1974. Simultaneous adsorption equilibria of organic solutes in dilute aqueous solutions on activated carbon. Chem. Eng. Sci. 29, 1279–1282.

Fritz, W., Merk, W., Schlunder, E.U., 1981. Competitive adsorption of two dissolved organics onto activated carbon, II adsorption kinetics in batch reactors. Chem. Eng. Sci. 36, 731–741.

Geankoplis, C.J., 1998. Procesos de Transporte y Operaciones Unitarias. Companı́a Editorial Continental, Ciudad de México.

Günay, A., Arslankaya, E., Tosun, İ., 2007. Lead removal from aqueous solution by natural and pretreated clinoptilolite: adsorption equilibrium and kinetics. J. Hazard. Mater. 146 (1-2), 362–371.

Guo, X., Wang, J., 2019. A general kinetic model for adsorption: theoretical analysis and modeling. J. Mol. Liq. 288, 111100.

Gupta, V.K., Srivastava, S.K., Mohan, D., Sharma, S., 1997. Design parameters for fixed bed reactors of activated carbon developed from fertilizer waste for the removal of some heavy metal ions. Waste Manag. 17 (8), 517–522.

Halsey, G., 1948. Physical adsorption on non-uniform surfaces. J. Chem. Phys. 16, 931–937.

Hill, A.V., 1910. The possible effects of the aggregation of the molecules of haemoglobin on its dissociation curves. J. Physiol. Lond. 40, 4–7.

Hill, T.L., 1952. Theory of physical adsorption. Adv. Catal. 4, 211–258.

Ho, Y.S., McKay, G., 1999. Pseudo-second order model for sorption processes. Process Biochem. 34 (5), 451–465.

Hsieh, J.S.C., Turian, R.M., Tien, C., 1977. Multicomponent liquid phase adsorption in fixed bed. AIChE J. 23 (31), 263.

Hu, Q., Zhang, Z., 2019. Application of Dubinin–Radushkevich isotherm model at the solid/solution interface: a theoretical analysis. J. Mol. Liq. 277, 646–648.

Huang, J.-C., Steffens, C.T., 1976. Competitive adsorption of organic materials by activated carbon. Proceedings of the 31st Industrial Waste Conference Purdue University, Ann Arbor Science, Michigan. pp. 107–121.

Hutchins, R., 1975. New method simplifies design of activated carbon. Chem. Eng. 80, 133.

Jain, J.S., Snoeyink, V.L., 1973. Adsorption from bisolute systems on active carbon. Water Pollut. Control Fed. 45 (12), 2463–2479.

Kim, Y., Kim, C., Choi, I., Rengaraj, S., Yi, J., 2004. Arsenic removal using mesoporous alumina prepared via a templating method. Environ. Sci. Technol. 38, 924–931.

Koble, R.A., Corrigan, T.E., 1952. Adsorption isotherms for pure hydrocarbons. Ind. Eng. Chem. 44 (2), 383–387.

Kumar, H., Patel, M., Mohan, D., 2019. Simplified batch and fixed-bed design system for efficient and sustainable fluoride removal from water using slow pyrolyzed okra stem and black gram straw biochars. ACS Omega 4 (22), 19513–19525.

Lagergren, S., 1898. Zur theorie der sogenannten adsorption geloester stoffe. Veternskapsakad. Handl. 24, 1–39.

Langmuir, I., 1916. The constitution and fundamental properties of solids and liquids. Part 1. Solids. J. Am. Chem. Soc. 38 (11), 2221–2295.

Le Roux, J., Bryson, A.W., Young, B., 1991. A comparison of several kinetic models for the adsorption of gold cyanide onto activated carbon. J. South. Afr. Inst. Min. Metall. 91 (3), 95–103.

Lee, V.K.C., Porter, J.F., McKay, G., 2000. Development of fixed-bed adsorber correlation models. Ind. Eng. Chem. 39, 2427–2433.

Lee, C.-G., Kim, J.-H., Kang, J.-K., Kim, S.-B., Park, S.-J., Lee, S.-H., Choi, J.-W., 2015. Comparative analysis of fixed-bed sorption models using phosphate breakthrough curves in slag filter media. Desalin. Water Treat. 55 (7), 1795–1805.

Limousin, G., Gaudet, J.-P., Charlet, L., Szenknect, S., Barthès, V., Krimissad, M., 2007. Sorption isotherms: a review on physical bases, modeling and measurement. Appl. Geochem. 22 (2), 249–275.

Liu, Y., Shen, L., 2008. From Langmuir kinetics to first- and second-order rate equations for adsorption. Langmuir 24 (20), 11625–11630.

Low, K.S., Lee, C.K., 1991. Cadmium uptake by the moss, Calymperes delessertii, Besch. Bioresour. Technol. 38 (1), 1–6.

Majd, M.M., Kordzadeh-Kermani, V., Ghalandari, V., Askari, A., Sillanpää, M., 2021. Adsorption isotherm models: A comprehensive and systematic review (2010−2020). Sci. Total Environ. 812, 151334.

Maliyekkal, S.M., Sharma, A.K., Philip, L., 2006. Manganese-oxide-coated alumina: a promising sorbent for defluoridation of water. Water Res. 40 (19), 3497–3506.

Marczewski, A.W., 2010. Application of mixed order rate equations to adsorption of methylene blue on mesoporous carbons. Appl. Surf. Sci. 256, 5145–5152.

Mathews, A., Weber, W.J., 1975. Mathematical Modeling of Muiticomponent Adsorption Kinetics. Presented at the 68th Annual Meets. American Institute of Chemical Engineers, Los Angeles.

McKay, G., 1995. Use of Adsorbents for the Removal of Pollutants From Wastewater. CRC Press.

McKay, G., Bino, M.J., 1985. Application of two resistance mass transfer models to adsorption systems. Chem. Eng. Res. Des. 63, 168.

McKay, G., Bino, M.J., 1990. Simplified optimisation procedure for fixed bed adsorption systems. Water Air Soil Pollut. 51, 33–41.

Mcmillan, W.G., Teller, E., 1951. The assumptions of the B.E.T. theory. J. Phys. Colloid Chem. 55 (1), 17–20.

Merle, T., Knappe, D.R.U., Pronk, W., Vogler, B., Hollender, J., von Gunten, U., 2020. Assessment of the breakthrough of micropollutants in full-scale granular activated carbon adsorbers by rapid small-scale column tests and a novel pilot-scale sampling approach. Environ. Sci. Water Res. Technol. 6 (10), 2742–2751.

Mohan, D., Chander, S., 2006. Removal and recovery of metal ions from acid mine drainage using lignite—A low cost sorbent. J. Hazard. Mater. B137, 1545–1553.

Mohan, D., Pittman Jr., C.U., 2006. Activated carbons and low cost adsorbents for remediation of tri- and hexavalent chromium from water. J. Hazard. Mater. 137 (2), 762–811.

Mohan, D., Pittman Jr., C.U., Steele, P.H., 2006. Single, binary and multi-component adsorption of copper and cadmium from aqueous solutions on Kraft lignin—a biosorbent. J. Colloid Interface Sci. 297 (2), 489–504.

Mohan, D., Pittman Jr., C.U., Bricka, M., Smith, F., Yancey, B., Mohammad, J., Steele, P.H., Alexandre-Franco, M.F., Gómez-Serrano, V., Gong, H., 2007. Sorption of arsenic, cadmium, and lead by chars produced from fast pyrolysis of wood and bark during bio-oil production. J. Colloid Interface Sci. 310 (1), 57–73.

Mohan, D., Singh, K.P., 2005. Granular activated carbon. In: Lehr, J., Keeley, J., Lehr, J. (Eds.), The Encyclopedia of Water: Domestic, Municipal, and Industrial Water Supply and Waste Disposal. John Wiley & Sons, New York, NY.

Myers, A.L., Prausnitz, J.M., 1965. Thermodynamics of mixed-gas adsorption. Am. Inst. Chem. Eng. J. 11 (1), 121–127.

Navarathna, C.M., Karunanayake, A.G., Gunatilake, S.R., Pittman Jr., C.U., Perez, F., Mohan, D., Mlsna, T., 2019. Removal of Arsenic(III) from water using magnetite precipitated onto Douglas fir biochar. J. Environ. Manag. 250, 109429.

Patel, M., Kumar, R., Kishor, K., Mlsna, T., Pittman Jr., C.U., Mohan, D., 2019. Pharmaceuticals of emerging concern in aquatic systems: chemistry, occurrence, effects, and removal methods. Chem. Rev. 119 (6), 3510–3673.

Patel, M., Kumar, R., Pittman Jr., C.U., Mohan, D., 2021. Ciprofloxacin and acetaminophen sorption onto banana peel biochars: environmental and process parameter influences. Environ. Res. 201, 111218.

Radke, C.J., Prausnitz, J.M., 1972. Adsorption of organic solutes from dilute aqueous solution on activated carbon. Ind. Eng. Chem. Fundam. 11, 445–451.

Ram, N.M., Christman, R.F., Cantor, K.P., 1990. Significance and Treatment of Volatile Organic Compounds in Water Supplies. CRC Press.

Redlich, O., Peterson, D.L., 1959. A useful adsorption isotherm. J. Phys. Chem. 63, 1024.

Revellame, E.D., Fortela, D.L., Sharp, W., Hernandez, R., Zappi, M.E., 2020. Adsorption kinetic modeling using pseudo-first order and pseudo-second order rate laws: A review. Cleaner Eng. Technol. 1, 100032.

Roginsky, S., Zeldovich, Y.B., 1934. The catalytic oxidation of carbon monoxide on manganese dioxide. Acta Phys. Chem. USSR 1 (554), 2019.

Sahel, M., Ferrandon-Dusart, O., 1993. Adsorption dynamique en phase liquide sur charbon actif: comparaison et simplification de différents modèles. Rev. Des Sci. De L'Eau (J. Water Sci.) 6 (1), 63–80.

Sarin, V., Singh, T.S., Pant, K.K., 2006. Thermodynamic and breakthrough column studies for the selective sorption of chromium from industrial effluent on activated eucalyptus bark. Bioresour. Technol. 97 (16), 1986–1993.

Sarswat, A., Mohan, D., 2016. Sustainable development of coconut shell activated carbon (CSAC) & a magnetic coconut shell activated carbon (MCSAC) for phenol (2-nitrophenol) removal. RSC Adv. 6, 85390–85410.

Sheindorf, C.H., Rebhun, M., Sheintuch, M., 1981. Freundlich-type multicomponent isotherm. J. Colloid Interface Sci. 79 (1), 136–142.

Simonin, J.-P., 2016. On the comparison of pseudo-first order and pseudo-second order rate laws in the modeling of adsorption kinetics. Chem. Eng. J. 300, 254–263.

Sing, K.S.W., Everett, D.H., Haul, R.A.W., Moscou, L., Pieroti, R.A., Rouquerol, J., Siemieniewska, T., 1985. Reporting physisorption data for gas/solid systems with special reference to the determination of surface area and porosit. Pure Appl. Chem. 57 (4), 603–619.

Sips, R., 1948. Combined form of Langmuir and Freundlich equations. J. Chem. Phys. 16, 490–495.

Soetaredjo, F.E., Kurniawan, A., Ong, L.K., Widagdyo, D.R., Ismadji, S., 1994. Investigation of the continuous flow sorption of heavy metals through biomass-packed column: Revisiting Thomas design model for correlation of binary component systems. RSC Adv. 4, 52856–52870.

Sontheimer, H., Summers, R.S., Crittenden, J.C., 1988. Activated Carbon for Water Treatment. DVGW-Forschungsstelle, Engler-Bunte-Institut, Universitat Karlsruhe (TH).

Tan, K., Hameed, B., 2017. Insight into the adsorption kinetics models for the removal of contaminants from aqueous solutions. J. Taiwan Inst. Chem. Eng. 74, 25–48.

Temkin, M.J., Pyzhev, V., 1940. Recent modifications to Langmuir isotherms. Acta Physicochim. URSS 12, 217–222.

Thomas, H.C., 1944. Heterogeneous ion exchange in a flowing system. J. Am. Chem. Soc. 66 (10), 1664–1666.

Toth, J., 1971. State equations of the solid gas interface layer. Acta Chem. Acad Hung. 69, 311–317.

Trgo, M., Medvidović, N.V., Peric, J., 2011. Application of mathematical empirical models to dynamic removal of lead on natural zeolite clinoptilolite in a fixed bed column. Indian J. Chem. Technol. 18 (2), 123–131.

Vimal, V., Patel, M., Mohan, D., 2019. Aqueous carbofuran removal using slow pyrolyzed sugarcane bagasse biochar: equilibrium and fixed-bed studies. RSC Adv. 9 (45), 26338–26350.

Walker, G.M., Weatherley, L.R., 1997. Adsorption of acid dyes on to granular activated carbon in fixed beds. Water Res. 31 (8), 2093–2101.

Wang, J., Guo, X., 2020. Adsorption kinetic models: physical meanings, applications, and solving methods. J. Hazard. Mater. 390, 122156.

Weber Jr., W.J., 1972. Physicochemical Processes for Water Quality Control. Wiley-Interscience, New York.

Weber, T.W., Chakravorti, R.K., 1974. Pore and solid diffusion models for fixed-bed adsorbers. AIChE J. 20 (2).

Weber Jr., W.J., Morris, J.C., 1963. Kinetics of adsorption on carbon from solution. J. Sanit. Eng. Div. 89 (2), 31–59.

Weber Jr., W.J., Smith, E.H., 1987. Simulation and design models for adsorption processes. Environ. Sci. Technol. 21 (1), 1040–1050.

Wolborska, A., 1989. Adsorption on activated carbon of p-nitrophenol from aqueous solution. Water Res. 23 (1), 85–91.

Worch, E., 2012. Adsorption Technology in Water Treatment-Fundamentals, Processes, and Modeling. Walter de Gruyter, GmbH & Co. KG, Berlin.

Wurster, D.E., Alkhamis, K.A., Matheson, L.E., 2000. Prediction of adsorption from multicomponent solutions by activated carbon using single-solute parameters part 1. AAPS PharmSciTech 1 (3), E25.

Xu, Z., Cai, J.-G., Pan, B.-C., 2013. Mathematically modeling fixed-bed adsorption in aqueous systems. J. Zhejiang Univ.-Sci. A (Appl. Phys. Eng.) 14 (13), 155–176.

Ye, N., Cimetiere, N., Heim, V., Fauchon, N., Feliers, C., Wolbert, D., 2019. Upscaling fixed bed adsorption behaviors towards emerging micropollutants in treated natural waters with aging activated carbon: model development and validation. Water Res. 148, 30–40.

Yoon, Y.H., Nelson, J.H., 1984. Application of gas adsorption kinetics I. A theoretical model for respirator cartridge service life. Am. Ind. Hyg. Assoc. J. 45 (8), 509–516.

Yuh-Shan, H., 2004. Citation review of Lagergren kinetic rate equation on adsorption reactions. Scientometrics 59, 171–177.

# Techno-economic analysis of biochar in wastewater treatment

**Arna Ganguly[a] and Mark Mba Wright[b]**
[a]Department of Mechanical Engineering, Iowa State University, Ames, IA, United States. [b]Department of Mechanical Engineering and Bioeconomy Institute, Iowa State University, Ames, IA, United States

## 1 Introduction

Techno-economic analysis (TEA) is a methodology for assessing the economic potential of novel technologies. It helps researchers to forecast and estimate the economic feasibility and profitability of a particular innovation. A TEA model thus consists of five fundamental components (Fig. 6.1), which are as follows:

1. Process model: The process model represents an engineering simulation of the system. The simulation quantifies the process's mass and energy balances depending on various inputs and assumptions, and it provides operating parameters and performance metrics for process streams and unit operations.
2. Equipment listing: This component consists of a list of the required process equipment. Each equipment entry contains the required information to size and cost the unit based on the process model's information. The equipment listing includes estimates for purchasing the equipment and for their installation. These two costs are often known as the free-on-board cost and installation costs. Free-on-board costs are the purchase price of the equipment at the manufacturer's facility. Installation costs include the direct costs required to transport and install the equipment like piping, wiring, building the facility, and indirect costs such as legal and accounting fees. The sum of equipment and installation costs is the fixed capital investment, which represents the total cost to build the

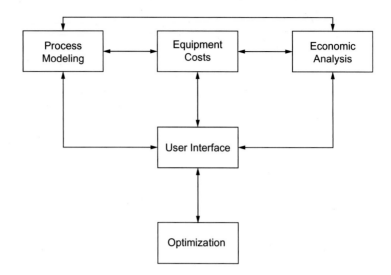

**Fig. 6.1** Techno-economic analysis framework for cost analysis. Reproduced from Burk, C., 2018. Techno-economic modeling for new technology development. Chem. Eng. Prog. 43–52.

facility (Burk, 2018). Typically, projects include a working capital charge of 10%–30% of the fixed capital investment to cover unexpected expenses.

3. Economics: The facility's economic performance can be determined by quantifying the system's costs and revenues. The operating costs are grouped into two sections: (i) variable costs, including material and energy costs, and (ii) fixed costs, which mainly constitute labor, maintenance, and insurance costs. Depreciation, financing interest charges, income taxes, and return on investment (ROI) are also part of fixed costs. To be profitable, a facility needs a positive ROI. A positive ROI is achieved when the revenue of selling the facility's products exceeds the operating costs.

4. User interface: Modern TEA models provide convenient interfaces for analysts to evaluate the impacts of key process and economic parameters. This is commonly accomplished by developing spreadsheets, but there are existing commercial and open-source software available for TEA.

5. Optimization analysis: The final step in TEA involves conducting optimization studies. A growing number of techniques have been developed for this purpose, including scenario, sensitivity, and uncertainty analysis. A robust optimization analysis reduces commercialization risk by evaluating the potential range of economic returns the facility could achieve based on its performance metrics.

In summary, TEA represents a combination of engineering and economic techniques to evaluate the commercialization potential of a technical innovation.

Scientists and engineers have employed TEA to investigate a wide range of technologies. Thus, it provides a helpful guide for evaluating biochar production economics in general and for wastewater treatment in particular.

Biochar is a long-established carbon-rich substance used as a combustion fuel in domestic surroundings throughout various parts of the world. Biochar is one of the primary by-products of biorenewable technologies. Due to its noteworthy advantage as a combustion and carbon sequestration material, scientists are taking a lot of interest in studying the energy efficiency and techno-economic potential of biochar. Biochar development addresses global environmental concerns like global warming and climate change, and has multifunctional uses in science and technology, which eventually improves the global economy.

The specific composition and structural characteristics of biochar have expanded the range of potential applications to various fields like soil amendment, catalysis, and adsorption agent in wastewater treatment plants. There are many TEA studies investigating technologies that generate biochar. However, most of these studies consider biochar as a by-product of secondary economic value. The development of high-value biochar applications and growing interest in carbon prices are leading toward more TEAs of biochar production. This chapter discusses key considerations in the TEA of biochar production for wastewater treatment systems and helps readers understand the potential cost factors and market drivers related to biochar use in treatment plants.

## 2  Biochar markets and economics

Biochar is of growing commercial interest. According to their application, online merchants list biochar prices, with most biochar being sold for horticulture and agricultural purposes. The biochar price depends on the biochar's quality, provenance, and consumer target. A few examples include biochar obtained from softwood forest residues having shallow ash content, biochar with lesser carbon and significant mineral content, biochar composts solely for horticulturists, etc. Prices vary according to the quality (related to carbon content) and volume of biochar blended with the mix, numbers being \$40–\$350/yard$^3$ (Pacific Biochar, 2020; Biochar Supreme (TM) & Black Owl Biochar (TM), n.d.).

Biochar has been considered for a wide range of applications, including agricultural land management, soil, water remediation, gardening and composting, filtration, livestock feeding, building materials, packaging materials, fuel cells, 3D printing, and carbon offsets. Estimated market values for biochar products range from $200 to $2k/tonne (Miles, 2014). Fig. 6.2 shows the wide ranges of biochar market size along with their varied prices. Large markets include agriculture, specialty crops, composting, and soil remediation. Smaller, specialty markets consist of soil and wastewater treatment, nurseries, medical, electrochemical, and food markets, among others.

The use of biochar in agriculture dates back centuries when land clearing with forest fires would improve soil productivity by increasing its organic carbon content. In recent years, biochar application results in significant yield improvements for carbon- and nutrient-depleted soils. Li et al. (2019) evaluated the economic benefit of increasing crop yields through biochar application and estimated the value of biochar to farmers to range between $100 and $400/tonne. Higher values are associated with either higher-valued crops or significantly depleted lands. Biochar's ability to improve soil nutrient availability and productivity is also useful for composting and land remediation, for example.

Biochar can selectively absorb heavy metals and minerals. Industrial intensification has led to an increasing concentration of pollutants in water and air streams. Biochar acts as a "win-win" substance when used as an adsorbent in wastewater treatment plants. It can purify wastewater streams containing organic and inorganic contaminants at an affordable price providing propitious performances comparable to activated carbon. It is used for sanitizing water as a biosorbent for effacing crude oil spills from synthetic seawater (AlAmeri et al., 2019). Contaminated gaseous mixtures can be treated on a large scale using biochar as a biofilter. A study by Das et al. (2019) utilized the adsorptive property of biochar to remove $H_2S$ (gaseous).

Biochar also has potential as a supercapacitor material to be used in energy storage devices. According to Bartoli et al. (2018), physically and chemically activated biochars can be utilized as supercapacitor electrodes mainly because of their elevated surface area, which aids in producing double ionic layers. Luo et al. (2014) pointed out that chemical activation of biochar using ammonia increased its capacitance to $40\,mF/m^2$. Biochar can also be employed as anodic materials for lithium-ion batteries, showing an adequate discharging capacity of $1169\,mAh/g$ (Dai et al., 2019). Biochar-based fuel cells have also gained prominence in the past few years. Titanate-based anodes in a biochar-carbon-based fuel cell converted chemical energy to electrical energy generating power of about $78\,mW/m^2$ (Ali et al., 2019). Huggins et al. (2014) studied biochar use as microbial fuel cell electrodes, generating power of around $532 \pm 18\,mW/m^2$, but also at reduced carbon footprints and

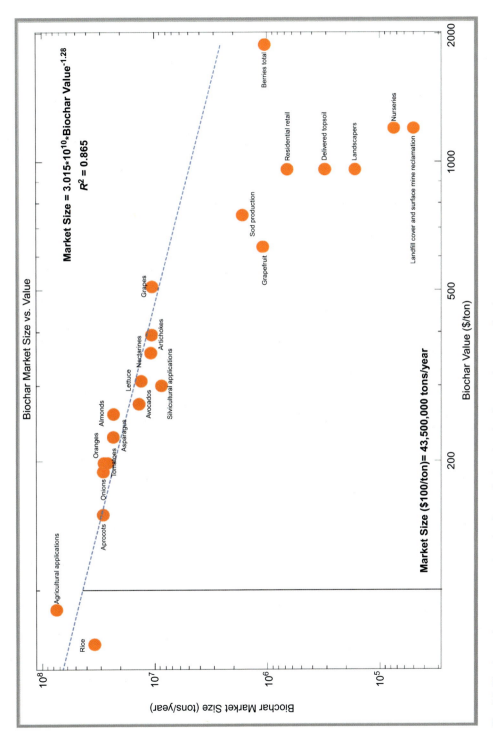

**Fig. 6.2** Biochar value versus market size for agricultural and soil amendment applications.

expenditure. Biochar has also been used as composites for building envelopes that provide good insulation and constitute increased water-retaining characteristics (Mu and Wang, 2019; Lee et al., 2019).

Biochar has shown adsorption capabilities in pharmaceutical and agrichemical industries along with personal care products, other than wastewater treatment plants or soil fertilization facilities. Biochar cost ranges between $350 and $1200/tonne (Shackley et al., 2015), which is less than powdered activated carbon with cost varying between $1100 and $1700/tonne (Alibaba, n.d.). A study by Fagbohungbe et al. (2017) demonstrates that it can adsorb inhibitors present in anaerobic digesters and aids in increasing the final biogas output. Biochar also can be used as a catalyst (Razzaghi et al., 2020) and in wastewater treatment plants because of the same characteristic features (Yao et al., 2019). Biochar, used as a substitute for activated carbon, is a fairly new technology. There is a lot of evidence that proves its efficiency being at par or sometimes more than activated carbon if engineered under some specific conditions. However, these conditions are determined by the system's targeted contaminants, which requires adsorption and separation from the wastewater being treated.

The key factors affecting the market price of biochar are its structure, porosity, ash content, and pH. Biochar structure and porosity are fundamental for applications that regulate the transport of materials. These applications rely on biochar's ability to selectively adsorb and desorb compounds based on their size relative to the biochar pores. High-value health and chemical applications require distinctive nanomaterials, and these industries are willing to pay high prices for suitable materials. Biochar can also regulate water and nutrient retention and release, in lower-valued markets. Ash content and pH are essential in applications where they influence chemical and biological interactions. Therefore, biochar can be tailored or optimized depending on the application.

Identifying the optimal biochar structure and henceforth its adsorption abilities require two interdependent steps: (a) knowing the contaminants present in the system and (b) determining the specific production conditions required for efficient adsorption. Thus, the economic factors of biochar production are very much dependent on how they are engineered. Biochar structures are governed by the raw material properties and the biochar production reaction conditions (Pan et al., 2021). Biochar can be obtained through various biorenewable technologies, namely torrefaction, pyrolysis (fast and slow), gasification, hydrothermal carbonization, and flash carbonization. Out of these processes, slow and fast pyrolysis achieve biochar yields of around 30 wt% (with 95% carbon content) and 12–26 wt% (with 74% carbon content), respectively (Meyer et al., 2011).

There are a few studies focused on the TEA of biochar production. According to Roberts et al. (2010), biochar produced from a slow pyrolysis plant with 10 tonnes/h of dry feedstock capacity earned a revenue of $16/tonne of dry biomass when being used for soil management and synthesis gas production purposes. Brown et al. (2011) reported a total annual operating cost, including coproduct credits of around $$_{2010}$ 71 million/year from a slow pyrolysis system with a biochar coproduction of approximately 262k tonnes each year. Slow pyrolysis has more biochar production and lesser fuel production. The production cost per tonne biochar from this facility was $$_{2010}$ 272. Harsono et al. (2013) observed slow pyrolysis of palm oil from empty fruit bunches resulting in a biochar revenue of $531.6/year per tonne of feedstock, the production cost for the same being $523.6/year for 1-tonne of input. Shabangu et al. (2014) reported an integrated system for biochar and methanol production by thermally decomposing pine in two stages. The results demonstrated that for the economic feasibility of the whole system, biochar market value was very pivotal. Sensitivity analysis of slow pyrolysis demonstrated that system profitability depends on the biochar selling price, with a break-even being attained at a biochar price of $220/tonne at 300°C and $280/tonne at 450°C.

## 3 Biochar costs in wastewater facilities

Rapid population and industrial growth have increased the need for wastewater treatment plants to avoid the concentration of heavy metals, minerals, and organic matter that threaten ecological systems' safety. Wastewater toxicity levels depend on the pollution sources and water flows in the regions. Wastewater can contain many contaminants resulting in many potential interactions with the environment and other materials. To efficiently absorb and neutralize these contaminants, wastewater treatment plants require absorbents tailor-made for the types and quantities of contaminants in their stream. This section discusses some of the TEA considerations related to using biochar in wastewater facilities.

### 3.1 Process design and capital costs

Wastewater decontamination generally includes five main steps: (a) pretreatment of the impure water, which provides for sedimentation, particle size reduction, etc., (b) primary treatment, which mainly includes processes like coagulation, precipitation, flocculation, etc., followed by (c) purification, which involves

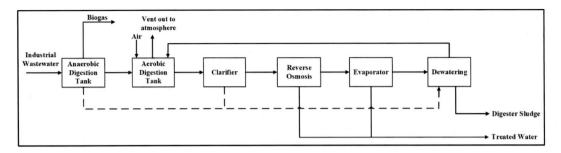

**Fig. 6.3** Block flow diagram of a wastewater system producing biogas, digester sludge, and treated water. Reproduced from Humbird, D., Davis, R., Tao, L., Kinchin, C., Hsu, D., Aden, A., et al., 2011. Process design and economics for biochemical conversion of lignocellulosic biomass to ethanol: dilute-acid pretreatment and enzymatic hydrolysis of corn stover. Nat. Renew. Energy Lab. 1–147. Available from: http://www.nrel.gov/docs/fy11osti/47764.pdf%5Cnpapers3://publication/uuid/A010A50F-E6E1-4445-88E3-B4DBDB2DF51C.

biodegradation, filtration, adsorption, etc., (d) final treatment which includes oxidation, membrane filtration, etc., and eventually (e) treating the sludge produced which involves supervised tipping, recycling, and incineration (Crini and Lichtfouse, 2019). The process design taken into account for understanding wastewater treatment equipment and facilities here is taken from Thomas Steinwinder et al. (2011) subcontract report. Fig. 6.3 is the block diagram of the wastewater treatment plant described below.

Traditionally, wastewater treatment plants include two consecutive treatments: anaerobic and aerobic digestion followed by filtration through reverse osmosis. The primary process design components used in wastewater treatment plants are as follows:

1. Anaerobic digestion process: This biological process converts organic compounds into mostly methane and carbon dioxide gases in an oxygen-free environment. The performance of this unit helps evaluate the potential for treating the wastewater. Biogas produced from this system could be reused for process heating purposes.
2. Aerobic digestion process: The wastewater leaving the anaerobic digestion system passes through an aerobically induced activated sludge process to process any remaining organic compounds. Usually, nitrification is required due to high ammonia concentrations in the anaerobic digestion effluent stream.
3. Secondary clarification or filtration: In this process, solid-liquid separation takes place via filtration. The filter typically employs a coal-based activated carbon material. The activated carbon could be replaced or combined with biochar depending on the contaminants present in the treated wastewater.

4. Sludge handling and dewatering: All waste materials in the sludge are pumped from the sludge storage tank and passed into a dewatering device to reduce the slurry's water content.
5. Salt removal: Reverse osmosis is done to remove inorganic salts before discharge to the environment.

The major equipment required in the whole procedure is as follows:

(i) Anaerobic treatment basins, (ii) iron sponge (only the equipment), (iii) aeration basins, (iv) membrane bioreactor, (v) -sludge-holding tank, (vi) centrifuge, (vii) reverse osmosis, (viii) evaporator, (ix) pumps, and (x) blowers. All these contribute toward the significant capital cost of the wastewater system plant.

## 3.2 Operation and maintenance costs

The operating and maintenance costs include costs associated with (i) operating the anaerobic system, (ii) power requirements inside the plant, (iii) replacement of the iron sponge, (iv) adding polymer for sludge dewatering and alkalis, usually NaOH, as a source of alkalinity, (v) labor cost for engineers and operators, and (vi) other maintenance costs such as repairs and replacement of minor equipment.

## 3.3 Techno-economic analysis of wastewater facilities

The traditional approach of treating wastewater demands high energy consumption and wastes organic nutrients. Proper upgrading of the facilities would not only make proper usage of the nutrients but also would be economically viable and would save energy. Some of the examined designs for treating wastewater are anaerobic digestion technologies which further produce biogas. However, the problem of treating biosolids prevails, which incurs a cost of around $0.4 million per year for landfilling.

The economic feasibility of retrofitting such systems has shown interesting results when compared to the above-mentioned system as reference. The first scenario constituted feeding the undigested slurry for hydrothermal liquefaction followed by an up-flow anaerobic sludge blanket (UASB) reactor. The output was bio-oil (used in refineries for biofuel production) and hydrochar (sold as soil fertilizer), and the remaining organic solubles are fed into the UASB reactor for biogas production.

The biosolids were used as an input for slow pyrolysis producing bio-oil (sold to refineries as done in the first scenario), biochar (used for soil amendment purposes), and noncondensable gases.

The biogas from the anaerobic digestion system and the UASB reactor were combusted in a combined heat and power (CHP) system for recovering some of the process heat and electricity required for pretreating the feedstock for slow pyrolysis. The third scenario, where the residual biosolids were thermochemically decomposed by fast pyrolysis, was similar to the second scenario. The only differences being increased reaction temperature and heating rate along with lesser residence time. The output, in this case, had more bio-oil (upgraded to gasoline and diesel range biofuels) and lesser biochar by-product.

TEA results showed an increased equipment cost for the three scenarios when compared to the reference design, the highest being for slow pyrolysis and the lowest for a hydrothermal liquefaction scenario. Fig. 6.4 depicts a comparison in operating costs of the three scenarios and the reference system. For all the scenarios, operation and maintenance costs were the primary contributors. The net present value of the reference system accounted for around $154.40 million. The same was $177.36 million, $101.15 million, and $103.35 million for hydrothermal liquefaction, slow, and fast pyrolysis retrofit, respectively. Hydrothermal retrofit design was most profitable primarily because of the production of biogas through the UASB reactor. The thermochemical systems were not that economically attractive, specifically due to higher capital costs and the undervalued cost of biochar. The sensitivity analysis for the three scenarios showed that biochar price (range assumed between $50 and $1900/tonne) could significantly influence the net present value (NPV) of the pyrolysis systems. Thus, from an economic point of view, the three retrofit designs have potential if aided by government policies providing incentives for biochar and hydrochar production (Tian et al., 2020).

### 3.4 Biochar properties and impacts on cost

As already discussed, pyrolytic feedstock, temperature, and conversion technology have a significant role to play in the efficiency of biochar as an adsorbent. In the case of temperature, it has been observed that with the increase in temperature, the contribution of biochar toward the adsorption of organic impurities increases. Due to increased temperature, the biomass organic matters are completely carbonized, which gives rise to more nanopores on the surface of the biochar, increasing the adsorption capacity of naphthalene (Chen et al., 2012). With the increase in pyrolytic temperature, the oxygen- and hydrogen-carrying functional groups get removed, resulting in highly hydrophobic biochars, which further aids in the adsorption of trichloroethylene

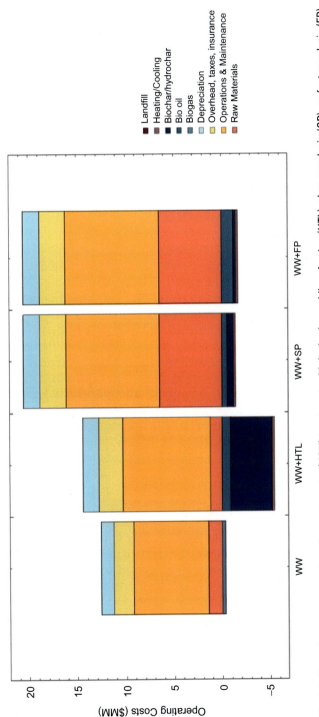

**Fig. 6.4** Operating cost comparison for wastewater (WW) treatment, with hydrothermal liquefaction (HTL), slow pyrolysis (SP), or fast pyrolysis (FP) biochar units. Revenue sources include biogas, bio-oil, biochar, heating/cooling, and landfill avoidance. Data from Tian, X., Richardson, R.E., Tester, J.W., Lozano, J.L., You, F., 2020. Retrofitting municipal wastewater treatment facilities toward a greener and circular economy by virtue of resource recovery: techno-economic analysis and life cycle assessment. ACS Sustain. Chem. Eng. 8(36), 13823–13837.

(Ahmad et al., 2013). Studies have also shown that biochar elemental, structural, and morphological properties change with the increase in temperature, which eventually brings in changes in the biochar's pH and surface area. Temperatures greater than 500°C help to increase cadmium adsorption (Kim et al., 2013). Increasing the pyrolysis temperature impacted the total capital investment due to the increased cost of heaters and heat exchangers required to meet high temperatures.

The natural composition of biomass feedstock also influences adsorption capability of biochar. Kou et al. (2011) reported that the adsorption capacity of methyl violet varied for biochar obtained from the pyrolysis of canola straw > peanut straw > soybean straw > rice hull (in order). Xu et al. (2013) observed increased efficiency of adsorbing lead, copper, zinc, and cadmium when rice husk biochar was being replaced with biochar obtained from dairy manure. Biomass feedstock cost is mostly dependent on the plant's location and the availability of the feedstock, which adds to the plant's operating cost.

Demineralization and deashing treatments of biochar increase the adsorbing ability by influencing the composition as well as the surface characteristics. Sun et al. (2013) observed changes in biochar hydrophobicity, which helped to increase adsorption of phenanthrene. Both treatments would demand more equipment which would increase the system capital investment. According to Tan et al. (2015), both the treatments still lack experimental evidence and are under further investigation.

One of the most influencing parameters of the adsorption process is solution pH, which influences target impurities and the type of biochar used for adsorption. It affects the biochar ionization properties by influencing the surface charge of the adsorbent. Biochar consists of many functional groups like carboxylic acid (—COOH), alcohol (—OH), etc., the behavior of which changes with the change in pH of the solution. At low pH, the biochar is positively charged, aiding in the adsorption of anions. However, electrostatic repulsion often takes place between cation-based impurities and positively charged biochar, leading to lower adsorption at low pH. As the solution pH increases, the concentration of $H_3O^+$ decreases and the surface of the biochar becomes more negatively charged often leading to an increase in adsorption of metal cations (Oh et al., 2012; Abdel-Fattah et al., 2014; Lu et al., 2012). Different studies have reported different optimal pH depending on the type of contaminant targeted; the pH values have ranged between 3.0 and 9.0 (Tan et al., 2015). Maintaining the solution's pH requires continuous operation and addition of acid or base whenever needed, which would increase the adsorption efficiency. Hence, maintaining pH increases the operating cost of the system.

Proper adsorbent dosage is crucial for efficient biochar adsorption and is also critical for making it a cost-effective application. Chen et al. (2011a) observed a decrease in adsorption efficiency with an increase in biochar content. Adsorption efficiency of biochar produced from swine manure decreased with increase in biochar dose. Proper dosage of biochar is crucial to keep the whole system economical and feasible.

Proper biochar management is very crucial once subjected to an active adsorption process. Since biochar adsorption mainly deals with the removal of toxic contaminants, biochar should be handled with care after processing. Desorption or regenerating properties are being investigated for reutilization of the biochar as adsorbent. Zhang et al. (2013a) observed that used biochar could be desorbed after going through four cycles of adsorption-desorption using 0.05 mol/L HCl. Another study demonstrated the desorption of dye-loaded biochar using ethanol. However, the economic feasibility of the desorption process still requires further research. Methods like incinerations and safe landfills could be utilized to treat the spent biochar (Tan et al., 2015).

## 4  Biochar production

Biochar can be produced via pyrolysis, gasification, flash and hydrothermal carbonization technologies, and torrefaction. Fig. 6.5 depicts a tree diagram of biochar production methods.

Pyrolysis involves thermal decomposition of the biomass feedstock at 450–500°C in an oxygen-depleted environment, leading to biofuel and biochar production. Pyrolysis processes are usually recognized by their temperature and residence time (either slow or fast pyrolysis). Pyrolysis differs from gasification by the absence of oxygen in the reaction chamber.

In the case of gasification, the feedstock is partially oxidized in the reactor at atmospheric pressure, maintaining about 800°C. The prime end product of this process is syngas. Biochar yield from this process is significantly less than that of other methods. Flash carbonization occurs at a pressure of 1–2 MPa, maintaining a temperature of around 300–600°C. In this process, a flash of fire ignites the bottom of packed bed biomass, flowing in an upward direction against the descending airflow. The main products of the process are syngas and solid biochar.

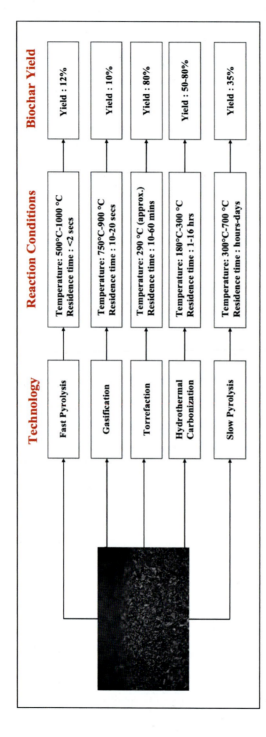

**Fig. 6.5** Biochar production pathways organized by technology, reaction conditions, and biochar yield. Based on Zhang, Z., Zhu, Z., Shen, B., Liu, L., 2019. Insights into biochar and hydrochar production and applications: a review. Energy 171, 581–598. https://doi.org/10.1016/j.energy.2019.01.035.

Hydrothermal carbonization, also known as "wet pyrolysis" (Libra et al., 2011), occurs at a much lower temperature than fast pyrolysis. In this process, the biomass is heated to a temperature of 180–220°C in the absence of oxygen, as a suspension in water under high pressure of 2–6 MPa for long hours. The primary products of this process are hydrochar and some gaseous phases mainly constituting of $CO_2$.

Torrefaction demands heating of biomass feedstock in the absence of oxygen at a temperature of 250–300°C. The primary output of this process is solid biochar.

## 4.1 Impacts of feedstock price

The pyrolysis process can convert a wide range of different biomass types. A plant-based feedstock usually results in high biochar yield due to higher lignin content; organic wastes like animal manure and sewage sludge have more biochar yield compared to agricultural waste and woody biomass. Organic wastes also lead to the mesoporous structure (uniform pore dimensions showcasing distinctive adsorption, separation, and catalytic characteristics) of biochar due to the presence of increased volatile substances in it (Pan et al., 2021).

According to the Billion-Ton Report of 2016 (Langholtz et al., 2016), primary group forest, agricultural, and waste resources used as biomass feedstock can be obtained at $60 or less at the roadside. This was part of the total supply, around 311 million dry tonnes in the near term, long-term base case, and long-term high-yield instances. The report indicated that delivery of 45%, 37%, and 54% of the supplies for the near-term, long-term base case, and long-term high-yield scenarios, respectively, is possible at approximately $84 per dry tonne, which would include all steps starting from producing, harvesting, transporting, and grinding. Quantification was also done in terms of weighted average prices, which depicted 70%, 69%, and 84% of the supplies for the near-term, long-term base case, and long-term high-yield scenarios, respectively, can be transported and delivered at a price up to $84/tonne. Moreover, the whole supply system of biomass feedstock follows specific canonical steps necessary to produce a substance that meets feed requirements and specifications for the raw biomass's thermochemical conversion.

Thus, although the feedstock range for biochar production is vast, the price varies since the cost of this kind of supply system varies with environmental, location, and regulatory factors (Meyer et al., 2020). Pyrolysis of food waste has gained significance because of its profitability, by bringing down the production cost (gate fees) to $50/tonne of food waste input when compared to $2.2k/tonne of conventional biomass input. Thus, biochar production proves profitable from this system (Grycová et al., 2016). Hence, the feedstock price used for pyrolysis has a significant role to play in biochar production to be used as an adsorbent in wastewater plants.

## 4.2 Impacts of biochar production technology

Efficient thermochemical decomposition of pyrolysis feedstock is pronouncedly dependent on the reactor type. One of the governing factors that impact the distribution of yields in final products is the bed reactors' height (fixed or moving). It has been observed that increasing the height of the reactor above 10 cm increases the biochar yield. A taller reactor bed increases the pyrolysis vapor's residence time and can lead to repolymerization of the vapors to yield more biochar (Yadav and Jagadevan, 2019). Primary factors that impact the reactor's cost include the area, the reactor diameter, and the thickness. With the increase in the height of the reactor, the volume of it increases, so does the reactor's cost, which furthermore increases the fixed capital investment for biochar production. Thus, the pyrolysis system reactor is one of the most significant cost factors to consider while going for biochar production (primarily) from pyrolysis.

Temperature, pressure, and residence time inside the pyrolysis reactor play a big role in controlling pyrolysis products' yield. Zhang et al. (2017) reported that biochar produced from wood chips of oak and pine showed a decrease in volatile matter content from 65% to 30% when the temperature increased from 350°C to 900°C. In the case of fixed matter content, an increase in pyrolysis temperature from 350°C to 900°C resulted in a biochar increase from 30% to 64% for woody biomass, 35% to 56% for sugarcane used as feedstock, and 28% to 42% for peanut shell feed.

Higher temperature requires an increase in the heating rate of the pyrolysis system, which requires increasing the size of the heat exchanger, adding cost to the system. The cost of the process furnaces and direct-fired heaters increases proportionally with heating duty depending on the type of heat exchanger material used. Carbon-steel tubes usually have the lowest price at higher heating duty compared to chrome or moly tubes, which incur a much higher cost, although both rise in a directly proportional manner with increasing heating duty (Peters, 1991). Hence, temperature plays a significant role as a cost factor of biochar production since an increased heating system area increases the fixed capital investment.

Pyrolysis, as already discussed in the introduction section, has a lot of potential to produce biochar. Sebastian Meyer et al. (2011) reported that the direct cost of producing biochar (not making or selling any composts) would aggregate to $50/tonne of dry biomass or around $172/tonne of biochar at a biochar yield of 29 wt%. On the other hand, gross income would amount to $112/tonne of dry biomass or $368/tonne of biochar, while not considering the avoided composting costs. The heat generated from the pyrolysis synthesis gas contributes around $35/tonne of dry biomass or $121/tonne of biochar produced within this revenue earned. The agricultural value of biochar sums up to $11/tonne of dry biomass or $38/tonne of produced biochar. Hence, revenue earned from GHG offset certificates is one of the essential factors driving the profitability of pyrolysis system.

Char availability from gasification is low, due to the produced char being reused in the system for heating purposes. The reason behind this is primarily because the focus of this system is to obtain energy technical challenges associated with it. Gasifier engineers will only sell the cogenerated biochar if the price compensates for the cost of an alternative heating fuel (fed to the system) for the whole gasification system. Jorapur and Rajvanshi (1997) indicated an HHV of 18.9 MJ for coproduced biochar in the gasification process, which corresponded to an energy value of around $_{2010}$ 380/tonne of biochar when replaced with wood pellets as the heating fuel.

Flash carbonization incurs a hefty investment, firstly on setting up and fabricating the carbonization reactor, which accounts for around $_{2010}$ 290k. Secondly, air compressor installations add to the investment cost. According to the reactor's pressure requirement, for 1.27 MPa, the amount is around $_{2010}$ 32.3k, while for 2.17 MPa, the price sums up to about $_{2010}$ 129k. Biochar production at 1.27 and 2.17 MPa air compressor-based flash carbonization system were 6.1 tonnes/day and 8.4 tonnes/day, respectively, operating 24 h. Although both the systems showed very short payback periods of 3.7 years (for 1.27 MPa system) and 1.3 years (for 2.17 MPa system), the overall profitability of the systems was not available. This was primarily due to lack of available financial data (Antal Jr. et al., 2007).

Hydrochar, the main product of hydrothermal carbonization, acts as a good adsorbent because it has similar characteristics as biochar. Both can be used to separate contaminants in wastewater treatment plants. In the case of monocyclic aromatic impurities, hydrochar surpasses biochar's performance in terms of sorption capacity, especially in adsorbing toluene. Polycyclic aromatic contaminants and halides can only be removed by using biochar. The sorption capacity of both is similar in terms of adsorbing phenolic compounds; on the other hand, adsorption capacity is higher for hydrochar when subjected to nitrobenzene, nitrophenol, and heterocyclic-based organic contaminants (Liu et al., 2021). Hydrochar cost is very much dependent on the feedstock used and the hydrothermal carbonization plant capacity. Hydrothermal carbonization of empty palm oil fruit bunches accounted for around $12.86 per $GJ_{HHV}$, which did not prove to be competitive to the fossil fuel market (Stemann et al., 2013).

On the other hand, the cost of producing pelletized hydrochar from a 20,000 tonne/year plant using grape marc as input was $177.41/tonne. The break-even value for hydrochar was around $226/tonne. In terms of energy produced, the price is estimated to be about $9.38 per $GJ_{HHV}$ (Lucian and Fiori, 2017). The minimum selling price of hydrochar was estimated to be around $30/tonne when processing 2000 tonne/day of food waste with a plant life of 20 years (Mahmood et al., 2016). Akbari et al. (2019) reported a TEA of producing hydrochar from yard waste from two different plants. Results depicted that plant capacity had a significant role to play regarding hydrochar production. Increasing the biomass input from 9.6 to 960 tonnes/day decreased the hydrochar production cost from $47.2/GJ to $4/GJ, respectively. Shao et al. (2019) quantified the input cost of hydrochar to be $50/tonne produced from green waste by considering hydrochar price and input energy price. Hydrothermal carbonization of sewage sludge is a field under research because of its profitability being observed widely. The estimated minimum cost of hydrochar from a 500 kW unit of sewage sludge-based plant was about $169/tonne (Zeymer et al., 2017). Sewage sludge treatments using hydrothermal carbonization indicate that the cost of the process decreases to $38/tonne while the traditional treatment of the same ranges between $38 and $92/tonne. The cost of the process is influenced by the production of electricity and steam in the plant utilizing the waste fractions (Child, 2014). Thus, similar to biochar, the economic efficacy of hydrochar is governed by the feedstock selection, capacity of the plant, supply logistics, process conditions (temperature, pressure, etc.), and the process design of the hydrothermal carbonization process.

Torrefaction operates under autothermal conditions except for the initial firing required for the thermal decomposition of the biomass input. Usually, pelletized biomass is preferred for a torrefaction process, because the output of torrefied biomass is similar to coal in terms of heating value and bulk density. Although biomass torrefaction is not competitive with the current fossil fuel market, it would gain significant prominence in a carbon-credit-based market and an increased plant capacity (Pirraglia et al., 2013; Svanberg et al., 2013; Batidzirai et al., 2013). According to Pirraglia et al. (2013), carbon credits would earn additional revenue of around $36–$72/tonne of torrefied wood pellets, which eventually would increase the internal rate of return (IRR) and NPV. The author also pointed out that scaling up the reactor would also notably bring down the capital investment costs. Batidzirai et al. (2013) pointed out that on a long-term basis, integration of technological optimizations and scaling up of the torrefaction plant would bring down around 50% of the total costs, which would further decrease the production costs of the biomass (woody) input to $2.1–$5.1 per $GJ_{HHV}$.

### 4.3 Impacts of biochar coproducts

Coproducing biochar along with biofuels increases the potential of the whole system having financial success. Comparing methanol and biochar coproduction by employing two-stage pyrolysis and gasification from forest biomass feedstock depicted that gasification successfully produced methanol being competitive to the fossil fuel market. On the contrary, the pyrolysis process required biochar to be sold between a break-even price of $220–$280/tonne. Zhang et al. (2013b) investigated the economic viability of producing monosaccharides, hydrogen, and transportation fuel via fast pyrolysis, where the biochar was assumed to be sold to the market as boiler fuel, the price being $18.21/tonne. Another study depicted a biochar price of $50/tonne (being sold as boiler fuel) when coproduced with purified levoglucosan obtained from the pyrolysis of lignocellulosic biomass. Results show that the biochar price reduced the minimum selling price of anhydrosugar by around 4% (Rover et al., 2019). Techno-economic analysis of fast pyrolysis of corn stover producing diesel and naphtha-range fuels depicted a coproduced biochar price of $22/tonne (Wright et al., 2010). Thus, it can be concluded that the price of biochar as a by-product is sensitive to the main output product of the biorenewable technology.

### 4.4 Impacts of environmental incentives

Biochar is a well-known $CO_2$ sequestration component when applied to the soil. To understand the influence of biochar use and production on the GHG emissions, it should be compared to an established system. This specific reference system would identify GHG emissions changes when it has been replaced with a biochar-based system. Taking yard waste as composting material, around 885 kg $CO_{2eqv}$/tonne dry biomass feedstock can be displaced when this yard waste is replaced with biochar for the same application. Similarly, if instead of producing bioenergy, corn stover is used solely for producing only biochar, around 793–864 kg $CO_{2eqv}$/tonne of dry corn stover can be displaced (Roberts et al., 2010). Libra et al. (2011) observed a reduction in both $N_2O$ and $CO_2$ emissions after four weeks of applying pyrochar in soil. However, GHG emissions were not lower if the energy crops to be used as feedstock are solely pyrolyzed to produce biochar. When only considering GHG emissions, biochar helps in $CO_2$ removal through sequestration. $CO_2$-activated biochar produces a material with higher adsorption capability and regenerability, and contributes toward reducing $CO_2$ emissions (Rind et al., 1990; Feijoo et al., 2019; Kellogg, 2019). Granular biochar used as an anode in a rice plant microbial fuel cell for producing bioelectricity also demonstrated the potential of biochar in reducing $CH_4$ emissions without decreasing biomass yield (Md Khudzari et al., 2019).

## 5 Biochar market drivers

Biochar current market drivers are mainly led by crop growers, water treatment facilities, remediation of mine areas and brownfields, stormwater, and turf management. Agriculture is the most commonly cited driver for biochar demand. However, there is increasing demand from a broad range of industrial sectors as companies develop a better understanding of biochar uses.

Biochar's market demand growth is largely driven by its ability to adsorb metals, chemicals, and toxins. Wastewater treatment plans employ biochar to reduce waste odors and adsorb different organic and inorganic contaminants. The green building and stormwater management sectors also employ adsorption materials like biochar. The animal husbandry sector employs biochar in bedding and sorption of phosphorous from dairy flush water.

Biochar also has value as a carbon sequestration agent. A future carbon market and revenues earned from carbon offsets could help biochar producers sell it at a much lower price (USBI, 2018). The biochar global market has reached a value of around $484 million in 2019. The market breakup of biochar in terms of application and technology type primarily includes farming, horticulture, water-air-soil treatment plants (application), slow/fast pyrolysis, hydrothermal carbonization, and gasification (technology), respectively. The market value of biochar is predicted to reach around $870.7 million by 2024, cataloging a compound annual growth rate of about 12% (APNews, n.d.).

The recovery of high-value metals is a market driver of increasing importance. In recent years, studies have identified its comparable abilities of adsorption with activated carbon. Biochar is fairly successful in removing heavy metals and other toxic contaminants from municipal and industrial wastewater (Chen et al., 2011a,b; Han et al., 2013; Karim et al., 2015). Alhashimi and Aktas (2017) quantified an average emission of $-0.9$ kg $CO_{2eqv}$/kg for biochar production while the same for activated carbon is 6.6 kg $CO_{2eqv}$/kg. In terms of the environmental impact of biochar as an adsorbent, the energy demand for heavy metals adsorption is strikingly lower than activated carbon. Similar patterns are observed in GHG emissions, where biochar adsorption achieves negative $CO_2$ emissions due to its capability to sequester carbon. Thus, biochar could play a role in increasing the sustainability of green supply chains for metal recovery.

There are a few niche biochar markets that could generate significant revenues for the biochar industry. Biochar can play a role in the human and animal feeding, pharmaceutical, and electronic industries. The high value of products in these industries and their strict quality demands can support high biochar market prices. The future growth of the biochar industry will therefore depend on balancing low- and high-volume market drivers with the supply of high-quality biochar.

## 6 Conclusions

Society has a growing demand for wastewater treatment plants that are technically efficient, cost-effective, and environmentally friendly. Biochar is a candidate for replacing adsorption materials and filters in wastewater treatment plants. It can effectively recover heavy metals, chemicals, and contaminants from wastewater streams. However, wastewater treatment plants must balance the costs of using biochar with its impact on the environment and performance of the facility. This chapter discussed the key techno-economic factors driving biochar adoption in wastewater treatment plants, biochar production, and global biochar markets.

The combination of a wastewater treatment and a biochar production facility can increase the facility's net return when there are markets for the by-products. A wastewater treatment plant combined with a pyrolysis, hydrothermal, or gasification unit generates additional revenues from the sale of bio-oil, biochar, and biogas. This combination can be more profitable than a conventional facility utilizing coal-based activated carbon for wastewater treatment. However, the economic returns will depend on the biochar performance and production costs.

This chapter discussed various techno-economic costs associated with biochar production. Feedstock, capital, and labor costs are among the key contributors to biochar costs. The use of low-case feedstock can significantly reduce biochar costs, but this could also lower biochar quality. Therefore, the biochar production process should consider the target market for the biochar when possible. However, current market prices primarily support the sale of biochar as a biorefinery by-product.

Market projections suggest that there is a growing demand for biochar. Market drivers include agricultural applications, wastewater treatment, and specialty markets such as pharmaceuticals and electronics. These markets are driven by innovations in the production of high-quality biochar, and the desire for replacing commercial materials with greener alternatives. Increasing awareness of the benefits of biochar use in existing markets will help improve the environmental footprint of various industrial processes (Pourhashem et al., 2019). In particular, the use of biochar in wastewater treatment appears to offer significant economic and environmental benefits.

## Questions and problems

1. What is techno-economic analysis and why is it important in the bio-based industry?
2. What is biochar? What role does it play in the bio-based market?
3. What is the use of biochar in a wastewater treatment plant? Does it serve a similar role in other industries? Name two.
4. When we already have activated carbon, what is the purpose of using biochar over activated carbon?
5. What are the primary building blocks of a wastewater treatment plant? Name each with small description. Where do you think we can use biochar in this process?
6. In a wastewater treatment plant, what are the major contributors of capital cost of the plant?
7. Wastewater treatment plants along with retrofit designs like slow and fast pyrolysis have potential if aided by good governmental policies. What do you think about this statement? Explain.
8. What are the bio-based methods of producing biochar? Which one do you think is the best process and why?

9. Do you think feedstock price is one of the impactful parameters toward biochar production and price? Why?
10. Name the primary market drivers of biochar. What will help the biochar industry grow in future?

# References

Abdel-Fattah, T.M., Mahmoud, M.E., Ahmed, S.B., Huff, M.D., Lee, J.W., Kumar, S., 2014. Biochar from woody biomass for removing metal contaminants and carbon sequestration. J. Ind. Eng. Chem. https://doi.org/10.1016/%0Dj.jiec.2014.06.030.

Ahmad, M., Lee, S.S., Rajapaksha, A.U., Vithanage, M., Zhang, M., Cho, J.S., et al., 2013. Trichloroethylene adsorption by pine needle biochars produced at various pyrolysis temperatures. Bioresour. Technol. 143, 615–622. https://doi.org/10.1016/j.biortech.2013.06.033.

Akbari, M., Oyedun, A.O., Kumar, A., 2019. Comparative energy and techno-economic analyses of two different configurations for hydrothermal carbonization of yard waste. Bioresour. Technol. Rep. 7, 100210.

AlAmeri, K., Giwa, A., Yousef, L., Alraeesi, A., Taher, H., 2019. Sorption and removal of crude oil spills from seawater using peat-derived biochar: an optimization study. J. Environ. Manag. 250. https://doi.org/10.1016/j.jenvman.2019.109465, 109465.

Alhashimi, H.A., Aktas, C.B., 2017. Life cycle environmental and economic performance of biochar compared with activated carbon: a meta-analysis. Resour. Conserv. Recycl. 118, 13–26. https://doi.org/10.1016/j.resconrec.2016.11.016.

Ali, A., Raza, R., Shakir, M.I., Iftikhar, A., Alvi, F., Ullah, M.K., et al., 2019. Promising electrochemical study of titanate based anodes in direct carbon fuel cell using walnut and almond shells biochar fuel. J. Power Sources 434. https://doi.org/10.1016/j.jpowsour.2019.05.085, 126679.

Alibaba, 325 Mesh Powdered Activated Carbon Factory Price. Shanghai. Available from: Alibaba.com.

Antal Jr., M.J., Wade, S.R., Nunoura, T., 2007. Biocarbon production from Hungarian sunflower shells. J. Anal. Appl. Pyrolysis 79 (1–2), 86–90. https://doi.org/10.1016/j.jaap.2006.09.005.

APNews, Biochar Market Report 2019-2024 | Global Industry Analysis, Current Trends, Market Share, Size, Growth and Opportunities. Wired Release. Available from: https://apnews.com/press-release/Wired%2520Release/10e8d6f0203bc607c7f1eec745c75f08.

Bartoli, M., Rosi, L., Frediani, M., Frediani, P., 2018. Challenges and Opportunities in the Field of Energy Storage: Supercapacitors and Activated Biochar. Nov Publ Hauppauge.

Batidzirai, B., Mignot, A.P.R., Schakel, W.B., Junginger, H.M., Faaij, A.P.C., 2013. Biomass torrefaction technology: techno-economic status and future prospects. Energy 62, 196–214. https://doi.org/10.1016/j.energy.2013.09.035.

Biochar Supreme (TM) & Black Owl Biochar (TM), Available from: https://www.biocharsupreme.com/products/bob-s-ag-hort.

Brown, T.R., Wright, M.M., Brown, R.C., 2011. Estimating profitability of two biochar production scenarios: slow pyrolysis vs fast pyrolysis. Biofuels Bioprod. Biorefin. 5 (1), 54–68. https://doi.org/10.1002/bbb.

Burk, C., 2018. Techno-economic modeling for new technology development. Chem. Eng. Prog., 43–52.

Chen, B., Chen, Z., Lv, S., 2011a. A novel magnetic biochar efficiently sorbs organic pollutants and phosphate. Bioresour. Technol. 102 (2), 716–723. https://doi.org/10.1016/j.biortech.2010.08.067.

Chen, X., Chen, G., Chen, L., Chen, Y., Lehmann, J., McBride, M.B., et al., 2011b. Adsorption of copper and zinc by biochars produced from pyrolysis of hardwood and corn straw in aqueous solution. Bioresour. Technol. 102 (19), 8877–8884. https://doi.org/10.1016/j.biortech.2011.06.078.

Chen, Z., Chen, B., Chiou, C.T., 2012. Fast and slow rates of naphthalene sorption to biochars produced at different temperatures. Environ. Sci. Technol. 46 (20), 11104–11111.

Child, M., 2014, June. Industrial-scale Hydrothermal Carbonization of Waste Sludge Materials for Fuel. Faculty of Technology Master's Degree Programme in Energy Technology,.

Crini, G., Lichtfouse, E., 2019. Advantages and disadvantages of techniques used for wastewater treatment. Environ. Chem. Lett. 17 (1), 145–155.

Dai, X.H., Fan, H.X., Zhang, J.J., Yuan, S.J., 2019. Sewage sludge-derived porous hollow carbon nanospheres as high-performance anode material for lithium ion batteries. Electrochim. Acta 319, 277–285. https://doi.org/10.1016/j.electacta.2019.07.006.

Das, J., Rene, E.R., Dupont, C., Dufourny, A., Blin, J., van Hullebusch, E.D., 2019. Performance of a compost and biochar packed biofilter for gas-phase hydrogen sulfide removal. Bioresour. Technol. 273, 581–591.

Fagbohungbe, M.O., Herbert, B.M.J., Hurst, L., Ibeto, C.N., Li, H., Usmani, S.Q., et al., 2017. The challenges of anaerobic digestion and the role of biochar in optimizing anaerobic digestion. Waste Manag. 61, 236–249. https://doi.org/10.1016/j.wasman.2016.11.028.

Feijoo, F., Mignone, B.K., Kheshgi, H.S., Hartin, C., McJeon, H., Edmonds, J., 2019. Climate and carbon budget implications of linked future changes in $CO_2$ and non-$CO_2$ forcing. Environ. Res. Lett. 14 (4).

Grycová, B., Koutník, I., Pryszcz, A., 2016. Pyrolysis process for the treatment of food waste. Bioresour. Technol. 218, 1203–1207.

Han, Y., Boateng, A.A., Qi, P.X., Lima, I.M., Chang, J., 2013. Heavy metal and phenol adsorptive properties of biochars from pyrolyzed switchgrass and woody biomass in correlation with surface properties. J. Environ. Manag. 118, 196–204. https://doi.org/10.1016/j.jenvman.2013.01.001.

Harsono, S.S., Grundman, P., Lau, L.H., Hansen, A., Salleh, M.A.M., Meyer-Aurich, A., et al., 2013. Energy balances, greenhouse gas emissions and economics of biochar production from palm oil empty fruit bunches. Resour. Conserv. Recycl. 77, 108–115. https://doi.org/10.1016/j.resconrec.2013.04.005.

Huggins, T., Wang, H., Kearns, J., Jenkins, P., Ren, Z.J., 2014. Biochar as a sustainable electrode material for electricity production in microbial fuel cells. Bioresour. Technol. 157, 114–119. https://doi.org/10.1016/j.biortech.2014.01.058.

Jorapur, R., Rajvanshi, A.K., 1997. Sugarcane leaf-bagasse gasifiers for industrial heating applications. Biomass Bioenergy 13 (3), 141–146.

Karim, A., Kumar, M., Mohapatra, S., Panda, C., Singh, A., 2015. Banana peduncle biochar: characteristics and adsorption of hexavalent chromium from aqueous solution. Int. Res. J. Pure Appl. Chem. 7 (1), 1–10.

Kellogg, W.W., 2019. Climate Change and Society: Consequences of Increasing Atmospheric Carbon Dioxide. Routledge, Abingdon-on-Thames, https://doi.org/10.4324/9780429048739.

Kim, W.K., Shim, T., Kim, Y.S., Hyun, S., Ryu, C., Park, Y.K., et al., 2013. Characterization of cadmium removal from aqueous solution by biochar produced from a giant Miscanthus at different pyrolytic temperatures. Bioresour. Technol. 138, 266–270. https://doi.org/10.1016/j.biortech.2013.03.186.

Kou, X.R., Cheng, X.S., Hua, Y.J., Zhen, Z.A., 2011. Adsorption of methyl violet from aqueous solutions by the biochars derived from crop residues. Bioresour. Technol. 102 (22), 10293–10298. https://doi.org/10.1016/j.biortech.2011.08.089.

Langholtz, M., Stokes, B., Eaton, L., 2016. 2016 Billion-ton report: advancing domestic resources for a thriving bioeconomy (executive summary). Ind. Biotechnol. 12 (5), 282–289.

Lee, H., Yang, S., Wi, S., Kim, S., 2019. Thermal transfer behavior of biochar-natural inorganic clay composite for building envelope insulation. Constr. Build. Mater. 223, 668–678. https://doi.org/10.1016/j.conbuildmat.2019.06.215.

Li, W., Dumortier, J., Dokoohaki, H., Miguez, F.E., Brown, R.C., Laird, D., et al., 2019. Regional techno-economic and life-cycle analysis of the pyrolysis-bioenergy-biochar platform for carbon-negative energy. Biofuels Bioprod. Biorefin. 13 (6), 1428–1438.

Libra, J.A., Ro, K.S., Kammann, C., Funke, A., Berge, N.D., Neubauer, Y., et al., 2011. Hydrothermal carbonization of biomass residuals: a comparative review of the chemistry, processes and applications of wet and dry pyrolysis. Biofuels 2 (1), 71–106.

Liu, Z., Wang, Z., Chen, H., Cai, T., Liu, Z., 2021. Hydrochar and pyrochar for sorption of pollutants in wastewater and exhaust gas: a critical review. Environ. Pollut. 268. https://doi.org/10.1016/j.envpol.2020.115910, 115910.

Lu, H., Zhang, W., Yang, Y., Huang, X., Wang, S., Qiu, R., 2012. Relative distribution of $Pb^{2+}$ sorption mechanisms by sludge-derived biochar. Water Res. 46 (3), 854–862. https://doi.org/10.1016/j.watres.2011.11.058.

Lucian, M., Fiori, L., 2017. Hydrothermal carbonization of waste biomass: process design, modeling, energy efficiency and cost analysis. Energies 10 (2).

Luo, W., Wang, B., Heron, C.G., Allen, M.J., Morre, J., Maier, C.S., et al., 2014. Pyrolysis of cellulose under ammonia leads to nitrogen-doped nanoporous carbon generated through methane formation. Nano Lett. 14 (4), 2225–2229.

Mahmood, R., Parshetti, G.K., Balasubramanian, R., 2016. Energy, exergy and techno-economic analyses of hydrothermal oxidation of food waste to produce hydro-char and bio-oil. Energy 102, 187–198. https://doi.org/10.1016/j.energy.2016.02.042.

Md Khudzari, J., Gariépy, Y., Kurian, J., Tartakovsky, B., Raghavan, G.S.V., 2019. Effects of biochar anodes in rice plant microbial fuel cells on the production of bioelectricity, biomass, and methane. Biochem. Eng. J. 141, 190–199. https://doi.org/10.1016/j.bej.2018.10.012.

Meyer, S., Glaser, B., Quicker, P., 2011. Technical, economical, and climate-related aspects of biochar production technologies: a literature review. Environ. Sci. Technol. 45 (22), 9473–9483.

Meyer, P.A., Snowden-Swan, L.J., Jones, S.B., Rappé, K.G., Hartley, D.S., 2020. The effect of feedstock composition on fast pyrolysis and upgrading to transportation fuels: techno-economic analysis and greenhouse gas life cycle analysis. Fuel 259. https://doi.org/10.1016/j.fuel.2019.116218, 116218.

Miles, T., 2014. Consultants TRMT. of the US Biochar Industry.

Mu, B., Wang, A., 2019. Fabrication and applications of carbon/clay mineral nanocomposites. In: Nanomaterials from Clay Minerals. Elsevier, pp. 537–587, https://doi.org/10.1016/B978-0-12-814533-3.00011-9.

Oh, T.K., Choi, B., Shinogi, Y., Chikushi, J., 2012. Effect of pH conditions on actual and apparent fluoride adsorption by biochar in aqueous phase. Water Air Soil Pollut. 223 (7), 3729–3738.

Pacific Biochar, 2020. Bulk Price Sheet for Fall 2020. Available from: https://pacificbiochar.com/biochar-price-sheet/.

Pan, X., Gu, Z., Chen, W., Li, Q., 2021. Preparation of biochar and biochar composites and their application in a Fenton-like process for wastewater decontamination: a review. Sci. Total Environ. 754. https://doi.org/10.1016/j.scitotenv.2020.142104, 142104.

Peters, M.S., 1991. In: Peters, M.S., Timmerhaus, K.D. (Eds.), Plant Design and Economics for Chemical Engineers, fourth ed. McGraw-Hill Chemical Engineering Series, McGraw-Hill, New York.

Pirraglia, A., Gonzalez, R., Saloni, D., Denig, J., 2013. Technical and economic assessment for the production of torrefied ligno-cellulosic biomass pellets in the US. Energy Convers. Manag. 66, 153–164. https://doi.org/10.1016/j.enconman.2012.09.024.

Pourhashem, G., Hung, S.Y., Medlock, K.B., Masiello, C.A., 2019. Policy support for biochar: review and recommendations. GCB Bioenergy 11 (2), 364–380.

Razzaghi, F., Obour, P.B., Arthur, E., 2020. Does biochar improve soil water retention? A systematic review and meta-analysis. Geoderma 361. https://doi.org/10.1016/j.geoderma.2019.114055, 114055.

Rind, D., Suozzo, R., Balachandran, N.K., Prather, M.J., 1990. Climate change and the middle atmosphere. Part I: the doubled $CO_2$ climate. J. Atmos. Sci. 4 (47), 475–494. https://doi.org/10.1175/1520-0469(1990)047%3C0475:CCATMA%3E2.0.CO;2.

Roberts, K.G., Gloy, B.A., Joseph, S., Scott, N.R., Lehmann, J., 2010. Life cycle assessment of biochar systems: estimating the energetic, economic, and climate change potential. Environ. Sci. Technol. 44 (2), 827–833.

Rover, M.R., Aui, A., Wright, M.M., Smith, R.G., Brown, R.C., 2019. Production and purification of crystallized levoglucosan from pyrolysis of lignocellulosic biomass. Green Chem. 21 (21), 5980–5989.

Shabangu, S., Woolf, D., Fisher, E.M., Angenent, L.T., Lehmann, J., 2014. Techno-economic assessment of biomass slow pyrolysis into different biochar and methanol concepts. Fuel 117 (Part A), 742–748. https://doi.org/10.1016/j.fuel.2013.08.053.

Shackley, S., Clare, A., Joseph, S., McCarl, B.A., Schmidt, H.P., 2015. Economic Evaluation of Biochar Systems: Current Evidence and Challenges.

Shao, Y., Long, Y., Wang, H., Liu, D., Shen, D., Chen, T., 2019. Hydrochar derived from green waste by microwave hydrothermal carbonization. Renew. Energy 135, 1327–1334. https://doi.org/10.1016/j.renene.2018.09.041.

Steinwinder, T., Gill, E., Gerhardt, M., 2011. Process Design of Wastewater Treatment for the NREL Cellulosic Ethanol Model. NREL. (September). Available from: http://www.brownandcaldwell.com/Tech_Papers/TP 1331 Process Design of Wastewater Treatment.pdf.

Stemann, J., Erlach, B., Ziegler, F., 2013. Hydrothermal carbonisation of empty palm oil fruit bunches: Laboratory trials, plant simulation, carbon avoidance, and economic feasibility. Waste Biomass Valorization 4 (3), 441–454. https://doi.org/10.1007/s12649-012-9190-y.

Sun, K., Kang, M., Zhang, Z., Jin, J., Wang, Z., Pan, Z., et al., 2013. Impact of deashing treatment on biochar structural properties and potential sorption mechanisms of phenanthrene. Environ. Sci. Technol. 47 (20), 11473–11481.

Svanberg, M., Olofsson, I., Flodén, J., Nordin, A., 2013. Analysing biomass torrefaction supply chain costs. Bioresour. Technol. 142, 287–296. https://doi.org/10.1016/j.biortech.2013.05.048.

Tan, X., Liu, Y., Zeng, G., Wang, X., Hu, X., Gu, Y., et al., 2015. Application of biochar for the removal of pollutants from aqueous solutions. Chemosphere 125, 70–85. https://doi.org/10.1016/j.chemosphere.2014.12.058.

Tian, X., Richardson, R.E., Tester, J.W., Lozano, J.L., You, F., 2020. Retrofitting municipal wastewater treatment facilities toward a greener and circular economy by virtue of resource recovery: techno-economic analysis and life cycle assessment. ACS Sustain. Chem. Eng. 8 (36), 13823–13837.

USBI, 2018. Council of Western State Foresters Biochar Market Analysis—Final Report.

Wright, M.M., Daugaard, D.E., Satrio, J.A., Brown, R.C., 2010. Techno-economic analysis of biomass fast pyrolysis to transportation fuels. Fuel 89 (Suppl. 1), S2–10. https://doi.org/10.1016/j.fuel.2010.07.029.

Xu, X., Cao, X., Zhao, L., 2013. Comparison of rice husk- and dairy manure-derived biochars for simultaneously removing heavy metals from aqueous solutions: role of mineral components in biochars. Chemosphere 92 (8), 955–961. https://doi.org/10.1016/j.chemosphere.2013.03.009.

Yadav, K., Jagadevan, S., 2019. Influence of process parameters on synthesis of biochar by pyrolysis of biomass: an alternative source of energy. In: Recent Advances in Pyrolysis. IntechOpen. Available from: https://www.intechopen.com/chapters/68186.

Yao, S., Li, X., Cheng, H., Zhang, C., Bian, Y., Jiang, X., et al., 2019. Resource utilization of a typical vegetable waste as biochars in removing phthalate acid esters from water: a sorption case study. Bioresour. Technol. 293 (71). https://doi.org/10.1016/j.biortech.2019.122081, 122081.

Zeymer, M., Meisel, K., Clemens, A., Klemm, M., 2017. Technical, economic, and environmental assessment of the hydrothermal carbonization of green waste. Chem. Eng. Technol. 40 (2), 260–269.

Zhang, Z.-B., Cao, X.-H., Liang, P., Liu, Y.-H., 2013a. Adsorption of uranium from aqueous solution using biochar produced by hydrothermal carbonization. J. Radioanal. Nucl. Chem. 295, 1201–1208.

Zhang, Y., Brown, T.R., Hu, G., Brown, R.C., 2013b. Techno-economic analysis of monosaccharide production via fast pyrolysis of lignocellulose. Bioresour. Technol. 127, 358–365. https://doi.org/10.1016/j.biortech.2012.09.070.

Zhang, H., Chen, C., Gray, E.M., Boyd, S.E., 2017. Effect of feedstock and pyrolysis temperature on properties of biochar governing end use efficacy. Biomass Bioenergy 105, 136–146. https://doi.org/10.1016/j.biombioe.2017.06.024.

# Retention of oxyanions on biochar surface

Santanu Bakshi[a], Rivka Fidel[b], Chumki Banik[a,d],
Deborah Aller[c,*], and Robert C. Brown[a]

[a]Bioeconomy Institute, Iowa State University, Ames, IA, United States,
[b]Department of Environmental Science, University of Arizona, Tucson, AZ, United States, [c]Agriculture Program, Cornell Cooperative Extension of Suffolk County, Riverhead, NY, United States, [d]Department of Agricultural and Biosystems Engineering, Iowa State University, Ames, IA, United States

## 1  Introduction

Biochars present a potential solution to the widespread contamination of wastewater with various oxyanions, reducing the threat to the water quality of both natural water bodies and drinking water (Xiang et al., 2020). Collectively, oxyanions present unique challenges to wastewater treatment, because (a) they are commonly repelled by negatively charged sorbents and natural porous media (e.g., soil and sediments), (b) they often persist after traditional biological wastewater treatments, and (c) oxyanions adsorb to media via unique mechanisms, making it difficult for one sorbent to remove them all (Shah et al., 2020; Villarín and Merel, 2020). Here, we explore the possibility of biochar retaining three particularly widespread and problematic oxyanions: arsenate ($AsO_4^{3-}$), phosphate ($PO_4^{3-}$), and nitrate ($NO_3^-$).

Biochars may prove potentially advantageous over other oxyanion sorbents due to: (a) lower cost and subsequent greater accessibility, (b) greater capacity for customization and optimization, (c) potential for reuse, and d) potential synergies with other sectors (Maroušek, 2014; Qambrani et al., 2017; Xiang et al., 2020; Yang et al., 2020). Because biochars can be produced from a wide variety of locally available feedstocks, producers can choose less expensive feedstock options, including waste materials (De Gisi et al., 2016; Inyang et al., 2016). This in turn enables the

---

*Present address: Section of Crop and Soil Sciences, School of Integrative Plant Science, Cornell University, Ithaca, NY, United States.

production of biochars specifically designed to retain particular oxyanions, using a combination of (1) inherent feedstock properties, (2) feedstock pretreatments, (3) pyrolysis conditions, and (4) postpyrolysis biochar treatments. This chapter examines the properties of biochars that enable their retention of the three oxyanions of interest and discusses how production and modification parameters affect biochars' retention of these oxyanions.

## 2 Motivation for removing oxyanions of interest

**Arsenate** ($AsO_4^{3-}$) contamination is of particular concern, as arsenic (As) is an acute metabolic poison of both geologic and anthropogenic origin. It is recognized as one of the world's greatest environmental hazards, representing a threat to the lives of hundreds of millions of people (Ravenscroft et al., 2009). Naturally occurring As, thought to originate from arsenite-bearing minerals, is found in natural bodies of water globally. Typically, the effect of underlying bedrock is dwarfed by the effect of biochemical surface processes that concentrate and mobilize As in soils, sediments, and consequently groundwater. Reducing and alkaline conditions, in particular, favor dissolution and hence greater mobility of As. Anthropogenic As contamination originates primarily from industrial processes, such as metal smelting; glassmaking; and the production of pigments, leather, textiles, paper, wood preservatives, and ammunition (Ravenscroft et al., 2009; Yadav et al., 2021). It is a widespread natural groundwater contaminant, which is particularly concerning since most of the global population relies on groundwater for drinking water and irrigation water. Groundwater As concentrations are known to exceed the World Health Organization's recommended drinking water limit of 10 μg/L in specific regions within many countries, including Argentina, Bangladesh, Chile, China, India, Mexico, Mongolia, Nepal, Pakistan, and the United States (Nordstrom, 2002; Yadav et al., 2021). The population of Bangladesh is thought to face the greatest threat from As contamination with about 48% of its population (~77 million people) exposed to As concentrations >10 μg/L (WHO limit), and about 21% (~35 million people) exposed to concentrations >50 μg/L (the Bangladesh government limit) (Uppal et al., 2019; Yadav et al., 2021). Within the USA, the southwest region has the greatest likelihood of groundwater As contamination, including Arizona, California, New Mexico, Nevada, and Utah (Ravenscroft et al., 2009). Thus, As contamination presents a truly widespread threat to human health, and developing improved methods for its removal is of critical importance.

The chemical form of As affects its toxicity, mobility, and approach to mitigation. Under oxic and suboxic conditions, the highly toxic arsenite (As[III], i.e., with an oxidation state of +3, found in the hydroxyl anion $As(OH)_3$) is oxidized to less toxic and less mobile arsenate (As[V], i.e., with an oxidation state of +5, found in the oxyanion $AsO_4^{3-}$) (Pepper et al., 2006; Yadav et al., 2021). Although various biochars could theoretically be used to remove As in multiple forms, here we focus on using biochar to remove As V in its various protonated and deprotonated forms, which shall be collectively referred to as arsenate or $AsO_4^{3-}$ ($AsO_4^{3-}$, $HAsO_4^{2-}$, $H_2AsO_4^-$, and $H_3AsO_4^0$). The many forms of As present a special challenge, as sorption mechanisms and hence optimal removal strategies differ between oxidation states and various protonated/deprotonated forms. Due to arsenate's greater adsorption and precipitation tendencies, many As removal methods rely first on oxidizing As(III) to As(V). Subsequently, the most common treatments for the resulting $AsO_4^{3-}$ are (a) precipitation with Fe and/or Mn and (b) adsorption and/or precipitation onto filtration media. Filtration media commonly include sorbents such as metal (Al, Fe, Mn, etc.) oxides, sulfur-modified iron, iron-activated alumina, modified zeolite, and metal-impregnated activated carbons (Hristovski and Markovski, 2017; Lal et al., 2020; Liu et al., 2020; Weerasundara et al., 2021). The cost of filtration media is commonly a limiting factor preventing the adoption of effective As removal techniques, especially in less affluent regions. Therefore, lower cost $AsO_4^{3-}$ sorbents such as biochar produced from relatively inexpensive organic waste materials are of special interest.

**Phosphate** ($PO_4^{3-}$) is chemically similar to arsenate, but its sources and impacts differ. Phosphorous-rich rocks, commonly containing apatite, are relatively rare. Hence, phosphate contamination is typical of anthropogenic origin. An essential component in ubiquitous biomolecules such as DNA and ATP, phosphate, and phosphate-bearing organic compounds are found in virtually all biomaterials, including biowastes. Biowastes such as municipal or household sewage and manure lagoons are common point sources of $PO_4^{3-}$ pollution. Phosphate and compounds that degrade into $PO_4^{3-}$ are also common in detergents and industrial by-products, which constitute another point source of $PO_4^{3-}$ pollution (Fig. 7.1) (Zhang et al., 2020). Municipal wastewater effluent tends to be the dominant source of P pollution in urban watersheds. The application of P-rich biowastes as well as $PO_4^{3-}$ fertilizers to soils is a common nonpoint source of $PO_4^{3-}$ pollution (Pepper et al., 2006). These nonpoint sources are much more difficult to detect and hence regulate and are therefore more likely to

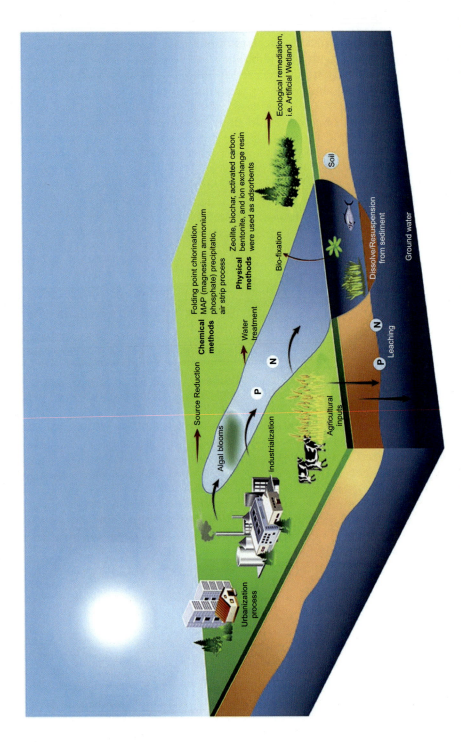

**Fig. 7.1** Source and control of nutrients (N and P) in the water environment. Reprinted from Zhang, M., Song, G., Gelardi, D.L., Huang, L., Khan, E., Mašek, O., Parikh, S.J., Ok, Y.S., 2020. Evaluating biochar and its modifications for the removal of ammonium, nitrate, and phosphate in water. Water Res. 186, 116303. https://doi.org/10.1016/j.watres.2020.116303 with permission of the publisher.

continue affecting water quality into the future, especially in agricultural watersheds. While $PO_4^{3-}$ is not very mobile in soil, its relatively low plant availability results in its frequent overapplication to the soil. Frequently, farmers also apply manure based on N needs, which results in overapplication of P. Once saturated with $PO_4^{3-}$ due to overapplication, agricultural soils become a net source of $PO_4^{3-}$. From 2002 to 2010, the total anthropogenic load to freshwater bodies was estimated at 1.5 Tg of P. Of this, 54% originated from the domestic sector, 38% from agriculture, and 8% from industry (Mekonnen and Hoekstra, 2018). Although not toxic to humans at typical wastewater concentrations, $PO_4^{3-}$ contamination is responsible for the widespread eutrophication of freshwater bodies. Eutrophication can be triggered at relatively low P concentrations, just 0.03 mg/L of dissolved P and 0.1 mg/L total P (Chambers et al., 2012). Soil amendment with biowastes should not stop entirely, however, as (a) P is necessary for crop growth and (b) rock P reserves are nonrenewable and may be exhausted as soon as 2033 (Rhodes, 2013). Consequently, there is a need to recycle and reuse P in a manner that does not exacerbate pollution. Capturing P from wastewater and applying it to the soil, for example, may constitute an important component of sustainable P management (Neset and Cordell, 2012). Therefore, the removal of P from wastewater is of utmost importance for protecting freshwater ecosystems and may enable the reuse of this scarce resource.

**Nitrate** ($NO_3^-$) is the most soluble of the aforementioned oxyanions and therefore presents a unique threat to water bodies and drinking water (Zhang et al., 2020, 2021). Nitrate is a contaminant concern because it can cause blue baby syndrome as well as eutrophication of saltwater bodies (Holland et al., 1999; Howarth, 2008). Nitrate can originate from biological as well as anthropogenic sources, the latter of which has increased dramatically since World War II and the subsequent green revolution (Fig. 7.1) (Holland et al., 1999). Ammonia ($NH_3$) fixed from the atmosphere through the Haber-Bosch process can be easily protonated to ammonium ($NH_4^+$), which can then be oxidized through industrial reactions into nitrate fertilizers, and/or applied directly to soil as $NH_4^+$. Once in the soil, $NH_4^+$ is oxidized under aerobic conditions into $NO_3^-$. Nitrate is ubiquitous in low concentrations in soils, sediments, and natural water bodies (Ascott et al., 2017). However, leaching and runoff of $NO_3^-$ from agricultural soils has dramatically increased the prevalence of $NO_3^-$ in natural water bodies, including streams and groundwater used as municipal and household drinking water sources in agricultural regions such as the central United States, China, and Pakistan. In addition to contaminating groundwater, $NO_3^-$ has caused eutrophication and hypoxia in coastal saltwater bodies such as the Louisiana coastline adjacent to the Gulf of Mexico (Holland et al., 1999; Howarth, 2008; US EPA, 2020). Unfortunately, $NO_3^-$ has proven particularly difficult to remove cost-effectively, mainly due to: (a) low sorption capacity of typical sorbents such as activated carbon for $NO_3^-$; (b) exchangeability of $NO_3^-$ with other anions such as sulfate and chloride on sorbents, resulting in competition for sorption sites and re-contamination of waste; and (c) biological removal methods require an anaerobic environment—either a constructed wetland or a separate, anaerobic treatment phase at treatment plants—adding to up-front costs (Archna et al., 2012; Du et al., 2019). Modified biochars, on the contrary, could potentially remove $NO_3^-$ in a more economical manner by alleviating these issues, especially the improvement of sorption capacity among sorbent-based approaches.

## 3 Chemical properties of biochar influencing oxyanion sorption

Biochars in their unmodified (also referred to as "untreated" or "pristine") state contain organic surface functional groups as well as inorganic compounds that may facilitate the retention of oxyanions from water (Fig. 7.2). The former enables oxyanion retention via ion exchange and/or ligand exchange, whereas the latter is more likely to facilitate precipitation of oxyanions onto biochar surfaces (Bakshi et al., 2021; Eduah et al., 2020; Zhou et al., 2017b). While most studies cite relatively low retention rates of oxyanions by unmodified biochars, understanding mechanisms of retention help inform the design of both modified and unmodified biochars with greater capacity for oxyanion retention, as well as filtration systems to house them in Wang et al. (2020a, b).

Biochar surface functional groups are predominately oxygen-containing weak acids and their conjugate bases. These groups carry a pH-dependent charge that becomes negative at high pHs, thereby repelling anions. When the pH exceeds the p$K_a$s of these groups, unmodified biochar has relatively low anion exchange capacity and thus low retention of oxyanions. Conversely, as pH decreases, the functional groups become protonated and develop a positive charge capable of attracting anions and thus increasing the anion exchange capacity (AEC) of the biochar (Lawrinenko and Laird, 2015). Indeed, Lawrinenko and Laird (2015) documented increasing AEC with decreasing pH. At any

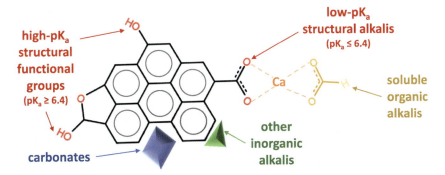

**Fig. 7.2** Hypothetical biochar structure, showing condensed aromatic framework, for example, functional groups of various $pK_a$s and inorganic ash components (carbonates and other inorganic alkalis). The soluble organic alkali is adsorbing to the biochar surface through cation bridging, a mechanism that inorganic oxyanions may also participate in. Reprinted from Fidel, R.B., Laird, D.A., Thompson, M.L., Lawrinenko, M., 2017. Characterization and quantification of biochar alkalinity. Chemosphere 167, 367–373 with permission of the publisher.

given pH, the AEC then depends on the $pK_a$ distribution of the surface functional groups. A greater concentration of low-$pK_a$ functional groups such as carboxylic acids leads to a lower PZNC (point of zero net charge) and hence lower AEC. A greater concentration of high-$pK_a$ functional groups, such as phenols or amines, would lead to a higher PZNC and AEC (Banik et al., 2018). This $pK_a$ distribution varies for biochar feedstock and pyrolysis conditions (Fig. 7.3). Another potential source of positive charge is O- and N-containing heterocycles, which carry a formal positive charge even without a proton adsorbed. Thus, it is expected that biochars with greater concentrations of high-$pK_a$ functional groups and heterocycles would adsorb greater quantities of oxyanions.

In addition to organic functional groups, the inorganic (ash) components of biochar are relevant to oxyanion retention, as this can (a) coprecipitate with oxyanions, (b) facilitate sorption through other means such as cation bridging, and (c) influence the solution pH adjacent to biochar (Fidel et al., 2017; Sanford et al., 2019). Coprecipitation is a likely mechanism for retention of anions that form salts of low solubility; $PO_4^{3-}$ and $AsO_4^{3-}$, for example, are likely to precipitate with $Al^{3+}$ as well as divalent base cations like $Ca^{2+}$ and $Mg^{2+}$, and transition metals such as $Fe^{3+}$ and $Mn^{2+}$ (Bakshi et al., 2018; Eduah et al., 2020; Li et al., 2019; Tan et al., 2016). In addition to precipitation mechanisms, multivalent cations could potentially facilitate bridging between negatively charged biochar surfaces and oxyanions (Sanford et al., 2019). Such metals and metalloids can commonly be found in biochar

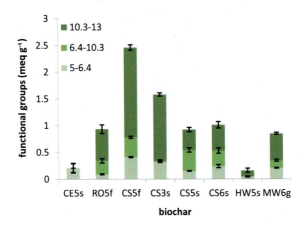

**Fig. 7.3** Surface organic functional group concentrations in discrete p$K_a$ ranges (5–6.4, 6.4–10.3, and 10.3–13), as quantified using the Boehm titration. Error bars represent the standard deviation of the mean ($n=3$). From left to right, biochar feedstocks with their corresponding temperatures (and pyrolysis rate/type) are cellulose at 500°C (slow), red oak at 500°C (fast), corn stover at 500°C (fast); corn stover at 300°C, 500°C, and 600°C (slow); hardwood at 500°C (slow); and mixed wood at 600°C. Reprinted from Fidel, R.B., Laird, D.A., Thompson, M.L., Lawrinenko, M., 2017. Characterization and quantification of biochar alkalinity. Chemosphere 167, 367–373 with permission of the publisher.

feedstocks or added to biochars to obtain desired properties. Furthermore, base cations in biochars are often associated with alkaline anions such as $CO_3^{2-}$, $OH^-$, and $O^{2-}$. These anions can raise the pH of biochar to about 8–8.5 when paired with divalent base cations, or upward of pH 12 when paired with monovalent base cations (Fidel et al., 2017). Such high pH conditions can then facilitate precipitation of $PO_4^{3-}$ and $AsO_4^{3-}$ with $Ca^{2+}$ and $Mg^{2+}$ and promote a more negative surface charge. This latter effect can cause repulsion of oxyanions by surface functional groups, potentially counteracting to a degree the positive effects of the aforementioned cations on oxyanion retention (Eduah et al., 2020; Fidel et al., 2018). Thus, inorganic components of biochar are expected to have interacted effects with each other and with biochar's organic components.

The interacting effects of p$K_a$ distribution, PZNC, and pH must be considered when comparing sorption capacities of biochar across studies, and when designing filtration systems for the removal of oxyanions. This will help identify biochar production and modification parameters that optimize concentrations of both organic surface functional groups and inorganic (ash) components of biochar, thereby enhancing the resulting biochars' capacity to remove oxyanions from water.

## 4 Feedstock and pyrolysis temperatures influence on surface chemistry and physical properties of biochar

Biomass feedstock and/or pyrolysis temperature (T) exerts a strong influence over biochar surface chemistry, which in turn impacts ion sorption, causing ion sorption capacities to vary widely with production parameters (Ascough et al., 2008; Banik et al., 2018; Keiluweit et al., 2010; Lawrinenko and Laird, 2015; Mukherjee et al., 2011; Uchimiya et al., 2011). Biomass feedstocks are composed of varying relative concentrations of cellulose, hemicellulose, lignin, and minerals, and these fractions undergo differential fragmentation as a function of pyrolysis T. As pyrolysis T rises, biomass cellulose, hemicellulose, and lignin are dehydrated and depolymerized (Keiluweit et al., 2010; Uchimiya et al., 2011), resulting in a loss of functional groups that otherwise could have contributed toward biochars' negative surface charge. Besides, the changes in biomass feedstock and pyrolysis T promote a wide variation in the yield of biochar by changing the molecular structure and pore size distribution, thus causing wide variation in biochar surface properties (Hassan et al., 2020). Since oxyanion sorption onto biochar surface occurs mainly through anion exchange, ligand exchange, and surface functional group sites, understanding the influence of feedstock and pyrolysis T is critical for optimizing specific oxyanion sorption and the specific application of biochar for water and wastewater treatment.

**Pyrolysis T:** For all other factors equal, increasing pyrolysis T strongly and consistently increases the degree of aromaticity of biochar C (Hassan et al., 2020), whereas H and O content decrease. This results in relatively high cation exchange capacity (CEC) for biochars produced at low pyrolysis T and high anion exchange capacity (AEC) for biochars produced at high pyrolysis T (Lawrinenko and Laird, 2015). The origin of positive surface charges contributing to high AEC biochars remains obscure, with some researchers attributing it to O or N heterocycles and/or $\pi$-$\pi$ $e^-$ interactions of the biochar aromatic carbon (Lawrinenko et al., 2016; Lawrinenko and Laird, 2015). This work has recently been supported by Banik et al. (2018) who reported a diverse number of functional groups to contribute to the biochar surface charge which are prevalent on biochar surfaces produced at low (400°C or low), moderate (500–700°C), and high (700–900°C) pyrolysis T. Biochar from three feedstocks (red oak, corn stover, and cellulose) at low pyrolysis T exhibited a high concentration of carboxyl

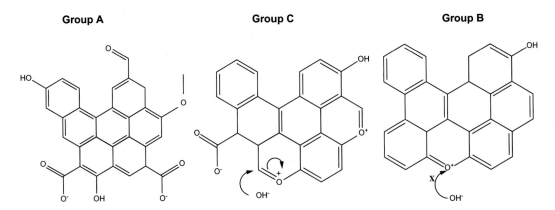

**Fig. 7.4** Different functional groups prevalent on biochar surface produced at different pyrolysis T, Group A (pyrolysis T 400°C or low: carboxyl or hydroxyl); Group B (pyrolysis T 700–900°C: bridging oxonium groups and hydroxyl); and Group C (pyrolysis T 500–700°C: nonbridging oxonium groups). Reprinted from Banik, C., Lawrinenko, M., Bakshi, S., Laird, D.A., 2018. Impact of pyrolysis temperature and feedstock on surface charge and functional group chemistry of biochars. J. Environ. Qual. 47, 452–461 with permission of the publisher.

groups (Fig. 7.4: Group A), whereas the high T biochars exhibited a relatively high concentration of bridging oxonium groups (Fig. 7.4: Group C), and biochar produced at moderate pyrolysis T (500–700°C) contains nonbridging oxonium groups (Fig. 7.4: Group B). The carboxyl groups contributed to CEC, whereas the bridging oxonium groups contributed to AEC. These observations indicate that the higher T biochars were more suitable for removing oxyanions. Notably, the moderate pyrolysis T biochars were shown to be capable of sorbing moderate amounts of both anions and cations, but not optimized for sorbing either. However, the trend of increasing AEC with increasing pyrolysis T does not continue indefinitely: a pyrolysis T of about 700°C appears to be optimal; beyond that, the extent of condensation of the aromatic C occurs, where a normal mode of structural vibration is challenging (Lawrinenko and Laird, 2015).

Since negatively charged biochar surfaces of low T biochars repel oxyanions, biochar produced at low pyrolysis T is a poor sorbent of oxyanions relative to that of high T biochars (Alsewaileh et al., 2019; Park et al., 2018). This trend appears relatively consistent in the literature, especially when sorption is measured at lower pHs. The effect of pH is likely an effect of sorption site acidity. As pH increases, acidic functional groups lose protons and develop more negative charges that repel oxyanions (Chintala et al., 2013; Fidel et al., 2018). Thus, as pyrolysis T increases and pH decreases, biochar becomes more effective at retaining the oxyanions of interest: nitrate ($NO_3^-$), phosphate ($PO_4^{3-}$), and arsenate ($AsO_4^{3-}$).

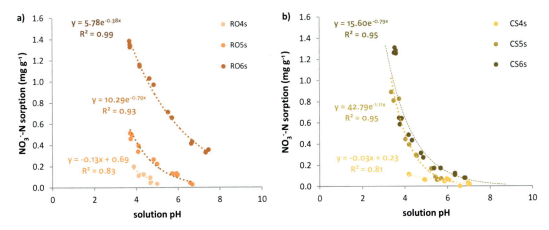

**Fig. 7.5** Sorption of $NO_3^-$ to acid-washed (A) red oak and (B) corn stover biochars from $10\,mg\,N\,L^{-1}$ solutions adjusted to various pHs using $Ca(OH)_2$. Reprinted from Fidel, R.B., Laird, D.A., Spokas, K.A., 2018. Sorption of ammonium and nitrate to biochars is electrostatic and pH-dependent. Sci. Rep. 8, 17627. https://doi.org/10.1038/s41598-018-35534-w.

Similar to $NO_3^-$ sorption on soil organic matter, sorption of $NO_3^-$ on biochar likely occurs through a predominately electrostatic exchange mechanism, which explains the effect of biochar pyrolysis T. As pyrolysis T increases, the shift from lower to higher $pK_a$ functional groups increases to the point of zero net charge (PZNC) and consequently developing a more positive (or less negative) charge at a given pH (Banik et al., 2018). All other factors remaining equal, an increase in pyrolysis T should increase $NO_3^-$ sorption. Indeed, this has been observed in several studies detailed later. However, because the functional groups to which $NO_3^-$ adsorbs are primarily acidic, this effect is pH-dependent. Indeed, both Chintala et al. (2013) and Fidel et al. (2018) document increased $NO_3^-$ sorption with decreasing pH (Fig. 7.5). This trend is explained by the protonation of functional groups with decreasing pH and a subsequent increase in positive surface charge (and a decrease in negative surface charge). These studies also documented the displacement of $NO_3^-$ by competing anions, supporting an electrostatic exchange mechanism with a high likelihood of pH sensitivity. Lastly, it should be noted that this pH effect means that the positive correlation between pyrolysis T and $NO_3^-$ sorption could be counteracted by the positive correlation between biochar pH and T. Thus, pH must be considered when examining the positive correlation between $NO_3^-$ sorption and pyrolysis T.

Research consistently finds a positive correlation between $NO_3^-$ sorption and pyrolysis T, especially when pH is controlled.

As an example, $NO_3^-$ was found to be sorbed more readily on date palm biochar produced at 700°C vs 300°C (Alsewaileh et al., 2019). The high surface area and net positive surface charge at relatively high basicity of the high T date palm biochar is the responsible factor. Additionally, biochars produced at 600°C from four different feedstocks (sugarcane bagasse, peanut hull, Brazilian pepperwood, and bamboo) were found to be capable of removing $NO_3^-$ from water, whereas biochar produced from the same feedstocks at low pyrolysis T (300°C and 450°C) released $NO_3^-$ (Yao et al., 2012). These results suggest that high pyrolysis T influences biochar surface area and increases anion sorption sites that affect $NO_3^-$ removal; however, low pyrolysis T biochar with more negatively charged surfaces may even release $NO_3^-$ from biochar (indicating that the pristine biochar itself contained nitrate). Thus, the effect of T on $NO_3^-$ sorption may be attributed to both surface functional groups and surface area. Another $NO_3^-$ sorption study reported increasing sorption rates for corn stover and red oak biochars as T increased from 400°C to 500°C and 600°C, this time controlling for pH (Fidel et al., 2018). In this study, adjusting the solution pH from 6 down to 3.7 more than doubled $NO_3^-$ sorption to all biochars (Fig. 7.5), and at a pH of 3.7, the effect of T became more pronounced in Fig. 7.5.

Among the corn stover biochars, $NO_3^-$ sorption at pH 3.7 doubled for each 100 °C increase in pyrolysis T (Fidel et al., 2018) (Fig. 7.6). These latter findings suggest that measuring $NO_3^-$ sorption at insufficiently acidic pHs may mask the effects of biochar pyrolysis

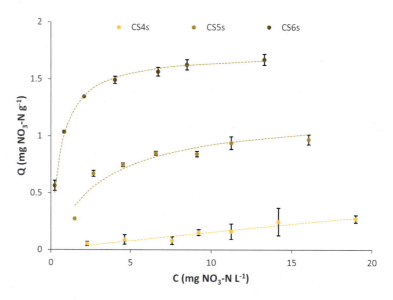

**Fig. 7.6** Sorption isotherms of $NO_3^-$ (pH 3.7) to acid-washed corn stover biochars, fit with the Langmuir model ($C$=final solution concentration; $Q$=final sorbed concentration). Error bars indicate ±s.d. ($n$=3). Reprinted from Fidel, R.B., Laird, D.A., Spokas, K.A., 2018. Sorption of ammonium and nitrate to biochars is electrostatic and pH-dependent. Sci. Rep. 8, 17627. https://doi.org/10.1038/s41598-018-35534-w.

parameters. Thus, $NO_3^-$ sorption, occurring via a pH-sensitive electrostatic exchange mechanism, is positively correlated with pyrolysis T, and this correlation is likely attributed to higher $pK_a$ functional groups and increasing surface area with increasing pyrolysis T.

Sorption of $PO_4^{3-}$ on biochar surfaces also can occur by electrostatic exchange; however, surface precipitation and ligand exchange reactions are likely stronger in this regard (Eduah et al., 2020). The positive correlation between the high pyrolysis T and the ash content promotes precipitation reactions and increases sorption of P on high T biochars more prominently than do biochars produced at low pyrolysis. For example, biochar produced from crawfish chitin at 800°C showed 70.9 mg g$^{-1}$ $PO_4^{3-}$ removal from wastewater compared to that of 9.5 mg g$^{-1}$ on biochar produced from the same feedstock at 200°C (Park et al., 2018). Enhanced inorganic element content (ash content) with high pyrolysis T specifically increased the calcium content in this biochar, which improved the $PO_4^{3-}$ removal by the precipitation reaction. Interestingly, Chintala et al. (2013) found that $PO_4^{3-}$ readily displaced adsorbed $NO_3^-$ at pH 4 after acid activation, which should have removed most of the ash, suggesting that $PO_4^{3-}$ adsorbs more strongly to biochar exchange sites than $NO_3^-$, and/or $PO_4^{3-}$ sorbs onto some of the same exchange sites as $NO_3^-$. These findings imply that $NO_3^-$ and $PO_4^{3-}$ compete for sorption sites, potentially making it difficult to optimize for the sorption of both. The negative charges of surface functional groups at higher pHs appear to affect both $NO_3^-$ and $PO_4^{3-}$; thus, a low pH solution is favorable for the sorption of both oxyanions.

The relationship between $AsO_4^{3-}$ (As(V)) removal capacity by pristine biochar and pyrolysis T appears nonlinear. As removal is low compared to activated, surface modified, and metal-impregnated biochars (Zoroufchi Benis et al., 2020). The sorption capacity of $AsO_4^{3-}$ on municipal solid waste-derived biochar (pH 8.0–9.0) increased with an increase in pyrolysis T from 400°C to 500°C; however, it dropped 1.4 times for biochar produced at 600°C (Jin et al., 2014). Metal impregnation on biochar can improve $AsO_4^{3-}$ sorption relative to pristine biochar; however, pyrolysis T still affects sorption capacity. For example, sorption of $AsO_4^{3-}$ onto Fe-impregnated water hyacinth biochar increased linearly with increasing the initial Fe content independent of pyrolysis T (Zhang et al., 2016); however, biochar produced at 250°C was more effective than biochar produced at 450°C due to high surface area and a larger number of active sites of amorphous Fe oxides produced at low pyrolysis T biochar surface and the high competition of negatively charged $AsO_4^{3-}$ and the hydrolysis product of the mineral component $OH^-$ of biochar produced at high pyrolysis T. Thus, nonlinearities in the relationship between $AsO_4^{3-}$ (As(V)) sorption on pristine biochar and pyrolysis T may be attributed to greater biochar alkalinity counteracting effects of greater sorption sites, coprecipitates, and/or surface area in higher T biochars. Pyrolysis T has a profound effect on oxyanion sorption, and it is important to optimize the T to ensure maximum removal of the particular oxyanion of interest.

**Feedstock type**: Besides pyrolysis T, feedstock properties also influence the ability of biochar surfaces to retain oxyanions. Biomass feedstock properties greatly vary with inorganic mineral (ash) content which is also inherently related to the biochar pH (Hassan et al., 2020), thereby influencing the oxyanion sorption. The ash content of woody biomass is relatively low compared to herbaceous biomass, wherein higher concentrations of alkali earth metals produce more of the ash components that can significantly increase oxyanion sorption by cation bridging or precipitation reactions. Moreover, herbaceous biochars contain more oxygen-containing functional groups relative to the wood-derived biochars, while wood-derived biochars are prone to have higher degrees of aromaticity. These inherent differences in biochars should be kept in mind in selecting feedstocks for removing specific oxyanions.

The chemical composition of biomass is an important factor in oxyanion sorption/release by specific biochar. For example, Cui et al. (2016b) compared 22 wetland plant-based biochars prepared at 500°C, among which only five biochars showed positive aqueous-phase sorption of phosphate, while the remaining biochars released phosphate due to their inherent high phosphate content. Low Mg or Ca content of biomass ash results in biochars with low P holding capability. Leaching of these alkaline earth metals from biochar can significantly influence the release of the biochar P into the water upon exposure. For an example, Jung et al. (2015a) reported that biochar derived from peanut shell at 700°C achieved $PO_4^{3-}$-P sorption of $3.8\,mg\,g^{-1}$ due to its high surface area and Mg/P and Ca/P ratios, whereas biochars produced at the same pyrolysis T from oak wood, bamboo wood, maize residue, and soybean stover were not as effective in P removal. Biochar from soybean stover and bamboo wood even released P. Similarly, biochars produced at 350–400°C from corn cob and cacao shell released $PO_4^{3-}$ and base cations when washed with water as reported by Hale et al. (2013). Chars with high alkali metal content in the ash fraction can sorb several times more $PO_4^{3-}$ than chars with little alkali metal content, as the $PO_4^{3-}$ minerals are a stable form of the divalent cations. The presence of colloidal and nanosized forms of periclase (MgO) in biochar produced at 600°C from the solid effluent of anaerobically digested sugar beet tailings increased $PO_4^{3-}$ removal significantly compared to biochar from nondigested sugar beet tailings despite having similar Mg content (Yao et al., 2011). These observations suggest that the sorption of $PO_4^{3-}$ by unmodified biochar is feedstock-specific, largely occurring via precipitation mechanisms. Retention may also occur via electrostatic attraction or divalent cation bridging (Takaya et al., 2016), but $PO_4^{3-}$ molecules sorbed by the latter two mechanisms are easily exchangeable with other similar charged ions.

Sorption of $NO_3^-$ is similarly feedstock-specific, as the sorption occurs mainly by electrostatic and divalent metal bridging. Chintala et al. (2013) reported that for three biochars produced at 650°C, switchgrass and corn stover biochars sorbed higher $NO_3^-$ than ponderosa pine biochar due to higher concentrations of volatile organic compounds and minerals. Cation bridging supported by the high concentration of divalent cations is thought to facilitate $NO_3^-$ sorption. Similarly, Yao et al. (2012) reported that among several biochars produced at 600°C, only sugarcane bagasse and bamboo biochar removed significant amounts of $NO_3^-$ (3.7% and 2.5%, respectively) compared to biochar produced from peanut hulls (0.2%) and Brazilian pepperwood (0.12%). In a recent study, irreversible sorption of $NO_3^-$ on six biochars derived under various pyrolysis conditions ranged from 82% to 89%. Sorption was as high as 15.2–15.9 mg g$^{-1}$ for biochars from six different biomass feedstocks (maize stover, sugarcane, grape pip, grape skin, and pinewood) and rubber tires (Aghoghovwia et al., 2020). The biochars promoted $NO_3^-$ sorption that followed a reduction-oxidation (redox) transformation of sorbed $NO_3^-$ resulting in the incorporation of that N to biochar structure was the reported explanation of the irreversible N sorption. Finally, Fidel et al. (2018) found mild acid-washing of biochars from high ash corn stover and low ash red oak to remove soluble ions eliminated differences in sorption capacity. This combined with demonstrated exchangeability of the adsorbed $NO_3^-$ supports the hypothesis that a significant portion of feedstock effects on $NO_3^-$ sorption can be attributed to both ash content and composition of the parent feedstock.

Sorption of $AsO_4^{3-}$ on unmodified biochar is generally low and the low stability of the sorbed oxyanion makes most biochars unsuitable for $AsO_4^{3-}$ removal. On the pristine biochar surface, biochar produced from pine wood, pine bark, oak wood, and oak bark feedstocks using fast pyrolysis at 400–450°C showed low $AsO_4^{3-}$ sorption (Mohan and Pittman, 2007). The presence of metals (Ca, Fe, and Al) in the biomass that contributes to the high ash content of biochar increases the sorption of $AsO_4^{3-}$ by precipitation reactions. For example, the presence of high Ca oxides in solid waste biochar sorbed significantly more $AsO_4^{3-}$ than the rice husk or sewage sludge biochar produced at the same T (Agrafioti et al., 2014a). Metal impregnation can improve $AsO_4^{3-}$ sorption several folds on biochars produced from several different feedstocks at the same pyrolysis T compared to unmodified biochars. Biochar produced from red oak (RO) and switchgrass (SG) at 900°C was only 1.42 and 1.15 mg g$^{-1}$, respectively. However, Fe-impregnated RO and SG biochar produced at the same T removed $AsO_4^{3-}$ from contaminated water by coprecipitation with Fe (Bakshi et al., 2018). However, the high ash content of SG impaired the stability of the Fe surface and RO biomass-derived biochar resulted in better removal of As than that of SG biomass-derived biochar.

It is clear from the discussion mentioned already that understanding the influences of feedstock type and pyrolysis T on biochar properties is critical to facilitate the biochar design for sustainable water and wastewater treatment. Generally, higher pyrolysis T, greater feedstock ash content, and lower pH result in greater oxyanion retention. However, exceptions do occur, especially when these factors interact. The synergistic and antagonistic effects of different biomass feedstocks and different pyrolysis T, as mediated by pH, on oxyanion sorption-desorption behavior of biochars require further exploration.

## 5 Physical properties of biochar influencing oxyanion sorption

At the most fundamental level, biochar structures and sorption properties depend upon biomass feedstock and pyrolysis conditions. Physical properties of the biochar have both direct and indirect impacts on the soil and environment. The skeletal structure of biochar is mainly comprised of carbon and minerals with different pore sizes (Warnock et al., 2007), consisting of various aromatic compounds and other functional groups derived from the original biomass feedstock. Therefore, biochar influences the porosity, particle size distribution, density, and packing arrangement of soils and other materials upon addition (Chia et al., 2015).

The highly porous structure of biochars, which is a function of pyrolysis temperature, is responsible for its unique properties. For example, pyrolysis above 900°C destroys carbonized cell walls, thus enlarging pores and increasing the adsorptive capacity of biochar (Qambrani et al., 2017). Scanning electron microscopy

(SEM) allows visualization of differences in pore structure, which are a combination of interconnected pores of different sizes (micro-, meso-, and macropores) (Brewer et al., 2009; Chia et al., 2015). The high specific surface area (SSA) and subsequent high cation/anion absorption capacity are the results of numerous micropores. Mesopores impact liquid-solid adsorption, whereas aeration, water movement, root growth, and bulk soil structure are controlled by macropores (Qambrani et al., 2017). A recent meta-analysis found that wood-based biochars have the highest SSA and when combined with certain pyrolysis conditions, they are likely to have the greatest impact on soil physical characteristics compared to other feedstocks produced by similar conditions (Ippolito et al., 2020).

Like its chemical and biological properties, the physical properties of biochar change in the environment with distinct behaviors for short (Brodowski et al., 2007) and long (Kuzyakov et al., 2009) timescales. Natural processes affecting these changes include oxidation, leaching, hydrolysis, hydration, wetting-drying, and freeze-thaw (Mia et al., 2017), although these weathering processes are beyond the scope of this chapter.

Most of the earliest research into biochar's physical characteristics focused on pristine biochar. More recently, however, desirable properties for environmental remediation and agricultural applications such as enlarged surface areas, unique pore structures, and enriched surface functional groups have been engineered into biochars, which are referred to as designer biochars (Novak et al., 2009) or modified or treated biochars. Biochars with enhanced porosity have higher SSA, measured by the Brunauer-Emmett-Teller (BET) method. Studies have measured SSA on pristine biochars ranging from less than $10 m^2/g$ to over $400 m^2/g$ (Brewer et al., 2009; Brown et al., 2006; Lehmann, 2007). Pre- and posttreatment in the production of biochar can increase its SSA to over $800 m^2/g$ (Angın et al., 2013; Qambrani et al., 2017). These pre- and posttreatments can make biochars highly effective for water pollution remediation, wastewater treatment, and oxyanion sorption.

Various physical and chemical posttreatments have been developed to increase surface area, pore volume, and surface functional groups (Dai et al., 2017; Tan et al., 2016). With changes in ion exchange capacity, surface change density; SFGs of biochars responsible for many of its favorable properties, however, from a physical property perspective pore structure and pore density (total volume and size distribution); and SSA are essential to the high oxyanion sorption capacity of biochar (Chia et al., 2015). Several of the physical modification methods commonly used to modify biochars include steam/gas activation, magnetic modification, microwave modification, and ball milling. All of these treatment methods increase the SSA by creating more micro- and mesopores for adsorption of various pollutants (Wang et al., 2017a, b).

Wet biomass can be hydrothermally carbonized to so-called hydrochars which have large adsorption capacity when produced at relatively high temperatures and long hold times (Saha et al., 2019; Shao et al., 2019; Zhou et al., 2017a, b). Ball milling is a simple, cost-effective, and efficient method for modifying biochar structures that is an alternative or complement to chemical methods for modifying biochar properties (Gao et al., 2015; Lyu et al., 2017; Richard et al., 2016; Spokas et al., 2009). Ball milling is a physical process that breaks biochar structure, enhancing SSA, pore volume, oxygen-containing functional groups, and absorption capacities (Lyu et al., 2018; Xiang et al., 2020). Oxidation treatments with chemicals such as hydrogen peroxide ($H_2O_2$) have been shown to increase oxygen and carboxyl functional groups on biochar surfaces (Wang and Liu, 2018). One study found that more biochar led to a greater reduction in the leaching of the oxyanion $NO_3^-$ from low SOC levels, but as native SOC levels declined $NO_3^-$ leaching increased (Kanthle et al., 2016). Another study determined that metal-blending prior to pyrolysis strongly affected the physical structure of a softwood biochar, increasing its SSA and pore volume as pyrolysis production temperature increased from 400°C to 700°C, which impacted the different biochars adsorption behavior of the oxyanions $PO_4^{3-}$ and $AsO_4^{3-}$ (Dieguez-Alonso et al., 2019). However, processes such as microwave-assisted pyrolysis (MAP) result in the loss of SSA through overheating that physically degrades micropore structure (Jimenez et al., 2017).

Overall, the enhanced physiochemical and adsorptive properties of biochars after specific treatments make them useful in a variety of environmental applications (Xiang et al., 2020).

## 6 Applications of biochar for removing oxyanions from water and wastewater

The ability of designer biochars to remove oxyanions makes them extremely promising in the control of water pollution and the treatment of wastewater. Due to the relative lack of positively charged surface sites of most unmodified biochars (Li et al., 2016b; Yao et al., 2013b), biochars typically exhibit limited anion exchange capacity (Lawrinenko and Laird, 2015; Zhang et al., 2016) and ability to sorb oxyanions (electrostatic repulsion).

Moreover, such as pristine, unmodified biochar typically lacks enough surface area and surface functional groups to strongly bind oxyanions and significantly reduces its capacities, as discussed in the previous section (Xiang et al., 2020). Therefore, the greatest area of potential improvement for biochars' environmental applications lies in enhancing their specific surface reactivity to ensure maximum retention of oxyanions. Based on this current understanding of biochar, surface modification is a growing research interest to implement biochar as the most effective biosorbents and to achieve its high oxyanion removal capacity (Agrafioti et al., 2014b; Chingombe et al., 2005).

## 6.1 Comparison of unmodified and modified biochar toward oxyanion sorption

Significant improvements in oxyanion sorption onto biochar surfaces have been reported in several studies in the literature as a result of surface modification. Table 7.1 summarizes changes in $AsO_4^{3-}$, $NO_3^-$, and $PO_4^{3-}$ removal by different biochars as a result of surface modification. Biochar surface modification by different chemical means is a promising technique to increase biochar oxyanion removal capacity several folds compared with their unmodified precursors.

It should be noted that biochar surface area and surface functionalities are manipulated by different surface modifications to enhance sorption capacities for specific oxyanion (Sizmur et al., 2017). Besides, surface modification offers several other advantages such as carbonization and activation in a single step and enhances catalytic oxidation capability (Liu et al., 2012). It is also evident from listed studies that evaluating biochar as the scavenger of oxyanions has been done in two different ways: (1) modifications of biochar surface to form exogenous functional groups for improved oxyanion sorption; and (2) modifications of biochar surface to increase positive charge, which would result in the reduction of the electrostatic repulsion of oxyanion species to the biochar surface (Benis et al., 2020). The addition of exogenous functional groups (such as metal phase onto biochar surface) plays an important role in attracting oxyanions and a promising way to impart specific oxyanion sorption sites to biochar. For the exogenous functional groups, biochar acts as a porous carbon scaffold onto which the neoformed exogenous groups gather higher surface area than its normal and attract oxyanions strongly (Sizmur et al., 2017). Though different mechanisms govern oxyanion sorption onto differently modified biochar surfaces, the

Table 7.1 Removal of different oxyanions from aqueous solution by biochars.

| Biomass feedstock | Pyrolysis temperature (°C) | Pre-/posttreatment | Oxyanion | Sorbent dosage (g L$^{-1}$) | Initial concentration (mg/L) | Sorption capacities (mg g$^{-1}$) | | Ref. |
|---|---|---|---|---|---|---|---|---|
| | | | | | | Unmodified | Modified | |
| Red oak | 900 | Pre; magnetite | $AsO_4^{3-}$ | 1 | 0–25 | 1.42 | 15.58 | Bakshi et al. (2018) |
| Switchgrass | 900 | Pre; magnetite | $AsO_4^{3-}$ | 1 | 0–25 | 1.15 | 7.92 | Bakshi et al. (2018) |
| Municipal solid waste | 500 | Post; 2 M KOH | $AsO_4^{3-}$ | 2 | 5–400 | 24.49 | 30.98 | Jin et al. (2014) |
| Pinewood | 600 | Pre; hematite | $AsO_4^{3-}$ | 2.5 | 1–50 | 0.265 | 0.43 | Wang et al. (2015a, b, c) |
| Loblolly pine | 600 | Pre; $MnCl_2$ | $AsO_4^{3-}$ | 2.5 | 1–20 | 0.2 | 0.59 | Wang et al. (2015b) |
| Hickory chips | 600 | Post; $Fe(NO_3)_3$ | $AsO_4^{3-}$ | 2 | 0.1–55 | Little or no sorption | 2.16 | Hu et al. (2015) |
| Empty fruit bunch | NR | Post; $FeCl_3$ | $AsO_4^{3-}$ | 5 | 3–300 | 5.5 | 15.2 | Samsuri et al. (2013) |
| Rice husk | NR | Post; $FeCl_3$ | $AsO_4^{3-}$ | 5 | 3–300 | 7.1 | 16 | Samsuri et al. (2013) |
| Sewage sludge digestate | 350 | Post; KOH or $H_2O_2$ | $AsO_4^{3-}$ | 4 | 1.39–41.68 | 0.22 | 1.18 | Wongrod et al. (2018) |
| Rice straw | 450 | Post; chitosan | $AsO_4^{3-}$ | 1 | 0.2–50 | 3.681 | 11.961 | Liu et al. (2017) |
| Sugarcane bagasse | 300 | Post; epichlorohydrin and dimethylformamide | $NO_3^-$ | 1 | 50 | 11.56 | 28.21 | Hafshejani et al. (2016) |
| Poplar wood | 550 | $AlCl_3$ | $NO_3^-$ | 2 | At pH3 50 At pH6 | ~5.5 | 89.58 | Yin et al. (2018a) |

| Feedstock | Temp | Modification | Ion | Col5 | Col6 | Col7 | Col8 | Reference |
|---|---|---|---|---|---|---|---|---|
| Soybean straw | 500 | Pre $MgCl_2$ | $NO_3^-$ | 2 | 50 | 2.26 At pH8.5 | 13.54 At pH8.8 | Yin et al. (2018b) |
| Soybean straw | 500 | Pre $AlCl_3$ | $NO_3^-$ | 2 | 50 | 2.26 At pH6.5 | 40.63 At pH8.8 | Yin et al. (2018b) |
| Soybean straw | 500 | Pre $MgCl_2+AlCl_3$ | $NO_3^-$ | 2 | 50 | 2.26 At pH7.5 | 24.86 At pH8.8 | Yin et al. (2018b) |
| Corn stover | 650 | Post; HCl | $NO_3^-$ | 25 | 100 at pH4 | 38.4 | 43.5 | Chintala et al. (2013) |
| Pinewood | 650 | Post; HCl | $NO_3^-$ | 25 | 100 at pH4 | 11.4 | 43.1 | Chintala et al. (2013) |
| Switchgrass | 650 | Post; HCl | $NO_3^-$ | 25 | 100 at pH4 | 38.7 | 49.6 | Chintala et al. (2013) |
| Rice straw | 400 | $FeCl_3$ Pre; HCl+KOH +$H_2O_2$+autoclave Post | $NO_3^-$ | 1.6 | 100 | 25.76 | 65.20 | Chandra et al. (2020) |
| Rice straw | 600 | $FeCl_3$ Pre; HCl+KOH +$H_2O_2$+autoclave Post | $NO_3^-$ | 1.6 | 100 | 32.92 | 67.65 | Chandra et al. (2020) |
| Rice straw | 400 | $FeCl_3$ Pre; HCl+KOH +$H_2O_2$+autoclave Post | $PO_4^{3-}$ | 1.6 | 100 | 38.62 | 72.63 | Chandra et al. (2020) |
| Rice straw | 600 | $FeCl_3$ Pre; HCl+KOH +$H_2O_2$+autoclave Post | $PO_4^{3-}$ | 1.6 | 100 | 43.52 | 75.22 | Chandra et al. (2020) |
| Soybean straw | 500 | Pre $MgCl_2$ | $PO_4^{3-}$ | 2 | 50 | 74.47 At pH9.8 | 1.73 At pH8.2 | Yin et al. (2018b) |
| Soybean straw | 500 | Pre $AlCl_3$ | $PO_4^{3-}$ | 2 | 50 | 44.98 At pH6.7 | 1.73 At pH8.2 | Yin et al. (2018b) |
| Soybean straw | 500 | Pre $MgCl_2+AlCl_3$ | $PO_4^{3-}$ | 2 | 50 | 36.33 At pH7.9 | 1.73 At pH8.2 | Yin et al. (2018b) |
| Maplewood | 500 | Post; poly-DADMAC | $PO_4^{3-}$ | 5 | 0–300 | 3.71 | 22.8 | Wang et al. (2020a, b) |
| Maplewood | 500 | Pre; Mg-acetate | $PO_4^{3-}$ | 5 | 0–300 | 3.71 | 46.1 | Wang et al. (2020a, b) |
| Oak sawdust | 500 | Pre; $LaCl_3$ | $PO_4^{3-}$ | 2 | 3–3000 | 32 | 142.7 | Wang et al. (2015a, b, c) |
| Orange peel | 700 | Pre; $FeCl_3+FeCl_2$ | $PO_4^{3-}$ | NR | 0–36.77 | 0.477 | 1.24 | Chen et al. (2011) |

*Continued*

Table 7.1 Removal of different oxyanions from aqueous solution by biochars—cont'd

| Biomass feedstock | Pyrolysis temperature (°C) | Pre-/posttreatment | Oxyanion | Sorbent dosage (g L$^{-1}$) | Initial concentration (mg/L) | Sorption capacities (mg g$^{-1}$) Unmodified | Sorption capacities (mg g$^{-1}$) Modified | Ref. |
|---|---|---|---|---|---|---|---|---|
| Sugarcane | 550 | Pre; FeCl$_2$+MgCl$_2$ | PO$_4^{3-}$ | 2.5 | 50 | <0.05 | 121.25 | Li et al. (2016b) |
| Cotton stalk | 350 | Post; Fe$_3$O$_4$ | PO$_4^{3-}$ | 20 | 20 | 0 | 0.963 | Ren et al. (2015) |
| Ground corn | 450 | Pre; MgCl$_2$ | PO$_4^{3-}$ | 10 | 84–2600 | 201 | 233 | Fang et al. (2014) |
| Ramie stem | 500 | Pre; CaCl$_2$ | PO$_4^{3-}$ | 2 | 10–400 | 40 | 105.4 | Liu et al. (2016) |
| Sewage sludge | 800 | Pre; dolomite | PO$_4^{3-}$ | 1.2 | 50 | ~0 | 29.18 | Li et al. (2019) |
| Coconut shell | NR | Post; FeCl$_3$ | PO$_4^{3-}$ | 2.5 | 2–100 | 13.23 | 36 | Zhong et al. (2019) |
| Walnut shell | 600 | Pre; FeCl$_2$+MgCl$_2$ | PO$_4^{3-}$ | 2 | 5–400 | 2.136 | 6.945 | Tao et al. (2020) |
| Corn stover | 500 | Pre; FeSO$_4$ | PO$_4^{3-}$ | 5 | 0–400 | 3.76 | 46.3 | Bakshi et al. (2021) |

overall changes in enhanced oxyanion sorption are related to the enlarged surface area, optimizing the pore size distribution, increasing positive surface charge, and increasing ligand-binding density.

## 6.2 Surface modification of biochar for water and wastewater treatment

As discussed earlier, biochar surfaces must be modified to achieve high oxyanion sorption capacities. Innovative methods have been developed for modifying biochar, either by treating the biomass feedstock pretreatments or by biochar posttreatments (Yang et al., 2019). Both pre- and posttreatments include physical (washing, sieving, crushing, etc.), and chemical (treating with functional or corrosive agents) alteration methods (Usman et al., 2015; Xiang et al., 2020). Fig. 7.7 demonstrates the most common biochar surface modification methods that have been published in the literature. The oxyanion removal efficacy results from both modification method and biochar production parameters, most notably pyrolysis T and feedstock type.

The primary objective of these modification methods is to enhance the biochar efficiencies to remove oxyanion removal from water and wastewater by changing biochar physical and chemical properties. Though physical modification methods are effective in increasing biochar surface area and pore size distribution (Lou et al., 2016; Shim et al., 2015; Xiao and Pignatello, 2016), their effects are minimal so long as biochar maintains a delocalized cloud of electrons associated with aromatic groups on the surface of carbonaceous biochar. Thus, physical modifications are most effective when augmenting chemical modifications to reduce the negative charge of biochar. Chemical modification mainly includes reaction with chemical additives to increase biochar oxyanion sorption capacity creating either by new functional groups or by opening up blocked pores and enlarging the diameter of small pores of biochar (Goswami et al., 2016; Jin et al., 2014).

**Pretreatment technologies**: Pretreatment chemical modification involves a chemical reaction with fresh biomass feedstocks before pyrolysis (Fig. 7.7). The most common way to treat the biomass feedstock is to wash or soak the feedstock with a solution of the chemical agent followed by drying and pyrolysis. As a result of the chemical pretreatment and subsequent pyrolysis, the exogenous functional groups are created onto the biochar surface to form the biochar-based nanocomposite, stabilized on the pores of the carbon surface (Xiao and Pignatello, 2016). Several chemical

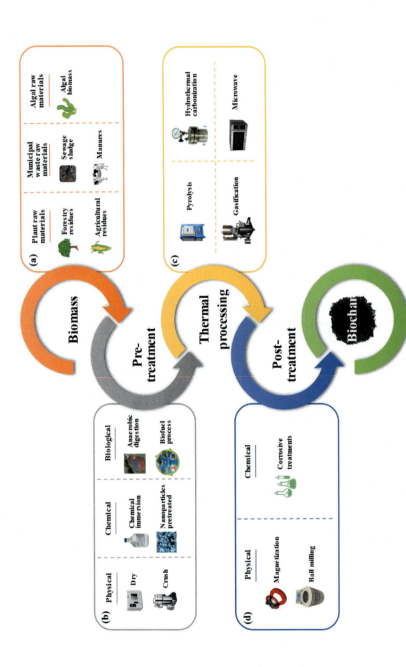

**Fig. 7.7** Most commonly used surface modification to activate biochar physically and chemically. Reprinted from Xiang, W., Zhang, X., Chen, J., Zou, W., He, F., Hu, X., Tsang, D.C.W., Ok, Y.S., Gao, B., 2020. Biochar technology in wastewater treatment: a critical review. Chemosphere 252, 126539. https://doi.org/10.1016/j.chemosphere.2020.126539 with permission of the publisher.

compounds have been used to pretreat the biomass feedstock for producing the biochar-based nanocomposite, the most common is the metal-based (Fe, Al, Ca, Mg, Mn, etc.) both in forms of salt solutions or natural colloid forms (Bakshi et al., 2018; Li et al., 2016b; Wang et al., 2020a, b; Zhang and Gao, 2013). The resultant metal (oxy)hydroxides are hypothesized to be stable onto biochar surface and offer excellent functional sites for oxyanion removal, by a combined effect of the high surface area of biochar and the positive surface charge of the metal (oxy)hydroxides (Sizmur et al., 2017).

Agrafioti et al. (2014b) demonstrated that soaking the rice husk biomass with CaO and $FeCl_3$ produced biochars that have 95% $AsO_4^{3-}$ removal capacities by precipitating out with the neoformed Ca or Fe (oxy)hydroxides. Wang et al. (2015a) explored $MnCl_2$ pretreatment of loblolly pine to improve biochar's $AsO_4^{3-}$ sorption ability and found four times greater sorption than its control biochar; however, they did not find any association of $AsO_4^{3-}$ and neoformed MnO nanoparticles. Zhang and Gao (2013) demonstrated that AlOOH nanoflake biochar composite is a multifunctional adsorbent to remove both $AsO_4^{3-}$ and $PO_4^{3-}$ from aqueous solution, by soaking and pyrolysis of cottonwood and $AlCl_3$. This finding is supported by Lawrinenko et al. (2017) who revealed the formation of Al—O—C or Fe—O—C organometallic moieties onto the biochar surface as a result of soaking biomass with $AlCl_3$ or $FeCl_3$. Wang et al. (2020b) reported that MgO-doped hardwood and softwood biochars, produced by copyrolysis of biomass feedstock and Mg-acetate, remove $PO_4^{3-}$ by forming amorphous Mg-phosphate species from simulated wastewater. Bakshi et al. (2021) showed that Fe-impregnated corn stover biochar, produced by copyrolysis of $FeSO_4$ and biomass feedstock, has high adsorption and low desorption capacity of $PO_4^{3-}$ in simulated wastewater.

The incorporation of Mg ($MgCl_2$) and Ca ($CaCl_2$) salts in biomass before production of biochar increases the adsorption of $PO_4^{3-}$ from wastewater, attributed to the formation of MgO and CaO nanoparticles during pyrolysis (Fang et al., 2014; Liu et al., 2016). Growth studies of tomatoes in soils amended with biochars produced from Mg pretreatment show enhanced $PO_4^{3-}$ retention in an aqueous solution (Yao et al., 2013b). Struvite (a Mg-P mineral) formation has also been reported as the dominant mechanism for $PO_4^{3-}$ removal from wastewater (Xu et al., 2018). Moreover, the so-called magnetic biochar, characterized by the presence of magnetite ($Fe^{II}+Fe^{III}$ mineral) formed during the pyrolysis of orange peel biomass treated with a solution of $FeCl_3+FeCl_2$, is reported to increase $PO_4^{3-}$ sorption threefold

(Chen et al., 2011). It should be noted that impregnating biomass with metal salts can reduce the overall surface area of biochar by clogging pores with metal oxides (Rajapaksha et al., 2016); nevertheless, the creation of pH-dependent positively charged functional groups can increase $PO_4^{3-}$ sorption by a factor of 12–50 (Micháleková-Richveisová et al., 2017; Sizmur et al., 2017).

Fang et al. (2015) have shown a metal oxide nanoparticle-biochar composite can increase $PO_4^{3-}$ removal from wastewater. The authors utilized $CaCl_2$ and $MgCl_2$ for the pretreatment of corncob biomass to produce the $CaO+MgO$ nanoparticle-biochar composite which removed $326.63\,mg\,g^{-1}$ $PO_4^{3-}$ from wastewater and reduced the negative interaction effects of competitive ions. Jung et al. (2015b) reported the highest so far $PO_4^{3-}$ sorption capacity ($887\,mg\,g^{-1}$) by Mg-Al oxides nanoparticle-biochar composite produced through electric-field-assisted pyrolysis. Additionally, Yin et al. (2018b) found that pretreatment of soybean straw with $Mg^{2+}$, $Al^{3+}$, or mixture of the two cations increased $PO_4^{3-}$ sorption by 43, 26, and 21 times, respectively. This increased sorption was attributed to the formation of Mg and Al oxide nanoparticles, which may have facilitated complexation and/or precipitation with $PO_4^{3-}$.

Pretreatment with metals and metalloids can enhance $NO_3^-$ adsorption by biochars. For example, Zhang et al. (2012) found that soaking various biomass feedstocks in $MgCl_2$ enabled $NO_3^-$ sorption capacity to rise as high as $95\,mg\,g^{-1}$ N. This increase was attributed in part to the formation of MgO nanoparticles onto which $NO_3^-$ likely adsorbed. These findings were supported by Yin et al. (2018b), who found that soaking feedstock (soybean straw) in $MgCl_2$ increased the $NO_3^-$ adsorption sixfold, to about $14\,mg\,g^{-1}$ $NO_3^-$ relative to the control. In the case of metalloids, Yin et al. (2018a) found that soaking popular woodchips with $AlCl_3$ solution and subsequent pyrolysis at 550°C resulted in biochar with 15% Al that increases $NO_3^-$ adsorption 18-fold. Additionally, Yin et al. (2018b) found that soaking soybean straw with $AlCl_3$ solution before pyrolysis at 500°C increased $NO_3^-$ adsorption 18-fold. Interestingly, combining both $MgCl_2$ and $AlCl_3$ in the soaking solution increased $NO_3^-$ adsorption only 11-fold relative to the control. The increased sorption rates of Mg and Al-treated biochars were both attributed to the increased surface area, precipitation of Mg+Al minerals (MgO, AlOOH, and $MgAl_2O_4$), and an increase in pH point of zero net charge (PZNC). The higher sorption of $NO_3^-$ by the Al-treated biochar compared with Mg-treated biochar was partly attributed to its lower pH, but also to the formation of AlOOH, which has a PZNC of 8.2.

Besides metal salts, natural mineral colloids have been used to pretreat biomass to produce biochar for application in oxyanion removal. Using magnetite as the modification agent with red oak and switchgrass, Bakshi et al. (2018) demonstrated the formation of $Fe^0$ nanoparticles and the removal of $AsO_4^{3-}$ onto neoformed Fe (oxy)hydroxides phase from contaminated drinking water. The $\gamma$-$Fe_2O_3$ enriched biochar, produced from pyrolyzing a mixture of pine wood and hematite, a naturally occurring Fe mineral, is effective in removing $AsO_4^{3-}$ from simulated wastewater (Wang et al., 2015b). Dolomite, a Ca-Mg-enriched naturally occurring mineral, has been used as the modification agent with dewatered sewage sludge to achieve 96.8% $PO_4^{3-}$ removal in the presence of other competing ions via the formation of inner- and outer-sphere surface complex (Li et al., 2019). The natural clay mineral montmorillonite mixed with bamboo produced a biochar composite with enhanced surface area and pore volume. This montmorillonite-biochar composite has been shown to sorb 105.28 mg g$^{-1}$ of $PO_4^{3-}$ by electrostatic attraction and ionic bonding (Chen et al., 2017). Wang et al. (2019) mixed attapulgite (an Mg-Al containing clay mineral) and $ZnCl_2$ with rice straw biomass to produce a functionalized clay-biochar composite which is effective to reduce the $AsO_4^{3-}$ bioavailability.

Besides metal salts and naturally occurring colloids, pretreatments based on acid or alkaline solutions have been reported to increase oxyanion sorption. Acid pretreatment adds surface carboxylic groups to biochar that repels negatively charged oxyanions. Borah et al. (2008) demonstrated that the pretreatment of commercial carbon black with 20% $H_2SO_4$ can effectively remove 90% $AsO_4^{3-}$ from an aqueous medium. Sulfonic acid groups and ligand exchange reactions are thought to be responsible for the adsorption of $AsO_4^{3-}$.

**Posttreatment technologies**: Like pretreatment, posttreatment chemical modification involves chemical reactions that change the physicochemical properties of biochar including surface area, pore volume, surface chemistry, and functional agents (Tan et al., 2016; Wang et al., 2020b). The most common way to treat the biochar with chemical additives is to soak or immerse the biochar with the chemical solution of interest, and either grafting or insertion of functional groups (Wang et al., 2020b; Zhong et al., 2019). As a result of chemical posttreatment, exogenous functional groups are grafted or inserted onto the biochar surface to form a biochar-based nanocomposite stabilized on the pores of the carbon surface (Tan et al., 2016).

Although pretreatment is simpler than posttreatment (Li et al., 2018), posttreatment is effective for producing biochar-based nanocomposites that are stable onto biochar surface and offer excellent functional sites for oxyanion removal (Sizmur et al., 2017). The most common is the metal-based (Fe and Mn) salt solutions (Wang et al. (2015a, b); Zhong et al., 2019), of which the resultant metal (oxy)hydroxides/$Fe^0$ is hypothesized to be stable onto biochar surface. Other chemicals used in posttreatments include layered double hydroxides (LDHs) (Li et al., 2016a; Zhang et al., 2013a, b), organic compounds (Cui et al., 2016a; Liu et al., 2017; Wang et al., 2020a, b), a combination of Fe salt/$Fe^0$ and organic compound (Zhou et al., 2014, 2017a, b), and alkaline solutions (Jin et al., 2014). As a result of these posttreatments, the oxyanions are stabilized onto grafted functional groups in the biochars by either electrostatic attraction or π-π electron donor-acceptor interaction.

Different Fe salts have been used as posttreatment of biochar to facilitate the oxyanion sorption. Hu et al. (2015) introduced a facile method of goethite (α-FeOOH) production by direct $Fe(NO_3)_3$ hydrolysis onto hickory chip biochar that facilitated 2.16 mg g$^{-1}$ $AsO_4^{3-}$ sorption through the chemisorption process. Samsuri et al. (2013) generated Fe-(oxy)hydroxide precipitation onto rice husk and empty fruit bunch biochars after treating with $FeCl_3$ solution which facilitated sorption of $AsO_4^{3-}$ via terminal FeOH and $FeOH_2^+$ groups. In this case, the author acknowledged almost three times more sorption than the unmodified counterpart. $MnCl_2$ has been explored as a posttreatment of biochar to form the crystalline phase of MnO which promoted three times more $AsO_4^{3-}$ sorption than the control biochar (Wang et al., 2015a). Wang et al. (2017b) embedded $Fe^0$ onto pinewood biochar to form biochar-$Fe^0$ composite by an in situ adsorption-reduction steps using $FeCl_3$ solution which facilitated 100% $AsO_4^{3-}$ removal from aqueous solutions. Similar to $AsO_4^{3-}$, $FeCl_3$ solution has been used to modify biochar surface to facilitate $PO_4^{3-}$ sorption. As an example, Zhong et al. (2019) demonstrated the formation of amorphous 2-line ferrihydrite onto coconut shell biochar which facilitated about 2.5 more $PO_4^{3-}$ sorption than the control biochar. In this case, the author acknowledged ligand exchange, electrostatic attraction, chemical precipitation, and inner-sphere surface complexation as the relevant mechanisms for sorption. Besides $FeCl_3$, $Fe_3O_4$ has been used to trap $PO_4^{3-}$ (from 0 to 0.963 mg g$^{-1}$) from aqueous solution due to the granular nature of Fe-oxides which has increased the biochar surface area by 340% (Ren et al., 2015).

Moreover, Fe salts have been used to prepare magnetic biochar, which has been demonstrated as an excellent sorbent for oxyanions and with the advantage that the sorbent is easy to separate after treatment (Baig et al., 2014; Tan et al., 2016; Zhang et al., 2013a). Nanosized magnetic Fe-oxides including $Fe_3O_4$, $\gamma$-$Fe_2O_3$, and $CoFe_2O_4$ particles are loaded onto biochar surfaces by these methods. Apart from metal oxides, LDHs are effective to strip oxyanions in water treatment. Zhang et al. (2013b) demonstrated a biochar-LDH composite with Mg and Al combination with a 3:1 M ratio which facilitated very high $PO_4^{3-}$ sorption (410 mg g$^{-1}$) by LDH colloidal dispersion. This finding is supported by Li et al. (2016a) who demonstrated the formation of different LDHs with Mg and Al by changing the molar ratio of 2:1, 3:1, and 4:1, which achieved $PO_4^{3-}$ sorption of 164, 221, and 251 mg g$^{-1}$, respectively, compared to only 3 mg g$^{-1}$ for pristine biochar. It should also be noted that the type of LDH strongly affected oxyanion sorption: an Mg-Fe LDH-wheat straw biochar composite was shown to be effective in removing $NO_3^-$ (25 mg g$^{-1}$) despite the presence of $PO_4^{3-}$ in the system (Xue et al., 2016).

Besides Fe and Mn posttreatment, organic compounds have been used to modify biochar surfaces which facilitated oxyanion sorption, which forms strong bonds with both the biochar surface and the oxyanion (Sizmur et al., 2017). Wang et al. (2020b) explored the grafting of polydiallyldimethylammonium chloride (poly-DADMAC) onto hardwood and softwood biochars, which promoted $PO_4^{3-}$ sorption by increasing anion exchange reaction with $Cl^-$. Chitosan, a natural biopolymer, has been used to synthesize chitosan-modified biochars characterized by networks of biochar porosities well as high biochar surface area. Chitosan has been applied to improve Fe oxide insertion onto biochar surfaces. As an example, Zhou et al. (2014) modified bamboo biochar with chitosan and $Fe^0$ particles to achieve high $AsO_4^{3-}$ (23%–95%) and $PO_4^{3-}$ (40%–96%) removal efficiency. The author acknowledged that chitosan is only effective in adsorbing $AsO_4^{3-}$ and $PO_4^{3-}$ in the presence of $Fe^0$ due to the electrostatic interactions with $Fe^0$ particles. In a similar study, Liu et al. (2017) combined chitosan and Fe solution (molar ratio of $Fe^{II}$:$Fe^{III}$=2:3) with rice straw biochar to form a magnetic chitosan/biochar composite. This composite sorbed about three times more $AsO_4^{3-}$ than the pristine biochar in an aqueous solution containing competing ions. Besides Fe salts, Mg salt has also been used with chitosan to modify *Thalia dealbata* biochar surface (Cui et al., 2016a). The resultant Mg-alginate/chitosan biochar removed 98.3% of added $PO_4^{3-}$ in the aqueous solution through precipitation and ligand exchange reactions.

Apart from organic compounds, corrosive alkaline solutions have also been used as in posttreatment. For example, Jin et al. (2014) activated biochar from municipal solid waste with 2 M

KOH, which increased surface area from 29 to 49 m$^2$/g although $AsO_4^{3-}$ sorption only increased 26% (from 24.49 to 30.98 mg g$^{-1}$) in an aqueous medium. This finding is supported by Wongrod et al. (2018) who demonstrated 2 M KOH activation of sewage sludge digestate biochar to increase $AsO_4^{3-}$ sorption; however, they found about a fivefold increase in $AsO_4^{3-}$ sorption (from 0.22 to 1.18 mg g$^{-1}$) compared to the control biochar. The author acknowledged that KOH activation opened up partially blocked pores that resulted in a tremendous increase in Brunauer-Emmett-Teller surface area from 0.4 to 7.9 m$^2$ g$^{-1}$.

Acid activation has also been used to increase $NO_3^-$ sorption. For example, Chintala et al. (2013) found that activation with HCl at 200°C increased $NO_3^-$ sorption 2–4.4 times relative to unmodified controls. This increase was attributed to an increase in surface area and a decrease in pH. Indeed, $NO_3^-$ sorption decreased with increasing pH, among both activated and nonactivated biochars.

It is clear from the aforementioned discussion that pre- and posttreatment biochar surface modifications are robust and a hot topic in current research for water and wastewater treatment. The modifications of biochar surface for oxyanion removal, especially with metal salts, are relatively cheap since the metals are abundant in nature (Li et al., 2017a). Existing wastewater treatment technologies to remove oxyanions are expensive and require the disposal of waste material (Shannon et al., 2008). The surface-modified biochar could be a cost-effective, sustainable, and eco-friendly solution for this disposal problem since the spent sorbent can be used as a soil amendment. To reduce the cost of the production, one should consider the lowest amount of the use of chemicals while maintaining the maximum sorption efficiency of the given oxyanion (Sizmur et al., 2017).

## 6.3 Mechanisms of oxyanion removal by biochars

As discussed in the previous section, sorption capacities of modified biochars for removal of $AsO_4^{3-}$, $NO_3^-$, and $PO_4^{3-}$ from water and wastewater varied considerably, depending on the type of biomass feedstock, nature of the treatment, and the difference in properties among biochar-chemical composites and their interactions with specific oxyanions. There are several mechanisms responsible for the high sorption capacities of the modified biochars. In general, the mechanism of oxyanion sorption onto modified biochar surface can be summarized into three main categories: (1) chemical sorption, (2) physical sorption, and (3) precipitation (Fig. 7.8). The relative contribution of each mechanism

Chapter 7 Retention of oxyanions on biochar surface 263

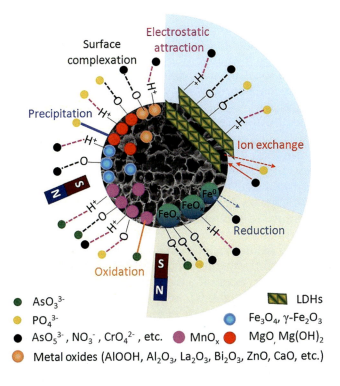

Fig. 7.8 Removal mechanism of oxyanions from wastewater by different metal-biochar composites. Reprinted from Li, R., Wang, J., Gaston, L.A., Zhou, B., Li, M., Xiao, R., Wang, Q., Zhang, Z., Huang, H., Liang, W., Huang, H., Zhang, X., 2018. An overview of carbothermal synthesis of metalebiochar composites for the removal of oxyanion contaminants from aqueous solution. Carbon 129, 674–687 with permission of the publisher.

depends upon the specific oxyanions and the physicochemical characteristics of the biochar-chemical composites (Li et al., 2018). Fig. 7.8 represents the interactive removal mechanisms of oxyanion from wastewater by different metal-biochar composites that have been published in the literature.

**Chemical sorption** is accomplished through ion exchange, ligand exchange, oxidation–reduction, and surface complexation of oxyanions. It should be noted that both inner- and outer-sphere sorption are entirely stoichiometric for the available sorption sites, and the sorption site availability depends on the pH of the medium (Sizmur et al., 2017). In general, the chemical properties of biochar and/or modified biochar depend on the surface charge, surface functional groups, mineral contents, and surface area which affect the type of governing mechanism for oxyanion removal (Li et al., 2018). For $PO_4^{3-}$ removal, several studies have shown that the ligand exchange, surface complexation, and precipitation mechanisms are dominant. As an example, Bakshi et al. (2021) demonstrated the formation of inner-sphere surface complexation of Fe-$PO_4^{3-}$ onto hematite particles due to the intercalation of $PO_4^{3-}$. Wang et al. (2020b) showed evidence of ion exchange of $Cl^-$ ion with $PO_4^{3-}$ using poly-DADMAC as a modifying agent of biochar. Furthermore, for LDH-biochar composite, different molar ratios of Mg and Al resulted in ion exchange of $PO_4^{3-}$ with the exchange of $NO_3^-$ (Li et al., 2017b). Cui et al. (2016a) demonstrated the ligand exchange mechanism between unprotonated oxygen atoms of $PO_4^{3-}$ and Mg-OH in the surface of biochar. In the study of Zhong et al. (2019), the presence of three asymmetric vibration peaks of 970, 1066, and 1167 $cm^{-1}$ in the FTIR reveals the presence of formation of monodentate and bidentate inner-sphere complexes of Fe—O and Al—O with $PO_4^{3-}$. Also, Li et al. (2016b) showed that the evidence of the disappearance of 870 and 770 $cm^{-1}$ peaks in FTIR corresponds to Fe—OH vibrations and appearance of 1050 $cm^{-1}$ peak due to P—O bond that suggests the formation of surface inner-sphere complexation of $PO_4^{3-}$ with Fe oxide. Besides $PO_4^{3-}$, the removal of $AsO_4^{3-}$ from wastewater mainly involves redox and surface complexation mechanisms. For example, Wang et al. (2017b) showed the extraordinary ability of $AsO_4^{3-}$ sorption by $Fe^0$-biochar composite by the surface inner-sphere complexation with surface Fe-OH groups. The author acknowledged the possibility of $As^{5+}$ reduction to $As^{3+}$ as part of the $AsO_4^{3-}$ removal mechanism. This finding is supported by Bakshi et al. (2018) who demonstrated the simultaneous oxidation of $Fe^0$ to $Fe^{3+}$ and reduction of $As^{5+}$ to $As^{3+}$ and subsequent isomorphous substitution of $As^{3+}$ for $Fe^{3+}$ in FeOOH. In another study, Hu et al. (2015) explained the surface complexation of $AsO_4^{3-}$ onto goethite surface of Fe-impregnated biochar by providing the FTIR evidence of weakening the 881 $cm^{-1}$ As—O stretching vibrations.

**Physical sorption** or electrostatic sorption for oxyanion occurs when a layer of water remains between the adsorbing anion and positively charged surface. This may first occur between positively charged hydroxylated metal ion surface and the delocalized cloud of electrons associated with the oxyanions, creating an electron donor-acceptor relationship. Unlike chemical sorption, physical sorption does not require the stoichiometric release of the proton from the biochar (Sizmur et al., 2017). It should be noted that this type of generalized electrostatic sorption is an initial stage of the surface complexation or chemical precipitation mechanism for oxyanion retention onto biochar. Several studies have reported the electrostatic sorption of $PO_4^{3-}$ (Li et al., 2016b; Li et al., 2017b; Ren et al., 2015), $AsO_4^{3-}$ (Wang et al., 2015a, b, c; Wang et al., 2016, 2017a), and $NO_3^-$. Should the oxyanion get sufficiently close to the biochar surface, it may also adsorb through the electrostatic exchange.

**Precipitation** or coprecipitation of oxyanions occurs particularly with metal ions as amorphous or crystalline salts on the surface of biochar. For example, precipitation of $MgHPO_4$ and $Mg(H_2PO_4)_2$ minerals have been demonstrated in x-ray diffraction analysis by Yao et al. (2013a), where MgO-biochar composite derived from anaerobically digested biomass applied for $PO_4^{3-}$ removal in water. This result is supported by Li et al. (2016b) who also demonstrated the formation of new $MgHPO_4$ and $Mg(H_2PO_4)_2$ in the x-ray diffraction analysis when the MgO decorated sugarcane harvest residues, biochar is challenged with $PO_4^{3-}$ aqueous solution. However, Wang et al. (2020b) did not find any crystalline mineral-phase precipitation instead of an amorphous-phase $Mg$-$PO_4^{3-}$ species. Similarly, Bakshi et al. (2021) found the formation of amorphous Fe-O-P film precipitation onto the biochar surface. On the contrary, coprecipitation of $Fe^0$ and $AsO_4^{3-}$ after oxidation-reduction onto reed oak and switchgrass biochar surface has been reported by Bakshi et al. (2018) by using magnetite as the biochar modifying agent. However, Wang et al. (2015b) did not find any x-ray diffraction evidence of precipitation or coprecipitation by using hematite as the biochar modifying agent.

So, it is clear from the earlier discussion that the formation of precipitation of oxyanion and metal ion onto biochar surface primarily depends on the biomass feedstock, biochar modifying agent and modification technique, and solution chemistry. It should be noted that oxyanion sorption onto biochar surface might not be limited to only one mechanism, and indeed, evidence for more than one mechanism was documented in most cases.

## 7 Conclusions and future perspectives

Biochar is an effective low-cost adsorbent that can be produced from a variety of biomass feedstocks including agricultural residues, forestry residues, sewage sludge, and municipal wastes. The capacity of biochar for the removal of oxyanions in an aqueous solution is dependent on the physicochemical properties of biochar which are resulted from the nature of the biomass feedstock, production technique, and modification of the resultant biochar surface. It is very important to recognize that different pyrolysis conditions, biomass feedstock, and modification techniques alter biochar properties and suitability as oxyanion sorbent, but control of these factors can produce a suitable sorbent. Since different oxyanion removal requires different mechanisms, attention must be paid to the biochar modification method to improve its performance that aims to improve the surface area, porosity, and surface sorption sites of biochar. Water and wastewater with high levels of oxyanions are adsorbed onto biochar surfaces, as it provides an excellent medium for their management. Overall, the application of biochar undoubtedly offers several positive economic and environmental benefits, and its efficiency for the removal of different oxyanions operated on the laboratory scale has been widely reported in the literature. So, there is a need for future research on large-scale biochar production, and subsequent in situ experiments in a large-scale application should be initiated to test the biochar efficacy in real effluents and to determine the economic and environmental viability of biochar. Additional studies are still needed to evaluate: (1) low-cost and high-efficiency biochar modification technology, (2) reuse and regeneration strategies of the used biochar, (3) practical application of biochar in water and wastewater treatment, and (4) long-term adsorption efficiency of biochar toward different oxyanions.

## Questions and problems

1. What are the types of biochar surface modifications and how can they be achieved?
2. What are the different oxyanion sorption mechanisms of biochar?
3. What are the main differences between chemical sorption and physical sorption mechanisms of oxyanions?
4. Which of the following methods are a form of physical modification to biochar surfaces? (select all that apply)
   A. Steam activation
   B. Saturation in a $H_2O_2$ solution
   C. Microwave modification
   D. Ball milling
5. A change in which biochar physical property would lead to the greatest increase in oxyanion sorption capabilities of a biochar?
   A. macropore structure
   B. specific surface area
   C. surface charge density
   D. ion exchange capacity
6. In general, biochars produced at what pyrolysis temperature have the greatest surface area for physical sorption of oxyanions and other pollutants?

    A. Biochars produced at temperatures $\leq 350°C$
    B. Pyrolysis temperature does not impact biochar physical properties
    C. Biochars produced at temperatures $\geq 500°C$
    D. Biochars produced at temperatures between 350°C and 500°C
7. State "True" or "False"

    Phosphate tends to cause eutrophication to occur in saltwater bodies, whereas nitrate tends to cause this in freshwater bodies.
8. Which arsenic form is easier to remove from water-arsenite ($As^{3+}$, as in the hydroxyl anion $As(OH)_3$) or arsenate ($As^{5+}$, as in the oxyanion $AsO_4^{3-}$)?
9. Which of the following scenario is true of the surface charge and anion exchange capacity of biochar as pH increases?
    A. Becomes more positive and increases
    B. Becomes more negative and decreases
    C. Becomes more positive and decreases
    D. Remains same

10. State "True" or "False"

    As pyrolysis temperature increases up to about 700°C, ash content tends to increase, pH increases, and oxyanion sorption is decreased by surface functional groups.

11. State "True" or "False"

    Herbaceous feedstocks have higher ash content and more oxygenated functional groups than the woody feedstocks.

12. Fill up the blanks

    Biochar derived from _____ (high ash/low ash) feedstock is effective for removing $NO_3^-$ whereas biochar derived from _____ (high ash/low ash) feedstock is effective for removing $PO_4^{3-}$ from wastewater.

# References

Aghoghovwia, M.P., Hardie, A.G., Rozanov, A.B., 2020. Characterisation, adsorption and desorption of ammonium and nitrate of biochar derived from different feedstocks. Environ. Technol. 0, 1–14. https://doi.org/10.1080/09593330.2020.1804466.

Agrafioti, E., Kalderis, D., Diamadopoulos, E., 2014a. Arsenic and chromium removal from water using biochars derived from rice husk, organic solid wastes and sewage sludge. J. Environ. Manage. 133, 309–314. https://doi.org/10.1016/j.jenvman.2013.12.007.

Agrafioti, E., Kalderis, D., Diamadopoulos, E., 2014b. Ca and Fe modified biochars as adsorbents of arsenic and chromium in aqueous solutions. J. Environ. Manag. 146, 444–450. https://doi.org/10.1016/j.jenvman.2014.07.029.

Alsewaileh, A., Usman, A., Al-Wabel, M., 2019. Effects of pyrolysis temperature on nitrate-nitrogen ($NO_3^-$-N) and bromate ($BrO_3^-$) adsorption onto date palm biochar. J. Environ. Manag. 237, 289–296. https://doi.org/10.1016/j.jenvman.2019.02.045.

Angın, D., Altintig, E., Köse, T.E., 2013. Influence of process parameters on the surface and chemical properties of activated carbon obtained from biochar by chemical activation. Bioresour. Technol. 148, 542–549. https://doi.org/10.1016/j.biortech.2013.08.164.

Archna, A., Sharma, S.K., Sobti, R.C., 2012. Nitrate removal from ground water: a review. E J. Chem. https://doi.org/10.1155/2012/154616.

Ascott, M.J., Gooddy, D.C., Wang, L., Stuart, M.E., Lewis, M.A., Ward, R.S., Binley, A.M., 2017. Global patterns of nitrate storage in the vadose zone. Nat. Commun. 8, 1–7. https://doi.org/10.1038/s41467-017-01321-w.

Ascough, P.L., Bird, M.I., Wormald, P., Snape, C.E., Apperley, D., 2008. Influence of production variables and starting material on charcoal stable isotopic and molecular characteristics. Geochim. Cosmochim. Acta 72, 6090–6102. https://doi.org/10.1016/j.gca.2008.10.009.

Baig, S.A., Zhu, J., Muhammad, N., Sheng, T., Xu, X., 2014. Effect of synthesis methods on magnetic Kans grass biochar for enhanced As(III, V) adsorption from aqueous solutions. Biomass Bioenergy 71, 299–310. https://doi.org/10.1016/j.biombioe.2014.09.027.

Bakshi, S., Banik, C., Rathke, S.J., Laird, D.A., 2018. Arsenic sorption on zero-valent iron-biochar complexes. Water Res. 137, 153–163. https://doi.org/10.1016/j.watres.2018.03.021.

Bakshi, S., Laird, D.A., Smith, R.G., Brown, R.C., 2021. Capture and release of orthophosphate by Fe-modified biochars: mechanisms and environmental applications. ACS Sustain. Chem. Eng. 9, 658–668. https://doi.org/10.1021/acssuschemeng.0c06108.

Banik, C., Lawrinenko, M., Bakshi, S., Laird, D.A., 2018. Impact of pyrolysis temperature and feedstock on surface charge and functional group chemistry of biochars. J. Environ. Qual. 47, 452–461.

Benis, K.Z., Damuchali, A.M., Soltan, J., McPhedran, K.N., 2020. Treatment of aqueous arsenic—a review of biochar modification methods. Sci. Total Environ. 739. https://doi.org/10.1016/j.scitotenv.2020.139750, 139750.

Borah, D., Satokawa, S., Kato, S., Kojima, T., 2008. Surface-modified carbon black for As(V) removal. J. Colloid Interface Sci. 319, 53–62.

Brewer, C.E., Schmidt-Rohr, K., Satrio, J.A., Brown, R.C., 2009. Characterization of biochar from fast pyrolysis and gasification systems. Environ. Prog. Sustain. Energy 28, 386–396.

Brodowski, S., Amelung, W., Haumaier, L., Zech, W., 2007. Black carbon contribution to stable humus in German arable soils. Geoderma 139, 220–228. https://doi.org/10.1016/j.geoderma.2007.02.004.

Brown, R.A., Kercher, A.K., Nguyen, T.H., Nagle, D.C., Ball, W.P., 2006. Production and characterization of synthetic wood chars for use as surrogates for natural sorbents. Org. Geochem. 37, 321–333. https://doi.org/10.1016/j.orggeochem.2005.10.008.

Chambers, P.A., McGoldrick, D.J., Brua, R.B., Vis, C., Culp, J.M., Benoy, G.A., 2012. Development of environmental thresholds for nitrogen and phosphorus in streams. J. Environ. Qual. 41, 7–20. https://doi.org/10.2134/jeq2010.0273.

Chandra, S., Medha, I., Bhattacharya, J., 2020. Potassium-iron rice straw biochar composite for sorption of nitrate, phosphate, and ammonium in soil for timely and controlled release. Sci. Total Environ. 712. https://doi.org/10.1016/j.scitotenv.2019.136337, 136337.

Chen, B.L., Chen, Z.M., Lv, S.F., 2011. A novel magnetic biochar efficiently sorbs organic pollutants and phosphate. Bioresour. Technol. 102 (2), 716–723.

Chen, L., Chen, X.L., Zhou, C.H., Yang, H.M., Ji, S.F., Tong, D.S., Zhong, Z.K., Yu, W.H., Chu, M.Q., 2017. Environmental-friendly montmorillonite-biochar composites: facile production and tunable adsorption-release of ammonium and phosphate. J. Clean. Prod. 156, 648–659. https://doi.org/10.1016/j.jclepro.2017.04.050.

Chia, C.H., Downie, A., Munroe, P., 2015. Characteristics of biochar: Physical and structural properties. In: Lehmann, J., Joseph, S. (Eds.), Biochar for Environmental Management. Earthscan Books Ltd., London, pp. 89–109.

Chingombe, P., Saha, B., Wakeman, R.J., 2005. Surface modification and characterisation of a coal-based activated carbon. Carbon 43, 3132–3143. https://doi.org/10.1016/j.carbon.2005.06.021.

Chintala, R., Mollinedo, J., Schumacher, T.E., Papiernik, S.K., Malo, D.D., Clay, D.E., Kumar, S., Gulbrandson, D.W., 2013. Nitrate sorption and desorption in biochars from fast pyrolysis. Microporous Mesoporous Mater. 179, 250–257. https://doi.org/10.1016/j.micromeso.2013.05.023.

Cui, X., Dai, X., Khan, K.Y., Li, T., Yang, X., He, Z., 2016a. Removal of phosphate from aqueous solution using magnesium-alginate/chitosan modified biochar microspheres derived from Thalia dealbata. Bioresour. Technol. 218, 1123–1132. https://doi.org/10.1016/j.biortech.2016.07.072.

Cui, X., Hao, H., He, Z., Stoffella, P.J., Yang, X., 2016b. Pyrolysis of wetland biomass waste: potential for carbon sequestration and water remediation. J. Environ. Manag. 173, 95–104. https://doi.org/10.1016/j.jenvman.2016.02.049.

Dai, L., Fan, L., Liu, Y., Ruan, R., Wang, Y., Zhou, Y., Zhao, Y., Yu, Z., 2017. Production of bio-oil and biochar from soapstock via microwave-assisted co-catalytic fast pyrolysis. Bioresour. Technol. 225, 1–8. https://doi.org/10.1016/j.biortech.2016.11.017.

De Gisi, S., Lofrano, G., Grassi, M., Notarnicola, M., 2016. Characteristics and adsorption capacities of low-cost sorbents for wastewater treatment: a review. Sustain. Mater. Technol. https://doi.org/10.1016/j.susmat.2016.06.002.

Dieguez-Alonso, A., Anca-Couce, A., Frišták, V., Moreno-Jiménez, E., Bacher, M., Bucheli, T.D., Cimò, G., Conte, P., Hagemann, N., Haller, A., Hilber, I., Husson, O., Kammann, C.I., Kienzl, N., Leifeld, J., Rosenau, T., Soja, G., Schmidt, H.P., 2019. Designing biochar properties through the blending of biomass feedstock with metals: impact on oxyanions adsorption behavior. Chemosphere 214, 743–753. https://doi.org/10.1016/j.chemosphere.2018.09.091.

Du, R., Cao, S., Peng, Y., Zhang, H., Wang, S., 2019. Combined partial denitrification (PD)-anammox: a method for high nitrate wastewater treatment. Environ. Int. 126, 707–716. https://doi.org/10.1016/j.envint.2019.03.007.

Eduah, J.O., Nartey, E.K., Abekoe, M.K., Henriksen, S.W., Andersen, M.N., 2020. Mechanism of orthophosphate ($PO_4$-P) adsorption onto different biochars. Environ. Technol. Innov. 17. https://doi.org/10.1016/j.eti.2019.100572, 100572.

Fang, C., Zhang, T., Li, P., Jiang, R.F., Wang, Y.C., 2014. Application of magnesium modified corn biochar for phosphorus removal and recovery from swine wastewater. Int. J. Environ. Res. Public Health 11, 9217–9237.

Fang, C., Zhang, T., Li, P., Jiang, R., Wu, S., Nie, H., Wang, Y., 2015. Phosphorus recovery from biogas fermentation liquid by Ca–Mg loaded biochar. J. Environ. Sci. 29, 106–114. https://doi.org/10.1016/j.jes.2014.08.019.

Fidel, R.B., Laird, D.A., Thompson, M.L., Lawrinenko, M., 2017. Characterization and quantification of biochar alkalinity. Chemosphere 167, 367–373.

Fidel, R.B., Laird, D.A., Spokas, K.A., 2018. Sorption of ammonium and nitrate to biochars is electrostatic and pH-dependent. Sci. Rep. 8, 17627. https://doi.org/10.1038/s41598-018-35534-w.

Gao, J., Wang, W., Rondinone, A.J., He, F., Liang, L., 2015. Degradation of trichloroethene with a novel Ball milled Fe-C nanocomposite. J. Hazard. Mater. 300, 443–450. https://doi.org/10.1016/j.jhazmat.2015.07.038.

Goswami, R., Shim, J., Deka, S., Kumari, D., Kataki, R., Kumar, M., 2016. Characterization of cadmium removal from aqueous solution by biochar produced from *Ipomoea fistulosa* at different pyrolytic temperatures. Ecol. Eng. 97, 444–451. https://doi.org/10.1016/j.ecoleng.2016.10.007.

Hafshejani, L.D., Hooshmand, A., Naseri, A.A., Mohammadi, A.S., Abbasi, F., Bhatnagar, A., 2016. Removal of nitrate from aqueous solution by modified sugarcane bagasse biochar. Ecological Engineering 95, 101–111. https://doi.org/10.1016/j.ecoleng.2016.06.035.

Hale, S.E., Alling, V., Martinsen, V., Mulder, J., Breedveld, G.D., Cornelissen, G., 2013. The sorption and desorption of phosphate-P, ammonium-N and nitrate-N in cacao shell and corn cob biochars. Chemosphere 91, 1612–1619. https://doi.org/10.1016/j.chemosphere.2012.12.057.

Hassan, M., Liu, Y., Naidu, R., Parikh, S.J., Du, J., Qi, F., Willett, I.R., 2020. Influences of feedstock sources and pyrolysis temperature on the properties of biochar and functionality as adsorbents: a meta-analysis. Sci. Total Environ. 744. https://doi.org/10.1016/j.scitotenv.2020.140714, 140714.

Holland, E.A., Dentener, F.J., Braswell, B.H., Sulzman, J.M., 1999. Contemporary and pre-industrial global reactive nitrogen budgets. In: New Perspectives on Nitrogen Cycling in the Temperate and Tropical Americas. Springer, Netherlands, pp. 7–43, https://doi.org/10.1007/978-94-011-4645-6_2.

Howarth, R.W., 2008. Coastal nitrogen pollution: a review of sources and trends globally and regionally. Harmful Algae. https://doi.org/10.1016/j.hal.2008.08.015.

Hristovski, K.D., Markovski, J., 2017. Engineering metal (hydr)oxide sorbents for removal of arsenate and similar weak-acid oxyanion contaminants: a critical review with emphasis on factors governing sorption processes. Sci. Total Environ. 598, 258–271. https://doi.org/10.1016/j.scitotenv.2017.04.108.

Hu, X., Ding, Z., Zimmerman, A.R., Wang, S., Gao, B., 2015. Batch and column sorption of arsenic onto iron-impregnated biochar synthesized through hydrolysis. Water Res. 68, 206–216. https://doi.org/10.1016/j.watres.2014.10.009.

Inyang, M.I., Gao, B., Yao, Y., Xue, Y., Zimmerman, A., Mosa, A., Pullammanappallil, P., Ok, Y.S., Cao, X., 2016. A review of biochar as a low-cost adsorbent for aqueous heavy metal removal. Crit. Rev. Environ. Sci. Technol. 46, 406–433. https://doi.org/10.1080/10643389.2015.1096880.

Ippolito, J.A., Cui, L., Kammann, C., Wrage-Mönnig, N., Estavillo, J.M., Fuertes-Mendizabal, T., Cayuela, M.L., Sigua, G., Novak, J., Spokas, K., Borchard, N., 2020. Feedstock choice, pyrolysis temperature and type influence biochar characteristics: a comprehensive meta-data analysis review. Biochar 2, 421–438. https://doi.org/10.1007/s42773-020-00067-x.

Jimenez, G.D., Monti, T., Titman, J.J., Hernandez-Montoya, V., Kingman, S.W., Binner, E.R., 2017. New insights into microwave pyrolysis of biomass: preparation of carbon-based products from pecan nutshells and their application in wastewater treatment. J. Anal. Appl. Pyrolysis 124, 113–121. https://doi.org/10.1016/j.jaap.2017.02.013.

Jin, H., Capareda, S., Chang, Z., Gao, J., Xu, Y., Zhang, J., 2014. Biochar pyrolytically produced from municipal solid wastes for aqueous As(V) removal: adsorption property and its improvement with KOH activation. Bioresour. Technol. 169, 622–629. https://doi.org/10.1016/j.biortech.2014.06.103.

Jung, K.-W., Hwang, M.-J., Ahn, K.-H., Ok, Y.S., 2015a. Kinetic study on phosphate removal from aqueous solution by biochar derived from peanut shell as renewable adsorptive media. Int. J. Environ. Sci. Technol. 12. https://doi.org/10.1007/s13762-015-0766-5.

Jung, K.-W., Jeong, T.-U., Hwang, M.-J., Kim, K., Ahn, K.-H., 2015b. Phosphate adsorption ability of biochar/Mg–Al assembled nanocomposites prepared by aluminum-electrode based electro-assisted modification method with MgCl2 as electrolyte. Bioresour. Technol. 198, 603–610. https://doi.org/10.1016/j.biortech.2015.09.068.

Kanthle, A.K., Lenka, N.K., Lenka, S., Tedia, K., 2016. Biochar impact on nitrate leaching as influenced by native soil organic carbon in an inceptisol of central India. Soil Tillage Res. 157, 65–72. https://doi.org/10.1016/j.still.2015.11.009.

Keiluweit, M., Nico, P.S., Johnson, M.G., Kleber, M., 2010. Dynamic molecular structure of plant biomass-derived black carbon (biochar). Environ. Sci. Technol. 44, 1247–1253.

Kuzyakov, Y., Subbotina, I., Chen, H., Bogomolova, I., Xu, X., 2009. Black carbon decomposition and incorporation into soil microbial biomass estimated by 14C labeling. Soil Biol. Biochem. 41, 210–219.

Lal, S., Singhal, A., Kumari, P., 2020. Exploring carbonaceous nanomaterials for arsenic and chromium removal from wastewater. J. Water Process Eng. https://doi.org/10.1016/j.jwpe.2020.101276.

Lawrinenko, M., Laird, D.A., 2015. Anion exchange capacity of biochar. Green Chem. 17, 4628–4636.

Lawrinenko, M., Laird, D.A., Johnson, R.L., Jing, D., 2016. Accelerated aging of biochars; impact on anion exchange capacity. Carbon 103, 217–227.

Lawrinenko, M., Jing, D., Banik, C., Laird, D.A., 2017. Aluminum and iron biomass pretreatment impacts on biochar anion exchange capacity. Carbon 118, 422–430.

Lehmann, J., 2007. A handful of carbon. Nature 447, 143–144. https://doi.org/10.1038/447143a.

Li, R., Wang, J.J., Zhou, B., Awasthi, M.K., Ali, A., Zhang, Z., Gaston, L.A., Lahori, A.H., Mahar, A., 2016a. Enhancing phosphate adsorption by Mg/Al layered double hydroxide functionalized biochar with different Mg/Al ratios. Sci. Total Environ. 559, 121–129. https://doi.org/10.1016/j.scitotenv.2016.03.151.

Li, R., Wang, J.J., Zhou, B., Awasthi, M.K., Ali, A., Zhang, Z., Lahori, A.H., Mahar, A., 2016b. Recovery of phosphate from aqueous solution by magnesium oxide decorated magnetic biochar and its potential as phosphate-based fertilizer substitute. Bioresour. Technol. 215, 209–214. https://doi.org/10.1016/j.biortech.2016.02.125.

Li, H., Dong, X., da Silva, E.B., de Oliveira, L.M., Chen, Y., Ma, L.Q., 2017a. Mechanisms of metal sorption by biochars: biochar characteristics and modifications. Chemosphere 178, 466–478. https://doi.org/10.1016/j.chemosphere.2017.03.072.

Li, R., Wang, J.J., Zhou, B., Zhang, Z., Liu, S., Lei, S., Xiao, R., 2017b. Simultaneous capture removal of phosphate, ammonium and organic substances by MgO impregnated biochar and its potential use in swine wastewater treatment. J. Clean. Prod. 147, 96–107. https://doi.org/10.1016/j.jclepro.2017.01.069.

Li, R., Wang, J., Gaston, L.A., Zhou, B., Li, M., Xiao, R., Wang, Q., Zhang, Z., Huang, H., Liang, W., Huang, H., Zhang, X., 2018. An overview of carbothermal synthesis of metal‒biochar composites for the removal of oxyanion contaminants from aqueous solution. Carbon 129, 674–687.

Li, J., Li, B., Huang, H., Lv, X., Zhao, N., Guo, G., Zhang, D., 2019. Removal of phosphate from aqueous solution by dolomite-modified biochar derived from urban dewatered sewage sludge. Sci. Total Environ. 687, 460–469. https://doi.org/10.1016/j.scitotenv.2019.05.400.

Liu, P., Liu, W.-J., Jiang, H., Chen, J.-J., Li, W.-W., Yu, H.-Q., 2012. Modification of biochar derived from fast pyrolysis of biomass and its application in removal of tetracycline from aqueous solution. Bioresour. Technol. 121, 235–240. https://doi.org/10.1016/j.biortech.2012.06.085.

Liu, S., Tan, X., Liu, Y., Gu, Y., Zeng, G., Hu, X., Wang, H., Zhou, L., Jiang, L., Zhao, B., 2016. Production of biochars from Ca impregnated ramie biomass (*Boehmeria nivea* (L.) Gaud.) and their phosphate removal potential. RSC Adv. 6, 5871–5880. https://doi.org/10.1039/C5RA22142K.

Liu, S., Huang, B., Chai, L., Liu, Y., Zeng, G., Wang, X., Zeng, W., Shang, M., Deng, J., Zhou, Z., 2017. Enhancement of As(v) adsorption from aqueous solution by a magnetic chitosan/biochar composite. RSC Adv. 7, 10891–10900. https://doi.org/10.1039/C6RA27341F.

Liu, B., Kim, K.H., Kumar, V., Kim, S., 2020. A review of functional sorbents for adsorptive removal of arsenic ions in aqueous systems. J. Hazard. Mater. https://doi.org/10.1016/j.jhazmat.2019.121815.

Lou, K., Rajapaksha, A.U., Ok, Y.S., Chang, S.X., 2016. Pyrolysis temperature and steam activation effects on sorption of phosphate on pine sawdust biochars in aqueous solutions. Chem. Speciat. Bioavailab. 28, 42–50.

Lyu, H., Gao, B., He, F., Ding, C., Tang, J., Crittenden, J.C., 2017. Ball-milled carbon nanomaterials for energy and environmental applications. ACS Sustain. Chem. Eng. 5, 9568–9585. https://doi.org/10.1021/acssuschemeng.7b02170.

Lyu, H., Gao, B., He, F., Zimmerman, A.R., Ding, C., Huang, H., Tang, J., 2018. Effects of ball milling on the physicochemical and sorptive properties of biochar: experimental observations and governing mechanisms. Environ. Pollut. 233, 54–63. https://doi.org/10.1016/j.envpol.2017.10.037.

Maroušek, J., 2014. Significant breakthrough in biochar cost reduction. Clean Techn. Environ. Policy 16, 1821–1825. https://doi.org/10.1007/s10098-014-0730-y.

Mekonnen, M.M., Hoekstra, A.Y., 2018. Global anthropogenic phosphorus loads to freshwater and associated grey water footprints and water pollution levels: a high-resolution global study. Water Resour. Res. 54, 345–358. https://doi.org/10.1002/2017WR020448.

Mia, S., Dijkstra, F.A., Singh, B., 2017. Chapter one—Long-term aging of biochar: a molecular understanding with agricultural and environmental implications. In: Sparks, D.L. (Ed.), Advances in Agronomy. Academic Press, pp. 1–51, https://doi.org/10.1016/bs.agron.2016.10.001.

Micháleková-Richveisová, B., Frišták, V., Pipíška, M., Ďuriška, L., Moreno-Jimenez, E., Soja, G., 2017. Iron-impregnated biochars as effective phosphate sorption materials. Environ. Sci. Pollut. Res. 24, 463–475.

Mohan, D., Pittman Jr., C.U., 2007. Arsenic removal from water/wastewater using adsorbents—a critical review. J. Hazard. Mater. 142 (1–2), 1–53.

Mukherjee, A., Zimmerman, A.R., Harris, W.G., 2011. Surface chemistry variations among a series of laboratory-produced biochars. Geoderma 163, 247–255.

Neset, T.-S.S., Cordell, D., 2012. Global phosphorus scarcity: identifying synergies for a sustainable future. J. Sci. Food Agric. 92, 2–6. https://doi.org/10.1002/jsfa.4650.

Nordstrom, D.K., 2002. Worldwide occurrences of arsenic in ground water. Science. https://doi.org/10.1126/science.1072375.

Novak, J., Lima, I., Xing, B., Gaskin, J., Steiner, C., Das, K.C., Ahmedna, M., Rehrah, D., Watts, D.W., Busscher, W., Schomberg, H., 2009. Characterization of Designer Biochar Produced at Different Temperatures and Their Effects on a Loamy Sand. United States Dep. Agric. Agric. Res. Serv. Coast. Plains Res.

Lab, Florence, South Carolina. 29501, United States dep. Agric. Agric. Res. Serv. South. Reg. Res. Center, New Orle 3.

Park, J.-H., Wang, J.J., Xiao, R., Zhou, B., Delaune, R.D., Seo, D.-C., 2018. Effect of pyrolysis temperature on phosphate adsorption characteristics and mechanisms of crawfish char. J. Colloid Interface Sci. 525, 143–151. https://doi.org/10.1016/j.jcis.2018.04.078.

Pepper, I.L., Gerba, C.P., Brusseau, M.L., 2006. Environmental and Pollution Science, second ed. Elsevier Science & Technology.

Qambrani, N.A., Rahman, M.M., Won, S., Shim, S., Ra, C., 2017. Biochar properties and eco-friendly applications for climate change mitigation, waste management, and wastewater treatment: a review. Renew. Sustain. Energy Rev. 79, 255–273. https://doi.org/10.1016/j.rser.2017.05.057.

Rajapaksha, A.U., Chen, S.S., Tsang, D.C.W., Zhang, M., Vithanage, M., Mandal, S., Gao, B., Bolan, N.S., Ok, Y.S., 2016. Engineered/designer biochar for contaminant removal/immobilization from soil and water: potential and implication of biochar modification. Chemosphere 148, 276–291. https://doi.org/10.1016/j.chemosphere.2016.01.043.

Ravenscroft, P., Brammer, H., Richards, K., 2009. Arsenic Pollution: A Global Synthesis., https://doi.org/10.1002/9781444308785.

Ren, J., Li, N., Li, L., An, J.-K., Zhao, L., Ren, N.-Q., 2015. Granulation and ferric oxides loading enable biochar derived from cotton stalk to remove phosphate from water. Bioresour. Technol. 178, 119–125. https://doi.org/10.1016/j.biortech.2014.09.071.

Rhodes, C.J., 2013. Peak phosphorus—peak food? The need to close the phosphorus cycle. Sci. Prog. 96, 109–152. https://doi.org/10.3184/003685013X13677472447741.

Richard, S., Rajadurai, J.S., Manikandan, V., 2016. Influence of particle size and particle loading on mechanical and dielectric properties of biochar particulate-reinforced polymer nanocomposites. Int. J. Polym. Anal. Charact. 21, 462–477. https://doi.org/10.1080/1023666X.2016.1168602.

Saha, N., Saba, A., Reza, M.T., 2019. Effect of hydrothermal carbonization temperature on pH, dissociation constants, and acidic functional groups on hydrochar from cellulose and wood. J. Anal. Appl. Pyrolysis 137, 138–145. https://doi.org/10.1016/j.jaap.2018.11.018.

Samsuri, A.W., Sadegh-Zadeh, F., Seh-Bardan, B.J., 2013. Adsorption of As(III) and As(V) by Fe coated biochars and biochars produced from empty fruit bunch and rice husk. J. Environ. Chem. Eng. 1, 981–988. https://doi.org/10.1016/j.jece.2013.08.009.

Sanford, J.R., Larson, R.A., Runge, T., 2019. Nitrate sorption to biochar following chemical oxidation. Sci. Total Environ. 669, 938–947. https://doi.org/10.1016/j.scitotenv.2019.03.061.

Shah, A.I., Din Dar, M.U., Bhat, R.A., Singh, J.P., Singh, K., Bhat, S.A., 2020. Prospectives and challenges of wastewater treatment technologies to combat contaminants of emerging concerns. Ecol. Eng. https://doi.org/10.1016/j.ecoleng.2020.105882.

Shannon, M.A., Bohn, P.W., Elimelech, M., Georgiadis, J.G., Mariñas, B.J., Mayes, A.M., 2008. Science and technology for water purification in the coming decades. Nature 452, 301–310.

Shao, Y., Long, Y., Wang, H., Liu, D., Shen, D., Chen, T., 2019. Hydrochar derived from green waste by microwave hydrothermal carbonization. Renew. Energy 135, 1327–1334. https://doi.org/10.1016/j.renene.2018.09.041.

Shim, T., Yoo, J., Ryu, C., Park, Y.-K., Jung, J., 2015. Effect of steam activation of biochar produced from a giant Miscanthus on copper sorption and toxicity. Bioresour. Technol. 197, 85–90. https://doi.org/10.1016/j.biortech.2015.08.055.

Sizmur, T., Fresno, T., Akgül, G., Frost, H., Moreno-Jiménez, E., 2017. Biochar modification to enhance sorption of inorganics from water. Bioresour. Technol. 246, 34–47. https://doi.org/10.1016/j.biortech.2017.07.082.

Spokas, K.A., Koskinen, W.C., Baker, J.M., Reicosky, D.C., 2009. Impacts of woodchip biochar additions on greenhouse gas production and sorption/degradation of two herbicides in a Minnesota soil. Chemosphere 77, 574–581. https://doi.org/10.1016/j.chemosphere.2009.06.053.

Takaya, C.A., Fletcher, L.A., Singh, S., Anyikude, K.U., Ross, A.B., 2016. Phosphate and ammonium sorption capacity of biochar and hydrochar from different wastes. Chemosphere 145, 518–527. https://doi.org/10.1016/j.chemosphere.2015.11.052.

Tan, X., Liu, Y., Gu, Y., Xu, Y., Zeng, G., Hu, X., Liu, S., Wang, X., Liu, S., Li, J., 2016. Biochar-based nano-composites for the decontamination of wastewater: a review. Bioresour. Technol. https://doi.org/10.1016/j.biortech.2016.04.093.

Tao, X., Huang, T., Lv, B., 2020. Synthesis of Fe/Mg-biochar nanocomposites for phosphate removal. Materials 13, 816.

Uchimiya, M., Wartelle, L.H., Klasson, K.T., Fortier, C.A., Lima, I.M., 2011. Influence of pyrolysis temperature on biochar property and function as a heavy metal sorbent in soil. Agric. Food Chem. 59, 2501–2510.

Uppal, J.S., Zheng, Q., Le, X.C., 2019. Arsenic in drinking water—recent examples and updates from Southeast Asia. Curr. Opin. Environ. Sci. Health. https://doi.org/10.1016/j.coesh.2019.01.004.

US EPA, 2020. Estimated Nitrate Concentrations in Groundwater Used for Drinking | Nutrient Pollution Policy and Data. US EPA (WWW Document).

Usman, A., Ahmad, M., El-mahrouky, M., Alomran, A., Ok, Y.S., Sallam, A., El-Naggar, A., Al-Wabel, M., 2015. Chemically modified biochar produced from conocarpus waste increases $NO_3$ removal from aqueous solutions. Environ. Geochem. Health 38. https://doi.org/10.1007/s10653-015-9736-6.

Villarín, M.C., Merel, S., 2020. Paradigm shifts and current challenges in wastewater management. J. Hazard. Mater. 390. https://doi.org/10.1016/j.jhazmat.2020.122139, 122139.

Wang, Y., Liu, R., 2018. $H_2O_2$ treatment enhanced the heavy metals removal by manure biochar in aqueous solutions. Sci. Total Environ. 628–629, 1139–1148. https://doi.org/10.1016/j.scitotenv.2018.02.137.

Wang, S., Gao, B., Li, Y., Mosa, A., Zimmerman, A.R., Ma, L.Q., Harris, W.G., Migliaccio, K.W., 2015a. Manganese oxide-modified biochars: preparation, characterization, and sorption of arsenate and lead. Bioresour. Technol. 181, 13–17. https://doi.org/10.1016/j.biortech.2015.01.044.

Wang, S., Gao, B., Zimmerman, A.R., Li, Y., Ma, L.Q., Harris, W.G., Migliaccio, K.W., 2015b. Removal of arsenic by magnetic biochar prepared from pinewood and natural hematite. Bioresour. Technol. 175, 391–395.

Wang, Z., Guo, H., Shen, F., Yang, G., Zhang, Y., Zeng, Y., Wang, L., Xiao, H., Deng, S., 2015c. Biochar produced from oak sawdust by lanthanum (La)-involved pyrolysis for adsorption of ammonium ($NH_4^+$), nitrate ($NO_3^-$), and phosphate ($PO_4^{3-}$). Chemosphere 119, 646–653. https://doi.org/10.1016/j.chemosphere.2014.07.084.

Wang, S., Gao, B., Li, Y., 2016. Enhanced arsenic removal by biochar modified with nickel (Ni) and manganese (Mn) oxyhydroxides. J. Ind. Eng. Chem. 37, 361–365. https://doi.org/10.1016/j.jiec.2016.03.048.

Wang, B., Gao, B., Fang, J., 2017a. Recent advances in engineered biochar productions and applications. Crit. Rev. Environ. Sci. Technol. 47, 2158–2207. https://doi.org/10.1080/10643389.2017.1418580.

Wang, S., Gao, B., Li, Y., Creamer, A.E., He, F., 2017b. Adsorptive removal of arsenate from aqueous solutions by biochar supported zero-valent iron nanocomposite: batch and continuous flow tests. J. Hazard. Mater. 322, 172–181.

Wang, X., Gu, Y., Tan, X., Liu, Y., Zhou, Y., Hu, X., Cai, X., Xu, W., Zhang, C., Liu, S., 2019. Functionalized biochar/clay composites for reducing the bioavailable fraction of arsenic and cadmium in river sediment. Environ. Toxicol. Chem. 38, 2337–2347. https://doi.org/10.1002/etc.4542.

Wang, X., Guo, Z., Hu, Z., Zhang, J., 2020a. Recent advances in biochar application for water and wastewater treatment: a review. PeerJ 8. https://doi.org/10.7717/peerj.9164, e9164.

Wang, Z., Bakshi, S., Li, C., Parikh, S.J., Hsieh, H.-S., Pignatello, J.J., 2020b. Modification of pyrogenic carbons for phosphate sorption through binding of a cationic polymer. J. Colloid Interface Sci. 579, 258–268. https://doi.org/10.1016/j.jcis.2020.06.054.

Warnock, D.D., Lehmann, J., Kuyper, T.W., Rillig, M.C., 2007. Mycorrhizal responses to biochar in soil—concepts and mechanisms. Plant Soil 300, 9–20. https://doi.org/10.1007/s11104-007-9391-5.

Weerasundara, L., Ok, Y.S., Bundschuh, J., 2021. Selective removal of arsenic in water: a critical review. Environ. Pollut. https://doi.org/10.1016/j.envpol.2020.115668.

Wongrod, S., Simon, S., van Hullebusch, E.D., Lens, P.N.L., Guibaud, G., 2018. Changes of sewage sludge digestate-derived biochar properties after chemical treatments and influence on As(III and V) and Cd(II) sorption. Int. Biodeterior. Biodegrad. 135, 96–102. https://doi.org/10.1016/j.ibiod.2018.10.001.

Xiang, W., Zhang, X., Chen, J., Zou, W., He, F., Hu, X., Tsang, D.C.W., Ok, Y.S., Gao, B., 2020. Biochar technology in wastewater treatment: a critical review. Chemosphere 252. https://doi.org/10.1016/j.chemosphere.2020.126539, 126539.

Xiao, F., Pignatello, J.J., 2016. Effects of post-pyrolysis air oxidation of biomass chars on adsorption of neutral and ionizable compounds. Environ. Sci. Technol. 50, 6276–6283.

Xu, K., Lin, F., Dou, X., Zheng, M., Tan, W., Wang, C., 2018. Recovery of ammonium and phosphate from urine as value-added fertilizer using wood waste biochar loaded with magnesium oxides. J. Clean. Prod. 187, 205–214. https://doi.org/10.1016/j.jclepro.2018.03.206.

Xue, L., Gao, B., Wan, Y., Fang, J., Wang, S., Li, Y., Muñoz-Carpena, R., Yang, L., 2016. High efficiency and selectivity of MgFe-LDH modified wheat-straw biochar in the removal of nitrate from aqueous solutions. J. Taiwan Inst. Chem. Eng. 63, 312–317. https://doi.org/10.1016/j.jtice.2016.03.021.

Yadav, M.K., Saidulu, D., Gupta, A.K., Ghosal, P.S., Mukherjee, A., 2021. Status and management of arsenic pollution in groundwater: a comprehensive appraisal of recent global scenario, human health impacts, sustainable field-scale treatment technologies. J. Environ. Chem. Eng. https://doi.org/10.1016/j.jece.2021.105203.

Yang, X., Wan, Y., Zheng, Y., He, F., Yu, Z., Huang, J., Wang, H., Ok, Y.S., Jiang, Y., Gao, B., 2019. Surface functional groups of carbon-based adsorbents and their roles in the removal of heavy metals from aqueous solutions: a critical review. Chem. Eng. J. 366, 608–621. https://doi.org/10.1016/j.cej.2019.02.119.

Yang, H., Ye, S., Zeng, Z., Zeng, G., Tan, X., Xiao, R., Wang, J., Song, B., Du, L., Qin, M., Yang, Y., Xu, F., 2020. Utilization of biochar for resource recovery from water: a review. Chem. Eng. J. https://doi.org/10.1016/j.cej.2020.125502.

Yao, Y., Gao, B., Inyang, M., Zimmerman, A.R., Cao, X., Pullammanappallil, P., Yang, L., 2011. Biochar derived from anaerobically digested sugar beet tailings:

characterization and phosphate removal potential. Bioresour. Technol. 102, 6273–6278. https://doi.org/10.1016/j.biortech.2011.03.006.

Yao, Y., Gao, B., Zhang, M., Inyang, M., Zimmerman, A.R., 2012. Effect of biochar amendment on sorption and leaching of nitrate, ammonium, and phosphate in a sandy soil. Chemosphere 89, 1467–1471.

Yao, Y., Gao, B., Chen, J., Yang, L., 2013a. Engineered biochar reclaiming phosphate from aqueous solutions: mechanisms and potential application as a slow-release fertilizer. Environ. Sci. Technol. 47, 8700–8708.

Yao, Y., Gao, B., Chen, J., Zhang, M., Inyang, M., Li, Y., Alva, A., Yang, L., 2013b. Engineered carbon (biochar) prepared by direct pyrolysis of Mg-accumulated tomato tissues: characterization and phosphate removal potential. Bioresour. Technol. 138, 8–13. https://doi.org/10.1016/j.biortech.2013.03.057.

Yin, Q., Ren, H., Wang, R., Zhao, Z., 2018a. Evaluation of nitrate and phosphate adsorption on Al-modified biochar: influence of Al content. Sci. Total Environ. 631–632, 895–903. https://doi.org/10.1016/j.scitotenv.2018.03.091.

Yin, Q., Wang, R., Zhao, Z., 2018b. Application of Mg–Al-modified biochar for simultaneous removal of ammonium, nitrate, and phosphate from eutrophic water. J. Clean. Prod. 176, 230–240. https://doi.org/10.1016/j.jclepro.2017.12.117.

Zhang, M., Gao, B., 2013. Removal of arsenic, methylene blue, and phosphate by biochar/AlOOH nanocomposite. Chem. Eng. J. 226, 286–292. https://doi.org/10.1016/j.cej.2013.04.077.

Zhang, M., Gao, B., Yao, Y., Xue, Y., Inyang, M., 2012. Synthesis of porous MgO-biochar nanocomposites for removal of phosphate and nitrate from aqueous solutions. Chem. Eng. J. 210, 26–32. https://doi.org/10.1016/j.cej.2012.08.052.

Zhang, M., Gao, B., Varnoosfaderani, S., Hebard, A., Yao, Y., Inyang, M., 2013a. Preparation and characterization of a novel magnetic biochar for arsenic removal. Bioresour. Technol. 130, 457–462.

Zhang, M., Gao, B., Yao, Y., Inyang, M., 2013b. Phosphate removal ability of biochar/MgAl-LDH ultra-fine composites prepared by liquid-phase deposition. Chemosphere 92, 1042–1047. https://doi.org/10.1016/j.chemosphere.2013.02.050.

Zhang, F., Wang, X., Xionghui, J., Ma, L., 2016. Efficient arsenate removal by magnetite-modified water hyacinth biochar. Environ. Pollut. 216, 575–583. https://doi.org/10.1016/j.envpol.2016.06.013. Epub 2016 Jul 2. PMID 27376988.

Zhang, M., Song, G., Gelardi, D.L., Huang, L., Khan, E., Mašek, O., Parikh, S.J., Ok, Y.-S., 2020. Evaluating biochar and its modifications for the removal of ammonium, nitrate, and phosphate in water. Water Res. 186. https://doi.org/10.1016/j.watres.2020.116303, 116303.

Zhang, X., Zhang, Y., Shi, P., Bi, Z., Shan, Z., Ren, L., 2021. The deep challenge of nitrate pollution in river water of China. Sci. Total Environ. https://doi.org/10.1016/j.scitotenv.2020.144674.

Zhong, Z., Yu, G., Mo, W., Zhang, C., Huang, H., Li, S., Gao, M., Lu, X., Zhang, B., Zhu, H., 2019. Enhanced phosphate sequestration by Fe(III) modified biochar derived from coconut shell. RSC Adv. 9, 10425–10436. https://doi.org/10.1039/C8RA10400J.

Zhou, Y., Gao, B., Zimmerman, A.R., Chen, H., Zhang, M., Cao, X., 2014. Biochar-supported zerovalent iron for removal of various contaminants from aqueous solutions. Bioresour. Technol. 152, 538–542.

Zhou, N., Chen, H., Xi, J., Yao, D., Zhou, Z., Tian, Y., Lu, X., 2017a. Biochars with excellent Pb(II) adsorption property produced from fresh and dehydrated banana peels via hydrothermal carbonization. Bioresour. Technol. 232, 204–210. https://doi.org/10.1016/j.biortech.2017.01.074.

Zhou, Z., Liu, Y., Liu, S., Liu, H., Zeng, G., Tan, X., Yang, C., Ding, Y., Yan, Z., Cai, X., 2017b. Sorption performance and mechanisms of arsenic(V) removal by magnetic gelatin-modified biochar. Chem. Eng. J. 314, 223–231. https://doi.org/10.1016/j.cej.2016.12.113.

# Arsenic removal from household drinking water by biochar and biochar composites: A focus on scale-up

Jacinta Alchouron[a], Amalia L. Bursztyn Fuentes[b], Abigail Musser[c], Andrea S. Vega[a], Dinesh Mohan[d], Charles U. Pittman, Jr.[e], Todd E. Mlsna[e], and Chanaka Navarathna[e]

[a]Facultad de Agronomía, Departamento de Recursos Naturales y Ambiente, Cátedra de Botánica General, Universidad de Buenos Aires, Buenos Aires, Argentina. [b]Technological Center of Mineral Resources and Ceramics (CETMIC-CIC/CONICET/UNLP), M.B. Gonnet, Argentina. [c]Department of Civil and Environmental Engineering, Mississippi State University, Starkville, MS, United States. [d]School of Environmental Sciences, Jawaharlal Nehru University, New Delhi, India. [e]Department of Chemistry, Mississippi State University, Mississippi State, MS, United States

## 1 Introduction

Arsenic (As) is a metalloid present in environmental matrixes worldwide. Arsenic concentrations vary in natural waters and depend on local factors such as geology, hydrology, and geochemical characteristics of the aquifer (Akter et al., 2005). The WHO recommends a maximum arsenic concentration $10\,\mu g\,L^{-1}$ for drinking water or less, but natural groundwater concentrations of $>10\,\mu g\,L^{-1}$ have been reported in over 120 countries, with the highest concentrations in Argentina, Bangladesh, Chile, China, Hungary, India, Mexico, Nepal, Romania, Taiwan, Vietnam, and USA. The occurrence, distribution, and origin of As in water have received a lot of attention in the last two decades, including comprehensive books and papers (Kumar et al., 2019; Litter et al., 2019a; Mohan and Pittman Jr., 2007; Ravenscroft et al., 2011).

Arsenic originates from both natural processes and anthropogenic sources. Anthropogenic activities include smelting, mining of As-bearing minerals, semiconductor manufacturing, exploitation and burning of coal, disposal and incineration of municipal and industrial wastes, use of As-containing products in wood preservatives, arsenical pesticides, herbicides, pharmaceuticals, etc. (Chowdhury and Mulligan, 2011; Li et al., 2018). However, except in specific cases, most groundwater and surface water contamination is natural, related to geological processes and the associated hydrothermal activity of mountain ranges. The main natural environmental sources are volcanic rocks, marine sedimentary rocks, hydrothermal mineral deposits, and associated geothermal waters. In addition, there are more than 250 reported types of As-bearing minerals; it is frequently found combined with sulfur (AsS, $As_2S_3$) or metals such as arsenopyrite (FeAsS) (Alkurdi et al., 2019). Water-solid phase geochemical interactions, such as sorption-desorption reactions and precipitation-dissolution from the solid phase, control As mobilization into water.

In water, arsenic is found mainly in inorganic forms as dissolved arsenite (+3) and arsenate (+5) oxyanions. The main factors that regulate the speciation and concentration of the respective oxoanions in water are their redox potentials ($E_h$) and the pH of the medium (Darling, 2016; Shakoor et al., 2017; Shakoor et al., 2016) (Fig. 8.1A, B and C). As(III) will predominate in environments with low redox potentials and oxygen-starved conditions (ground water), while As(V) is the most important species in environments with high redox potentials. However, their conjugate dissociation patterns are completely different. In reducing conditions, $H_3AsO_3$ predominates in solution through pH 9. $H_2AsO_3^-$ begins to form at ~pH 8 and predominates from ~pH 9.2 to 12. $HAsO_3^{2-}$ is first formed at ~pH 10 and is the predominant species from ~pH 12.2 to 13 (Fig. 8.1A) (Smedley and Kinniburgh, 2002). $H_3AsO_3$ can also be protonated ($H_4AsO_3^+$) but only in strongly acidic solution. In oxidizing conditions, $H_2AsO_4^-$ dominates at lower pH (<6.9) and $HAsO_4^{2-}$ is dominant at higher pH (>8). $H_3AsO_4$ and $AsO_4^{3-}$ may be present in strong acid or base conditions, respectively (Fig. 8.1B) (Smedley and Kinniburgh, 2002). In addition, the arsenic concentration in natural waters is controlled by a complex set of solid-liquid interactions with the surfaces of the minerals present and, in many cases, with organic matter (Dong et al., 2014).

Arsenic is one of the most toxic elements. Arsenic can enter the human body through food, water, soil, and air. Today, arsenic poisoning due to the consumption of contaminated drinking water is

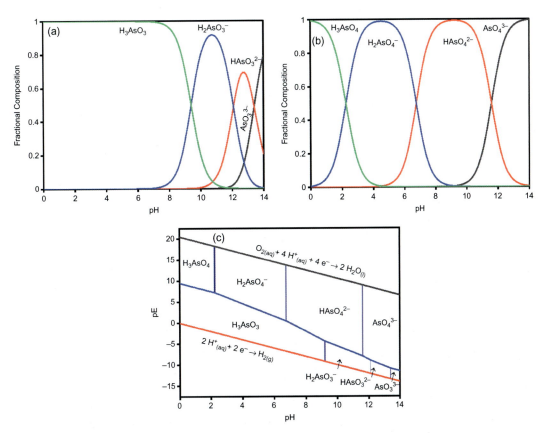

**Fig. 8.1** pH dependent (A) arsenite, (B) arsenate, and (C) speciation diagram of As as a function of pH and $E_h$ at 25°C and 101.3 kPa.

a major health concern worldwide (Huang et al., 2015). The number of exposed people is estimated at around 14 million in Latina America alone (Kumar et al., 2019). The presence of relatively high As concentrations in water is a problem due to the harmful side effects it can produce. Chronic ingestion of As-bearing water produces arsenicosis or Chronic Regional Endemic Hydroarsenicism (HACRE, for its Spanish acronym) (Astolfi et al., 1981). This pathology is associated with skin lesions (palm-plantar hyperkeratosis), gastrointestinal problems, hepatic damage, disturbances of cardiovascular and nervous system functions, as well as lung, liver, kidney, bladder, and skin cancer (Bardach et al., 2015; Naujokas et al., 2013). Arsenic is detrimental to those exposed, causing severe health problems and sometimes even death.

Proposed arsenic removal techniques for water and wastewater include coagulation-filtration, (Sancha and Fuentealba, 2009; Sancha Fernández, 2006) including using a preoxidation step (Bordoloi et al., 2013), electrocoagulation (Kumar and Goel, 2010), ion exchange (Wang et al., 2000), and membrane/reverse osmosis (Ning, 2002; Shih, 2005). Most of these conventional approaches require expensive reagents and instruments. Moreover, these systems require some specific technical skills, including the manipulation and mixture of chemicals and production of considerable amounts of hazardous sludge (Baig et al., 2014).

Adsorption is an attractive alternative to conventional techniques, and it is widely employed because it is cost-effective and simple to operate. This is ideal for rural areas, and it generates lower environmental impacts than conventional technologies (Asere et al., 2019; Baig et al., 2014). Traditional As adsorbents include activated alumina, activated carbon, and iron oxide-based sorbents. Recent attention has been given to low-cost sorbents produced from accessible and abundant natural sources such as clays, iron oxides, zeolites (Asere et al., 2019; Uddin, 2017), soils (Deschamps et al., 2003; Rajapaksha et al., 2011; Yazdani et al., 2016), biosorbents or nonliving biomass (Michalak et al., 2013; Shakoor et al., 2016), and low-cost carbonaceous materials or biochars (Amen et al., 2020; Bursztyn Fuentes et al., 2021; Mohan and Pittman, 2007).

Biochars and biochar-derived composites have gained significant recent attention, especially arsenic mitigation with biochars. In contrast to the raw feedstock or biomass adsorbents (Michalak et al., 2013), biochar can be stored long term. As removal was achieved with high efficiencies and sorption capacities, particularly with modified biochars. These biochar and biochar-derived composite capacities were obtained under controlled conditions using synthetic solutions without competing ions at optimal pH. Trials are usually only performed in batch systems on a laboratory scale, and column tests are rarely conducted. Thus, major knowledge gaps exist in engineered adsorbent real-world performance on larger working scales. Additionally, no updated document exists compiling the recent advances on biochar applications to scaled-up household water treatments to provide As-safe drinking water.

This chapter presents an overview of biochar-based adsorbents for arsenic removal, based on a literature search (2010–2020 period) and the authors' own experiences. An understanding of the ongoing scale-up experiences is essential to further develop economical methods effective for water treatment to meet the urgent need for safe drinking water.

## 2 Biochar-based adsorbent synthesis

### 2.1 Biochar production and its relevance in arsenic removal

Biochar is low cost, sustainable, and environmentally friendly. Also it can be synthesized from almost any residual feedstock or by-product locally available. Biochars for As removal have been produced from different biomass residues and agricultural wastes including bamboo culms (Alchouron et al., 2020), Japanese oak wood (Niazi et al., 2018b), red oak, switchgrass (Bakshi et al., 2018), corn straw (Fan et al., 2018; He et al., 2018), cotton stalks (Hussain et al., 2019), banana pith (Lata et al., 2019), rice straw (Nguyen et al., 2019), white birch (Braghiroli et al., 2020), spent coffee grounds (Cho et al., 2017), rice and wheat husks (Singh et al., 2020), etc. Additionally, nonconventional feedstocks such as bones (Liu et al., 2016), algae (Johansson et al., 2016), tires (Mouzourakis et al., 2017), municipal solid waste and sewage sludge (Agrafioti et al., 2014; Jin et al., 2014; Tavares et al., 2012) have been used.

The feedstock type defines many of the initial biochar textural properties, while pyrolysis temperature determines the final physicochemical properties (Weber and Quicker, 2018). Higher temperatures remove greater amounts of volatile compounds, promoting a more graphitic-like structure and higher pore volumes, pore distributions, and surface areas. Additionally, composition and surface chemistry vary. When ash content increases, C content decreases, oxygen-containing surface groups decrease, and $pH_{PZC}$ increases as pyrolysis temperature rises. The O/C ratio decreases with temperature as aromaticity rises, and the surface becomes more hydrophilic. Biochars produced at higher pyrolysis temperatures (600–800°C) are usually more effective than those prepared at low temperatures (300–500°C) (Amen et al., 2020; Benis et al., 2020). For instance, Niazi et al. (2018a) found that perilla leaf biochar produced at 700°C had better As(III) and As(V) sorption capacities than when produced at 300°C. The former had a higher surface area, smaller pore sizes, and higher aromaticity. However, some biochars produced at lower pyrolysis temperatures can be more effective (Vithanage et al., 2017) when considering the crucial role of polar functional groups with a higher As affinity (mainly –OH, –COOH, –SH) or those which promote As removal through electrostatic attraction, ion exchange, and surface complexation (Amen et al., 2020). There is a consensus that the maximum As sorption capacity is achieved when biochar has an optimal net surface charge (Benis et al., 2020). This is related to the abundance and type of biochar surface functionalities present. Consequently, FTIR spectroscopy and $pH_{PZC}$ have been widely used to provide biochar functional group information and surface net charge, both before and after As sorption. Further research is needed to better understand the effect of low and high pyrolysis temperatures on biochar's physicochemical properties and its correlation with As sorption capacities (Amen et al., 2020).

The reported maximum As(III) and As(V) sorption capacities of pristine biochars vary over a narrow range ($<13\,mg\,g^{-1}$), dependent on both the type of feedstock and the pyrolysis conditions. Mohan et al. (2007) achieved 1.2, 5.85, 12.15, and $7.4\,mg\,g^{-1}$ maximum As(III) sorption capacities for byproduct pinewood biochar, oak wood biochar, pine bark biochar, and oak bark biochar, respectively, produced at 400–450°C by a relatively fast pyrolysis in a bio-oil reactor. Van Vinh et al. (2015) achieved a maximum adsorption capacity of $5.7 \times 10^{-3}\,mg\,g^{-1}$ for As(III) with pine cone biochar pyrolyzed at 500°C. Agrafioti et al. (2014) achieved $0.013\,mg\,g^{-1}$ for As(V) with sewage sludge biochar pyrolyzed at 300°C, and Norazlina et al. (2014) achieved As(V) maximum sorption capacities of 0.42 and $0.35\,mg\,g^{-1}$ with empty fruit bunch and rice rusk biochars produced at 300–350°C and 500°C, respectively. More recently, Niazi et al. (2018b) achieved As(III) and As(V) maximum sorption capacities of 3.89 and $3.16\,mg\,g^{-1}$, respectively, for Japanese oak wood biochar produced at 500°C, and Ali et al. (2020) achieved As(III) and As(V) maximum sorption capacities of 12.86 and $5.65\,mg\,g^{-1}$, respectively, for almond shell biochar produced at 450°C.

It has been well accepted that pristine biochars are less effective than modified or the so-called engineered biochars for arsenic removal. Indeed, currently few articles report the sorption capacities of the pristine material, and the modified counterparts are directly characterized and used. The enhanced As sorption capacities of modified biochars have been attributed to improvements in both textural properties, surface chemistry, higher porosity and surface areas, along with an increase in desired functional groups or compounds with higher affinity to As (Amen et al., 2020).

## 2.2 Engineering strategies to enhance arsenic removal

Because pristine biochars have shown relatively low As removal efficiencies, several engineering strategies have been studied to enhance As sorption capacities (Fig. 8.2A). Some of

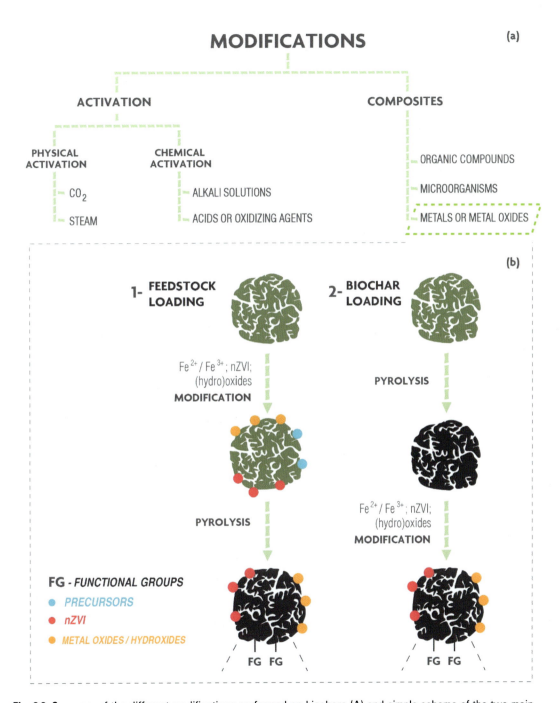

Fig. 8.2 Summary of the different modifications performed on biochars (A) and simple scheme of the two main methodological paths to achieve Fe-based composites (B). Adapted from Sizmur, T., Fresno, T., Akgül, G., Frost, H. and Moreno-Jiménez, E. (2017) Biochar modification to enhance sorption of inorganics from water. Bioresour. Technol. 246, 34-47 and Amen, R., Bashir, H., Bibi, I., Shaheen, S.M., Niazi, N.K., Shahid, M., Hussain, M.M., Antoniadis, V., Shakoor, M.B. and Al-Solaimani, S.G. (2020) A critical review on arsenic removal from water using biochar-based sorbents: the significance of modification and redox reactions. Chem. Eng. J., 125195.

the more frequent pre- or postpyrolysis modifications (Sizmur et al., 2017) are treatments with alkali or acidic solutions such as KOH (Wongrod et al., 2018) and $H_3PO_4$ (Hussain et al., 2019), oxidation with $HNO_3$ (Budinova et al., 2006; Frišták et al., 2018), functionalization with particular surface groups (e.g., thiols) (Roy et al., 2013), Impregnation with metals or metal oxide/hydroxides nanoparticles produces composites, containing Zn (Cruz et al., 2020), Mn (Liu et al., 2016), Ti (Luo et al., 2019), Ca (Mondal et al., 2010), Al (Ding et al., 2018; Qian et al., 2013), Fe (Hao et al., 2018), or mixtures (Lin et al., 2017; Lin et al., 2019; Liu et al., 2019).

Chemically and physically modified biochars have exhibited a wide range of As sorption capacities both for As(III) and As(V), depending mainly on the feedstock and treatment characteristics. Sorption capacities after chemical activation have been reported between 0.001 and 30.9 $mg\,g^{-1}$. Steam activation was the only reviewed physical treatment and had sorption capacities between 0.001 and 1.8 $mg\,g^{-1}$ for both As(III) and As(V) (Benis et al., 2020). The higher sorption capacities of chemically and physically modified biochars contrast with their pristine counterparts. Activating agents have positive effects on textural parameters such as surface area, pore volume, and pore size distribution, which facilitate diffusion of As into the adsorbents pores. Interestingly, low As adsorption values have been found for some materials with high specific surfaces and porosity, indicating that many factors can govern the adsorption process besides textural parameters: biochar surface functional groups being among these (Vithanage et al., 2017).

Biochar-based composite sorbents are increasingly popular (Benis et al., 2020) since the surface chemical and physical treatments offer technical and environmental benefits (Premarathna et al., 2019). Most of them have affinity for metals and metallic compounds or phases (Bowell et al., 2014; Hao et al., 2018). Supporting and dispersing metallic or oxide/hydroxide nanoparticles with affinity to As onto biochar reduces the self-agglomeration of these particles when used alone, enhancing adsorption capacities (Zhang et al., 2013a). Moreover, the porous carbonaceous skeleton provides mechanical strength which is essential for column-based adsorption processes. Also, the biochar carbonaceous phase can adsorb additional sorbates. Biochar composites can be endowed with a magnetic response (e.g., magnetite phase), enabling easy recovery of the As-loaded adsorbents from solution using an external magnetic field (Wang et al., 2017). Hence, these easy-to-produce nanodispersions present reusability potential after several application cycles. Additionally, fewer hazardous chemicals and less energy are needed for biochar-based adsorption versus some other methods. Finally, nanoparticles are regarded as an emerging class of pollutants, so their direct incorporation in water is highly discouraged (Hao et al., 2018). Tiny adsorbent particles also lead to high pressure drops in columns. Therefore, supporting high surface area nanoparticles on much larger biochar particles allows easier flow through columns while still exposing the polluted water to nanoparticle surfaces.

Fe-based biochar composites are most studied materials in the past decade. Several methods have been presented for their synthesis (Fig. 8.2B). Different iron species can be obtained, from zero-valent iron (ZVI) to iron oxide/hydroxide nanoparticles such as magnetite, maghemite, and hematite, prepared both ex situ and in situ (Amen et al., 2020; Li et al., 2018). Biochar loading is performed traditionally by wetness impregnation and coprecipitation. Briefly, the general procedure consists of a soaking step with impregnating solutions containing Fe, must frequently $FeCl_3$, $Fe(NO_3)_3$ and $Fe_2(SO_4)_3$ followed by a drying step. In the coprecipitation method, a further reducing ($NaBH_4$) or precipitating agent (NaOH, KOH, $NH_4OH$) are added to the suspension before the drying step. A thermal conversion to promote nZVI or oxide particle nucleation and growth on the surface of the biochar can also be performed (Wang et al., 2019; Wen et al., 2017). Additionally, iron oxides or their nanoparticles can be produced separately and then physically mixed together with the feedstock or biochar in order to achieve the Fe-based composite after copyrolysis (Wang et al., 2019).

Recent focus has been placed on simple procedures or "one-pot synthesis" to obtain composites. For instance, Wang et al. (2015) produced an iron oxide biochar by direct pyrolysis of natural hematite-treated biomass. Bakshi et al. (2018) produced low-cost ZVI-biochar composites by copyrolysis of high and low ash biomass feedstocks and magnetite. Duan et al. (2017) used an iron oxide-modified biochar prepared by impregnation of walnut shells with $FeCl_3$ solution and subsequent microwave-assisted pyrolysis (800 W, 20 min, and $N_2$ atmosphere). Zubrik et al. (2018) proposed a one-step microwave synthesis of magnetic biochar hybrids from wheat straw mixed with ferrofluid, followed by thermal treatment. Fan et al. (2018) reported an one-pot synthesis of nZVI-embedded biochar to remove As from polluted soils. Some adsorbents have been proposed for other pollutants with potential for arsenic removal. Chen et al. (2019) reported a single-step synthesis of magnetic activated carbon by direct pyrolysis of eucalyptus sawdust with $FeCl_3$ to remove methylene blue (MB). Yin et al. (2018) documented a one-step synthesis of magnetic activated carbon to remove 17β-estradiol and $Cu^{2+}$ with rice straw impregnated with $ZnCl_2$ and $FeCl_3$. Liu et al. (2020) synthesized a magnetic biochar from *Salix mongolica* impregnated with $ZnCl_2$ and $FeCl_3$ for antibiotics removal. Bursztyn Fuentes et al. (2020) obtained biochar from *Eucalyptus pruning* residues with a nonconventional device named Kon-Tiki kiln to remove paracetamol.

Overall, the reported As sorption capacities of Fe oxide composites are higher than those of pristine biochars for both As(III) and As(V). In addition, these sorption capacities tend to be higher than those reported for physical and chemical treatments (Benis et al., 2020). Table 8.1 summarizes the efficiency of the Fe-oxide-modified biochar over pristine biochar for the removal

Table 8.1 Summary of As removal efficiencies of pristine and Fe-modified biochars.

| Feedstock | Pyrolysis (°C) | Speciation | pH | Langmuir $q_m$ (mg g$^{-1}$) | $S_{BET}$ (m$^2$ g$^{-1}$) | Reference |
|---|---|---|---|---|---|---|
| **Pristine biochars** | | | | | | |
| Pine wood | 400 | As(III) | 3.5 | 2.6 | 3 | (Mohan et al., 2007) |
| Pine bark | 400 | As(III) | 3.5 | 13.0 | 2 | (Mohan et al., 2007) |
| Oak wood | 400–450 | As(III) | 3.5 | 4.0 | 2 | (Mohan et al., 2007) |
| Oak bark | 400 | As(III) | 3.5 | 3.0 | 25 | (Mohan et al., 2007) |
| Almond shell | 450 | As(III) | 6.2 | 12.86 | NA | (Ali et al., 2020) |
| | 450 | As(V) | 7.2 | 5.65 | NA | (Ali et al., 2020) |
| Rice husks | 300 | As(V) | 9 | NA | NA | (Agrafioti et al., 2014) |
| Solid waste | 300 | As(V) | 9 | NA | NA | (Agrafioti et al., 2014) |
| Sewage sludge | 300 | As(V) | 9 | 0.013 | NA | (Agrafioti et al., 2014) |
| Oak wood | 500 | As(III) | 6 | 3.16 | 475 | (Niazi et al., 2018b) |
| Oak wood | 500 | As(V) | 7 | 3.89 | 475 | (Niazi et al., 2018b) |
| Perilla leaf | 300 | As(III) | 9.7 | 4.7 | 473 | (Niazi et al., 2018a) |
| | 300 | As(V) | 9.7 | 3.8 | 473 | (Niazi et al., 2018a) |
| | 700 | As(III) | 10.6 | 11.0 | 3 | (Niazi et al., 2018a) |
| | 700 | As(V) | 10.6 | 7.2 | 3 | (Niazi et al., 2018a) |
| *Tectona sp.* leaves | 800 | As(III) | NA | 0.66 | 35 | (Verma and Singh, 2019) |
| | 800 | As(V) | NA | 1.3 | 35 | (Verma and Singh, 2019) |

Table 8.1 Summary of As removal efficiencies of pristine and Fe-modified biochars—cont'd

| Feedstock | Pyrolysis (°C) | Speciation | pH | Langmuir $q_m$ (mg g$^{-1}$) | $S_{BET}$ (m$^2$ g$^{-1}$) | Reference |
|---|---|---|---|---|---|---|
| *Lagerstroemia sp.* leaves | 800 | As(III) | NA | 0.45 | 6 | (Verma and Singh, 2019) |
| | 800 | As(V) | NA | 0.71 | 6 | (Verma and Singh, 2019) |
| Sewage sludge | 500 | As(III) | 3 | 0.07 | 70 | (Tavares et al., 2012) |
| **Pristine and modified counterparts** | | | | | | |
| Empty fruit bunch | CAC | As(III) | 9 | BC: 18.9; MBC: 31.4 | 2/NA | (Sadegh-Zadeh and Seh-Bardan, 2013) |
| | CCA | As(V) | 5 | BC: 5.5; MBC: 15.2 | 2/NA | (Sadegh-Zadeh and Seh-Bardan, 2013) |
| Rice husks | CAC | As(III) | 9 | BC: 19.3; MBC: 30.7 | 25/NA | (Sadegh-Zadeh and Seh-Bardan, 2013) |
| | CAC | As(V) | 5 | BC: 7.1; MBC: 16.9 | 25/NA | (Sadegh-Zadeh and Seh-Bardan, 2013) |
| Grape seeds | 600 | As(V) | 6 | BC: 5.08 | 405 | (Frišták et al., 2018) |
| | 600 | As(V) | 6 | MBC: 39.44 | 416 | (Frišták et al., 2018) |
| | 600 | As(V) | 6 | MBC: 34.9 | 348 | (Frišták et al., 2018) |
| Bamboo (*Guadua chacoensis*) culms | 700 | As(V) | 7 | BC: 256 | 7 | (Alchouron et al., 2020) |
| | 700 | As(V) | 7 | MBC: 457 | 29 | (Alchouron et al., 2020) |
| | 700 | As(V) | 7 | MBC: 868 | 482 | (Alchouron et al., 2020) |
| Wheat straw | 600 | As(III) | 4 | BC: 1.04 | 3 | (Zhu et al., 2020) |
| | 600 | As(III) | 4 | MCB: 78.3 | 215 | (Zhu et al., 2020) |
| Chestnut shell | 450 | As(V) | 7 | BC: 17.5 | NA | (Zhou et al., 2017) |
| | 450 | As(V) | 4 | MBC: 45.8 | NA | (Zhou et al., 2017) |
| Hickory chips | 600 | As(V) | 5.8 | BC: nd | NA | (Hu et al., 2015) |
| | 600 | As(V) | 5.8 | MBC: 2.16 | 16 | (Hu et al., 2015) |
| Corn straw | 600 | As(V) | 6 | BC: 0.017 | 71 | (He et al., 2018) |
| | 600 | As(V) | 6 | MBC: 6.8 | 297 | (He et al., 2018) |
| Pinewood | 600 | As(V) | 7 | BC: 0.265 | 210 | (Wang et al., 2015) |
| | 600 | As(V) | 7 | MBC: 0.429 | 193 | (Wang et al., 2015) |
| **Modified biochars** | | | | | | |
| Banana pith | 400 | As(V) | 6.5 | 0.12 | 32 | (Lata et al., 2019) |
| Maize straw | 500 | As(III) | 3 | 168 | 94 | (Huang et al., 2020) |
| | 500 | As(V) | 3 | 182 | 94 | (Huang et al., 2020) |
| Rice straw | 500 | As(V) | 5.3 | 7.41 | 69 | (Zhang et al., 2016) |
| Spent-coffee ground | 700 | As(V) | 5 | 12.6 | 8 | (Cho et al., 2017) |
| | 700 | As(V) | 5 | 12.1 | 512 | (Cho et al., 2017) |

*Continued*

Table 8.1 Summary of As removal efficiencies of pristine and Fe-modified biochars—cont'd

| Feedstock | Pyrolysis (°C) | Speciation | pH | Langmuir $q_m$ (mg g$^{-1}$) | $S_{BET}$ (m$^2$ g$^{-1}$) | Reference |
|---|---|---|---|---|---|---|
| Rice husks | 600 | As(III) | 7 | 0.096 | 300 | (Singh et al., 2020) |
| Wheat husks | 600 | As(III) | 7 | 0.113 | 339 | (Singh et al., 2020) |
| Cotton | 800 | As(III) | 7 | 70.22 | 9 | (Wei et al., 2019) |
|  | 800 | As(V) | 7 | 93.94 | 9 | (Wei et al., 2019) |
| Red oak | 900 | As(V) | 7 | 15.6 | NA | (Bakshi et al., 2018) |
| Switchgrass | 900 | As(V) | 7 | 6.48 | NA | (Bakshi et al., 2018) |
| Tea waste | 300 | As(V) | 5 | 38.03 | 63 | (Wen et al., 2017) |
|  | 400 | As(V) | 5 | 31.67 | 34 | (Wen et al., 2017) |
|  | 500 | As(V) | 5 | 27.81 | 31 | (Wen et al., 2017) |
| Waste wood | 900 | As(III) | 7 | 5.06 | 320 | (Navarathna et al., 2019) |

*CAC*, commercial activated carbon; *BC*, biochar; *MCB*, iron-modified biochar; *nd*, nondetected; *NA*, not available data.
Adapted from Amen, R., Bashir, H., Bibi, I., Shaheen, S.M., Niazi, N.K., Shahid, M., Hussain, M.M., Antoniadis, V., Shakoor, M.B. and Al-Solaimani, S.G. (2020) A critical review on arsenic removal from water using biochar-based sorbents: the significance of modification and redox reactions. Chem. Eng. J., 125195 and Benis, K.Z., Damuchali, A.M., Soltan, J. and McPhedran, K. (2020) Treatment of aqueous arsenic—a review of biochar modification methods. Sci. Total Environ., 139750.

of As from the liquid phase. Sadegh-Zadeh and Seh-Bardan (2013) found that the maximum sorption capacities for As(III) with empty fruit biochar and rice husks biochar were 18.9 mg g$^{-1}$ and 19.3 mg g$^{-1}$, respectively. Capacities increased to 31.4 mg g$^{-1}$ and 30.7 mg g$^{-1}$ for the Fe-oxide-loaded. As(V) capacities of the pristine biochars were 5.5 mg g$^{-1}$ and 7.1 mg g$^{-1}$, respectively, while the Fe-oxide-loaded biochar capacities increased to 15.2 mg g$^{-1}$ and 16.0 mg g$^{-1}$. Wang et al. (2015) produced pinewood biochar and a hematite-modified counterpart, achieving As(V) maximum sorption capacities of 0.265 and 0.429 mg g$^{-1}$, respectively. Hematite modification almost doubled the still small As sorption ability of the pristine biochar. Zhou et al. (2017) produced an Fe-oxide-modified biochar derived from chestnut shells and a magnetic gelatin. As(V) sorption capacity of this modified biochar was three times higher than unmodified biochar: rising from 17.5 to 45.8 mg g$^{-1}$. He et al. (2018) observed a 400-fold increase in the sorption capacity of an iron-oxide-impregnated biochar in contrast to that of the pristine counterpart.

Various iron compounds have different As sorption capacities. This has been related to the iron compound phase crystallinity (Wen et al., 2017; Zhang et al., 2016). Low crystallinity and more amorphous Fe oxide phases will have higher surface areas with more surface active As sorption sites (Kumar et al., 2014; Zhang et al.,

2007). As(V) sorption capacities progressively decreased for magnetic tea waste hybrids pyrolyzed at 300, 400, and 500 °C because their iron oxide nanoparticles produced at higher temperatures had higher crystallinity and larger diameters. X-ray diffraction (XRD), X-ray photoelectron spectroscopy (XPS), and Mossbauer spectroscopy, obtained before and after As sorption, are used to identify the specific oxide/hydroxide causing As removal. They help describe the structures of the adsorbed As species. More recently, X-ray absorption near-edge structure analysis (XANES) has also been used as a complementary tool to study the structure of As bound to the adsorbent (Niazi et al., 2018b; Xu et al., 2020).

Lately, other engineered biochar-derived composites have been gaining attention for As removal. For instance, layered double hydroxides supported on biochar (LDH-biochar) have been proposed as a "win-win" material with improved chemical properties to overcome the limitations of pristine biochar in pollutant removal (Vithanage et al., 2017). Wang et al. (2016) studied As(V) removal with a Ni/Fe LDH-biochar composite derived from pine wood. The maximum adsorption capacity of this composite more than doubled that of the pristine biochar (1.5 to 4.4 mg g$^{-1}$) at a low pH. However, studies involving LDH-biochar composites for As removal are still scarce (Wang et al., 2016). Good phosphate sorption capacities have been achieved. With LDH-biochar hybrids Zhang et al. (2013b) achieved a maximum phosphate sorption capacity of 410 mg g$^{-1}$ with a Mg/Al LDH-cotton wood biochar composite. Jiang et al. (2019) achieved a 185 mg g$^{-1}$ phosphate capacity with a Zn/Al LDH-biochar composite derived from banana straw versus the 149 mg g$^{-1}$ with a Mg/Al LDH-rice husk biochar composite (Lee et al., 2019). These mineral-biochar composites could be efficiently used for As removal given the structural similarities of phosphate and arsenate anions. This will be described in Section 3.2.1.

It is worth mentioning that a lot of focus has been placed lately on describing As removal mechanisms from aqueous solutions using biochars and modified biochars. These mechanisms include ion exchange, electrostatic interactions, hydrogen bonding, surface complexation, precipitation, pore-filling mechanisms, and π-π electron-donor-acceptor interactions (Fig. 8.3). Many papers and thorough reviews have been published recently describing different mechanisms involved in As removal with modified biochars, especially for iron oxide-based composites (Benis et al., 2020; Navarathna et al., 2020; Vithanage et al., 2017).

Although analysis of the different mechanisms for As removal is beyond the aim of this chapter, it is important to highlight the importance of their understanding to better predict the removal

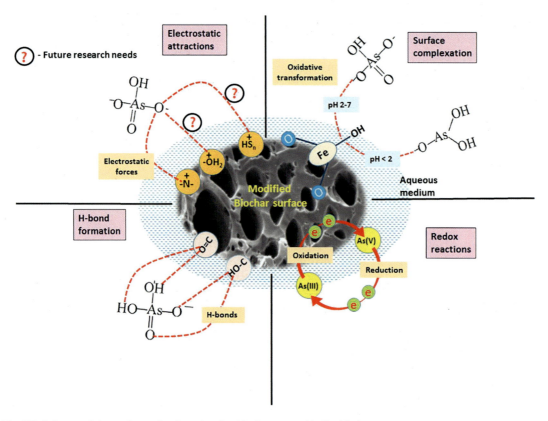

**Fig. 8.3** Scheme of the main mechanisms involved in As removal in liquid phase. Published with permission of Alkurdi, S.S.A., Herath, I., Bundschuh, J., Al-Juboori, R.A., Vithanage, M. and Mohan, D. (2019) Biochar versus bone char for a sustainable inorganic arsenic mitigation in water: what needs to be done in future research? Environ. Int. 127, 52-69.

performance of adsorbents and further optimize their synthesis. For example, nZVI-biochar has a redox ability which enables the oxidation of As(III) to As(V), a less toxic species (Navarathna et al., 2020). Dissolved oxygen in aqueous media corrodes the Fe(0), and a mixture of iron compounds—including oxides like magnetite or maghemite and hydroxides—is formed on the surface of the nanoparticles (Bang et al., 2005). Consequently, As(III) is oxidized to As(V) by $Fe^{3+}/O_2$ and As-Fe compounds coprecipitate. Indeed, Yan et al. (2012) studied redox transformations with XPS and found that the mechanism of As removal with nZVI is related to surface-mediated redox transformations instead of adsorption, which differs from that of other Fe oxide-based adsorbents (Yan et al., 2012).

It is evident so far that the physicochemical properties of both pristine biochars and biochar-derived adsorbents have a strong impact on As adsorption capacity. However, there are other factors to consider when designing an adsorption application. These include amount of adsorbent, initial concentrations or surges adsorbate, solution pH, redox conditions, temperature, interfering substances, etc. (Premarathna et al., 2019; Vithanage et al., 2017). Many studies report high removal efficiencies at low pH, but these are unrealistic conditions in natural waters. Since arsenite and arsenate ion behavior is different, the uptake efficiency of As(III) is not the same as with As(V). Differences in experimental variables and their effects on As removal need to be further developed.

## 3 Adsorbent performance testing for scaled-up drinking water arsenic treatment

This chapter focuses on the ability of recently developed pristine biochars and their composites to provide As-free household drinking water. Technology transfer begins with the synthesis and selection of a suitable adsorbent for the task (Section 2), followed by performance screening and optimization for As decontamination. The capability to provide As-safe drinking water (As < 10 µg L$^{-1}$) is determined last, but this result will define if this adsorbent can be scaled-up.

Biochars are eco-friendly, biocompatible, widely available, and cost-efficient, ensuring its technology transfer potential. However, none of these features will be useful if there is no proof that it can produce safe drinking water. Most studies usually follow a well-established scheme based mainly on pH, kinetics, isotherms, and thermodynamic experiments. Afterward, a series of other studies can be performed. Their design and analysis ultimately depend on the aim of the project. Here, we have systematized an approach that follows lab experimentation that has provided comprehensive evidence of the adsorbent´s capability to provide As-safe drinking water. The experimentation path is displayed in Fig. 8.4. We have separated it into three successive early performance screenings: a 1st backbone screening, a 2nd spinoff screening, and a 3rd on-site screening.

We named the first experimental stage "backbone screening." This comprises the necessary basic battery of experiments that are carried out to assess the adsorbents' ability to eliminate aqueous arsenic. These experiments are carried out in batch scenarios using single-component systems (As + deionized water). To saturate the adsorbent/adsorbate sites, the arsenic tests

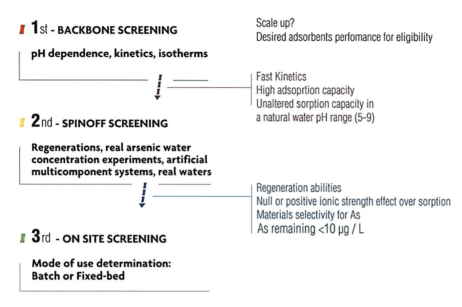

**Fig. 8.4** Experimental optimization path for biochar or biochar composite technology transfer from lab experimentation to provide safe household As drinking water.

concentrations employed are much higher than naturally occurring concentrations. The influence of several operational parameters is estimated during this first backbone screening related to the effect of particle size, solution pH, contact time, initial As concentration, and temperature. This backbone screening is what is most commonly addressed in literature. Such studies will be further developed in Section 3.1. The desired parameters that justify the material's eligibility to be further pursued toward a scale-up for household water provision (second screening) have to do with its ability to maintain an adequate and unaltered sorption capacity in a natural water pH range (5–9), relatively fast kinetics, and a high arsenic adsorption capacity with the major competitive ions and neutrals (organic+inorganic) present in those media.

Solutions with only one solute almost never exist in natural environments. Thus, once the material's performance in monocomponent systems is well understood and optimized through the first screening, a second experimental stage considering technology spinoff should be carried out. This second screening is mainly performed in aqueous heterogeneous media. Experiments in artificial multicomponent systems and in naturally contaminated samples are carried out. In an iterative process, these results serve to elaborate aqueous matrices which are representative of the ones where the technology transfer operation will take place, eventually. Simultaneously, the adsorbent's regeneration, reusability, and adsorption performance with real arsenic concentrations should also be assessed. At this point, the desired parameters that justify eligibility to continue the technological scale-up include the sorbent's regeneration abilities, the negative, null, or positive effect of ionic strength over sorption, the material selectivity or preference for arsenic over other substances dissolved in water, and the ability to decrease the arsenic concentration to below a safe drinking level (As $< 10\,\mu g\,L^{-1}$) with a reasonable adsorbent dose.

A third and last screening is introduced next. This is directly related to how this adsorbent's technology will be transferred. There are several shapes/designs that the adsorption process can adopt in the field. These depend on the scale and other site-specific considerations. Small-scale filters for family use or medium-scale filters for a communal water treatment system be used, batch or continuous systems, bulk or fixed-bed configurations, with or without stirring (mechanical or electrical), among others. Defining the type of device or configuration is essential because the desired mode of use will ultimately determine the many additional parameters that need to be studied and optimized in accordance with site-specific technical, economic, and social constraints.

Fig. 8.5 generated by classifying all published aqueous As/biochar sorption studies from 2010 to 2020 ($n=54$) via the 3-stage experimental optimization path (Fig. 8.4). Some studies performed only one type of experiment (e.g., =1st Screening) while others performed two (e.g., =1st and 2nd Screenings) or three (1st/2nd and 3rd Screenings). The total percentages of the individual screenings performed are displayed in the pie chart (Fig. 8.5). The bulk of previous research has been focused on As removal only from synthetic As-rich solutions (Total 1st screening=56%). Remarkably, only a few studies have devoted efforts to using biochar-based sorbents for remediating natural drinking water polluted with As (Total 3rd screening=8%).

## 3.1 1st screening: Adsorbents' performance in monocomponent aqueous systems

Optimizing an adsorbent for aqueous pollutant remediation primarily requires screenings in monocomponent aqueous systems. This screening involves studying adsorbent dose (solid-

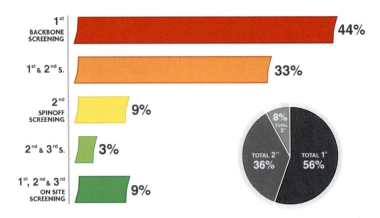

**Fig. 8.5** Percent of studied reporting data for 1st, 2nd, and 3rd screenings (S.). Some studied report data for two so-called screenings. Hence, the inset pie chart is the total % of each individual screening (% was normalized to 100). As a reminder: 1st S = 1st backbone screening: Adsorbents' performance in monocomponent aqueous systems (dose, pH, kinetics and isotherms); 2nd S = 2nd adsorbents *spinoff* screening: multicomponent testing, naturally occurring [As] concentrations testing, real water tests, large-scale column tests; 3rd S = 3rd screening: utilization of biochar and biochar composites for drinking water treatment and ongoing scaled-up experiences.

to-liquid ratio) influence, pH dependence, kinetics, isotherms, thermodynamics, and column performance in monocomponent systems. Fast kinetics and/or high adsorption capacities in natural water pH range are the basic characteristics sought. At first glance, they manifest the potential of an adsorbent for the screened purpose (Tan et al., 2015). This first backbone screening is routinely performed in most papers dealing with As, and very recent reviews have comprehensively compiled biochar and modified biochar sorption parameters (Amen et al., 2020; Asere et al., 2019; Navarathna et al., 2020; Vithanage et al., 2017).

We have reviewed papers on 70 biochars and biochar composites for aqueous As(III) and As(V) uptake sorption performance. Histograms in Fig. 8.6 display the reported studies for different value ranges of (a) optimum sorption pH's, (b) equilibrium sorption time's, and (c) Langmuir's adsorption capacities.

Changes in the adsorbent's performance versus pH variations are one of the key limitations when assessing surface interactions. A pH of 5–9 covers the range of most underground As-contaminated waters (Singh et al., 2015). Many authors have reported good removal efficiencies up to a pH of 7–8 and decreasing removal efficiencies with increasing pH (Amen et al., 2020).

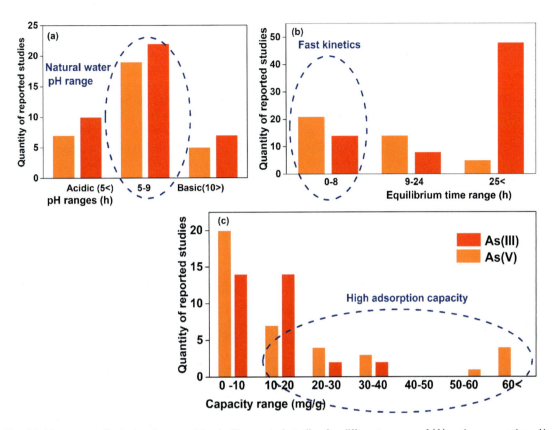

**Fig. 8.6** Histograms displaying the quantities in 70 reported studies for different ranges of (A) optimum sorption pH, (B) equilibrium sorption time, hours, and (C) Langmuir adsorption capacities, mg/g. The best ranges for As-safe drinking water eligibility are circled with dashed lines.

Hence, the ability to present mostly unaltered sorption capacities in a wide natural water pH range is a robust characteristic that permits the scale-up of the adsorbent independent of the pH, avoiding extra steps needed to efficiently remove the pollutant. Most screening studies report adsorbents for which a pH 5 to 9 variation has a negligible effect on arsenic removal. These are represented by a wide variety of adsorbents such as aluminum biochar composites (Ding et al., 2018), Fe-modified iron rice husk biochar (Sadegh-Zadeh and Seh-Bardan, 2013), thioglycolate sugarcane carbon (Roy et al., 2013), Leonardite char (Chammui et al., 2014), etc.

Estimation and optimization of contact time and adsorption capacity are of utmost importance when designing a practical and large-scale adsorption-based treatment method. "Fast kinetics" and "high adsorption capacities" are the two targeted features. To define these qualities, arbitrary assumptions were necessary and are presented here. We defined that adsorbents presented "fast kinetics" when equilibrium was reached before 8 h. This time range represents a full work day. Though an 8 h equilibrium time may be too much for an in-house water supply, it is suitable for water supply to a network of houses. A total of 35 materials [14 for As(III) and 21 for As(V)] presented "fast kinetics." These consist of different biochar and biochar composite sources such as *Tectona sp.* biochar (Verma and Singh, 2019) or almond shell biochar (Ali et al., 2020) for As(III), presenting 6 h and 2 h equilibrium times, respectively. Moreover, KOH-activated cotton stalk biochar (Hussain et al., 2019) and bismuth-activated carbon (Zhu et al., 2016) displayed 3 h and 2 h equilibrium times, respectively, for As(V) sorption. The average Langmuir adsorption capacities for all of the 70 reviewed materials for As(III) and As(V) were $10.2\,\mathrm{mg\,g^{-1}}$ and $32\,\mathrm{mg\,g^{-1}}$, respectively. For As(III) sorption, values ranged from $5.70\mathrm{E}\text{-}03\,\mathrm{mg\,g^{-1}}$ (Van Vinh et al., 2015) to $48.57\,\mathrm{mg\,g^{-1}}$ (Xu et al., 2020) for raw pine cone biochar and Fe-modified Populus L. biochar, respectively. Furthermore, As(V) sorption values ranged from a low of $0.0258\,\mathrm{mg\,g^{-1}}$ (Singh et al., 2015) to a high of $457\,\mathrm{mg\,g^{-1}}$ (Alchouron et al., 2020) for fly ash Bagasse iron-coated and $Fe_3O_4$-bamboo biochar, respectively. We have used these values and the histogram frequencies to define "high adsorption capacities." Thus, all the materials that present a Langmuir sorption capacity over $10\,\mathrm{mg\,g^{-1}}$ are considered to have "high arsenic adsorption capacity." Considerations of the concentration ranges used in most studies to calculate these capacities are further discussed in Section 3.2.

Table 8.2 displays all the materials that exhibited these three characteristics: 5–9 optimum sorption pH, fast kinetics (equilibrium time less than 8 h), and high arsenic adsorption capacities (Langmuir sorption capacity over $10\,\mathrm{mg\,g^{-1}}$) after the first backbone screening. In other words, these materials are available candidates for advancing continue to the second screening with the chance to successfully provide As-free household drinking water.

The fact that other materials do not precisely combine these features does not mean that such adsorbents will not accomplish the targeted aim. An adsorbents' adequacy can be specifically designed to provide As-safe drinking water. The first features that can be altered are particle size and the material's synthesis. Both are unique features that determine arsenic sorption performance, and their modification seeks achieving more adsorption of arsenic per unit of surface area. Gupta et al. (2011) found that the amount

Table 8.2 Eligible readily optimized materials for 2nd scale-up screening.

| Adsorbent | Speciation | Langmuir $q_m$ (mg g$^{-1}$) | Optimum pH | Equilibrium time (h) | Reference |
| --- | --- | --- | --- | --- | --- |
| Iron-modified corn straw BC[a] | As(V) | 14.8 | 2 to 8 | 0.3 | (Fan et al., 2018) |
| Iron oxide-modified bamboo BC | As(V) | 457.0 | 5 to 9 | 1.0 | (Alchouron et al., 2020) |
| Titanium-modified ultrasonic BC | As(V) | 118.6 | 4 | 8.0 | (Luo et al., 2019) |
| Iron-modified rice straw BC | As(V) | 28.5 | 6 | 2.0 | (Nguyen et al., 2019) |
| Iron-modified tea waste BC | As(V) | 38.0 | 5 | 1.5 | (Wen et al., 2017) |
| Fe-Mn-La-impregnated BC | As(III) | 14.9 | 4 | 6.0 | (Lin et al., 2019) |
| Pyrolytic-tire BC | As(III) | 31.2 | 4 to 8.5 | 1.0 | (Mouzourakis et al., 2017) |

[a]BC stands for biochar.

of adsorbed acid blue raised when decreasing activated biochar particle size. Additionally, Yean et al. (2005) proved that increases in maximum adsorption capacities for As(III) and As(V) were observed with decreasing magnetite particle size. To the best of our knowledge, no studies varying biochar synthesis or biochar composite particle size and further rescreening As(III) and As(V) sorption are published. After completing particle size modification, the first screening needs to be carried out as an iterative process. Interestingly, Verma and Singh (2019) compared the As(III) and As(V) sorption performance of biochars derived from two different species of *Tectona* and *Lagerstroemia*. These adsorbents were synthetized following the same protocol, yet the *Tectona sp.* adsorbent presented a higher size particle (250–450 nm) than the *Lagerstroemia sp.*-derived adsorbent (60–90 nm). However, the *Lagerstroemia sp.*-derived adsorbent displayed less adsorption capacity for both As(III) and As(V). Hence, the biochar's synthesis also plays an important role. Broadly, a materials' synthesis adequacy has to do with the pyrolysis temperature, surface area increase, and engineered coatings, factors which were comprehensively discussed in Section 2.

## 3.2 2nd screening: Multicomponent testing, naturally occurring As concentration testing, real water tests, large-scale column tests, statistical analysis

Water chemistry dictates the adsorbent's performance. To position biochar or biochar composites as an alternative to other well-established and available technologies for household As-safe drinking water provision, multicomponent system optimization needs to be performed. Thus, once the performance of the material in monocomponent systems is well understood via optimized through the first screening, a second experimental stage should be carried out considering the *spinoff* of technology. Regeneration abilities, behavior in the presence of other aqueous elements, ionic strength effects on sorption, and the capability to provide As-safe drinking water ($<10 \mu g\, L^{-1}$) will be the targeted variables. This will enable the elaboration of representative aqueous matrices to optimize the adsorbent's use while considering site-specific factors.

Some studies have assessed aspects of this second screening (competitive ions, ionic strength, and regeneration studies) as an expansion of the first screening. Other articles directly faced this second screening, focusing on column scale-up studies, naturally occurring As concentrations, and real As-contaminated groundwaters (Tabassum et al., 2019). Roy et al. (2013) presented a full first and second screening of sorbent in the same paper, thoroughly exploring column parameters.

### 3.2.1 Multicomponent testing

In natural water or soils, solutions of only one solute usually do not occur. This emphasizes that it is crucial to consider the effects of other cations and anions. While some studies approach this issue by studying the effect of variable ionic strength in batch experiments, others present dual or triple coadsorption studies or utilizing multiple-ion competitive experiments with severed known ions to prove their particular effect.

The ionic composition of the adsorption media will condition the adsorption of As(III) and As(V) (Bolan et al., 1986). In a $10\, mg\, L^{-1}$ As(V) solution, an increase in ionic strength, from null to 0.1 M with $NaNO_3$ as the background electrolyte, reduced the $FeCl_3$-modified corn-straw biochar's pollutant removal from 86.48% to 84.82% (He et al., 2018). A 0–0.32 mM (0–20 $mg\, L^{-1}$) $NO_3^-$ ionic strength increase in a $100\, mg\, L^{-1}$ As(V) solution slightly decreased ($\sim 1\, mg\, g^{-1}$) As(V) sorption capacity of ultrasonically modified $TiO_2$-biochar's from $\sim 72$ to $\sim 71\, mg\, g^{-1}$ (Luo et al.,

2019). It is important to highlight that these multicomponent studies are usually performed in high arsenic concentrations ($>5\,mg\,L^{-1}$). Thus, increases in ionic strength (0–0.1 M) display nonsignificant effects over other factors and consequently, As uptake. At these As oversaturated conditions, the effect of electrolyte concentration on anion adsorption is minor. This trend is not always followed in natural arsenic aqueous waters, where arsenic concentrations are very low compared with other naturally occurring ions. This aspect is further discussed in Section 3.2.3.

Dual coadsorption studies explore the performance of an adsorbent in the presence of arsenic and another pollutant (most commonly a cationic heavy metal). A $Fe_3O_4/CaCO_3$-modified rice straw biochar displayed competitive and synergistic effects during simultaneous Cd(II) and As(III) adsorption. The addition of Cd(II) to the solution suppressed As(III) adsorption by 15%–33%. Cocontamination with Cd(II) lowered As(III) adsorption because of the electrostatic interaction and the formation of type B ternary surface complexes (Wu et al., 2018). Simultaneous As (V) and Cd(II) sorption was assessed by ultrasonicated titanium-modified biochar. Once again, the addition of Cd(II) decreased the adsorption capacity of As(V), but the ability to adsorb both anions and cations this biochar-based material is displayed (Luo et al., 2019).

Triple-component coadsorption was assessed by Johansson et al. (2016). After performing a complete first screening for Se, As, and Mo onto two Fe-modified algal derived biochars, triple simultaneous Se, As, Mo sorption was analyzed from ash disposal waters (Power Station, Tarong, Australia). The *Oedogonium sp.* algae-derived biochar was the best at sequestering the metalloids, displaying simultaneous removal percentages of 64%, 80%, and 88% for Se, As, and Mo, respectively.

Though simultaneous coadsorption is not assessed, some studies screen material adsorption performance for multiple pollutants individually in the same project. Conclusions concerning the adsorbent's synthesis and characterization are mainly obtained from these studies. Additionally, since some adsorption capacity is shown, ideas related to enhancing the material usability as a "broad-spec adsorbent" are mentioned. Sewage sludge digestate and its KOH- or $H_2O_2$- activated analogues were studied for As(III), (V), and Cd(II) sorption (Wongrod et al., 2018). Although a slower As(V) kinetic sorption was found after KOH treatment compared to the raw biochar, no significant difference was noted for the $H_2O_2$-modified biochar. Cd(II) sorption kinetics was unaffected by the activation treatment. Cruz et al. (2020) compared ZnO-modified coffee husk and corn cob biochars for Pb(II) and As(V) sorption. The kinetics and the isotherm data performance revealed that the corncob-derived biochar could be a better broad-spec adsorbent. As(V) and $F^-$ sorption were screened for $FeCl_2$-modified yak dung biochar (Chunhui et al., 2018) and KOH-activated perennial grass (*Saccharum ravennae* (L.) L.) (Saikia et al., 2017). Both studies found higher As(V) sorption abilities compared with $F^-$. Chunhui et al. (2018) reported 3.9 mg g$^{-1}$ As(V) and 2.9 mg g$^{-1}$ $F^-$ removal. Saikia et al. (2017) reported an optimized 72.1% As(V) and $F^-$ (24.8%) uptake. Dose/volume, temperatures, pollutant concentration, and equilibrium times were independently optimized for each uptake (As(V) or $F^-$).

Organic matter can play an important role in As removal (Dong et al., 2014). The effect of humic acid on As(V) sorption, assessed for $FeCl_3$-modified tea waste biochar, exhibited little effect over the pH range from 2 to 10 (Wen et al., 2017). Interestingly, $FeCl_3$-modified tea waste biochar's adsorption capacity for this humic acid was very high, 95.92 mg g$^{-1}$. Studies quantifying both humic acid removal and arsenic sorption might help clarify organic matter and As interactions.

In a multicomponent system of seven competitive ions, As(V) sorption was most affected by 0.01 M $SO_4^{2-}$ (Wen et al., 2017). Hence, these authors recommended $SO_4^{2-}$ removal before sorbing As(V). The effect of $SO_4^{2-}$, $NO_3^-$, and $H_2PO_4^-$ on As(III) uptake was assessed for a Fe-Mn-La-modified biochar composite. As the $H_2PO_4^-$ concentration rose, As(III) sorption capacity dropped significantly. $SO_4^{2-}$ and $NO_3^-$ had no effect. Sorption of As(III) and As(V) was most influenced in the presence of $HCO_3^-$, $Cl^-$, and $PO_4^{3-}$ on $Fe^{3+}/Fe^{2+}$-modified Kans grass biochar (Baig et al., 2014). Chloride had the least binding attraction to the adsorbent. Yan et al. (2018) also observed a $PO_4^{3-} > SO_4^{2-} > HCO_3^- > Cl^-$ order of influence over As(V)'s sorption. The electrostatic repulsion between the positively charged surface of the employed $ZnCl_2$-modified crawfish shell biochar and the anions may be related to these phenomena.

Overall, $PO_4^{3-}$ is an ion that seriously competes with As(V) and As(III) for adsorption sites. These findings can be clarified by considering the similar chemical structure of arsenates and phosphates. Inner sphere oxygen exchange chemisorption ($\equiv$M-OH + $PO_4^{2-}$ → $\equiv$M-O-$PO_3^{2-}$ + 2$H_2O$) with the adsorbent's surface functional groups occurs with both As(V) and As(III) anions (Karunanayake et al., 2019). The presence of phosphate in natural water is related to eutrophic streams of water from soil erosion (Golterman, 1973) and from domestic wastewater containing detergents and cleaning preparations, fertilizer-containing farm runoff, fertilizer, detergent, and soap industrial effluents

(Kundu et al., 2015). Knowing that a near 2.7-fold rise in phosphorus-driven eutrophication of terrestrial, freshwater, and marine habitats is expected to occur in the near future (Kundu et al., 2015), this As/P interaction is critical for scaling up As adsorption processes. Providing As-safe drinking water in the presence of $PO_4^{3-}$ was achieved by an aluminum-enriched tetrapak-derived biochar Ding et al. (2018).

As(V) uptake was quantified in the presence of As(V) with individual $SO_4^{2-}$, $MoO_4^{2-}$, $PO_4^{3-}$, $CO_3^{2-}$, $NO_3^-$, $Cr_2O_7^{2-}$, $F^-$, $SeO_4^{2-}$, and $Cl^-$ competing ions in Alchouron et al. (2021). The potential sorption interactions and pathways for arsenate and competitive contaminant ions over the multiple surfaces in a $Fe_3O_4$-modified bamboo biochar are shown in Fig. 8.7. All in all, interactions may occur through hydrogen bonding, electrostatic attractions, and weak chemisorption with the phenolics present on the biochar phase of the adsorbent. Strong $PO_4^{3-}$, $MoO_4^{2-}$, $SeO_4^{2-}$, $SO_4^{2-}$, and $Cr_2O_7^{2-}$ chemisorption dominate on the nano-$Fe_3O_4$ face of the adsorbent. Electrostatic attractions and hydrogen bonding in $CO_3^{2-}$, $NO_3^-$, $F^-$, and $Cl^-$ adsorption play an important role. Considerations must be given to stoichiometric precipitation of insoluble metal compounds formed from metal cation ($Mg^{2+}$, $Ca^{2+}$, and $Fe^{3+}/Fe^{2+}$)-oxyanion ($PO_4^{3-}$, $MoO_4^{2-}$, $SeO_4^{2-}$, and $Cr_2O_7^{2-}$) reactions (Alchouron et al., 2020).

### 3.2.2 Tests with naturally occurring As concentrations

Most arsenic remediation studies are carried out at very high, unrealistic As(III) or (V) concentrations ($>5000\,\mu g\,L^{-1}$). Certain adsorbents capable of remediating arsenic at high concentrations have failed to function in the low concentration range. Adsorption mechanisms which were comprehensively described and parameterized at high aqueous arsenic concentrations may not be the dominant mechanisms at lower concentrations. Very recently, Singh et al. (2020) conducted a full first screening study at low As(III) concentrations (50–1000 $\mu g\,L^{-1}$) to show the removal of As(III) at concentrations normally present in real water bodies. A 111 $\mu g\,g^{-1}$ adsorption capacity was obtained by a $FeCl_3$-modified wheat husk biochar. The arsenic sorption rate-limiting step at 50 $\mu g\,L^{-1}$ As was related to particle diffusion, while at 100 and 200 $\mu g\,L^{-1}$, it was controlled by film diffusion rate (Singh et al., 2020). Braghiroli et al. (2020) studied $FeCl_3$-modified white birch biochar's sorption efficiency in a 300–1000 $\mu g\,L^{-1}$ As(V) concentration range. While an 80.1% As sorption was achieved in a monocompositional 1000 $\mu g\,L^{-1}$ As-contaminated water, 95.4% (270 $\mu g\,g^{-1}$) As sorption was achieved in a real mine

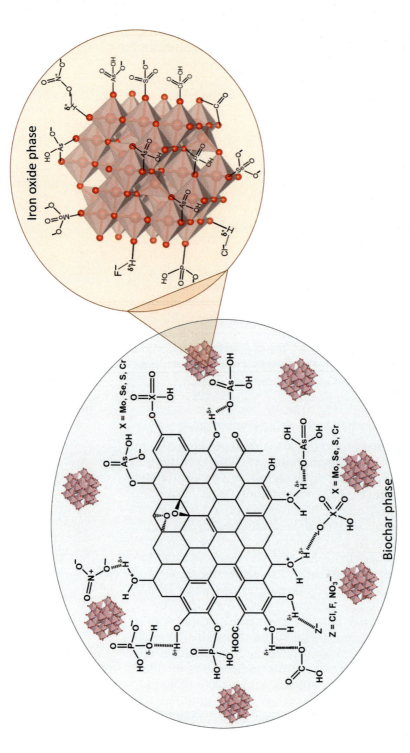

**Fig. 8.7** Schematic representation of potential interactions between arsenate and competitive contaminant ions on the biochar and the iron oxide phase of the $Fe_3O_4$-modified bamboo (Guadua chacoensis) biochar. Modified from Alchouron, J., Navarathna, C., Rodrigo, P.M., Snyder, A., Chludil, H. D., Vega, A.S., Bosi, G., Perez, F., Mohan, D., Pittman, C.U., Jr., 2021. Household arsenic contaminated water treatment employing iron oxide/bamboo biochar composite: an approach to technology transfer. J. Colloid Interface Sci. 587.

effluent (initial As = 300 µg L$^{-1}$). Capacity was only reported for the arsenic absorbed from mine effluent. These capacities may seem low if compared with those carried out at high concentrations. However, the high sorption capacity of many adsorbents is due to the high As concentration ranges used in batch experiments. Interestingly, Feng et al. (2020) chose a 0 to 5000 µg L$^{-1}$ As concentration range to evaluate FeCl$_3$/FeSO$_4$ modified wood (*Populus adenopoda*)-based biochar arsenic sorption based on groundwater concentrations from Datong Basin, China (2680 µg L$^{-1}$) (Pi et al., 2015). Though the focus of the paper was placed on material synthesis and characterization, we highlight the selection criteria employed for the tests. A 99.9% As(V) and a 99.9% As(III) removal were achieved for a 10 g L$^{-1}$ dose at a 48 h equilibration time.

Testing naturally occurring concentrations is a practical way to bridge the technology spinoff. Still, obtaining the arsenic capacity at saturating concentrations allows for making consistent predictions regarding the usability of the adsorbent. Evidence of this statement is observed in Ding et al. (2018). Competitive sorption of As(III) and (V) with PO$_4^{3-}$ was studied on aluminum-enriched tetrapak-derived biochar (TP-BC), at two sorbate concentration levels: high concentrations (7.5–75 mg L$^{-1}$) and trace concentrations (75–750 µg L$^{-1}$). Since the TP-BC displayed a higher (~400 mmol kg$^{-1}$) adsorption capacity for As(V) than for As(III) (~175 mmol kg$^{-1}$) at high/saturated As concentrations, its performance in the presence of phosphate followed the same trend. At trace As concentrations, TP-BC gave 100% As(V) removal in both monocomponent and bi-component (arsenate and phosphate) solutions. As(III) removal in single-element solutions was 95%, but only 70% when coexisting with phosphate (using a 2 g L$^{-1}$ dose and a 24 h equilibration). The main reason why As(III) removal at low concentrations was more inhibited by PO$_4^{3-}$.

### 3.2.3 As sorption in natural water samples

Additional information on the ionic strength effect, practical optimization of parameters, the material selectivity, preference for arsenic over other substances dissolved in water, and the adsorbent´s ability to reduce arsenic concentration to a healthy drinking level (As < 10 µg L$^{-1}$) is given by performing arsenic sorption in natural As-polluted aqueous environments. The majority of the available arsenic removal studies using pristine and modified biochars were performed with synthetic effluents. Very few study the use of biochars with natural water samples. These second screening experiments are vital. The ability to reduce arsenic concentrations to a healthy drinking level from natural water sources must be corroborated.

Mondal et al. (2010) found that an $8\,g\,L^{-1}$ dose of $Ca^{2+}$-impregnated granular activated biochar was enough to provide As-safe drinking water from a $188\,\mu g\,L^{-1}$ as containing groundwater source from West Bengal at a pH 7.1 and 24 h equilibration time at 30°C (Mondal et al., 2010). Also, safe As drinking water was obtained from a $91\,\mu g\,L^{-1}$ [As] Vietnamese water sample decontaminated by $FeCl_3/FeSO_4$ modified biochar at a $0.007\,g\,L^{-1}$ dose (Nguyen et al., 2019). A $1\,g\,L^{-1}$ dose Oak wood-derived biochar decontaminated groundwaters with As spanning concentrations of $27–144\,\mu g\,L^{-1}$ (2 h equilibration, 20°C) (Niazi et al., 2018b). Broad usability predictions can be done by correlations between dose and arsenic concentrations.

Arsenic removal from industrial wastewater samples has also been assessed. A heavy metal wastewater from the Mississippi State Chemical Laboratory (HMW) with $4\,mg\,L^{-1}$ of added As(III) was remediated by $Fe_3O_4$-modified Douglas fir biochar. This industrial disposal effluent conditions included a $0.2\,mg\,L^{-1}$ As(III) ($4\,g\,L^{-1}$ dose, 1 h equilibration, 200 rpm, 25°C) (Navarathna et al., 2019). All the As(III) levels dropped below the WHO safe limits. In a column study, Braghiroli et al. (2020) assessed these experimental conditions (52–75 g of adsorbent, 20 cm column height, 1.5 cm column diameter, of $1.5\,mL\,min^{-1}$ flow), $FeCl_3$-modified $CO_2$-activated white birch biochar. Calculations revealed the column could treat for 63 consecutive days (outlet concentration $<0.2\,mg\,L^{-1}$) an $\sim 842\,\mu g\,L^{-1}$ As-contaminated Québec's mine effluent.

### 3.2.4 Adsorbent recycling, regeneration, and/or safe disposal

A major concern when dealing with adsorbents is their lifespan or the number of cycles they can be used without significantly altering their original removal efficiency. Regeneration studies are critical to provide information related to the durability of the adsorbent. Additionally, leaching tests are a complementary tool to study the stability of the adsorbent under real operational conditions. For instance, iron ions may leach when Fe-based composites are subjected to long-term periods in an aqueous solution or to low pH operating conditions (Zubrik et al., 2018). Also, the recycling and regeneration potential are important factors when analyzing the technological application of an adsorbent and the overall treatment cost. However, few papers deal with this stage of adsorption.

Various methods have been used for adsorbent regeneration, such as using acidic and alkaline solutions at different concentrations or purging gas, and changing the partial pressure (Ahmed et al., 2016). While acidic solutions are not very efficient for the regeneration of arsenic-laden biochar, the highest desorption rates have been achieved with alkaline solutions (Alkurdi et al., 2019; Frišták et al., 2018). Hence, different alkaline reagents, such as NaOH, $Na_2CO_3$, $NaHCO_3$, and $K_3PO_4$, have been used to strip the As-loaded biochar, such as NaOH, $Na_2CO_3$, $NaHCO_3$, and $K_3PO_4$. Verma and Singh (2019) performed the regeneration study of pristine biochars derived from waste leaf litter with a solution of NaOH 0.1 M as the desorbing agent. Good results were achieved for both As(III) and As(V), up to four regeneration cycles. For their best biochar, the removal percentages were found to be 76.4%, 68.8%, 65.1%, and 56.3% for As(III) and 95.6%, 90.6%, 83.85%, and 63.85% for As(V) (Verma and Singh, 2019).

The regeneration of biochar-based composites often seems more challenging because the stripping agent might have some effect high-affinity sorbates or functional groups on the surface of the modified biochars. Baig et al. (2014) performed arsenic desorption studies on magnetic Kans grass biochar using NaOH 0.5 M and achieved ~89% As(V) and ~68% As(III) removal at pH 13.5. The regenerated biochar was used in sand filters, and significant arsenic removal efficiencies were achieved for three cycles (Baig et al., 2014). Alchouron et al. (2020) performed desorption studies using $K_3PO_4$ 1 M and $NaHCO_3$ 1 M. With $K_3PO_4$, the most effective stripping agent, the $Fe_3O_4$-loaded biochar underwent almost 70% of desorption of As(V) while the pristine biochar lost only ~25% of its sorbed As(V). Complementary leaching tests showed higher initial arsenate concentrations and higher temperatures caused more iron to be leached (Alchouron et al., 2020). This was also observed by other authors (Karunanayake et al., 2019). Navarathna et al. (2019) stripped a $Fe_3O_4$-biochar composite colum laden with As(III) with $KH_2PO_4$ and achieved 87% As(III) desorption. The efficiency decreased to 76% in the third cycle. Wang et al. (2017) achieved successful desorption of the As-loaded nZVI-biochar [98.2% As(V)] with 0.1 M NaOH for 24 h and found that the regenerated adsorbent could remove almost 4.22 mg g$^{-1}$ As(V), a higher value than that achieved with a composite based on activated carbon.

The scarce results reported on this subject show that it is possible to reuse biochar-derived sorbents to achieve several operating cycles. This is a key factor when designing a remediation scheme, and it increases the worth of a sorbent for potential As treatments. However, some important operational considerations exist. First, alkaline stripping agents raise pH to levels which are not always compatible with water treatment facilities (Alkurdi et al., 2019) and force the introduction of new steps in the water treatment system (i.e., washing step to reach neutral pH). Second, Fe leaching from composites has been detected. Despite the fact that Fe in drinking water is not hazardous (Colter and Mahler, 2006), concentrations above $0.3\,mg\,L^{-1}$ may affect some organoleptic properties (taste and color) (Alchouron et al., 2020). The treated water may lack social acceptance. Third, the regenerated materials do lose their initial As removal efficiency (Amen et al., 2020), so these adsorbents need reasonable disposition alternatives, which will also impact the overall cost-effectiveness. Hence, emphasis must be placed on these further adsorption stages and their limitations. More research is needed to fully understand the applicability of the envisioned adsorption technology. Ultimately, the economic, environmental, and social aspects are crucial factors to position biochar as a promising As removal adsorbent.

The final disposition of the As-loaded biochar must be planned when no more regeneration cycles are possible. Since As is a toxic, carcinogenic pollutant, the disposal alternatives are an important issue to consider (Tan et al., 2015). Techniques that have been suggested include land filling, using them as pore-forming agents in the preparation of many building construction materials and mixing them with livestock waste (Sullivan et al., 2010). On the other hand, the As-loaded solution from adsorbent regeneration is a new waste. Its final disposal must also be considered. Kim et al. (2019) studied the effect of these stripping solutions on plankton, which is present in natural water sources and commonly used as a biomarker. Higher As concentrations produced enhanced toxicity on *Daphnia magna*. For this, acute toxicity tests with neonate *D. magna* were carried when the As(III) concentrations were increased from 0 to $16\,mg\,L^{-1}$. *Daphnia magna* immobilization percentages increased with higher aqueous As(III) concentrations. Still, a $2.02\,mg\,L^{-1}$ concentration (corresponding to the As stripped from the $Fe(NO_3)_3$-modified *Miscanthus sp.* biochar) had a negligible effect on the plankton (Kim et al., 2019).

## 3.3 3rd screening: Utilization of biochar and biochar composites for drinking water treatment and ongoing scale-up experiences

Water and wastewater treatment systems based on conventional technologies to achieve As-safe water have been implemented in many centralized facilities worldwide. Some comprehensive articles on this topic have been written (Cortina et al., 2016; Hering et al., 2017). However, field experiences with point-of-use (POU) treatment systems for arsenic removal are rare. Research reports using biochar and biochar-based composites are even more rare. Furthermore, reviews screening low-cost point-of-use water treatment systems for developing communities consider other readily available technologies such as coagulation-adsorption-filtration processes, reverse osmosis, and membrane filtration (Litter et al., 2019b; Pooi and Ng, 2018). Biochar and biochar composites for arsenic adsorptive treatment are simply not mentioned. Hence, household level POU biochar and biochar composite arsenic removal systems need to be studied and optimized since they can significantly contribute to safe drinking water, especially for rural, isolated populations in developing countries.

Indeed, there are few reported field trials, including biochar or biochar-derived composites and arsenic. In developing countries, different NGOs have been carrying out projects to implement low-cost biochar-based technologies to provide As-safe drinking water for household scale. Biochar-based sorbents were studied for water treatment system optimization in vulnerable contexts in thesis research projects performed in collaboration between academic institutions and a local NGO from Mexico. Eikelboom (2017) produced biochar from residual wood in a homemade gasifier at temperatures above 850 °C and modified it with iron precursors ($Fe_2O_3$, $FeCl_3$, and $Fe(OH)_3$). The materials were tested in batch systems to explore kinetics and sorption capacities. The preselected material with the best qualities was tested in a fixed-double cartridge filter system (25 L $day^{-1}$ flow, 150 g adsorbent, 450 mL Bv). A schematic chart of the experimental set up is displayed in Fig. 8.8-I. The water used for tests belonged to a well from a rural area near San Miguel de Allende (Mexico) and contained about 80 $\mu g L^{-1}$. The best iron-modified biochar gave a batch Langmuir sorption capacity of 55.52 $\mu g g^{-1}$, and a 300 $\mu g g^{-1}$ cartridge capacity. An 8-day provision of As-safe drinking water was projected in the column configuration. Instead, only 50 Bv (bed volumes) (35 liters; ~1.5 days) of As-safe water provision were achieved. Although it did not reach the NGO's objective (300 $\mu g g^{-1}$), these results served as a solid base to continue with optimization. In general, the potential of adsorption from fixed bed column studies is lower than that of batch studies (Navarathna et al., 2020). The residence time of the sorbate in the column is not long enough for adsorption equilibrium to be achieved. Thus, arsenic will exhaust the column before reaching equilibrium.

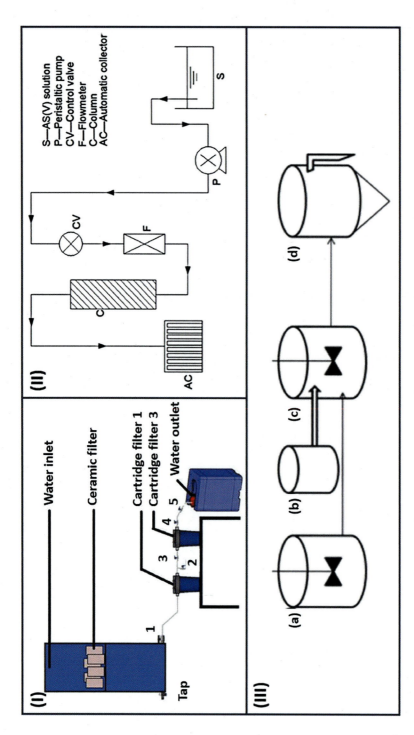

**Fig. 8.8** Schematic charts of different scale-up systems. (I) Technical drawing of the set-up used in Eikelboom (2017); (II) Experimental setup of dynamic column adsorption (Yan et al., 2018); (III) Completely mixed reactors for As(V) removal: (A) agitation stirrer, (B) reservoir of As(V) stock solutions, (C) reactor where adsorbent and adsorbent contact, (D) up flow clarified where treated water is collected and spent adsorbents settle (Wang et al., 2017). Published with permission from (I) Eikelboom, M., 2017. Developing a Low-Cost Filter Medium for Remediation of Arsenic in Contaminated Groundwater. San Miguel de Allende: sn; (II) Yan, J., Xue, Y., Long, L., Zeng, Y., Hu, X., 2018. Adsorptive removal of As(V) by crawfish shell biochar: batch and column tests. Environ. Sci. Pollut. Res. Int. 25 (34), 34674–34683; (III) Wang, F., Liu, L.-Y., Liu, F., Wang, L.-G., Ouyang, T., Chang, C.-T., 2017. Facile one-step synthesis of magnetically modified biochar with enhanced removal capacity for hexavalent chromium from aqueous solution. J. Taiwan Inst. Chem. Eng. 81, 414–418.

A theoretical basis for the practical design of biochar-based filters was derived and displayed by Yan et al. (2018). A PVC column setup (7.5 × ø1.5 cm) was installed, and the effect of flow rate (5 to 10 mL min$^{-1}$), inlet As (V) concentration (20–80 mg L$^{-1}$), and the mass of packed adsorbent (1–3 g) for As (V) adsorption by ZnCl$_2$-modified crawfish shell biochar was studied and analyzed. A schematic chart of the experiment is displayed in Fig. 8.8-II. Based on all the defined parameters, a combination of the mathematics of the breakthrough models of Thomas, Yoon-Nelson, and Adams-Bohart was expressed in a unified EXY model (Equation 8.1) which can ultimately be used to design biochar-based filters (Yan et al., 2018).

$$EXY\ model : \frac{C_t}{C_0} = \frac{1}{1 + \exp(T - K_y\ t)} \qquad (8.1)$$

where $t$ is the breakthrough time (min), $C_0$ is the influent concentration (mg L$^{-1}$), $C_t$ is the concentration at the breakthrough time (mg L$^{-1}$), $T$ is a dimensionless constant and $K_y$ is the EXY rate constant (min$^{-1}$).

Some studies have investigated large scaled-up fixed-bed column systems. For example, FeCl$_2$-modified commercial (James Cummins P/L, Australia) granular activated biochar columns managed to provide 1494 Bv (441 μg g$^{-1}$) of safe As drinking water at a 100 μg L$^{-1}$ inlet As solution (30 × ø2 cm containing 32 g adsorbent) in a 2.5 m h$^{-1}$ flow rate. Doubling the flow rate (5 m h$^{-1}$) reduced it to 892 Bv (306 μg g$^{-1}$) (Kalaruban et al., 2019). FeCl$_3$-modified sun-dried water hyacinth biochar managed to provide ~1000 Bv of As-safe drinking water at a 402 μg L$^{-1}$ (3.5 × 0.5 cm containing 2.5 g adsorbent) in ~0.55 mL min$^{-1}$. The choice of the inlet concentration was based on the average As level of Huangshui River in the Shimen realgar mining district in Hunan, China. Before the effluent concentration hit 50 μg L$^{-1}$, a total of 400 more Bv was handled (Zhang et al., 2016). Though WHO recommends a 10 μg L$^{-1}$ As limit for drinking water, some countries such as Argentina, Peru, Uruguay, Bolivia, and Vietnam haven't adopted this recommendation in their local regulations (in all these mentioned countries, the maximum permitted limit of As in drinking water is 50 μg L$^{-1}$) (Kumar et al., 2019).

Completely mixed reactors (CMRs) are widely used as a decontaminant process for pollutants (Wang et al., 2017). Wang et al. (2017) scaled up results from batch studies to bench-scale continuous CMR experiments in order to provide guidelines for As(V) remediation employing nZVI-modified pinewood-derived biochars. The whole CMR system is displayed in Fig. 8.8-III, and it comprises a (1) 1L reactor (a), where $2\,gL^{-1}$ nZVI-biochar and distilled water have been constantly agitated to ensure homogeneous adsorbent concentrations with an overhead stirrer, (2) a 250-mL reactor (b) with concentrated As(V) solutions, (3) a 400-mL reactor (c), where an overhead stirrer has been paired with the adsorbent and sorbate, and (4) a 350-mL reactor (d) where the As(V) solution treated was gathered and the adsorbent settled out of the solution. The residence time on the reactor was optimized based on kinetics data from its first screening (30 min equilibrium time). At a constant rate and adsorbent concentration ($1.9\,gL^{-1}$), 90% and 100% of the initial As(V) concentration were removed 60 min after the first effluent. In contrast, the initial As(V) concentrations were 5.5 and $2.1\,mgL^{-1}$, respectively (Wang et al., 2017). The CMR design provides an effective, easy, and convenient approach to reducing the concentrations of As(V) to meet the common requirements for the As-safe drinking water provision. An in-depth analysis of the first, second, and third screening for this nZVI-biochar material will assure a successful technology transfer for safe drinking water provision.

## 4 Conclusions

From the analysis presented, it is evident that much is known about arsenic removal in controlled systems and with different biochar-derived adsorbents. Most articles published so far study batch configuration with monocomponent solutions. However, some have lately been including the performance of adsorbents in multicomponent systems, trying to understand the effect of the competitive ions and the different interactions involved. These controlled experimental designs have clear limitations and are essential to overcome if the technology is to be implemented in the field. These limitations become even more challenging when scales for real-world applications are larger. However, there seems to be only scarce efforts to transfer these so-called efficient materials from the academic area to concrete field applications.

Indeed, there are several operational limitations when designing a water treatment system which must be considered. However, given that most studies reach the first and second screening, there are few experiences that reach this last stage (3rd screening) in order to identify the actual bottlenecks of this technology. Site-specific considerations play a significant role in adapting the operational parameters to the particularities of a community and its water. In this vein, for scale-up outcomes to become a practical household solution, a transdisciplinary approach is vital. Articulation between academic institutions, field experts, social organizations, NGOs, and communities should be promoted.

Where centralized water supply systems are unavailable, a biochar-based system for As removal at a household scale is a feasible, short-term solution. However, efforts should be put into developing appropriate water and sanitation policies to ensure access to potable drinking water sources for informal settlements or rural, isolated communities. This will require the contribution of different disciplines as well as top-down and bottom-up approaches.

## Questions and problems

(1) Why does the World Health Organization recommend such a low maximum value for As in drinking water ($10\,\mu g\,L^{-1}$)?
(2) What are the main factors that regulate arsenic speciation and concentration in water? Describe for each arsenate and arsenite oxyanion.
(3) Enumerate traditional arsenic removal techniques for water and wastewater treatment.
(4) Why is adsorption an attractive technique for arsenic removal?
(5) What are the benefits of using biochar-based composite sorbents for As treatment? Enumerate.
(6) Why are biochar adsorption capacities for As(V) sorption superior compared to those of As(III)?
(7) Briefly, describe the 3-stage methodological approach presented in the chapter aimed at providing a comprehensive idea of adsorbent's capability to provide As-safe drinking water.
(8) What does the first experimental stage or "backbone screening" involve?
(9) Given the following graph (Alchouron et al., 2020, 2021) for BC (raw biochar), BCA (activated biochar), BC-Fe (raw biochar impregnated with iron), and BCA-Fe (activated biochar impregnated with iron), describe the influence of pH in As(V) sorption (the other parameters were kept constant: 25 mL of $10\,mg\,L^{-1}$ As(V) solution in contact with 50 mg of biochar).

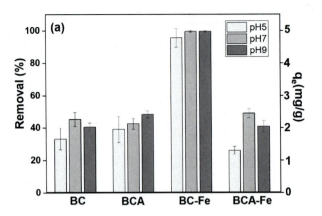

(10) Given the following As(V) adsorption isotherms for four different materials: BC (raw biochar), BCA (activated biochar), BC-Fe (raw biochar impregnated with iron), and BCA-Fe (activated biochar impregnated with iron) (Alchouron et al., 2020, 2021), what is the temperature effect?

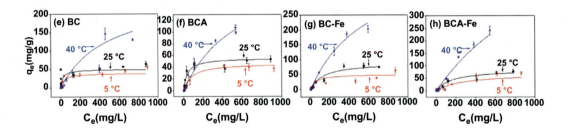

(11) What does the second experimental stage or "spinoff screening" involve?
(12) Why is it necessary to promote on-site screening?
(13) Why have the bulk of papers on the 1st screening stage, while the 3rd screening results are scarce?
(14) What information does a "breakthrough curve" give and what is its relevance in column tests?
(15) What is the difference between determining a removal percentage or removal efficiency compared to an adsorption capacity?

**(16)** "Where centralized water supply systems are unavailable, a biochar-based system for As removal at a household scale is a feasible, short-term solution." Based on this statement, we encourage readers of this chapter to search for updates on biochar-based systems for As removal at a household level: Are there new patents or new 3rd on-site screening papers? Are there new online/market releases?

# References

Agrafioti, E., Kalderis, D., Diamadopoulos, E., 2014. Arsenic and chromium removal from water using biochars derived from rice husk, organic solid wastes and sewage sludge. J. Environ. Manag. 133, 309–314.

Ahmed, M.B., Zhou, J.L., Ngo, H.H., Guo, W., Chen, M., 2016. Progress in the preparation and application of modified biochar for improved contaminant removal from water and wastewater. Bioresour. Technol. 214, 836–851.

Akter, K.F., Owens, G., Davey, D.E., Naidu, R., 2005. Reviews of Environmental Contamination and Toxicology. Springer, pp. 97–149.

Alchouron, J., Navarathna, C., Chludil, H.D., Dewage, N.B., Perez, F., Hassan, E.B., Pittman Jr., C.U., Vega, A.S., Mlsna, T.E., 2020. Assessing South American Guadua chacoensis bamboo biochar and Fe3O4 nanoparticle dispersed analogues for aqueous arsenic(V) remediation. Sci. Total Environ. 706, 135943.

Alchouron, J., Navarathna, C., Rodrigo, P.M., Snyder, A., Chludil, H.D., Vega, A.S., Bosi, G., Perez, F., Mohan, D., Pittman Jr., C.U., 2021. Household arsenic contaminated water treatment employing iron oxide/bamboo biochar composite: An approach to technology transfer. J. Colloid Interface Sci. 587.

Ali, S., Rizwan, M., Shakoor, M.B., Jilani, A., Anjum, R., 2020. High sorption efficiency for As (III) and As (V) from aqueous solutions using novel almond shell biochar. Chemosphere 243, 125330.

Alkurdi, S.S.A., Herath, I., Bundschuh, J., Al-Juboori, R.A., Vithanage, M., Mohan, D., 2019. Biochar versus bone char for a sustainable inorganic arsenic mitigation in water: what needs to be done in future research? Environ. Int. 127, 52–69.

Amen, R., Bashir, H., Bibi, I., Shaheen, S.M., Niazi, N.K., Shahid, M., Hussain, M.M., Antoniadis, V., Shakoor, M.B., Al-Solaimani, S.G., 2020. A critical review on arsenic removal from water using biochar-based sorbents: the significance of modification and redox reactions. Chem. Eng. J., 125195.

Asere, T.G., Stevens, C.V., Du Laing, G., 2019. Use of (modified) natural adsorbents for arsenic remediation: a review. Sci. Total Environ. 676, 706–720.

Astolfi, E., Maccagno, A., Fernández, J.G., Vaccaro, R., Stimola, R., 1981. Relation between arsenic in drinking water and skin cancer. Biol. Trace Elem. Res. 3 (2), 133–143.

Baig, S.A., Zhu, J., Muhammad, N., Sheng, T., Xu, X., 2014. Effect of synthesis methods on magnetic Kans grass biochar for enhanced As (III, V) adsorption from aqueous solutions. Biomass Bioenergy 71, 299–310.

Bakshi, S., Banik, C., Rathke, S.J., Laird, D.A., 2018. Arsenic sorption on zero-valent iron-biochar complexes. Water Res. 137, 153–163.

Bang, S., Johnson, M.D., Korfiatis, G.P., Meng, X., 2005. Chemical reactions between arsenic and zero-valent iron in water. Water Res. 39 (5), 763–770.

Bardach, A.E., Ciapponi, A., Soto, N., Chaparro, M.R., Calderon, M., Briatore, A., Cadoppi, N., Tassara, R., Litter, M.I., 2015. Epidemiology of chronic disease related to arsenic in Argentina: a systematic review. Sci. Total Environ. 538, 802–816.

Benis, K.Z., Damuchali, A.M., Soltan, J., McPhedran, K., 2020. Treatment of aqueous arsenic—a review of biochar modification methods. Sci. Total Environ., 139750.

Bolan, N., Syers, J., Tillman, R., 1986. Ionic strength effects on surface charge and adsorption of phosphate and sulphate by soils. J. Soil Sci. 37 (3), 379–388.

Bordoloi, S., Nath, S.K., Gogoi, S., Dutta, R.K., 2013. Arsenic and iron removal from groundwater by oxidation–coagulation at optimized pH: laboratory and field studies. J. Hazard. Mater. 260, 618–626.

Bowell, R.J., Alpers, C.N., Jamieson, H.E., Nordstrom, D.K., Majzlan, J., 2014. The environmental geochemistry of arsenic—an overview. Rev. Mineral. Geochem. 79 (1), 1–16.

Braghiroli, F.L., Calugaru, I.L., Gonzalez-Merchan, C., Neculita, C.M., Bouafif, H., Koubaa, A., 2020. Efficiency of eight modified materials for As (V) removal from synthetic and real mine effluents. Miner. Eng. 151, 106310.

Budinova, T., Petrov, N., Razvigorova, M., Parra, J., Galiatsatou, P., 2006. Removal of arsenic (III) from aqueous solution by activated carbons prepared from solvent extracted olive pulp and olive stones. Ind. Eng. Chem. Res. 45 (6), 1896–1901.

Bursztyn Fuentes, A.L., Barraqué, F., Mercader, R.C., Scian, A.N., Montes, M.L., 2021. Efficient low-cost magnetic composite based on eucalyptus wood biochar for arsenic removal from groundwater. Groundwater for Sustainable Development 14, 100585.

Bursztyn Fuentes, A.L., Canevesi, R.L.S., Gadonneix, P., Mathieu, S., Celzard, A., Fierro, V., 2020. Paracetamol removal by Kon-Tiki kiln-derived biochar and activated carbons. Ind. Crops Prod. 155.

Chammui, Y., Sooksamiti, P., Naksata, W., Arqueropanyo, O.-A., 2014. Kinetic and mechanism of arsenic ions removal by adsorption on leonardite char as low cost adsorbent material. J. Chil. Chem. Soc. 59 (1), 2378–2381.

Chen, C., Mi, S., Lao, D., Shi, P., Tong, Z., Li, Z., Hu, H., 2019. Single-step synthesis of eucalyptus sawdust magnetic activated carbon and its adsorption behavior for methylene blue. RSC Adv. 9 (39), 22248–22262.

Cho, D.-W., Yoon, K., Kwon, E.E., Biswas, J.K., Song, H., 2017. Fabrication of magnetic biochar as a treatment medium for As (V) via pyrolysis of $FeCl_3$-pretreated spent coffee ground. Environ. Pollut. 229, 942–949.

Chowdhury, M.R.I., Mulligan, C.N., 2011. Biosorption of arsenic from contaminated water by anaerobic biomass. J. Hazard. Mater. 190 (1-3), 486–492.

Chunhui, L., Jin, T., Puli, Z., Bin, Z., Duo, B., Xuebin, L., 2018. Simultaneous removal of fluoride and arsenic in geothermal water in Tibet using modified yak dung biochar as an adsorbent. R. Soc. Open Sci. 5 (11), 181266.

Colter, A., Mahler, R.L., 2006. Iron in Drinking Water. University of Idaho, Moscow.

Cortina, J.L., Litter, M.I., Gibert, O., Travesset, M., Ingallinella, A., Fernández, R., 2016. Latin American experiences in arsenic removal from drinking water and mining effluents. J. Total Environ.

Cruz, G.J., Mondal, D., Rimaycuna, J., Soukup, K., Gómez, M.M., Solis, J.L., Lang, J., 2020. Agrowaste derived biochars impregnated with ZnO for removal of arsenic and lead in water. J. Environ. Chem. Eng. 8 (3), 103800.

Darling, B.K., 2016. Geochemical factors controlling the mobilization of arsenic at an artificial recharge site, Clearwater, Florida. J. Contemp. Water Res. Educ. 159 (1), 105–116.

Deschamps, E., Ciminelli, V.S., Weidler, P.G., Ramos, A.Y., 2003. Arsenic sorption onto soils enriched in Mn and Fe minerals. Clay Clay Miner. 51 (2), 197–204.

Ding, Z., Xu, X., Phan, T., Hu, X., Nie, G., 2018. High adsorption performance for As (III) and As (V) onto novel aluminum-enriched biochar derived from abandoned Tetra Paks. Chemosphere 208, 800–807.

Dong, X., Ma, L.Q., Gress, J., Harris, W., Li, Y., 2014. Enhanced Cr (VI) reduction and As (III) oxidation in ice phase: important role of dissolved organic matter from biochar. J. Hazard. Mater. 267, 62–70.

Duan, X., Zhang, C., Srinivasakannan, C., Wang, X., 2017. Waste walnut shell valorization to iron loaded biochar and its application to arsenic removal. Resour. Technol. 3 (1), 29–36.

Eikelboom, M., 2017. Developing a Low-Cost Filter Medium for Remediation of Arsenic in Contaminated Groundwater. San Miguel de Allende: sn.

Fan, J., Xu, X., Ni, Q., Lin, Q., Fang, J., Chen, Q., Shen, X., Lou, L., 2018. Enhanced As (V) removal from aqueous solution by biochar prepared from iron-impregnated corn straw. J. Chem. 2018.

Feng, Y., Liu, P., Wang, Y., Finfrock, Y.Z., Xie, X., Su, C., Liu, N., Yang, Y., Xu, Y., 2020. Distribution and speciation of iron in Fe-modified biochars and its application in removal of As (V), As (III), Cr (VI), and Hg (II): an X-ray absorption study. J. Hazard. Mater. 384, 121342.

Frišták, V., Moreno-Jimenéz, E., Fresno, T., Diaz, E., 2018. Effect of physical and chemical activation on arsenic sorption separation by grape seeds-derived biochar. Separations 5 (4), 59.

Golterman, H., 1973. Natural phosphate sources in relation to phosphate budgets: a contribution to the understanding of eutrophication. Phosphorus in fresh water and the marine environment. Progr. Water Technol. 2, 3–17p.

Gupta, V., Gupta, B., Rastogi, A., Agarwal, S., Nayak, A., 2011. A comparative investigation on adsorption performances of mesoporous activated carbon prepared from waste rubber tire and activated carbon for a hazardous azo dye—Acid Blue 113. J. Hazard. Mater. 186 (1), 891–901.

Hao, L., Liu, M., Wang, N., Li, G., 2018. A critical review on arsenic removal from water using iron-based adsorbents. RSC Adv. 8 (69), 39545–39560.

He, R., Peng, Z., Lyu, H., Huang, H., Nan, Q., Tang, J., 2018. Synthesis and characterization of an iron-impregnated biochar for aqueous arsenic removal. Sci. Total Environ. 612, 1177–1186.

Hering, J.G., Katsoyiannis, I.A., Theoduloz, G.A., Berg, M., Hug, S.J., 2017. Arsenic Removal From Drinking Water: Experiences With Technologies and Constraints in Practice. American Society of Civil Engineers.

Hu, X., Ding, Z., Zimmerman, A.R., Wang, S., Gao, B., 2015. Batch and column sorption of arsenic onto iron-impregnated biochar synthesized through hydrolysis. Water Res. 68, 206–216.

Huang, Y., Gao, M., Deng, Y., Khan, Z.H., Liu, X., Song, Z., Qiu, W., 2020. Efficient oxidation and adsorption of As (III) and As (V) in water using a Fenton-like reagent, (ferrihydrite)-loaded biochar. Sci. Total Environ. 715, 136957.

Huang, L., Wu, H., van der Kuijp, T.J., 2015. The health effects of exposure to arsenic-contaminated drinking water: a review by global geographical distribution. Int. J. Environ. Health Res. 25 (4), 432–452.

Hussain, M., Imran, M., Abbas, G., Shahid, M., Iqbal, M., Naeem, M.A., Murtaza, B., Amjad, M., Shah, N.S., Khan, Z.U.H., 2019. A new biochar from cotton stalks for As (V) removal from aqueous solutions: its improvement with $H_3PO_4$ and KOH. Environ. Geochem. Health, 1–16.

Jiang, Y.-H., Li, A.-Y., Deng, H., Ye, C.-H., Li, Y., 2019. Phosphate adsorption from wastewater using ZnAl-LDO-loaded modified banana straw biochar. Environ. Sci. Pollut. Res. 26 (18), 18343–18353.

Jin, H., Capareda, S., Chang, Z., Gao, J., Xu, Y., Zhang, J., 2014. Biochar pyrolytically produced from municipal solid wastes for aqueous As (V) removal: adsorption property and its improvement with KOH activation. Bioresour. Technol. 169, 622–629.

Johansson, C.L., Paul, N.A., de Nys, R., Roberts, D.A., 2016. Simultaneous biosorption of selenium, arsenic and molybdenum with modified algal-based biochars. J. Environ. Manag. 165, 117–123.

Kalaruban, M., Loganathan, P., Nguyen, T.V., Nur, T., Hasan Johir, M.A., Nguyen, T.-H., Trinh, M.V., Vigneswaran, S., 2019. Iron-impregnated granular activated carbon for arsenic removal: application to practical column filters. J. Environ. Manag. 239, 235–243.

Karunanayake, A.G., Navarathna, C.M., Gunatilake, S.R., Crowley, M., Anderson, R., Mohan, D., Perez, F., Pittman Jr., C.U., Mlsna, T., 2019. Fe3O4 nanoparticles dispersed on douglas Fir biochar for phosphate sorption. ACS Appl. Nano Mater. 2 (6), 3467–3479.

Kim, J., Song, J., Lee, S.-M., Jung, J., 2019. Application of iron-modified biochar for arsenite removal and toxicity reduction. J. Ind. Eng. Chem. 80, 17–22.

Kumar, E., Bhatnagar, A., Hogland, W., Marques, M., Sillanpää, M., 2014. Interaction of inorganic anions with iron-mineral adsorbents in aqueous media—a review. Adv. Colloid Interf. Sci. 203, 11–21.

Kumar, N.S., Goel, S., 2010. Factors influencing arsenic and nitrate removal from drinking water in a continuous flow electrocoagulation (EC) process. J. Hazard. Mater. 173 (1-3), 528–533.

Kumar, R., Patel, M., Singh, P., Bundschuh, J., Pittman Jr., C.U., Trakal, L., Mohan, D., 2019. Emerging technologies for arsenic removal from drinking water in rural and peri-urban areas: methods, experience from, and options for Latin America. Sci. Total Environ. 694, 133427.

Kundu, S., Coumar, M.V., Rajendiran, S., Rao, A., Rao, A.S., 2015. Phosphates from detergents and eutrophication of surface water ecosystem in India. Curr. Sci., 1320–1325.

Lata, S., Prabhakar, R., Adak, A., Samadder, S.R., 2019. As (V) removal using biochar produced from an agricultural waste and prediction of removal efficiency using multiple regression analysis. Environ. Sci. Pollut. Res. 26 (31), 32175–32188.

Lee, S.Y., Choi, J.-W., Song, K.G., Choi, K., Lee, Y.J., Jung, K.-W., 2019. Adsorption and mechanistic study for phosphate removal by rice husk-derived biochar functionalized with Mg/Al-calcined layered double hydroxides via co-pyrolysis. Compos. Part B 176, 107209.

Li, R., Wang, J.J., Gaston, L.A., Zhou, B., Li, M., Xiao, R., Wang, Q., Zhang, Z., Huang, H., Liang, W., 2018. An overview of carbothermal synthesis of metal–biochar composites for the removal of oxyanion contaminants from aqueous solution. Carbon 129, 674–687.

Lin, L., Qiu, W., Wang, D., Huang, Q., Song, Z., Chau, H.W., 2017. Arsenic removal in aqueous solution by a novel Fe-Mn modified biochar composite: characterization and mechanism. Ecotoxicol. Environ. Saf. 144, 514–521.

Lin, L., Song, Z., Huang, Y., Khan, Z.H., Qiu, W., 2019. Removal and oxidation of arsenic from aqueous solution by biochar impregnated with Fe-Mn oxides. Water Air Soil Pollut. 230 (5), 105.

Litter, M.I., Ingallinella, A.M., Olmos, V., Savio, M., Difeo, G., Botto, L., Torres, E.M.F., Taylor, S., Frangie, S., Herkovits, J., 2019a. Arsenic in Argentina: occurrence, human health, legislation and determination. Sci. Total Environ. 676, 756–766.

Litter, M.I., Ingallinella, A.M., Olmos, V., Savio, M., Difeo, G., Botto, L., Torres, E.M.F., Taylor, S., Frangie, S., Herkovits, J., Schalamuk, I., Gonzalez, M.J., Berardozzi, E., Garcia Einschlag, F.S., Bhattacharya, P., Ahmad, A., 2019b. Arsenic in Argentina: technologies for arsenic removal from groundwater sources, investment costs and waste management practices. Sci. Total Environ. 690, 778–789.

Liu, X., Gao, M., Qiu, W., Khan, Z.H., Liu, N., Lin, L., Song, Z., 2019. Fe–Mn–Ce oxide-modified biochar composites as efficient adsorbents for removing As (III) from water: adsorption performance and mechanisms. Environ. Sci. Pollut. Res. 26 (17), 17373–17382.

Liu, J., He, L., Dong, F., Hudson-Edwards, K.A., 2016. The role of nano-sized manganese coatings on bone char in removing arsenic (V) from solution: implications for permeable reactive barrier technologies. Chemosphere 153, 146–154.

Liu, P., Li, H., Liu, X., Wan, Y., Han, X., Zou, W., 2020. Preparation of magnetic biochar obtained from one-step pyrolysis of salix mongolica and investigation into adsorption behavior of sulfadimidine sodium and norfloxacin in aqueous solution. J. Dispers. Sci. Technol. 41 (2), 214–226.

Luo, M., Lin, H., He, Y., Li, B., Dong, Y., Wang, L., 2019. Efficient simultaneous removal of cadmium and arsenic in aqueous solution by titanium-modified ultrasonic biochar. Bioresour. Technol. 284, 333–339.

Michalak, I., Chojnacka, K., Witek-Krowiak, A., 2013. State of the art for the biosorption process—a review. Appl. Biochem. Biotechnol. 170 (6), 1389–1416.

Mohan, D., Pittman Jr., C.U., Bricka, M., Smith, F., Yancey, B., Mohammad, J., Steele, P.H., Alexandre-Franco, M.F., Gómez-Serrano, V., Gong, H., 2007. Sorption of arsenic, cadmium, and lead by chars produced from fast pyrolysis of wood and bark during bio-oil production. J. Colloid Interface Sci. 310 (1), 57–73.

Mohan, D., Pittman Jr., C.U., 2007. Arsenic removal from water/wastewater using adsorbents—a critical review. J. Hazard. Mater. 142 (1-2), 1–53.

Mondal, P., Mohanty, B., Balomajumder, C., 2010. Treatment of arsenic contaminated groundwater using calcium impregnated granular activated carbon in a batch reactor: optimization of process parameters. CLEAN–Soil Air Water 38 (2), 129–139.

Mouzourakis, E., Georgiou, Y., Louloudi, M., Konstantinou, I., Deligiannakis, Y., 2017. Recycled-tire pyrolytic carbon made functional: a high-arsenite [As (III)] uptake material PyrC350®. J. Hazard. Mater. 326, 177–186.

Naujokas, M.F., Anderson, B., Ahsan, H., Aposhian, H.V., Graziano, J.H., Thompson, C., Suk, W.A., 2013. The broad scope of health effects from chronic arsenic exposure: update on a worldwide public health problem. Environ. Health Perspect. 121 (3), 295–302.

Navarathna, C., Alchouron, J., Liyanage, A., Herath, A., Wathudura, P., Nawalage, S., Rodrigo, P., Gunatilake, S., Mohan, D., Pittman Jr., C.U., 2020. Contaminants in Our Water: Identification and Remediation Methods. ACS Publications, pp. 197–251.

Navarathna, C.M., Karunanayake, A.G., Gunatilake, S.R., Pittman Jr., C.U., Perez, F., Mohan, D., Mlsna, T., 2019. Removal of Arsenic(III) from water using magnetite precipitated onto Douglas fir biochar. J. Environ. Manag. 250, 109429.

Nguyen, T.H., Pham, T.H., Nguyen Thi, H.T., Nguyen, T.N., Nguyen, M.-V., Tran Dinh, T., Nguyen, M.P., Do, T.Q., Phuong, T., Hoang, T.T., 2019. Synthesis of iron-modified biochar derived from rice straw and its application to arsenic removal. J. Chem. 2019.

Niazi, N.K., Bibi, I., Shahid, M., Ok, Y.S., Burton, E.D., Wang, H., Shaheen, S.M., Rinklebe, J., Lüttge, A., 2018a. Arsenic removal by perilla leaf biochar in aqueous solutions and groundwater: an integrated spectroscopic and microscopic examination. Environ. Pollut. 232, 31–41.

Niazi, N.K., Bibi, I., Shahid, M., Ok, Y.S., Shaheen, S.M., Rinklebe, J., Wang, H., Murtaza, B., Islam, E., Nawaz, M.F., 2018b. Arsenic removal by Japanese oak wood biochar in aqueous solutions and well water: investigating arsenic fate using integrated spectroscopic and microscopic techniques. Sci. Total Environ. 621, 1642–1651.

Ning, R.Y., 2002. Arsenic removal by reverse osmosis. Desalination 143 (3), 237–241.

Norazlina, A.S., Che, F.I., Rosenani, A.B., 2014. Characterization of oil palm empty fruit bunch and rice husk biochars and their potential to adsorb arsenic and cadmium. Am. J. Agric. Biol. Sci. 9 (3), 450–456.

Pi, K., Wang, Y., Xie, X., Su, C., Ma, T., Li, J., Liu, Y., 2015. Hydrogeochemistry of co-occurring geogenic arsenic, fluoride and iodine in groundwater at Datong Basin, northern China. J. Hazard. Mater. 300, 652–661.

Pooi, C.K., Ng, H.Y., 2018. Review of low-cost point-of-use water treatment systems for developing communities. Npj Clean Water 1 (1), 1–8.

Premarathna, K., Rajapaksha, A.U., Sarkar, B., Kwon, E.E., Bhatnagar, A., Ok, Y.S., Vithanage, M., 2019. Biochar-based engineered composites for sorptive decontamination of water: a review. Chem. Eng. J. 372, 536–550.

Qian, W., Zhao, A.-Z., Xu, R.-K., 2013. Sorption of As (V) by aluminum-modified crop straw-derived biochars. Water Air Soil Pollut. 224 (7), 1610.

Rajapaksha, A.U., Vithanage, M., Jayarathna, L., Kumara, C.K., 2011. Natural Red Earth as a low cost material for arsenic removal: kinetics and the effect of competing ions. Appl. Geochem. 26 (4), 648–654.

Ravenscroft, P., Brammer, H., Richards, K., 2011. Arsenic Pollution: A Global Synthesis. John Wiley & Sons.

Roy, P., Mondal, N.K., Bhattacharya, S., Das, B., Das, K., 2013. Removal of arsenic(III) and arsenic(V) on chemically modified low-cost adsorbent: batch and column operations. Appl Water Sci 3 (1), 293–309.

Sadegh-Zadeh, F., Seh-Bardan, B.J., 2013. Adsorption of As (III) and As (V) by Fe coated biochars and biochars produced from empty fruit bunch and rice husk. J. Environ. Chem. Eng. 1 (4), 981–988.

Saikia, R., Goswami, R., Bordoloi, N., Senapati, K.K., Pant, K.K., Kumar, M., Kataki, R., 2017. Removal of arsenic and fluoride from aqueous solution by biomass based activated biochar: optimization through response surface methodology. J. Environ. Chem. Eng. 5 (6), 5528–5539.

Sancha Fernández, A.M., 2006. Review of coagulation technology for removal of arsenic: case of Chile. J. Health Popul. Nutr. 24.

Sancha, A., Fuentealba, C., 2009. Application of coagulation–filtration processes to remove arsenic from low-turbidity waters. In: Bundschuh, J., Bhattacharya, P. (Eds.), Natural Arsenic in Groundwater of Latin America. Arsenic in the Environment, vol. 1, pp. 687–697.

Shakoor, M.B., Nawaz, R., Hussain, F., Raza, M., Ali, S., Rizwan, M., Oh, S.-E., Ahmad, S., 2017. Human health implications, risk assessment and remediation of As-contaminated water: a critical review. Sci. Total Environ. 601, 756–769.

Shakoor, M.B., Niazi, N.K., Bibi, I., Murtaza, G., Kunhikrishnan, A., Seshadri, B., Shahid, M., Ali, S., Bolan, N.S., Ok, Y.S., 2016. Remediation of arsenic-contaminated water using agricultural wastes as biosorbents. Crit. Rev. Environ. Sci. Technol. 46 (5), 467–499.

Shih, M.-C., 2005. An overview of arsenic removal by pressure-driven membrane processes. Desalination 172 (1), 85–97.

Singh, P., Sarswat, A., Pittman Jr., C.U., Mlsna, T., Mohan, D., 2020. Sustainable low-concentration arsenite [As (III)] removal in single and multicomponent systems using hybrid iron oxide-biochar nanocomposite adsorbents—a mechanistic study. ACS Omega 5 (6), 2575–2593.

Singh, R., Singh, S., Parihar, P., Singh, V.P., Prasad, S.M., 2015. Arsenic contamination, consequences and remediation techniques: a review. Ecotoxicol. Environ. Saf. 112, 247–270.

Sizmur, T., Fresno, T., Akgül, G., Frost, H., Moreno-Jiménez, E., 2017. Biochar modification to enhance sorption of inorganics from water. Bioresour. Technol. 246, 34–47.

Smedley, P.L., Kinniburgh, D.G., 2002. A review of the source, behaviour and distribution of arsenic in natural waters. Appl. Geochem. 17, 517–568.

Sullivan, C., Tyrer, M., Cheeseman, C.R., Graham, N.J., 2010. Disposal of water treatment wastes containing arsenic—a review. Sci. Total Environ. 408 (8), 1770–1778.

Tabassum, R.A., Shahid, M., Niazi, N.K., Dumat, C., Zhang, Y., Imran, M., Bakhat, H.-F., Hussain, I., Khalid, S., 2019. Arsenic removal from aqueous solutions and groundwater using agricultural biowastes-derived biosorbents and biochar: a column-scale investigation. Int. J. Phytoremed. 21 (6), 509–518.

Tan, X., Liu, Y., Zeng, G., Wang, X., Hu, X., Gu, Y., Yang, Z., 2015. Application of biochar for the removal of pollutants from aqueous solutions. Chemosphere 125, 70–85.

Tavares, D.S., Lopes, C.B., Coelho, J.P., Sánchez, M.E., Garcia, A.I., Duarte, A.C., Otero, M., Pereira, E., 2012. Removal of arsenic from aqueous solutions by sorption onto sewage sludge-based sorbent. Water Air Soil Pollut. 223 (5), 2311–2321.

Uddin, M.K., 2017. A review on the adsorption of heavy metals by clay minerals, with special focus on the past decade. Chem. Eng. J. 308, 438–462.

Van Vinh, N., Zafar, M., Behera, S., Park, H.-S., 2015. Arsenic (III) removal from aqueous solution by raw and zinc-loaded pine cone biochar: equilibrium, kinetics, and thermodynamics studies. Int. J. Environ. Sci. Technol. 12 (4), 1283–1294.

Verma, L., Singh, J., 2019. Synthesis of novel biochar from waste plant litter biomass for the removal of Arsenic (III and V) from aqueous solution: a mechanism characterization, kinetics and thermodynamics. J. Environ. Manag. 248, 109235.

Vithanage, M., Herath, I., Joseph, S., Bundschuh, J., Bolan, N., Ok, Y.S., Kirkham, M., Rinklebe, J., 2017. Interaction of arsenic with biochar in soil and water: a critical review. Carbon 113, 219–230.

Wang, H.-Y., Chen, P., Zhu, Y.-G., Cen, K., Sun, G.-X., 2019. Simultaneous adsorption and immobilization of As and Cd by birnessite-loaded biochar in water and soil. Environ. Sci. Pollut. Res. 26 (9), 8575–8584.

Wang, L., Fields, K.A., Chen, A.S., 2000. Arsenic Removal From Drinking Water by Ion Exchange and Activated Alumina Plants. National Risk Management Research Laboratory.

Wang, S., Gao, B., Li, Y., Zimmerman, A.R., Cao, X., 2016. Sorption of arsenic onto Ni/Fe layered double hydroxide (LDH)-biochar composites. RSC Adv. 6 (22), 17792–17799.

Wang, S., Gao, B., Zimmerman, A.R., Li, Y., Ma, L., Harris, W.G., Migliaccio, K.W., 2015. Removal of arsenic by magnetic biochar prepared from pinewood and natural hematite. Bioresour. Technol. 175, 391–395.

Wang, F., Liu, L.-Y., Liu, F., Wang, L.-G., Ouyang, T., Chang, C.-T., 2017. Facile one-step synthesis of magnetically modified biochar with enhanced removal capacity for hexavalent chromium from aqueous solution. J. Taiwan Inst. Chem. Eng. 81, 414–418.

Weber, K., Quicker, P., 2018. Properties of biochar. Fuel 217, 240–261.

Wei, Y., Wei, S., Liu, C., Chen, T., Tang, Y., Ma, J., Yin, K., Luo, S., 2019. Efficient removal of arsenic from groundwater using iron oxide nanoneedle array-decorated biochar fibers with high Fe utilization and fast adsorption kinetics. Water Res. 167, 115107.

Wen, T., Wang, J., Yu, S., Chen, Z., Hayat, T., Wang, X., 2017. Magnetic porous carbonaceous material produced from tea waste for efficient removal of As(V), Cr(VI), humic acid, and dyes. ACS Sustain. Chem. Eng. 5 (5), 4371–4380.

Wongrod, S., Simon, S., Van Hullebusch, E.D., Lens, P.N., Guibaud, G., 2018. Changes of sewage sludge digestate-derived biochar properties after chemical treatments and influence on As (III and V) and Cd (II) sorption. Int. Biodeterior. Biodegradation 135, 96–102.

Wu, J., Huang, D., Liu, X., Meng, J., Tang, C., Xu, J., 2018. Remediation of As (III) and Cd (II) co-contamination and its mechanism in aqueous systems by a novel calcium-based magnetic biochar. J. Hazard. Mater. 348, 10–19.

Xu, Y., Xie, X., Feng, Y., Ashraf, M.A., Liu, Y., Su, C., Qian, K., Liu, P., 2020. As (III) and As (V) removal mechanisms by Fe-modified biochar characterized using synchrotron-based X-ray absorption spectroscopy and confocal micro-X-ray fluorescence imaging. Bioresour. Technol. 304, 122978.

Yan, W., Ramos, M.A., Koel, B.E., Zhang, W.-X., 2012. As (III) sequestration by iron nanoparticles: study of solid-phase redox transformations with X-ray photoelectron spectroscopy. J. Phys. Chem. C 116 (9), 5303–5311.

Yan, J., Xue, Y., Long, L., Zeng, Y., Hu, X., 2018. Adsorptive removal of As(V) by crawfish shell biochar: batch and column tests. Environ. Sci. Pollut. Res. Int. 25 (34), 34674–34683.

Yazdani, M.R., Tuutijärvi, T., Bhatnagar, A., Vahala, R., 2016. Adsorptive removal of arsenic (V) from aqueous phase by feldspars: kinetics, mechanism, and thermodynamic aspects of adsorption. J. Mol. Liq. 214, 149–156.

Yean, S., Cong, L., Yavuz, C.T., Mayo, J., Yu, W., Kan, A., Colvin, V., Tomson, M., 2005. Effect of magnetite particle size on adsorption and desorption of arsenite and arsenate. J. Mater. Res. 20 (12), 3255–3264.

Yin, Z., Liu, Y., Liu, S., Jiang, L., Tan, X., Zeng, G., Li, M., Liu, S., Tian, S., Fang, Y., 2018. Activated magnetic biochar by one-step synthesis: enhanced adsorption and coadsorption for 17β-estradiol and copper. Sci. Total Environ. 639, 1530–1542.

Zhang, M., Gao, B., Varnoosfaderani, S., Hebard, A., Yao, Y., Inyang, M., 2013a. Preparation and characterization of a novel magnetic biochar for arsenic removal. Bioresour. Technol. 130, 457–462.

Zhang, M., Gao, B., Yao, Y., Inyang, M., 2013b. Phosphate removal ability of biochar/MgAl-LDH ultra-fine composites prepared by liquid-phase deposition. Chemosphere 92 (8), 1042–1047.

Zhang, G., Qu, J., Liu, H., Liu, R., Wu, R., 2007. Preparation and evaluation of a novel Fe–Mn binary oxide adsorbent for effective arsenite removal. Water Res. 41 (9), 1921–1928.

Zhang, F., Wang, X., Xionghui, J., Ma, L., 2016. Efficient arsenate removal by magnetite-modified water hyacinth biochar. Environ. Pollut. 216, 575–583.

Zhou, Z., Liu, Y.-G., Liu, S.-B., Liu, H.-Y., Zeng, G.-M., Tan, X.-F., Yang, C.-P., Ding, Y., Yan, Z.-L., Cai, X.-X., 2017. Sorption performance and mechanisms of arsenic (V) removal by magnetic gelatin-modified biochar. Chem. Eng. J. 314, 223–231.

Zhu, S., Qu, T., Irshad, M.K., Shang, J., 2020. Simultaneous removal of Cd (II) and As (III) from co-contaminated aqueous solution by α-FeOOH modified biochar. Biochar, 1–12.

Zhu, N., Yan, T., Qiao, J., Cao, H., 2016. Adsorption of arsenic, phosphorus and chromium by bismuth impregnated biochar: adsorption mechanism and depleted adsorbent utilization. Chemosphere 164, 32–40.

Zubrik, A., Matik, M., Lovás, M., Štefušová, K., Danková, Z., Hredzák, S., Václavíková, M., Bendek, F., Briančin, J., Machala, L., 2018. One-step microwave synthesis of magnetic biochars with sorption properties. Carbon Lett. 26, 31–42.

# Application of biochar for the removal of actinides and lanthanides from aqueous solutions

Amalia L. Bursztyn Fuentes[a], Beatrice Arwenyo[b,c], Andie L.M. Nanney[b], Arissa Ramirez[b,d], Hailey Jamison[b,d], Beverly Venson[b,e], Dinesh Mohan[f], Todd E. Mlsna[b], and Chanaka Navarathna[b]

[a]Technological Center of Mineral Resources and Ceramics (CETMIC - CIC/CONICET/UNLP), M.B. Gonnet, Argentina. [b]Department of Chemistry, Mississippi State University, Mississippi State, MS, United States. [c]Department of Chemistry, Gulu University, Gulu, Uganda. [d]Department of Biochemistry, Molecular Biology, Entomology and Plant Pathology, Mississippi State University, Mississippi State, MS, United States. [e]Department of Biological Sciences, Mississippi State University, Mississippi State, MS, United States. [f]School of Environmental Sciences, Jawaharlal Nehru University, New Delhi, India

## 1 Introduction

Actinides and lanthanides are elements with unfilled f orbitals and are collectively labeled as the inner transition elements (ITEs). The actinide series includes 15 radioactive metallic elements with atomic numbers 89–103 (Cooper, 2000). The actinide elements are classified as light (Ac, Th, Pa, U, Np, Pu, Am) or heavy (Cm, Bk, Cf, Es, Fm, Md, No, Lr) based on their atomic numbers. The actinides fill their 5f sublevels progressively and exhibit characteristics of both the d-block and the f-block elements. Elements of the actinide series can have oxidation states from +2 to plus +7. In addition to being radioactive, all actinides are paramagnetic and pyrophoric. Except for actinium with one oxidation state of +3, the other actinides are known to show variable oxidation states and more than one crystalline phase. To date, only the first four elements in the actinide series have been found to occur naturally.

Lanthanide series is made up of 15 elements with atomic numbers 57–71 (Herrmann et al., 2016). The elements in this series are chemically similar to scandium and yttrium. They are usually together referred to as the rare earth elements (REEs). Like the actinides, the lanthanides series fill their 4f sublevels progressively. They are classified as light REEs (La, Ce, Pr, Nd, Pm, Sm, Eu, Gd) or heavy REEs (Tb, Dy, Ho, Er, Tm Yb, Lu). Unlike the actinides, lanthanides' properties differ from those of transition metals and the main group elements. Their 4f orbitals are shielded from the atom's environment by 4d and 5p electrons. Accordingly, the lanthanides' properties primarily depend on their size, which declines gradually with rising atomic number. In addition to trivalent compounds, some elements of the lanthanide series can form divalent or tetravalent compounds. Whereas the actinides are mainly used where their radioactive nature is applicable, such as in cardiac pacemakers, nuclear reactors, and nuclear medicine, the lanthanides are widely used in optical devices, petroleum refining, and alloys (Kusrini et al., 2018). These elements have been reported to be relevant in the "clean energy" economy (Smith et al., 2016) and are widely used for various cutting-edge technological applications in energy, medicine, optics, and in electronic, chemical, automotive, and nuclear industries.

Water is the primary transport medium for most elements in the environment. Environmental ITEs differ in their oxidation state and complexation (Maher et al., 2013). In general, the trivalent state is the most thermodynamically stable form for lanthanides in aqueous environments (Elsalamouny et al., 2017). Usually, at a given oxidation state, the stability of actinide complexes increases with rising atomic number (Cooper, 2000). For example, the actinide uranium typically forms compounds in the +4 and +6 oxidation states in geological environments (Ervanne, 2004), with compounds in the former oxidation state being more toxic than the latter. Like most contaminants, the origin of the ITEs in water can be natural or anthropogenic. For example, uranium and thorium are naturally abundant in the Earth's crust and can contaminate groundwater through leaching (Kumar et al., 2011). Different anthropogenic activities contribute to incorporation and accumulation of ITEs in various environmental matrixes (Kołodynska et al., 2018). The primary man-made sources of ITEs in the environment are nuclear power generation, nuclear weapons manufacturing, testing and accidental releases (Maher et al., 2013), unregulated direct discharges and/or improper disposal of mine tailings and effluents containing ITEs, and the exploitation of ITEs for their extraction, concentration, and purification (Maher et al., 2013; Tan, 2019).

ITEs in water are reported to cause detrimental effects on ecosystems and human health (Srivastava et al., 2010). Some ITEs are radionuclides, emitting high energy gamma, beta, and alpha rays. Consequently, many ITEs have been classified as carcinogenic and genotoxic because long-term human exposure can damage tissue and DNA (NCR, 2012). However, little is known of the toxicity of accumulated ITEs in aquatic biota. A sustainable and cost-effective approach for the separation and recovery of ITEs from wastewater is both inevitable and vital to reduce ecological risk (Iftekhar et al., 2018). Moreover, from an economic point of view, the expense of these elements demands their recovery from generated waste solutions (Wang et al., 2020a).

Many different techniques have been used for the removal and recovery of ITEs from wastewater. Some of the conventional methods include precipitation (Djedidi et al., 2009; Royer-Lavallée et al., 2020; Souza et al., 2011), ion exchange (Hershey and Keliher, 1989), solvent extraction (Tong et al., 2009), and adsorption (Huang et al., 2018). However, some of these traditional techniques are costly, energy intensive and require large quantities of chemical reagents (Wang et al., 2016). Among these conventional remediation methods, adsorption stands out as a promising technique due to its operational simplicity, relatively low cost, and environmental friendliness (Anastopoulos et al., 2016a). Many adsorbents have been investigated for nuclear waste management: biosorbents (Pinto seaweeds) (Kumar et al., 2018), oxides (Das et al., 2010; Yu et al., 2014), resins (Lee et al., 2009; Serkan et al., 2020), silica (Liang and Fa, 2005; Wu et al., 2013), clays (Beall et al., 1979; Sylwester et al., 2000), nanoparticles (Aytas et al., 2004), activated carbon (Taskaev and Apostolov, 1978), carbon nanotubes (Sengupta and Gupta, 2017), and ordered mesoporous carbons (Lefrancois Perreault et al., 2017). However, some of the adsorbents remain cost-ineffective compared to biosorbents (Huang et al., 2018). Rare earth ion adsorbents require high selectivity over matrix metal ions for practical use, including high adsorption and desorption rates and capacities, durability, low cost, and high mechanical resistance. Aside from their selectivity, high adsorption rate, and removal efficiencies, biosorbents are stable toward the radiation damage (da Costa et al., 2020).

In the last decades, biochar and biochar-derived composites have received significant attention for water and wastewater remediation. Biochar is a carbon-rich material produced by pyrolysis of carbonaceous biomass such as crop residues, wood, and manure under oxygen-reduced conditions Wang et al., 2016. Biochar and biochar-derived materials can be easily synthesized and modified to produce efficient and inexpensive adsorbents. However, there is little information on the use of biochar and biochar-derived composites to recover ITEs from aqueous media (Mahmoud et al., 2019). Therefore, this chapter aims to compile available information from the literature on aqueous phase ITEs' sequestration using biochar and biochar-derived composites. Additionally, the adsorption mechanisms, knowledge gaps, and priority areas for future research shall be highlighted.

## 2  Biochar and engineered biochar

Over the last decade, biochar has been widely used for the removal of aqueous pollutant (Amen et al., 2020; Oliveira et al., 2017), including inorganic pollutants such as nitrate and phosphate (Li et al., 2014), heavy metals (Sizmur et al., 2017), organic compounds (Bursztyn Fuentes et al., 2020; Fan et al., 2017; Jang and Kan, 2019; Liu et al., 2012), and radioactive nuclides (Wang et al., 2016). Unlike activated carbon, which is costly and presents regeneration challenges, biochar is an inexpensive, eco-friendly, and effective adsorption material that can easily be synthesized from almost any organic residue. Moreover, biochar's sorption capacity can be enhanced by modification with chemical agents such as oxidants, reductants, metal salts, and ammonium citrate (Wang et al., 2016). Some of the conventional and nonconventional feedstock include pinewood sawdust (Komnitsas, 2017), rice straw (Dong et al., 2017a), corn cobs (Frišták et al., 2017), pine needles (Zhang et al., 2013a); banana peels (Kusrini et al., 2018), switchgrass (Kumar et al., 2011), *Liquidambar styraciflua* fruit (Mahmoud et al., 2019), bamboo (CITA Hu et al., 2018b), coffee residues and olive kernels (Paschalidou et al., 2020), algae (Wang et al., 2016), cyanobacterium (Wang et al., 2020a), invasive plant species (Li et al., 2020), recycled tires. (Smith et al., 2016), and municipal sewage sludge (Fristak et al., 2018; Zeng et al., 2020).

However, it has been found that biochar's final physicochemical properties and sorption performance depend entirely on feedstock type and pyrolysis temperature (Weber and Quicker, 2018). For instance, Frišták et al. (2017) achieved $0.11\,mg\,g^{-1}$ maximum sorption capacity for Eu(III) using corncob-derived biochar produced by slow pyrolysis at 500°C. In a separate study for Eu(III) sequestration, Fristak et al. (2018) obtained maximum sorption capacities of 0.375, 0.795, and $1.169\,mg\,g^{-1}$, using biochar produced by slow pyrolysis at 430°C from microcrystalline cellulose, organic cotton, and sewage sludge, respectively. They observed that the acquired sorption capacities were comparable to those of commercial titanium dioxide and zeolite. Similarly, Komnitsas (2017) produced a wood-derived biochar at 350°C. They obtained an Nd(III) removal efficiency comparable to that of a commercial activated carbon. Furthermore, Kumar et al. (2011) made switchgrass-derived biochar at 300°C through hydrothermal carbonization and found a maximum U(VI) sorption capacity of $2.12\,mg\,g^{-1}$. This biochar's adsorption ability has been attributed to surface functional groups (i.e., -COOH, C=O, and —OH), which represent active sites to which different pollutants can selectively interact with Wang et al. (2018). For example, to assess the effect of pyrolysis temperature on biochar removal capacities for ITEs, Wang et al., 2016 synthesized biochar from algal biomass (*Sargassum fusiforme*) by slow pyrolysis at different maximum temperatures (300, 500, and 700°C) and attained maximum La (III) sorption capacities of 170.36, 185.53, and $275.48\,mg\,g^{-1}$, respectively. The authors attributed the improved sorption capacity with increased pyrolysis temperature to the biochar's surface properties, including increased surface area and changes in their functional groups. IR band changes in the range of $1000–1400\,cm^{-1}$ and at $1604\,cm^{-1}$ were observed, corresponding to a reduction of negatively charged oxygen-containing functional groups in the biochar produced at 700°C. In contrast, Guilhen et al. (2019) synthesized palm tree-derived biochar at pyrolysis temperatures ranging from 250°C to 750°C. The highest U(VI) adsorption capacities corresponded to biochar produced at 250°C, followed by one obtained at 350°C. Their removal efficiencies were 86% and 80%, respectively. The authors observed that results from Fourier transform infrared spectroscopy (FTIR) and X-ray photoelectron spectrometry (XPS) indicated that oxygen-containing groups on biochar responsible for binding of U(VI) decreased with rising pyrolysis temperature. Zeng et al. (2020) studied U(VI) removal of sewage sludge-derived biochar produced at two pyrolysis temperatures (450°C and 600°C), and their corresponding maximum U(VI) adsorption capacities at pH=3 were 43.13 and $47.47\,mg\,g^{-1}$, respectively. For biochar pyrolyzed from 450°C to 600°C, the authors noticed increased pore size, pore volume, and specific surface area with rising pyrolysis temperature, whereas their volatile matter, C, H, O, and N, contents declined. They attributed the lower H/C and O/C associated with biochar pyrolyzed at 600°C to the decomposition of surface functional groups which, in turn, resulted in higher aromaticity and hydrophobicity of the biochar.

Although raw biochar can attain good sorption capacities for several pollutants compared to other biosorbents, better adsorptive capacities have been reported with modified or engineered biochar due to their improved surface functional characteristics (Shaheen et al., 2019). Wang et al. (2017) claimed that chemically treated and engineered biochar composites have an advantage over their raw counterparts because their sorption capacity can further be increased by introducing desirable properties required for enhancing adsorption. For example, specific compounds or functional groups with selectivity toward a targeted pollutant can be introduced to maximize the adsorption performance. Additionally, supporting nanoparticles onto the biochar enables the production of multiphase composites with affinities toward various pollutants without the problems of agglomeration of the particles used in isolation (Zhang et al., 2013b). For example, supporting magnetic nanoparticles onto the biochar provides a magnetic response that enables indirect manipulation of the adsorbent: it can be easily recovered from solution after metal sequestration with the aid of an external magnetic field, avoiding problems of clogging during the filtration step and allowing further recycling of the adsorbent and the adsorbate.

Traditional biochar modification techniques involve biochar or its feedstock treatments with chemicals such as acids, alkalis, and metal salts to improve the surface textural properties. For instance, Yakout (2016) studied the effect of different oxidizing agents (KOH, $HNO_3$, $H_2SO_4$, $H_2O_2$, and $KMnO_4$) on rice straw-derived biochar properties and U(VI) removal efficiency. Chemical activation induces significant effects on the abundance and distribution of basic and acidic surface sites of rice straw-based biochar. The magnitude relied on the strength of the chemical reagent used. According to the authors, biochar surface chemistry accounted for much of the U(VI) adsorption rather than textural modifications (i.e., changes in surface area and pore size distribution). Additionally, Alahabadi et al. (2020) performed a fractional design and an artificial neural network analysis to screen and optimize U(VI) removal efficiency of a large set of wood-derived biochar. Their biochar was treated with different acids (HCl, $HNO_3$, $H_2SO_4$, or $H_3PO_4$) in concentrations ranging from 2 to 12 M and was thermally treated (200–800 W fixed microwave power) for a predetermined time (3–15 min). The U(VI) removal efficiencies recorded with $H_3PO_4$, HCl, $H_2SO_4$, and $HNO_3$ treatments were 68.54%, 68.53%, 84.83%, and 94.14%, corresponding to sorption capacities of 342.70, 342.65, 424.15, and 470.4 mg g$^{-1}$, respectively.

Additional chemically activated biochar examples for the sequestration of ITEs have been reported in literature including NaOH-activated biochar (Kusrini et al., 2018), $HNO_3$-activated biochar (Alahabadi et al., 2020; Jin et al., 2018; Liatsou et al., 2017;

Paschalidou et al., 2020; Yakout, 2016), functionalized biochar (Lefrancois Perreault et al., 2017; Wang et al., 2016; Zhou et al., 2020), and composites, including magnetic composites (Hu et al., 2018a; Wang et al., 2020a, b; Wang et al., 2018; Zhu et al., 2018), zero-valent iron composites (ZVI-biochar) (Kołodynska et al., 2018; Ying et al., 2020), layered double hydroxide composites (LDH-biochar) (Li et al., 2020), and organo-biochar composites (Gao et al., 2020; Mahmoud et al., 2019). Such engineered biochar has been produced through different treatments and synthesis pathways. For instance, Gao et al. (2020) synthesized an iron oxide/persimmon tannin/graphene oxide nanocomposite ($Fe_3O_4$/PT/GO) by a green and facile hydrothermal method for the adsorption of Er(III) from aqueous solution. Yang et al. (2018) proposed a facile synthesis of a novel graphene oxide/manganese oxide composite through ultrasonic radiation for U(VI) sorption. Zhou et al. (2020) synthesized $PO_4$-functionalized biochar through ball-milling *Typha angustifolia*-derived biochar with phytic acid, a natural and nontoxic saturated cyclic organic acid extracted from plant tissues (Sizmur et al., 2017).

Like the case of raw biochar, contrasting results from chemically activated and/or engineered biochar's sorption capacity have been reported (Table 9.1). For example, Jin et al. (2018) observed a 40 times improved U(VI) adsorption capacity for a wheat straw-derived biochar chemically activated with $HNO_3$ (from 8.7 to 355.6 mg g$^{-1}$, pH = 4.5 at 298 K). They attributed their findings to the higher concentration of surface carboxylic groups and a more negative surface charge associated with the treated biochar. In contrast, Paschalidou et al. (2020) saw a decrease in the U(VI) adsorption capacity of three feedstocks after oxidation with $HNO_3$. They claimed that reduced oxidation of surface carboxylic acids functional groups during the reduction and precipitation of U(VI) to U(IV) could explain the apparent decrease in sorption capacity after modification.

Table 9.1 summarizes some of the recently published materials used for ITEs' sequestration and their corresponding sorption capacities. Generally, most studies reported enhanced sorption capacities for the pollutant under analysis with modified or engineered biochar compared to their untreated counterparts. For example, Wang et al. (2016) prepared modified algal-derived biochar (AC-BC) using ammonium citrate for remediation of La(III) from aqueous solution. They reported that the maximum La(III) adsorption capacity of the modified biochar doubled the capacity of the raw biochar produced at the same temperature (170.36 and 362.3 mg g$^{-1}$). FTIR analysis indicates that AC-BC300 functional groups were much more abundant compared to raw biochar

Table 9.1 Summary of the materials used in ITEs' removal from aqueous solution, their sorption capacities and experimental conditions.

| Adsorbent | Ref. | Pollutant | Reactor | Dosage (g L$^{-1}$) | pH | Ionic strength (mol L$^{-1}$) | Contact time (min) | $C_0$ (mg L$^{-1}$) | $q_e$ (mg g$^{-1}$) | Kinetics | Isotherms | Thermodynamic | Competitive ions |
|---|---|---|---|---|---|---|---|---|---|---|---|---|---|
| **Raw biochars** | | | | | | | | | | | | | |
| Microcrystalline cellulose biochar | Fristak et al. (2018) | Eu(III) | Batch | 31 | 6 | – | 1440 | 5 to 250 | 0.375 | $q_e = 0.40$, $k_2 = 0.11$, $R^2 = 0.98$ | $q_m = 0.38$, $R^2 = 0.88$ | – | – |
| Cotton biochar | | | | | | | | | 0.795 | $q_e = 0.69$, $k_2 = 0.02$, $R^2 = 0.97$ | $q_m = 0.80$, $R^2 = 0.88$ | | |
| Sewage sludge biochar | | | | | | | | | 1169 | $q_e = 1.41$, $k_2 = 0.00$, $R^2 = 0.98$ | $q_m = 1.17$, $R^2 = 0.90$ | | |
| Rice straw-derived biochar | Dong et al. (2017b) | U(VI) | Batch | 0.66 | 4.48, 5.26, 6.15 | 0.001; 0.01 M; 0.1 NaClO$_4$ | 180 | – | 2.80e$^{-5}$ mol/g | $q_e = 3.38e^{-5}$, $k_2 = 2.44$, $R^2 = 1.00$ | ($q_m = 2.70e^{-5}$, $R^2 = 1.00$); ($q_m = 3.01e^{-5}$, $R^2 = 1.00$); ($q_m = 3.38e^{-5}$, $R^2 = 1.00$) | 293 K, 313 K, 333 K | – |
| Sewage sludge-derived biochar @450°C | Zeng et al. (2020) | U(VI) | Batch | 0.25–4.0 | 2 to 7 | 0 to 50 mg L$^{-1}$ CaCl$_2$ | – | – | 43.13 | $q_e = 6.93$, $k_2 = 0.06$, $R^2 = 1.00$ | ($q_m = 40.51$, $R^2 = 0.99$);($q_m = 41.62$, $R^2 = 0.99$); ($q_m = 43.13$, $R^2 = 1.00$) | 293 K, 308 K, 318 K | – |
| Sewage sludge-derived biochar @600°C | | | | | | | | | 47.47 | $q_e = 6.98$, $k_2 = 0.08$, $R^2 = 1.00$ | ($q_m = 42.37$, $R^2 = 0.99$);($q_m = 44.84$, $R^2 = 1.00$); ($q_m = 47.47$, $R^2 = 1.00$) | | |
| Bamboo-derived biochar | Hu et al. (2018b) | U(VI) | Batch | 0.5–5 | 1–7 (4) | 0.1 to 1 NaCl | 600 | 20–100 | 35.2 | $q_e = 18.8$, $k_2 = 0.00$, $R^2 = 0.99$ | ($q_m = 32.30$, $R^2 = 1.00$);($q_m = 34.90$, $R^2 = 1.00$); ($q_m = 35.20$, $R^2 = 1.00$) | 298 K, 308 K, 318 K | – |
| Switchgrass-derived biochar HTC | Kumar et al. (2011) | U(VI) | Batch/Column | 1, 4 and 5 | 3.9 | 0.1 NaNO$_3$ | 2040 | 5 to 30 | 2.12 | – | $q_m = 2.12$ | – | – |
| Sawdust-derived biochar | Kommitsas (2017) | Nd(III) | Batch | 1 | 3 | – | 1440 | 20 | 8 | $q_e = 1.06$, $k_2 = 2.72$, $R^2 = 1.00$ | $R^2 = 0.97$ | – | – |
| | | | | 2 | | | | | 4.7 | $q_e = 1.54$, $k_2 = 13.80$, $R^2 = 1.00$ | $R^2 = 0.88$ | | |
| | | | | 5 | | | | | 2.5 | $q_e = 1.61$, $k_2 = 0.88$, $R^2 = 1.00$ | $R^2 = 0.89$ | | |
| | | | | 10 | | | | | 1.6 | $q_e = 1.77$, $k_2 = 2.26$, $R^2 = 1.00$ | $R^2 = 0.86$ | | |

| Material | Reference | Target | Mode | C1 | C2 | C3 | C4 | C5 | C6 | Capacity | Kinetics | Isotherm | Other |
|---|---|---|---|---|---|---|---|---|---|---|---|---|---|
| Palm tree-derived biochar @250°C | Guilhen et al. (2019) | U(VI) | Batch | 10 | 3 | 1440 | – | 5 | – | 417 | – | $q_m=417$ | – |
| Palm tree-derived biochar @350°C | | | | | | | | | | 408 | | $q_m=408$ | |
| Palm tree-derived biochar @450°C | | | | | | | | | | 372 | | $q_m=372$ | |
| Palm tree-derived biochar @550°C | | | | | | | | | | 88 | | $q_m=88$ | |
| Palm tree-derived biochar @650°C | | | | | | | | | | 56 | | $q_m=56$ | |
| Palm tree-derived biochar @750°C | | | | | | | | | | 52 | | $q_m=52$ | |
| Palm tree-derived biochar @350°C | Guilhen et al. (2018) | U(VI) | Batch | 10 | 3 | 180 | – | 5 | – | 400 | $q_e=454$, $k_2=9.17e^{-5}$, $R^2=0.99$ | $q_m=439$, $R^2=1.00$ | – |
| **Raw and engineered counterparts** | | | | | | | | | | | | | |
| Algae biochar @300°C | Wang et al., 2016 | La(III) | Batch | 1 | 3 to 9 (7) | – | – | – | – | 170.36 | $q_e=47.92$, $k_2=0.00$, $R^2=1.00$ | – | – |
| Algae biochar @500°C | | | | | | | | | | 185.5 | $q_e=5.72$, $k_2=49.80$, $R^2=1.00$ | | |
| Algae biochar @700°C | | | | | | | | | | 275.49 | $q_e=0.13$, $k_2=0.13$, $R^2=1.00$ | | |
| Ammonium citrate-modified biochar (@300°C) | | | | | | | | 25 to 500 | | 362.32 | $q_e=0.01$, $k_2=49.85$, $R^2=1.00$ | | $K^+$, $Ca^{2+}$, $Mg^{2+}$ |
| Hydrophyte biomass-derived biochar | Hu et al. (2018a) | U(VI) | Batch | 1 | 3 | 0.01 M NaClO4 | 2880 | – | – | 52.36 | – | $q_m=52.36$, $R^2>1.00$ 293 K | – |
| Magnetic biochar | | | | | | | | | | 54.35 | | $q_m=54.35$, $R^2>1.00$ | |
| Corncob-derived biochar | Frištak et al. (2017) | Eu(III) | Batch | 33.3 | 6 | – | 1440 | 2.5 to 25 | | 0.11 | $q_e=0.58$, $k_2=0.01$, $R^2=0.99$ | $q_m=0.89$, $R^2=1.00$ | – |
| Fe-modified biochar | | | | | | | | | | 2.26 | $q_e=0.64$, $k_2=0.02$, $R^2=0.88$ | $q_m=0.98$, $R^2=0.99$ | |
| Cyanobacteria biochar | Wang et al. (2020a, b) | U(VI) | Batch | 0.1 g L$^{-1}$ 0.67 g L$^{-1}$ | 2 to 10 (6) | 0.001 M, 0.01 M and 0.1 M NaNO3 | 480 | – | – | 54.46 | $k_2=0.02$ | $q_m=44.15$, $R^2=0.08$; 298 K, 308 K, 318 K $q_m=48.04$, $R^2=0.91$; $q_m=54.46$, 0.88) | – |
| Fe3O4-modified biochar | | | | | | 0.001 M, 0.01 M and 0.1 M NaNO3 | | | | 49.19 | $k_2=0.02$ | $q_m=28.24$, $R^2=0.91$; $q_m=36.09$, $R^2=0.86$; $q_m=49.19$, $R^2=0.93$ | |
| | | U(VI) | Batch | – | 4.5 | – | – | – | – | 8.7 | – | $q_m=8.70$, $R^2=0.98$ 298 K | – |

*Continued*

**Table 9.1** Summary of the materials used in ITEs' removal from aqueous solution, their sorption capacities and experimental conditions—cont'd

| Adsorbent | Ref. | Pollutant | Reactor | Dosage (g L$^{-1}$) | pH | Ionic strength (mol L$^{-1}$) | Contact time (min) | Co (mg L$^{-1}$) | q$_e$ (mg g$^{-1}$) | Kinetics | Isotherms | Thermodynamic | Competitive ions |
|---|---|---|---|---|---|---|---|---|---|---|---|---|---|
| Wheat straw biochar | Jin et al. (2018) | | | | | | | | 64.0 | $q_e = 26.06$, $k_2 = 1.8 e^{-3}$, $R^2 = 1.00$ | $q_m = 64.00$, $R^2 = 0.98$ | | |
| Cow manure biochar | | | | | | | | | | HNO$_3$-treated wheat straw biochar 355.6 | | | |
| HNO$_3$-treated cow manure biochar | | | | 73.3 | | $q_e = 27.90$, $k_2 = 2.6 e^{-3}$, $R^2 = 1.00$ | $q_m = 73.30$, $R^2 = 0.96$ | | | $q_m = 355.60$, $R^2 = 0.98$ | | | |
| Malt spent rootlets biochar | Paschalidou et al. (2020) | U(VI) | Batch | 0.33 (0.01 g in 30 mL) | | – | 1440 | – | 547 | – | $q_m = 2.30 \pm 0.1$, $R^2 = 0.99$ | | – |
| Coffee residues | | | | | | | | | 547 | | $q_m = 2.30 \pm 0.5$, $R^2 = 0.98$ | | |
| Olive kernels | | | | | | | | | 357 | | $q_m = 1.50 \pm 0.3$, $R^2 = 0.98$ | | |
| HNO$_3$-treated malt spent rootlets biochar | | | | | | | | | 500 | | $q_m = 1.50 \pm 0.2$, $R^2 = 0.98$ | | |
| HNO$_3$-treated coffee residues | | | | | | | | | 357 | | $q_m = 2.10 \pm 0.4$, $R^2 = 0.99$ | | |
| HNO$_3$-treated olive kernels | | | | | | | | | 381 | | $q_m = 1.60 \pm 0.2$, $R^2 = 0.99$ | | |
| Ficus microcarpa-derived biochar | Li et al. (2019) | U(VI) | Batch | 0.5 to 25 (1) | 2 to 10 (4) | – | <180 | 1 to 200 | 19.09 | $q_e = 46.49$, $k_2 = 1.82e^{-3}$, $R^2 = 1.00$ | ($q_m = 3.69e^{-4}$, $R^2 = 1.00$)($q_m = 4.02e^{-4}$, $R^2 = 1.00$)($q_m = 4.13e^{-4}$, $R^2 = 1.00$) | 296 K, 316 K, 336 K | – |
| KMnO4-modified biochar | | | | | | | | | 27.29 | – | – | – | – |

**Only engineered biochars**

| Adsorbent | Ref. | Pollutant | Reactor | Dosage (g L$^{-1}$) | pH | Ionic strength (mol L$^{-1}$) | Contact time (min) | Co (mg L$^{-1}$) | q$_e$ (mg g$^{-1}$) | Kinetics | Isotherms | Thermodynamic | Competitive ions |
|---|---|---|---|---|---|---|---|---|---|---|---|---|---|
| Fe3O4/PT/GO | Gao et al. (2020) | Er(III) | Batch | 30-100 | 2 to 8 (4) | 0.05 to 0.25 | 300 | 50 to 300 | 403.2 | – | – | – | – |

| Material | Reference | Target | Mode | Dose | pH (optimal) | Time (min) | Concentration | $q_m$ | Parameters | Notes |
|---|---|---|---|---|---|---|---|---|---|---|
| CaAl-LDH/PB | Li et al. (2020) | Eu(III) | Batch | 70 | 3 to 10 (7) | – | 120 | 10 to 80 | 120.4 | $q_e$ = 86.957, $k_2$ = 4.70 × 10$^{-4}$, $R^2$ = 0.999 | ($q_m$ = 76.923, $R^2$ = 0.882) 298K, 318K, 338K; ($q_m$ = 86.207, $R^2$ = 0.927); ($q_m$ = 94.340, $R^2$ = 0.955) | – |
| Magnetic biochar | Zhu et al. (2018) | Eu(III) | Batch | 1.2 | 2 to 10 (3) | – | – | 10 to 50 | 105.5 | $q_e$ = 6.786, $k_1$ = 0.015, $R^2$ 0.8524 | $q_m$ = 105.53, $R^2$ = 0.9993 | – |
| Oxidized OMC | Lefrancois Perreault et al. (2017) | La(III) | Batch | 1 | 2.6 | – | 240 | 0.3 | – | ($q_e$ = 1.17, $k_2$ = 368.25, $R^2$ = 0.9981); ($q_e$ = 12.55, $k_2$ = 2.89, $R^2$ = 0.9989); ($q_e$ = 9.12, $k_2$ = 9.26, $R^2$ = 0.9993) | ($q_m$ = 3.31, $R^2$ = 0.95); ($q_m$ = 22.30, $R^2$ = 0.97); ($q_m$ = 9.90, $R^2$ = 1.00) | – |
| NaOH-activated banana peels-derived biochar | Kusrini et al. (2018) | Y, La, Ce, Nd, and Sm | Batch | 2 | 4 | – | – | – | 362.3 | – | – | – |
| ZVI-biochar (MBC1) | Kolodynska et al. (2018) | La(III), Ce(III) and Nd(III) | Batch | 5 g | 2 to 5 | – | – | 50-200 mg L$^{-1}$ | – | – | – | – |
| ZVI-biochar (MBC2) | | | | | | | | | | ($q_e$ = 11.07, $k_2$ = 0.010, $R^2$ = 0.993); ($q_e$ = 12.54, $k_2$ = 0.013, $R^2$ = 0.998); ($q_e$ = 16.67, $k_2$ = 0.011, $R^2$ = 0.998) | – | – |
| ZVI-biochar (MBC3) | | | | | | | | | | – | – | – |
| Oxidized biochar fibers | Liatsou et al. (2017) | Sm(III) | Batch | 0.01 g | 3 | – | 1440 | – | – | – | $q_m$ = 2.4 | – |
| Glutaraldehyde-nanosilica-biochar | Mahmoud et al. (2019) | U(VI) | Batch | 0.04 to 0.6 (0.2) | 1 to 10 (4) | – | 10 | 15 to 75 (30) | – | – | – | X+tap water |
| Magnetic acid-modified biochar | Alahabadi et al. (2020) | U(VI) | Batch | 10 to 50 (20) | 2 to 10 (5) | – | 1200 | 12.5 to 250 | 688.03 | $q_e$ = 503.1, $k_2$ = 5.51E-04, $R^2$ = 0.9999 | $q_m$ = 688.03, $R^2$ = 0.997 | – |
| | | Th(IV) | | | | | | | 432.04 | | | |

*Continued*

Table 9.1 Summary of the materials used in ITEs' removal from aqueous solution, their sorption capacities and experimental conditions—cont'd

| Adsorbent | Ref. | Pollutant | Reactor | Dosage ($gL^{-1}$) | pH | Ionic strength ($mol L^{-1}$) | Contact time (min) | $C_0$ ($mgL^{-1}$) | $q_e$ ($mgg^{-1}$) | Kinetics | Isotherms | Thermodynamic | Competitive ions |
|---|---|---|---|---|---|---|---|---|---|---|---|---|---|
| Magnetic acid-modified biochar | | | | | 2 to 10 (5) | | | | | | $q_m=432.04$; $R^2=0.9991$ | – | – |
| Pine needle-derived biochar in citric acid + HTC | Zhang et al. (2013a, b) | U(VI) | Batch | 0.1 to 3 | 3 to 8 (6) | 0.01 to 0.3 $KNO_3$ | 50 | 10 to 110 (50) | 62.7 | $q_e=380.78$; $k_2=6.17E-04$, $R^2=0.9999$ | $q_m=62.7$, $R^2=0.99$ | – | – |
| Dictyophora indusiata-derived biochar supported sulfide NZVI | Pang et al. (2019) | U(VI) | Batch | 0.05 | 2 to 11 (5) | 0.1 M and 0.01 M $NaNO_3$ | 180 | – | 427.9 | $q_e=56.4$, $k_2=0.0054$, $R^2=0.993$ | ($q_m=427.9$, $R^2=0.992$ 298 K, 313 K, 328 K ($q_m=483$, $R^2=0.994$);($q_m=530.1$, $R^2=0.988$) | | X+seawater |
| Chemically modified rice straw-based biochars | Yakout (2016) | U(VI) | Batch | – | 2 to 10 | – | – | – | – | $q_e=85.47$, $k_2=0.0011$, $R^2=0.994$ | $q_m=100$ | – | – |
| P-functionalized biochar | Zhou et al. (2020) | U(VI) | Batch | – | – | – | – | – | 128.5 | – | 128.5 | – | – |

BC300. The 3512 cm$^{-1}$ band in AC-BC300 is due to —OH stretching vibrations and 1573 cm$^{-1}$ band was attributed to —NH stretching vibration in the amide bond. Peaks at 1461 and 1376 cm$^{-1}$ result due to —CN stretching. Peaks at 2995, 2956, and 1054 cm$^{-1}$ originate from alkyl —CH$_2$ and —CO stretching, respectively. A new strong peak at 1768 cm$^{-1}$ (carboxyl group) was detected, suggesting carboxyl groups were attached to the biochar surface. These results suggest the successive AC grafting to the biochar surface providing more binding sites for metal adsorption. Frišták et al. (2017) produced Fe-modified corncob-derived biochar. They achieved 2.26 mg g$^{-1}$ maximum Eu(III) sorption capacity, 20 times higher than that of the raw biochar. Zhu et al. (2018) realized the highest Eu(III) sorption capacity with the magnetic composite (105.53 mg g$^{-1}$). In contrast, the raw biochar and the magnetite particles in isolation exhibited 88.45 and 97.95 mg g$^{-1}$, respectively, at pH=3 and 298 K.

## 3 Sorption performance and dynamics

Numerous studies are required to optimize experimental conditions under which a given adsorbent material performs best. The basic requirements for a promising adsorbent are high sorption capacity and fast removal rates (i.e., remove as much as possible in the shortest time). In most cases, kinetic and isotherm studies are performed to gain an insight into these aspects. Other vital features needed to assess the sorption capacity are the adsorbent dosage, the working pH (which must be compatible with the available facilities), and the material's ability to effectively adsorb a specific contaminant in the presence of competing ions or other pollutants. Studies such as pH dependency, effect of adsorbent dosage, and multicomponent studies are commonly carried out to evaluate an adsorbent's performance concerning these aspects. Additionally, the adsorbent regeneration and reusability, through several cycles, is a crucial characteristic to consider for the cost-effectiveness of the overall process. A suitable adsorbent will satisfy most of the above aspects and other location-related factors. Whereas many studies have been done to evaluate many (bio)sorbents (Royer-Lavallée et al., 2020), only work done for biochar and biochar composites has been considered in the next sections.

### 3.1 pH dependency

The removal efficiency of ITEs by different sorption materials is often affected by solution pH alterations. The solution pH governs both the adsorbent's surface charge and the pollutant's speciation and ionization in the solution. For example, Kumar et al. (2011) studied U(VI) removal using switchgrass-derived biochar. They found that the U(VI) sorption capacity was 2.12 mg g$^{-1}$ at pH = 3.9. They noted that the sorption process was highly pH dependent, and that sorption capacity almost doubled at near neutral pH. Zhang et al. (2013b) used pine needle-derived biochar produced by hydrothermal carbonization and found a similar trend. In their study, U(VI) adsorption increased from pH 3 to 6, with a maximum adsorption capacity of 55.91 mg g$^{-1}$. Similar results were reported with rice straw-derived biochar (Dong et al., 2017b). The U(VI) removal efficiency increased from 31.67% to 99.17% at a pH range of 3.0–6.8, respectively. However, the adsorption capacity reduced in the pH range of 7.3–9.2. Additionally, Alahabadi et al. (2020) studied the effect of pH on U(VI) and Th(IV) removal with a magnetic HNO$_3$-activated biochar. They obtained maximum U(VI) and Th(IV) removal efficiencies of 87.96% and 97.94%, respectively, corresponding to pH ranges of 4.5–5.0 and 4.5–5.5. At pH > 6.0, a smooth decline in sorption capacity was recorded.

Dong et al. (2017b) ascribed increased U(VI) sorption with rising pH values to reduced competition between protons (H$^+$) and U(VI) for surface sites of biochar at higher pH. They argued that at low pH, the dominant species in solution are $UO_2^{2+}$ or $(UO_2)_3(OH)_5^+$ and $(UO_2)_4(OH)$. Therefore, the presence of both H$^+$ and the positive U(VI) ions causes electrostatic repulsion on the positive biochar surface and thus less adsorption. The greater U(VI) removal of biochar at pH 6.8, however, is due to adsorption and successive precipitation which resulted in the formation of schoepite. The reduced U(VI) adsorption at pH 7.3 can partly be explained by the existence of dissolved inorganic carbon or the formation of negatively charged ions including $(UO_2)_3(OH)_7^-$, $UO_2(CO_3)_2^{2-}$, $UO_2(CO_3)_3^{4-}$ which are repelled from the negatively charged sorbent surface by electrostatic forces.

The impact of solution pH on the adsorbent's surface charge is usually assessed by determining the pH point of zero charge (pH$_{PZC}$). The pH$_{PZC}$ is the pH at which the net charge on the adsorbent is zero. Above the pH$_{PZC}$ value, the sorbent surface will be negatively charged, promoting electrostatic attraction between the sorbent and positively charged radionuclides. On the contrary, at pH values lower than the pH$_{PZC}$, the sorbent's surface will be positively charged, promoting electrostatic repulsion between the surface and the positively charged radionuclides (Alam et al., 2018). The surface net charge and the pH$_{PZC}$ within biochars vary and depend on the adsorbent's properties defined by the

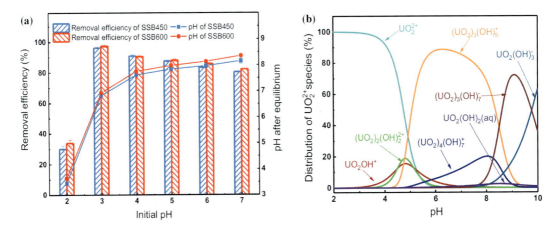

**Fig. 9.1** (A) Variation of the U(VI) removal efficiency of the sewage sludge-derived biochar and the final pH after equilibrium with different initial pH and (B) the U(VI) species distribution percentages in aquatic solution with variable pH (temperature = 30°C and an initial U(VI) concentration = 7.2 mg L$^{-1}$). Reprinted by permission from Springer Nature Customer Service Centre GmbH: Springer, Zeng, T., Mo, G., Zhang, X., Liu, J., Liu, H., Xie, S., 2020. U(VI) removal efficiency and mechanism of biochars derived from sewage sludge at two pyrolysis temperatures. J. Radioanal. Nucl. Chem. https://doi.org/10.1007/s10967-020-07423-y. Copyright (2020).

feedstock, thermal treatment, posterior modifications, and the particular surface functional groups (Komnitsas, 2017).

Furthermore, U(VI)'s adsorption mechanism onto biochar is strongly related to its aqueous speciation. U(VI) has a complex aqueous speciation (Alam et al., 2018). Generally, at pH < 6, $UO_2^{2+}$ is the predominant species, while at pH > 6, hydroxy complexes such as $(UO_2)_3(OH)_5^+$ $(UO_2)_3(OH)_7^+$ predominate (Fig. 9.1) (Zeng et al., 2020).

Research studies should be done using processing industries' real working conditions for a better understanding of a materials industrial applicability in addition to finding the optimal pH to maximize an adsorbent's sequestration capacity. Lefrancois Perreault et al. (2017) studied lanthanide separation with a modified, ordered mesoporous carbon under acidic conditions (pH = 4) to mimic industrial settings' real conditions.

## 3.2 Adsorbent dosage

Optimizing the adsorbent dose is also required to maximize an adsorbent sorption performance. Generally, increasing the adsorbent's dose improves the adsorbent sorption capacity by increasing active sites available for sorption. Nevertheless, since sorption capacity is expressed in per unit weight (usually per gram of material), the overall adsorption value can decrease. Increasing the amount of adsorbent can increase the pollutant removal; however, it can increase cost, which could be a constraint, especially if the adsorbent is expensive.

Mahmoud et al. (2019) studied U(VI) sequestration varying the adsorbent mass between 2.0 and 28.0 mg while maintaining other experimental conditions at their optimum values (pH = 4.0; contact time = 10 min). The adsorbent's removal efficiency increased up to 89.7% at 10 mg. Higher doses did not increase the removal efficiency. Probably, increased adsorbent dose causes buildup of adsorbent, and so, the available adsorption sites could reduce due to the adsorption density. Komnitsas (2017) studied Nd(III) removal with sawdust-derived biochar. While removal efficiencies increased from 40.2% to 78.6%, adsorption capacities decreased from 8 to 1.6 mg g$^{-1}$ in 1 and 10 mg L$^{-1}$ solutions, respectively (initial concentration = 20 mg L$^{-1}$; pH 3.0; contact time = 24 h; 298 K). Alahabadi et al. (2020) studied the effect of the dose on U(VI) and Th(VI) removal efficiency by varying the adsorbent's dose between 1 and 5 g L$^{-1}$ (initial concentration = 100 mg L$^{-1}$; pH 5.0; contact time = 1 h; 298 K). With the increasing doses, U(VI) and Th(VI) removal efficiencies improved from 93.91% to 98.29% and from 71.11% to 92.27%, respectively. Zeng et al. (2020) varied the dose between 0.25 and 4.0 g L$^{-1}$ to study the sewage sludge-derived biochar U(VI) removal efficiency. An increase from 0.25 to 1.0 g L$^{-1}$ enhanced the removal efficiency from 56.80% to 97.14% and from 63.92% to 98.05% for the material produced at 450 °C and 600 °C respectively, and the removal efficiency did not change significantly at higher doses.

## 3.3 Contact time and kinetic studies

Kinetic studies are of utmost importance when determining the effectiveness of an adsorbent. Adsorption kinetics involves studying the adsorbate removal over time. It provides information on the time required to reach equilibrium, which is useful for designing and optimizing full-scale treatment plants. Fast reaction kinetics reduces the required reactor volume and ensures high process efficiency (Zhang et al., 2016). Additionally, information on reaction kinetics helps determine the adsorption mechanism and the rate-limiting steps (Anastopoulos et al., 2016b). Many different models have been used to study kinetic data including pseudo-first-order (PFO), pseudo-second-order (PSO), intraparticle diffusion (IPD), Elovich (EM), and mass transfer (MT) models (Iftekhar et al., 2018). Each model's distinctive features have been comprehensively studied (Inyang et al., 2015;

Nguyen et al., 2017) and go beyond this chapter's scope. Here, the reviewed papers indicate that the time required to reach equilibrium depends on the adsorbent, the adsorbate, and the experimental conditions. In almost every case, the pseudo-second-order model best fits the experimental data, providing insight on the rate of chemical adsorption of ITEs on most bio/adsorbents reported in the literature.

PFO and PSO models are the most widely employed (da Costa et al., 2020). For example, Wang et al. (2020a, b) studied the kinetics of a $Fe_3O_4$-modified biochar ($Fe_3O_4$/MB) and its raw counterpart for U(VI) removal. Equilibrium was reached in 8 h for both materials ($C_0 = 50$ mg L$^{-1}$, dose = 0.5 g L$^{-1}$, pH = 6.0, ionic strength = 0.01 M $NaNO_3$ at $T = 289$ K). PFO and PSO models were fitted to the experimental data. The PSO model had the best correlation coefficient, indicating that chemical processes drove sorption on the materials. The PSO rate constants ($k_2$) for the raw biochar and $Fe_3O_4$/MB were 0.016 g mg$^{-1}$ h$^{-1}$ and 0.018 g mg$^{-1}$ h$^{-1}$, respectively. Statistical analysis confirmed no significant difference between the U(VI) sorption capacities of these materials. In this case, the magnetic composite did not increase the speed and rate of the adsorption process but enabled easy recovery of the adsorbent from the solution, an important advantage.

Kołodynska et al. (2018) studied the kinetic behavior of ZVI-biochar composites on La(III), Ce(III), and Nd(III) simultaneous sorption. The path to equilibrium can be explained by the adsorption kinetics. During the beginning of the adsorption process, the adsorbate attaches to the widely available adsorbent pores quickly. However, as time goes on, these pores fill up, and unadsorbed chemicals begin to run into more filled pores than unfilled ones, slowing down the process entirely. Furthermore, the electrostatic interactions of the filled pores repulse unadsorbed ions away from the adsorbent.

There were two stages noted during kinetic studies: a relatively fast initial adsorption process (1 h) followed by a slow increase in adsorption until equilibrium is reached (6 h). The two-stage sorption process can be attributed to the vacant site accessibility on the modified biochar composites particles for La(III), Ce(III), and Nd(III) ions loading. Initially, the number of accessible sites is vast; hence, the sharp curve is rapid sorption process. Over time, the number of accessible sites reduces, making it more difficult to adsorb additional ions. The curve thus flattens and reaches a plateau. When approaching equilibrium, adsorbent and adsorbate molecule repulsive force slows down ions' uptake in the next stages. Four kinetic models were fitted to the experimental data:

PFO, PSO, IPD, and Elovich models. The PSO model exhibited the best fit, with a correlation coefficient ($R^2 > 0.993$) greater than the PFO model ($R^2 < 0.930$). Further analysis of the diffusion-based model (i.e., IPD) showed the adsorption process was controlled by intraparticle diffusion.

Jin et al. (2018) studied the kinetic behavior of cow-manure-derived biochar and an $HNO_3$-oxidized counterpart ($C_0 = 10.0\,mg\,L^{-1}$, pH = 4.5, ionic strength = 0.01 M $NaNO_3$ at $T = 298\,K$) on U(VI) sorption. Equilibrium was reached after 12 h, but during the first 4 h, there was a relatively fast adsorption process, which then slowed. The PSO model had a better correlation coefficient ($R^2 = 0.999$) than the PFO model ($R^2 \leq 0.950$), indicating that the U(VI) adsorption process by the biochars was primarily attributed to chemical reaction. Guilhen et al. (2018) studied the kinetic behavior of a palm tree-derived biochar in U(VI) sorption. The equilibrium was reached in 3 h ($C_0 = 5.0\,mg\,L^{-1}$, dose = $10\,g\,L^{-1}$, pH = 3), achieving 80% of U(VI) removal efficiency. Several kinetic models were fitted (PFO, PSO, IPD, and Elovich) in their linearized and nonlinearized versions when possible. The best correlation coefficient was achieved with the nonlinear PFO model.

Initial pollutant concentration effect can also provide information on adsorbent performance. Research has shown that the initial concentration can impact adsorption kinetics because it affects U(VI) mass transfer resistance between aqueous and the solid phases. Zhang et al. (2013b) studied the effect of initial U(VI) concentration from 10 to $110\,mg\,L^{-1}$ at 298 K. Up to $50\,mg\,L^1$, the adsorption capacity increased with the increasing initial U(VI) concentration and then remained constant. Wang et al. (2016) studied the effect of initial La(III) concentration with an ammonium citrate-modified biochar between 25 and $500\,mg\,L^{-1}$. The adsorption capacity increased with the initial La(III) concentration up to $500\,mg\,L^{-1}$; however, adsorption rates slowed as the initial concentration increased (Fig. 9.2).

### 3.4 Equilibrium studies: Isotherms and the effects of temperature

Equilibrium adsorption isotherms provide valuable information that help gain insight into the system's mechanics needed for the proper design and application of an adsorption technology (Zhang et al., 2016). To fit the experimental data, often two-parameter isotherm models have been used, with Langmuir and Freundlich equations being the most frequently applied (Iftekhar et al., 2018; Inyang et al., 2015). In most cases, the maximum

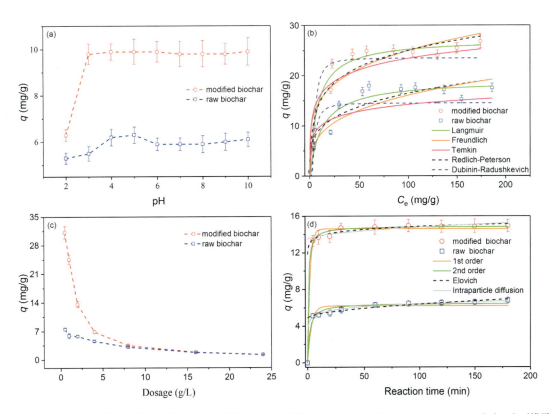

**Fig. 9.2** Preliminary (A) pH effect, (B) isotherm, (C) dose, and (D) kinetic studies in batch systems to optimize the U(VI) adsorption performance. Reprinted from Li, N., Yin, M., Tsang, D.C.W., Yang, S., Liu, J., Li, X., Song, G., Wang, J., 2019. Mechanisms of U(VI) removal by biochar derived from *Ficus microcarpa* aerial root: a comparison between raw and modified biochar. Sci. Total Environ. 697. https://doi.org/10.1016/j.scitotenv.2019.134115, Copyright (2019), with permission from Elsevier.

sorption capacity ($q_m$) calculated through these models is the typical performance parameter used to compare different adsorbent materials. In their review of REEs' sorbents, Anastopoulos et al. (2016a, b) found that out of 69 sorption materials, 78.3% of the isotherms were best described by Langmuir models and 21.7% by Freundlich equation. Their analysis included reported data on sorption isotherms for Ce(III), Dy(III), Eu(III), La(III), Gd(III), Nd(III), Sm(III), Pr(III), Sc(III), and Y(III).

Although most studies were performed at room temperature (298 K), temperature changes on sorption can be examined by building different isotherms by varying the working temperatures. Changing the working temperature enables the calculation of thermodynamic parameters, useful for determining the adsorption process nature and the underlying mechanisms. The important thermodynamic parameters are Gibbs free energy ($\Delta G^0$), enthalpy change ($\Delta H^0$), and entropy change ($\Delta S^0$) values. Negative $\Delta G^0$ values indicate spontaneity of the adsorption process. Positive values of $\Delta H^0$ signifies an endothermic process, whereas a negative value shows an exothermic process. The magnitude of $\Delta H^0$ can offer information related to the type of sorption (physisorption <50 kJ mol$^{-1}$; chemisorption >50 kJ mol$^{-1}$) (Iftekhar et al., 2018). Furthermore, the $\Delta S^0$ value can provide information about the affinity of the adsorbent toward the adsorbate. Values of $\Delta S^0 > 0$ imply high affinity, while values of $\Delta S^0 < 0$ denote low affinity (da Costa et al., 2020). Table 9.2 shows some thermodynamic parameters reported in the literature for sorption of ITEs with different biochar and engineered biochar. Most adsorption isotherm studies reported in literature for ITEs indicated that the adsorption processes were spontaneous and endothermic.

Dong et al. (2017b) studied sorption isotherms at 293, 313, and 333 K using rice-straw-based biochar. Their $\Delta H^0$ values were positive, with negative $\Delta G^0$ values which increased as the temperature rose, signifying the endothermicity, spontaneity, and greater efficiency of the process at higher temperatures. These authors attributed their findings to the U(VI) desolvation, which favors sorption.

One explanation for this phenomenon suggests that ions are stripped of their hydration sheath during adsorption in an endothermic process. Despite the ions attaching to the surface of the adsorbent being exothermic, the magnitude of the aforementioned endothermic process may exceed that of the exothermic one, making the net process endothermic. As a result, the adsorption process is favored at higher temperatures. The increase in the disorders at the biochar-solution interface may result from some structural changes in the U(VI) and biochar, as suggested by the positive $\Delta S^0$ value. This would lead to an increase in the disorders at the biochar-solution interface. The thermodynamic process described supports the conclusion that U(VI)'s sorption process on biochar is spontaneous and endothermic.

A similar trend was reported for sorption with magnetic biochar (Wang et al., 2020a) and $MnO_2$/orange peel biochar composite (Ying et al., 2020). Alahabadi et al. (2020) calculated thermodynamic parameters for a magnetic biochar at 288, 298, 308, and 328 K. They found $\Delta G°$ values of 27.45, 29.22, 31.08, and 32.98 kJ mol$^{-1}$ for U(VI) and 24.32, 26.28, 28.21, and 30.26 kJ mol$^{-1}$ for Th(VI), respectively. Their results indicated that the sorption processes were, in both cases, spontaneous, feasible, and favorable at higher temperatures. Their calculated $\Delta H^0$ values for uranium and thorium, respectively, were 25.697 and

Table 9.2 Reported thermodynamic state functions for adsorption.

| Element | Sorbent | $\Delta H^0$ (kJ mol$^{-1}$) | $\Delta S^0$ (J mol$^{-1}$ K$^{-1}$) | $\Delta G^0$ (kJ mol$^{-1}$) | Temp. (K) | References |
|---|---|---|---|---|---|---|
| U | Magnetic acid-treated biochar | 25.69 | 184.41 | 27.45 | 288 | Alahabadi et al. (2020) |
|   |   |   |   | 29.22 | 298 |   |
|   |   |   |   | 31.08 | 308 |   |
|   |   |   |   | 32.98 | 328 |   |
| Th | Magnetic acid-treated biochar | 32.57 | 197.49 | 24.32 | 288 | Alahabadi et al. (2020) |
|   |   |   |   | 26.28 | 298 |   |
|   |   |   |   | 28.21 | 308 |   |
|   |   |   |   | 30.26 | 328 |   |
| U | MnO$_2$/orange peel biochar composite | 57.67 | 246.97 | −13.49 | 288 | Ying et al. (2020) |
|   |   |   |   | −14.73 | 293 |   |
|   |   |   |   | −15.96 | 298 |   |
|   |   |   |   | −17.2 | 303 |   |
|   |   |   |   | −19.67 | 313 |   |
| U | Orange peel-derived biochar | 33.33 | 173.43 | −16.64 | 288 | Ying et al. (2020) |
|   |   |   |   | −17.51 | 293 |   |
|   |   |   |   | −18.38 | 298 |   |
|   |   |   |   | −19.25 | 303 |   |
|   |   |   |   | −20.98 | 313 |   |
| U | Cyanobacterium-derived magnetic biochar | 14.25 | 68.23 | −6.08 | 298 | Wang et al. (2020a, b) |
|   |   |   |   | −6.76 | 308 |   |
|   |   |   |   | −7.44 | 318 |   |
| U | Cyanobacterium-derived biochar | 44.86 | 173.78 | −6.92 | 298 | Wang et al. (2020a, b) |
|   |   |   |   | −8.66 | 308 |   |
|   |   |   |   | −10.4 | 318 |   |

*Continued*

Table 9.2 Reported thermodynamic state functions for adsorption—cont'd

| Element | Sorbent | $\Delta H^0$ (kJ mol$^{-1}$) | $\Delta S^0$ (J mol$^{-1}$ K$^{-1}$) | $\Delta G^0$ (kJ mol$^{-1}$) | Temp. (K) | References |
|---|---|---|---|---|---|---|
| U | Rice straw-derived biochar | 7.48 | 107.12 | −23.9<br>−25.77<br>−28.38 | 293<br>313<br>333 | Dong et al. (2017b) |
| La | Tire-derived biochar | 4.91 | 52.2 | −9.76<br>−12.2<br>−13.7<br>−12.4 | 298<br>313<br>333<br>353 | Smith et al. (2016) |
| Ce | Tire-derived biochar | 22.0 | 106 | −8.44<br>−13.0<br>−13.1<br>−14.9 | 298<br>313<br>333<br>353 | Smith et al. (2016) |
| Nd | Tire-derived biochar | 58.8 | 220 | −7.27<br>−8.81<br>−14.7<br>−14.7 | 298<br>313<br>333<br>353 | Smith et al. (2016) |
| Sm | Tire-derived biochar | 79.9 | 289 | −7.34<br>−10.6<br>−13.8<br>−24.4 | 298<br>313<br>333<br>353 | Smith et al. (2016) |
| Y | Tire-derived biochar | 17.5 | 89.9 | −9.37<br>−10.1<br>−13.4<br>−13.8 | 298<br>313<br>333<br>353 | Smith et al. (2016) |
| La | Bamboo biochar | 6.61 | 84.6 | −17.8<br>−18.6<br>−19.4 | 288<br>298<br>308 | Chen (2010) |

32.57 kJ mol$^{-1}$, indicating that the sorption process of both radionuclides was endothermic.

Due to the hardness of the Th(IV) ion compared to U(VI) ions present as uranyl ions in the solution, as well as the higher measured $\Delta H^0$ magnitude of thorium sorption (32.57 kJ mol$^{-1}$ for thorium and 25.70 kJ mol$^{-1}$ for uranium), the conclusions can be drawn that chelating complexation plays a significant role in the sorption of these radionuclides onto new sorbents and thorium sorption proceeds via stronger interactions. The $\Delta S^0$ values were positive, which supports the theory that hydration is liberated at the solid/solution interface during the sorption process, and were measured to be 184.41 and 197.49 J/mol K for U(VI) and Th(IV), respectively.

Adsorption is usually a spontaneous and exothermic process that results in a decrease in entropy. Adsorption/bioadsorption cases in which enthalpy and entropy are both positive are an indication that the interaction is endothermic and has increasing randomness in the solution/solution interface during the adsorption process. This endothermicity might be explained by the endothermic process of ions or molecules losing part of their hydration sheath just as they are about to be adsorbed, having a magnitude higher than that of the exothermic process of ions attaching to the surface. In the case of the adsorption process of REEs, positive values of $\Delta S^0$ could be due to a metal ion being displaced by more than one water molecule.

Smith et al. (2016) studied the thermodynamics of biochar derived from recycled tires in the adsorption of Y, La, Ce, Nd, and Sm. They found that the REEs' adsorption process was endothermic, as indicated by their positive enthalpy values (Table 9.2). The authors claimed that enthalpy values corresponding to Nd and Sm sorption were $\Delta H° > 50$ kJ mol$^{-1}$, suggesting chemisorption through covalent bonding. In contrast, enthalpy values for La, Ce, and Y were $\Delta H° < 50$ kJ mol$^{-1}$ indicating physisorption via van der Waals forces. The negative $\Delta G^0$ values suggested a spontaneous adsorption process.

## 3.5 Multicomponent systems: Ionic strength and competitive ions

In most cases, studies on sorption performance are done in single-component systems (i.e., distilled water + pollutant). However, nuclear waste solutions often exist as multicomponent systems or mixtures of several pollutants. Therefore, studying the effect of competing ions for the adsorbent's surface-active sites is vital to mimic industrial systems. These conditions are usually evaluated by (1) varying the ionic strength to assess if the presence of other solutes can affect removal efficiency and (2) employing a variable matrix, resembling the natural environment where the adsorption technology is performed, with known concentrations of potential competing species.

Zhang et al. (2013b) studied the effect of ionic strength (0.01–0.3 M) on U(VI) removal with pine needle-derived biochar, using $KNO_3$ as a background electrolyte. They recorded a marked decrease in U(VI) sorption capacity, from 42.25 to 33.20 mg g$^{-1}$ for 0.01-0.1 M solutions. However, there was no significant effect on ionic strength. Presence of $KNO_3$ in the solution screens the electrostatic interaction between the charges on sorbent surface and the U(VI) ions in solution, which competed with the U(VI) ions for surface adsorption sites. Additionally, the ionic strength of the solution affected the U(VI) activity coefficient, which restricted their transfer to the sorbent's surface. Zeng et al. (2020) studied the effect of ionic strength on U(VI) removal with sewage-sludge-derived biochar, varying from 0 to 50 mg L$^{-1}$ with $CaCl_2$ as a background electrolyte (dosage = 1 g L$^{-1}$ and pH = 3). Their findings showed that increasing ionic strength reduced U(VI) removal efficiencies from 96.45% to 76.07% and 97.33% to 78.68% for biochar produced at 450°C and 600°C, respectively. The authors supposed that $Ca^{2+}$ ions could form ternary complexes, which reduces U(VI) removal efficiency. Hu et al. (2018b) studied the effect of ionic strength on U(VI) removal using bamboo-derived biochar from 0.01 to 1 M, using NaCl as a background electrolyte (dose = 2 g L$^{-1}$, $C_0$ = 50 mg L$^{-1}$, pH = 4, at 298 K). Removal efficiency slightly decreased to 86% at the highest NaCl concentration, indicating the adsorbent's high selectivity toward U(VI).

In more complex systems, Wang et al. (2016) studied the effect of various cations ($K^+$, $Ca^{2+}$, and $Mg^{2+}$) on La(III) uptake by an ammonium citrate-modified algal-derived biochar. They found that at the tested concentrations (0.01 M), the effect was low. However, the divalent ions had a higher effect than the monovalent ion. They attributed their results to the stability of the M-OH$_2$ bonds in an aqueous solution. Fristak et al. (2018) studied the competitive effect of $Na^+$, $K^+$, $Mg^{2+}$, $Ca^{2+}$, and $Al^{3+}$ on Eu(III) uptake by biochar derived from microcrystalline cellulose, organic cotton, and sewage sludge in comparable concentrations (100 mg L$^{-1}$). They found that the ions significantly affected the removal efficiency ($Na^+ < K^+ < Ca^{2+} < Mg^{2+} < Al^{3+}$), with $Al^{3+}$ being the most competitive ion due to its chemical similarity with Eu(III). Furthermore, Mahmoud et al. (2019) studied the effect of $K^+$, $Na^+$, $Ca^{2+}$, $Mg^{2+}$, and $Zn^{2+}$ on U(VI) uptake by a glutaraldehyde-nanosilica-biochar composite. They found that $K^+$, $Na^+$, $Ca^{2+}$ showed slight or no interference in the U(VI) removal. However, $Mg^{2+}$ and $Zn^{2+}$ exhibited high interference, decreasing removal efficiency from 81.3% to 25.5% and 32.7%, respectively (for an initial U concentration of 30 mg g$^{-1}$) and from 87.5% to 32.7% and 29.0% (for an initial U concentration of 60 mg g$^{-1}$), respectively. They ascribed the reduction in removal efficiency to the potential complex formation on the surface of the composite.

Pang et al. (2019) studied the *Dictyophora indusiate*-derived biochar supported sulfide NZVI (DI-SNZVI) performance in simulated wastewater (pH = 5.0) and real seawater (pH = 7.6) to better represent the complexity of a real-world aqueous matrix. The maximum U(VI) sorption capacity was 427.9 mg g$^{-1}$ in simulated wastewater and 150.0 mg g$^{-1}$ in seawater. They argued that the marked decrease in the U(VI) uptake in the seawater system was related to the complexity of the real matrix, including the coexisting ions, high salinity, and microorganisms that could compete for the active sites of the adsorbent and change the redox environment. However, the reported sorption capacity in the more realistic scenario was still adequate.

Kusrini et al. (2018) studied the simultaneous sequestration of various lanthanides in a synthetic solution, including Y, La, Ce, Nd, and Sm, with a NaOH-activated banana peel-derived biochar. The maximum Y, La, Ce, Nd, and Sm removal efficiencies were 31.1%, 40.8%, 33.2%, 7.6%, and 26.2, respectively, at pH = 4. A similar trend was observed with a ZVI-biochar composite for La(III), Ce(III), and Nd(III) simultaneous sorption (dose = 0.1 g, pH = 4, and contact time of 360 min (Kołodynska et al., 2018). Smith et al. (2016) also studied simultaneous sequestration of those same REEs (Y, La, Ce, Nd, and Sm) with a tire-derived biochar. Hu et al. (2020) studied U(VI) sorption of a phosphate-functionalized biochar in the presence of competitive metal ions, including Cs(I), Sr(II), Co(II), Cu(II), Fe(III), Ni(II), and Eu(III) ($C_0 = 2.0 \times 10^{-4}$ M, pH = 4.0; $T$ = 298 K). The U(VI) sorption capacities were barely influenced by Cs(I), Sr(II), Co(II), Cu(II), and Ni(II), in the range of 134.2–137.1 mg g$^{-1}$. However, the U(VI) removal efficiency decreased by 40.3% with coexisting Fe(III) and 35.2% with Eu(III). Authors attributed the reduced removal efficiency to the valence and ionic radius of the interfering elements (i.e., higher valence and smaller ionic radius are more competitive for coordination on the surface).

Phosphate biochar (PBs) may be considered a promising sorbent for effective reclamation of U(VI) from wastewater with the ability to act as a coseparation material for the actinides (Hu et al., 2020). Perhaps due to the higher affinity of modified -PO$_4$ with U(VI) than other competitive metal ions, PBs overall exhibited a favorable sorption selectivity toward U(VI). Similar to the structure of UO$_2$(NO$_3$)$_2$(trimethyl phosphate)$_2$, the U—O—P bond in

U(VI) loaded PBs is stable. $UO_2^{2+}$ is also a strong Lewis acid which can in turn interact with -$PO_4$, a Lewis base, effectively.

### 3.6 Regeneration, recycling, and reuse

The adsorbent's regeneration is another essential feature because it impacts the feasibility and cost-effectiveness of the adsorption technology in a real-world application. Different stripping agents such as $HNO_3$, HCl, NaCl, thiourea, and $CaCl_2$ (Iftekhar et al., 2018) can be used to treat saturated biochar and desorb the attached pollutant. However, acids have shown the best desorption and regeneration efficiencies (Smith et al., 2016). Kołodynska et al. (2018) used three different stripping acids ($HNO_3$, HCl, and $H_2SO_4$) at 1 M. They found that $HNO_3$ was more effective for lanthanide's desorption. Further desorption kinetics showed that, after 6 h, desorption efficiency of La(III), Ce(III), and Nd(III) was 81.8%, 94.4%, and 86.6%, respectively. Several cycles can be performed to assess the regeneration-recyclability dynamics and study how the removal and desorption capacity change through sequential operation cycles. Gao et al. (2020) studied the effect of HCl 18% on a magnetic composite ($Fe_3O_4$/PT/GO) used for Er(III) removal. After the first regeneration, the adsorption efficiency was 92.3%, and after 5 operative cycles, it decreased to 83.3%. Wang et al. (2016) used HCl 0.2 M as the stripping agent for the modified algal-derived biochar. They found that the removal efficiency decreased, but La(III) removal remained above 85% even in the sixth cycle. Among the different stripping agents tested, Smith et al. (2016) achieved between 68 and 98% desorption efficiencies for all REEs tested (Y, La, Ce, Nd, and Sm) with HCl 0.15 M. Both adsorption and desorption linearly decreased with each cycle (at rates of 20%/cycle and 16%/cycle, respectively), except for Sm and Ce, which exhibited almost complete desorption in each cycle.

The adsorption's reduction after each cycle may be due to the decrease in adsorption sites—for every adsorption site filled, there is one less available site for unadsorbed compound to go to, leading to a reduction in adsorption. However, perhaps due to their higher solubility in HCl over La, Nd, and Y, this trend was not observed for Sm and Ce, making their desorption nearly constant.

Zhang et al. (2013b) studied U(VI) desorption from pine needle-derived biochar, particularly for uranium. They found that it was completed using HCl 0.05 M. The adsorption capacity, removal efficiency, and desorption efficiency decreased from 50.17 to 45.41 mg g$^{-1}$, 99.11% to 88.37%, and 96.03% to 87.41% in the fourth cycle, respectively. Zeng et al. (2020) used HCl 0.1 M to regenerate sewage sludge-derived biochar. After 5 operating cycles, they achieved U(VI) removal efficiencies of 82.33% for biochar produced at 450°C and 83.88% for biochar pyrolyzed at 600°C with corresponding regeneration efficiencies of 86.52% and 87.61%, respectively. The authors attributed the decreasing removal efficiencies to the loss of specific surface area, pore volume, or adsorbent material loss.

Pang et al. (2019) put emphasis on studying the stability and recyclability of the NZVI-biochar composite, given that these materials can be easily oxidized and become unstable, restricting their applicability. This study exposed DI-NZVI/SNZVI suspensions (0.5 g L$^{-1}$) in aerobic conditions for several days. After 3 days, the DI-NZVI became yellow. After 5 days, DI-SNZVI appeared to rust. After 20 days, the DI-NZVI suspension turned yellow (due to Fe$^0$ oxidation to iron oxides). This means the DI-SNZVI suspension was more limpid than DI-NZVI and, in turn, demonstrates the reusability and stability of DI-SNZVI.

## 3.7 Sorption mechanisms

Many techniques can have been used to study and understand the different mechanisms involved in aqueous ITEs' removal. Characterization of adsorbent using FTIR, XRD, and XPS before and after removal experiments helps to visualize changes occurring on the adsorbent's surface and provides support for the mechanisms hypothesized via equilibrium studies. ITEs' adsorption typically arises through either surface complexation or electrostatic interaction with different functional groups on the adsorbent surface. The involvement of the different functional groups is often shown by the shift of FTIR, XRD, or XPS peaks (Iftekhar et al., 2018).

For instance, Gao et al. (2020) characterized the Fe$_3$O$_4$/PT/GO biosorbent with SEM, FTIR, VSM, XRD, and XPS techniques before and after sorption of Er(III). They observed that the intensity of phenolic -OH stretching (at 3423 cm$^{-1}$) and the peak at 1635 cm$^{-1}$ (sp$^2$ carbon skeleton vibration) deteriorated, and that the peak at 1215 cm$^{-1}$ disappeared upon Er(III) adsorption. These changes indicated that chemical reactions happen between the adsorbent and Er(III). This is possibly due to a redox reaction between Er(III) ions and the phenolic groups and their oxidation to quinones. Additionally, -OH bending at 1400 cm$^{-1}$ and the peak at 1090 cm$^{-1}$ were shifted to 1385 and 1085 cm$^{-1}$, respectively. The authors assumed that under weakly acidic conditions, the Er(III) ions formed a cyclic metal-ligand chelate with the phenolic hydroxyl groups on the adsorbent. Another mechanism associated with Er(III) sorption on Fe$_3$O$_4$/PT/GO proposed by the authors which involved electrostatic interaction indicated by the influence of pH on the adsorption capacity was described previously in Section 3.1.

Li et al. (2020) studied five CaAl-LDH/IPB composites. They found that the dominant sorption mechanism relied on the mass ratio of LDH and biochar. They recorded improved sorption capacity with increasing biochar content (e.g., in CaAl-LDH/IPB-13) due to ion exchange, surface complexes, and precipitation reactions on the adsorbent surfaces. In contrast, an increase of LDH content (e.g., in CaAl-LDH/IPB-31) lowered the sorption capacity because ion-exchange reactions control the sorption mechanism.

XPS analysis on magnetic biochar showed abundant oxygen-containing functional groups responsible for Eu(III) adsorption (Zhu et al., 2018). They observed that Eu(III) adsorption on magnetic biochar at low and high pH was dominated by inner-sphere surface complexation and surface coprecipitation, respectively. Zeng et al. (2020) reported that based on FTIR, XRD, and XPS analysis, the mechanisms of U(VI) adsorption on adsorbents were primarily ion exchange with M—X (M-metal) and complexation with —CO, —C=O, and —COO groups.

Pang et al. (2019) found that the dispersion of SNZVI could be improved through the addition of Dictyophora indusiata (DI)-derived biochar by undertaking an adsorptive role for U(VI) capture. The structure of DI-SNZVI allowed for this as the $Fe^0$ inner core, through redox reactions, accelerated U(VI) elimination, and the surface functional groups (iron/sulfur/oxygen-containing groups) primarily participated in adsorption. Table 9.1 compares DI-SNZVI with other comparable adsorbents in U(VI) removal. They concluded that DI-SNZVI showed promise as a material for U(VI) removal from aqueous solutions due to its surface properties and adsorption-reduction processes.

Wang et al. (2020a, b) performed XPS and noted that U(VI) sorption on both MB and $Fe_3O_4$/MB was primarily attributed to surface complexation between U(VI) and oxygen-containing functional groups on the MB surface. They also found that while $Fe_3O_4$ particles on the surface of MB did not offer different active sites for U(VI) sorption, it enabled magnetic separation of the adsorbent.

Alam et al. (2018) performed U(VI) batch adsorption experiments coupled to surface complexation modeling (SCM) and isothermal titration calorimetry (ITC), supported by synchrotron-based X-ray absorption spectroscopy (XAS) analyses in order to study U(VI) adsorption mechanisms on biochar. Results from FT-IR and XPS studies, along with XAS analyses, show that U(VI) adsorption takes place at the surface carboxyl (—COOH) and hydroxyl (—OH) groups, whereas SCM results indicated solution pH strongly influenced U adsorption by aqueous U(VI) speciation.

ITC measurements showed that U(VI) adsorption on biochar was due to both inner- and outer-sphere complexation.

Hu et al. (2018a, b) studied uranium removal by magnetic biochar. XRD, XPS, and XAS indicated a reductive coprecipitation of U(VI) to U(IV) by magnetic biochar. The authors alleged that U—Fe and U—U shells' existence signified that at low pH, the removal of uranium was driven by inner-sphere coordination and reductive coprecipitation.

Fig. 9.3 summarizes selected mechanism pathways presented for U(VI) and Th(IV) sorption. These involve hydrogen bonding, amide bonding, acetal linkage, imine bonding (Wang et al., 2020a, b), coordination (Li et al., 2019), and ion exchange onto biochar or metal oxide surface functionalities (Mishra et al., 2017; Hadjittofi and Pashalidis, 2016; Alam et al., 2018; Philippou et al., 2019; Liatsou et al., 2018; Philippou et al., 2018). In addition, the biochar-supported ZVI adsorbs via the formation of inner-sphere surface complexes and U(VI) reduction to U(IV) (Pang et al., 2019).

## 4 Scaling up: From batch to column configuration

Batch experiments are a useful strategy for preliminary studies, particularly to gain insight into the adsorbent's performance under different experimental conditions and to optimize the potential design in an iterative process. However, batch tests have several limitations since they involve treating small volumes of solution, and efficiencies tend to be higher than those in large-scale systems. Besides, continuous or dynamic processes are usually more suitable configurations in industrial applications. In columns and fixed-bed reactors, the adsorbent is packaged and immobilized in a container through which the solution to be treated is drained, and the adsorbate is removed until the adsorbent reaches a certain state of saturation. Breakthrough curves are typically constructed to assess the adsorption profiles for a given adsorbate, adsorbent, and column size, and several mathematical models can be applied to make predictions on performance in industrial scale (da Costa et al., 2020).

For most articles reviewed, the removal of ITEs by biochar and biochar composites was on a batch scale. However, for REEs' biosorption, 11 studies involving dynamic systems were performed (da Costa et al., 2020). Only two articles were found that studied the performance of biochar in columns or fixed-bed reactors. Kumar et al. (2011) designed a permeable reactive barrier with switchgrass-

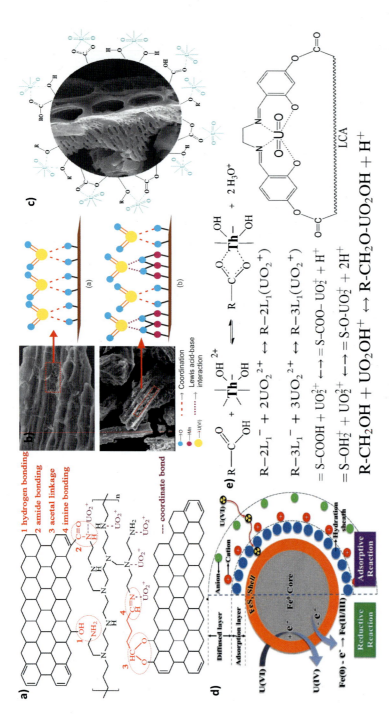

**Fig. 9.3** Summary of sorption mechanisms. Adapted from Wang, X., Feng, J., Cai, Y., Fang, M., Kong, M., Alsaedi, A., Hayat, T., Tan, X., 2020b. Porous biochar modified with polyethyleneimine (PEI) for effective enrichment of U(VI) in aqueous solution. Sci. Total Environ. 708. https://doi.org/10.1016/j.scitotenv.2019.134575; Li, N., Yin, M., Tsang, D.C.W., Yang, S., Liu, J., Li, X., Song, G., Wang, J., 2019. Mechanisms of U(VI) removal by biochar derived from *Ficus microcarpa* aerial root: a comparison between raw and modified biochar. Sci. Total Environ. 697. https://doi.org/10.1016/j.scitotenv.2019.134115; Mishra, V., Sureshkumar, M.K., Gupta, N., Kaushik, C.P., 2017. Study on sorption characteristics of uranium onto biochar derived from eucalyptus wood. Water Air Soil Pollut. 228. https://doi.org/10.1007/s11270-017-3480-8; Pang, H., Diao, Z., Wang, Xiangxue, Ma, Y., Yu, S., Zhu, H., Chen, Z., Hu, B., Chen, J., Wang, Xiangke, 2019. Adsorptive and reductive removal of U(VI) by *Dictyophora indusiate*-derived biochar supported sulfide NZVI from wastewater. Chem. Eng. J. 366, 368–377. https://doi.org/10.1016/j.cej.2019.02.098; Hadjittofi, L., Pashalidis, I., 2016. Thorium removal from acidic aqueous solutions by activated biochar derived from cactus fibers. Desalin. Water Treat. 57, 27864–27868. https://doi.org/10.1080/19443994.2016.1168580; Alam, S., Gorman-Lewis, D., Chen, N., Safari, S., Baek, K., Konhauser, K.O., Alessi, D.S., 2018. Mechanisms of the removal of U(VI) from aqueous solution using biochar: a combined spectroscopic and modeling approach. Environ. Sci. Technol. 52, 13057–13067. https://doi.org/10.1021/acs.est.8b01715; Philippou, K., Anastopoulos, I., Dosche, C., Pashalidis, I., 2019. Synthesis and characterization of a novel $Fe_3O_4$-loaded oxidized biochar from pine needles and its application for uranium removal. Kinetic, thermodynamic, and mechanistic analysis. J. Environ. Manag. 252. https://doi.org/10.1016/j.jenvman.2019.109677; Liatsou, I., Pashalidis, I., Nicolaides, A., 2018. Triggering selective uranium separation from aqueous solutions by using salophen-modified biochar fibers. J. Radioanal. Nucl. Chem. 318, 2199–2203. https://doi.org/10.1007/s10967-018-6186-5; Philippou, K., Savva, I., Pashalidis, I., 2018. Uranium(VI) binding by pine needles prior and after chemical modification. J. Radioanal. Nucl. Chem. 318, 2205–2211. https://doi.org/10.1007/s10967-018-6145-1.

derived biochar. U(VI) breakthrough from the PRB column gave an adsorption capacity of $0.52\,mg\,g^{-1}$ (473× higher than that onto quartz). This biochar U(VI) adsorption capacity in the PRB column configuration was about 25% of batch systems, likely due to preferential flow channels in the column resulting in less contact with the adsorbent's reactive surface than expected from batch trials.

Chen (2010) performed dynamic experiments with bamboo biochar. The La(III) breakthrough curve was built with the following operational parameters: biochar mass = 150.0 mg, $C_0 = 41.0\,mg\,L^{-1}$, dimensions = 23.5 × 0.45 × 7.4 cm; flow rate = 0.076 mL min$^{-1}$, pH = 7.20 at 25°C. The experimental data are better fitted by the model of Thomas with a correlation coefficient of 0.9798. The Thomas equation coefficients for La(III) adsorption were $K_T = 2.07 \times 10^{-2}\,mL\,min^{-1}\,mg^{-1}$ and $Q = 128\,mg\,g^{-1}$. The theoretical result predicted by the model agreed with the experimental Q value.

## 5  Conclusions, gaps, and future perspectives

ITEs have gained considerable attention in recent years due to their application in various cutting-edge industries. However, safe and sustainable methods for their removal and recovery from aqueous solutions remain a challenge. Although biochar has gained relevance as an effective adsorbent, there are not many articles that deal with real operating conditions. Interestingly, nearly every study reported in the literature has been performed on a batch scale with synthetic solutions. There were only two papers in which column studies were reported. Therefore, for biochar to become a promising technique, further studies should be done on an industrial scale to meet actual industrial application needs.

This literature survey has shown that, to date, most research available on bio/adsorption of ITEs using low-cost adsorbents is limited to batch systems (kinetics and adsorption equilibrium). Very few studies have been reported on continuous fixed-bed adsorption and regeneration studies, which are crucial for predicting and simulating dynamic performance. The studies reported in the literature are also limited to a laboratory scale to examine the adsorption mechanism and capacity. There are few or no studies on practical applications, mainly industrial adsorption processes to recover ITEs.

In conclusion, while the number of studies reported on low-cost nonconventional bio/adsorbents for removing ITEs increased, there is still a need for real application on an industrial scale. Moreover, studies on regeneration of bio/adsorbents, fixed-bed adsorption, and cost evaluation are needed. More studies using natural systems as desirable since most recent studies use synthetic and mono-component solutions at higher concentrations than those found in an industrial environment. Future studies should also involve multicomponent bio/adsorption, regeneration of bio/adsorbent, treatment of real effluents, continuous adsorption, and the recovery of ITEs in sequential adsorption-desorption cycles.

To our knowledge, future research on ITEs' removal from polluted water and their recapture from secondary sources should emphasize (1) the advance of a low-cost and efficient process for the removal and recovery of ITEs from aqueous media, (2) the purification of ITEs from the recovered secondary source, and (3) the assessment of the remaining waste's long-term stability in the environment.

## Questions and problems

1. Name the elements included in actinide series. How are actinide elements classified?
2. Name the elements included in lanthanide series. How are lanthanide elements classified?
3. What are the primary anthropogenic sources of inner transition elements (ITEs) in the environment?
4. How do actinides and lanthanides behave in aqueous media?
5. Why is it important to provide a sustainable approach for the separation and recovery of ITEs from water?
6. Enumerate the different methods used for the removal and recovery of ITEs from wastewater.
7. How can biochar be a sustainable approach to remove ITEs from wastewater?

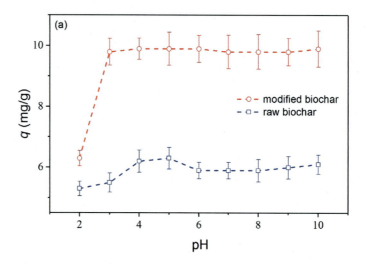

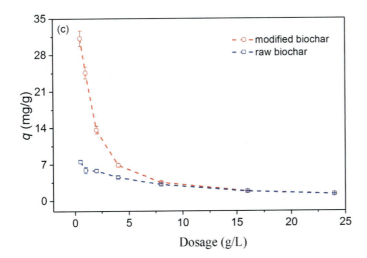

8. In the following figure (Li et al., 2019), describe roughly the effect of pH in U(V) adsorption capacity for raw and modified biochars.
9. In the following figure (Li et al., 2019), describe the effect of raw biochar and modified biochar dosages on U(V) adsorption.
10. How is sorption kinetics useful in material optimization?
11. What are the most common models used for fitting equilibrium adsorption data?
12. Why are multicomponent adsorption studies necessary in process optimization?
13. Why are adsorbent regeneration studies necessary?
14. Name the most effective stripping agents reported.
15. Name the characterization techniques usually employed to describe sorption mechanisms.

# References

Alahabadi, A., Singh, P., Raizada, P., Anastopoulos, I., Sivamani, S., Dotto, G.L., Landarani, M., Ivanets, A., Kyzas, G.Z., Hosseini-bandegharaei, A., 2020. Activated carbon from wood wastes for the removal of uranium and thorium ions through modification with mineral acid. Colloids Surfaces A 607. https://doi.org/10.1016/j.colsurfa.2020.125516, 125516.

Alam, S., Gorman-lewis, D., Chen, N., Safari, S., Baek, K., Konhauser, K.O., Alessi, D.-S., 2018. Mechanisms of the removal of U(VI) from aqueous solution using biochar: a combined spectroscopic and modeling approach. Environ. Sci. Technol. 52, 13057–13067. https://doi.org/10.1021/acs.est.8b01715.

Amen, R., Bashir, H., Bibi, I., Shaheen, S.M., Khan, N., Shahid, M., Mahroz, M., Antoniadis, V., Bilal, M., Al-solaimani, S.G., Wang, H., Bundschuh, J., 2020. A critical review on arsenic removal from water using biochar-based sorbents: the significance of modification and redox reactions. Chem. Eng. J. 396, 1–17. https://doi.org/10.1016/j.cej.2020.125195.

Anastopoulos, I., Bhatnagar, A., Lima, E.C., 2016a. Adsorption of rare earth metals: a review of recent literature. J. Mol. Liq. 221, 954–962. https://doi.org/10.1016/j.molliq.2016.06.076.

Anastopoulos, I., Karamesouti, M., Mitropoulos, A.C., Kyzas, G.Z., 2016b. A review for coffee adsorbents. J. Mol. Liq. 229, 555–565. https://doi.org/10.1016/j.molliq.2016.12.096.

Aytas, S.O., Akyil, S., Eral, M., 2004. Adsorption and thermodynamic behavior of uranium on natural zeolite. J. Radioanal. Nucl. Chem. 260, 119–125.

Beall, G.W., Ketelle, B.H., Haire, R.G., Kelley, G.D.O., 1979. Sorption behavior of trivalent actinides and rare earths on clay minerals. In: Fried, S. (Ed.), Radioactive Waste in Geologic Storage. American Chemical Society, https://doi.org/10.1021/bk-1979-0100.

Bursztyn Fuentes, A.L., Canevesi, R.L.S., Gadonneix, P., Mathieu, S., Celzard, A., Fierro, V., 2020. Paracetamol removal by Kon-tiki kiln-derived biochar and activated carbons. Ind. Crop. Prod. 155, 1–10. https://doi.org/10.1016/j.indcrop.2020.112740.

Chen, Q., 2010. Study on the adsorption of lanthanum(III) from aqueous solution by bamboo charcoal. J. Rare Earths 28 (1), 125–131. https://doi.org/10.1016/S1002-0721(10)60272-4.

Cooper, N.G., 2000. The chemical interactions of actinides in the environment. Los Alamos Sci. Challenges Plutonium Sci. 26, 392–411.

da Costa, T.B., da Silva, M.G.C., Vieira, M.G.A., 2020. Recovery of rare-earth metals from aqueous solutions by bio/adsorption using non-conventional materials: a review with recent studies and promising approaches in column applications. J. Rare Earths 38, 339–355. https://doi.org/10.1016/j.jre.2019.06.001.

Das, D., Sureshkumar, M.K., Koley, S., Mithal, N., Pillai, C.G.S., 2010. Sorption of uranium on magnetite nanoparticles. J. Radioanal. Nucl. Chem. 285, 447–454. https://doi.org/10.1007/s10967-010-0627-0.

Djedidi, Z., Bouda, M., Aly Souissi, M., Ben Cheikh, R., Mercier, G., Dayal Tyagi, R., Blais, J.-F., 2009. Metals removal from soil, fly ash and sewage sludge leachates by precipitation and dewatering properties of the generated sludge. J. Hazard. Mater. 172, 1372–1382. https://doi.org/10.1016/j.jhazmat.2009.07.144.

Dong, L., Chang, K., Wang, L., Linghu, W., Zhao, D., Asiri, A.M., Alamry, K.A., Alsaedi, A., Hayat, T., Li, X., 2017a. Application of biochar derived from rice straw for the removal of Th(IV) from aqueous solution. Sep. Sci. Technol. 00, 1–11. https://doi.org/10.1080/01496395.2017.1411363.

Dong, L., Yang, J., Mou, Y., Sheng, G., Wang, L., 2017b. Effect of various environmental factors on the adsorption of U(VI) onto biochar derived from rice straw. J. Radioanal. Nucl. Chem. 314, 377–386. https://doi.org/10.1007/s10967-017-5414-8.

Elsalamouny, A.R., Desouky, O.A., Mohamed, S.A., Galhoum, A.A., Guibal, E., 2017. Uranium and neodymium biosorption using novel chelating polysaccharide. Int. J. Biol. Macromol. https://doi.org/10.1016/j.ijbiomac.2017.06.081.

Ervanne, H., 2004. Oxidation State Analysis of Uranium with Emphasis on Chemical Speciation. University of Helsinki.

Fan, S., Wang, Y., Wang, Z., Tang, J., Tang, J., Li, X., 2017. Removal of methylene blue from aqueous solution by sewage sludge-derived biochar: adsorption kinetics, equilibrium, thermodynamics and mechanism. J. Environ. Chem. Eng. 5, 601–611. https://doi.org/10.1016/j.jece.2016.12.019.

Frišták, V., Miháleková-Richveisová, B., Víglašová, E., Duriska, L., Galambos, M., Moreno-jiménez, E., Pipiska, M., Soja, G., 2017. Sorption separation of Eu and As from single-component systems by Fe-modified biochar: kinetic and equilibrium study. J. Iran. Chem. Soc. 14, 521–530. https://doi.org/10.1007/s13738-016-1000-1.

Fristak, V., Pipíska, M., Hubenak, M., Kadleciková, M., Galambos, M., Soja, G., 2018. Pyrogenic materials-induced immobilization of Eu in aquatic and soil systems: comparative study. Water Air Soil Pollut. 229.

Gao, L., Wang, Z., Qin, C., Chen, Z., Gao, M., He, N., Qian, X., Zhou, Z., Li, G., 2020. Preparation and application of iron oxide/persimmon tannin/graphene oxide nanocomposites for efficient adsorption of erbium from aqueous solution. J. Rare Earths. https://doi.org/10.1016/j.jre.2020.02.003.

Guilhen, S.N., Rovani, S., Filho, L.P., Alves, D., 2018. Kinetic study of uranium removal from aqueous solutions by macaúba biochar. Chem. Eng. Commun. 0, 1–13. https://doi.org/10.1080/00986445.2018.1533467.

Guilhen, S.N., Masek, O., Ortiz, N., Izidoro, J.C., Fungaro, D.A., 2019. Pyrolytic temperature evaluation of macauba biochar for uranium adsorption from aqueous solutions. Biomass Bioenergy 122, 381–390. https://doi.org/10.1016/j.biombioe.2019.01.008.

Hadjittofi, L., Pashalidis, I., 2016. Thorium removal from acidic aqueous solutions by activated biochar derived from cactus fibers. Desalin. Water Treat. 57, 27864–27868. https://doi.org/10.1080/19443994.2016.1168580.

Herrmann, H., Nolde, J., Berger, S., Heise, S., 2016. Aquatic ecotoxicity of lanthanum—a review and an attempt to derive water and sediment quality criteria. Ecotoxicol. Environ. Saf. 124, 213–238. https://doi.org/10.1016/j.ecoenv.2015.09.033.

Hershey, J.W., Keliher, P.N., 1989. Some atomic absorption hydride generation inter-element interference reduction studies utilizing ion exchange resins. Spectrochim. Acta 44, 329–337.

Hu, H., Zhang, X., Wang, T., Sun, L., Wu, H., Chen, X., 2018a. Bamboo (*Acidosasa longiligula*) shoot shell biochar: its potential application to isolation of uranium(VI) from aqueous solution. J. Radioanal. Nucl. Chem. 316, 349–362. https://doi.org/10.1007/s10967-018-5731-6.

Hu, Q., Zhu, Y., Hu, B., Lu, S., Sheng, G., 2018b. Mechanistic insights into sequestration of U(VI) toward magnetic biochar: batch, XPS and EXAFS techniques. J. Environ. Sci. 70, 217–225. https://doi.org/10.1016/j.jes.2018.01.013.

Hu, R., Xiao, J., Wang, T., Chen, G., Chen, L., Tian, X., 2020. Engineering of phosphate-functionalized biochars with highly developed surface area and porosity for efficient and selective extraction of uranium. Chem. Eng. J. 379. https://doi.org/10.1016/j.cej.2019.122388, 122388.

Huang, Y., Hu, Y., Chen, L., Yang, T., Huang, H., Shi, R., Lu, P., Zhong, C., 2018. Selective biosorption of thorium(IV) from aqueous solutions by ginkgo leaf. PLoS ONE 13, 1–25. https://doi.org/10.1371/journal.pone.0193659.

Iftekhar, S., Lakshmi, D., Srivastava, V., Bilal Asif, M., Sillanpaa, M., 2018. Understanding the factors affecting the adsorption of lanthanum using different adsorbents: a critical review. Chemosphere 204, 413–430. https://doi.org/10.1016/j.chemosphere.2018.04.053.

Inyang, M.I., Gao, B., Yao, Y., Xue, Y., Zimmerman, A., Mosa, A., Pullammanappallil, P., Ok, Y.S., Cao, X., 2015. A review of biochar as a low-cost adsorbent for aqueous heavy metal removal. Crit. Rev. Environ. Sci. Technol. https://doi.org/10.1080/10643389.2015.1096880.

Jang, H.M., Kan, E., 2019. A novel hay-derived biochar for removal of tetracyclines in water. Bioresour. Technol. 274, 162–172. https://doi.org/10.1016/j.biortech.2018.11.081.

Jin, J., Li, S., Peng, X., Liu, W., Zhang, C., Yang, Y., Han, L., Du, Z., Sun, K., Wang, X., 2018. $HNO_3$ modified biochars for uranium(VI) removal from aqueous solution. Bioresour. Technol. 256, 247–253. https://doi.org/10.1016/j.biortech.2018.02.022.

Kołodynska, D., Bak, J., Majdanska, M., Fila, D., 2018. Sorption of lanthanide ions on biochar composites. J. Rare Earths 36, 1212–1220. https://doi.org/10.1016/j.jre.2018.03.027.

Komnitsas, K., 2017. Adsorption of scandium and neodymium on biochar derived after low-temperature pyrolysis of sawdust. Fortschr. Mineral. 7, 1–18. https://doi.org/10.3390/min7100200.

Kumar, S., Loganathan, V.A., Gupta, R.B., Barnett, M.O., 2011. An assessment of U(VI) removal from groundwater using biochar produced from hydrothermal carbonization. J. Environ. Manag. 92, 2504–2512. https://doi.org/10.1016/j.jenvman.2011.05.013.

Kumar, N., Sengupta, A., Gupta, A., Ravindra Sonwane, J., Sahoo, H., 2018. Biosorption—an alternative method for nuclear waste management: a critical review. J. Environ. Chem. Eng. 6, 2159–2175. https://doi.org/10.1016/j.jece.2018.03.021.

Kusrini, E., Kinastiti, D.D., Wilson, L.D., Usman, A., Rahman, A., 2018. Adsorption on lanthanide ions from an aqueous solution in multicomponent systems using activated carbon from banana peels (*Musa paradisiaca* L.). Int. J. Technol. 6, 1132–1139.

Lee, G.S., Uchikoshi, M., Mimura, K., Isshiki, M., 2009. Distribution coefficients of La, Ce, Pr, Nd, and Sm on Cyanex 923-, D2EHPA-, and PC88A-impregnated resins. Sep. Purif. Technol. 67, 79–85. https://doi.org/10.1016/j.seppur.2009.03.033.

Lefrancois Perreault, L., Giret, S., Gagnon, M., Florek, J., Larivie, D., Kleitz, F., 2017. Functionalization of mesoporous carbon materials for selective separation of lanthanides under acidic conditions. ACS Appl. Mater. Interfaces Interfaces 9, 12003–12012. https://doi.org/10.1021/acsami.6b16650.

Li, J., Lv, G., Bai, W., Liu, Q., Zhang, Y., Song, J., 2014. Modification and use of biochar from wheat straw (Triticum aestivum L.) for nitrate and phosphate removal from water. Desalin. Water Treat. 1–13. https://doi.org/10.1080/19443994.2014.994104.

Li, N., Yin, M., Tsang, D.C.W., Yang, S., Liu, J., Li, X., Song, G., Wang, J., 2019. Mechanisms of U(VI) removal by biochar derived from *Ficus microcarpa* aerial root: a comparison between raw and modified biochar. Sci. Total Environ. 697. https://doi.org/10.1016/j.scitotenv.2019.134115.

Li, S., Dong, L., Wei, Z., Sheng, G., Du, K., Hu, B., 2020. Adsorption and mechanistic study of the invasive plant-derived biochar functionalized with CaAl-LDH for Eu(III) in water. J. Environ. Sci. 96, 127–137. https://doi.org/10.1016/j.jes.2020.05.001.

Liang, P., Fa, W., 2005. Determination of La, Eu and Yb in water samples by inductively coupled plasma atomic emission spectrometry after solid phase extraction of their 1-phenyl-3-methyl-4-benzoylpyrazol-5-one complexes on silica gel column. Microchim. Acta 150, 15–19. https://doi.org/10.1007/s00604-005-0340-9.

Liatsou, I., Pashalidis, I., Oezaslan, M., Dosche, C., 2017. Surface characterization of oxidized biochar fibers derived from *Luffa Cylindrica* and lanthanide binding. J. Environ. Chem. Eng. 5, 4069–4074. https://doi.org/10.1016/j.jece.2017.07.040.

Liatsou, I., Pashalidis, I., Nicolaides, A., 2018. Triggering selective uranium separation from aqueous solutions by using salophen-modified biochar fibers. J. Radioanal. Nucl. Chem. 318, 2199–2203. https://doi.org/10.1007/s10967-018-6186-5.

Liu, P., Liu, W., Jiang, H., Chen, J., Li, W., Yu, H., 2012. Modification of bio-char derived from fast pyrolysis of biomass and its application in removal of tetracycline from aqueous solution. Bioresour. Technol. 121, 235–240. https://doi.org/10.1016/j.biortech.2012.06.085.

Maher, K., Bargar, J.R., Brown, G.E., 2013. Environmental speciation of actinides. Inorg. Chem. 52, 3510–3532. https://doi.org/10.1021/ic301686d.

Mahmoud, M.E., Khalifa, M.A., El, Y.M., Header, M.S., El-sharkawy, R.M., Kumar, S., Abdel-fattah, T.M., 2019. A novel nanocomposite of *Liquidambar styraciflua* fruit biochar-crosslinked-nanosilica for uranyl removal from water. Bioresour. Technol. 278, 124–129. https://doi.org/10.1016/j.biortech.2019.01.052.

Mishra, V., Sureshkumar, M.K., Gupta, N., Kaushik, C.P., 2017. Study on sorption characteristics of uranium onto biochar derived from eucalyptus wood. Water Air Soil Pollut. 228. https://doi.org/10.1007/s11270-017-3480-8.

NCR, 2012. Analysis of Cancer Risks in Populations Near Nuclear Facilities: Phase I. Washington (DC).

Nguyen, H., You, S., Hosseini-bandegharaei, A., Chao, H.-P., 2017. Mistakes and inconsistencies regarding adsorption of contaminants from aqueous solutions: a critical review. Water Res. 120, 88–116. https://doi.org/10.1016/j.watres.2017.04.014.

Oliveira, F.R., Patel, A.K., Jaisi, D.P., Adhikari, S., Lu, H., Khanal, S.K., 2017. Environmental application of biochar: current status and perspectives. Bioresour. Technol. 246, 110–122. https://doi.org/10.1016/j.biortech.2017.08.122.

Pang, H., Diao, Z., Wang, X., Ma, Y., Yu, S., Zhu, H., Chen, Z., Hu, B., Chen, J., Wang, X., 2019. Adsorptive and reductive removal of U(VI) by Dictyophora indusiate-derived biochar supported sulfide NZVI from wastewater. Chem. Eng. J. 366, 368–377. https://doi.org/10.1016/j.cej.2019.02.098.

Paschalidou, P., Pashalidis, I., Manariotis, I.D., Karapanagioti, H.K., 2020. Hyper sorption capacity of raw and oxidized biochars from various feedstocks for U(VI). J. Environ. Chem. Eng. 8. https://doi.org/10.1016/j.jece.2020.103932, 103932.

Philippou, K., Savva, I., Pashalidis, I., 2018. Uranium(VI) binding by pine needles prior and after chemical modification. J. Radioanal. Nucl. Chem. 318, 2205–2211. https://doi.org/10.1007/s10967-018-6145-1.

Philippou, K., Anastopoulos, I., Dosche, C., Pashalidis, I., 2019. Synthesis and characterization of a novel $Fe_3O_4$-loaded oxidized biochar from pine needles and its application for uranium removal. Kinetic, thermodynamic, and mechanistic analysis. J. Environ. Manag. 252. https://doi.org/10.1016/j.jenvman.2019.109677.

Royer-Lavallée, A., Neculita, C.M., Coudert, L., 2020. Removal and potential recovery of rare earth elements from mine water. J. Ind. Eng. Chem. 89, 47–57. https://doi.org/10.1016/j.jiec.2020.06.010.

Sengupta, A., Gupta, N.K., 2017. MWCNTs based sorbents for nuclear waste management: a review. J. Environ. Chem. Eng. 5, 5099–5114. https://doi.org/10.1016/j.jece.2017.09.054.

Serkan, Ş., Sahin, U., Hazer, O., Ülgen, A., 2020. An automated system for preconcentration on imprinted polymer and laser diode spectrometric determination of uranium in mineral water samples. Spectrochim. Acta Part A Mol. Biomol. Spectrosc. 240, 1–7. https://doi.org/10.1016/j.saa.2020.118572.

Shaheen, S.M., Niazi, N.K., Hassan, N.E.E., Bibi, I., Wang, H., Tsang, D.C.W., Ok, Y.S., Bolan, N., Rinklebe, J., 2019. Wood-based biochar for the removal of potentially toxic elements in water and wastewater: a critical review. Int. Mater. Rev. 6608. https://doi.org/10.1080/09506608.2018.1473096.

Sizmur, T., Fresno, T., Akgül, G., Frost, H., Moreno-jiménez, E., 2017. Bioresource technology biochar modification to enhance sorption of inorganics from water. Bioresour. Technol. 246, 34–47. https://doi.org/10.1016/j.biortech.2017.07.082.

Smith, Y.R., Bhattacharyya, D., Willhard, T., Misra, M., 2016. Adsorption of aqueous rare earth elements using carbon black derived from recycled tires. Chem. Eng. J. 296, 102–111. https://doi.org/10.1016/j.cej.2016.03.082.

Souza, A.L., Lemos, S.G., Oliveira, P.V., 2011. A method for Ca, Fe, Ga, Na, Si and Zn determination in alumina by inductively coupled plasma optical emission spectrometry after aluminum precipitation. Spectrochim. Acta Part B At. Spectrosc. 66, 383–388. https://doi.org/10.1016/j.sab.2011.03.001.

Srivastava, S., Bhainsa, K.C., Souza, S.F.D., 2010. Investigation of uranium accumulation potential and biochemical responses of an aquatic weed Hydrilla verticillata (L. f.) Royle. Bioresour. Technol. 101, 2573–2579. https://doi.org/10.1016/j.biortech.2009.10.054.

Sylwester, E.R., Hudson, E.A., Allen, P.G., 2000. The structure of uranium(VI) sorption complexes on silica, alumina, and montmorillonite. Geochim. Cosmochim. Acta 64, 2431–2438.

Tan, W., 2019. Effects of phosphorus modified bio-char on metals in uranium-containing soil. Water Air Soil Pollut. 230, 1–10.

Taskaev, E., Apostolov, D., 1978. On uranium(VI) adsorption on activated carbon. J. Radioanal. Chem. 45, 65–71.

Tong, S., Song, N., Jia, Q., Zhou, W., Liao, W., 2009. Solvent extraction of rare earths from chloride medium with mixtures of 1-phenyl-3-methyl-4-benzoyl-pyrazalone-5 and sec-octylphenoxyacetic acid. Sep. Purif. Technol. 69, 97–101. https://doi.org/10.1016/j.seppur.2009.07.003.

Wang, Y., Lu, H., Liu, Y., Yang, S., 2016. Ammonium citrate-modified biochar: an adsorbent for La(III) ions from aqueous solution. Colloids Surfaces A Physicochem. Eng. Asp. 509, 550–563. https://doi.org/10.1016/j.colsurfa.2016.09.060.

Wang, B., Gao, B., Fang, J., 2017. Recent advances in engineered biochar productions and applications. Crit. Rev. Environ. Sci. Technol. 3389. https://doi.org/10.1080/10643389.2017.1418580.

Wang, S., Guo, W., Gao, F., Wang, Y., Gao, Y., 2018. Lead and uranium sorptive removal from aqueous solution using magnetic and nonmagnetic fast. RSC Adv. 8, 13205–13217. https://doi.org/10.1039/c7ra13540h.

Wang, B., Li, Y., Zheng, J., Hu, Y., Wang, X., Hu, B., 2020a. Efficient removal of U(VI) from aqueous solutions using the magnetic biochar derived from the biomass of a bloom-forming cyanobacterium (*Microcystis aeruginosa*). Chemosphere 254. https://doi.org/10.1016/j.chemosphere.2020.126898, 126898.

Wang, X., Feng, J., Cai, Y., Fang, M., Kong, M., Alsaedi, A., Hayat, T., Tan, X., 2020b. Porous biochar modified with polyethyleneimine (PEI) for effective enrichment of U(VI) in aqueous solution. Sci. Total Environ. 708. https://doi.org/10.1016/j.scitotenv.2019.134575.

Weber, K., Quicker, P., 2018. Properties of biochar. Fuel 217, 240–261. https://doi.org/10.1016/j.fuel.2017.12.054.

Wu, D., Sun, Y., Wang, Q., 2013. Adsorption of lanthanum(III) from aqueous solution using 2-ethylhexyl phosphonic acid mono-2-ethylhexyl ester-grafted magnetic silica nanocomposites. J. Hazard. Mater. 260, 409–419. https://doi.org/10.1016/j.jhazmat.2013.05.042.

Yakout, S.M., 2016. Effect of porosity and surface chemistry on the adsorption-desorption of uranium(VI) from aqueous solution and groundwater. J. Radioanal. Nucl. Chem. 308, 555–565. https://doi.org/10.1007/s10967-015-4408-7.

Yang, A., Zhu, Y., Huang, C.P., 2018. Facile preparation and adsorption performance of graphene oxide- manganese oxide composite for uranium. Sci. Rep. 1–10. https://doi.org/10.1038/s41598-018-27111-y.

Ying, D., Hong, P., Jiali, F., Qinqin, T., Yuhui, L., Youqun, W., Zhibin, Z., Xiaohong, C., Yunhai, L., 2020. Removal of uranium using $MnO_2$/orange peel biochar composite prepared by activation and in-situ deposit in a single step. Biomass Bioenergy 142. https://doi.org/10.1016/j.biombioe.2020.105772, 105772.

Yu, S., Yao, Q., Zhou, G., Fu, S., 2014. Preparation of hollow core/shell microspheres of hematite and its adsorption ability for samarium. ASC Appl. Mater. Interfaces 6, 1–11. https://doi.org/10.1021/am502166p.

Zeng, T., Mo, G., Zhang, X., Liu, J., Liu, H., Xie, S., 2020. U(VI) removal efficiency and mechanism of biochars derived from sewage sludge at two pyrolysis temperatures. J. Radioanal. Nucl. Chem. https://doi.org/10.1007/s10967-020-07423-y.

Zhang, Z., Cao, X., Liang, P., Liu, Y., 2013a. Adsorption of uranium from aqueous solution using biochar produced by hydrothermal carbonization. J. Radioanal. Nucl. Chem. 295, 1201–1208. https://doi.org/10.1007/s10967-012-2017-2.

Zhang, M., Gao, B., Varnoosfaderani, S., Hebard, A., Yao, Y., Inyang, M., 2013b. Preparation and characterization of a novel magnetic biochar for arsenic removal. Bioresour. Technol. 130, 457–462. https://doi.org/10.1016/j.biortech.2012.11.132.

Zhang, F., Wang, X., Xionghui, J., Ma, L., 2016. Efficient arsenate removal by magnetite-modified water hyacinth biochar. Environ. Pollut., 575–583. https://doi.org/10.1016/j.envpol.2016.06.013.

Zhou, Y., Xiao, J., Hu, R., Wang, T., Shao, X., Chen, G., Chen, L., Tian, X., 2020. Engineered phosphorous-functionalized biochar with enhanced porosity using phytic acid-assisted ball milling for efficient and selective uptake of aquatic uranium. J. Mol. Liq. 303. https://doi.org/10.1016/j.molliq.2020.112659, 112659.

Zhu, Y., Zheng, C., Wu, S., Song, Y., Hu, B., 2018. Interaction of Eu(III) on magnetic biochar investigated by batch, spectroscopic and modeling techniques. J. Radioanal. Nucl. Chem. 316, 1337–1346. https://doi.org/10.1007/s10967-018-5839-8.

# 10

# Microplastic removal from water and wastewater by carbon-supported materials

Virpi Siipola[a], Henrik Romar[b], and Ulla Lassi[b]
[a]VTT Technical Research Centre of Finland Ltd, Espoo, Finland. [b]University of Oulu, Research Unit of Sustainable Chemistry, Oulu, Finland

## 1 Introduction

Annual global production of plastics has grown from 1.7 million tonnes up to 335 million tonnes since 1950. Due to the high production capacity and durability of plastics, huge amounts of plastic waste end up in the environment. In recent years, microplastics (MPs) in the aquatic environment have attracted a lot of attention and new research, even they were found in the environment and water bodies as early as the 1970s. Overall, 80%–85% of all marine waste is plastic-based. In this chapter, sources and adverse effects of microplastics in water bodies are considered. In addition, various methods of quantitative and qualitative analysis of microplastics are presented. The main focus is on the removal methods of microplastics over carbon-supported materials.

## 2 Microplastics—Sources and effects in the aquatic environment

Plastics are produced to be durable with a long lifetime, and they are typically nonbiodegradable. The decomposition of plastics can take decades making them persistent pollutants in the environment, especially in water. The most common sources of microplastics are the breakdown of larger pieces of plastic into small microplastics, microbeads contained in hygiene products, and microplastic fibers detached from man-made clothing.

Microplastics contained in wastewater can be identified, and analysis of the quality and quantity of microplastics by various methods clarifies the situation. Wastewater treatment plants use a variety of advanced equipment and wastewater treatment methods to reduce microplastic concentration in the treated effluent.

## 2.1 Definition of microplastics

Microplastics are defined as plastic pieces with a diameter of less than 5 mm. Great variations exist in the physical and chemical properties of microplastics, such as particle shape, size, and chemical ingredients. The shape of microplastics varies depending on whether they consist of fibers, microbeads, or fragments. Microplastics can be composed of several polymers. The most typical polymers include polyethylene, polypropylene, polyvinyl chloride, and polystyrene (Jaikumar et al., 2018). Microplastics are classified into large (1–5 mm) and small microplastics (20 μm–1 mm) and nanoplastics (<20 μm) according to the particle diameter distribution (Wagner et al. 2014).

Microplastics can also be classified into primary and secondary microplastics depending on the formation mechanism (Fig. 10.1). Primary microplastics are intentionally produced into very small pellets or powder for commercial applications such as personal care products (Jaikumar et al., 2018). Secondary microplastics are formed as larger plastic particles break down into smaller particles. Decomposition can occur in materials such as textiles, paints, and car tires (Talvitie et al., 2017a). The environmental effects causing microplastics to break down are mainly abrasion, ultraviolet B (UV-B) radiation, and temperature changes (Jaikumar et al., 2018).

## 2.2 Sources of microplastics

The two largest sources of microplastics are the disintegration of larger plastic pieces into small plastic particles and hygiene and plastic peelers contained in cleaning products, called microbeads (Chang, 2015). Microplastics are used in cosmetics to replace natural peeling agents such as pumice, oatmeal, apricot, and walnut skin. Cosmetics containing microplastics include hand sanitizers, soaps, toothpaste, shaving foams, bath foams, and sunscreens (Napper et al., 2015).

The use of abrasive and exfoliating materials in personal hygiene products has increased. Abrasive products containing natural materials, such as ground fruit stones, as well as synthetic

Fig. 10.1 Primary and secondary microplastics. Reproduced with permission from Elsevier. Talvitie, J., Mikola, A., Koistinen, A., Setälä, O., 2017. Solutions to microplastic pollution—removal of microplastics from wastewater effluent with advanced wastewater treatment technologies. Water Res. 123, 401–407. https://doi.org/10.1016/j.watres.2017.07.005.

materials, polyethylene beads, alumina, and sodium decahydrate granules, change the abrasiveness of the product (Chang, 2015). About 93% of the microbeads used in cosmetics are produced from polyethylene (PE), but microbeads can also be made of polypropylene (PP), polyethylene terephthalate (PET), polymethyl methacrylate, or acrylic and nylon (Napper et al., 2015). Polyethylene beads are used in abrasive products because their softness reduces redness of the skin and damages the skin less than most other materials. Polyethylene beads are 4 μm–1 mm in diameter, so they are classified as small microplastics. Most skin cleansing products contain polyethylene beads, so the consumer is most likely to use cleaning products that contain small microplastics (Chang, 2015).

One important source of microplastics originates in clothing and other textiles as washing releases synthetic fibers, i.e., microplastic fibers. Such textile fibers include polyester, acrylic, and nylon. The yarn used in textiles affects the detachment of fibers as loosely structured yarns release more fibers than yarns with a solid structure. In addition, the more worn the fabric is, the more fibers come off during washing (Carney Almroth et al., 2018).

Transportation is a major source of microplastics, especially in large cities. In addition to rubber, there are also many particulate microplastic emissions from traffic. Particles are released especially from car tires as well as road marking masses used for the permanent marking of traffic, such as guardrails and directional arrows. Microplastic is also generated in the plastics industry through the manufacture and use of plastics in processing processes (Mahon et al., 2017). Microplastics can end up in the environment from the plastic industries through leaks and transport accidents (Lares et al., 2018).

## 2.3 Effects of microplastics in the aquatic environment

A large amount of microplastics enters water bodies from hygiene products, washing of fiber-containing clothes, shipping, and stormwater. The main sources of migration of microplastics into water bodies are municipal wastewater treatment plants, agricultural, tourist and industrial areas, and harbors (Wagner et al., 2014).

Macroplastics degrade into secondary microplastics in the environment and in water bodies as a result of photolytic, mechanical, and biodegradation. In photolytic or light scattering degradations, the sun's UV radiation oxidizes the chemical structure of the plastic causing the polymer bonds to break. The structure of the plastic becomes brittle and the plastic easily disintegrates into small particles. In addition to UV radiation, larger plastic debris can be decomposed into microplastics by sea currents and the force caused by the waves and the abrasion of sediment particles. Macroplastics can break down into smaller particles also by biological degradation by bacteria and fungi (Browne et al., 2007).

Industrial products, such as toilet and skin cleaners, contain small polyethylene and polystyrene particles of about 1 mm in diameter. Acrylic, melamine, and polyester particles with a diameter of about 0.25–1.7 mm are used to clean various machines, including boats (Browne et al., 2007). The microbeads are transported with the wastewater to the wastewater treatment plant, where some of the microplastics are recovered in oxidation tanks or wastewater sewage sludge. However, due to the small size of microplastics, a significant part passes through wastewater treatment and ends up in water bodies.

Synthetic fibers, such as polyester and acrylic fibers, transport with washing machine effluent to municipal sewage treatment plants and thereby possibly into water bodies. Water samples have shown that a single garment can release up to more than 1900 fibers during one wash. Much of the microfiber fibers found in water bodies are probably from washing textiles (Browne et al., 2011). A more sensible textile structure design, textile prewash at the production stage, and more efficient filter use in household washing machines could potentially reduce microfiber discharge into wastewater (Carney Almroth et al., 2018).

In traffic, microplastics are released into the environment from car tires as well as particles released from road marking masses. Microplastics are released into the air and on the road. Such particulate emissions are leached with stormwater through sewers into rivers and further into lakes and seas. The number of microplastics carried by stormwater can be managed through comprehensive and proactive design, taking into account piping well solutions. The efficiency of the stormwater treatment can be improved with new material and structural solutions. In wastewater treatment plants, a large part of the microplastics removed from wastewater passes to sewage sludge that is or can be used as a soil improver and as a fertilizer in agriculture (Wagner et al., 2014; Lares et al., 2018). Thus, sewage sludge can act as a significant release route of microplastics into water bodies, as microplastics can migrate from agricultural areas with runoff water to rivers, lakes, and eventually to the seas (Wagner et al., 2014).

## 2.4 Effects of microplastics in organisms

Due to the small size of microplastics, they are also available for organisms at the bottom of the food chain. Microplastics are the same size as plankton natural food, so in addition to large organisms, plankton and other microorganisms can also swallow microplastics (Desforges et al., 2014). Microplastics can become enriched and transported in the food chain when organisms feed on organisms that have ingested microplastics (Wright et al., 2013). In addition to the size of microplastics, the shape and density of the particles also affect which organisms eat them (Browne et al., 2007). Low-density microplastic particles, such as polyethylene, float in the surface layers of the water column (Wright et al.,

2013), whereby organisms living in the surface layers of the water eat the floating surface particles. High-density microplastics, such as polyvinyl chloride, sink to the bottom of the water, making them available to those living in the aquifers (Browne et al., 2007).

Ingestion of microplastics may cause mechanical stress on organisms and in addition to the risk of suffocation, microplastics also cause other disadvantages (Talvitie et al., 2017a). Microplastic particles swallowed by organisms may persist in the gastrointestinal tract or be absorbed in the intestine. If microplastic particles are absorbed by the organism's intestines, the migration of microplastics around the body is possible. A rodent study has shown that solid polystyrene microspheres can move from the intestine to the lymphatic system. The lymphatic system controls the transport of the substances into the bloodstream (Browne et al., 2007) so that microplastic particles can travel into the bloodstream involved and bioaccumulate, i.e., accumulate in individual tissues (Chang, 2015). Assuming that the gastrointestinal tract of rodents is similar to that of other organisms, the migration of microplastics from the intestines of aquatic organisms around the body is likely (Browne et al., 2007). Microplastics are particularly dangerous because of their good durability and because they can move from one organism to another in the food chain (Talvitie et al., 2017a).

Many plastics, such as polyvinyl chloride, polystyrene, and polycarbonate, have been shown to release toxic monomers that are associated with cancer and reproductive disorders in humans, rodents, and invertebrates. In addition, in the plastic manufacturing process, many chemical additives are incorporated into plastics, such as antioxidants, flame-retardants, as well as antimicrobials. Ingestion of microplastics can contribute to toxicity migration of monomers and chemical additives into organisms (Browne et al., 2007).

Due to the hydrophobic or water-repellent nature of microplastics, they can adsorb organic pollutants such as polychlorinated biphenyls (PCBs), dichlorodiphenyltrichloroethane (DDT), polycyclic aromatic hydrocarbons (PAH), and polybrominated diphenyl ethers (PBDE) on their surface from the marine environment (Desforges et al., 2014). Ingestion of microplastics can facilitate the transport of these hydrophobic contaminants into the body of organisms in the food chain. Microplastics have a higher reaction area than larger pieces of plastic, so impurities are transported especially well with microplastic particles (Browne et al., 2007).

Ingestion of microplastics can also have physiological effects. Monomer from polystyrene production (styrene) and polycarbonate containing the monomer, bisphenol A, can interfere with the endocrine system of organisms and thus affect hormonal function. The functioning of the endocrine system affects the organism's development, reproduction, and immune system (Rochman et al., 2014). Bisphenol A, for instance, has been found to interfere hormonal activity of invertebrates, fish, and amphibians (Ballent et al., 2016). The harm caused by microplastics affects not only organisms but also water bodies. Microplastics can transport the organic contaminants with the water flows from one area to another (Lobelle and Cunliffe, 2011). Toxic substances released from microplastics also decrease water quality (McCormick et al., 2014). In the aquatic environment, microplastics can also act as artificial growth media for microbes and bacteria (Ballent et al., 2016).

## 3 Analysis methods for microplastics

Microplastics found in water bodies have rather complex structures and physical and chemical properties as presented in Table 10.1. Microplastics appear in different shapes like spheres, filaments, and different-shaped fragments. They also have different chemical compositions like polyethylene (PE), low-density PE (LDPE), PE terephthalate (PET), and polyacrylates (PA) just to mention a few (Picó and Barcelo, 2019). Microplastics are found together with other particulate materials like sand and organic debris making the separation and identification of MPs a demanding task (Al-Azzawi et al., 2020). The life span of the MPs in waters can range from a few years to over 100 years. Due to this life span, the MPs can undergo changes, such as surface oxidation and formation of biofilms on the surface.

There are several analytical methods used for the quantitation and identification of microplastics in water bodies and sediments. Some summaries of the analytical methods including sample pretreatment procedures and laboratory demands have been published and can be found in published reviews (Hidalgo-Ruz et al., 2012; Huppertsberg and Knepper, 2018; Li et al., 2018; Rios Mendoza and Balcer, 2019).

Independent of the methods used for detection and quantitation, the measurements must be performed in a plastic-free or low plastic environment. This includes all steps from sampling to preparation and detection. During all analytical steps (sampling, preparation, and detection), "plastic-free" or low-plastic working conditions must be ensured. These include the avoidance of standard plastic products and the use of alternatives made of metal,

Table 10.1 Some properties of the most common polymers in microplastics.

| _____ Several characteristics of the most common plastics found in the MPs _____ | | | | |
|---|---|---|---|---|
| Name | Acronym | Products | Density (g/mL) | Life span (years) |
| Polyethylene terephthalate | PET | Water bottles | 1.37–1.45 | 20 |
| polyester | PES | Polyester clothes | 1.39 | >20 |
| Low-density polyethylene | LDPE | Plastic bags, squeeze bottles | 0.917–0.930 | |
| High-density polyethylene | HDPE | Detergent bottles | 0.93–0.97 | >28 |
| Polyvinylchloride | PVC | Pipes, electric cables, clothing | 1.20–1.45 | 140 |
| Polypropylene | PP | Clothing, stoppers | 0.89–0.94 | >100 |
| Polyamide | PA | Textile (nylon), tooth brush | 1.13–1.35/1.41 | >20 |
| Polystyrene | PS | Ready to-eat food | 1.04–1.11 | 50 |
| Acrylonitrile-butadiene-styrene | ABS | Pipe systems, musical instruments | 1.04–1.06 | |
| Polytetrafluoroethylene | PTFE | Plain bearings, gears, slide plates, seals, gaskets, bushings | 2.10–2.30 | >140 |

Reproduced with permission of ACS Author Choice License. Picó, Y., Barcelo, D., 2019. Analysis and prevention of microplastics pollution in water: current perspectives and future directions. ACS Omega 4, 6709–6719. Retrieved from https://pubs.acs.org/doi/pdf/10.1021/acsomega.9b00222.

glass, or silicone. An exception is the use of plastics that are not to be detected or evaluated. Because of the risk of getting false positives due to contaminated samples, the use of known standards and internal standard additions is highly recommended (Nguyen et al., 2019). The efficiency of the collection method is verified by spiking the samples with standard components (Nguyen et al., 2019). The whole analytical procedure includes steps like sampling, extraction and cleanup, identification, and confirmation as described in Fig. 10.2.

### 3.1 Sampling condition and volumes

Depending on the origin of the samples (soil, sludge, or water) different sampling and pretreatment methods are needed. Regarding samples taken from water (natural waters, effluents, or drinking waters) different sampling volumes are needed; some

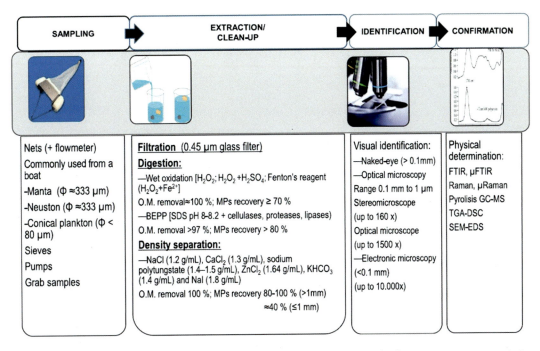

**Fig. 10.2** A summary of the steps included in the sampling and analysis of microplastics. Reproduced with permission of ACS AuthorChoice License. Picó, Y., Barcelo, D., 2019. Analysis and prevention of microplastics pollution in water: current perspectives and future directions. ACS Omega 4, 6709–6719. Retrieved from https://pubs.acs.org/doi/pdf/10.1021/acsomega.9b00222.

rules for sampling volumes are presented in Table 10.2 (Braun et al., 2018). For the sampling of MPs from waters, three major methods are used: sampling by nets, by sieves, and by pumping. The volumes needed for sampling are highly dependent on the content of solid particles in the water (Picó and Barcelo, 2019). After collection, the MPs must be concentrated and cleaned up.

## 3.2 Concentration and cleanup of collected samples

Microplastic will, after being in the water for a while, be covered by biofilms, and/or the surface is oxidized due to the stability of the particles, stable for a few up to 100 years. These changes in composition make identification much harder to perform and the plastics must be converted into their native state before analysis. Inorganic matter in the samples can be removed using differences in densities, saturated solutions of sodium chloride (NaCl), or sodium iodide (NaI) having densities of 1.2 and 1.8 can be used to separate inorganic and organic particles. Organic particles

**Table 10.2 Recommended sampling volumes.**

| | Very high solid content | Rich in solids | Low in solids | Nearly solid-free |
|---|---|---|---|---|
| Filterable substances/plankton | More than 500 mg/L | 100–500 mg/L | 1–100 mg/L | Less than 1 mg/L |
| Examples | Sewage plant intake | Street drainage | Sewage plant effluent, surface waters | Groundwater, mineral water, drinking water |
| Recommended sample volume for particle analysis of ~50–1 μm | 5 mL | 500 mL | 1 L | 500 L |

From Braun, U., Jekel, M., Gerdts, G., Ivleva, N. P., & Reiber, J. (2018). Microplastics analytics sampling, preparation and detection methods. Discussion paper published in the course of the research focus "Plastics in the Environment—Sources • Sinks • Solutions". Berlin: Ecologic Institute. https://www.ecologic.eu/sites/files/publication/2018/discussion_paper_mp_analytics_en.pdf.

are removed using oxidizing conditions as described in the following (Al-Azzawi et al., 2020).

So far, no standardized methods for the cleanup of MP samples have been published making the comparison of results mode difficult (Al-Azzawi et al., 2020). The two methods mostly used for the oxidative cleanup procedures are the Fenton method and the $KOH/H_2O_2$ method. The Fenton reagent consists of a mixture of $H_2O_2$ and $Fe(II)SO_4$. Both methods are described in Fig. 10.3, note that this setup is for wastewater, and natural waters might need other sample volumes and modified procedures (Al-Azzawi et al., 2020). Finally, the cleaned microplastics are concentrated by filtration over a filter with 0.45 μm pores.

## 3.3 Analytical methods for identification and verification of MPs

To investigate the negative effects of microplastics, there must be methods to do quantitative and qualitative measurements of the MPs. Several methods have been used to identify microplastics and also to make quantitative measurements of them. The methods can be divided into two main categories, fast identification, and instrumental verification methods.

The fast identification methods are mostly based on microscopic methods or even visual identification can be used. The

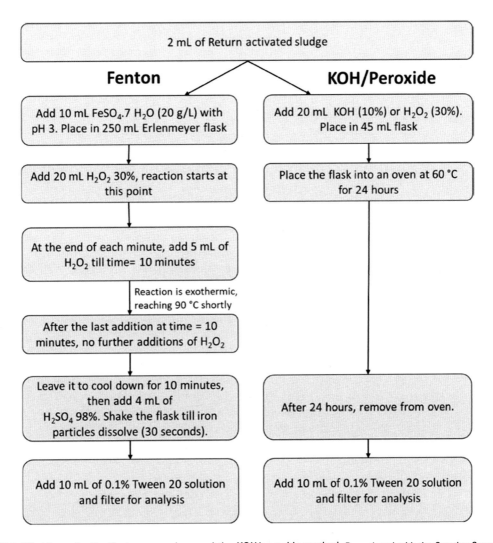

**Fig. 10.3** Workflows for the Fenton procedure and the KOH/peroxide method. Reproduced with the Creative Common Attribute License. Al-Azzawi, M.S.M., Kefer, S., Weißer, J., Reichel, J., Schwaller, C., Glas, K., et al., 2020. Validation of sample preparation methods for microplastic analysis in wastewater matrices—reproducibility and standardization. Water (Basel) 12 (9). https://doi.org/10.3390/w12092445.

microscopic methods (optical and electron microscopes) cover all microscopic techniques ranging from stereomicroscopes over light microscopes to scanning electron microscope (SEM) and transmission electron microscope (TEM). The method used is highly dependent on the particle sizes available and the complexity of the samples.

Another method used for fast identification is based on the particle's density. In Table 10.3 are the densities for the most common MP presented; the density ranges from 0.94 g/mL (LPDE) to 2.4 g/mL (PTFE). An estimation of the density is made by preparing a series of saturated salt solutions (Table 10.3) (Liu et al., 2020). The final determination is made by immersing the samples into the series of solutions.

### 3.4 Instrumental methods for confirmation tests

The instrumental methods include spectroscopic methods (FTIR, ATR and μATR FTIR, and Raman) (Araujo et al., 2018; Huppertsberg and Knepper, 2018), and gas-chromatographic methods including pyrolytic pretreatment (Py-GC-MS). The selection of the method depends on the information needed and the quantity of samples available. A summary of the most common methods used is presented in Table 10.4.

The most common methods are those based on Fourier-transform infrared (FTIR) spectroscopy and Raman spectroscopy, especially methods based on attenuated total reflectance (ATR) or μATR spectroscopy (Huppertsberg and Knepper, 2018).

An instrumental method based on fluorometry using fluorescent probes for the analysis of MPs has been published (Maes et al., 2017). In the fluorometric method presented by Maes et al., the microplastics are labeled with a fluorescent probe. The critical step in the method is the capacity of the probe to attach to the surface of the MPs. Several fluorescent probes were

**Table 10.3** A series of saturated salt solutions used for density tests.

| Salt | Density of saturated solution (g/mL) |
| --- | --- |
| NaCl | 1.19 |
| $CaCl_2$ | 1.42 |
| NaBr | 1.55 |
| $ZnCl_2$ | 1.68 |
| NaI | 1.89 |
| $H_2O$ | 1.0 |
| KBr | 1.58 |

From Liu, M., Lu, S., Chen, Y., Cao, C., Bigalke, M., He, et al., 2020. Analytical methods for microplastics in environments: current advances and challenges. In: He, D., Luo, Y. (Eds.), Microplastics in Terrestrial Environments. The Handbook of Environmental Chemistry, vol. 95. Springer, Cham.

Table 10.4 Information available using different instrumental methods.

| Characteristics | Spectroscopic | | | | | | Thermoanalytical | | | | Chemical |
|---|---|---|---|---|---|---|---|---|---|---|---|
| | μ-Raman | μ-FTIR (trans) | FPA FTIR (trans) | μ-ATR-FTIR | ATR-FTIR | NIR/hyperspectral Imaging | Py-GC-MS | Mod. Py-GC-MS | TED-GC-MS | DSC | ICP-MS |
| Type of polymer | Yes | Yes | Yes | Yes | Yes | Yes | Yes | Yes | Yes | Only PE, PP | Only tire abrasion |
| Detectable additives | Pigments | No | No | No | No | No | Yes | No | No | No | No |
| Particle surface (chemical) | Yes | No | No | No | Yes | Yes | No | No | No | No | No |
| State of degradation | Surface oxidation | No | No | Surface oxidation | Surface oxidation | No | Oxidation | No | No | Mol. weight | No |
| Particle number, particle size, particle shape, and particle surface morphology | Yes | Yes | Yes | Yes | Yes | No | No | Yes | No | No | No |
| Mass balance | No | No | No | No | No | No | No | Yes | Yes | Yes | Yes |

From Braun, U., Jekel, M., Gerdts, G., Ivleva, N.P., Reiber, J., 2018. Microplastics analytics sampling, preparation and detection methods. Discussion paper published in the course of the research focus "Plastics in the Environment—Sources • Sinks • Solutions". Ecologic Institute, Berlin. https://www.ecologic.eu/sites/files/publication/2018/discussion_paper_mp_analytics_en.pdf.

tested, but Nile Red, NR (9-diethylamino-5H-benzo[α]phenoxazine-5-one), proved to be the most effective one (Tamminga, 2017).

The MPs were initially separated using density solutions as described in Table 10.3, separated, and incubated with different concentrations of the probes. Finally, the samples were filtered onto a filter dish. The samples were analyzed using a fluorescence microscope using a blue LED as excitation light. So far, no mechanism has been proposed for the attachment of the probe molecule to MPs.

## 4 Microplastic retention by biochars and activated carbons

### 4.1 Current carbon-based methods for microplastic purification

Key facilities in MP removal from water are the wastewater treatment plants (WWTPs) and drinking water treatment plants (DWTPs). WWTPs can reduce the MP waste from ending up in ecosystems, whereas DWTPs reduce the risks of MP human consumption (Shen et al., 2020). The amounts, sizes, and shapes of MPs in raw waste and drinking water can vary depending on the plant location (Pivokonsky et al., 2018; Talvitie et al., 2017a). MP amounts in DWTP raw water have been reported to range from 0 to >4000 particles/L (Novotna et al., 2019). Extremely high MP amounts (up to >6000) have been found in bottled water contained in recyclable PET or glass bottles (Oßmann et al., 2018). Number of MPs in municipal sewage treatment plant influent ranged from 15.1 to 640 particles/L in 9 different countries (Kang et al. 2018). Fig. 10.4 presents the number of microplastics found in different water types (Koelmans et al., 2019).

Both WWTP and DWTP facilities normally use sequential water purification treatments, including coagulation/flocculation, sedimentation, filtration, aeration, and membrane bioreactors. Good MP recoveries have been achieved by each method, but results depend on a high number of variables, such as MP size, shape, concentration, filter pore size, and water type. Talvitie et al. (2017a), for instance, compared disk filtering (DF) (10 and 20 μm pore size), rapid sand filtering (RSF), dissolved air flotation, and membrane biofiltering (MBR) for their performance in >20 μm MP removal from primary and secondary effluents. The obtained removal efficiencies were 40% (10 μm disc filtering), 98.5% (20 μm DF), 97.1% (RSF), 95.0% (DAF), and 99.9% (MBR).

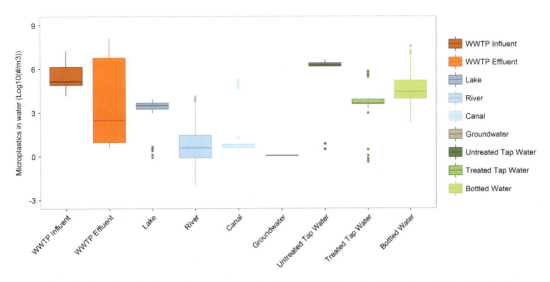

Fig. 10.4 Variation in microplastic number concentrations in individual samples taken from different water types. References included in the figure: (Estahbanati and Fahrenfeld, 2016; Faure et al., 2015; Fischer et al., 2016; Hoellein et al., 2017; Kosuth et al., 2018; Leslie et al., 2017; Magnusson and Noren, 2014; Mason et al., 2016; McCormick et al., 2016; Mintenig et al., 2019; Oßmann et al., 2018; Pivokonsky et al., 2018; Rodrigues et al., 2018; Schymanski et al., 2018; Simon et al., 2018; Talvitie et al., 2015, 2017a,b; Ziajahromi et al., 2017). Reproduced with Creative Commons CC-BY license. Koelmans, A.A., Mohamed Nor, N.H., Hermsen, E., Kooi, M., Mintenig, S.M., De France, J., 2019. Microplastics in freshwaters and drinking water: critical review and assessment of data quality. Water Res. 155, 410–422. Elsevier Ltd. https://doi.org/10.1016/j.watres.2019.02.054.

The poor result of 10 μm DF was caused by excessive polymer addition in the system due to disturbances in the WWTP treatment process, and the high MBR result again was affected by its smallest filter pore size (0.4 μm) and treatment of clarified wastewater with higher MP concentration compared to secondary effluent. High removal efficiency using MBR was verified by the study of Lares et al. (2018) who compared conventional activated sludge (CAS) process to MBR. The removal efficiency of MBR was 99.4% and 98.3% for CAS (conventional activated sludge) (Lares et al., 2018), whereas CAS treatment of wastewater in China had a much lower removal rate of 64.4% (Liu et al., 2019). Powdered activated carbons (PACs) have been tested also as part of MBR purification, where the benefit sought has been in mitigating the membrane fouling (Skouteris et al., 2015). Sludges formed during water treatments can have very high MP concentrations (Gies et al., 2018), which needs to be accounted for if sludge is planned to be mixed, for instance, in soil.

Granulated activated carbons (GACs) are normally used after primary and secondary steps (Fig. 10.5) to remove the remaining

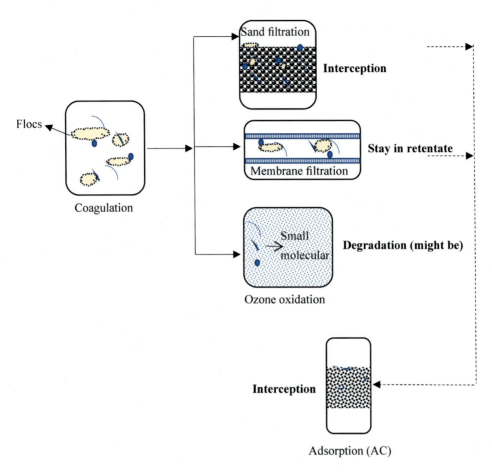

**Fig. 10.5** Example of MP removal including GAC filtration in tertiary water treatment. Reproduced with permission from Elsevier. Zhang, X., Chen, J., Li, J., 2020. The removal of microplastics in the wastewater treatment process and their potential impact on anaerobic digestion due to pollutants association. Chemosphere 251, 126360. Elsevier Ltd. https://doi.org/10.1016/j.chemosphere.2020.126360.

small concentrations of harmful compounds left before releasing the water back to nature or human consumption. Fossil-based GACs have been used for drinking water and wastewater purification for a long time, but only recently, attention has been focused on their ability to retain MPs. The purification steps preceding GAC filtering can remove a majority of the MPs, but adding of GAC step at the end can further enhance the water quality (Pivokonský et al., 2020). The effect is mainly seen in the size range <20 μm as larger MPs are more efficiently captured during previous steps and only the smallest particles remain in the effluent (Fig. 10.6) (Pivokonský et al., 2020). According to

(Talvitie et al., 2017a) the smallest <100 μm MP fraction may remain in the effluent after traditional purification processes. The benefits of using GAC purification are also in the possibility to remove many types of organic pollutants simultaneously. GAC has been found efficient in removing organic pollutants from landfill wastewater, including bisphenol-A which is a monomer used by the plastic industries and pharmaceutical residues (Baresel et al., 2019; Kalmykova et al., 2014). Baresel et al. 2019 studied the removal of micropollutants including pharmaceutical residues and microplastics from MBR-effluent using commercial GACs. They found 100% removal efficiency for microplastics and 90%–98% efficiency for the pharmaceutical compounds. Comparison to full-scale CAS was also performed, showing 90.7% removal efficiency for the MPs (Baresel et al., 2019). One disadvantage of using GAC filtration is the clogging of the pores with highly contaminated waters. The use of other efficient purification treatments, such as MBR, before GAC filtration, is beneficial for the system, as it can reduce the clogging and backwash frequency, thereby reducing operational costs (Baresel et al., 2019).

Terrestrial ecosystems are also affected by microplastics (Corradini et al., 2019; Huerta Lwanga et al., 2016; Nizzetto et al., 2016; Scheurer and Bigalke, 2018). The use of sewage sludge as fertilizer or soil amendment may, for instance, cause MP accumulation in soils (Corradini et al., 2019; Wagner et al., 2014). An estimated 63,000–43,000 and 44,000–300,000 tonnes of MP end up in farmland soils in Europe and North America, respectively (Nizzetto et al., 2016). As biochar has long been used as a soil amendment and adsorbent for various contaminants, MP retention using soil or sand as filter media has also been studied (Abit et al., 2014; Tong, et al., 2020a,b; Yan et al., 2020). Since the presence of MPs can have negative effects on soil biota (e.g., earthworms (Huerta Lwanga et al., 2016)), the topic is of high importance.

One concern regarding the existing studies is the use of mainly primary plastics and pristine filter media (such as quartz sand) which do not present natural conditions (Alimi et al., 2018; Yan et al., 2020). Different soil types and natural degradation may affect the MP retention capacity. Natural aging causes chemical changes in the MP particles, such as increased hydrophilicity, fragility, and changes in particle shape (Alimi et al., 2018; Celina, 2013; Hüffer et al., 2017; Yan et al., 2020). Reported MP values may also be underestimated due to the higher presence of nano-sized plastics that have not been detected with the currently used or available analysis methods (Alimi et al., 2018; Enfrin et al., 2019). Larger MPs can fragment into smaller nanoplastics due

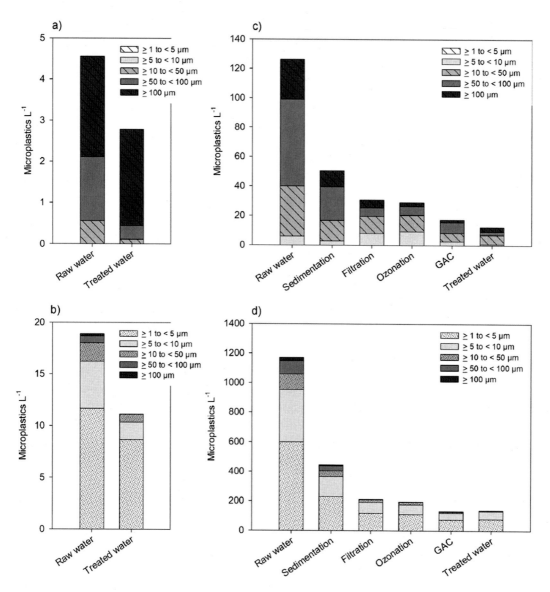

**Fig. 10.6** Size distributions and concentrations of differently shaped microplastics at the drinking water treatment plant Milence (A, B) and Plzen (C, D). (A) and (C) show results for microplastic fibers and (B) and (C) for microplastic fragments. Reproduced with permission from Elsevier. Pivokonský, M., Pivokonská, L., Novotná, K., Čermáková, L., Klimtová, M., 2020. Occurrence and fate of microplastics at two different drinking water treatment plants within a river catchment. Sci. Total Environ. 741, 140236. https://doi.org/10.1016/j.scitotenv.2020.140236.

to mechanical or chemical stress during the purification process, for example, mixing, pumping, or grinding by silica sand in rapid sand filtering (Enfrin et al., 2019, 2020; Wang et al., 2020a). This underlines the importance of analytic method development in addition to efficient purification methods.

## 4.2 Biochar for microplastic removal

Despite the wide use of fossil-based activated carbons for water treatment, only a few research outputs concentrating on MP using biochars or biobased ACs have been published to date, despite the high number of publications focusing on biochar and biobased AC purification applications. Biochars and activated biochars have been proven efficient in removing inorganic and organic contaminants from water and can be fairly easily modified to serve different adsorbates by creating, e.g., suitable porosities (Dias et al., 2007; Rivera-Utrilla et al., 2011; Siipola et al., 2018).

The existing studies where biochar has been used as filter material have shown promising results in preventing MP leaching. The research covered by this chapter includes the following research. Retention of 1 μm carboxylated polystyrene microspheres mixed in two soil types has been studied by (Abit et al., 2014). Bark-based activated carbon has been tested in small-scale column study for the retention of fleece fibers, 5 mm and 10 μm MP particles (Siipola et al., 2020). Cellulose-based biochar as such and $Fe_3O_4$-modified biochar mixed in quartz sand was tested for the purification of 0.02, 0.2, and 2 μm polystyrene particles in a column study using 5 and 25 mM NaCl solution elution (Tong et al., 2020a). Tong et al. (2020a) addressed also the recovery of the biochar from the sand utilizing the magnetic properties of the Fe-modified biochar. Another study by Tong et al. (2020b) addressed heteroaggregation, cotransport, and deposition of 0.02 μm nanoplastics and 0.2 and 2 μm MPs with biochar using saturated quartz sand matrix (Tong, et al., 2020b). Corn straw and hardwood biochar have been tested for 10 μm polystyrene microsphere removal in a column test setup where biochar has been placed between sand layers (Wang et al., 2020b).

### 4.2.1 Effect of biochar physical characteristics

Biochar can be produced from various lignocellulosic biomasses of different chemical and physical characteristics. Carbonization can retain some of the morphology of the biomass in the biochar, e.g., the cell structure, which affects the porosity of the produced biochar. Biochar porosity depends on the carbonization parameters, such as temperature and time, and further porosity can be produced using activation methods (González-García, 2018). The large pores in biochar combined with a rough surface can serve as an efficient barrier to MP movement through the biochar column (Figs. 10.7–10.10). Surface chemistry can enhance MP retention in the smallest colloidal-size range (Abit

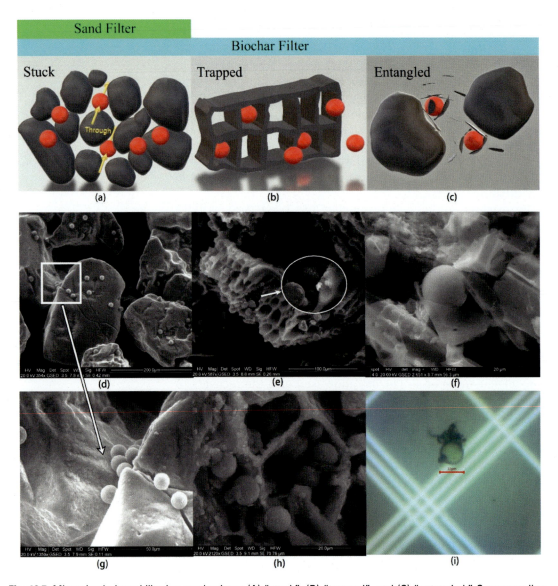

**Fig. 10.7** Microplastic immobilization mechanisms: (A) "stuck", (B) "trapped", and (C) "entangled." Corresponding ESEM images: (D and G) sand filter; biochar filter (E, F, H) and optical microscope image (I). Reproduced with permission from Elsevier. Wang, Z., Sedighi, M., Lea-Langton, A., 2020. Filtration of microplastic spheres by biochar: removal efficiency and immobilisation mechanisms. Water Res. 184, 116165. https://doi.org/10.1016/j.watres.2020.116165.

et al., 2014; Tong, et al., 2020a), but its meaning is reduced when MP particle size increases.

Different MP retention mechanisms have been studied by Wang et al. (2020b) using environmental scanning electron

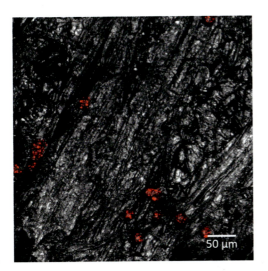
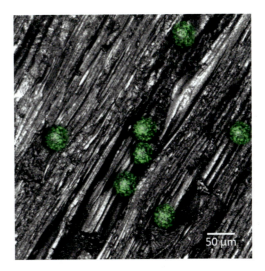

**Fig. 10.8** Microplastic particles attached to bark biochar. Used MP particles were in the size range from 10 to 22 μm (Cospheric UVPMS-BR-1.20). The picture was taken with a confocal laser scanning microscope (CLSM) using a diode laserline of 405 nm for reflection to visualize the surface of the sample and an argon laserline of 488 nm for the excitation of fluorescence of MP particles. Emission from the MP particles was detected at 552–678 nm. The final CLSM micrograph, in which MP particles appear *red*, was reconstructed by superimposing the reflection and emission images. Copyright VTT.

**Fig. 10.9** Microplastic particles attached to bark biochar. Used MP particles were in the size range from 38 to 45 μm (Cospheric UVPMS-BG-1.035). The picture was taken with a confocal laser scanning microscope (CLSM) using a diode laserline of 405 nm for reflection to visualize the surface of the sample and an argon laserline of 488 nm for the excitation of fluorescence of MP particles. Emission from the MP particles was detected at 503–532 nm. The final CLSM micrograph, in which MP particles appear *green*, was reconstructed by superimposing the reflection and emission images. Copyright VTT.

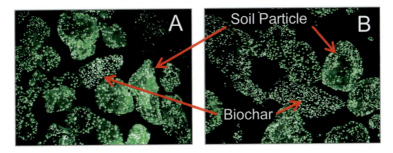

**Fig. 10.10** Fluorescent microspheres attached to biochar and soil particles in sand (A) and sandy loam (B). Reproduced with permission from Wiley. Abit, S.M., Bolster, C.H., Cantrell, K.B., Flores, J.Q., Walker, S.L., 2014. Transport of Escherichia coli, Salmonella typhimurium, and microspheres in biochar-amended soils with different textures. J. Environ. Qual. 43 (1), 371–388. https://doi.org/10.2134/jeq2013.06.0236.

microscopy (ESEM) imaging (Wang et al., 2020b). They studied 10 μm MP retention using silica sand, hardwood biochar, and corn straw biochar produced in 300°C, 400°C, or 500°C and 10 μm MP spheres. The MP size was selected to minimize the effect of ionic strength, zeta potential, and van der Waals forces on the retention. Three MP immobilization mechanisms were detected: *"stuck,"* *"trapped,"* and *"entangled"* (Fig. 10.7). Biochar samples exhibited all three immobilization mechanisms, but with silica sand, only "stuck" mechanism was present. In the "stuck" mechanism, sand or biochar acts as a sieve, where MP particles are trapped in the gaps between the filter particles. The "entangled" mechanism was described as MP immobilization by small flaky biochar particles, which may possess some colloidal surface properties affecting the MP attachment. In the "trapped" mechanism, MP particles are immobilized in pores slightly larger than the MP particle size. The plastic particle form plays a role as well, as spherical microbeads with a smooth surface are more likely to escape the filter (Ding et al., 2020).

Physical straining preventing the MP movement in the filter column is one of the main mechanisms keeping the particles in the filter material. For sand filters, immobilization in the porous system between the sand particles is one of the main retention mechanisms (Talvitie et al., 2017a; Wang et al., 2020b). According to Talvitie et al. (2017a), retention using rapid sand filtering (RSF) is immobilization between the sand grains or adherence on the sand grain surfaces. They also discuss the possibility of improving the result by using a coagulant in the process that could improve the adherence. Sand grains have a relatively smooth surface compared to biochar, for which reason adherence to the surface is smaller and mobility of the particles larger (Wang et al., 2020b). Biochar can again be highly porous with a coarse surface structure, which increases the possibilities of particle attachment.

Pore sizes and pore morphology of biochar can vary depending on the biochar raw material and production parameters. Biochar can retain small, colloidal-size plastic particles inside the porous system (Abit et al., 2014). Larger MP particles attach at the surface or are captured by the cracks or large surface pores, whereas the micro- and mesoporosity (<2 nm and 2–50 nm, respectively) probably have a minimal contribution to the retention. This can be seen in the studies of Siipola et al., where pine and spruce bark ACs and biochars have been tested for MP retention. Activated carbon (AC) produced from spruce bark has been found best performing for 10 μm MP compared to pine bark activated carbon (Siipola et al., 2020). The reason for the result was seen to be the higher amount of macroporosity (>50 nm) in the spruce bark

AC. Retention of fleece fibers (5 mm) was 100% in the same study. Comparison of pine and spruce bark AC to pine bark biochar revealed superior performance of biochar for <1 mm MP retention (Siipola et al., 2019, Project report, unpublished data (in Finnish)). The pine bark biochar had 22% macropore content, whereas the ACs had only 1%–2%.

The biochars described in the study by Wang et al. (2020b) were evaluated for their MP retardation efficiency as C500 > C400 > Hardwood > C300 > silica sand. The ESEM images showed biochars produced in 500°C (C500) or 400°C (C400) to have "chipped" or "loofah" structure and hardwood biochar had "honeycomb" structure. Silica sand and biochar produced in 300°C (C300) had relatively smooth surfaces. These surface morphologies of C500, C400, and hardwood along with surface roughness were seen to cause better MP retention compared to silica sand and C300. MP attachment due to surface roughness and large surface cracks or pores is visually presented in Figs. 10.8 and 10.9, where 10–22 μm MP particles (red) or 38–45 μm particles (green) have been retained by biochar produced from a mixture of pine and spruce barks (Siipola et al., 2021, unpublished). The microspheres were injected into a biochar column (d 3 cm, height approximately 28 cm) using tap water after which the column was flushed with 1.5–2 L of tap water.

### 4.2.2 Effect of biochar and solution chemistry

In addition to acting as a physical barrier, the surface chemistry of the biochar and plastic particles and the ionic strength of the eluent affect MP retention with colloidal-scale nanoplastics (Mitropoulou et al., 2013; Tong, et al., 2020a,b). The MP particles can flocculate and biofilms can form on the particle surfaces (Carr et al., 2016; Zhang et al., 2020). Flocculation may affect the tendency of the particles to float or sink as well as the adherence to the filtering material. Biofilm formation may affect the surface properties of the particles (Carr et al., 2016) that can affect retention to filter material. Biochar as larger particles do not have colloidal properties, but colloidal-size biochar chips can be formed (Wang et al., 2020b). These chips have colloidal surface properties and may attach to the plastic particles, a mechanism referred to as "entangled" (Wang et al., 2020b).

The ionic strength of the influent solution is one parameter affecting all adsorption phenomena, where charge-dependent adsorption occurs. In a study by Mitropoulou et al. (2013), retention of colloidal-size nanoplastics (0.075, 0.3, and 2.1 μm) to fine (0.181 mm) and medium (0.513 mm) grain-size quartz sand was addressed using varying ionic strengths (0.1–1000 mM NaCl) and unsaturated water media. The results showed a significant increase in retention with increasing ionic strength. The grain size of the sand had also some effect on the retention. Retention using deionized water was slightly higher with the medium grain size than with the fine grain size (Mitropoulou et al., 2013). Similar results using biochar-amended quartz sand as filter media have been obtained by Tong et al. (2020a). They used 5 and 25 mM NaCl ionic strengths for the eluents and 0.02, 0.2, and 2 µm polystyrene particles. The higher 25 mM ionic strength increased the retention of all plastic particle sizes with clean quartz sand, but especially with the biochar addition.

Zeta-potential and van der Waals forces of the particles (both plastic and biochar) affect the retention capacity along with the solution ionic strength. In the study of Tong et al. (2020a), biochar particles were treated with $FeCl_3$ and $FeSO_4$ creating $Fe_3O_4$-modified biochar. As $Fe_3O_4$ nanoparticles contain positive surface charge, this method was used to add positive charge on the biochar particles. The $Fe_3O_4$-modified biochars experienced increased retention of the tested MPs compared to unmodified biochars. Increased surface roughness caused by the $Fe_3O_4$ modification was seen as one reason for the better performance, but the $Fe_3O_4$-modified biochars had also better performance in varying ionic strengths compared to unmodified biochar. The results were seen to indicate reduced electrostatic repulsion caused by less negative zeta potential of the quartz sand induced by the biochar addition. Surface charges can also lead to heteroaggregation between the biochar and plastic particles. Tong et al. 2020b studied cotransport of nano- and microplastics and biochar in saturated quartz sand. Heteroaggregation of biochar with 0.02 and 0.2 µm plastic particles was detected which caused changes in the aggregate zeta potentials and thereby in the retention capacities. Steric repulsion caused by the changes in particle surface charge through aggregate formation was seen to increase the movement of particles through filtering material (Tong et al., 2020b). Similar results were obtained by Yan et al. (2020) using loamy sand as filter material. They used simulated rainwater and MPs derived from waste plastic seedling ropes collected from farmland soil. The MPs were washed with water (water-washed rope MP, RMP), or with diluted HCl, ethanol, and $H_2O_2$ (further cleaned RMP, CMP). The RMPs had therefore some soil colloids adsorbed, whereas CMPs were very clean. Their results showed CMP movement through the column to be negligible, whereas RMP movement was higher. Mixing humic acid into the rainwater produced the most extensive RMP movement. Heteroaggregation of soil colloids with MPs increasing the MP zeta potential was seen as the driving mechanism for the downward movement, which was further enhanced by humic acid addition. Natural aging of the MPs may also change the surface charge and thereby the retention capability (Yan et al., 2020).

### 4.2.3 Mixing biochar with soil/sand

Biochar as such has been shown to retain substantial amounts of the MPs in all size fractions (Wang et al., 2020b; Siipola et al., 2019). Mixing biochar with sand or soil appears also to increase the adsorption capacity compared to sand or soil alone. In addition, quite small biochar additions (0.5%–5%) seem to be enough for increased MP retention. Tong et al. (2020a) studied the retention of nano- and microscale plastics (0.02, 0.2, and 2 µm polystyrene MPs) using sand filtering with or without biochar addition. They found that sand filtering without biochar addition exhibited high breakthrough curves with more than 75% of the plastic particles escaping from the sand using 5 mM NaCl elution and 58%–70% escape with 25 mM NaCl elution in all MP sizes. Mixing of 0.5% of biochar or $Fe_3O_4$-modified biochar with the sand increased the retention substantially in all studied size fractions, especially high retention was obtained using the Fe-modified biochar, the MP effluent recoveries (5 mM NaCl solution) being 0.5%–2.6%. Abit et al. (2014) used 2% biochar loading in the soil samples and the recoveries from the effluent decreased from 18% to 0.05% with fine sand and from 5.6% to 0.02% with sandy loam. Respective values without biochar additions were 18% and 5.6%. Attachment of microspheres in both soil and biochar particles is visualized in Fig. 10.10.

Studies that have addressed the aggregation of biochar and plastic particles have, however, showed also some negative effects. Increased biochar transport through a quartz sand column was observed in the presence of nanoplastics (0.02 and 0.2 µm) when biochar and plastic particles were injected using 5 and 25 mM NaCl solutions (Tong et al., 2020b). With the larger 2 µm MPs, increased transport was detected only with 5 mM NaCl. Increased biochar particle size due to plastic adsorption and decreased steric repulsion with the higher (25 mM) NaCl solution were seen as the reason for the behavior. The presence of biochar in the sand column did increase the deposition of all tested MP sizes, but the results suggest that copresence and aggregation of biochar and MP in a solution could increase colloidal biochar particle transport through the sandy soil column if especially nano-sized MPs adsorb on their surfaces. Similar results have been obtained in the studies of Wang et al. (2013) and Yang et al. (2019), where natural organic compounds, such as humic acid, increased biochar transport through saturated quartz sand media. Increased biochar particle transport can potentially cause groundwater contamination (Tong et al., 2020b) as adsorbed harmful chemical compounds can be transported through the soil.

## 5 Conclusions

In this chapter, a review of microplastics in aquatic/water environment, analytics and methods for microplastics removal was conducted. Studies related to microplastic treatment methods are quite new, even though the existence and effects of microplastics on the environment and organisms were known for a long time. Of course, some long-term effects might still be partly unknown. Analytics, quantitative, and qualitative determination of microplastics have also improved during the past years. Based on this, microplastic characterization has become more detailed, enabling also the use of more efficient treatment methods. Among those, mechanical methods are often applied. One potential method for microplastics removal is adsorption over biochar or activated carbon adsorbents.

The retention of microplastics using carbon-based materials is a combination of many factors. To date, laboratory studies have reached promising results showing biochars as an efficient filter material for MP particles. The subject is still only superficially touched and further research is still needed regarding the retention parameters. The effect of natural aging of real-world MP particles, which changes the surface chemistry of the MP particles through, for instance, photooxidation (Yan et al., 2020), changes the retention characteristics. Pristine MP particles used in laboratory studies cannot, therefore, fully predict the behavior of actual MPs in wastewater.

Along with the changes in MP surface chemistry, the differences in MP shape and size can change the MP retention capacity by different filtering materials. The smallest colloidal-size-range plastic particles are the ones most likely to escape filters. On a colloidal-size scale, the surface charge of the plastic particles can have a more pronounced role in the retention than the filter pore structure (Tong et al., 2020b). The porosity and surface charge of biochar on colloidal MP retention needs to be addressed for a better understanding of the retention mechanisms, as well as the particle or grain size and surface roughness of biochar.

The degree of water saturation in the filter column and the ionic strength of the purified water are also factors for consideration. The thickness of the water film on filter particles affects the adherence of the plastic particles if the water film thickness is smaller than the plastic particle size (Mitropoulou et al., 2013). The ionic strength of water has been shown to be an affecting parameter for nanoplastics, for which the surface charges and steric repulsion of the particles have an impact on the retention. Still, despite being in the early stages of the research in the field, biochar has already shown high potential to serve as an effective and low-cost filter material for microplastic removal from water.

## Questions and problems

1. What are the key sources of microplastics in water environment?
2. Briefly describe the steps used for the identification of microplastics according to Picó and Barcelo?
3. What are the risks and sources of contamination in each step and how to prevent them?
4. What are the advantages of carbon-based methods for microplastic remediation?
5. What are the disadvantages of carbon-based methods for microplastic remediation?
6. How biochar physical properties affect the microplastic removal?
7. How does biochar surface chemistry affect the microplastic removal?
8. How does microplastic particle size affect its retention in different adsorptive materials?

## References

Abit, S.M., Bolster, C.H., Cantrell, K.B., Flores, J.Q., Walker, S.L., 2014. Transport of Escherichia coli, Salmonella typhimurium, and microspheres in biochar-amended soils with different textures. J. Environ. Qual. 43 (1), 371–388. https://doi.org/10.2134/jeq2013.06.0236.

Al-Azzawi, M.S.M., Kefer, S., Weißer, J., Reichel, J., Schwaller, C., Glas, K., et al., 2020. Validation of sample preparation methods for microplastic analysis in wastewater matrices—reproducibility and standardization. Water (Basel) 12 (9). https://doi.org/10.3390/w12092445.

Alimi, O.S., Farner Budarz, J., Hernandez, L.M., Tufenkji, N., 2018. Microplastics and nanoplastics in aquatic environments: aggregation, deposition, and enhanced contaminant transport. Environ. Sci. Technol. 52 (4), 1704–1724. https://doi.org/10.1021/acs.est.7b05559.

Araujo, C.F., Nolasco, M.M., Ribeiro, A.M.P., Ribeiro-Claro, P.J.A., 2018. Identification of microplastics using Raman spectroscopy: latest developments and future prospects. Water Res. (Oxford) 142, 426–440. https://doi.org/10.1016/j.watres.2018.05.060.

Ballent, A., Corcoran, P.L., Madden, O., Helm, P.A., Longstaffe, F.J., 2016. Sources and sinks of microplastics in Canadian Lake Ontario nearshore, tributary and beach sediments. Mar. Pollut. Bull. 110 (1), 383–395. https://doi.org/10.1016/j.marpolbul.2016.06.037.

Baresel, C., Harding, M., Fång, J., 2019. Ultrafiltration/granulated active carbon-biofilter: efficient removal of a broad range of micropollutants. Appl. Sci. 9 (4). https://doi.org/10.3390/app9040710. p. 710. MDPI AG.

Braun, U., Jekel, M., Gerdts, G., Ivleva, N.P., Reiber, J., 2018. Microplastics analytics sampling, preparation and detection methods. Discussion paper published in the course of the research focus "Plastics in the Environment – Sources • Sinks • Solutions. Ecologic Institute, Berlin. https://www.ecologic.eu/sites/files/publication/2018/discussion_paper_mp_analytics_en.pdf.

Browne, M.A., Galloway, T., Thompson, R., 2007. Microplastic—an emerging contaminant of potential concern? Integr. Environ. Assess. Manag. 3 (4), 559–561. https://doi.org/10.1002/ieam.5630030412.

Browne, M.A., Crump, P., Niven, S.J., Teuten, E., Tonkin, A., Galloway, T., Thompson, R., 2011. Accumulation of microplastic on shorelines worldwide: sources and sinks. Environ. Sci. Technol. 45 (21), 9175–9179. https://doi.org/10.1021/es201811s.

Carney Almroth, B.M., Åström, L., Roslund, S., Petersson, H., Johansson, M., Persson, N.-K., 2018. Quantifying shedding of synthetic fibers from textiles; a source of microplastics released into the environment. Environ. Sci. Pollut. Res. 25 (2), 1191–1199. https://doi.org/10.1007/s11356-017-0528-7.

Carr, S.A., Liu, J., Tesoro, A.G., 2016. Transport and fate of microplastic particles in wastewater treatment plants. Water Res. 91, 174–182. https://doi.org/10.1016/j.watres.2016.01.002.

Celina, M.C., 2013. Review of polymer oxidation and its relationship with materials performance and lifetime prediction. Polym. Degrad. Stab. 98 (12), 2419–2429. https://doi.org/10.1016/j.polymdegradstab.2013.06.024.

Chang, M., 2015. Reducing microplastics from facial exfoliating cleansers in wastewater through treatment versus consumer product decisions. Mar. Pollut. Bull. 101 (1), 330–333. https://doi.org/10.1016/j.marpolbul.2015.10.074.

Corradini, F., Meza, P., Eguiluz, R., Casado, F., Huerta-Lwanga, E., Geissen, V., 2019. Evidence of microplastic accumulation in agricultural soils from sewage sludge disposal. Sci. Total Environ. 671, 411–420. https://doi.org/10.1016/j.scitotenv.2019.03.368.

Desforges, J.-P.W., Galbraith, M., Dangerfield, N., Ross, P.S., 2014. Widespread distribution of microplastics in subsurface seawater in the NE Pacific Ocean. Mar. Pollut. Bull. 79 (1–2), 94–99. https://doi.org/10.1016/j.marpolbul.2013.12.035.

Dias, J.M., Alvim-Ferraz, M.C.M., Almeida, M.F., Rivera-Utrilla, J., Sánchez-Polo, M., 2007. Waste materials for activated carbon preparation and its use in aqueous-phase treatment: a review. J. Environ. Manag. 85 (4), 833–846. https://doi.org/10.1016/J.JENVMAN.2007.07.031.

Ding, N., An, D., Yin, X., Sun, Y., 2020. Detection and evaluation of microbeads and other microplastics in wastewater treatment plant samples. Environ. Sci. Pollut. Res. 27 (13), 15878–15887. https://doi.org/10.1007/s11356-020-08127-2.

Enfrin, M., Dumée, L.F., Lee, J., 2019. Nano/microplastics in water and wastewater treatment processes – Origin, impact and potential solutions. Water Res. 161, 621–638. Elsevier Ltd https://doi.org/10.1016/j.watres.2019.06.049.

Enfrin, M., Lee, J., Gibert, Y., Basheer, F., Kong, L., Dumée, L.F., 2020. Release of hazardous nanoplastic contaminants due to microplastics fragmentation under shear stress forces. J. Hazard. Mater. 384. https://doi.org/10.1016/j.jhazmat.2019.121393, 121393.

Estahbanati, S., Fahrenfeld, N.L., 2016. Influence of wastewater treatment plant discharges on microplastic concentrations in surface water. Chemosphere 162, 277–284. https://doi.org/10.1016/j.chemosphere.2016.07.083.

Faure, F., Demars, C., Wieser, O., Kunz, M., De Alencastro, L.F., 2015. Plastic pollution in Swiss surface waters: nature and concentrations, interaction with pollutants. Environ. Chem. 12 (5), 582–591. https://doi.org/10.1071/EN14218.

Fischer, E.K., Paglialonga, L., Czech, E., Tamminga, M., 2016. Microplastic pollution in lakes and lake shoreline sediments—a case study on Lake Bolsena and Lake Chiusi (central Italy). Environ. Pollut. 213, 648–657. https://doi.org/10.1016/j.envpol.2016.03.012.

Gies, E.A., LeNoble, J.L., Noël, M., Etemadifar, A., Bishay, F., Hall, E.R., Ross, P.S., 2018. Retention of microplastics in a major secondary wastewater treatment plant in Vancouver, Canada. Mar. Pollut. Bull. 133, 553–561. https://doi.org/10.1016/j.marpolbul.2018.06.006.

González-García, P., 2018. Activated carbon from lignocellulosics precursors: a review of the synthesis methods, characterization techniques and applications. Renew. Sust. Energ. Rev. 82, 1393–1414. https://doi.org/10.1016/J.RSER.2017.04.117.

Hidalgo-Ruz, V., Gutow, L., Thompson, R.C., Thiel, M., 2012. Microplastics in the marine environment: A review of the methods used for identification and quantification. Environ. Sci. Technol. 46 (6), 3060–3075. https://doi.org/10.1021/es2031505.

Hoellein, T.J., McCormick, A.R., Hittie, J., London, M.G., Scott, J.W., Kelly, J.J., 2017. Longitudinal patterns of microplastic concentration and bacterial assemblages in surface and benthic habitats of an urban river. Freshw. Sci. 36 (3).

Huerta Lwanga, E., Gertsen, H., Gooren, H., Peters, P., Salánki, T., van der Ploeg, M., Besseling, E., Koelmans, A.A., Geissen, V., 2016. Microplastics in the terrestrial ecosystem: implications for Lumbricus terrestris (Oligochaeta, Lumbricidae). Environ. Sci. Technol. 50 (5), 2685–2691. https://doi.org/10.1021/acs.est.5b05478.

Hüffer, T., Praetorius, A., Wagner, S., von der Kammer, F., Hofmann, T., 2017. Microplastic exposure assessment in aquatic environments: learning from similarities and differences to engineered nanoparticles. Environ. Sci. Technol. 51 (5), 2499–2507. https://doi.org/10.1021/acs.est.6b04054.

Huppertsberg, S., Knepper, T.P., 2018. Instrumental analysis of microplastics—benefits and challenges. Anal. Bioanal. Chem. 410, 6343–6352. Retrieved from https://link.springer.com/article/10.1007/s00216-018-1210-8.

Jaikumar, G., Baas, J., Brun, N.R., Vijver, M.G., Bosker, T., 2018. Acute sensitivity of three Cladoceran species to different types of microplastics in combination with thermal stress. Environ. Pollut. 239, 733–740. https://doi.org/10.1016/j.envpol.2018.04.069.

Kalmykova, Y., Moona, N., Strömvall, A.-M., Björklund, K., 2014. Sorption and degradation of petroleum hydrocarbons, polycyclic aromatic hydrocarbons, alkylphenols, bisphenol A and phthalates in landfill leachate using sand, activated carbon and peat filters. Water Res. 56, 246–257. https://doi.org/10.1016/j.watres.2014.03.011.

Kang, H.-J., Park, H.-J., Kwon, O.-K., Lee, W.-S., Jeong, D.-H., Ju, B.-K., Kwon, J.-H., 2018. Occurrence of microplastics in municipal sewage treatment plants: a review. Environ. Health Toxicol. 33 (3). https://doi.org/10.5620/eht.e2018013, e2018013.

Koelmans, A.A., Mohamed Nor, N.H., Hermsen, E., Kooi, M., Mintenig, S.M., De France, J., 2019. Microplastics in freshwaters and drinking water: critical review and assessment of data quality. Water Res. 155, 410–422. Elsevier Ltd https://doi.org/10.1016/j.watres.2019.02.054.

Kosuth, M., Mason, S.A., Wattenberg, E., 2018. Anthropogenic contamination of tap water, beer, and sea salt. PLoS One 13 (4). https://doi.org/10.1371/journal.pone.0194970.

Lares, M., Ncibi, M.C., Sillanpää, M., Sillanpää, M., 2018. Occurrence, identification and removal of microplastic particles and fibers in conventional activated sludge process and advanced MBR technology. Water Res. 133, 236–246. https://doi.org/10.1016/J.WATRES.2018.01.049.

Leslie, H.A., Brandsma, S.H., van Velzen, M.J.M., Vethaak, A.D., 2017. Microplastics en route: field measurements in the Dutch river delta and Amsterdam canals, wastewater treatment plants, north sea sediments and biota. Environ. Int. 101, 133–142. https://doi.org/10.1016/j.envint.2017.01.018.

Li, J., Liu, H., Paul Chen, J., 2018. Microplastics in freshwater systems: a review on occurrence, environmental effects, and methods for microplastics detection., https://doi.org/10.1016/j.watres.2017.12.056.

Liu, X., Yuan, W., Di, M., Li, Z., Wang, J., 2019. Transfer and fate of microplastics during the conventional activated sludge process in one wastewater treatment plant of China. Chem. Eng. J. 362, 176–182. https://doi.org/10.1016/j.cej.2019.01.033.

Liu, M., Lu, S., Chen, Y., Cao, C., Bigalke, M., He, D., 2020. Analytical methods for microplastics in environments: current advances and challenges. In: He, D., Luo, Y. (Eds.), Microplastics in Terrestrial Environments—Emerging Contaminants and Major Challenges. The Handbook of Environmental Chemistry. 95. Springer, Berlin, Heidelberg, pp. 3–24.

Lobelle, D., Cunliffe, M., 2011. Early microbial biofilm formation on marine plastic debris. Mar. Pollut. Bull. 62 (1), 197–200. https://doi.org/10.1016/j.marpolbul.2010.10.013.

Maes, T., Jessop, R., Wellner, N., Haupt, K., Mayes, A.G., 2017. A rapid-screening approach to detect and quantify microplastics based on fluorescent tagging with Nile Red. Nat. Publ. Group. https://doi.org/10.1038/srep44501.

Magnusson, K., Noren, F., 2014. Screening of microplastic particles in and downstream a wastewater treatment plant. IVL Swedish Environmental Research Institute Report B, 2208, pp. 1–31.

Mahon, A.M., O'Connell, B., Healy, M.G., O'Connor, I., Officer, R., Nash, R., Morrison, L., 2017. Microplastics in sewage sludge: effects of treatment. Environ. Sci. Technol. 51 (2), 810–818. https://doi.org/10.1021/acs.est.6b04048.

Mason, S.A., Garneau, D., Sutton, R., Chu, Y., Ehmann, K., Barnes, J., Fink, P., Papazissimos, D., Rogers, D.L., 2016. Microplastic pollution is widely detected in US municipal wastewater treatment plant effluent. Environ. Pollut. 218, 1045–1054. https://doi.org/10.1016/j.envpol.2016.08.056.

McCormick, A., Hoellein, T.J., Mason, S.A., Schluep, J., Kelly, J.J., 2014. Microplastic is an abundant and distinct microbial habitat in an urban river. Environ. Sci. Technol. 48 (20), 11863–11871. https://doi.org/10.1021/es503610r.

McCormick, A.R., Hoellein, T.J., London, M.G., Hittie, J., Scott, J.W., Kelly, J.J., 2016. Microplastic in surface waters of urban rivers: concentration, sources, and associated bacterial assemblages. Ecosphere 7 (11).

Mintenig, S.M., Löder, M.G.J., Primpke, S., Gerdts, G., 2019. Low numbers of microplastics detected in drinking water from ground water sources. Sci. Total Environ. 648, 631–635. https://doi.org/10.1016/j.scitotenv.2018.08.178.

Mitropoulou, P.N., Syngouna, V.I., Chrysikopoulos, C.V., 2013. Transport of colloids in unsaturated packed columns: role of ionic strength and sand grain size. Chem. Eng. J. 232, 237–248. https://doi.org/10.1016/j.cej.2013.07.093.

Napper, I.E., Bakir, A., Rowland, S.J., Thompson, R.C., 2015. Characterisation, quantity and sorptive properties of microplastics extracted from cosmetics. Mar. Pollut. Bull. 99 (1–2), 178–185. https://doi.org/10.1016/j.marpolbul.2015.07.029.

Nguyen, B., Claveau-Mallet, D., Hernandez, L.M., Xu, E.G., Farner, J.M., Nathalie Tufenkji, N., 2019. Separation and analysis of microplastics and nanoplastics in complex environmental samples. ACC Chem. Res. 52 (4), 858–866. https://doi.org/10.1021/acs.accounts.8b00602.

Nizzetto, L., Langaas, S., Futter, M., 2016. Pollution: do microplastics spill on to farm soils? Nature 537 (7621), 488. https://doi.org/10.1038/537488b.

Novotna, K., Cermakova, L., Pivokonska, L., Cajthaml, T., Pivokonsky, M., 2019. Microplastics in drinking water treatment—current knowledge and research needs. Sci. Total Environ. 667, 730–740. https://doi.org/10.1016/j.scitotenv.2019.02.431.

Oßmann, B.E., Sarau, G., Holtmannspötter, H., Pischetsrieder, M., Christiansen, S.-H., Dicke, W., 2018. Small-sized microplastics and pigmented particles in bottled mineral water. Water Res. 141, 307–316. https://doi.org/10.1016/j.watres.2018.05.027.

Picó, Y., Barcelo, D., 2019. Analysis and prevention of microplastics pollution in water: current perspectives and future directions. ACS Omega 4, 6709–6719. Retrieved from https://pubs.acs.org/doi/pdf/10.1021/acsomega.9b00222.

Pivokonsky, M., Cermakova, L., Novotna, K., Peer, P., Cajthaml, T., Janda, V., 2018. Occurrence of microplastics in raw and treated drinking water. Sci. Total Environ. 643, 1644–1651. https://doi.org/10.1016/j.scitotenv.2018.08.102.

Pivokonský, M., Pivokonská, L., Novotná, K., Čermáková, L., Klimtová, M., 2020. Occurrence and fate of microplastics at two different drinking water treatment plants within a river catchment. Sci. Total Environ. 741. https://doi.org/10.1016/j.scitotenv.2020.140236, 140236.

Rios Mendoza, R.L.M., Balcer, M., 2019. Microplastics in freshwater environments: a review of quantification assessment. TrAC, Trends Anal. Chem. (Regular Ed.) 113, 402–408. https://doi.org/10.1016/j.trac.2018.10.020.

Rivera-Utrilla, J., Sánchez-Polo, M., Gómez-Serrano, V., Álvarez, P.M., Alvim-Ferraz, M.C.M., Dias, J.M., 2011. Activated carbon modifications to enhance its water treatment applications. An overview. J. Hazard. Mater. 187 (1–3), 1–23. https://doi.org/10.1016/j.jhazmat.2011.01.033.

Rochman, C.M., Kurobe, T., Flores, I., Teh, S.J., 2014. Early warning signs of endocrine disruption in adult fish from the ingestion of polyethylene with and without sorbed chemical pollutants from the marine environment. Sci. Total Environ. 493, 656–661. https://doi.org/10.1016/j.scitotenv.2014.06.051.

Rodrigues, M.O., Abrantes, N., Gonçalves, F.J.M., Nogueira, H., Marques, J.C., Gonçalves, A.M.M., 2018. Spatial and temporal distribution of microplastics in water and sediments of a freshwater system (Antuã River, Portugal). Sci. Total Environ. 633, 1549–1559. https://doi.org/10.1016/j.scitotenv.2018.03.233.

Scheurer, M., Bigalke, M., 2018. Microplastics in Swiss floodplain soils. Environ. Sci. Technol. 52 (6), 3591–3598. https://doi.org/10.1021/acs.est.7b06003.

Schymanski, D., Goldbeck, C., Humpf, H.U., Fürst, P., 2018. Analysis of microplastics in water by micro-Raman spectroscopy: release of plastic particles from different packaging into mineral water. Water Res. 129, 154–162. https://doi.org/10.1016/j.watres.2017.11.011.

Shen, M., Song, B., Zhu, Y., Zeng, G., Zhang, Y., Yang, Y., Wen, X., Chen, M., Yi, H., 2020. Removal of microplastics via drinking water treatment: current knowledge and future directions. Chemosphere 251, 126612. Elsevier Ltd https://doi.org/10.1016/j.chemosphere.2020.126612.

Siipola, V., Tamminen, T., Källi, A., Lahti, R., Romar, H., Rasa, K., Keskinen, R., Hyväluoma, J., Hannula, M., Wikberg, H., 2018. Effects of biomass type, carbonization process, and activation method on the properties of bio-based activated carbons. Bioresources 13 (3), 5976–6002.

Siipola, V., Sorsamäki, L., Koukkari, P., Karlsson, M., Björnsröm, M., 2019. Metsäteollisuuden sivuvirroista valmistetun biohiilen lisäarvo ja liiketoimintapotentiaali-BioCar. VTT research report: VTT-R-01290-19.

Siipola, V., Pflugmacher, S., Romar, H., Wendling, L., Koukkari, P., 2020. Low-cost biochar adsorbents for water purification including microplastics removal. Appl. Sci. 10 (3), 788. https://doi.org/10.3390/app10030788.

Simon, M., van Alst, N., Vollertsen, J., 2018. Quantification of microplastic mass and removal rates at wastewater treatment plants applying focal plane array (FPA)-based Fourier transform infrared (FT-IR) imaging. Water Res. 142, 1–9. https://doi.org/10.1016/j.watres.2018.05.019.

Skouteris, G., Saroj, D., Melidis, P., Hai, F.I., Ouki, S., 2015. The effect of activated carbon addition on membrane bioreactor processes for wastewater treatment and reclamation—a critical review. Bioresour. Technol. 185, 399–410. https://doi.org/10.1016/j.biortech.2015.03.010.

Talvitie, J., Heinonen, M., Pääkkönen, J.-P., Vahtera, E., Mikola, A., Setälä, O., Vahala, R., 2015. Do wastewater treatment plants act as a potential point source of microplastics? Preliminary study in the coastal Gulf of Finland, Baltic Sea. Water Sci. Technol. 72 (9), 1495–1504. https://doi.org/10.2166/wst.2015.360.

Talvitie, J., Mikola, A., Koistinen, A., Setälä, O., 2017a. Solutions to microplastic pollution—removal of microplastics from wastewater effluent with advanced wastewater treatment technologies. Water Res. 123, 401–407. https://doi.org/10.1016/j.watres.2017.07.005.

Talvitie, J., Mikola, A., Setälä, O., Heinonen, M., Koistinen, A., 2017b. How well is microlitter purified from wastewater?—a detailed study on the stepwise removal of microlitter in a tertiary level wastewater treatment plant. Water Res. 109, 164–172. https://doi.org/10.1016/j.watres.2016.11.046.

Tamminga, M., 2017. Nile red staining as a subsidiary method for microplastic quantification: a comparison of three solvents and factors influencing application reliability. SDRP J. Earth Sci. Environ. Stud. 2 (2). https://doi.org/10.25177/JESES.2.2.1.

Tong, M., He, L., Rong, H., Li, M., Kim, H., 2020a. Transport behaviors of plastic particles in saturated quartz sand without and with biochar/$Fe_3O_4$-biochar amendment. Water Res. 169. https://doi.org/10.1016/j.watres.2019.115284, 115284.

Tong, M., Li, T., Li, M., He, L., Ma, Z., 2020b. Cotransport and deposition of biochar with different sized-plastic particles in saturated porous media. Sci. Total Environ. 713. https://doi.org/10.1016/j.scitotenv.2019.136387, 136387.

Wagner, M., Scherer, C., Alvarez-Muñoz, D., Brennholt, N., Bourrain, X., Buchinger, S., Fries, E., Grosbois, C., Klasmeier, J., Marti, T., Rodriguez-Mozaz, S., Urbatzka, R., Vethaak, A.D., Winther-Nielsen, M., Reifferscheid, G., 2014. Microplastics in freshwater ecosystems: what we know and what we need to know. Environ. Sci. Eur. 26 (1), 12. https://doi.org/10.1186/s12302-014-0012-7.

Wang, D., Zhang, W., Zhou, D., 2013. Antagonistic effects of humic acid and iron oxyhydroxide grain-coating on biochar nanoparticle transport in saturated sand. Environ. Sci. Technol. 47 (10), 5154–5161. https://doi.org/10.1021/es305337r.

Wang, Z., Lin, T., Chen, W., 2020a. Occurrence and removal of microplastics in an advanced drinking water treatment plant (ADWTP). Sci. Total Environ. 700. https://doi.org/10.1016/j.scitotenv.2019.134520, 134520.

Wang, Z., Sedighi, M., Lea-Langton, A., 2020b. Filtration of microplastic spheres by biochar: removal efficiency and immobilisation mechanisms. Water Res. 184. https://doi.org/10.1016/j.watres.2020.116165, 116165.

Wright, S.L., Thompson, R.C., Galloway, T.S., 2013. The physical impacts of microplastics on marine organisms: a review. Environ. Pollu. (Barking, Essex : 1987) 178, 483–492. https://doi.org/10.1016/j.envpol.2013.02.031.

Yan, X., Yang, X., Tang, Z., Fu, J., Chen, F., Zhao, Y., Ruan, L., Yang, Y., 2020. Downward transport of naturally-aged light microplastics in natural loamy sand and the implication to the dissemination of antibiotic resistance genes. Environ. Pollut. 262. https://doi.org/10.1016/j.envpol.2020.114270, 114270.

Yang, W., Bradford, S.A., Wang, Y., Sharma, P., Shang, J., Li, B., 2019. Transport of biochar colloids in saturated porous media in the presence of humic substances or proteins. Environ. Pollut. 246, 855–863. https://doi.org/10.1016/j.envpol.2018.12.075.

Zhang, X., Chen, J., Li, J., 2020. The removal of microplastics in the wastewater treatment process and their potential impact on anaerobic digestion due to pollutants association. Chemosphere 251, 126360. Elsevier Ltd https://doi.org/10.1016/j.chemosphere.2020.126360.

Ziajahromi, S., Neale, P.A., Rintoul, L., Leusch, F.D.L., 2017. Wastewater treatment plants as a pathway for microplastics: development of a new approach to sample wastewater-based microplastics. Water Res. 112, 93–99. https://doi.org/10.1016/j.watres.2017.01.042.

# Sorptive removal of pharmaceuticals using sustainable biochars

**Manvendra Patel[a], Abhishek Kumar Chaubey[a], Chanaka Navarathna[b], Todd E. Mlsna[b], Charles U. Pittman, Jr.[b], and Dinesh Mohan[a]**

[a]School of Environmental Sciences, Jawaharlal Nehru University, New Delhi, India. [b]Department of Chemistry, Mississippi State University, Mississippi State, MS, United States

## 1 Introduction

Rising medical needs have intensified the release of pharmaceuticals and their residues in the environment (Lapworth et al., 2012; Patel et al., 2019; Wilkinson et al., 2017). Pharmaceuticals were reported in aquatic water bodies of 71 countries (aus der Beek et al., 2016). Pharmaceuticals were commonly detected in ng/L to µg/L concentration ranges (Larsson, 2014; Mestre et al., 2007; Richardson and Ternes, 2011). However, manufacturing industry units release pharmaceuticals in their effluents in up to mg/L concentration levels (Kleywegt et al., 2019; Larsson et al., 2007). Pharmaceutical's universal presence causes both environmental toxicity and human health concerns (Ebele et al., 2017; Patel et al., 2019). The inability of wastewater treatment plants to completely remove pharmaceuticals raises more concerns (Stackelberg et al., 2004; Ternes et al., 2002). Therefore, aqueous pharmaceutical removal technologies are becoming necessary.

Advanced oxidation processes, nanofiltration, ultrafiltration, membrane filtrations, reverse osmosis, photocatalysis, and adsorption are important pharmaceutical removal processes that have been tested so far (Patel et al., 2019). Easy operability and scalability for both small- and large-scale operations, and no additional chemical requirements make adsorption a successful option for water treatment (Homem and Santos, 2011; Kumar et al., 2019a; Rivera-Utrilla et al., 2013). Activated carbon, biological sorbents, biochars, mineral oxides, polymer resins, nanomaterials, and metal-organic frameworks are adsorbents commonly used for aqueous pharmaceutical removal (de Andrade et al., 2018; Rivera-Utrilla et al., 2013; Tahar et al., 2013). Activated carbons are the most commonly used and tested adsorbent for contaminant removal (de Andrade et al., 2018). However, high costs and regeneration difficulties make it unfeasible for developing and undeveloped economies (Crini, 2006; de Andrade et al., 2018). Biochars produced from locally available biomass serve as a low-cost sustainable option (Mohan et al., 2014). Several studies have applied biochars for pharmaceutical removal (Choudhary et al., 2020; Essandoh et al., 2015; Peiris et al., 2017; Rajapaksha et al., 2014). However, studies summarizing pharmaceutical sorption onto biochars still lack the needed understanding of several aspects versus sorption on activated carbons (Tong et al., 2019). Thus, this review summarizes various biochars for pharmaceutical removal. Biochar preparation processes can alter biochar surface properties (Ahmad et al., 2012). Thus, impact of biochar properties on pharmaceutical sorption is also summarized. The role of environmental factors on pharmaceutical sorption has been emphasized. Biochar-pharmaceutical interactions are explained by adsorption mechanisms based on adsorbent and adsorbate physicochemical properties in the bulk solution. Thus, understanding of sorption processes is assisted through attempts to understand the sorption mechanisms.

## 2 Pharmaceuticals: Emerging contaminants

Pharmaceuticals are designed for diagnostic, preventive, and active treatments of human and veterinary ailments. Extensive pharmaceutical usage for human and veterinary purposes has led to their presence in aquatic systems (Patel et al., 2019). Since the discovery of pharmaceuticals in aquatic systems in the 1980s, numerous pharmaceuticals have been detected in aquatic systems around the world (aus der Beek et al., 2016; Patel et al., 2019). Classifying pharmaceuticals as homogeneous compounds like chlorofluorocarbons, polychlorinated biphenyls, or polyaromatic hydrocarbons are impractical; they exist in wide structural variations with different physical, chemical, or biological properties (Taylor and Senac, 2014). Pharmaceuticals are commonly grouped as (i) anti-inflammatories and analgesics (diclofenac, ibuprofen, and paracetamol), (ii) antibiotics (β-lactams, fluoroquinolones, imidazoles, macrolides, penicillins, sulfonamides, and tetracyclines), (iii) antiepileptics (carbamazepine), (iv) antidepressants (benzodiazepines), (v) lipid-lowering agents (fibrates), (vi) antihistamines (famotidine and ranitidine), (vii) β-blockers (metoprolol, atenolol, and propranolol), and (viii) other substances (barbiturates, narcotics, antiseptics, and contrast media) (Bush, 1997; Rivera-Utrilla et al., 2013).

Global antibiotics consumption reached 34.8 billion defined daily doses in 2015, with a staggering 65% increase from 21.1 billion defined daily doses in 2000 (Klein et al., 2018). Global medicine usage reached 4.5 trillion doses/year by 2020. ~4000 active pharmaceutical ingredients are currently available in the global market with an approximate consumption of ~100,000 tons/year (Lindim et al., 2017). Environmental pharmaceutical detection has appeared in hospital effluents; sewage treatment plant effluents and influents; and ground, surface, and drinking water (aus der Beek et al., 2016; Hughes et al., 2013). Seasonal, spatial, and temporal pharmaceutical variations have also been evaluated (aus der Beek et al., 2016; Hughes et al., 2013). Based on the available literature, the concentration of pharmaceutical contamination follows the general order: Industrial effluents > hospital effluents > wastewater treatment plant effluents > surface water > groundwater > drinking water (Patel et al., 2019).

Overall, antibiotics are most frequently detected and are followed by analgesics (Patel et al., 2019). However, widely varied results can be observed. These variations depend upon country, region, area consumption pattern, and manufacturing industry locations (aus der Beek et al., 2016; Hughes et al., 2013). For example, globally the most frequently reported pharmaceuticals are painkillers, but antibiotics have this distinction in Asia. On the basis of concentration, Europe reported the highest painkiller concentrations, whereas Asia reported the highest antibiotic concentrations (Hughes et al., 2013). Several pharmaceutical sources and pathways were reported and summarized as shown in Fig. 11.1 (Patel et al., 2019). Manufacturing industries and human and veterinary usage are the main pharmaceutical sources of effluents and wastes from industries, hospitals, domestic usage, and animals (Fig. 11.1). The presence of environmental pharmaceuticals in low concentrations (ng/L to µg/L) poses both chronic and acute effects on flora, fauna, and microorganisms (Patel et al., 2019). Complex pharmaceutical mixtures can cause serious toxicity, despite concentrations below threshold levels (Patel et al., 2019). Pharmaceuticals in the environment can even affect organisms of higher trophic levels in a similar manner as they are designed to have impacts on humans (Kümmerer, 2009). Thus, aquatic pharmaceuticals' removal becomes a necessity (Patel et al., 2019).

Photocatalysis, oxidation, advanced oxidation, reverse osmosis, nanofiltration, and adsorption were applied for aqueous

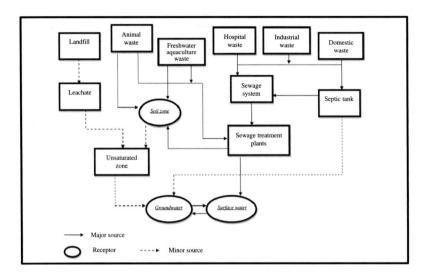

**Fig. 11.1** Sources, pathways, and receptors of pharmaceuticals in the environment. Adapted with permission from Li, W.C., 2014. Occurrence, sources, and fate of pharmaceuticals in aquatic environment and soil. *Environ. Pollut.* 187, 193–201. Copyright @ Elsevier.

pharmaceutical removal (Patel et al., 2019). These techniques can achieve high pharmaceutical removal (up to 90%). Many of these techniques have several disadvantages (Patel et al., 2019). Catalysis and oxidation processes generate toxic by-products and require high-cost inputs and skilled laborers. Similarly, high-cost, skilled laborer requirements and membrane fouling are disadvantages of nanofiltrations and reverse osmosis (Patel et al., 2019). Instead, adsorption overcomes these disadvantages as it does not require high-cost inputs and skilled laborers. Ease of application and adsorbent regeneration also enhance the use of adsorption (Kumar et al., 2019a; Patel et al., 2019). Clays, activated carbons, graphene, carbon nanotubes, and biochars are the adsorbents reported for pharmaceutical removal (de Andrade et al., 2018; Patel et al., 2019).

## 3 Biochar as an adsorbent

So far, a wide range of biochars developed from grasses, wood, agricultural residues, food wastes, manure, vegetable residues, algae, and garden wastes have been applied for removing pharmaceuticals (Table 11.1). High surface area, porosity, surface oxygenated groups, and the presence of inorganic ash content make biochar an excellent adsorbent (Alkurdi et al., 2019; Kumar et al., 2019a, b; Mohan et al., 2007, 2014; Vimal et al., 2019).

Table 11.1 Pharmaceutical adsorption performance of reported adsorbents.

| Adsorbent | Adsorbate | pH | Temp. (°C) | Conc. range (mg/L) | Surface area (m²/g) | Models used | Langmuir sorption capacity (mg/g) | Recovery/ regeneration | Provided sorption mechanism | References |
|---|---|---|---|---|---|---|---|---|---|---|
| Alfalfa biochar 500 | Tetracycline | 5 | – | 10–100 | 31.1 | Langmuir Freundlich Temkin | 372.31 | Yes | H-bonding, electrostatic interactions, surface complexation | (Jang and Kan, 2019b) |
| Alfalfa biochar 300 ($BC_R$) | Tetracycline | 5 | 20 | 10–100 | 0.68 | Langmuir Freundlich Temkin | 30.70 | No | Hydrophobic cation-π and π-π EDA interactions | (Jang and Kan, 2019a) |
| Alfalfa biochar 800 ($BC_{800}$) | | | | | 50.58 | Dubinin-Radushkevich | 55.62 | | | |
| Ball-milled hickory chips biochar | Sulfamethoxazole (water) | 6 | 25 | – | – | Langmuir Freundlich | 100.30 | No | Hydrophobic, π-π interactions, H-bonding, electrostatic interaction | (Huang et al., 2020) |
| | Sulfamethoxazole (waste water) | 7.6 | | | | | 24.30 | | | |
| Ball-milled bamboo biochar | Sulfapyridine (water) | 6 | | | | | 57.90 | | | |
| | Sulfapyridine (waste water) | 7.6 | | | | | 53.70 | | | |
| Bamboo biochar | Sulfamethoxazole | 7.0 ± 0.1 | 25 | 1–80 | 24.56 | Langmuir Freundlich | 128.2 | Yes | H-bonding, hydrophobic, and π-π electron donor-acceptor interactions | (Heo et al., 2019) |

*Continued*

Table 11.1 Pharmaceutical adsorption performance of reported adsorbents—cont'd

| Adsorbent | Adsorbate | pH | Temp. (°C) | Conc. range (mg/L) | Surface area (m$^2$/g) | Models used | Langmuir sorption capacity (mg/g) | Recovery/ regeneration | Provided sorption mechanism | References |
|---|---|---|---|---|---|---|---|---|---|---|
| Bamboo charcoal (BC) | Chloramphenicol | – | 25 | 5–100 | <1 | Langmuir Freundlich | | No | Enhanced π-π interaction | (Fan et al., 2010) |
| Bamboo charcoal | Tetracycline Chloramphenicol | 7.0 | 30 | 5–100 | 67.8 | Langmuir Freundlich Dubinin-Radushkevich | 22.7 8.1 | No | π-π electron donor-acceptor (EDA), cation-π bond in conjunction with H-bonding | (Liao et al., 2013) |
| Barley straw biochar | Salicylic acid | 3 | 25 35 45 | 50–250 | 435.5 | Langmuir Freundlich | 189.21 197.09 210.53 | No | Not available | (Ahmed and Hameed, 2018) |
| Bermuda grass biochar | Tetracycline | 5 | | 10–100 | 34.7 | Langmuir Freundlich Temkin | 44.24 | Yes | H-bonding, electrostatic interactions, and surface complexation | (Jang and Kan, 2019b) |
| Beetle kill pine timber biochar | Acetaminophen | – | 20 | 0–10 | – | Langmuir Freundlich | 24.63 | – | – | (Clurman et al., 2020) |
| Chili seeds Chili seed biochar 450 | Ibuprofen | 7 7 | 25 | 50–1000 | 0.19 0.52 | Langmuir Freundlich | 0.496 12.78 | No | Hydrophobic, π-acceptor, and attractive | (Ocampo-Perez et al., 2019) |

| Biochar | Pollutant | pH | Temperature | Concentration range | Isotherm model | $q_{max}$ | Reusability | Mechanism | Reference |
|---|---|---|---|---|---|---|---|---|---|
| Chili seed biochar 550 | | 7 | | 0.24 | | 16.64 | | electrostatic interactions | |
| Chili seed biochar 600 | | 7 | | 0.18 | | 26.13 | | | |
| | | 9 | | 0.18 | | 19.78 | | | |
| | | 11 | | 0.18 | | 18.15 | | | |
| | | 7 | | 0.18 | | 25.69 | | | |
| Cotton gin waste biochar (CG-700) | Sulfapyridine | – | 35 | 50–1000 | Langmuir Freundlich | 1.22 | No | H-bonding, hydrophobic, π-π interactions, and diffusion | (Ndoun et al., 2020) |
| | Docusate | | RT | 2–50 | | 19.68 | | | |
| | Erythromycin | | | 16.33 | | 17.12 | | | |
| Food and garden waste biochar | Tetracycline | 7.0 | RT | 2–100 | Langmuir Freundlich Sips | 15.5 | Yes | H-bonding and π-π EDA interactions | (Hoslett et al., 2020) |
| Garden waste biochar | Diclofenac | 7 | 20 | 0.1–20 | Langmuir Freundlich | 7.25 | No | H-bonding, electrostatic interactions, and π-electrons dispersions | (Li et al., 2019) |
| Garden waste biochar | Trimethoprim | | | 0.1–400 8.89 | Langmuir Freundlich | 2.08 | | | |
| Invasive plant-derived biomass | Sulfamethazine | 5 | 25 | 2.5–50 2.28 | Langmuir Freundlich Temkin | 6.688 | No | Hydrophobic, H-bonding, electrostatic interaction, and π-π interactions | (Rajapaksha et al., 2015) |
| Invasive plant-derived biochar 300 | | 5 | | 0.85 | Dubinin-Radushkevich | 15.656 | | | |
| Invasive plant-derived 700 | | 3 | | 2.31 | | 20.559 | | | |
| | | 5 | | 2.31 | | 18.776 | | | |
| | | 7 | | 2.31 | | 18.591 | | | |
| | | 9 | | 2.31 | | 10.451 | | | |

*Continued*

Table 11.1 Pharmaceutical adsorption performance of reported adsorbents—cont'd

| Adsorbent | Adsorbate | pH | Temp. (°C) | Conc. range (mg/L) | Surface area (m²/g) | Models used | Langmuir sorption capacity (mg/g) | Recovery/ regeneration | Provided sorption mechanism | References |
|---|---|---|---|---|---|---|---|---|---|---|
| Macroalgae | Diclofenac Trimethoprim | 7 | 20 | 0.1–20 0.1–400 | 0.29 | Langmuir Freundlich | 0.022 71.4 | No | H-bonding, electrostatic interactions, and π-electrons dispersions | (Li et al., 2019) |
| Maize straw biochar (300) Maize straw biochar (600) | Sulfamethazine | 6.0 | 25 | 0.5–50 | — — | Langmuir Freundlich | 5.75 4.32 | No | π-π EDA interaction, van der Waals force | (Jia et al., 2018) |
| Oak wood chippings | Diclofenac Trimethoprim | 7 | 20 | 0.1–20 0.1–400 | 1.77 | Langmuir Freundlich | 0.033 8.33 | No | H-bonding, electrostatic interactions, π-electrons dispersions | (Li et al., 2019) |
| Peanut shell-derived biochar (800-PSB) | Naproxen | 7.0 | 25 40 55 | 5–1000 | 571 | Langmuir Freundlich Redlich-Peterson | 105 96.1 88.7 | Yes | Pore filling, H-bonding, n-π and π-π interactions, electrostatic attraction, and van der Waals force | (Tomul et al., 2020) |
| Peanut shell-derived biochar (190–800-PSB) | Naproxen | 7.0 | 25 40 55 | 5–1000 | 496 | Langmuir Freundlich Redlich-Peterson | 215 174 130 | | | |
| Peanut shell-derived biochar (800–800-PSB) | Naproxen | 7.0 | 25 40 55 | 5–1000 | 596 | Langmuir Freundlich Redlich-Peterson | 324 287 215 | | | |

| Biochar | Pollutant | pH | Temp (°C) | Conc. range | $q_{max}$ | Isotherm models | Value | Reusability | Mechanism | Reference |
|---|---|---|---|---|---|---|---|---|---|---|
| Pig manure biochar | Diclofenac | 6.5 | 25 | 0.1–10 | 43.5 | Langmuir Freundlich Temkin Dubinin-Radushkevich | 12.5 | No | H-bonding, van der Waals force, and electrostatic interactions | (Lonappan et al., 2018) |
| Pine sawdust (PS) biochar | Daptomycin | 4 | 25 | 0.5–100 | 602.066 | Langmuir Freundlich Temkin Dubinin-Radushkevich | 55.56 | No | Physisorption and multilayer adsorption | (Ai et al., 2019) |
| Pinewood biochar | Diclofenac | 6.5 | 25 | 0.1–10 | 13.3 | Langmuir Freundlich Temkin Dubinin-Radushkevich | 0.5263 | No | H-bonding, van der Waals force, and electrostatic interactions | (Lonappan et al., 2018) |
| Pinewood biochar | Salicylic acid | 2.5 | 25<br>35<br>45 | 25–100 | 1.35 | Langmuir Freundlich | 7.56<br>16.84<br>22.70 | Yes | H-bonding, van der Waals force, and π-π interactions | (Essandoh et al., 2015) |
| | Ibuprofen | 3.0 | 35 | | | | 10.74 | | | |
| Pomelo peel spherical biochar | Paracetamol | 7.0 | 25 | 0–700 | 1292 | Langmuir Freundlich Redlich-Peterson Dubinin-Radushkevich | 286 | Yes | H-bonding and n-π interactions Pore filling | (Tran et al., 2020) |
| Pomelo peel nonspherical biochar | | | | | 1033 | | 147 | | | |
| Potato stems and leaves biochar | Ciprofloxacin | — | 15<br>25<br>35 | 2–16 | 0.99 | Langmuir Freundlich | 10.13<br>6.27<br>8.48 | No | Hydrophobic, H-bonding, electrostatic interaction, and π-π interactions | (Li et al., 2018) |
| Rice husk biochar | Tetracycline | — | 30 | | 34.4 | Langmuir Freundlich | 16.95 | No | π-π interaction | (Liu et al., 2012) |

*Continued*

Table 11.1 Pharmaceutical adsorption performance of reported adsorbents—cont'd

| Adsorbent | Adsorbate | pH | Temp. (°C) | Conc. range (mg/L) | Surface area (m²/g) | Models used | Langmuir sorption capacity (mg/g) | Recovery/ regeneration | Provided sorption mechanism | References |
|---|---|---|---|---|---|---|---|---|---|---|
| Rice straw biochar 300 | Tetracycline | – | 25 | 0–40 | 20.202 | Langmuir Freundlich | 4.07 | No | EDA (electron donor-acceptor) π-π interactions and electrostatic interactions | (Fan et al., 2018) |
| Rice straw biochar 500 | | | 25 | | 50.107 | | 20.37 | | | |
| Rice straw biochar 700 | | | 15 | | 288.341 | | 26.48 | | | |
| | | | 25 | | | | 34.60 | | | |
| | | | 35 | | | | 50.72 | | | |
| Rice straw biochar 700 | Doxycycline | 6 | 25 | 5–60 | 20.55 | Langmuir Freundlich Temkin BET | 170.36 | No | π-π EDA interaction and H-bonding | (Zeng et al., 2018) |
| | | | 35 | | | | 207.90 | | | |
| | | | 45 | | | | 432.90 | | | |
| | Ciprofloxacin | | 25 | | | | 48.80 | | | |
| | | | 35 | | | | 76.69 | | | |
| | | | 45 | | | | 131.58 | | | |
| Spirulina biochar 350 | Tetracycline | 6 | 20 | 10–100 | 0.31 | Langmuir Freundlich Temkin | 18 | Yes | Hydrophobic interaction, functional groups, and metal complexation | (Choi et al., 2020) |
| Spirulina biochar 550 | | | | | 1.55 | | 80 | | | |
| Spirulina biochar 750 | | | | | 2.63 | | 147.9 | | | |
| Tea waste biomass (BM) | Sulfamethazine | 5.0 | 25 | 0–50 | 2.23 | Langmuir Freundlich | 0.537 | No | π-π electron donor-acceptor, cation-π interaction. | (Rajapaksha et al., 2014) |
| Tea waste biochars (TWBC-300) | | 5.0 | | | 2.28 | | 2.43 | | | |

| Adsorbent | Adsorbate | Temp | Conc. range | pH | Capacity | Isotherm | Value | Reusability | Mechanism | Reference |
|---|---|---|---|---|---|---|---|---|---|---|
| Tea waste biochars (TWBC-300N) | | | | 5.0 | 0.90 | | 2.79 | | and cation exchange | |
| Tea waste biochars (TWBC-300S) | | | | 5.0 | 1.46 | | 1.88 | | | |
| Tea waste biochars (TWBC-700) | | | | 3.0<br>5.0<br>7.0<br>9.0 | 342.22 | | 7.12<br>6.58<br>8.25<br>2.70 | | | |
| Tea waste biochars (TWBC-700N) | | | | 3.0<br>5.0<br>7.0<br>9.0 | 421.31 | | 30.06<br>26.76<br>26.49<br>16.72 | | | |
| Tea waste biochars (TWBC-700S) | | | | 3.0<br>5.0<br>7.0<br>9.0 | 576.09 | | 33.81<br>30.95<br>27.47<br>24.62 | | | |
| Water hyacinth biochar (WHC 350) | Caffeine<br>Ciprofloxacin | 25 | 2–10 | — | — | Langmuir<br>Freundlich | 2.488<br>2.717 | No | Hydrophobic interactions and other noncoulombic interactions | (Ngeno et al., 2016) |
| Wheat straw biochar (300)<br>Wheat straw biochar (600) | Sulfamethazine | 25 | 0.5–50 | 6.0 | — | Langmuir<br>Freundlich | 1.39<br>3.06 | No | $\pi$-$\pi$ EDA interaction and van der Waals force | (Jia et al., 2018) |

Biochar modifications made for specific applications also enhanced biochar's applicability for pharmaceutical removal (Ahmed et al., 2017a,b; Chakraborty et al., 2018; Huang et al., 2020). Compared to activated carbons, biochar production requires less energy and chemicals (Hagemann et al., 2018; Thompson et al., 2016). Activated carbon and biochar also share similarities in adsorption functionalities, e.g., $\pi$-$\pi$ electron donor-acceptor sites within carbon/biochar's polycyclic aromatic network as a major adsorption contributor (Zhu and Pignatello, 2005). Much lower average production costs (0.2–0.5 US$/Kg) as compared to activated carbons (up to 5 US$/Kg) often make biochar suitable for activated carbon replacement (Ahmed et al., 2015). Thus, biochar can be a low-cost environmental-friendly alternative for more costly and chemical-intensive adsorbents such as activated carbons (de Andrade et al., 2018; Mohan et al., 2014).

## 4 Pharmaceutical sorption onto biochar

Several studies have employed biochar for pharmaceutical sorption (Ahmed and Hameed, 2018; Ahmed et al., 2017a; Choi et al., 2020; Essandoh et al., 2015; Krasucka et al., 2020; Li et al., 2018; Lian et al., 2015; Ndoun et al., 2020; Ngeno et al., 2016; Patel et al., 2021; Xie et al., 2014; Zeng et al., 2018). Biochars produced from activated sludge, alfalfa, hickory wood, bamboo, barley straw, and many other feedstocks (Table 11.1) have been applied for removing pharmaceuticals. Our previous study compiles the removal of sulfonamides and tetracyclines from aquatic environments using biochar-based adsorbents (Peiris et al., 2017). In our previous major review, biochars have shown equivalent or higher sorption capacities than many other adsorbents for several pharmaceuticals (Patel et al., 2019). For example, pinewood biochar (11 mg/g) shows equivalent or higher ibuprofen sorption capacity with SBA-15 (0.41 mg/g), mesoporus activated carbon (17 mg/g), and olive waste activated carbon (13 mg/g) (Patel et al., 2019). Similar results have also been reported for sulfamethazine and tetracycline (Patel et al., 2019).

Organic compound sorption onto adsorbents is highly influenced by both adsorbates' and adsorbents' physicochemical properties. A pharmaceutical's physicochemical properties such as polarity and ionic forms (neutral, cationic, anionic, or zwitterionic) in aqueous solution play an important role in its sorption (Patel et al., 2019; Peiris et al., 2017; Tong et al., 2019). Elemental composition, aromaticity, polarity, surface charge, surface area, pore diameter distribution, and porosity are important biochar properties affecting pharmaceutical sorption (Tong et al., 2019). For example, alteration in banana peel biochar properties with pyrolysis temperature caused massive changes in acetaminophen and ciprofloxacin sorption (Patel et al., 2021). Direct relations exist between Brunauer-Emmett-Teller (BET) surface area, total pore volume, % carbon, ash content, and C/N ratio for ciprofloxacin and acetaminophen sorption in banana peel biochars (prepared at 450°C, 550°C, 650°C, and 750°C) (Patel et al., 2021). Aqueous pH, temperature, hardness, salinity, ionic strength, dissolved ions, and organic matter can also affect pharmaceutical sorption. For example, increases in ionic strength, divalent cations ($Ca^{2+}$, $Mg^{2+}$), and bicarbonate ions increased acetaminophen sorption, while these factors decreased ciprofloxacin sorption (Patel et al., 2021). Altogether, these factors can affect both sorption capacity and rate of pharmaceutical sorption by altering pharmaceutical's sorption mechanism (Patel et al., 2019; Peiris et al., 2017). The role of all these parameters is summarized in detail in separate subsections.

## 4.1 Role of biochar's physicochemical properties

Understanding sorption onto biochar requires the knowledge of biochar physicochemical properties. Biochars possess a wide range of physicochemical and surface properties as explained in chapters previously. Surface properties depend upon the precursor biomass and production parameters including pyrolysis technique, residence time, pyrolysis temperature, and atmosphere (Mohan et al., 2014; Tong et al., 2019). Pretreatments of biomass can vary biochar properties. Altogether, these properties play important roles in pharmaceutical sorption. For example, linear pharmaceutical sorption correlations were observed with a rise in biochar surface area and porosity. For example, the surface area of tea residue biochar increases from 2.28 to 342.22 $m^2/g$ as the pyrolysis temperature increases from 300°C to 700°C, and the sulfamethazine sorption capacity also increases from 2.43 to 6.58 mg/g (Rajapaksha et al., 2014).

Biochar produced from different biomass feedstocks under the same conditions has different sorptive capacities. For example, maximum Langmuir capacity for sulfamethoxazole sorption by wheat straw biochar (WBC made at 600°C) and maize straw biochar (MBC made at 600°C) were 3.06 and 4.32 mg/g, respectively (Jia et al., 2018). Different pyrolysis techniques can produce biochars with different properties, thus affecting sorption properties (Mohan et al., 2014). For example, fast pyrolysis biochars have low surface area generally (Essandoh et al., 2015). Pinewood fast pyrolysis biochar ($S_{BET} = 1.35 \, m^2/g$) produced in an auger-fed reactor at 425°C was applied for salicylic acid and ibuprofen removal (Essandoh et al., 2015). When its adsorption per unit of surface area ($m^2/g$) was compared, fast pyrolysis pinewood biochar absorbs up to 100 times more salicylic acid and ibuprofen than other adsorbents (Essandoh et al., 2015). This enhanced sorption capability of fast pyrolysis pine biochar is due to the permeation of adsorbates onto solid somewhat hydrophobic biochar structure (Essandoh et al., 2015; Patel et al., 2019). Fast pyrolysis biochars of this type have a high oxygen content (10%–26%) with plenty of oxygenated functional groups on the surface and below their pore surfaces (Mohan et al., 2014). Thus, diffusion of water and dissolved adsorbates can occur beyond the surface, enhancing the sorptive potential of biochars (Patel et al., 2019). Thus, this biochar has excellent sorptive potential due to its ability to imbibe water and adsorbate into regions of the solid with sufficient oxygen to allow swelling.

Biochar's aromaticity and polarity are represented by H/C and (O+N)/C molar ratios, respectively. Both these parameters are influenced by changes in biomass feedstock and biochar production conditions (Ahmad et al., 2014). Biochar aromaticity increases with an increase in pyrolysis temperature. An inverse relation between biochar's aromaticity and H/C is observed (Chen et al., 2008). Biochar's aromaticity shows a negative correlation with the Freundlich parameter ($n$) of organic contaminant sorption (Chen et al., 2008; Sun et al., 2011). Biochars with lower H/C ratios and higher aromaticity have stronger interactions with sulfamethoxazole (Shimabuku et al., 2016). A previous review assessed 107 organic contaminants and suggested the importance of the surface area and H/C in biochar selection for their sorption (Hale et al., 2016). Biochar ash content (ranging from 1 to 80% depending upon precursor feedstock) is the residual mass that remained after high-temperature combustion (Enders et al., 2012; Shimabuku et al., 2016). Organic contaminant sorption showed a positive linear relationship with biochar production temperature and surface area, while declines in O/C and H/C ratios show an increment in organic contaminant sorption (Hale et al., 2016). Biochar surface charge versus pH results from base ionizing dissociable functional groups including –COOH, –OH, and nitrogen-containing functional groups that can be protonated in acidic media. Surface charge plays an important role in polar and ionic compound sorption (Mukherjee et al., 2011). The amount of these polar functional groups can decline with an increase in biochar pyrolysis temperature (Ahmad et al., 2014). Biochar's polar groups show a higher sorption affinity for polar or ionic pharmaceuticals, while hydrophobic organic compounds exhibit affinity toward aromatic carbon structure (Lian and Xing, 2017). Sulfamethoxazole sorption shows higher affinity with biochars having more polar functional groups (Lian et al., 2014).

Biochar preparations are immensely influenced by such biochar properties as surface area, pore size, aromaticity, polarity, ash content, cellulose, hemicellulose, lignin content, and functional groups. These properties determine an organic contaminant's sorption potential, equilibrium between two phases, adsorption rate, thermodynamic parameters, and possible sorption mechanisms. Thus, assessment of these parameters during sorption studies becomes essential.

## 4.2 Effect of solution pH

Solution pH is a key parameter for both inorganic and organic contaminants' uptake. Alteration of solution pH can significantly affect sorption capacities (Peiris et al., 2017). Change in pH can alter both biochar's surface charge and a pharmaceutical's ionic form in solution (Essandoh et al., 2015; Patel et al., 2019). Depending upon $pH_{pzc}$, a biochar's surface can be positive ($pH < pH_{pzc}$), negative ($pH > pH_{pzc}$), or neutral ($pH = pH_{pzc}$) in an aqueous solution. Similarly, a pharmaceutical's ionic form (cationic, anionic, neutral, or zwitterionic) in aqueous solution is dependent upon its $pK_a$ value (Fig. 11.2). For example, acetaminophen ($pK_a = 9.4$) exists in both neutral and deprotonated (anion) forms (Fig. 11.2A), whereas ciprofloxacin ($pK_{a1} = 6.09$ and $pK_{a2} = 8.64$) exists as protonated (pH below $pK_{a1}$), deprotonated (pH above $pK_{a2}$), and neutral or zwitterionic in a pH between $pK_{a1}$ and $pK_{a2}$ as shown in Fig. 11.2B (Patel et al., 2021). Therefore, pH is a controlling factor for both pharmaceutical sorption capacity and sorption mechanism (Patel et al., 2021).

Most pharmaceuticals show maximum adsorption in the acidic (∼3) to neutral pH range (Table 11.1). Maximum ibuprofen and salicylic acid adsorptions were reported at pH = 3 and pH = 2, respectively (Essandoh et al., 2015). Maximum sulfamethoxazole sorption on bamboo biochar ($pH_{pzc} = 2.5$) is achieved at pH = 3.25. Sorption decreases with both decrease and increase in pH (Ahmed et al., 2017b). The higher sulfamethoxazole sorption at pH = 3.25 is supported by an electron donor-acceptor mechanism (Ahmed et al., 2017b). Sulfamethoxazole sorption on rice straw biochar ($pH_{pzc} = 2$) also follows a similar pattern (Li et al., 2015). However, maximum sulfamethoxazole sorption on alligator flag biochar ($pH_{pzc} = 3.5$) was achieved at pH = 7. Sorption at pH = 7 is favored by the many oxygenated surface functional group contents on alligator flag biochar,

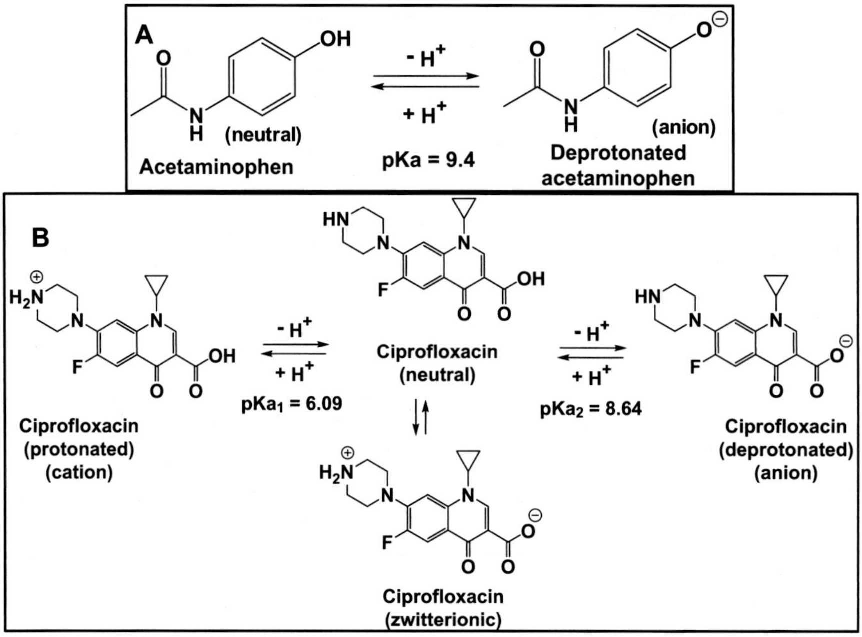

**Fig. 11.2** Acetaminophen (A) and ciprofloxacin (B) speciation showing different ionic forms. Adopted with permission from Patel M, Kumar R, Pittman Jr CU, Mohan D. Ciprofloxacin and acetaminophen sorption onto banana peel biochars: environmental and process parameter influences. Environ. Res. 2021; 201: 111218., Copyright @ Elsevier.

which enhances sulfamethoxazole sorption at neutral pH (Li et al., 2015). Higher tetracycline sorption was achieved at pH < 8 on sludge-based chars (Rivera-Utrilla et al., 2013). This enhanced sorption was attributed to π-π electron donor-acceptor interactions between the sorbent's surface arene rings and zwitterionic tetracycline (Peiris et al., 2017; Rivera-Utrilla et al., 2013). However, the decline in sorption at pH above 8 is due to repulsions between negatively charged tetracycline and sludge-based char which have a $pH_{pzc} = 8.7$–10.3 (Ocampo-Pérez et al., 2012).

## 4.3 Pharmaceutical removal: Isotherm, kinetics, and thermodynamics

Obtaining equilibrium data is a crucial step for understanding adsorption. Dynamic adsorption equilibrium is established when adsorption and desorption rates become equal and the total mass transfer of adsorbate onto and off the adsorbent is zero. Maximum adsorption capacity is also obtained at this point of equilibrium. Isotherm models are used to define maximum adsorption capacity using isotherm experimental data (Patel et al., 2019). Several isotherm models were applied for fitting pharmaceutical isotherm data and to assist with explanations. However, Langmuir and Freundlich isotherm models are most commonly applied for pharmaceutical sorption onto biochar (Table 11.1). Table 11.1 compares the performance of the reported biochars for pharmaceuticals removal. Table 11.1 compared the conditions applied during the sorption experiment as well as surface area of the applied biochars. Table 11.1 also compiled the reported pharmaceutical sorption mechanisms.

Salicylic acid sorption on barley straw biochar better fits the Langmuir isotherm ($R^2$ values >0.99) than the Freundlich model ($R^2$ values 0.97–0.98) (Ahmed and Hameed, 2018). Thus, a monolayer of salicylic acid predominated during the adsorption on the homogeneous barley straw biochar surface. Maximum Langmuir adsorption capacities at 25°C, 35°C, and 45°C were 189, 197, and 211 mg/g, respectively. Thus, an endothermic sorption process occurred (Ahmed and Hameed, 2018). The ranges of pH from acidic (3.0) to neutral (7.0) are the most commonly used pH ranges for pharmaceutical removal using biochar (Table 11.1). Antibiotics followed by NSAIDs are the two most investigated pharmaceutical classes for adsorptive removal (Table 11.1).

Tetracycline has been adsorbed by biochars of alfalfa (Jang and Kan, 2019a,b), bamboo (Liao et al., 2013), bermuda grass (Jang and Kan, 2019a), rice husk (Liu et al., 2012), rice straw (Fan et al., 2018), and spirulina (Choi et al., 2020). The Freundlich model best fits the tetracycline sorption on alfalfa biochar (prepared at 800°C), favoring multilayer tetracycline sorption (Jang and Kan, 2019b). Tetracycline uptake increased with a rise in temperature on rice straw biochar, showing sorption, was endothermic (Fan et al., 2018). The Dubinin-Radushkevich model provided approximate binding energy (Jang and Kan, 2019b). The binding energies for alfalfa biochars prepared at 300°C and 800°C were 8.29 kJ/mol and 9.60 kJ/mol, respectively. Binding energy values >8 kJ/mol suggest the dominance of chemisorption over physisorption (Jang and Kan, 2019b). Similar, endothermic trends were also achieved for salicylic acid sorption on pinewood biochar (Essandoh et al., 2015) and doxycycline and ciprofloxacin sorption on rice straw biochar (Zeng et al., 2018). The Langmuir capacity of salicylic acid sorption on pinewood biochar increases from 7.56 to 22.70 mg/g as the temperature rises from 25°C to 45°C (Essandoh et al., 2015). However, ibuprofen sorption on pinewood biochar is athermic in nature (Essandoh et al., 2015).

Kinetic studies evaluate the adsorption rate. Pseudo-first order and pseudo-second order rate expressions are the most common descriptions found for pharmaceutical sorption on biochars (Table 11.2). Ibuprofen, salicylic acid, sulfonamides, and

**Table 11.2 Adsorption kinetic data for pharmaceutical removal from aqueous solution using biochar.**

| Adsorbent | Adsorbate | pH | Temp (°C) | Conc. (mg/L) | Pseudo-first order rate constant (min$^{-1}$) | Pseudo-second order rate constant | References |
|---|---|---|---|---|---|---|---|
| Alfalfa biochar | Tetracycline | 5 | – | 20 | $1.55 \times 10^{-2}$ | $2.95 \times 10^{-4}$ (g mg$^{-1}$ min$^{-1}$) | (Jang and Kan, 2019b) |
|  |  |  |  | 100 | $5 \times 10^{-3}$ | $2.42 \times 10^{-5}$ (g mg$^{-1}$ min$^{-1}$) |  |
| Alfalfa-derived raw biochar 300 |  | 5 | 20 | 20 | $9.49 \times 10^{-3}$ | $1.31 \times 10^{-3}$ (g mg$^{-1}$ min$^{-1}$) | (Jang and Kan, 2019a) |
|  |  |  |  | 100 | $4.76 \times 10^{-3}$ | $3.31 \times 10^{-4}$ (g mg$^{-1}$ min$^{-1}$) |  |
| Alfalfa-derived raw biochar 800 |  |  |  | 20 | $3.31 \times 10^{-2}$ | $3.04 \times 10^{-3}$ (g mg$^{-1}$ min$^{-1}$) |  |
|  |  |  |  | 100 | $4.83 \times 10^{-3}$ | $1.69 \times 10^{-4}$ (g mg$^{-1}$ min$^{-1}$) |  |
| Ball-milled hickory chips biochar | Sulfamethoxazole (water) | 6 | 25 | 10 | 3.430 (1/h) | 0.136 (g/mg/h) | (Huang et al., 2020) |
| Ball-milled bamboo biochar | Sulfapyridine (water) | 6 |  |  | 1.314 (1/h) | 0.052 (g/mg/h) |  |
| Ball-milled hickory chips biochar | Sulfamethoxazole (waste water) | 7.6 |  |  | 1.947 (1/h) | 0.127 (g/mg/h) |  |
| Ball-milled bamboo biochar | Sulfapyridine (waste water) | 7.6 |  |  | 3.414 (1/h) | 0.125 (g/mg/h) |  |
| Bamboo biochar | Sulfamethoxazole | 7.0 | 25 | 20 | 0.043 | $0.432 \times 10^{-3}$ (g mg$^{-1}$ min$^{-1}$) | (Heo et al., 2019) |
| Banana pseudostem biochar | Furazolidone | 7 | 25 | 100 | $1.43 \times 10^{-2}$ | $4.82 \times 10^{-4}$ (g mg$^{-1}$ min$^{-1}$) | (Gurav et al., 2020) |
| Barley straw biochar | Salicylic acid | 3 | 25 | 50 | 0.9167 (1/h) | 0.0149 (g/mg.h) | (Ahmed and Hameed, 2018) |
|  |  |  |  | 150 | 29.629 (1/h) | 0.0091 (g/mg.h) |  |
|  |  |  |  | 250 | 31.261 (1/h) | 0.0090 (g/mg.h) |  |
| Bermuda grass biochar | Tetracycline | 5 |  | 20 | $1.55 \times 10^{-2}$ | $1.48 \times 10^{-3}$ (g mg$^{-1}$ min$^{-1}$) | (Jang and Kan, 2019b) |
|  |  |  |  | 100 | $5.11 \times 10^{-3}$ | $2.48 \times 10^{-4}$ (g mg$^{-1}$ min$^{-1}$) |  |

| Biochar | Compound | pH | Temp | Conc. | $k_1$ | $k_2$ | References |
|---|---|---|---|---|---|---|---|
| Garden waste biochar | Diclofenac | 7 | 20 | 0.50 | $9.21 \times 10^{-4}$ | $6.99 \times 10^{-6}$ (g μg$^{-1}$ min$^{-1}$) | (Li et al., 2019) |
| | Trimethoprim | | | | $4.61 \times 10^{-4}$ | $3.59 \times 10^{-5}$ (g μg$^{-1}$ min$^{-1}$) | |
| Macroalgae | Diclofenac | 7 | 20 | 0.50 | $7.14 \times 10^{-3}$ | $4.99 \times 10^{-3}$ (g μg$^{-1}$ min$^{-1}$) | |
| Macroalgae | Trimethoprim | 7 | 20 | 0.50 | $1.93 \times 10^{-2}$ | $1.36 \times 10^{-3}$ (g μg$^{-1}$ min$^{-1}$) | |
| Maize-straw-derived biochar | Oxytetracycline | 5.5 | 25 | 460.44 | | $2 \times 10^{4}$ (g mg$^{-1}$ min$^{-1}$) | (Jia et al., 2018) |
| Novel wasted sludge-based biochar | Tetracycline | — | 25 | 500 | 0.0095 | $2.81 \times 10^{-4}$ (g mg$^{-1}$ min$^{-1}$) | (Liu et al., 2019) |
| | | | 35 | | 0.0124 | $2.84 \times 10^{-4}$ (g mg$^{-1}$ min$^{-1}$) | |
| | | | 45 | | 0.0182 | $3.08 \times 10^{-4}$ (g mg$^{-1}$ min$^{-1}$) | |
| Oak wood chippings | Diclofenac | 7 | 20 | 0.50 | $3.69 \times 10^{-3}$ | $2.05 \times 10^{-3}$ (g mg$^{-1}$ min$^{-1}$) | (Li et al., 2019) |
| | Trimethoprim | | | | $1.24 \times 10^{-2}$ | $8.03 \times 10^{-4}$ (g mg$^{-1}$ min$^{-1}$) | |
| Pig manure biochar | Diclofenac | 6.5 | 25 | 0.5 | 0.0175 | $4.40 \times 10^{-5}$ (g μg$^{-1}$ min$^{-1}$) | (Lonappan et al., 2018) |
| Pine sawdust (PS) biochar | Daptomycin | 4 | 25 | 25 | 0.0283 | $16.31 \times 10^{-4}$ (g mg$^{-1}$ min$^{-1}$) | (Ai et al., 2019) |
| Pinewood biochar | Diclofenac | 6.5 | 25 | 0.5 | 0.0133 | $3.82 \times 10^{-5}$ (g μg$^{-1}$ min$^{-1}$) | (Lonappan et al., 2018) |
| Pinewood biochar | Salicylic acid | 2.5 | 25 | 25 | — | 0.36 (g mg$^{-1}$ min$^{-1}$) | (Essandoh et al., 2015) |
| | | | | 50 | | 0.21 (g mg$^{-1}$ min$^{-1}$) | |
| | | | | 100 | | 0.10 (g mg$^{-1}$ min$^{-1}$) | |
| | | | 35 | 25 | | 0.41 (g mg$^{-1}$ min$^{-1}$) | |
| | | | | 50 | | 0.15 (g mg$^{-1}$ min$^{-1}$) | |
| | | | | 100 | | 0.09 (g mg$^{-1}$ min$^{-1}$) | |
| | | | 45 | 25 | | 0.43 (g mg$^{-1}$ min$^{-1}$) | |
| | | | | 50 | | 0.23 (g mg$^{-1}$ min$^{-1}$) | |
| | | | | 100 | | 0.08 (g mg$^{-1}$ min$^{-1}$) | |

Table 11.2 Adsorption kinetic data for pharmaceutical removal from aqueous solution using biochar—cont'd

| Adsorbent | Adsorbate | pH | Temp (°C) | Conc. (mg/L) | Pseudo-first order rate constant (min$^{-1}$) | Pseudo-second order rate constant | References |
|---|---|---|---|---|---|---|---|
| | Ibuprofen | 3 | 25 | 25 | | 0.24 (g mg$^{-1}$ min$^{-1}$) | |
| | | | | 50 | | 0.07 (g mg$^{-1}$ min$^{-1}$) | |
| | | | | 100 | | 0.09 (g mg$^{-1}$ min$^{-1}$) | |
| | | | 35 | 25 | | 0.47 (g mg$^{-1}$ min$^{-1}$) | |
| | | | | 50 | | 0.11 (g mg$^{-1}$ min$^{-1}$) | |
| | | | | 100 | | 0.18 (g mg$^{-1}$ min$^{-1}$) | |
| | | | 45 | 25 | | 0.33 (g mg$^{-1}$ min$^{-1}$) | |
| | | | | 50 | | 0.06 (g mg$^{-1}$ min$^{-1}$) | |
| | | | | 100 | | 0.06 (g mg$^{-1}$ min$^{-1}$) | |
| Potato stems and leaves biochar | Ciprofloxacin | | 25 | 10 | 0.06 (/h) | 0.79 g·(mg·min)$^{-1}$ | (Li et al., 2018) |
| Rice husk biochar | Tetracycline | | 30 | | 0.0002 | 5.42 × 10$^{-4}$ (g mg$^{-1}$ min$^{-1}$) | (Liu et al., 2012) |
| Rice husk biochar 300 | Levofloxacin | 8 | 30 | 40 | 0.0108 | 0.0252 (g mg$^{-1}$ min$^{-1}$) | (Yi et al., 2016) |
| Rice husk biochar 600 | | 8 | 30 | 75 | 0.00599 | 0.00451 (g mg$^{-1}$ min$^{-1}$) | |
| Rice straw biochar 500 | Tetracycline | | 25 | 50 | 0.07407 | 0.00596 (g mg$^{-1}$ min$^{-1}$) | (Fan et al., 2018) |
| Rice straw biochar 700 | | | 25 | 50 | 0.2110 | 0.01214 (g mg$^{-1}$ min$^{-1}$) | |
| Rice straw biochar 700 | Doxycycline Ciprofloxacin | 6 | 25 | 40 | 4.35 (1 min$^{-1}$) 2.75 (1 min$^{-1}$) | 0.026 (g/mg min) 0.054 (g/mg min) | (Zeng et al., 2018) |
| Water hyacinth biochar (WHC 350) | Caffeine Ciprofloxacin | – | 25 | 5 | – | 0.352 (g mg$^{-1}$ min$^{-1}$) –0.736 (g mg$^{-1}$ min$^{-1}$) | (Ngeno et al., 2016) |
| Willow sawdust (WS) biochar | Daptomycin | 4 | 25 | 25 | 0.0150 | 6.98 × 10$^{-4}$ (g mg$^{-1}$ min$^{-1}$) | (Ai et al., 2019) |
| Wood apple biochar | Ibuprofen | 3 | 25 | 15 | 0.048 | 0.0361 (g mg$^{-1}$ min$^{-1}$) | (Chakraborty et al., 2018) |
| Wood-chip biochar 300 | Levofloxacin | 6.5 | 30 | 50 | 0.0198 | 0.0201 (g mg$^{-1}$ min$^{-1}$) | (Yi et al., 2016) |
| Wood-chip biochar 600 | Levofloxacin | 6.5 | 30 | 150 | 0.0168 | 0.00757 (g mg$^{-1}$ min$^{-1}$) | |

tetracyclines show best fits to the pseudo-second order (Essandoh et al., 2015; Peiris et al., 2017). Conversely, daptomycin sorption on willow sawdust biochar and pine sawdust biochar favors pseudo-first order kinetics with $R^2$ values greater than 0.99 (Ai et al., 2019).

Banana pseudostem biochar and magnetic iron oxide banana pseudostem biochar hybrids prepared at 600°C removed 44.5% and 54.9% furazolidone (conc. 100 mg/L) in initial 60-min contact time (Gurav et al., 2020). The high furazolidone removal rate in the initial stage is achieved because of its high concentration and the availability of various biochar surface functional groups and active sites providing hydrophobic interactions and H-bonding. Later, the increasing drop in available unoccupied adsorption sites on biochar's surface resulted in a drop in the adsorption rate. The pseudo-second order kinetic model best fits the data with high correlation coefficients ($R^2 \geq 0.99$) (Gurav et al., 2020).

## 4.4 Pharmaceutical sorption mechanisms onto biochar

Contaminant adsorption involves four main steps: (i) bulk solute transportation, (ii) adsorbate film diffusion, (iii) pore diffusion of the adsorbate, and (iv) interactions between the adsorbent's porous structure and adsorbate (Ahmed et al., 2015). Organic contaminant adsorption is a complex phenomenon because multiple mechanisms can operate simultaneously (Patel et al., 2019). Several factors including adsorbate and adsorbent physicochemical properties, pH, temperature, and the composition of the aqueous matrix (interfering chemicals and organic matter) contribute to pharmaceutical sorption onto biochar. Based on previously reported literature, major pharmaceutical adsorption interaction with biochar includes:

(I) H-bonding
(II) Electrostatic attractions
(III) Electron donor-acceptor interactions
(IV) π-π donor-acceptor and dispersion interactions
(V) Hydrophobic interactions
(VI) Charge-charge repulsions [(+)(+) and (−)(−)].

These interactions can occur individually or simultaneously during pharmaceutical sorption. Pore diffusion is another important feature of the sorption process of organic contaminant sorption. Adsorbents having a high surface area and porosity promote pore diffusion. Biochars can possess low to high surface area and their original plant-derived morphology assists in inducing plenty of micro-, meso- and macropores during pyrolysis (Ahmad et al.,

2014; Mohan et al., 2014). Despite the fact some biochars have a low surface area, these examples can possess high contaminant sorption ability. This enhanced sorption ability is due to large amounts of oxygenated surface functional groups, which allows swelling of some biochar regions with substantial O/C ratios (Essandoh et al., 2015, 2017; Mohan et al., 2011). Such biochar swelling in aqueous solution enhances contaminant diffusion into subsurface regions of the solid as well as enhances polar functional group availability (Patel et al., 2019).

Several sorption mechanisms can occur simultaneously during pharmaceutical sorption onto biochar (Patel et al., 2019, 2021). Different sulfonamide and tetracycline adsorption interactions with biochars were reviewed recently (Peiris et al., 2017). Mechanisms including electrostatic interactions, charge-assisted hydrogen bondings, surface complexation, π-π electron acceptor-acceptor, cation exchange, cation bridging, and nonspecific van der Waals interactions were advanced (Fig. 11.3) (Peiris et al., 2017). Fig. 11.3 illustrates a few types of attractive interactions acting individually. In reality, more than one such attractive interaction will jointly act to hold large pharmaceutical molecules to sorption sites. Electron donor-acceptor attractions, cation exchange, and cation-π interactions were also discussed for sulfamethazine sorption on tea waste steam-activated biochar (Rajapaksha et al., 2015, 2014). Biochar's π-electron-rich graphene surface and the protonated aniline ring (electron deficient) of sulfamethazine form π-π electron donor-acceptor (EDA) interactions (Rajapaksha et al., 2015). Solution pH also affects the number of donor or acceptor interactions. In an acidic environment, stoichiometric protonation of sulfamethazine mainly governed sorption, whereas at high pH, sulfamethazine forms anionic species and electrostatic repulsion between the negative sites on biochar surfaces and sulfamethazine results in lowered adsorption (Rajapaksha et al., 2015). A number of protonated nitrogen and other moieties in sulfamethazine also can participate in hydrogen bonding to biochar sites (Rajapaksha et al., 2015).

Both nitrogen-containing functions in pharmaceuticals (amide, amine, imine, piperazinyl, and pyridinium) and oxygen-containing functional groups in both biochars and pharmaceuticals (carbonyl, carboxyl, ester, hydroxyl, lactone, and pyrene) can play a role in pharmaceutical sorption (Patel et al., 2019). Solution pH affects the surface of adsorbent and pharmaceutical's functional group speciation, thus playing a central role in sorption (Patel et al., 2019, 2021). For example, ibuprofen's ($pK_a = 4.91$) carboxylic group deprotonates to a carboxylate ion, facilitating both attractive H-bonding and charge-charge repulsions with phenolic, hydroxyl, and carboxyl groups on biochar's surface (Fig. 11.4). However, at

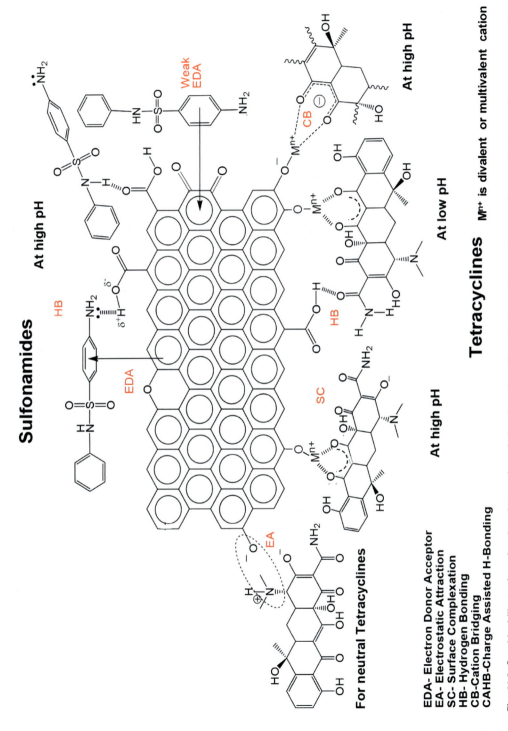

Fig. 11.3 Graphical illustration of various interactions during the adsorption of sulfonamides and tetracyclines on biochar surfaces. Adapted with permission from Peiris C, Gunatilake SR, Mlsna TE, Mohan D, Vithanage M. Biochar based removal of antibiotic sulfonamides and tetracyclines in aquatic environments: a critical review. Bioresour. Technol. 2017; 246: 150–159. Copyright 2017, Elsevier.

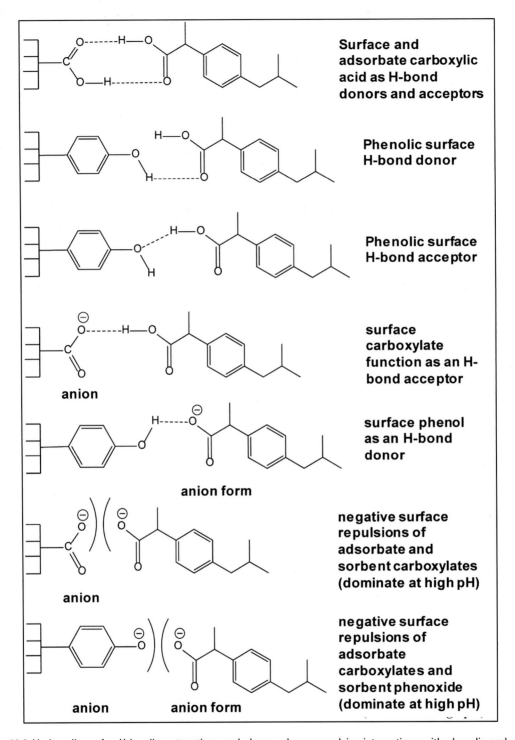

**Fig. 11.4** Various ibuprofen H-bonding attractions and charge-charge repulsive interactions with phenolic and carboxylic acid biochar functions. Both adsorbent surface functional groups and adsorbate can be H-donors and H-acceptors. Adapted with permission from Essandoh M, Kunwar B, Pittman Jr CU, Mohan D, Mlsna T. Sorptive removal of salicylic acid and ibuprofen from aqueous solutions using pine wood fast pyrolysis biochar. Chem. Eng. J. 2015; 265: 219–227. Copyright 2015, Elsevier.

higher pH (pH > p$K_a$ and pH$_{pzc}$) both biochar and ibuprofen's carboxylic group become negatively charged, resulting in repulsion and the reduction in the sorption (Fig. 11.4). Similar H-bondings were also reported for salicylic acid sorption onto fast pyrolysis pinewood biochar (Essandoh et al., 2015).

Ash and mineral phases in biochar (e.g., calcium carbonate, calcite, silicate, hydroxyapatite, and magnesium oxide) can also contribute to the overall sorption mechanism by interacting with pharmaceutical functions (Jang and Kan, 2019b). However, these effects can vary widely for different pharmaceuticals and biochars. For example, the ash content in biochar did not influence sulfamethoxazole sorption due to insignificant interactions between ash fraction and sulfamethoxazole (Shimabuku et al., 2016). Conversely, tetracycline sorption can be highly influenced by ash content (Jang and Kan, 2019b). Presence of hydroxyapatite and calcite in alfalfa biochar contributes to tetracycline sorption via electrostatic interactions, H-bonding, and surface complexations as shown in Fig. 11.4 (Jang and Kan, 2019b). This is supported by higher tetracycline sorption capacity by alfalfa ash (528.56 mg/g) as compared to alfalfa biochar (372.31 mg/g) (Jang and Kan, 2019b).

## 4.5 Matrix effects on pharmaceutical sorption onto biochar

Biochars can remove both inorganic and organic contaminants in single and multicontaminated solutions (Mohan et al., 2014). Pharmaceutical adsorption from environmental samples containing other coexisting contaminants can be hindered or promoted. Metal ions, inorganic anions, ionic strength variations, organic constituents, and organic matter are important matrix contributors posing various effects on pharmaceutical sorption (Patel et al., 2019; Peiris et al., 2017). The presence of metal ions can affect pharmaceutical sorption. Sulfamethoxazole sorption on rice straw biochar was enhanced considerably in the presence of $Cd^{2+}$ ions (Han et al., 2013). Enhancement in sulfamethoxazole sorption is due to cadmium binding with the negatively charged surface of biochar, facilitating neutral and anionic sulfamethoxazole sorption on biochar (Han et al., 2013). However, $Cu^{2+}$ ions inhibit sulfamethoxazole and sulfapyridine sorption by competing for available sorption sites (Xie et al., 2014). $Cu^{2+}$ ions form strong H-bonds with surface oxygen functionalities of biochar ($K_d$ values $10^3$–$10^5$ L/kg) and host hydration shells of dense water. Thus, $Cu^{2+}$ ions compete directly with sulfonamides by the formation of innersphere complexes on the surface of biochar (Xie et al., 2014). The presence of $Pb^{2+}$ ions does not cause any significant impact on sulfapyridine sorption onto carbon nanotube-modified biochar (Inyang et al., 2015).

Reduced, enhanced, and insignificant matrix effects were reported for tetracycline sorption on biochar (Peiris et al., 2017). Tetracycline sorption can increase with increasing ionic strength (NaCl), because tetracycline's solubility drops by a salting out effect (He et al., 2016). The presence of $Na^+$ and $Cl^-$ ions can have an electrostatic screening effect, thus causing a decrease in tetracycline sorption on carbonaceous adsorbents (Rivera-Utrilla et al., 2013; Tan et al., 2016). Interfering effects of $Na^+$ and $Cl^-$ ions can also reduce tetracycline sorption by competing with sorption sites (Tan et al., 2016). Oxytetracycline adsorption on maize-straw-derived biochar is influenced by metal ions (Jia et al., 2013). Oxytetracycline adsorption in presence of $Pb^{2+}$ ions (0.2 mM) is slightly reduced in acidic solution. However, $Cu^{2+}$ ions (0.2 mM) increased oxytetracycline sorption by 16%, 66%, and 54% at pH 3.5, 5.5, and 7.5 respectively. $K^+$, $Ca^{2+}$, $Zn^{2+}$, and $Cd^{2+}$ did not cause any significant impact on oxytetracycline adsorption (Jia et al., 2013).

Humic acids, a representative type of organic matter in natural waters, contain many polar moieties including carbonyl, carboxyl, amino, methoxy, and phenolic groups (Lian et al., 2015). Biochar's polar functions strongly interact with humic acid through H-bonding and electron donor-acceptor interactions to its hydrophobic fraction (Lian et al., 2015). Humic acid affects sulfonamide sorption on biochar (Lian et al., 2015). This sorption is generally higher in the single contaminant mode than in a multicontaminated system. For example, sulfamethazine, sulfamethoxazole, and sulfathiazole adsorption was 3 times lower than the singular adsorption mode (Ahmed et al., 2017a,b). Variations in contaminant type drive the adsorption mechanism in a copolluted environment. Insufficient data regarding matrix effects on pharmaceutical adsorbates are available so far. Thus, further investigations are required in the future.

## 5 Conclusions and recommendations

To date, many biochars have been developed and applied to remove several pharmaceuticals. Pharmaceutical sorption onto biochar shows enormous potential. Several biochars have provided comparable pharmaceutical removal to that achieved with activated carbons and other commercial adsorbents. Pharmaceutical's physicochemical nature and biochar's surface properties play crucial roles in the sorption process. Pharmaceutical polarity and its speciation to ionic forms (neutral, cationic, anionic, or zwitterionic) in aqueous solution influence uptake. Biochar's surface area, porosity, aromaticity, elemental composition, surface charge, and polarity are major parameters affecting sorption. H-bonding, π-π electron donor-acceptor attraction, and pore diffusion are important pharmaceutical sorption contributors. Only a few available studies have summarized the effects of biochar properties on pharmaceutical sorption in detail. Thus, further studies assessing the role of biochar properties on pharmaceutical sorption need to be conducted.

Solution pH and temperature affect both pharmaceutical sorption and its mechanism. pH can affect both a biochar's surface charge and a pharmaceutical's speciation in solution. Thus, pH plays one of the most important roles in pharmaceutical sorption. Exothermic, endothermic, and athermic pharmaceutical sorption responses have been reported. Still, a clear picture of the temperature's role on pharmaceutical sorption is lacking. While raising the temperature will increase the sorption rate, the effect of temperature on the free energy change for sorption will depend both on the entropy change and enthalpy going from solution to the solid phase: $\Delta G = \Delta H - T\Delta S$. Role of solution matrix (cations, anions, ionic strength, and organic matter) upon pharmaceutical sorption onto biochar has also been summarized. However, few studies evaluating matrix effects on pharmaceutical sorption exist. The role of complex solution matrices is scarce. Binary and multicomponent studies summarizing inhibitory or enhancing effects of different pharmaceuticals on the other's sorption on biochar are largely absent. Thus, an intense focus is still required to establish a clear picture of pharmaceutical sorption on biochar.

## Acknowledgments

Author (DM) is thankful to PSA, GOI for financial assistance under the project "Delhi Cluster-Delhi Research Implementation and Innovation (DRRIV)." Financial support [DST/TM/WTI/2K15/121 (C) dated: 19.09.2016] in the project entitled "Removal and recovery of pharmaceuticals from water using sustainable magnetic and nonmagnetic biochars" from Department of Science and Technology, New Delhi, India is thankfully acknowledged.

## Questions and problems

1. What are emerging contaminants? Why are pharmaceuticals categorized as emerging contaminants?
2. What kind of threat do pharmaceuticals pose for flora and fauna?

3. What are the steps or decisions taken by the WHO on pharmaceutical pollution?
4. What are the major sources of pharmaceuticals that get into natural water bodies?
5. What is the fate of aqueous pharmaceuticals in the environment?
6. What kind of problems are associated with the presence of pharmaceuticals in water bodies?
7. What are the methods available to treat/remove pharmaceuticals from aqueous media?
8. How can one decide the suitability of methods applied for removal of pharmaceuticals from water?
9. What is adsorption? How it is useful for the pharmaceutical removal?
10. How can the best adsorbent be selected for a particular pharmaceutical removal?
11. How do biochar properties influence the pharmaceutical sorption?
12. Explain the role of solution pH on pharmaceutical sorption.
13. What are the main attractive interactions involved in pharmaceutical sorption?
14. How do various ions affect pharmaceutical sorption?
15. Is it possible to reuse the spent biochar again?

## References

Ahmad, M., Lee, S.S., Dou, X., Mohan, D., Sung, J.-K., Yang, J.E., et al., 2012. Effects of pyrolysis temperature on soybean stover-and peanut shell-derived biochar properties and TCE adsorption in water. Bioresour. Technol. 118, 536–544.

Ahmad, M., Rajapaksha, A.U., Lim, J.E., Zhang, M., Bolan, N., Mohan, D., et al., 2014. Biochar as a sorbent for contaminant management in soil and water: a review. Chemosphere 99, 19–33.

Ahmed, M., Hameed, B., 2018. Adsorption behavior of salicylic acid on biochar as derived from the thermal pyrolysis of barley straws. J. Clean. Prod. 195, 1162–1169.

Ahmed, M.B., Zhou, J.L., Ngo, H.H., Guo, W., 2015. Adsorptive removal of antibiotics from water and wastewater: progress and challenges. Sci. Total Environ. 532, 112–126.

Ahmed, M.B., Zhou, J.L., Ngo, H.H., Guo, W., Johir, M.A.H., Belhaj, D., 2017a. Competitive sorption affinity of sulfonamides and chloramphenicol antibiotics toward functionalized biochar for water and wastewater treatment. Bioresour. Technol. 238, 306–312.

Ahmed, M.B., Zhou, J.L., Ngo, H.H., Guo, W., Johir, M.A.H., Sornalingam, K., 2017b. Single and competitive sorption properties and mechanism of functionalized biochar for removing sulfonamide antibiotics from water. Chem. Eng. J. 311, 348–358.

Ai, T., Jiang, X., Liu, Q., Lv, L., Wu, H., 2019. Daptomycin adsorption on magnetic ultra-fine wood-based biochars from water: kinetics, isotherms, and mechanism studies. Bioresour. Technol. 273, 8–15.

Alkurdi, S.S., Herath, I., Bundschuh, J., Al-Juboori, R.A., Vithanage, M., Mohan, D., 2019. Biochar versus bone char for a sustainable inorganic arsenic mitigation in water: what needs to be done in future research? Environ. Int. 127, 52–69.

de Andrade, J.L.R., Oliveira, M.F., da Silva, M.G., Vieira, M.G., 2018. Adsorption of pharmaceuticals from water and wastewater using nonconventional low-cost materials: a review. Ind. Eng. Chem. Res. 57, 3103–3127.

aus der Beek, T., Weber, F.A., Bergmann, A., Hickmann, S., Ebert, I., Hein, A., et al., 2016. Pharmaceuticals in the environment—global occurrences and perspectives. Environ. Toxicol. Chem. 35, 823–835.

Bush, K., 1997. Antimicrobial agents. Curr. Opin. Chem. Biol. 1, 169–175.

Chakraborty, P., Banerjee, S., Kumar, S., Sadhukhan, S., Halder, G., 2018. Elucidation of ibuprofen uptake capability of raw and steam activated biochar of *Aegle marmelos* shell: isotherm, kinetics, thermodynamics and cost estimation. Process Saf. Environ. Prot. 118, 10–23.

Chen, B., Zhou, D., Zhu, L., 2008. Transitional adsorption and partition of nonpolar and polar aromatic contaminants by biochars of pine needles with different pyrolytic temperatures. Environ. Sci. Technol. 42, 5137–5143.

Choi, Y.-K., Choi, T.-R., Gurav, R., Bhatia, S.K., Park, Y.-L., Kim, H.J., et al., 2020. Adsorption behavior of tetracycline onto Spirulina sp.(microalgae)-derived biochars produced at different temperatures. Sci. Total Environ. 710, 136282.

Choudhary, V., Patel, M., Pittman Jr., C.U., Mohan, D., 2020. Batch and continuous fixed-bed lead removal using Himalayan pine needle biochar: isotherm and kinetic studies. ACS Omega 5, 16366–16378.

Clurman, A.M., Rodríguez-Narvaez, O.M., Jayarathne, A., De Silva, G., Ranasinghe, M.I., Goonetilleke, A., et al., 2020. Influence of surface hydrophobicity/hydrophilicity of biochar on the removal of emerging contaminants. Chem. Eng. J. 402, 126277.

Crini, G., 2006. Non-conventional low-cost adsorbents for dye removal: a review. Bioresour. Technol. 97, 1061–1085.

Ebele, A.J., Abdallah, M.A.-E., Harrad, S., 2017. Pharmaceuticals and personal care products (PPCPs) in the freshwater aquatic environment. Emerg. Contam. 3, 1–16.

Enders, A., Hanley, K., Whitman, T., Joseph, S., Lehmann, J., 2012. Characterization of biochars to evaluate recalcitrance and agronomic performance. Bioresour. Technol. 114, 644–653.

Essandoh, M., Kunwar, B., Pittman Jr., C.U., Mohan, D., Mlsna, T., 2015. Sorptive removal of salicylic acid and ibuprofen from aqueous solutions using pine wood fast pyrolysis biochar. Chem. Eng. J. 265, 219–227.

Essandoh, M., Wolgemuth, D., Pittman Jr., C.U., Mohan, D., Mlsna, T., 2017. Phenoxy herbicide removal from aqueous solutions using fast pyrolysis switchgrass biochar. Chemosphere 174, 49–57.

Fan, S., Wang, Y., Li, Y., Wang, Z., Xie, Z., Tang, J., 2018. Removal of tetracycline from aqueous solution by biochar derived from rice straw. Environ. Sci. Pollut. Res. 25, 29529–29540.

Fan, Y., Wang, B., Yuan, S., Wu, X., Chen, J., Wang, L., 2010. Adsorptive removal of chloramphenicol from wastewater by NaOH modified bamboo charcoal. Bioresour. Technol. 101, 7661–7664.

Gurav, R., Bhatia, S.K., Choi, T.-R., Park, Y.-L., Park, J.Y., Han, Y.-H., et al., 2020. Treatment of furazolidone contaminated water using banana pseudostem biochar engineered with facile synthesized magnetic nanocomposites. Bioresour. Technol. 297, 122472.

Hagemann, N., Spokas, K., Schmidt, H.-P., Kägi, R., Böhler, M.A., Bucheli, T.D., 2018. Activated carbon, biochar and charcoal: linkages and synergies across pyrogenic carbon's ABCs. Water 10, 182.

Hale, S.E., Arp, H.P.H., Kupryianchyk, D., Cornelissen, G., 2016. A synthesis of parameters related to the binding of neutral organic compounds to charcoal. Chemosphere 144, 65–74.

Han, X., Liang, C.-f., Li, T.-q., Wang, K., Huang, H.-g., Yang, X.-e., 2013. Simultaneous removal of cadmium and sulfamethoxazole from aqueous solution by rice straw biochar. J. Zhejiang Univ. Sci. B 14, 640–649.

He, J., Dai, J., Zhang, T., Sun, J., Xie, A., Tian, S., et al., 2016. Preparation of highly porous carbon from sustainable α-cellulose for superior removal performance of tetracycline and sulfamethazine from water. RSC Adv. 6, 28023–28033.

Heo, J., Yoon, Y., Lee, G., Kim, Y., Han, J., Park, C.M., 2019. Enhanced adsorption of bisphenol A and sulfamethoxazole by a novel magnetic CuZnFe2O4–biochar composite. Bioresour. Technol. 281, 179–187.

Homem, V., Santos, L., 2011. Degradation and removal methods of antibiotics from aqueous matrices—a review. J. Environ. Manage. 92, 2304–2347.

Hoslett, J., Ghazal, H., Katsou, E., Jouhara, H., 2020. The removal of tetracycline from water using biochar produced from agricultural discarded material. Sci. Total Environ. 751, 141755.

Huang, J., Zimmerman, A.R., Chen, H., Gao, B., 2020. Ball milled biochar effectively removes sulfamethoxazole and sulfapyridine antibiotics from water and wastewater. Environ. Pollut. 258, 113809.

Hughes, S.R., Kay, P., Brown, L.E., 2013. Global synthesis and critical evaluation of pharmaceutical data sets collected from river systems. Environ. Sci. Technol. 47, 661–677.

Inyang, M., Gao, B., Zimmerman, A., Zhou, Y., Cao, X., 2015. Sorption and cosorption of lead and sulfapyridine on carbon nanotube-modified biochars. Environ. Sci. Pollut. Res. 22, 1868–1876.

Jang, H.M., Kan, E., 2019a. Engineered biochar from agricultural waste for removal of tetracycline in water. Bioresour. Technol. 284, 437–447.

Jang, H.M., Kan, E., 2019b. A novel hay-derived biochar for removal of tetracyclines in water. Bioresour. Technol. 274, 162–172.

Jia, M., Wang, F., Bian, Y., Jin, X., Song, Y., Kengara, F.O., et al., 2013. Effects of pH and metal ions on oxytetracycline sorption to maize-straw-derived biochar. Bioresour. Technol. 136, 87–93.

Jia, M., Wang, F., Bian, Y., Stedtfeld, R.D., Liu, G., Yu, J., et al., 2018. Sorption of sulfamethazine to biochars as affected by dissolved organic matters of different origin. Bioresour. Technol. 248, 36–43.

Klein, E.Y., Van Boeckel, T.P., Martinez, E.M., Pant, S., Gandra, S., Levin, S.A., et al., 2018. Global increase and geographic convergence in antibiotic consumption between 2000 and 2015. Proc. Natl. Acad. Sci. 115, E3463–E3470.

Kleywegt, S., Payne, M., Ng, F., Fletcher, T., 2019. Environmental loadings of active pharmaceutical ingredients from manufacturing facilities in Canada. Sci. Total Environ. 646, 257–264.

Krasucka, P., Pan, B., Ok, Y.S., Mohan, D., Sarkar, B., Oleszczuk, P., 2020. Engineered biochar-a sustainable solution for the removal of antibiotics from water. Chem. Eng. J. 405, 126926.

Kumar, H., Patel, M., Mohan, D., 2019b. Simplified batch and fixed-bed design system for efficient and sustainable fluoride removal from water using slow Pyrolyzed okra stem and black gram straw biochars. ACS Omega 4, 19513–19525.

Kumar, R., Patel, M., Singh, P., Bundschuh, J., Pittman Jr., C.U., Trakal, L., et al., 2019a. Emerging technologies for arsenic removal from drinking water in rural and peri-urban areas: methods, experience from, and options for Latin America. Sci. Total Environ. 694, 133427.

Kümmerer, K., 2009. The presence of pharmaceuticals in the environment due to human use–present knowledge and future challenges. J. Environ. Manage. 90, 2354–2366.

Lapworth, D., Baran, N., Stuart, M., Ward, R., 2012. Emerging organic contaminants in groundwater: a review of sources, fate and occurrence. Environ. Pollut. 163, 287–303.

Larsson, D.J., 2014. Pollution from drug manufacturing: review and perspectives. Philos. Trans. R. Soc. B 369, 20130571.

Larsson, D.J., de Pedro, C., Paxeus, N., 2007. Effluent from drug manufactures contains extremely high levels of pharmaceuticals. J. Hazard. Mater. 148, 751–755.

Li, T., Han, X., Liang, C., Shohag, M., Yang, X., 2015. Sorption of sulphamethoxazole by the biochars derived from rice straw and alligator flag. Environ. Technol. 36, 245–253.

Li, Y., Taggart, M.A., McKenzie, C., Zhang, Z., Lu, Y., Pap, S., et al., 2019. Utilizing low-cost natural waste for the removal of pharmaceuticals from water: mechanisms, isotherms and kinetics at low concentrations. J. Clean. Prod. 227, 88–97.

Li, R., Wang, Z., Guo, J., Li, Y., Zhang, H., Zhu, J., et al., 2018. Enhanced adsorption of ciprofloxacin by KOH modified biochar derived from potato stems and leaves. Water Sci. Technol. 77, 1127–1136.

Lian, F., Sun, B., Chen, X., Zhu, L., Liu, Z., Xing, B., 2015. Effect of humic acid (HA) on sulfonamide sorption by biochars. Environ. Pollut. 204, 306–312.

Lian, F., Sun, B., Song, Z., Zhu, L., Qi, X., Xing, B., 2014. Physicochemical properties of herb-residue biochar and its sorption to ionizable antibiotic sulfamethoxazole. Chem. Eng. J. 248, 128–134.

Lian, F., Xing, B., 2017. Black carbon (biochar) in water/soil environments: molecular structure, sorption, stability, and potential risk. Environ. Sci. Technol. 51, 13517–13532.

Liao, P., Zhan, Z., Dai, J., Wu, X., Zhang, W., Wang, K., et al., 2013. Adsorption of tetracycline and chloramphenicol in aqueous solutions by bamboo charcoal: a batch and fixed-bed column study. Chem. Eng. J. 228, 496–505.

Lindim, C., van Gils, J., Cousins, I., Kühne, R., Georgieva, D., Kutsarova, S., et al., 2017. Model-predicted occurrence of multiple pharmaceuticals in Swedish surface waters and their flushing to the Baltic Sea. Environ. Pollut. 223, 595–604.

Liu, P., Liu, W.-J., Jiang, H., Chen, J.-J., Li, W.-W., Yu, H.-Q., 2012. Modification of biochar derived from fast pyrolysis of biomass and its application in removal of tetracycline from aqueous solution. Bioresour. Technol. 121, 235–240.

Liu, J., Zhou, B., Zhang, H., Ma, J., Mu, B., Zhang, W., 2019. A novel Biochar modified by Chitosan-Fe/S for tetracycline adsorption and studies on site energy distribution. Bioresour. Technol. 294, 122152.

Lonappan, L., Rouissi, T., Brar, S.K., Verma, M., Surampalli, R.Y., 2018. An insight into the adsorption of diclofenac on different biochars: mechanisms, surface chemistry, and thermodynamics. Bioresour. Technol. 249, 386–394.

Mestre, A., Pires, J., Nogueira, J., Carvalho, A., 2007. Activated carbons for the adsorption of ibuprofen. Carbon 45, 1979–1988.

Mohan, D., Pittman Jr., C.U., Bricka, M., Smith, F., Yancey, B., Mohammad, J., et al., 2007. Sorption of arsenic, cadmium, and lead by chars produced from fast pyrolysis of wood and bark during bio-oil production. J. Colloid Interface Sci. 310, 57–73.

Mohan, D., Rajput, S., Singh, V.K., Steele, P.H., Pittman Jr., C.U., 2011. Modeling and evaluation of chromium remediation from water using low cost bio-char, a green adsorbent. J. Hazard. Mater. 188, 319–333.

Mohan, D., Sarswat, A., Ok, Y.S., Pittman Jr., C.U., 2014. Organic and inorganic contaminants removal from water with biochar, a renewable, low cost and sustainable adsorbent–a critical review. Bioresour. Technol. 160, 191–202.

Mukherjee, A., Zimmerman, A., Harris, W., 2011. Surface chemistry variations among a series of laboratory-produced biochars. Geoderma 163, 247–255.

Ndoun, M.C., Elliott, H.A., Preisendanz, H.E., Williams, C.F., Knopf, A., Watson, J.E., 2020. Adsorption of pharmaceuticals from aqueous solutions using biochar derived from cotton gin waste and guayule bagasse. Biochar, 1–16.

Ngeno, E., Orata, F., Lilechi, D., Shikuku, V.O., Kimosop, S., 2016. Adsorption of caffeine and ciprofloxacin onto pyrolytically derived water hyacinth biochar: isothermal, kinetics and thermodynamics. J. Chem. Chem. Eng. 10, 185–194.

Ocampo-Perez, R., Padilla-Ortega, E., Medellin-Castillo, N., Coronado-Oyarvide, P., Aguilar-Madera, C., Segovia-Sandoval, S., et al., 2019. Synthesis of biochar from chili seeds and its application to remove ibuprofen from water. Equilibrium and 3D modeling. Sci. Total Environ. 655, 1397–1408.

Ocampo-Pérez, R., Rivera-Utrilla, J., Gómez-Pacheco, C., Sánchez-Polo, M., López-Peñalver, J., 2012. Kinetic study of tetracycline adsorption on sludge-derived adsorbents in aqueous phase. Chem. Eng. J. 213, 88–96.

Patel, M., Kumar, R., Kishor, K., Mlsna, T., Pittman Jr., C.U., Mohan, D., 2019. Pharmaceuticals of emerging concern in aquatic systems: chemistry, occurrence, effects, and removal methods. Chem. Rev. 119, 3510–3673.

Patel, M., Kumar, R., Pittman Jr., C.U., Mohan, D., 2021. Ciprofloxacin and acetaminophen sorption onto banana peel biochars: environmental and process parameter influences. Environ. Res. 201, 111218.

Peiris, C., Gunatilake, S.R., Mlsna, T.E., Mohan, D., Vithanage, M., 2017. Biochar based removal of antibiotic sulfonamides and tetracyclines in aquatic environments: a critical review. Bioresour. Technol. 246, 150–159.

Rajapaksha, A.U., Vithanage, M., Ahmad, M., Seo, D.-C., Cho, J.-S., Lee, S.-E., et al., 2015. Enhanced sulfamethazine removal by steam-activated invasive plant-derived biochar. J. Hazard. Mater. 290, 43–50.

Rajapaksha, A.U., Vithanage, M., Zhang, M., Ahmad, M., Mohan, D., Chang, S.X., et al., 2014. Pyrolysis condition affected sulfamethazine sorption by tea waste biochars. Bioresour. Technol. 166, 303–308.

Richardson, S.D., Ternes, T.A., 2011. Water analysis: emerging contaminants and current issues. Anal. Chem. 83, 4614–4648.

Rivera-Utrilla, J., Sánchez-Polo, M., Ferro-García, M.Á., Prados-Joya, G., Ocampo-Pérez, R., 2013. Pharmaceuticals as emerging contaminants and their removal from water. A review. Chemosphere 93, 1268–1287.

Shimabuku, K.K., Kearns, J.P., Martinez, J.E., Mahoney, R.B., Moreno-Vasquez, L., Summers, R.S., 2016. Biochar sorbents for sulfamethoxazole removal from surface water, stormwater, and wastewater effluent. Water Res. 96, 236–245.

Stackelberg, P.E., Furlong, E.T., Meyer, M.T., Zaugg, S.D., Henderson, A.K., Reissman, D.B., 2004. Persistence of pharmaceutical compounds and other organic wastewater contaminants in a conventional drinking-water-treatment plant. Sci. Total Environ. 329, 99–113.

Sun, K., Keiluweit, M., Kleber, M., Pan, Z., Xing, B., 2011. Sorption of fluorinated herbicides to plant biomass-derived biochars as a function of molecular structure. Bioresour. Technol. 102, 9897–9903.

Tahar, A., Choubert, J.-M., Coquery, M., 2013. Xenobiotics removal by adsorption in the context of tertiary treatment: a mini review. Environ. Sci. Pollut. Res. 20, 5085–5095.

Tan, X., Liu, S., Liu, Y., Gu, Y., Zeng, G., Cai, X., et al., 2016. One-pot synthesis of carbon supported calcined-Mg/Al layered double hydroxides for antibiotic removal by slow pyrolysis of biomass waste. Sci. Rep. 6, 1–12.

Taylor, D., Senac, T., 2014. Human pharmaceutical products in the environment–the "problem" in perspective. Chemosphere 115, 95–99.

Ternes, T.A., Meisenheimer, M., McDowell, D., Sacher, F., Brauch, H.-J., Haist-Gulde, B., et al., 2002. Removal of pharmaceuticals during drinking water treatment. Environ. Sci. Technol. 36, 3855–3863.

Thompson, K.A., Shimabuku, K.K., Kearns, J.P., Knappe, D.R., Summers, R.S., Cook, S.M., 2016. Environmental comparison of biochar and activated carbon for tertiary wastewater treatment. Environ. Sci. Technol. 50, 11253–11262.

Tomul, F., Arslan, Y., Kabak, B., Trak, D., Kendüzler, E., Lima, E.C., et al., 2020. Peanut shells-derived biochars prepared from different carbonization processes: comparison of characterization and mechanism of naproxen adsorption in water. Sci. Total Environ. 726, 137828.

Tong, Y., McNamara, P.J., Mayer, B.K., 2019. Adsorption of organic micropollutants onto biochar: a review of relevant kinetics, mechanisms and equilibrium. Environ. Sci.: Water Res. Technol. 5, 821–838.

Tran, H.N., Tomul, F., Nguyen, H.T.H., Nguyen, D.T., Lima, E.C., Le, G.T., et al., 2020. Innovative spherical biochar for pharmaceutical removal from water: insight into adsorption mechanism. J. Hazard. Mater. 394, 122255.

Vimal, V., Patel, M., Mohan, D., 2019. Aqueous carbofuran removal using slow pyrolyzed sugarcane bagasse biochar: equilibrium and fixed-bed studies. RSC Adv. 9, 26338–26350.

Wilkinson, J., Hooda, P.S., Barker, J., Barton, S., Swinden, J., 2017. Occurrence, fate and transformation of emerging contaminants in water: an overarching review of the field. Environ. Pollut. 231, 954–970.

Xie, M., Chen, W., Xu, Z., Zheng, S., Zhu, D., 2014. Adsorption of sulfonamides to demineralized pine wood biochars prepared under different thermochemical conditions. Environ. Pollut. 186, 187–194.

Yi, S., Gao, B., Sun, Y., Wu, J., Shi, X., Wu, B., et al., 2016. Removal of levofloxacin from aqueous solution using rice-husk and wood-chip biochars. Chemosphere 150, 694–701.

Zeng, Z.-w., Tan, X.-f., Liu, Y.-g., Tian, S.-r., Zeng, G.-m., Jiang, L.-h., et al., 2018. Comprehensive adsorption studies of doxycycline and ciprofloxacin antibiotics by biochars prepared at different temperatures. Front. Chem. 6, 80.

Zhu, D., Pignatello, J.J., 2005. Characterization of aromatic compound sorptive interactions with black carbon (charcoal) assisted by graphite as a model. Environ. Sci. Technol. 39, 2033–2041.

# Dye removal using biochars

Gordon McKay[a], Prakash Parthasarathy[a], Samra Sajjad[b], Junaid Saleem[a], and Mohammad Alherbawi[a]

[a]Division of Sustainable Development, College of Science and Engineering, Hamad Bin Khalifa University, Qatar Foundation, Doha, Qatar. [b]Centre for Advanced Materials, Qatar University, Doha, Qatar

## 1 Introduction

The pollution of receiving waters, streams, and rivers, from colored dyestuffs arises due to incomplete effluent discharge treatment from many industries including: cosmetics, food and drink, paint, paper and pulp, pharmaceuticals, printing, textiles, and dyeing (Zhang et al., 2019). Due to their chemical properties, many dyes impose a serious threat to the water environmental ecosystem with serious adverse impacts on the health of humans, animals, and plants (Gupta and Saleh, 2013). Apart from several reports on the toxicity of certain dyestuffs (Mani et al., 2019; Khan and Malik, 2018), dyes in receiving waters reduce photosynthesis activity by preventing the passage of light. They reduce the dissolved oxygen content of the water during the degradation process, thus lowering the water quality for aquatic life. This yields a negative visual esthetic impact and can cause reproductive and other health impact problems in fish (Dil et al., 2016; Hernández-Zamora and Martínez-Jerónimo, 2019). In humans, certain dyestuffs will adversely affect the liver, heart, kidneys, reproductive system, skin, brain, nervous system and some could potentially be carcinogens or cause mutagenicity.

Quantitative information on dyestuffs production and their discharge volumes into the environment via receiving waters around the world is not well established or reported. For example, one production figure stated (Yagub et al., 2014) that $700 \times 10^3$ tons dyes are produced annually based on 100,000 commercial dyes. In another report (Research and Markets, 2020), the annual production of dyestuffs worldwide is in the range of 1.8–1.9 million metric tons comprising over 10,000 dyes and pigments, which are mainly used in the textiles, paper, food, leather, cosmetics, and plastics industries.

Depending on the nature of the dye and the processing technology adopted, approximately 1%–10% of dye is lost during manufacturing and utilization, revealing substantial amounts of dyes are released to the environment via different pathways (Forgacs et al., 2004).

Most dyes and pigments have targeted properties including chemical stability, photodegradation resistance for light stability and they reduce photosynthesis in rivers and streams. They inhibit a range of complex chemical functionalities, depending on which substance or material they will be applied to and with regard to which color they will impart (Fig. 12.1). All these features are beneficial to the dyestuff users and are promoted by the dye manufacturing companies. However, their presence in the large effluent volumes renders the treatment of these compounds to meet environmental discharge standards extremely difficult. Furthermore, a dye concentration as low as 1 ppm can impart color to water. Most of the dyeing applications use large volumes of water for the dyeing, rinsing, and washing stages (Research and Markets, 2020; Afroze and Sen, 2018).

In order to perform the removal of color from discharged waters containing dyes with a wide array of complicated and stable structures, as described in Section 2, several treatment technologies have been applied and these are discussed in Section 3.

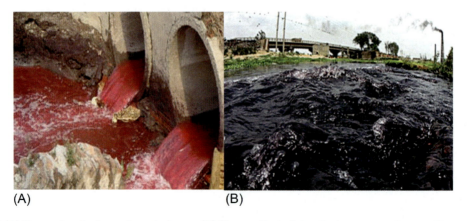

**Fig. 12.1** (A) Shows dye discharge from dyehouse. (B) Shows effect of dye discharge on a river quality.

## 2 Classification of dyestuffs

Dyes can be defined as colored compounds or substances that may be applied in aqueous solutions or dispersions to color a substrate, including textiles, polymers, rubbers, plastics, food, drinks, and many others. Water solubility is a key property of most dyes. In fact, many water-soluble dyestuffs owe their solubility to the presence of at least one sulfonic acid group, normally applied as the sodium salt. Dyes are colored because they absorb light of a specific wavelength within the visible spectrum. This key characterizing property is promoted by the presence of one or more unsaturated chromophores. The more common ones are shown in Fig. 12.2:

In addition to these chromophoric species, other groups to improve water solubility and enhance a dye's adsorption potential on the substrate material, such as amino, substituted amino, hydroxyl, sulfonic, or carbonyl groups, referred to as auxochromes.

Dyes may be classified according to their chemical composition or according to their application. Consequently, the chemical nature and type of dye should be the major consideration in deciding which dye wastewater treatment should be used for effluent purification and specially to evaluate what are the required adsorbent properties to be used for dye adsorption.

### 2.1 Reactive dyes

Reactive dyes have been applied widely to cellulose-based fibers, such as cotton and flax. They are also applicable for dyeing linen, viscose, and silk due to their strong ability to adhere to a substrate (Al-Degs et al., 2000, 2008). These dyes can create a covalent chemical linkage with fibers. The mechanism is shown in the mechanistic scheme for a dichlorotriazine-based reactive dye, becoming fixed to a cellulosic fiber with the displacement of chloride groups. One or both chlorides may be involved:

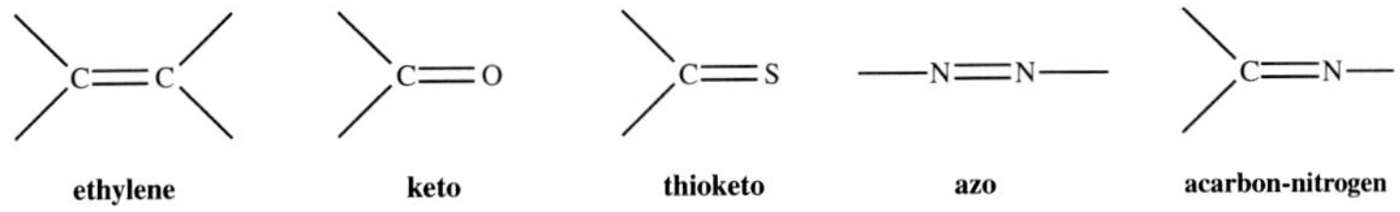

**Fig. 12.2** Color imparting chemical groups or chromophores.

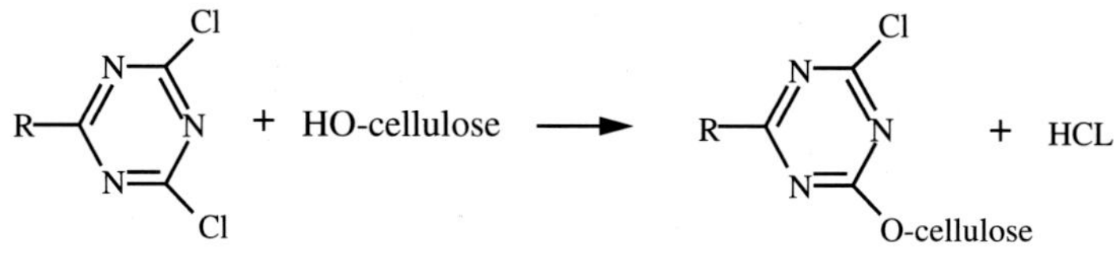

**Fig. 12.3**

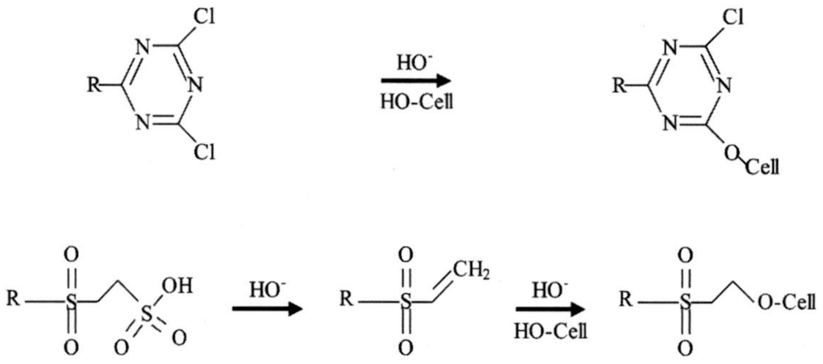

**Fig. 12.4** Mechanism of reactive dye adsorption onto cellulose.

Another very common mechanism for dye adsorption onto cellulose is presented in Figs. 12.3 and 12.4.

Due to the strong bonding affinity of reactive dyes for cellulose, biosorbents with hydroxyl groups have shown very positive adsorption results for removing reactive dyes from textile effluents (Chiou et al., 2003).

## 2.2 Disperse dyes

Disperse dyes are nonionic compounds widely used for polyester and also applicable for acetate and nylon fibers. They are soluble in water and applied to the fibers/fabrics from dispersions by diffusing into fibers at high temperature. Since there is an absence of basic surface groups, no attachment sites for acid dyes were present although there is a small affinity for basic dyes. The dyeing mechanism results from dipole–dipole attractions and the weaker van der Waals forces, suggesting a similar mechanism may be responsible for the disperse dye removal using biochars (Clark, 2011). One example of this class is Disperse Blue 7 as shown in Fig. 12.5.

The disperse dye is usually a very fine suspension which could be removed by biochars operating in a filtration mode.

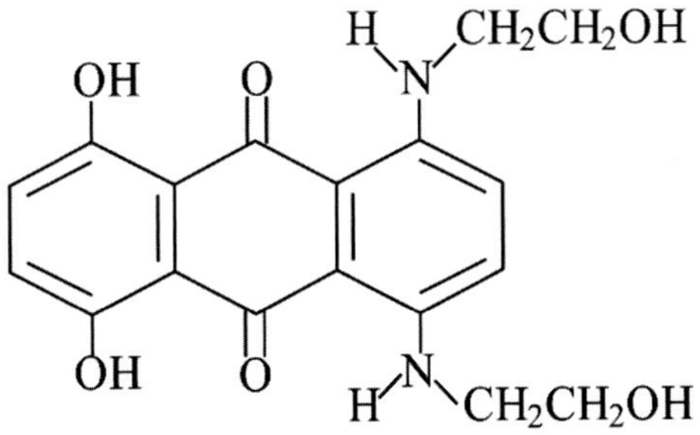

**Fig. 12.5** The Disperse Blue 7 dye molecule.

## 2.3 Vat dyes

Most vat dyes contain the ketone-containing chromophore and they are used for dyeing cellulosics, such as cotton, linen, and viscose fibers/materials. This is a large class of dyes, containing the following groups: anthraquinones, indanthrones, benzanthrones, carbazoles, acridones, and polycyclic quinones. Two structures of typical vat dyes are shown in Fig. 12.6A and B, for CI vat yellow 26 and CI vat green 1 (Yusuf, 2020).

The large anthraquinone groups indicate that the adsorption mechanism might occur by the anthraquinone electron clouds adsorbing by electrostatic attraction to the biochar positive surface group structures and adequate pore size.

## 2.4 Direct dyes

Direct dyes are sometimes referred to as substantive dyes because they do not require a fixation stage, and they are used for dyeing viscose and cotton yarn, fabrics, or loose cotton (Hunger et al., 2005). Some direct dyes, but not all, use mordants (chemical species, sometimes chromium salts, which complex with the colored chromophore to form an insoluble color) in order for the dye fixation and improve color fastness. This technique has

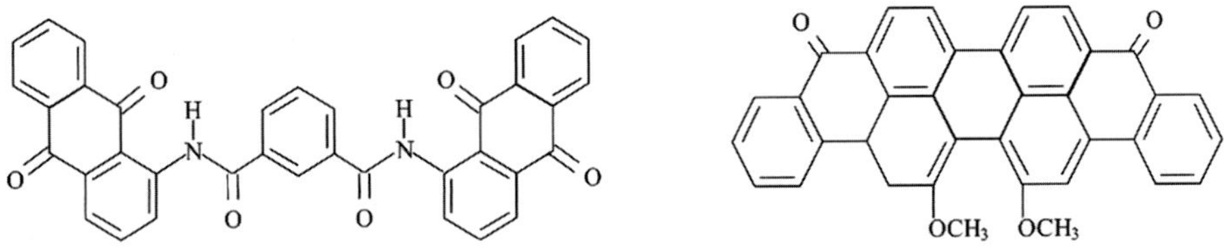

*CI Vat Yellow 26 (Anthraquinone Vat Dye Structure)*   *CI Vat Green 1*

**Fig. 12.6** Typical structures of two vat dyes: vat yellow 26 and vat green 1.

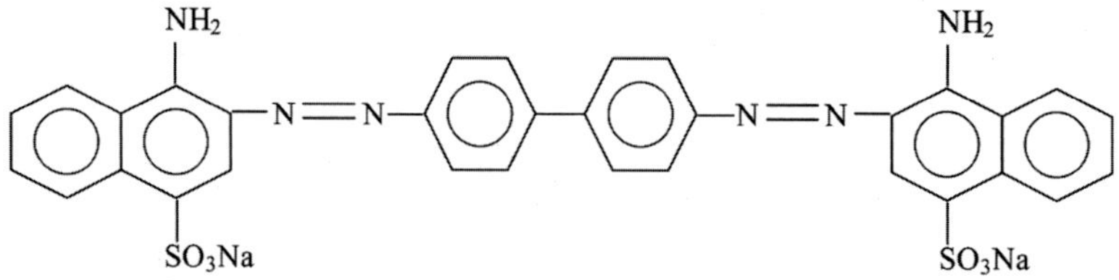

**Fig. 12.7** Structure of Congo red dye.

proven economical in achieving high color fastness when producing dark dye shades involving navy and black colors. Recent health, safety, and environmental awareness has restricted their application. The direct dyeing mechanism involves attaching the dye to the substrate, such as textiles and fibers, by establishing nonionic forces (Tahir et al., 2016). Fig. 12.7 shows the structure of Congo red dye.

In solution, these dyes produce negative ions and will adsorb onto positive biochar sites.

## 2.5 Basic dye

Congo red is one of many basic dyes distinguished by their brilliance in color and high tinctorial strength. Basic dyes are mainly applicable for acrylic fibers and can be applied to other fibrous textiles by using mordants. In addition, the basic dyes are soluble in acid and insoluble in basic solutions. These dyes comprise primarily amino or imino groups linked to xanthene or triarylmethane: their uses also include inks, carbon paper, and typewriter ribbon (Gupta et al., 2000). There are three main subclasses, namely monoazo, methane, and oxazine. A general formula for the basic dye structure is shown in Fig. 12.8.

The basic dye has been named because the dye ionizes in water and the colored ion component is the positively charged cation or basic species; consequently, these dyes are often referred to as cationic dyes (Allen et al., 2004) and will best adsorb onto negatively charged biochar sites.

## 2.6 Acid dye

Acid dyes have a strong color fastness quality and a great affinity for proteins making them very suitable for wool and also silk and nylon. Most of the acid dyes are in the form of the sodium salts of organic sulfonic acids (Fan et al., 2004). They are characterized by different combinations of unsaturated aromatic groups

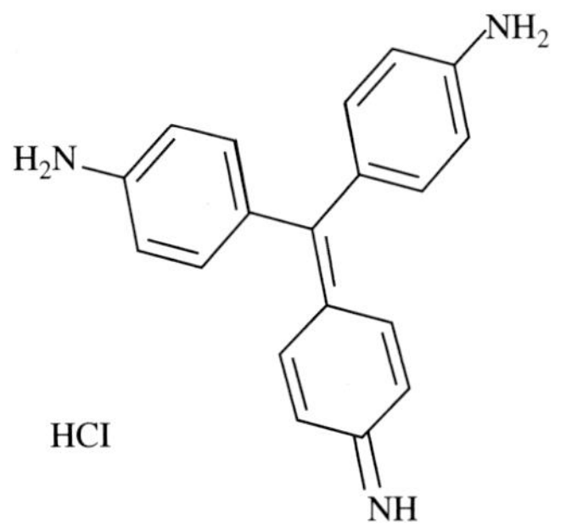

**Fig. 12.8** Typical basic dye structure for Basic Red 9.

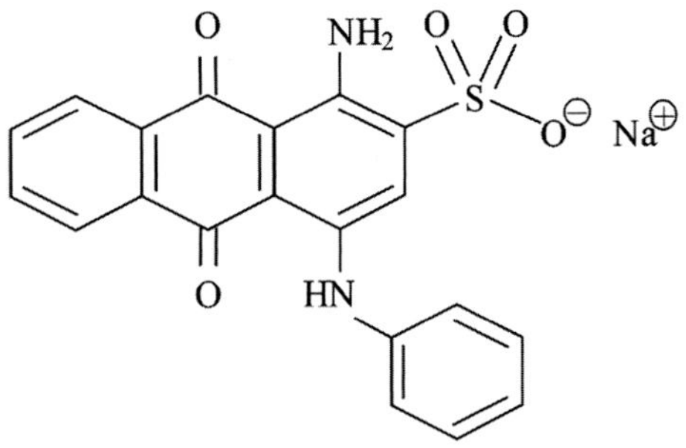

**Fig. 12.9** Structure of the anthraquinone dye Acid Blue 25.

called chromophores (Ho and Chiang, 2001). The acid dye containing discharge not only increases the overall COD loading to the receiving water but also results an esthetic nuisance to the environment at concentrations as low as 1 ppm (Valix et al., 2004). A few dyes, such as benzidine, benzidine-congener, and arylamine-based types, are of particular interest, because they are toxic or even carcinogenic substances (Yu et al., 2002). One major group of acid dyes is the anthraquinone group. One example is Acid Blue 25, as shown in Fig. 12.9.

The colored unit of the sodium salt is the anion giving rise to the frequent description of acid dyes as anionic dyes and the colored ions, sulfonic acid groups (sometimes carboxylic acid groups), will adsorb onto positive biochar sites.

## 2.7 Azo dyes

Azo dyes represent more than 50% of the annual production of dyes worldwide. They are extensively utilized in large industrial sectors such as cosmetic, food, leather, pharmaceutical, and textile

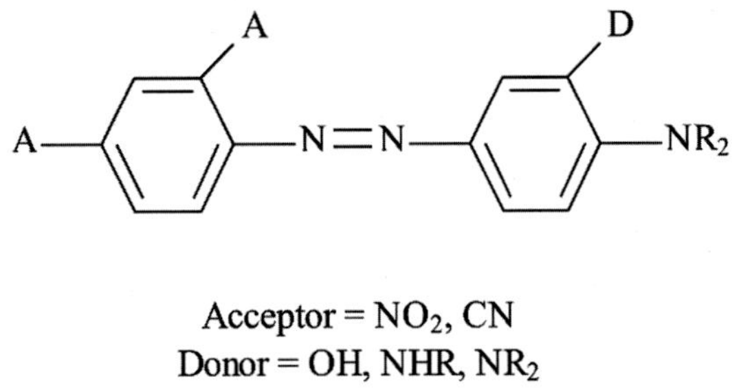

Acceptor = NO₂, CN
Donor = OH, NHR, NR₂

**Fig. 12.10** Typical structure of a monoazo dye.

coloring applications. These dyestuffs contain one or more azo (—N=N—) functional groups as the chromophore and are found widely in synthetic dyes containing an aromatic ring structure (Rovina et al., 2016), as shown in Fig. 12.10. If azo dyes are systemically absorbed, they can be metabolized by the intestinal microflora in liver cells producing aromatic amines that can be hazardous and the azo dye itself, if water soluble, can be carcinogenic.

Several countries have imposed legislation on the production, handling, and usage of azo dyes. Implementation legislation of azo dyes has been categorized into two groups (Nikfar and Jaberidoost, 2014):

(i) Those which are capable of generating a carcinogenic metabolite, and
(ii) Those that are not.

## 2.8 Sulfur dyes

Sulfur dyes have been extensively manufactured and have been widely applied in the cotton-based industries as they are easy to apply, have strong color fastness, and are not expensive. The dyes are insoluble in water at room temperature, but at alkaline pH with the addition of a reducing agent and elevated temperatures, the dyes become soluble in water enabling them to adsorb onto and hence dye the fabric. From an environmental perspective, the sulfur dyes are highly polluting in the dyehouse effluent discharge and consequently have been steadily losing popularity (Goswami and Basak, 2001). A typical sulfur dye is shown in Fig. 12.11.

The presence of so many lone pairs of electrons (sulfur, nitrogen, and oxygen) enables sulfur dyes to adsorb onto surface positive sites particularly chelating sites looking for lone pairs.

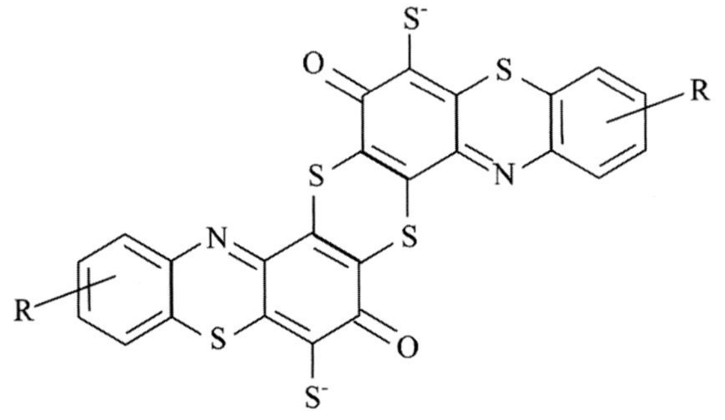

**Fig. 12.11** Structure of the sulfur dye—Sulfur Black 1.

## 2.9 Aniline dyes

This group of synthetic dyes is produced mostly from coal tar containing one or more phenyl groups. A whole new dye manufacturing industry sprang up around them producing some of the most extensively used and famous dyestuff names, like malachite green and crystal violet. The dyestuffs occur in the form of a fine powder. The aniline dyes are water soluble and come in a wide range of colors applicable to material, such as fabric, leather, and wood (Thakur and Qanungo, 2021). From a health perspective, some of the aniline dyes have been found to increase the risk of cancer. A typical aniline dye is shown in Fig. 12.12.

These dyes are cationic (Allen et al., 2004) and will best adsorb onto negatively charged biochar sites. Furthermore, the ring electron clouds and the nitrogen lone pair electrons are also attracted to the positive biochar sites.

## 2.10 Metal complex dyes

Metal complex dyes, often termed premetallized dyes, have a low water solubility. These dyes are mostly monoazo species applied to dye polyamides, wool, nylon, and silk. The basic

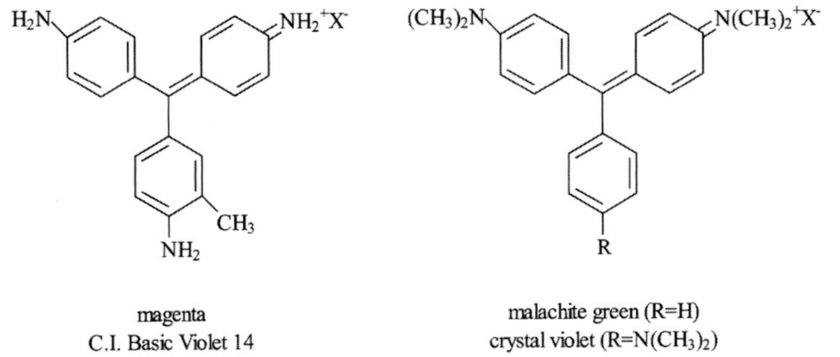

magenta
C.I. Basic Violet 14

malachite green (R=H)
crystal violet (R=N(CH$_3$)$_2$)

**Fig. 12.12** Structure of the aniline dyes: basic violet 14 and malachite green.

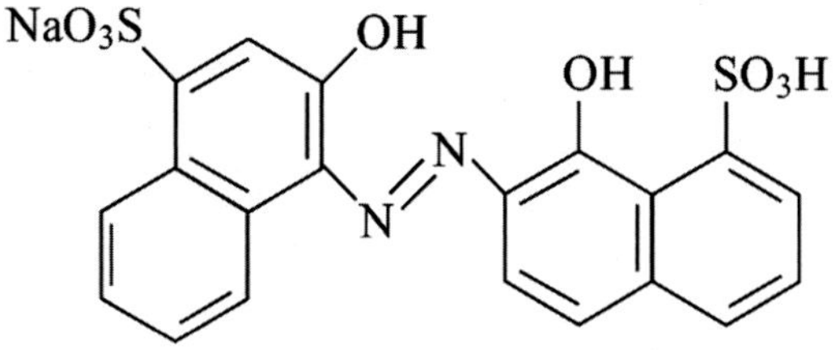

**Fig. 12.13** Structure of the metal complex dye acid blue 158.

structure has monoazo functions with some attached amino, carboxyl, or hydroxyl groups providing these dyes with a strong bond formation potential and the additional capability to form coordination complex compounds with transition metals, including chromium, cobalt, copper, and nickel. But these metal complex dyes are causing safety, health, and environmental concerns, since the effluent discharges from their use contain unfixed dye and the associated metal species (Chakraborty, 2011). The structure of the metal complex dye, acid blue 158, is shown in Fig. 12.13.

The dyes ionize in water producing colored anions which can adsorb on the positive biochar sites, but the removal of the toxic heavy metal counterion requiring negative adsorption sites is a greater concern.

## 2.11 Mordant dyes

The majority of natural dyes are unable to attach/fix very strongly to textile fibers and require the application of a mordant substance, which binds to the dye and the substrate. This can be applied to both the substrate fiber and to the dyestuff and applied to nylon and wood. This mordant chemical generates a bond between the dye and the fiber (Baker, 1958), but in recent years direct dyes are beginning to replace mordants. Fig. 12.14 shows the structure of mordant brown 35.

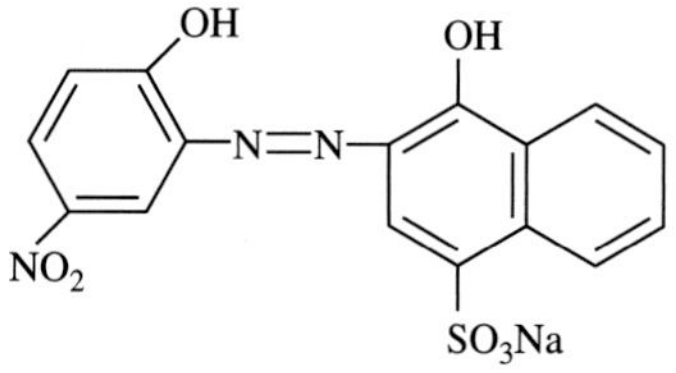

**Fig. 12.14** Structure of mordant brown 35.

In solution, these dyes ionize forming colored anions, therefore requiring treatment by biochars with positive sites.

## 3 Dye removal technologies

Typical dyehouse effluent treatment technologies involve a wide range of treatment processes. These can be broadly divided into three categories, namely physical (including adsorption), chemical, and biological processes. In practice, these technologies may be used solely or in a cocktail mix in order to obtain an optimal balance between economic benefit and technical efficiency during the course of water pollutants removal.

### 3.1 Physical processes

The mechanisms of adsorption (Gupta, 2009) and biosorption (Ip et al., 2010) are based on the attraction of dye molecules or dye ions to the surface of solid materials in powder or granular form by physical bonding by van der Waals forces or by stronger chemical bonding or ion exchange with surface functional groups on the adsorbent surface by higher energy chemical bonds or chemisorption forces. More detail on the adsorption and biosorption processes is explained in other parts of this book. Also, Section 5 of this chapter discusses adsorption equilibrium capacity and equilibrium kinetics so it will not be discussed further here.

Coagulation and flocculation have been widely used for the removal of various contaminants and dyes from wastewaters (Beluci et al., 2019). This involves the addition of one or more chemicals, frequently with agitation to the polluted wastewater, followed by a settling stage allowing the pollutant-loaded flocculant/coagulant to settle to the bottom of the vessel. The clarified water overflow for discharge into receiving water may undergo further tertiary processing and/or recycling. As in adsorption, the dye adsorbs on the coagulant/flocculant surface without any chemical degradation. Common materials used are alum and ferrous sulfate (Mcyotto et al., 2021). The disadvantages include the long settling time for the water to be clarified, the large volume of dye-loaded coagulant/flocculant generated and needing disposal and the cost of drying the very wet slurry prior to disposal in the landfill site.

Other options include electrokinetic coagulation (Bayramoglu et al., 2007) and irradiation. Aluminum electrodes have been widely studied for use in electrocoagulation. They liberate activated aluminum species in solution (Hasani et al., 2019) providing more active adsorption sites in the solution, but the disadvantages associated with coagulation discussed above also apply. Irradiation/catalytic irradiation for color removal has been investigated in a few cases. The removal efficiency was examined for methylene blue (MB). After 120 min of irradiation, the removal efficiency reached around 86.4% for the highest contribution of Pd catalyst (Al-Ahmed et al., 2020) representing a fivefold increase in the rate of removal. Both processes exhibited good removal efficiency for certain dye species but less effective in others (Robinson et al., 2001). Further in-depth studies of these methods need to be carried out in an attempt to verify their effectiveness.

Membrane filtration (Samsami et al., 2020) involves the physical separation of dissolved dye molecules from effluent through permeable membranes under pressure. Extensive interest now exists in membrane applications, nanofiltration (Yang et al., 2017), and ultrafiltration membranes (Tang et al., 2019). These commonly used processes are similar to membrane technology. Nanofiltration/ultrafiltration (Tang et al., 2019; Ellouze et al., 2010; Ammar et al., 2015) and forward/reverse osmosis (Ellouze et al., 2010; Ammar et al., 2015) both exhibit good ability to separate solutes and particles larger than 1 nm. The major issues with membranes are the pumping energy requirement cost and the potential for membrane pore blockage. If the dye effluent discharge was used previously for dyeing textiles/clothing, fine particulate textile threads are frequently discharged with the spent dye solution. A prefiltration stage may be beneficial in these cases to minimize early membrane fouling and blockage (Balcik-Canbolat et al., 2017). If a concentrated dye layer builds up, it gradually reduces the overall removal efficiency with a lower flux rate. Also, the cost of filter media becomes important especially in larger-scale production.

A number of studies (Chu et al., 2020; Anantha et al., 2020) have recently applied ultrasonics and photocatalysis to enhance the adsorption of dyes onto solid particles especially by the application of magnetic biochars and nanoparticles.

## 3.2 Chemical processes

Dye molecules are usually highly resistant to biodegradation or oxidation with the aid of ultraviolet light. Most chemical techniques for dye effluent treatment aim to break down the complex dye molecules into relatively simple and less toxic substances such as carbon dioxide or water via a series of oxidative reactions. Examples include chlorination, ozonation, wet air oxidation (WAO), and other advanced oxidation processes (AOPs).

Wet air oxidation (Cheremisinoff, 1994) has been proved effective in dye effluent treatment; however, it has been proposed that catalysts could enhance performance (Chang et al., 2003; Santos et al., 2007). Ozone is widely applied in many commercial-scale advanced oxidation systems for the destruction of color and organics (Boudissa et al., 2019). The application of catalysts in ozonation is being developed (Faria et al., 2009). Other strong oxidizing chemical reagents have been used commercially, including strong oxidizers such as chlorine, sodium hypochlorite, and hydrogen peroxide (Chaudhuri and Sur, 2000; Christie, 2007) oxidation of dye-containing effluent. These oxidants degrade dyes to simple substances like carbon dioxide and water (Raghavacharya, 1997). However, there is increasing evidence that the advantages of using chlorinated substances in water treatment could be outweighed by the generation of certain potentially toxic side products such as chloroanilines, chlorophenols, chloronitrobenzenes, and even down to chloromethanes in the oxidation pathway (Sarasa et al., 1998). In the widely applied field of ozonation, there is also concern regarding oxygenated by-products and bromoforms (Farzaneh et al., 2020) and a number of studies have looked at toxicity in such systems (Oliveira et al., 2010).

As an alternative to the standard oxidizers in advanced oxidation process (AOP) using Fenton's reagents, titanium oxides or ferrous ions as catalysts have been adopted in the degradation of organic pollutants such as dye and phenolic compounds (Martinez et al., 2005). Excellent removal efficiencies coupled with UV light due to the generation of high reduction potential intermediate radicals have been reported (Hislop and Bolton, 1999). However, one should be aware that the posttreatment of ferrous sludge associated with the reaction could be problematic in some cases (Jeong and Yoon, 2005). Several AOP processes are available. Achieving color removal of 80%–90% is common but careful pH control and sludge treatment after the process have to be taken into consideration. Photochemical degradation (Houas et al., 2001) is an ongoing area of interest for the removal of dye colors using the traditional titanium dioxide but being restricted by process design in the form of scale-up issues. Electrochemical treatment has been tested on an industrial azo dye effluent (Vaghela et al., 2005).

## 3.3 Biological processes

Conventional aerobic methods such as activated sludge have been successfully applied in domestic effluent treatment but are inadequate in the case of dyestuffs removal. Dye molecules are usually hydrophilic and have little affinity to the biomass, leading to early breakthrough in practice (Bell and Buckley, 2003). Toxic and stable dye molecular structures enable them to be highly resistant to microbial degradation. Anaerobic digestion involves a series of microbial reactions where dye molecules are degraded to fatty acids and soluble organics, then to methane and hydrogen sulfide in the absence of oxygen. Satisfactory results on azo, diazo, and reactive dye decoloration have been reported (Shoukat et al., 2019), but toxic aromatic amines in the effluent remained nonsensitive to the treatment (Banat et al., 1996).

# 4 Adsorption technology

Adsorption has long been used as a purification and separation process on an industrial scale. The process is one in which pollutants dissolved in water (or entrained in gas phase streams) are removed from the fluid stream by attaching themselves to the surface of a solid material, termed the adsorbent, leaving the solution/fluid phase pollutant free—ideally.

## 4.1 Adsorbent properties and adsorption mechanism

Highly porous adsorbents with good selectivity are required. Activated carbon has shown excellent ability to remove organic compounds such as dyestuffs (Gupta, 2009; McKay, 1980; Saleem et al., 2019). The adsorption of dye onto adsorbents can be a physical or a chemical process. In the physical adsorption mechanism, the dye molecules attach onto the adsorbent surface under the influence of van der Waals forces and hydrogen bonding (Cooney, 1998). During chemisorption, the dye molecule or ion attaches itself to a specific surface functional group or site by a chemical bond. Section 2 illustrates many different dye functional groups that can participate in multiple adsorptive attachments over a wide range of available adsorbents. Dye removal by adsorption onto biochars is reviewed in detail in Section 5.

## 4.2 Adsorbents

Adsorbent selection to use for adsorption depends primarily on the adsorbent's uptake capacity for the specific adsorbate, as well as the design of the contact system. Ideally, the adsorbent should satisfy several requirements (Saleem et al., 2019; Dotto and McKay, 2020):
1. Reasonably high surface area and pore volume
2. Suitable pore size distribution and pore network

3. Correct type and surface charge of adsorbent functional groups
4. Type of functional groups and charge on the colored dye group/ions
5. Suitable solution pH for uptake.

In the first two categories, it is very important to compare the dimensions of the dye molecule with the biochar pore size distribution to ensure that the dye can easily diffuse through the pores as many dyestuffs are quite large molecules. Also dye molecules often associate together into larger groupings or in micelles. Therefore, mesoporous adsorbents might be preferred over those with a very high surface area with a highly microporous structure. Nevertheless, larger surface areas are usually advantageous. A large mesoporous network allows a higher diffusion rate that is critical to adsorption kinetics and process design.

Common examples of adsorbents in commercial use are activated carbon, zeolites, silica gel, and activated alumina (Yang, 2003). Others include bone char (Ko et al., 2005), agricultural wastes or by-products (e.g., wood meals, bagasse, nutshells, rice husks, fruit stones, maize cobs, lignocellulosics) (Crini, 2006; Abdolali et al., 2014; Rodriguez-Reinoso and Molina-Sabio, 1992), inorganic minerals (e.g., bentonite, Fuller's earth, or clay), peat, lignite, chitosan, and ion-exchange resins (Allen et al., 1989). Novel adsorbents such as carbon nanotubes, MCM-41 (Rouquerol et al., 2013; Lam et al., 2006) or related structures, and aluminophosphate molecular sieves (Rouquerol et al., 2013) are found in recent literature.

Since large water volumes and relatively low dye concentrations are found, adsorption offers an attractive method for color removal, if appropriate adsorbents with high adsorption capacities are available at an attractive cost. Consequently, there is a great opportunity for the application of biochars for the removal of dyes in effluents. The biochars, for adsorption, are produced by pyrolysis at relatively low temperatures, namely 350–600°C, compared to activated carbons and using relatively low-cost biomass raw materials. The temperature selected and the feedstock type will control the properties. For highly charged surface properties, modifying agents are incorporated into the preparation phase. Section 5 presents many examples of both modified and unmodified biochars.

## 5 Dye removal using adsorption onto biochars

### 5.1 Biochar as a dye removal adsorbent

The term "activated carbon" defines a category of amorphous carbonaceous materials with high porosity and internal surface area. Nearly all organic materials with relatively high carbon content can serve as its feedstock, ranging from conventional materials like wood, coconut shell, or coal to natural or synthetic polymers. For commercial grade carbons, surface areas typically vary in the range between 500 and 1500 $m^2/g$ or even as high as 3000 $m^2/g$. Carbons can be further separated into two subcategories in accordance with their use in removing pollutants from the gas phase or liquid phase.

The former are usually microporous (pore diameter < 2 nm) in granular form (2.36 mm – 0.833 mm, or 8/20 in mesh size) and the latter are mesoporous (pore diameter lies between 2 and 50 nm) in powdered form (0.150 mm–0.043 mm, or 100/325) (Cooney, 1998). Both granular and powdered activated carbons are effective in wastewater treatment, where they play an important role in decolorizing, odor removal, metal recovery, as well as organic adsorption.

The adsorption capacity of activated carbon is proportional to the internal surface area, pore volume, pore size distribution, and surface chemistry (Cooney, 1998; Rouquerol et al., 2013). Organics are reported to adsorb in pores that are barely large enough to admit the adsorbate molecule (Cheremisinoff, 1994). Examples include dyes and humic acids with dimensions (1.5–3.0 nm) that favor their adsorption in mesopores (Mui et al., 2010a). Hence, the difference in pore size distribution of the biochar affects the adsorption capacity for molecules of different size and shape. The removal efficiency can be further enhanced by the electric force between the carbon surface and the adsorbate. Adsorption can depend upon the dissociation equilibria and functionality of particular functional groups on the biochar surface, for example, phenolic and alcoholic hydroxyl groups; carboxylic and lactonic groups; ketone and carbonyl groups; aromatic and heterocyclic carbons; pyridinic N, pyrrolic N, and quaternary N types of nitrogen species; and many others. This array of potential biochar surface groups depends on a number of factors, including biomass type, pyrolysis temperature, heating rate, and heating time, and the type of pyrolysis, for example, batch, continuous, or fluidized.

Although activated carbons and modified carbons are widely used for an adsorbent material, this is quite expensive on a large scale because of combined raw material, high energy, and high chemical production costs. For this reason, many researchers have targeted developing novel sorbents derived from biomass and carbonaceous wastes, which are of high capacity and low cost. More recently, higher capacity adsorbents have been developed using nanoadsorbent materials and metal organic frameworks, but the processing cost makes these materials currently more expensive.

Consequently, in the past several years, several cheap adsorbents are now being produced by biomass pyrolysis and described as biochars; these are being used for contaminated water clean-up applications. This process is feasible and economical if the adsorbent material is cheap and easily available (Mohan et al., 2014). The conversion of waste biomass into product biochars using the process of fast or slow pyrolysis provides an efficient and effective reaction operation with highly valuable by-products of biofuels like bio-syngas and bio-oil. The thermal process does require energy, but this can be provided from the side reaction by-products, and the biochars have a significantly greater surface area and porosity, coupled with chemically functional moieties providing a much more capable and powerful adsorbent material than the original raw biomaterial for water treatment and color removal (Mui et al., 2010b; Patel et al., 2021; Vimal et al., 2019; Gupta et al., 2016). There are well over 1000 papers in the literature on color removal and over 100 of these are based on dye removal using biochars, including biochars obtained from a worm-derived vermicomposting process, waste cabbage, algae residues, and animal manures (Yang et al., 2016; Sewu et al., 2017; Nautiyal et al., 2016; Khan et al., 2020; Huang et al., 2020). Furthermore, over 100 journal papers have appeared on the production and application of modified biochars alone for dye color removal. This will be discussed in Section 5.2.

The dye capacities for unmodified biochars and several others are presented in Tables 12.1 and 12.2. The tables are divided according to the type/category of dye since the chemical groups on the dye molecule will be one of the major influencing factors on the amount of dye uptake, as described in Section 2.

The key properties governing the adsorption potential of the biochars are presented in Tables 12.1 and 12.2 and also later in Tables 12.3 and 12.4 for the modified biochars. These parameters were listed in Section 4. The values and their influence on the dye adsorption capacity can be observed in the tables. The available surface area is an important sorbent property and has a strong dependence on the pyrolysis temperature as well as the type of biochar raw material. The pore volume and particle size distribution are similarly dependent on these two factors. Process factors, such as solution pH and temperature, can significantly influence adsorption capacities.

The adsorption properties for the uptake of cationic dyes onto the unmodified biochars are shown in Table 12.1. The data include values for methylene blue (Zubair et al., 2020a; Yu et al., 2021; Yin et al., 2019; Santos et al., 2019; Park et al., 2019; Biswas et al., 2020), malachite green (Vigneshwaran et al., 2021a; Ganguly et al., 2020;

Table 12.1 Adsorption of basic (cationic) dyes on biochars.

| Biochar feedstock | Dye name/type | Pyrolysis conditions | | | BET SA/ pore vol. (cm³/g) | Capacity (mg/g)/ isotherm type | Kinetic model | Parameters | | | Mechanism | Reference |
| --- | --- | --- | --- | --- | --- | --- | --- | --- | --- | --- | --- | --- |
| | | Temp. (°C) | Heating rate (°C/min) | Time | | | | Dose (mg/L) | pH | Equil. time (min) | | |
| Date palm fronds | Methylene blue/ (cationic) | 700 | – | 4h | 432 m²/g 0.134 | 207 mg/g | – | 0.5 g | 6 | 36 | – | Zubair et al. (2020a) |
| Tapioca peel | Malachite green/ cationic | 800 | 10 | 3h | – | 32% L, F | PFO, PSO | 0–0.16 g | 2–10 | 0–180 | – | Vigneshwaran et al. (2021a) |
| Tapioca peel | Rhodamine B/cationic | 800 | 10 | 3h | – | 66% L, F | PFO, PSO | 0–0.16 g | 2–10 | 0–180 | – | Vigneshwaran et al. (2021a) |
| Chlorella sp. microalgal | Methylene blue/ (cationic) | MW heating (2450 MHz, 800 W) | – | – | 2.66 m²/g | 113 mg/g F, Te | PFO, PSO, E, I | 0.1, 1, 2, 3, and 5 g/L | 2–10 | 7200 | B, I | Yu et al. (2021) |
| Rice husk | Malachite green/ cationic | 400–600 | – | 1h | – | 66 mg/g L, F | PFO, PSO, E, I | 50, 100, 150, 200 mg | 2, 4, 6, 8 | 1440 | – | Ganguly et al. (2020) |
| Crab shell | Malachite green/ cationic | 800 | – | 2h | 81.6 m²/g, 0.0861 | 12,502 mg/g L | PSO | 0.5 g/L | 7 | 2 | EA, HB, π-π | Dai et al. (2018) |
| Areca leaf | Methylene blue/ cationic | 200 | 5 | 1h | 20.9 m²/g | 123 mg/g L, F | PFO, PSO | 0.5 g/L | 7 | 720 | EA | Yin et al. (2019) |
| Wodyetia bifurcata | Methylene blue/ cationic | 700 | 10 | 30 | – | 149 S | PFO/ PSO | 10 | – | 30 | – | Santos et al. (2019) |
| Waste wheat straw/wheat bran | Malachite green/ cationic | 800 | 15 | 90 | – | 1739 L | PSO | 0.02 g | 2, 4, 6, 8, 10 | – | EI, C | Yang et al. (2019) |
| Waste wheat straw/wheat bran | Crystal violet/ (cationic) | 800 | 15 | 90 | – | 176, L | PSO | 0.02 g | 2, 4, 6, 8, 10 | – | EI, C | Yang et al. (2019) |

| Biomass | Dye/type | Temp | Heating rate | Time | BET SA / Pore volume | Value | Kinetic | Dose | pH | Eq. time | Mechanism | Reference |
|---|---|---|---|---|---|---|---|---|---|---|---|---|
| Switchgrass-biochar | Methylene blue/cationic | 600 | — | 1 h | 256 m²/g 0.029 | 37.6 L | PSO | 2 g/L | 6 | — | I | Park et al. (2019) |
| Switchgrass-biochar | Methylene blue/cationic | 900 | — | 1 h | 642 m²/g 0.058 | 196 L | PSO | 2 g/L | 6 | — | I | Park et al. (2019) |
| Mango leaves | Crystal violet/cationic | 800 | — | 1 h | 168 m²/g | 176 | — | 0.5 g | 8 | 48 | — | Vyavahare et al. (2019) |
| Ulothrix zonata algae | Malachite green/cationic | 800 | 15 | 90 | 133 m²/g | 5306 F | PSO | 5 mg | 2, 4, 6, 10 | 840 | C | Chen et al. (2018) |
| Ulothrix zonata algae | Crystal violet/cationic | 800 | 15 | 90 | 133 m²/g | 1222 F | PSO | 5 mg | 2, 4, 6, 10 | 840 | C | Chen et al. (2018) |
| Bovine bones | Basic Red 9/cationic | 800 | 10 | 1 h | 90.3 m²/g 0.271 | 49.5 F/L | PSO | 2.5 g/L | 7 | 180 | C | Côrtes et al. (2019) |
| Bovine bones | Basic Red 9/cationic | 800 | 10 | 3 h | 94.5 m²/g, 0.193 | 52.3 | PFO | 2.5 g/L | 7 | 180 | C | Côrtes et al. (2019) |
| Sugarcane bagasse | Methylene blue/cationic | 500 | 10 | 1.5 h | 264 m²/g | 71.2 L, F | PFO PSO | 0.05–0.25 g/50 mL | 7.4 | 180 | E, I | Biswas et al. (2020) |

Temp., temperature (°C); Heating rate, °C/minute; Time, minutes (min); BET SA: nitrogen surface area (m²/g); Pore volume: cm³/g; Chemicals: TETRA, triethylenetetramine; Technologies: HC, hydrothermal carbonization; Isotherm type: L, Langmuir; F, Freundlich; L-F, Langmuir–Freundlich (Sips); MW, microwave; R-P, Redlich-Peterson; T, Toth; Te, Temkin, H, Hill; Kinetic model: PFO, pseudo-first order; PSO, pseudo-second order; E, Elovich; A, Avrami, D-R, Dubinin-Radushkevich; Adsorption mechanism: I, intraparticle diffusion; EMT, external mass transfer, B, Boyd; EI, electrostatic interaction; IE, Ion exchange; MC, metal complexation; HB, hydrogen bonding; EA, electrostatic attraction; π-π, π-π interactions; SA, surface adsorption; PF, pore filling; ER, electrostatic repulsion, C, chemisorption; P, physisorption; PC, physiochemical; OB, outer boundary; Equilibrium time: minutes.

Table 12.2 Adsorption of acid (anionic) dyes on biochars.

| Biocharfeed material | Dye name/type | Pyrolysis conditions ||| | BET SA/ pore vol. | Capacity (mg/g) (or) removal efficiency (%)/ isotherm type | Adsorbent parameters |||| Mechanism | Reference |
| --- | --- | --- | --- | --- | --- | --- | --- | --- | --- | --- | --- | --- |
| | | Temp. (°C) | Heat rate (C/min) | Time | | | Kinet model | Dosage | pH | Equil. time (min) | | |
| *Chlorella sp.* microalgal | Congo red (CR)/ anionic | MW heating (2450 MHz, 800 W) | – | – | 2.66 | 164 mg/g/ L, F, Te | PFO, PSO, I, E | 0.1, 1, 2, 3.5 g/L | 2–10 | 240 | B, I | Yu et al. (2021) |
| Rice husk | Congo red (CR)/ anionic | 500 | 5 | 3h | – | 66.8–96.9%/L, F | – | 0.5, 1.0, 1.5, 2.5, 5.0 g/100 mL | 2, 4, 6, 7, 9, 11 | 5760 | – | Khan et al. (2020) |
| *Eucheuma spinosum* | Reactive red RR120/ anionic | 300–600 | 10 | 2h | – | 332 mg/g, L, F, Te | PFO, PSO, E | 0.250 g/L | 3–9 | 20 | EI, IE, MC, HB | Gurav et al. (2021) |
| *Phoenix dactylifera* leaves | Congo red/ anionic | 400 | – | – | 1.07 | 25.3 mg/g/L, F | PFO, PSO | 20 mg/L | 5.8 | 120 | – | Iqbal et al. (2021a) |
| Cotton stalks | Congo red/ anionic (acid) | 400 | 8 | 1.5h | – | 250 mg/g/ L, F, Te, D-R | PFO, PSO, I | 0.5–2 g/L | 2–10 | 180 | EA | Iqbal et al. (2021b) |
| Orange Peel | Congo red anionic | 800 | 15 | 15 | – | 17 mg/g | – | – | – | – | – | Yek et al. (2020) |
| Green marine algae (*Caulerpa scalpelliformis*) | Remazol brilliant violet 5R/ anionic | 300, 350, 400, 450, 500 | 5 | 2h | – | 70%/L, F, L-F, T | PFO, PSO | – | 2–5 | – | – | Gokulan et al. (2019) |
| Green marine algae (*Caulerpa scalpelliformis*) | Remazol brilliant orange 3R/ anionic | 300, 350, 400, 450, 500 | 5 | 2h | – | 77%/L, F, L-F, T | PFO, PSO | – | 2–5 | – | – | Gokulan et al. (2019) |
| Green marine algae | | 300, 350, 400, 450, 500 | 5 | 2h | – | 75%/L, F, L-F, T | PFO, PSO | – | 2–5 | – | – | Gokulan et al. (2019) |

| Material | Dye/type | Temp (°C) | Heating rate | Time | BET SA / Pore volume | Capacity | Kinetic | Concentration | pH | Equilibrium time | Mechanism | Reference |
|---|---|---|---|---|---|---|---|---|---|---|---|---|
| Crab shell (*Caulerpa scalpelliformis*) | | 800 | – | 2h | 81.6, 0.0861 | 20,317 mg/g, L | PFO, PSO | 0.5 g/L | 4 | 2 | EA, HB, π-π | Dai et al. (2018) |
| Activated carbon biochar | Congo red/anionic | 450 | 20 | 2h | – | 234 mg/g F | – | 0.2–1 g/100 mL | 2–10 | 120 | – | Das et al. (2021) |
| Spirulina platensis algae | Congo red/anionic | 450 | 20 | 2h | – | F | – | 0.2–1 g/100 mL | 2–10 | 120 | – | Nautiyal et al. (2016) |
| Waste wheat straw/wheat bran | Congo red/anionic | 800 | 15 | 90 | – | 86.9 mg/g, L | PSO | 0.02 g | 2, 4, 6, 8, 10 | – | C, EI | Yang et al. (2019) |
| Switchgrass-biochar | Orange G/anionic (acid) | 600 | – | 1 h | 256 m²/g 0.029 | 8.2 mg/g, L | PSO | 2 g/L | 6 | – | OB | Park et al. (2019) |
| Switchgrass-biochar | Congo red/anionic | 600 | – | 1 h | 256 m²/g 0.029 | 8.0 mg/g, L | PSO | 2 g/L | 6 | – | OB | Park et al. (2019) |
| Switchgrass-biochar | Congo red/anionic | 900 | – | 1 h | 642 m²/g 0.058 | 22.6 mg/g, L | PSO | 2 g/L | 6 | – | OB | Park et al. (2019) |
| *Ulothrix zonata* algae | Congo red/anionic | 800 | 15 | 90 min | 133 m²/g | 345 mg/g, F | PSO | 5 mg | 2, 4, 6, 10 | 840 | C | Chen et al. (2018) |
| Corn cob | Methyl orange/anionic | 600 | 15 | 120 min | 469 m²/g | 86.4 mg/g, F | PSO | 1.0 g/L | 5.6 | – | PCI | Zhang et al. (2020) |

Temp., temperature (°C); Heating rate, °C/minute; Time, minutes (min); BET SA: nitrogen surface area ($m^2/g$); Pore volume: $cm^3/g$; Chemicals: TETRA, triethylenetetramine; Technologies: HC, hydrothermal carbonization; Isotherm type: L, Langmuir; F, Freundlich; L-F, Langmuir-Freundlich (Sips); MW, microwave; R-P, Redlich-Peterson; T, Toth; Te, Temkin, H, Hill; Kinetic model: PFO, pseudo-first order; PSO, pseudo-second order; E, Elovich; A, Avrami, D-R, Dubinin-Radushkevich; Adsorption mechanism: I, intraparticle diffusion; EMT, external mass transfer, B, Boyd; El, electrostatic interaction; IE, ion exchange; MC, metal complexation; HB, hydrogen bonding; EA, electrostatic attraction; π-π, π-π interactions; SA, surface adsorption; PF, pore filling; ER, electrostatic repulsion; C, chemisorption; P, physisorption; PC, physiochemical; OB, outer boundary; Equilibrium time: minutes.

Table 12.3 Adsorption of basic (anionic) dyes on modified biochars.

| Biochar modification | Dye name/ type | Pyrolysis conditions | | | BET SA/ pore vol. | Capacity (mg/g) (or) removal efficiency (%)/isotherm type | Kinet model | Adsorbent parameters | | | | Reference |
|---|---|---|---|---|---|---|---|---|---|---|---|---|
| | | Temp. | Heat rate | Time | | | | Dosage | pH/ pH (opt.) | Equil. time | Mechanism | |
| Date palm fronds | Methylene blue cationic | 800 | 20 | 4h | 66.5 m²/g | 210 mg/g | – | 2 g/L | 7 | 180 | – | Hoda and Ghany (2019) |
| Tapioca peel + S doped | Malachite green cationic | 800 | 10 | 3h | 146 m²/g | 30.2 mg/g/ 74% (180 min)/ L, F | PFO, PSO | 0–0.16 g | 2–10 | 1080 | – | Vigneshwaran et al. (2021a) |
| Tapioca peel + S doped | Rhodamine B/cationic | 800 | 10 | 3h | 146 m²/g | 33.1 mg/g/ 92% (180 min)/ L, F | PFO, PSO | 0–0.16 g | 2–10 | 1080 | – | Vigneshwaran et al. (2021a) |
| Areca leaf + $K_2FeO_4^-$ | Methylene blue/ cationic | 200 | 5 | 1h | 20.9 m²/g | 252 mg/g L, F | PFO, PSO | 0.5 g/L | 7 | 720 | EA | Yin et al. (2019) |
| Chitosan-tapioca peel + S doped | Malachite green cationic | 600 | – | 2h | 121 m²/g | 53.3 mg/g / L, F | PFO, PSO | 25–200 mg | 2–12 | 160 | EA, HB | Vigneshwaran et al. (2021b) |
| Chitosan-tapioca peel + S doped | Rhodamine B/cationic | 600 | – | 2h | 120 m²/g | 40.86 mg/g / L, F | PFO, PSO | 25–200 mg | 2–12 | 160 | EA, HB, π-π | Vigneshwaran et al. (2021b) |
| Sugarcane bagasse + steam | Methylene blue cationic | 800 | 10 | 2h | 570 m²/g 0.356 | 5220 mg/g L, F | – | 0.05–0.25 g/50 mL | 7.4 | 180 | – | Carrier et al. (2012) |
| Date palm fronds with Fe/Mn | Methylene blue cationic | 700 | 3 | 4h | 432 m²/g | 303 mg/g/ L, F | PFO, PSO, ID, E | 2–50 mg | 4–10 | 240 | SA, π-π, IE, PF | Zubair et al. (2020b) |
| Wakame Undaria pinnatifida leaves with calcination | Methylene blue cationic | 800 | 10 | 2h | 1156 m²/g | 842 mg/g/ L, F | PFO, PSO | 0.01–0.07 g | 2–12 | 300 | SA, HB, π-π, PF | Yao et al. (2020) |

| Biomass | Dye | Temp (°C) | Heating rate | Time | BET SA / Pore volume | $q_{max}$ / Isotherm | Kinetic | Dose | pH | Equilibrium time | Mechanism | Reference |
|---|---|---|---|---|---|---|---|---|---|---|---|---|
| Wakame Undaria pinnatifida leaves with calcination | Rhodamine B Cationic | 800 | 10 | 2h | 1156.25 m²/g | 534 mg/g, L, F | PFO, PSO | 0.01–0.07 g | 2–12 | 300 | SA, HB, π-π, PF | Yao et al. (2020) |
| Wakame Undaria pinnatifida leaves with calcination | Malachite green cationic | 800 | 10 | 2h | 1156.25 m²/g | 4067 mg/g, L, F | PFO, PSO | 0.01–0.07 g | 2–12 | 300 | SA, HB, π-π, PF | Yao et al. (2020) |
| Corn straw | Malachite green cationic | 500 | – | 3h | 36.3 m²/g | 516 mg/g, L, F, Te | PFO, PSO, I | 0.005–0.020 mg | 2–9 | 20 | | Yang (2003) |
| Rice husk + Cu+Al | Malachite green cationic | 80 | – | 1h | 201 m²/g, 0.350 | 471 mg/g | L, F | 1 g/L | 9 | 200 | PF, π-π | Palapa et al. (2020) |
| Litchi Peel + HC | Malachite green cationic | 850 | | 1h | 1006 m²/g, 0.588 | 2468 mg/g, F | E | 0.02 g | 8 | 720 | HB, π-π, PF, EI | Wu et al. (2020) |
| Sugarcane bagasse + $ZnCl_2$ | Malachite green cationic | 800 | – | 2h | 51.7 m²/g, 0.0235 | 90.1 mg/g, F | PSO | 4 g/L | 8 | – | B | Das et al. (2021) |

Temp., temperature (°C); Heating rate, °C/minute; Time, minutes (min); BET SA: nitrogen surface area (m²/g); Pore volume: cm³/g; Chemicals: TETRA, triethylenetetramine; Technologies: HC, hydrothermal carbonization; Isotherm type: L, Langmuir; F, Freundlich; L-F, Langmuir–Freundlich (Sips); MW, microwave; R-P, Redlich-Peterson; T, Toth; Te, Temkin; H, Hill; Kinetic model: PFO, pseudo-first order; PSO, pseudo-second order; E, Elovich; A, Avrami; D-R, Dubinin-Radushkevich; Adsorption mechanism: I, intraparticle diffusion; EMT, external mass transfer; B = Boyd; EI, electrostatic interaction; IE, ion exchange; MC, metal complexation; HB, hydrogen bonding; EA, electrostatic attraction; π-π, π-π interactions; SA, surface adsorption; PF, pore filling; ER, electrostatic repulsion; C, chemisorption; P, physisorption; PC, physiochemical; OB, outer boundary; Equilibrium time: minutes.

Table 12.4 Adsorption of acid (cationic) dyes on modified biochars.

| Biochar modification | Dye name/type | Pyrolysis conditions | | | BET SA/ pore vol. | Capacity (mg/g) (or) removal efficiency (%)/isotherm type | Kinet model | Adsorbent parameters | | | Mechanism | Reference |
|---|---|---|---|---|---|---|---|---|---|---|---|---|
| | | Temp. (°C) | Heat rate | Time | | | | Dose | pH | Equil. Time min | | |
| Rubber seeds + NaOH | Congo red | 800 | – | 6h | – | 458 mg/g L, F, D-R | – | 0.01–0.08g | 6–7 | 120 | – | Nizam et al. (2021) |
| Shorea robusta leaf extract +AgNPs | Congo red (CR)/ anionic | 300 | – | 3h | 20.90 m²/g | 23.5 mg/g/ L, F, Te, D-R | PFO, PSO, I, E | 0.1–1.0 g/L | 2–10 | 90 | EA, HB | Shaikh et al. (2021) |
| Shorea robusta leaf extract | Congo red (CR)/ anionic | 300 | – | 3h | 1.5 m²/g | 1.68 mg/g | – | – | 2–10 | 60 | EA | Shaikh et al. (2021) |
| Food waste +ultrasound + $H_2O_2$ | Methylene orange/ (anionic) | 300 | 5 | 7h | – | 68.69%/ – | – | 50 mg/L | 7 | 60 | – | Chu et al. (2020) |
| Phoenix dactylifera leaves + Mn | Congo red/ anionic | 400 | – | – | 57.74 cm²/g | 118 mg/g/ L, F | PFO, PSO | 200, 500, 1000 mg/L | 2.5, 4.5, 5.8, 8.2, 11.2 | 120 | EI | Iqbal et al. (2021a) |
| Marine Chlorella + ultrasound | Reactive yellow/ anionic | 450, 550, 650 | 10 | 1 | 351 m²/g | 48.3 mg/g/ BC550°C L, F, Te | PFO, PSO | 0.1–0.5 g/L | 2.0–10.0 | 2 | EI, ER | Hernández-Zamora et al. (2015) |
| Sludge rice husk composite | Reactive blue 19/ acid Orange II Direct Red | 500 | 7 | 2h | 29.2 m²/g 0.058 | 38.5 mg/g 42.1 mg/g 59.8 mg/g L, F, Te, D-R | PFO PSO I, E | 0.1 g/L | 7 | 1440 | – | Chen et al. (2019) |
| Arjuna (Terminalia Arjuna) seeds | Congo red/ anionic | – | – | – | 170 m²/g | 92.0 ± 5.0%/ - | – | – | – | – | – | Goswami et al. (2020) |
| Cotton stalks | Congo red/ anionic | 400 | 8 | 90 min | – | 557 mg/g/L, F, Te, D-R | PFO, PSO, I | 0.5–2 g/L | 2–10 | 180 | EA | Iqbal et al. (2021b) |
| Orange peel + $CO_2$ + steam | Congo red/ anionic | 700 | – | 10 min | 305 m²/g | 136 mg/g F | PSO | 3 g/L | 2–3 | 1440 | EI | Yek et al. (2020) |

| | | | | | | | | | |
|---|---|---|---|---|---|---|---|---|---|
| Litchi peel + hydrothermal | Congo red/ anionic | 850 | | 1h | 1006 m²/g 0.588 | | E | 0.02 g | 4 | 720 | HB, $\pi$-$\pi$, PF, EI | Wu et al. (2020) |
| Spirulina/ alginate/ paper | Congo red/ anionic | 450 | 20 | 2h | — | 41.1 mg/g L, F, Te, D-R | PFO PSO | 0.2 g/100 mL | 6–8 | 0–120 | EA, HB, $\pi$-$\pi$ | Fawzy and Gomaa (2020) |
| Pine nutshell | Acid chrome blue Methyl orange | 700 | 10 | 180 | — | 27.2 mg/g 11.0 mg/g L, F | PFO PSO | 0.010 g | 3 | 1200 | EI, $\pi$-$\pi$ | Wang et al. (2020) |

Temp., temperature (°C); Heating rate, °C/minute; Time, minutes (min); BET SA: nitrogen surface area (m²/g); Pore volume: cm³/g; Chemicals: TETRA, triethylenetetramine; Technologies: HC, hydrothermal carbonization; Isotherm type: L, Langmuir; F = Freundlich; L-F, Langmuir–Freundlich (Sips); MW = microwave; R-P, Redlich-Peterson; T, Toth; Te, Temkin, H, Hill; Kinetic model: PFO, pseudo-first order; PSO, pseudo-second order; E, Elovich; A, Avrami; D-R, Dubinin-Radushkevich; Adsorption mechanism: I, intraparticle diffusion; EMT, external mass transfer, B, Boyd; EI, electrostatic interaction; IE, ion exchange; MC, metal complexation; HB, hydrogen bonding; EA, electrostatic attraction; $\pi$-$\pi$, $\pi$-$\pi$ interactions; SA, surface adsorption; PF, pore filling; ER, electrostatic repulsion; C, chemisorption; P, physisorption; PC, physiochemical; OB, outer boundary; Equilibrium time: minutes.

Dai et al., 2018; Yang et al., 2019; Chen et al., 2018), rhodamine B (Vigneshwaran et al., 2021a), crystal violet (Chen et al., 2018; Vyavahare et al., 2019), and basic red 9. Several references only report the percentage of dye removed (Zubair et al., 2020a; Yin et al., 2019; Vigneshwaran et al., 2021a; Ganguly et al., 2020; Vyavahare et al., 2019; Côrtes et al., 2019), which is useful, but this value is a variable depending on the adsorbent mass, the solution volume, and the dye concentration. In contrast, the maximum adsorption capacity, $q_{max}$ in mg dye/g biochar, is a fixed value. For all other equilibrium dye concentrations, at a fixed temperature, a single isotherm equation provides the value of the adsorption capacity for a fixed effluent discharge concentration.

The adsorption capacities of methylene blue (Table 12.1) are 113, 149, 37.6, and 196 mg/g for biochars from microalgae (Yu et al., 2021), Wodyetia (Santos et al., 2019), switchgrass prepared at two temperatures, namely 600°C and 900°C (Cheremisinoff, 1994; Park et al., 2019), respectively. The effect of temperature on switchgrass biochar has a major effect on the adsorption capacity of biochar. Some literature capacity values for methylene blue are 112, 137, and 454 mg/g for bagasse (Biswas et al., 2020), phosphoric acid-activated olive seed carbon (Lafi, 2001), and the high-capacity bamboo cane active carbon (Hameed et al., 2007). Biochar capacities for malachite green dye are exceptionally high, namely 12,502, 1739, and 5306 mg/g on crab shell (Dai et al., 2018), wheat/bran straw-fed larvae biochar (Yang et al., 2019), and Ulothrix algae-derived biochar (Chen et al., 2018). Some literature capacity values are 667, 318, and 1428 mg/g, respectively, from grape processing waste-activated carbon (Sayğılı and Güzel, 2015), shrimp shell (Salamat et al., 2019), and activated carbon derived from plastic waste (Li et al., 2021). The amounts of rhodamine B adsorbed on the tapioca shell biochar is quite low at 33.1 mg/g (Vigneshwaran et al., 2021a), while literature values include 76.7 using *Acacia mangium* wood-derived carbon (Danish et al., 2018) and 30–40 mg/g on activated carbons from carnauba, macauba, and pine nut wastes, activated using phosphoric acid and calcium chloride (Lacerda et al., 2013). Two capacities are listed for the removal of crystal violet dye, namely 176 mg/g and another very high value of 1222 mg/g on biochar from mango leaves (Yang et al., 2019) and from *Ulothrix zonata* algae (Chen et al., 2018), respectively. Literature capacity values vary widely from 602 mg/g for a bentonite-alginate composite (Fabryanty et al., 2017) to 76 mg/g for chitosan hydrogel beads (Pal et al., 2013). The final values in Table 12.1 are both for the adsorption of basic red 9 (BR9) onto biochar from animal waste (Côrtes et al., 2019) heated for 1 and 3 h. The BR9 capacities were 49.5 and 52.3 mg/g at 1 (surface area 90 m$^2$/g) and 3 h (surface area 94.5 m$^2$/g). These capacities followed the general trend in pyrolysis times and surface areas. Literature values are slightly lower but of a similar order of magnitude, for example, 28.8 and 15.0 mg/g for sepiolite (Osman et al., 2017) and fish bone (Kizilkaya, 2012).

The adsorption properties of anionic dyes onto the unmodified biochars are shown in Table 12.2. They include acid orange 7 (Zubair et al., 2020a), Congo red (Nautiyal et al., 2016; Khan et al., 2020; Yu et al., 2021; Park et al., 2019; Dai et al., 2018; Yang et al., 2019; Chen et al., 2018; Iqbal et al., 2021a,b; Yek et al., 2020; Das et al., 2021), reactive red, RR120 (Gurav et al., 2021), remazol violet 5R, remazol orange 3R, remazol blue R (Gokulan et al., 2019), orange G (Park et al., 2019), and methyl orange (Zhang et al., 2020). Several of these references only reported the percentage (Nautiyal et al., 2016; Khan et al., 2020; Huang et al., 2020; Zubair et al., 2020a; Gokulan et al., 2019) of dye removed, while this value varies depending on the adsorbent mass, the solution volume, and the dye concentration.

The adsorption properties of acid orange 7 (AO7) using peanut shell biochar are shown in Table 12.2, and only dye removal percentages are cited. The literature gives only a few examples of AO7 adsorption capacities which vary widely, including 83, 180, and 50 mg/g on fly ash (Janos et al., 2003), chemically activated sawdust (Janos et al., 2009) and oxihumolite (Janos et al., 2005), respectively.

Congo red is one of the most studied anionic dyes, so the several examples in Table 12.2 represent only a small fraction of the literature studies. The adsorption capacities are 164, 25.3, 250, 91, 20,317, 234, 86.9, 8.0, 22.6, and 345 mg/g on biochars prepared from chlorella microalgae (Yu et al., 2021), dactylifera (Iqbal et al., 2021a), cotton stalk (Iqbal et al., 2021b), orange peel (Yek et al., 2020), crab shell (Dai et al., 2018), activated carbon biochar (Das et al., 2021), spirulina algae (Nautiyal et al., 2016), wheat straw/wheat bran larvae (Yang et al., 2019), switchgrass (pyrolyzed at temperatures 600°C and 900°C) (Park et al., 2019), and Ulothrix algae (Chen et al., 2018). The switchgrass study indicated a higher biochar capacity is produced at the higher pyrolysis temperature.

Most studies have reported the adsorption capacity of Congo red dye as less than 100 mg/g with three or four exceptions going up to 350 mg/g and the very exceptional value of 20,317 mg/g on crab shell biochar (Dai et al., 2018). This highest value was obtained from crab shell pyrolyzed at 800°C, having a surface area of 81.57 m$^2$/g, using a volume to mass ratio of 2 and a pH value of 4 but requiring a very high initial Congo red concentration of over 20 g/L. Date stone-derived activated carbon has a capacity of 35.2 mg/g, which is low. These low capacities may be attributed to (i) the large dye molecular size (molecular mass of 697 g/mol) and (ii) and a total pore volume of 0.086. Congo red dye is a direct dye, lacking the strong anionic bonding properties of other anionic acid dyes (Bouchemal and Addoun, 2009; Bouchemal et al., 2012). This phenomenon emphasizes the importance of studying pore size distribution of adsorbents. A mesoporous activated carbon was produced (Li et al., 2016a) using phosphoric acid activation followed by microwave. This carbon achieved a higher surface area and an adsorption capacity of 351 mg/g.

Several bamboo biochars were produced between 240 and 280°C and reaction times from 0.5–6.0 h (Li et al., 2016b). Their Congo red uptake capacities were in the range 30–100 mg/g with the highest values on biochars produced at the two highest temperatures—240°C and 280°C, and the longer thermal processing periods of 5 and 6 h. An activated carbon derived using apricot stones (Abbas and Trari, 2015) at only 250°C for 4 h gave a low Congo red capacity of 32.9 mg/g. This was probably due to the low pyrolysis temperature and the low specific surface area. The last two studies had adsorption capacities in the same range as the date stone carbon. More studies should be tried to produce biochars using microwave and possibly plasma technology.

Anionic reactive red 120 was absorbed on *Eucheuma spinosum*-derived biochar (Gurav et al., 2021) in a high (332 mg/g) capacity. Literature values are high for activated carbon (Oueslati et al., 2020) and $Fe_3O_4$ magnetic nanoparticles (Absalan et al., 2011) with capacities of 255 and 167 mg/g, respectively. Green marine algae biochar (Gokulan et al., 2019) has been used for remazol dyes, brilliant violet 5R, and brilliant orange 3R, although no capacities were stated the dye removal percentages reported were over 70%.

Literature values for the remazol dyes are quite low at 16.0 and 66.8 mg/g for brilliant violet 5R and brilliant orange 3R, respectively, on calcined eggshell (Rapo et al., 2019) and coffee husk-derived activated carbon (Ahmad and Rahman, 2011), respectively.

A low orange G adsorption capacity of 8.2 mg/g on switchgrass biochar (Park et al., 2019) produced at 600°C is shown in Table 12.2. This was largely due to the low surface of 256 $m^2$/g. Other literature values include 18.8 and 9.1 for nanoporous activated carbon (Kundu et al., 2017) and activated carbon produced from *Thespesia populnea* (Arulkumar et al., 2011), respectively. All these values suggest orange G is a problematic dye to treat.

A biochar produced from corn cob (Zhang et al., 2020) adsorbed methyl orange with a capacity of 86.4 mg/g. An adsorbent derived from amidoxime (Rahman et al., 2019) has a capacity of 142 mg/g.

## 5.2 Dye removal using adsorption onto modified biochars

Many studies now modify biochars to enhance their adsorption performance. Several hundred papers have been published in the last few years applying various biochar modification techniques, for example, biochar treatment using various acids and bases, coating/encapsulation processes, chemical impregnation, and size modification. This section will also exhibit two tables showing the characteristic and performance properties of cationic dye adsorption in Table 12.3 and anionic dye adsorption in Table 12.4. Due to the very large number of existing references, the tables will mostly be confined to modifications of the biochar raw materials already appearing in Tables 12.1 and 12.2 for ease of comparison. Furthermore, since general literature values of adsorption capacities have already been cited, there is no further need to add to these. Table 12.3 is presented showing the adsorption characteristics of modified biochars for the adsorption of cationic dyes.

The date palm frond biochar (Zubair et al., 2020a) in Table 12.1 had a capacity of 207 mg/g for methylene blue, whereas the modified date frond biochars had capacities of 210 and 303 mg/g on 800°C thermally activated fronds (Hoda and Ghany, 2019) and Fe/Mn impregnated fronds activated at 700°C (Zubair et al., 2020b), respectively. The tapioca peel-derived biochar (Vigneshwaran et al., 2021a) presented in Table 12.1 had a malachite green removal percentage of 32% and a rhodamine B removal percentage of 66%. When prepared identically, sulfur-doped tapioca peel biochar (Vigneshwaran et al., 2021a) has 74% and 92% removal percentages, respectively, corresponding to adsorption capacities of 30.2 and 33.1 mg/g rhodamine B, respectively. Further modification was undertaken to coat the sulfur doped tapioca peel with chitosan (Vigneshwaran et al., 2021b) at 600°C, which further enhanced the dye uptake capacities to 53.3 mg/g for malachite green and 40.9 mg/g for rhodamine B, respectively. The unmodified areca plant biochar has a methylene blue capacity of 189 mg/g (Yin et al., 2019). After modification with $K_2FeO_4$, the wakame biochar capacity increased to 252 mg/g (Yin et al., 2019). In Table 12.1, the chlorella microalgae biochar has a methylene blue capacity of 113 mg/g and is one of several seaweed/algae species biochars with adsorption capacities in the range of 25–130 mg/g. The adsorption of malachite green on chlorella biochar gave a capacity of about around 80 mg/g (Tsai and Chen, 2010). Rhodamine B uptake on seaweed biochar was only 11 mg/g capacity. By contrast very large adsorption capacities of 842, 534, and 4067 mg/g for methylene blue, rhodamine B, and malachite green were obtained on calcined wakame algal biochar (Yao et al., 2020). The rice husk biochar capacity for malachite green in Table 12.1 was 66 mg/g (Ganguly et al., 2020); after modification with Cu+Al (Palapa et al., 2020), the capacity became 471 mg/g. Sugarcane bagasse biochar had a capacity of 71.2 mg/g (Biswas et al., 2020), while the $ZnCl_2$-modified bagasse biochars and steam-activated bagasse biochar (Carrier et al., 2012) had capacities of 90.1 and 5220 mg/g, respectively. The two modified biochars (Carrier et al., 2012) had surface areas 120 and 570 $m^2$/g. Table 12.3 shows other high-capacity adsorbents, for example, malachite green on corn straw (Yang, 2003) and litchi peel (Wu et al., 2020) with super capacities of 516 and 2468 mg/g, respectively.

The adsorption properties of anionic dyes onto a selection of the modified biochars are shown in Table 12.4. Biochars from rubber seeds were prepared at 800°C and activated with NaOH. The capacity for Congo red of 227 mg/g rose to 458 mg/g after activation (Nizam et al., 2021). The leaf extract of *Shorea robusta* provided a removal capacity of only 1.98 mg/g for Congo red dye with an extremely low surface area of 1.5 $m^2$/g after heating at 300°C (Shaikh et al., 2021). Modification of this biochar using Ag nanoparticles resulted in an increased capacity of 23.5 mg/g and a surface area of 20.9 $m^2$/g (Shaikh et al., 2021).

A magnetic food waste-derived biochar has been developed for use in a Fenton style type effluent treatment process (Chu et al., 2020) achieving a removal capacity of 23%. On ultrasonic treatment, the biochar capacity increased to 33%. Finally with ultrasonics + $H_2O_2$ + biochar, a 97% removal was achieved in 3 h. In Table 12.2, the unmodified *dactylifera* leaf biochar had an adsorption capacity of 25.3 mg/g for Congo red, but after Mn activation, the capacity became 118 mg/g (Iqbal et al., 2021a). *Marine chlorella vulgaris* algae was converted to biochar (Hernández-Zamora et al., 2015) at temperatures of 450, 550, and 650°C, yielding surface areas of 266, 351, and 151 $m^2$/g, respectively. This adsorbent's capacities for the anionic reactive yellow 45 dye were 48.3 mg/g (experimental) and 57.8 mg/g dye (Langmuir) for a synthesis temperature of 550°C.

Sludge biochars have been studied extensively (Hadi et al., 2015). The anionic remazol brilliant blue R dye has a Langmuir capacity of 127 mg/g (Raj et al., 2021). Acid orange II and the two other anionic dyes, direct red 4BS and reactive blue 19, were successfully removed from solution by a sludge-rice husk composite biochar prepared at 500°C with capacities of 42.1, 59.8, and 38.5 mg/g, respectively (Chen et al., 2019). A hybrid process developed using Arjuna seeds + microorganisms and ozonation (Goswami et al., 2020) enabled the removal of 92% of Congo red from the dye solution. Cotton stalk agricultural waste is in great abundance. Its biochar has been used without further modification to remove Congo red with a 250 mg/g capacity. Modifying this biochar using ZnO nanoparticles enhanced Congo red's adsorption capacity to 557 mg/g (Iqbal et al., 2021b). Orange peel's biochar had an adsorption capacity for Congo red dye of about 17 mg/g but two modified biochars were produced using carbon dioxide and steam gave improved capacities of 91 and 136 mg/g, respectively. An 800°C hydrothermal carbonization of litchi peel for 1 h produced a modified biochar having a surface area of 1006 m$^2$/g, leading to a high Congo red capacity of 404 mg/g (Goswami et al., 2020).

A modified biochar has been developed from spirulina and paper and alginate from seaweed (Fawzy and Gomaa, 2020). This study investigated the use of algae biorefinery waste and wastepaper in the preparation of cost-effective and eco-friendly xerogels for the removal of Congo red. The developed biosorbents exhibited a light and porous network structure and were characterized by a fast uptake of dye at the optimum pH value of 6–8. The adsorption capacity was 41.15 mg/g for Congo red.

Fast pyrolysis has been used to produce biochar from raw pine nut shells at 700°C for 3 h (Wang et al., 2020). Then, this biochar was tested for dye adsorption characteristics with anionic methyl orange and acid chrome blue K dye sorbates. The unmodified biochar was then subjected to further treatment using cetyl trimethyl ammonium bromide-modified magnetic biochar and subjected to a final $FeCl_3$ modification. The adsorption capacities on the unmodified biochar are 1.0 and 16 mg/g for methyl orange and acid chrome blue K, respectively. After the modification, the capacities were 11 and 27 mg/g for methyl orange and acid chrome blue K, respectively.

# 6 Summary

A discussion of the application of biochars and modified biochars for the removal of dyes from effluents has been performed. Several hundred papers were identified in the literature on this topic and certain ones have been cited and described in this review. Dyestuffs are represented by a wide range of chemical species, chemical groups, and properties, which are presented in Section 2. A brief overview of treatment technologies is presented in Section 3 followed by the detailed discussion of the benefits of applying adsorption technology in Section 4 and finally a detailed review and tables highlighting the major properties and color removal applications of the biochars (Section 5.1) and the modified biochars (in Section 5.2) to color removal.

In general, certain performance trends can be observed and related to the properties of the biochar, for example, for the same raw material source:
- The higher the pyrolysis temperature, the higher the dye adsorption capacity, which is strongly linked to the surface area;
- The higher the temperature, the lower the biochar yield but up to about 800 °C; higher surface areas and pore volumes are available for adsorption;
- Above 800–850°C, the pore volume continues to increase but the surface area starts to decrease—due to the micropore and small mesopore walls burning out, creating fewer but larger pores;
- Pore diameter size distribution is important and depends on both the raw material type and the pyrolysis temperature—dye molecules vary enormously in size and even small dye molecules are relatively large compared to many chemical molecules; therefore, it is important to consider setting the reaction conditions of time, temperature, and heating rate, depending on the nature of the dyestuff;
- The type of surface sites on the biochar is very significant since many dyes are either cationic or anionic in solution and are attracted to oppositely charged sites—the nature of the surface sites depends on the raw material and the pyrolysis temperature;
- Slow pyrolysis is the best for biochar production and controlling the char properties as it provides high biochar yields, better pore control development, and therefore higher surface areas with a narrower spectrum of pore size distribution.

Modified biochars exist in so many variations, it is impossible to review all the different types, and only a few brief examples are mentioned:
- acid or alkali chemical treatment produces chars with positive and negative surface groups; at pyrolysis temperatures above 550 °C, these chars are usually described as activated carbons;
- sulfur doping of the raw material prior to pyrolysis will produce chars/activated carbons with a strong affinity for the toxic heavy metal ions;
- iron oxide-doped modified chars have proved very attractive adsorbents for both anions, such as arsenate and chromate, and metal cations by ion exchange;

- coating the biochars has proved enormously successful, for example, coating with chitosan has produced capacities of 5-fold and 20-fold w/w adsorption capacities putting these modified biochars into the superadsorbent category.

Overall, the prospects for biochar technology looks very promising. Many biochars are produced from waste biomass resources and therefore are counted as carbon neutral, creating both an economic and environmentally attractive product with diverse applications. There are many gaps in the cited literature, and much of the missing data will be required for designing the dyehouse effluent biochar treatment plants. Some recent articles (Dotto and McKay, 2020; Mohan et al., 2014; Wang et al., 2020; Mudhoo et al., 2021) have highlighted the "way forward" that needs to be followed in future adsorption studies.

## Questions and problems

1. List the main problems with dye containing effluents.
2. What is the difference between an anionic dye and a cationic dye?
3. Give 2 examples, with formulae, of anionic dyes.
4. Show, with structures, how these dyes ionize in water.
5. What are the main treatment technologies for dye effluents?
6. List 4 key characteristic properties for a biochar adsorbent.
7. Name 3 advantages of biochar adsorption of dyes over the other dye treatment technologies.
8. Which biochar property is important for adsorbing a very large dye molecule?
9. 9. What surface functional properties would be important for a "modified" biochar to enhance Its adsorption capacity for (a) cationic dyes and (b) anionic dyes?
10. Give four biochar properties which are influenced by the pyrolysis temperature.
11. If you have a mixed dye effluent requiring treatment, for example, two dyes, one anionic and one cationic, what type of biochar adsorbent treatment system would you select?
12. Name 2 biochar activities/characteristics that need to be studied more.

## Acknowledgment

The authors would like to thank Qatar National Research Fund for supporting this research under the National Priorities Research Program grant number NPRP11S-0117-180328. Any opinions, findings and conclusions, or recommendations expressed in this material are those of the author(s) and do not necessarily reflect the views of Hamad Bin Khalifa University or Qatar Foundation.

# References

Abbas, M., Trari, M., 2015. Kinetic, equilibrium and thermodynamic study on the removal of Congo red from aqueous solutions by adsorption onto apricot stone. Process. Saf. Environ. Prot. 98, 424–432.

Abdolali, A., Guo, W.S., Ngo, H.H., Chen, S.S., Nguyen, N.C., Tung, K.L., 2014. Typical lignocellulosic wastes and by-products for biosorption process in water and wastewater treatment: a critical review. Bioresour. Technol. 160, 57–66. https://doi.org/10.1016/j.biortech.2013.12.037.

Absalan, G., Asadi, M., Kamran, S., Sheikhian, L., Goltz, D.M., 2011. Removal of reactive red-120 and 4-(2-pyridylazo) resorcinol from aqueous samples by $Fe_3O_4$ magnetic nanoparticles using ionic liquid as modifier. J. Hazard. Mater. 192 (2), 476–484. https://doi.org/10.1016/j.jhazmat.2011.05.046.

Afroze, S., Sen, T.K., 2018. A review on heavy metal ions and dye adsorption from water by agricultural solid waste adsorbents. Water Air Soil Pollut. 229, 225–232.

Ahmad, M.A., Rahman, N.K., 2011. Equilibrium, kinetics and thermodynamic of Remazol brilliant Orange 3R dye adsorption on coffee husk-based activated carbon. Chem. Eng. J. 170 (1), 154–161. https://doi.org/10.1016/j.cej.2011.03.045.

Al-Ahmed, Z.A., Al-Radadi, N.S., Ahmed, M.K., Shoueir, K., El-Kemary, M., 2020. Dye removal, antibacterial properties, and morphological behavior of hydroxyapatite doped with Pd ions. Arab. J. Chem. 13 (12), 8626–8637. https://doi.org/10.1016/j.arabjc.2020.09.049.

Al-Degs, Y., Khraisheh, A.M., Allen, S.J., Ahmad, M.N., 2000. Effect of carbon surface chemistry on the removal of reactive dyes from textile effluent. Water Res. 34 (3), 927–935.

Al-Degs, Y., El-Barghouthi, M.I., El-Sheikh, A.H., Walker, G.M., 2008. Effect of solution pH, ionic strength, and temperature on adsorption behavior of reactive dyes on activated carbon. Dyes Pigments 77, 16–23.

Allen, S.J., McKay, G., Khader, K.Y.H., 1989. Equilibrium adsorption isotherms for basic dyes onto lignite. J. Chem. Technol. Biotechnol. 45, 291–302.

Allen, S.J., McKay, G., Porter, J.F., 2004. Adsorption isotherm models for basic dye adsorption by peat in single and binary component systems. J. Colloid Interface Sci. 280 (2), 322–333. https://doi.org/10.1016/j.jcis.2004.08.078.

Ammar, A., Dofan, I., Jegatheesan, V., Muthukumaran, S., Shu, L., 2015. Comparison between nanofiltration and forward osmosis in the treatment of dye solutions. Desalin. Water Treat. 54 (4–5), 853–861. https://doi.org/10.1080/19443994.2014.908419.

Anantha, M.S., Olivera, S., Chunyan, H., Jayanna, B.K., Reddy, N., Krishna Venkatesh, H.B., Muralidhara, R.N., 2020. Comparison of the photocatalytic, adsorption and electrochemical methods for the removal of cationic dyes from aqueous solutions. Environ. Technol. Innov. 17. https://doi.org/10.1016/j.eti.2020.100612, 100612.

Arulkumar, M., Sathishkumar, P., Palvannan, T., 2011. Optimization of Orange G dye adsorption by activated carbon of *Thespesia populnea* pods using response surface methodology. J. Hazard. Mater. 186, 827–834. 2011.

Baker, J.R., 1958. Principles of Biological Microtechnique. Pub. Methuen, London, UK.

Balcik-Canbolat, C., Sengezer, C., Sakar, H., Karagunduz, A., Keskinler, B., 2017. Recovery of real dye bath wastewater using integrated membrane process: considering water recovery, membrane fouling and reuse potential of membranes. Environ. Technol. 38 (21), 2668–2676. https://doi.org/10.1080/09593330.2016.1272641.

Banat, I.M., Nigam, P., Singh, D., Marchant, R., 1996. Microbial decolorization of textile dye containing effluents: a review. Bioresour. Technol. 58, 217–227.

Bayramoglu, M., Eyvaz, M., Kobya, M., 2007. Treatment of textile wastewater by electrocoagulation economic evaluation. Chem. Eng. J. 128 (2–3), 155–161.

Bell, C.A., Buckley, C.A., 2003. Treatment of a textile dye in the anaerobic baffled reactor. Water 29 (2), 129–134.

Beluci, N., Mateus, G., Miyashiro, C., Homem, N., Gomes, R., Fagundes-Klen, M., Bergamasco, R., Vieira, A., 2019. Hybrid treatment of coagulation/flocculation process followed by ultrafiltration in TIO2-modified membranes to improve the removal of reactive black 5 dye. Sci. Total Environ. 664, 222–229. https://doi.org/10.1016/j.scitotenv.2019.01.199.

Biswas, S., Mohapatra, S.S., Kumari, U., Meikap, B.C., Sen, T.K., 2020. Batch and continuous closed circuit semi-fluidized bed operation: removal of MB dye using sugarcane bagasse biochar and alginate composite adsorbents. J. Environ. Chem. Eng. 8 (1). https://doi.org/10.1016/j.jece.2019.103637, 103637.

Bouchemal, N., Addoun, F., 2009. Adsorption of dyes from aqueous solution onto activated carbons prepared from date pits : the effect of adsorbents pore size distribution. Desalin. Water Treat. 7, 242–250.

Bouchemal, N., Azoudj, Y., Merzougui, Z., Addoun, F., 2012. Adsorption modeling of Orange G dye on mesoporous activated carbon prepared from Algerian date pits using experimental designs. Desalin. Water Treat. 45, 284–290.

Boudissa, F., Mirilà, D., Arus, V.-A., Terkmani, T., Semaan, S., Proulx, M., Nistor, I.-D., Roy, R., Azzouz, A., 2019. Acid-treated clay catalysts for organic dye ozonation–thorough mineralization through optimum catalyst basicity and hydrophilic character. J. Hazard. Mater. 364, 356–366.

Carrier, M., Hardie, A.G., Uras, U., Gorgens, J., Knoetze, J., 2012. Production of char from vacuum pyrolysis of south-African sugar cane bagasse and its characterization as activated carbon and biochar. J. Anal. Appl. Pyrolysis 96, 24–32. https://doi.org/10.1016/j.jaap.2012.02.016.

Chakraborty, J.N., 2011. Metal complex dyes. In: Clark, M. (Ed.), Handbook of Textile and Industrial Dyeing. Principles, Processes and Types of Dyes, vol. 1. Woodhead Publishing Limited, Cambridge, pp. 446–463.

Chang, D.J., Chen, I.P., Chen, M.T., Lin, S.S., 2003. Wet air oxidation of a reactive dye solution using $CoAlPO_4^{5-}$ and $CeO_2$ catalysts. Chemosphere 52 (6), 943–949.

Chaudhuri, S.K., Sur, B., 2000. Oxidative decolourization of reactive dye solution using fly ash as catalyst. J. Environ. Eng. 126, 583–594.

Chen, Y.-D., Lin, Y.-C., Ho, S.-H., Zhou, Y., Nan-qi, R., 2018. Highly efficient adsorption of dyes by biochar derived from pigments-extracted macroalgae pyrolyzed at different temperature. Bioresour. Technol. 259, 104–110. https://doi.org/10.1016/j.biortech.2018.02.094.

Chen, S., Qin, C., Wang, T., Chen, F., Li, X., Hou, H., Zhou, M., 2019. Study on the adsorption of dyestuffs with different properties by sludge-rice husk biochar: adsorption capacity, isotherm, kinetic, thermodynamics and mechanism. J. Mol. Liq. 285, 62–74. https://doi.org/10.1016/j.molliq.2019.04.035.

Cheremisinoff, P.N., 1994. Biomanagement of Wastewater and Wastes. Pub. PTRice Prentice Hall, Englewood Cliffs, NJ.

Chiou, M.S., Kuo, W.S., Li, H.Y., 2003. Removal of reactive dye from wastewater by adsorption using ECH cross-linked chitosan beads as medium. J. Environ. Sci. Health. Part A 38 (11), 2621–2631.

Christie, R.M., 2007. Handbook of Textile and Industrial Dyeing. Woodhead Publishing Ltd., ISBN: 978-1-84569-115-8.

Chu, J.-H., Kang, J.-K., Park, S.-J., Lee, C.-G., 2020. Application of magnetic biochar derived from food waste in heterogeneous sono-Fenton-like process for removal of organic dyes from aqueous solution. J. Water Proc. Eng. 37. https://doi.org/10.1016/j.jwpe.2020.101455, 101455.

Clark, M., 2011. Handbook of Textile and Industrial Dyeing: Principles, Processes and Types of Dyes. Elsevier, ISBN: 978-0-85709-397-4, p. 366.

Cooney, D.O., 1998. Adsorption Design for Wastewater Treatment. Lewis Publishers, Boca Raton, Florida.

Côrtes, L.N., Druzian, S.P., Streit, A.F.M., Godinho, M., Perondi, D., Collazzo, G.C., Oliveir, M.L.S., Cadaval Jr., T.R.S., Dotto, G.L., 2019. Biochars from animal wastes as alternative materials to treat colored effluents containing basic red 9. J. Environ. Chem. Eng. 7 (6). https://doi.org/10.1016/j.jece.2019.103446, 103446.

Crini, G., 2006. Non-conventional low-cost adsorbents for dye removal: a review. Bioresour. Technol. 97, 1061–1085.

Dai, L., Zhu, W., He, L., Tan, F., Zhu, N., Zhou, Q., He, M., Hu, G., 2018. Calcium-rich biochar from crab shell: an unexpected super adsorbent for dye removal. Bioresour. Technol. 267, 510–516. https://doi.org/10.1016/j.biortech.2018.07.090.

Danish, M., Ahmad, T., Hashim, R., Said, N., Akhtar, M.N., Mohamad-Saleh, J., Sulaiman, O., 2018. Comparison of surface properties of wood biomass activated carbons and their application against rhodamine B and methylene blue dye. Surf. Interfaces 11, 1–13.

Das, L., Sengupta, S., Das, P., Bhowal, A., Bhattacharjee, C., 2021. Experimental and numerical modeling on dye adsorption using pyrolyzed mesoporous biochar in batch and fixed-bed column reactor: isotherm, thermodynamics, mass transfer, kinetic analysis. Surf. Interfaces. https://doi.org/10.1016/j.surfin.2021.100985, 100985.

Dil, E.A., Ghaedi, M., Ghaedi, A., Asfaram, A., Goudarzi, A., Hajati, S., Soylak, M., Agarwal, S., Gupta, V.K., 2016. Modeling of quaternary dyes adsorption onto ZnO–NR–AC artificial neural network: analysis by derivative spectrophotometry. J. Ind. Eng. Chem. 34, 186–197.

Dotto, G.L., McKay, G., 2020. Current scenario and challenges in adsorption for water treatment. J. Environ. Chem. Eng., 103988. https://doi.org/10.1016/j.jece.2020.103988.

Ellouze, E., Souissi, S., Jrad, A., Amar, R.B., Salah, A.B., 2010. Performances of nanofiltration and reverse osmosis in textile industry waste water treatment. Desalin. Water Treat. 22 (1–3), 182–186. https://doi.org/10.5004/dwt.2010.1432.

Fabryanty, R., Valencia, C., Soetaredjo, F.E., Putro, J.N., Santoso, S.P., Kurniawan, A., Ju, Y.-H., Ismadji, S., 2017. Removal of crystal violet dye by adsorption using bentonite—alginate composite. J. Environ. Chem. Eng. 5 (6), 5677–5687. https://doi.org/10.1016/j.jece.2017.10.057.

Fan, Q., Hoskote, S., Hou, Y., 2004. Reduction of colorants in nylon flock dyeing effluent. J. Hazard. Mater. 112 (1–2), 123–131.

Faria, P.C.C., Órfão, J.J.M., Pereira, M., 2009. Activated carbon and ceria catalysts applied to the catalytic ozonation of dyes and textile effluents. Appl. Catal. B Environ. 88, 341–350. https://doi.org/10.1016/j.apcatb.2008.11.002.

Farzaneh, H., Loganathan, K., Saththasivam, J., McKay, G., 2020. Ozone and ozone/hydrogen peroxide treatment to remove gemfibrozil and ibuprofen from treated sewage effluent: factors influencing bromate formation. Emerg. Contam. 6, 225–234. https://doi.org/10.1016/j.emcon.2020.06.002.

Fawzy, M.A., Gomaa, M., 2020. Use of algal biorefinery waste and waste office paper in the development of xerogels: a low cost and eco-friendly biosorbent for the effective removal of Congo red and Fe (II) from aqueous solutions. J. Environ. Manag. 262. https://doi.org/10.1016/j.jenvman.2020.110380, 110380.

Forgacs, E., Cserhati, T., Oros, G., 2004. Removal of synthetic dyes from wastewaters: a review. Environ. Int. 30, 953–971.

Ganguly, P., Sarkhel, R., Das, P., 2020. Synthesis of pyrolyzed biochar and its application for dye removal: batch, kinetic and isotherm with linear and non-linear mathematical analysis. Surf. Interfaces 20. https://doi.org/10.1016/j.surfin.2020.100616, 100616.

Gokulan, R., Avinash, A., Jegan, J., 2019. Remediation of remazol dyes by biochar derived from *Caulerpa scalpelliformis*—an eco-friendly approach. J. Environ. Chem. Eng. 7 (5). https://doi.org/10.1016/j.jece.2019.103297, 103297.

Goswami, P., Basak, M., 2001. "Sulfur Dyes" in Kirk-Othmer Encyclopedia of Chemical Technology. John Wiley & Sons, https://doi.org/10.1002/0471238961.1921120619051409.a01.pub2.

Goswami, M., Chaturvedi, P., Sonwani, R.K., Gupta, A.D., Giri, B.S., Rai, B.N., Singh, H., Yadav, S., Singh, R.S., 2020. Application of Arjuna (Terminalia arjuna) seed biochar in hybrid treatment system for the bioremediation of Congo red dye. Bioresour. Technol. 307. https://doi.org/10.1016/j.biortech.2020.123203, 123203.

Gupta, V., 2009. Application of low-cost adsorbents for dye removal–a review. J. Environ. Manag. 90, 2313–2342.

Gupta, V.K., Saleh, T.A., 2013. Sorption of pollutants by porous carbon, carbon nanotubes and fullerene-an overview. Environ. Sci. Pollut. Res. 20, 2828–2843.

Gupta, V.K., Mohan, D., Sharma, S., Sharma, M., 2000. Removal of basic dyes (rhodamine B and methylene blue) from aqueous solutions using bagasse fly ash. Sep. Sci. Technol. 35, 2097–2113.

Gupta, V.K., Tyagi, I., Agarwal, S., Singh, R., Chaudhary, M., Harit, A., Kushwaha, S., 2016. Column operation studies for the removal of dyes and phenols using a low cost adsorbent. Glob. J. Environ. Sci. Manag. 2, 1–10.

Gurav, R., Bhatia, S.K., Choi, T., Choi, Y.-K., Kim, H.J., Song, H.-S., Lee, S.M., Park, S.-L., Lee, H.S., Koh, J., Jeon, J.-M., Yoon, J.-J., Yang, Y.-H., 2021. Application of macroalgal biomass derived biochar and bioelectrochemical system with *Shewanella* for the adsorptive removal and biodegradation of toxic azo dye. Chemosphere 264 (2). https://doi.org/10.1016/j.chemosphere.2020.128539, 128539.

Hadi, P., Xu, M., Ning, C., McKay, G., 2015. A critical review on preparation, characterization and utilization of sludge-derived activated carbons for wastewater treatment. Chem. Eng. J. 260, 895–906. https://doi.org/10.1016/j.cej.2014.08.088.

Hameed, B., Din, A., Ahmad, A., 2007. Adsorption of methylene blue onto bamboo-based activated carbon: kinetics and equilibrium studies. J. Hazard. Mater. 141, 819–825. https://doi.org/10.1016/j.jhazmat.2006.07.049.

Hasani, G., Maleki, A., Daraei, H., Ghanbari, R., Safari, M., McKay, G., Yetilmezsoy, K., Ilham, F., Marzban, N., 2019. A comparative optimization and performance analysis of four different electrocoagulation- flotation processes for humic acid removal from aqueous solutions. Process. Saf. Environ. Prot. 121, 103–117.

Hernández-Zamora, M., Martínez-Jerónimo, F., 2019. Congo red dye diversely affects organisms of different trophic levels: a comparative study with microalgae, cladocerans, and zebrafish embryos. Environ. Sci. Pollut. Res. 26, 11743–11755. https://doi.org/10.1007/s11356-019-04589-1.

Hernández-Zamora, M., Cristiani-Urbina, E., Martínez-Jerónimo, F., Perales-Vela, H., Ponce-Noyola, T., Montes-Horcasitas, M., Cañizares-Villanueva, R., 2015. Bioremoval of the azo dye Congo red by the microalga Chlorella vulgaris. Environ. Sci. Pollut. Res. Int. 22. https://doi.org/10.1007/s11356-015-4277-1.

Hislop, K.A., Bolton, J.R., 1999. The photochemical generation of hydroxyl radicals in the UV-vis/ferrioxalate/$H_2O_2$ system. Environ. Sci. Technol. 33 (18), 3119–3126.

Ho, Y., Chiang, C., 2001. Sorption studies of acid dye by mixed sorbents. Adsorption 7, 139–147. https://doi.org/10.1023/A:1011652224816.

Hoda, M., Ghany, A., 2019. Production of a new activated carbon prepared from palm fronds by thermal activation. Int. J. Eng. Technol. Manage. Res. 6 (4), 34–43. https://doi.org/10.29121/ijetmr.v6.i4.2019.368.

Houas, A., Lachheb, H., Ksibi, M., Elaloui, E., Guillard, C., Hermann, H.M., 2001. Photocatalytic degradation pathway of methylene blue in water. Appl. Catal. B Environ. 31 (2), 145–157.

Huang, M., Zhang, Y., Wang, J., Chen, J., Zhang, J., 2020. Biochars prepared from rabbit manure for the adsorption of rhodamine B and Congo red: characterization, kinetics, isotherms, and thermodynamic studies. Water Sci. Technol. 81 (3), 436–444.

Hunger, K., Mischke, P., Rieper, W., Raue, R., Kunde, K., Engel, A., 2005. "Azo Dyes" in Ullmann's Encyclopedia of Industrial Chemistry. Wiley-VCH, Weinheim, https://doi.org/10.1002/14356007.a03_245.

Ip, A.W.M., Barford, J.P., McKay, G., 2010. Biodegradation of reactive black 5 and bioregeneration in upflow fixed bed bioreactors packed with different adsorbents. J. Chem. Technol. Biotechnol. 85 (5), 658–667.

Iqbal, J., Shah, N.S., Sayed, M., Niazi, N.K., Imran, M., Khan, J.A., Khan, Z.U.H., Hussien, A.G.S., Polychronopoulou, K., Howari, F., 2021a. Nano-zerovalent manganese/biochar composite for the adsorptive and oxidative removal of Congo-red dye from aqueous solutions. J. Hazard. Mater. 403. https://doi.org/10.1016/j.jhazmat.2020.123854, 123854.

Iqbal, M.M., Imran, M., Hussain, T., Naeem, M.A., Al-Kahtani, A.A., Shah, G.M., Ahmad, S., Farooq, A., Rizwan, M., Majeed, A., Khan, A.R., Ali, S., 2021b. Effective sequestration of Congo red dye with ZnO/cotton stalks biochar nanocomposite: modelling, reusability and stability. J. Saudi Chem. Soc. 25 (2), 101176.

Janos, P., Buchtova, H., Ryznarova, M., 2003. Sorption of dyes from aqueous solutions onto fly ash. Water Res. 37, 4938–4944.

Janos, P., Sedivy, P., Ryznarova, M., Grotschelova, S., 2005. Sorption of basic and acid dyes from aqueous solutions onto oxihumolite. Chemosphere 59, 881–886.

Janos, P., Coskun, S., Pilarova, V., Rejnek, J., 2009. Removal of basic (methylene blue) and acid (Eg. acid orange) dyes from waters by sorption on chemically treated wood shavings. Bioresour. Technol. 100, 1450–1453.

Jeong, J., Yoon, J., 2005. pH effect on OH radical production in photo/ferrioxalate system. Water Res. 39 (13), 2893–2900.

Khan, S., Malik, A., 2018. Toxicity evaluation of textile effluents and role of native soil bacterium in biodegradation of a textile dye. Environ. Sci. Pollut. Res. 25, 4446–4458. https://doi.org/10.1007/s11356-017-0783-7.

Khan, N., Chowdhary, P., Ahmad, A., Giri, B.S., Chaturvedi, P., 2020. Hydrothermal liquefaction of rice husk and cow dung in Mixed-Bed-Rotating Pyrolyzer and application of biochar for dye removal. Bioresour. Technol. 309, 123294.

Kizilkaya, B., 2012. Usage of biogenic apatite (fish bones) on removal of basic Fuchsin dye from aqueous solution. J. Dispers. Sci. Technol. 33 (11), 1596–1602. https://doi.org/10.1080/01932691.2011.629497.

Ko, D.C.K., Porter, J.F., McKay, G., 2005. Application of the concentration-dependent surface diffusion model on the multicomponent fixed-bed adsorption systems. Chem. Eng. Sci. 60, 5472–5479.

Kundu, S., Chowdhury, I.H., Naskar, M., 2017. Synthesis of hexagonal shaped nanoporous carbon for efficient adsorption of methyl orange dye. J. Mol. Liq. 234, 417–423.

Lacerda, V.S., López-Sotelo, J.B., Correa-Guimarães, A., Hernández-Navarro, S., Sánchez-Báscones, M., Navas-Gracia, L.M., Martín-Gil, J., 2013. Rhodamine B removal with activated carbons obtained from lignocellulosic waste. J. Environ. Manag. 155, 67–76.

Lafi, W.K., 2001. Production of activated carbon from acorns and olive seeds. Biomass Bioenergy 20, 57–62. https://doi.org/10.1016/S0961-9534(00)00062-3.

Lam, K.F., Yeung, K.L., McKay, G., 2006. A rational approach in the design of mesoporous adsorbents. Langmuir 22 (23), 9632–9641.

Li, Y., Meas, A., Shan, S., Yang, R., Gai, X., 2016a. Production and optimization of bamboo hydrochars for adsorption of Congo red and 2-naphthol. Bioresour. Technol. 207, 379–386.

Li, C., Zhang, L., Xia, H., Peng, J., Zhang, S., Cheng, S., Shu, J., 2016b. Kinetics and isotherms studies for Congo red adsorption on mesoporous Eupatorium adenophorum-based activated carbon via microwave-induced H3PO4 activation. J. Mol. Liq. 224, 737–744.

Li, Z., Chen, K., Chen, Z., Li, W., Biney, B.W., Guo, A., Liu, D., 2021. Removal of malachite green dye from aqueous solution by adsorbents derived from polyurethane plastic waste. J. Environ. Chem. Eng. 9 (1). https://doi.org/10.1016/j.jece.2020.104704, 104704.

Mani, S., Chowdhary, P., Bharagava, R.N., 2019. Textile wastewater dyes: Toxicity profile and treatment approaches. In: Bharagava, R., Chowdhary, P. (Eds.), Emerging and Eco-Friendly Approaches for Waste Management. Springer, Singapore, https://doi.org/10.1007/978-981-10-8669-4_11.

Martinez, F., Calleja, G., Melero, J.A., Molina, R., 2005. Heterogeneous photo-Fenton degradation of phenolic aqueous solutions over iron-containing SBA-15 catalyst. Appl. Catal. B Environ. 60 (3–4), 181–190.

McKay, G., 1980. Waste Colour Removal from Textile Effluents. vol. 68 American Dyestuff Reporter, pp. 29–36.

Mcyotto, F., Wei, Q., Macharia, D.K., Huang, M., Shen, C., Chow, C.W.K., 2021. Effect of dye structure on color removal efficiency by coagulation. Chem. Eng. J. 405. https://doi.org/10.1016/j.cej.2020.126674, 126674.

Mohan, D., Sarswat, A., Ok, Y.S., Pittman Jr., C.U., 2014. Organic and inorganic contaminants removal from water with biochar, a renewable, low cost and sustainable adsorbent – a critical review. Bioresour. Technol. 160, 191–202. https://doi.org/10.1016/j.biortech.2014.01.120.

Mudhoo, A., Mohan, D., Pittman Jr., C.U., Sharma, G., Sillanpää, M., 2021. Adsorbents for real-scale water remediation: gaps and the road forward. J. Environ. Chem. Eng. 9 (4). https://doi.org/10.1016/j.jece.2021.105380, 105380.

Mui, E.L., Cheung, W.H., Valix, M., McKay, G., 2010a. Dye adsorption onto activated carbons from tyre rubber waste using surface coverage analysis. J. Colloid Interface Sci. 347 (2), 290–300. https://doi.org/10.1016/j.jcis.2010.03.061.

Mui, E.L.K., Cheung, W.H., Valix, M., McKay, G., 2010b. Dye adsorption onto char from bamboo. J. Hazard. Mater. 177, 1001–1005.

Nautiyal, P., Subramanian, K., Dastidar, M., 2016. Adsorptive removal of dye using biochar derived from residual algae after in-situ transesterification: alternate use of waste of biodiesel industry. J. Environ. Manag. 182, 187–197.

Nikfar, S., Jaberidoost, M., 2014. In: Wexler, P. (Ed.), Dyes and Colourants in Encyclopedia of Toxicology, third ed. Elsevier.

Nizam, N.U.M., Hanafiah, M.M., Mahmoudi, E., et al., 2021. The removal of anionic and cationic dyes from an aqueous solution using biomass-based activated carbon. Sci. Rep. 11, 8623. https://doi.org/10.1038/s41598-021-88084-z.

Oliveira, G.A.R., Ferraz, E.R.A., Chequer, F.M.D., Grando, M.D., Angeli, J.P.F., Tsuboy, M.S., Marcarini, J.C., Mantovani, M.S., Osugi, M.E., Lizier, T.M., Zanoni, M.V.B., Oliveira, D.P., 2010. Chlorination treatment of aqueous samples reduces, but does not eliminate, the mutagenic effect of the azo dyes disperse red 1, disperse red 13 and disperse orange 1. Mutat. Res. 703 (2), 200–208. https://doi.org/10.1016/j.mrgentox.2010.09.001.

Osman, D., Sibel, T., Tulin, G.P., 2017. Adsorptive removal of triarylmethane dye (Basic Red 9) from aqueous solution by sepiolite as effective and low-cost adsorbent. Microporous Mesoporous Mater. 210. https://doi.org/10.1016/j.micromeso.2015.02.040.

Oueslati, K., Lima, E.C., Ayachi, F., Cunha, M.R., Lamine, A.B., 2020. Modeling the removal of reactive red 120 dye from aqueous effluents by activated carbon. Water Sci. Technol. 82 (4), 651–662. https://doi.org/10.2166/wst.2020.347.

Pal, A., Pan, S., Saha, S., 2013. Synergistically improved adsorption of anionic surfactant and crystal violet on chitosan hydrogel beads. Chem. Eng. J. 217, 426–434. https://doi.org/10.1016/j.cej.2012.11.120.

Palapa, N.R., Taher, T., Rahayu, B.R., Mohadi, R., Rachmat, A., Lesbani, A., 2020. CuAl LDH/Rice husk biochar composite for enhanced adsorptive removal of cationic dye from aqueous solution. Bull. Chem. React. Eng. Catal. 15 (2), 525–537. https://doi.org/10.9767/bcrec.15.2.7828.525.

Park, J.-H., Wang, J.J., Meng, Y., Wei, Z., DeLaune, R.D., Seo, D.-C., 2019. Adsorption/desorption behavior of cationic and anionic dyes by biochars prepared at normal and high pyrolysis temperatures. Colloids Surf. A. Physiochem. Eng. Asp. 572, 274–282.

Patel, M., Mohan, D., Kumar, R., Pitman Jr., C.U., 2021. Ciprofloxacin and acetaminophen sorption onto banana peel biochars: environmental and process parameter influences. Environ. Res. https://doi.org/10.1016/j.envres.2021.111218.

Raghavacharya, C., 1997. Colour removal from industrial effluents. Chem. Eng. World 32 (7), 53–54.

Rahman, N., Dafader, N.C., Miah, A.R., Shahnaz, S., 2019. Efficient removal of methyl orange from aqueous solution using amidoxime adsorbent. Int. J. Environ. Stud. 76 (4), 594–607. https://doi.org/10.1080/00207233.2018.1494930.

Raj, A., Yadav, A., Rawat, A., Singh, A., Kumar, S., Kumar, A., Sirohi, R., Pandey, A., 2021. Kinetic and thermodynamic investigations of sewage sludge biochar in removal of Remazol Brilliant Blue R dye from aqueous solution and evaluation of residual dyes cytotoxicity. Environ. Technol. Innov. 23. https://doi.org/10.1016/j.eti.2021.101556, 101556.

Rapo, E., Posta, K., Suciu, M., Robert, S., 2019. Adsorptive removal of remazol brilliant violet-5R dye from aqueous solutions using calcined eggshell as biosorbent. Acta Chim. Slov. 66 (3), 648–658. https://doi.org/10.17344/acsi.2019.5079.

Research and Markets, 2020. Global Textile Dyes Market Research Report—Industry Analysis, Size, Share, Growth, Trends and Forecast Till 2026.

Robinson, T., McMullan, G., Marchant, R., Nigam, P., 2001. Remediation of dyes in textile effluent: a critical review on current treatment technologies with a proposed alternative. Bioresour. Technol. 77 (3), 247–255. https://doi.org/10.1016/S0960-8524(00)00080-8.

Rodriguez-Reinoso, F., Molina-Sabio, M., 1992. Activated carbons from lignocellulosic materials by chemical and or physical activation—an overview. Carbon N. Y. 30, 1111–1118.

Rouquerol, J., Rouquerol, F., Llewellyn, P., Maurin, G., Sing, K., 2013. Adsorption by Powders and Porous Solids, 2nd Edition: Principles, Methodology and Applications. Pub. Academic Press.

Rovina, K., Prabakaran, P.P., Siddiquee, S., Shaarani, S.M., 2016. Methods for the analysis of sunset yellow FCF (E110) in food and beverage products- a review. TrAC Trends Anal. Chem. 85 (Part B), 47–56. https://doi.org/10.1016/j.trac.2016.05.009.

Salamat, S., Hadavifar, M., Rezaei, H., 2019. Preparation of nanochitosan-STP from shrimp shell and its application in removing of malachite green from aqueous solutions. J. Environ. Chem. Eng. 7 (5). https://doi.org/10.1016/j.jece.2019.103328, 103328.

Saleem, J., Hijab, M., Mackey, H., McKay, G., 2019. Production and applications of activated carbons as adsorbents from olive stones-a review. Biomass Convers. Biorefin. https://doi.org/10.1007/s13399-019-00473-7.

Samsami, S., Mohamadi, M., Sarrafzadeh, M.H., Rene, E.R., Firoozbahr, M., 2020. Recent advances in the treatment of dye-containing wastewater from textile industries: overview and perspectives. Process. Saf. Environ. Prot. 143, 138–163. https://doi.org/10.1016/j.psep.2020.05.034.

Santos, A., Yustos, P., Rodriguez, S., Garcia-Ochoa, F., de Gracia, M., 2007. Decolorization of textile dyes by wet oxidation using activated carbon as catalyst. Ind. Eng. Chem. Res. 46, 2423–2427.

Santos, K.J.L., de Souzados Santos, G.E., Sá, Í.M.G.L., Ide, A.H., da Silva Duarte, J.L., Carvalho, S.H.V., Soletti, J.I., Meili, L., 2019. *Wodyetia bifurcata* biochar for methylene blue removal from aqueous matrix. Bioresour. Technol. 293. https://doi.org/10.1016/j.biortech.2019.122093, 122093.

Sarasa, J., Roche, M.P., Ormad, M.P., Gimeno, E., Puig, A., Ovelleiro, J.L., 1998. Treatment of a wastewater resulting from dyes manufacturing with ozone and chemical coagulation. Water Res. 32 (9), 2721–2727.

Sayğılı, H., Güzel, F., 2015. Performance of new mesoporous carbon sorbent prepared from grape industrial processing wastes for malachite green and congo red removal. Chem. Eng. Res. Des. 100, 27–38.

Sewu, D.D., Boakye, P., Woo, S.H., 2017. Highly efficient adsorption of cationic dye by biochar produced with Korean cabbage waste. Bioresour. Technol. 224, 206–213.

Shaikh, W.A., Islam, R.U., Chakraborty, S., 2021. Stable silver nanoparticle doped mesoporous biochar-based nanocomposite for efficient removal of toxic dyes. J. Environ. Chem. Eng. 9 (1). https://doi.org/10.1016/j.jece.2020.104982.

Shoukat, R., Khan, S.J., Jamal, Y., 2019. Hybrid anaerobic-aerobic biological treatment for real textile wastewater. J. Water Proc. Eng. 29. https://doi.org/10.1016/j.jwpe.2019.100804, 100804.

Tahir, M.A., Bhatti, H.N., Iqbal, M., 2016. Solar red and brittle blue direct dyes adsorption onto Eucalyptus angophoroides bark: equilibrium, kinetics and thermodynamic studies. J. Environ. Chem. Eng. 4 (2), 2431–2439. https://doi.org/10.1016/j.jece.2016.04.020.

Tang, L., Xiao, F., Wei, Q., Liu, Y., Zou, Y., Liu, J., Sand, W., Chow, C., 2019. Removal of active dyes by ultrafiltration membrane pre-deposited with a PSFM coagulant: performance and mechanism. Chemosphere 223, 204–210. https://doi.org/10.1016/j.chemosphere.2019.02.034.

Thakur, S., Qanungo, K., 2021. Removal of aniline blue from aqueous solution using adsorption: a mini review. Mater. Today Proc. 37 (2), 2290–2293. https://doi.org/10.1016/j.matpr.2020.07.725.

Tsai, W.-T., Chen, H.-R., 2010. Removal of malachite green from aqueous solution using low-cost chlorella-based biomass. J. Hazard. Mater. 175 (1–3), 844–849. https://doi.org/10.1016/j.jhazmat.2009.10.087.

Vaghela, S.S., Jethva, A.D., Mehta, B.B., Dave, S.P., Adimurthy, S., Ramachandraiah, G., 2005. Laboratory studies of electrochemical treatment of industrial azo dye effluent. Environ. Sci. Technol. 39 (8), 2848–2855.

Valix, M., Cheung, W.H., McKay, G., 2004. Preparation of activated carbon using low temperature carbonization and physical activation of high ash raw bagasse for acid dye adsorption. Chemosphere 56 (5), 493–501.

Vigneshwaran, S., Sirajudheen, P., Karthikeyan, P., Meenakshi, S., 2021a. Fabrication of sulfur-doped biochar derived from tapioca peel waste with superior adsorption performance for the removal of Malachite green and Rhodamine B dyes. Surf. Interfaces 23. https://doi.org/10.1016/j.surfin.2020.100920, 100920.

Vigneshwaran, S., Sirajudheen, P., Nikitha, M., Ramkumar, K., Meenaksh, S., 2021b. Facile synthesis of sulfur-doped chitosan/biochar derived from tapioca peel for the removal of organic dyes: isotherm, kinetics and mechanisms. J. Mol. Liq. 326. https://doi.org/10.1016/j.molliq.2021.115303, 115303.

Vimal, V., Patel, M., Mohan, D., 2019. Aqueous carbofuran removal using slow pyrolyzed sugarcane bagasse biochar: equilibrium and fixed- bed studies. RSC Adv. 9, 26338. https://doi.org/10.1039/c9ra01628g.

Vyavahare, G., Jadhav, P., Jadhav, J., Patil, R., Aware, C., Patil, D., Gophane, A., Yang, Y.-H., Gurav, R., 2019. Strategies for crystal violet dye sorption on biochar derived from mango leaves and evaluation of residual dye toxicity. J. Clean. Prod. 207, 296–305. https://doi.org/10.1016/j.jclepro.2018.09.193.

Wang, H., Wang, S., Gao, Y., 2020. Cetyl trimethyl ammonium bromide modified magnetic biochar from pine nut shells for efficient removal of acid chrome blue K. Bioresour. Technol. 312. https://doi.org/10.1016/j.biortech.2020.123564, 123564.

Wu, J., Yang, J., Feng, P., Huang, G., Xu, C., Lin, B., 2020. High-efficiency removal of dyes from wastewater by fully recycling litchi peel biochar. Chemosphere 246. https://doi.org/10.1016/j.chemosphere.2019.125734, 125734.

Yagub, M.T., Sen, T.K., Afroze, S., Ang, H.M., 2014. Dye and its removal from aqueous solution by adsorption: a review. Adv. Colloid Interf. Sci. 209, 172–184.

Yang, R.T., 2003. Adsorbents: Fundamentals and Applications. Wiley and Sons, Inc., Hoboken, NJ, p. 424.

Yang, G., Wu, L., Xian, Q., Shen, F., Wu, J., Zhang, Y., 2016. Removal of Congo red and methylene blue from aqueous solutions by vermicompost-derived biochars. PLoS One 11.

Yang, L., Wang, Z., Zhang, J., 2017. Zeolite imidazolate framework hybrid nanofiltration (NF) membranes with enhanced permselectivity for dye removal. J. Membr. Sci. 532, 76–86.

Yang, S.-S., Chen, Y.-D., Kang, J.-H., Xie, T.-R., He, L., Xing, D.-F., Ren, N.-Q., Ho, S.-H., Wu, W.-M., 2019. Generation of high-efficient biochar for dye adsorption using frass of yellow mealworms (larvae of *Tenebrio molitor* Linnaeus) fed with wheat straw for insect biomass production. J. Clean. Prod. 227, 33–47. https://doi.org/10.1016/j.jclepro.2019.04.005.

Yao, X., Ji, L., Guo, J., Ge, S., Lu, W., Chen, Y., Cai, L., Wang, Y., Song, W., 2020. An abundant porous biochar material derived from wakame (*Undaria pinnatifida*) with high adsorption performance for three organic dyes. Bioresour. Technol. 318. https://doi.org/10.1016/j.biortech.2020.124082, 124082.

Yek, P.N.Y., Peng, W., Wong, C.C., Liew, R.K., Ho, Y.L., Mahari, W.A.W., Azwar, E., Yuan, T.Q., Tabatabaei, M., Aghbashlo, M., Sonne, C., Lam, S.S., 2020. Engineered biochar via microwave $CO_2$ and steam pyrolysis to treat carcinogenic Congo red dye. J. Hazard. Mater. 395. https://doi.org/10.1016/j.jhazmat.2020.122636, 122636.

Yin, Z., Liu, N., Bian, S., Li, J., Xu, S., Zhang, Y., 2019. Enhancing the adsorption capability of areca leaf biochar for methylene blue by $K_2FeO_4$-catalyzed oxidative pyrolysis at low temperature. RSC Adv. 9, 42343–42350. https://doi.org/10.1039/C9RA06592J.

Yu, M.C., Skipper, P.L., Tannenbaum, S.R., Chan, K.K., Ross, R.K., 2002. Arylamine exposures and bladder cancer risk. Mutat. Res. Fundam. Mol. Mech. Mutagen. 506–507, 21–28.

Yu, K.L., Lee, X.J., Ong, H.C., Chen, W.-H., Chang, J.-S., Lin, C.-S., Show, P.L., Ling, T.-C., 2021. Adsorptive removal of cationic methylene blue and anionic Congo red dyes using wet-torrefied microalgal biochar: equilibrium, kinetic and mechanism modeling. Environ. Pollut. 272. https://doi.org/10.1016/j.envpol.2020.115986, 115986.

Yusuf, A., 2020. Vat dyes, properties, dyeing mechanism—a comprehensive look. https://textiletuts.com/vat-dyes/. (Accessed 31 March 2021).

Zhang, Y., Fan, R., Zhang, Q., Chen, Y., Sharifi, O., Leszczynska, D., Zhang, R., Dai, Q., 2019. Synthesis of CaWO4-biochar nanocomposites for organic dye removal. Mater. Res. Bull. 110, 169–173.

Zhang, Z., Wang, G., Li, W., Zhang, L., Chen, T., Ding, L., 2020. Degradation of methyl orange through hydroxyl radical generated by optically excited biochar: performance and mechanism. Colloids Surf. A. Physiochem. Eng. Asp. 601. https://doi.org/10.1016/j.colsurfa.2020.125034, 125034.

Zubair, M., Mu'azu, N.D., Jarrah, N., et al., 2020a. Adsorption behavior and mechanism of methylene blue, crystal violet, eriochrome black T, and methyl orange dyes onto biochar-derived date palm fronds waste produced at different pyrolysis conditions. Water Air Soil Pollut. 231, 240–248. https://doi.org/10.1007/s11270-020-04595-x.

Zubair, M., Manzar, M.S., Mu'azu, N.D., Anil, I., Blaisi, N.I., Al-Harthi, M.A., 2020b. Functionalized MgAl-layered hydroxide intercalated date-palm biochar for enhanced uptake of cationic dye: kinetics, isotherm and thermodynamic studies. Appl. Clay Sci. 190. https://doi.org/10.1016/j.clay.2020.105587, 105587.

# Further reading

McKay, G., 1984. Analytical solution using a pore diffusion model for a pseudo-irreversible isotherm for the adsorption of basic dye on silica. AICHE J. 30 (4), 692–697.

Selvakumar, A., Rangabhashiyam, S., 2019. Biosorption of rhodamine B onto novel biosorbents from Kappaphycus alvarezii, Gracilaria salicornia and Gracilaria edulis. Environ. Pollut. 255 (2). https://doi.org/10.1016/j.envpol.2019.113291, 113291.

# Biochar as a potential agent for the remediation of microbial contaminated water

Jayani J. Wewalwela[a], Prasad Sanjeewa[b], and Sameera R. Gunatilake[b]

[a]Department of Agricultural Technology, Faculty of Technology, University of Colombo, Colombo, Sri Lanka. [b]College of Chemical Sciences, Institute of Chemistry Ceylon, Rajagiriya, Sri Lanka

## 1 Introduction

Water is found in oceans, groundwater, glaciers, lakes, swamps, rivers, and the atmosphere (Gleick, 1993). An increase in population and intensive development of industries and agriculture cause a significant impact on the natural water bodies in the world. Water contamination can render water unusable for consumption and other human utilization (Arnone and Perdek Walling, 2007). Water can be contaminated through pathogenic microorganisms, toxic inorganic materials, synthetic organic contaminants, and radionuclides. Contamination of drinking water sources by pathogenic organisms is a major problem in many developing and developed countries as is the use of polluted water for agricultural purposes (Chandrasena et al., 2014; Mohanty and Boehm, 2014).

The utilization of water contaminated by pathogenic microorganisms may cause severe diseases in humans and animals. These pathogenic microorganisms can be removed by low-cost water management systems to obtain purified water for consumption. Common decontamination methods including ceramic filtration, solar water disinfection, and chlorination are found to have several limitations (Trenberth and Asrar, 2012; James and Joyce, 2004) during the decontamination process. Therefore, biochar, carbonaceous organic materials comprised of pyrolyzed biomass, is being considered as part of a remediation method for microbial contaminated water (Mohanty and Boehm, 2014). Several laboratory-scale studies have demonstrated that pathogenic microorganisms such as *Escherichia coli, Salmonella enterica, Typhimurium,* and *Staphylococcus aureus* can be removed using biochar-amended columns (World Health Organization; International Water Association, 2009). Additionally, improved methods utilizing biochar have been extensively applied for the remediation of microbial contaminated drinking water (Inyang and Dickenson, 2015), stormwater (Petterson et al., 2016), and polluted water used for agriculture (Ahmed et al., 2019).

## 2 Water sources and microbial contamination

### 2.1 Microbial contamination in water

Water is essential for life and has an enormous effect on public health. Therefore, ensuring an adequate supply of water to improve sanitation is a major goal of communities worldwide. Three-fourth of the Earth's surface is covered by water, primarily by oceans and inland surface water bodies. Water sources are affected by the continuous circulation of water in the water/hydrological cycle. Surface water is in contact with groundwater and contamination of surface water leads to groundwater contamination. In general, there is less contamination of groundwater compared to surface water, since groundwater typically has reduced exposure to contaminants. However, in some cases, groundwater has been found to have significant levels of contaminants. Stormwater and drain water channels are related to the amount of surface water (Trenberth and Asrar, 2012) and these entities can be affected by microorganisms that are pathogenic or innocuous to humans. Pathogenic microbes cause a variety of diseases such as cholera, diarrhea, dysentery, and various other infections and intoxications. Therefore, increasing awareness regarding the need for novel treatment methods for contamination is vital. Hence, there are explicit water quality assurance protocols that are internationally defined by the World Health Organization (WHO) and other state organizations. All these impediments are implemented to ensure that the human population has access to water free of pathogenic microbial contaminants (James and Joyce, 2004).

These standards ensure the protection of consumers against outbreaks of illnesses associated with the consumption of contaminated water. In addition to testing for fecal indicator bacteria

(FIB), a few more methods have been introduced by scientists during the past years. Most of the important criteria introduced by the WHO are to maintain standards all over the world while compiling water quality data (World Health Organization; International Water Association, 2009).

Macropollutants can be observed and treated with the traditional purification systems. Modified modern systems should also treat microscopic contaminants such as viruses, bacteria, and protozoa. As a result, there is a need for the further development of techniques for the treatment of microscale contaminants. Accidental introduction of microbes leads to contamination known as microbial stream water contamination. According to researchers, the microbial concentration is higher in inland water due to a range of seasonal effects. Stormwater quality is affected by certain bacterial species such as *E. coli*, fecal *Streptococci* (*Streptococcus faecalis*), and *Clostridium perfringens* which are inhabitants of the animals' large intestines (Inyang and Dickenson, 2015; Petterson et al., 2016). Therefore, these bacteria are constantly present in feces. The presence of any of the aforementioned bacterial species in aqueous systems is an indication of fecal pollution of water. Therefore, there is a higher tendency that these microbes contaminate stormwater in highly populated areas. Most of the available studies are based on total coliform or total fecal coliform count. Other microbial contaminants are waterborne and introduced by viruses (adenovirus, rotavirus, and hepatitis A) and protozoa (*Cryptosporidium* and *Giardia*). Protozoans are most commonly present in marine habitats and freshwater (Ahmed et al., 2019).

## 2.2 Methods of contamination

There are some particle and colloidal interactions of bacteria in the aquatic environment. Bacteria present in karstic water bodies, groundwater, recreational water, and lakes have bacterial groups present in the colloidal matrix. Previous studies have shown that microbe contamination caused by humans is much greater than the impact caused by contamination sources that are not utilized regularly by humans. Urban stormwater, combined sewers, sanitary sewer outflows, wastewater from animal feeding operations, decentralized wastewater treatment systems, and effluents of wastewater treatment plants could be identified as water sources rich in pollutants. The large number of microorganisms present within such water sources help

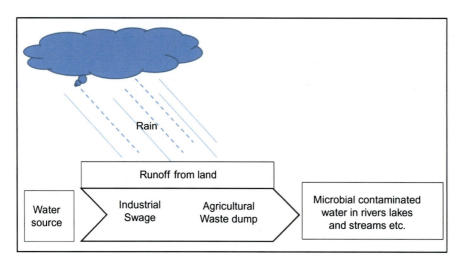

**Fig. 13.1** Schematic diagram of the effect of rainfall for microbial contamination in water bodies.

pathogenic microorganisms rapidly populate from the use of nutrients available in these systems.

Due to the limited availability of land and the high population density of animals in urban areas, a large amount of waste is accumulated daily. These environments are ideal for the growth of microbial pollutants. During rainfall, microorganisms can be transported by runoffs to rivers, lakes, and streams located nearby (Fig. 13.1). It is also noteworthy that depending on the amount of rainfall, the concentration of microorganisms reaching water bodies could vary. Higher rainfall results in water with a low concentration of microorganisms reaching water bodies and vice versa. In urban areas, effluent bathwater is regarded as a major source of pathogenic contaminants as it is rich in vast varieties of microbes including a higher percentage of FIBs. In rural areas and forests, humans and animals use rivers and water streams leading to a possibility of water being contaminated. In most cases, pollution is measured with the number of FIBs or coliform bacteria (CBs); however, there are many other species such as trace amounts of pathogenic bacterial species, viruses, as well as protozoa. Even in trace amounts, certain pathogens can cause harmful effects.

Decentralized wastewater treatment systems also contaminate the water supply chain due to their inherent poor treatment processes. These systems contain microbes such as bacteria (*Salmonella* spp., *Shigella* spp., *Vibrio cholerae*), viruses (rotavirus, hepatitis A virus, reovirus), and protozoa (*Cryptosporidium*,

*Balantidium coli*). The poor treatment processes associated with decentralized wastewater treatment systems can cause the leakage of pathogenic microorganisms causing communicable diseases such as salmonellosis, diarrhea, and giardiasis. Therefore, these harmful microbial pollutants should be documented and identified prior to the use of a particular water source for commercial water distribution.

Water canals contain sludge that gives protection for microbes against harmful UV light. It also contains submerged aquatic vegetation and sediments that support bacteria to survive in the aquatic environment. This protection is a result of the type of sediments, their particle sizes, and the high amount of organic carbon content present in the media. When heavy rainfalls occur, these systems get disturbed causing microorganisms to be suspended as part of a particle matrix and washed away along with the water streams. The same is observed in the water treatment plants containing sludge; however, in this case runoff is with the effluent water streams instead of rainfall runoffs. Microbes such as viruses are most prominently transported with the help of attached minerals. The reversible interactions between the microbes and minerals are affected by environmental conditions such as temperature, solar radiation, and pH. When the flow volume exceeds the system capacity, it may cause overflow of combined sewers (CS) and sanitary sewers (SS), which could transport pathogens to reservoirs. This can lead to a considerable impact on the water quality of the reservoir if the necessary treatment protocols are not adhered to.

Water contamination by livestock and poultry manure causes the release of zoonotic pathogens. This process is affected by factors such as temperature, pH value, and hydrophobicity. Toxic microbes that enter the soil from animal farms will be present in the soil until they are eventually transferred to the water. Exposure to these microbes imparts a high risk to human health. This can be avoided by subjecting the farm effluent water to a water treatment process. In 1972, the United States (US) implemented the Clean Water Act to reduce the transmission of contaminants to water bodies. Through the act, the United States forbids the release of feedstock water without the National Permit Discharge Elimination System (NPDES) permit. Studies on manure-related epidemics have been conducted since the 1990s and it has been found in many cases that manure runoffs cause the release of viruses, bacteria, protozoan parasites, helminth parasites, *E. coli*, *Campylobacter*, *Cryptosporidium*, etc. (Table 13.1).

Table 13.1 Common pathogenic microorganisms that contaminate aqueous systems from animal farms (James and Joyce, 2004)

| Livestock and poultry | Pathogenic microorganisms |
|---|---|
| Cattle | Salmonella spp., Pathogenic Escherichia coli, Campylobacter spp., Erysipelothrix rhusiopathiae, Cryptosporidium parvum, Giardia lamblia, Listeria monocytogenes |
| Chicken | Salmonella spp., Brucella spp. |
| Pigs | Yersinia enterocolitica, Leptospira spp., E. rhusiopathiae |
| Swine | Salmonella spp. |
| Deer | L. monocytogenes |

## 3 Mechanisms of microbial decontamination using biochar

The most commonly used water purification methods are decantation, filtration, boiling, and adding chemicals. The methods that are widely used for the removal of pathogenic microorganisms are facultative lagoons/storage, air drying, composting, anaerobic digestion, aerobic digestion, lime stabilization, and biofiltration (Chandrasena et al., 2014). Among these methods, filters utilizing biochar have been found to be extremely effective. Scientists are currently working on the development of wastewater and stream water treatment methods with the aid of biochar to get a potent filtration system with a distinct output result.

Previous studies have shown that biochar can be used to increase microbial populations due to the increase in the availability of micronutrients and soil cation exchange capacity (Peiris et al., 2019). Biochar is used in filtration systems to remove pathogenic microorganisms from contaminated water. The biochar feedstock and production method, the particle size, and percent amendment of biochar with other materials, all affect the efficiency of removal of microorganisms in both soil and aqueous media (Mohanty and Boehm, 2014; Kranner et al., 2019).

The effect of biochar amended with sand to remove E. coli in water was studied. Stormwater has a very complex matrix; however, the biochar-augmented sand columns had a strong potential to remove waterborne bacteria due to several characteristics.

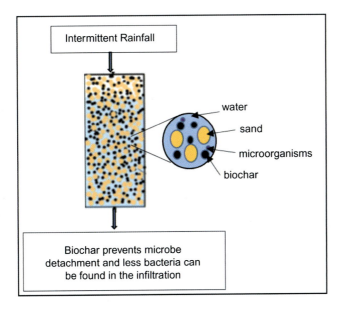

Fig. 13.2 Diagram of biochar-augmented sand filter (Mohanty et al., 2014).

During the experiment, researchers identified that sand alone columns had a higher infiltration rate for pathogenic microorganisms compared with biochar-amended columns with an increased number of *E. coli* in the filtration media when using stormwater (Fig. 13.2). Under the various flow conditions of *E. coli* contaminated water, removal was high in the biochar-amended columns (Mohanty and Boehm, 2014).

Conventional biofilters showed inadequate and inconsistent characteristics when removing FIBs in stormwater systems (Perez-Mercado et al., 2019). This led to the promotion of low-impact development (LID), including natural treatment systems to capture, treat, and recycle urban stormwater to restore the urban hydrologic regime to the pre-urbanization stage.

Biochar is the nonactivated charcoal product produced during pyrolysis. The characteristics of biochar can be altered by changing the conditions during the production process (Fig. 13.3). Microbial removal is mainly enhanced through an adsorption mechanism which also depends on material characteristics such as surface area, pore size, and distribution of pores.

Many studies have been carried out to investigate the efficiency of biochar against microbes. *E. coli* is one of the most analyzed bacterial species. This infers that the effect of biochar on the removal of microorganisms is appreciable and that it can be developed further to obtain maximum output as a filter material. The physicochemical and electrochemical properties of biochar

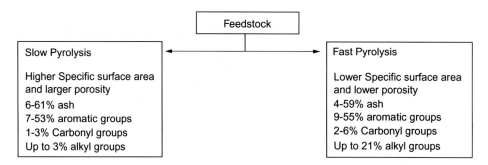

**Fig. 13.3** Diagram of biochar production and its properties (Mukherjee et al., 2011).

show that it can be modified to remove pathogens effectively and efficiently. Biochar filters have limitations in their microbial removal capacity related to particle size and flow rate (Dai and Hozalski, 2002). The common types of interactions between microbes and biochar material are electrostatic, hydrophobic, and weak van der Waal forces. Sometimes the interactions are affected by the natural organic matter (NOM) due to saturation of the surface and electric repulsion.

Biochar-assisted filters have a considerable impact on the removal of microbial contaminants in water sources. Yet, modified biofilter development studies are inadequate when considering their productivity and utility. The adsorption mechanism is controlled by biochar surface characteristics and other external conditions such as temperature, pH, and aqueous system flow rate. Bacteria and viruses are suspended in water bodies as tiny particles; when they come in contact with biofilms, adsorption onto the filter material is facilitated. Microbial attachments to biofilters are enhanced by rough surfaces (Bozorg et al., 2015). The interactions between other particle matrices and a biofilm will lead to occupation at surface binding sites and a change in the physiological properties of the filter material. This will alter the efficiency of pollutant filtration and may block the water flow. The movement and adsorption of microbes onto saturated biofilms are studied to determine (Hornberger et al., 1992):

(i) Transport—governed by convection/dispersion mechanisms
(ii) Exchange— between liquid suspension and solid substratum (e.g.: bacterial adhesion).

Additionally, studies are also based on the irreversible transport and adsorption of microbes into biofilms (Hornberger et al., 1992).

$$\frac{\partial C_{pb}}{\partial t} + \lambda_{att} C_{pb} = D\nabla^2 . C_{pb} - u\nabla . C_{pb}$$

$C_{pb}$ = Microbial concentration in suspension at time $t$, $D$ = Hydrodynamic dispersion coefficient, $\lambda_{att}$ = Pseudo-first-order attachment rate coefficient, $u$ = Average bacterial velocity in the pore space (i.e., interstitial pore velocity).

Changes in microbial concentration in the biofilters are due to convection, dispersion, and adsorption. Furthermore, to emphasize the effect of the colloidal nature of the bacterial suspension, classical colloid filtration theory (CFT) is effectively used by scientists. This gives a calculation for bacterial transportation and adsorption onto porous media such as biochar filters. In this theory, first-order kinetics with a constant rate of adsorption is used to explain the behavior of colloidal concentration in the suspension. Hence, to determine the attachment rate coefficient ($\lambda_{att}$), the packed bed bacterial removal efficacy can be used as a measure. The attachment rate coefficient is directly proportional to the single-collector removal efficiency ($\eta_r$) (Schijven and Hassanizadeh, 2000).

$$\lambda_{att} = \frac{a}{4} u \eta_r$$

The equation below is used to find "$\alpha$".

$$a = \frac{6(1-\varphi)}{d_c}$$

Where,
$a$ = Wetted area.
$\varphi$ = Porosity in a packed-bed of spherical collectors.
$d_c$ = Average diameter of the spherical collector.

The collision efficiency of a single collector ($\eta_c$) must be calculated to find the $\eta_r$ value. The $\eta_c$ value depends on geohydrological properties of porous media and probability of collision (strike probability of particles suspended in the flow toward the filter) (Tufenkji and Elimelech, 2004). Hence, $\eta_c$ stands for efficiency of all possibilities with the transport of bacteria in the flow and the following equation can be used to determine the value.

$$\eta_c = 2.44 A_s^{\frac{1}{3}} N_R^{-0.081} N_{pe}^{-0.715} N_{vdW}^{0.052} + 0.55 A_s N_R^{1.675} N_A^{0.125} + 0.22 N_R^{-0.24} N_G^{1.11} N_{vdW}^{0.053}$$

Where, $A_s$, $N_R$, $N_{pe}$, $N_{vdW}$, $N_A$, and $N_G$ are numbers used in the calculation. In addition to these values, diameters of the bacterial cells are also used. Phase-contrast microscopy and AxioVision image processing are systems useful in analysis.

Colloidal interactions between suspended particles and filter material are not the only possible collisions related to adsorption.

In most aquatic systems, $\eta_r$ is lower than $\eta_c$ and it can be calculated by:

$$\Omega = -\frac{2}{3}\frac{d_c}{(1-\varphi)\eta_c L}\ln RE_{pb}$$

In this case, $L$ is the bed depth and $RE_{pb}$ represents the bacterial recovery in the column effluent.

Moreover, $RE_{pb}$ is given by:

$$RE_{pb} = \frac{\sum_{i=0}^{t} C_{Pbi}\Delta t_i}{C_{Pb_o} t}$$

$C_{Pb_o}$ = initial bacterial concentration.
$t$ = the injection period of the bacterial suspension.

It has been found that extra polysaccharide secretion can alter the surface characteristics of porous biochar media. Such variations are not included in the filtration models. Hence, to evaluate the impact that biofilms have on bacterial deposition in porous biochar media, the biofilm interaction coefficient ($\sigma_B$) is calculated by:

$$\sigma_B = \frac{\Omega_{Biofilm} - \Omega_{Clean}}{\Omega_{Clean}}$$

$\Omega_{Clean}$ and $\Omega_{Biofilm}$ are the sticking efficiencies of clean and biofilm coated collectors, respectively.

The model described earlier is a reactive flow process in porous media that was solved numerically. Several other theories are also available to explain the process of removal of pathogenic microorganisms in biochar-amended media. Highly porous structure and increased attachment sites of biochar are some of the characteristics that make biochar an effective pathogen remover. Fine biochar particles have an increased surface area, and this favors the attachment of microbes; however, the pores are smaller than the size of a bacterium, which leads to enhanced removal of bacteria.

Derjaguin-Landau-Verwey-Overbeek (DLVO) theory (Hermansson, 1999) explains the combination of attractive van der Waal forces and repulsive electrostatic forces which are responsible for the removal of *E. coli* from stormwater. These results contrast with the fact that the biofilm formation of bacteria will reduce the effective adsorption of microorganisms.

Another theory focuses on the favorable characteristics of biochar such as rough surfaces and the presence of irregular shapes leading to the formation of the electrostatic repulsions and nonforces (hydrophobic attractions and steric interactions).

This improves bacterial attachment on biochar-based filtering systems and is known as non-DLVO forces. For example, bacterial retention in the biochar columns was greater than in the sand columns, suggesting the attachment of bacteria on biochar may occur as a result of the non-DLVO forces. However, the increase in overall removal of *E. coli* by biochar is attributed to the strong attachment of *E. coli* on biochar surfaces and the higher water-holding capacity of biochar-amended sand. It was determined that biochar composed of low volatile matter and polarity was most effective in removing *E. coli*.

This combined-complex relationship can be simplified; there are some equations used to obtain direct relationships such as the log removal value of effluent samples, which is a parameter used to measure the efficacy and saturation limits of the biofilters.

$$\log_{10} Removal = -\log_{10}\frac{C}{C_0} = -\log\frac{C}{C_0}$$

The $C$ stands for effluent fecal indicator concentration (MPN/100 mL or PFU/100 mL) and $C_0$ is the influent fecal indicator concentration (MPN/100 mL or PFU/100 mL). The removal of microorganisms in the conditioning phase is a function of the media type. Therefore, multiple linear regression is used to evaluate the effect of media type, saturated zone, biofilter age, and influent microbial concentration with respect to the removal of microorganisms.

$$\log_{10} Removal = \beta_0 + \beta_1(\text{Media Type}) + \beta_2(\text{Saturated Zone}) + \beta_3(\text{Age}) + \beta_4\left(\log_{10}\left(\frac{C_{0,event}}{C_{0,GM}}\right)\right)$$

In this equation,

$C_{0,\ event}$—influent *E. coli* concentration for a given sampling event

$C_{0,\ GM}$—geometric mean influent *E. coli* concentration for all sampling events

$(\beta_0)$—the model intercept

This model represents the expected *E. coli* log removal for an un-aged, unsaturated, amended biofilter when $C_{0,\ event}$ and $C_{0,\ GM}$ are identical (Kranner et al., 2019).

Furthermore, the formation of the biofilms alters the adsorption process and hence the removal efficiency of the microbes (Bhattacharjee et al., 1998) .Geohydrological properties of the filter surface interfere with the bacterial transport, attachment and the growth of the biofilm, and the maturation of microorganisms. This results in an alteration of filter media properties.

The formation of the thick biofilm on adsorbent surfaces will decrease the available area for adsorption of microorganisms and the porosity and the permeability of the biochar media according to the aforementioned equations. Biofilm formation is empowered by bacterial mass transport toward a surface, reversible bacterial adhesion, transition to irreversible adhesion, and cell wall deformation. The physicochemical properties of the substrate surface prescribe the initial reaction of the microbial adhesion and adhesion to other biofilm residents who appear to react differently (Carniello et al., 2018).

## 4   Recommendation and conclusion

Biochar is considered to be a good remediation material for the removal of microbes from water. Removal efficiency depends on biochar and microbial polluted water properties. When organic matter increases in stormwater, bacteria tend to adsorb to sites of biochar and increase their growth. Physical weathering of biochar also affects the bacterial removal capacity of the biochar. The fluctuations in soil moisture also influence microbial survival and have a negative impact on remediation.

Many studies have been conducted to increase microbial population by amending biochar to soil. However, less attention is given to the biochar-amended studies in aquatic systems. In few studies, pathogenic microorganisms were removed from stormwater using biochar-based filters. Thus, more studies are needed to better understand the biochar use in microbial decontamination of water.

## Questions and problems

1. Name three (03) water sources that are subject to microbial contamination.
2. What bacterial species present in an aqueous system is an indication of fecal water pollution?
3. Other than bacterial species, do other microorganisms influence contamination of water? If any, state specific names.
4. Name the sources of common pathogenic microorganisms originated from animal farms and responsible for aqueous system contamination?
5. List three water purification methods used for microbial decontamination.
6. How soil amendments with biochar affect the soil microbial population.

7. How biochar amendments affect the microbial population in water?
8. What properties of biochar-based materials are considered to be important in pathogen removal from water?
9. Explain the DLVO theory in *E. coli* removal.

# References

Ahmed, W., Hamilton, K., Toze, S., Cook, S., Page, D., 2019. A review on microbial contaminants in stormwater runoff and outfalls: potential health risks and mitigation strategies. Sci. Total Environ. 692, 1304–1321.

Arnone, R.D., Perdek Walling, J., 2007. Waterborne pathogens in urban watersheds. J. Water Health 5 (1), 149–162.

Bhattacharjee, S., Elimelech, M., Borkovec, M., 1998. DLVO interaction between colloidal particles: beyond Derjaguin's approximation. Croat. Chem. Acta 71, 883–903.

Bozorg, A., Gates, I.D., Sen, A., 2015. Impact of biofilm on bacterial transport and deposition in porous media. J. Contam. Hydrol. 183, 109–120.

Carniello, V., Peterson, B.W., van der Mei, H.C., Busscher, H.J., 2018. Physicochemistry from initial bacterial adhesion to surface-programmed biofilm growth. Adv. Colloid Interf. Sci. 261, 1–14.

Chandrasena, G.I., Pham, T., Payne, E.G., Deletic, A., McCarthy, D.T., 2014. E. coli removal in laboratory scale stormwater biofilters: influence of vegetation and submerged zone. J. Hydrol. 519, 814–822.

Dai, X., Hozalski, R.M., 2002. Effect of NOM and biofilm on the removal of *Cryptosporidium parvum* oocysts in rapid filters. Water Res. 36 (14), 3523–3532.

Gleick, P.H., 1993. Water in crisis. In: Pacific Institute for Studies in Development, Environment, and Security.; Stockholm Environment Institute. Oxford University Press, p. 473 (9, 1051–0761).

Hermansson, M., 1999. The DLVO theory in microbial adhesion. Colloids Surf. B: Biointerfaces 14 (1–4), 105–119.

Hornberger, G.M., Mills, A.L., Herman, J.S., 1992. Bacterial transport in porous media: evaluation of a model using laboratory observations. Water Resour. Res. 28 (3), 915–923.

Inyang, M., Dickenson, E., 2015. The potential role of biochar in the removal of organic and microbial contaminants from potable and reuse water: a review. Chemosphere 134, 232–240.

James, E., Joyce, M., 2004. Assessment and management of watershed microbial contaminants. Crit. Rev. Environ. Sci. Technol. 34 (2), 109–139.

Kranner, B.P., Afrooz, A.R.M.N., Fitzgerald, N.J.M., Boehm, A.B., 2019. Fecal indicator bacteria and virus removal in stormwater biofilters: effects of biochar, media saturation, and field conditioning. PLoS One 14 (9), e0222719.

Mohanty, S.K., Boehm, A.B., 2014. Escherichia coli removal in biochar-augmented biofilter: effect of infiltration rate, initial bacterial concentration, biochar particle size, and presence of compost. Environ. Sci. Technol. 48 (19), 11535–11542.

Mohanty, S.K., Cantrell, K.B., Nelson, K.L., Boehm, A.B., 2014. Efficacy of biochar to remove *Escherichia coli* from stormwater under steady and intermittent flow. Water Res. 61, 288–296.

Mukherjee, A., Zimmerman, A.R., Harris, W., 2011. Surface chemistry variations among a series of laboratory-produced biochars. Geoderma 163 (3), 247–255.

Peiris, C., Gunatilake, S.R., Wewalwela, J.J., Vithanage, M., 2019. Biochar for sustainable agriculture: nutrient dynamics, soil enzymes, and crop growth. In: Biochar from Biomass and Waste. Elsevier, pp. 211–224.

Perez-Mercado, L.F., Lalander, C., Joel, A., Ottoson, J., Dalahmeh, S., Vinnerås, B., 2019. Biochar filters as an on-farm treatment to reduce pathogens when irrigating with wastewater-polluted sources. J. Environ. Manag. 248, 109295.

Petterson, S.R., Mitchell, V.G., Davies, C.M., O'Connor, J., Kaucner, C., Roser, D., Ashbolt, N., 2016. Evaluation of three full-scale stormwater treatment systems with respect to water yield, pathogen removal efficacy and human health risk from faecal pathogens. Sci. Total Environ. 543 (Pt A), 691–702.

Schijven, J.F., Hassanizadeh, S.M., 2000. Removal of viruses by soil passage: overview of modeling, processes, and parameters. Crit. Rev. Environ. Sci. Technol. 30 (1), 49–127.

Trenberth, K.E., Asrar, G.R., 2012. Challenges and opportunities in water cycle research: WCRP contributions. Surv. Geophys., 515–532.

Tufenkji, N., Elimelech, M., 2004. Correlation equation for predicting single-collector efficiency in physicochemical filtration in saturated porous media. Environ. Sci. Technol. 38 (2), 529–536.

World Health Organization; International Water Association, 2009. Water Safety Plan Manual: Step-by-Step Risk Management for Drinking-Water Suppliers. World Health Organization, Geneva.

# 14

# Biochar-based constructed wetland for contaminants removal from manure wastewater

Valentine Nzengung[a] and Stefanie Gugolz[b]
[a]Department of Geology, College of Arts and Sciences, University of Georgia, Athens, GA, United States. [b]GEI Consultants, Inc., Atlanta, GA, United States

## 1 Introduction

### 1.1 Nutrient contamination

Agricultural runoff and urban stormwater not only represent the largest sources of surface water (e.g., especially lakes, rivers, reservoirs, estuaries, and ocean shoreline waters) and groundwater pollution in the world, but also exemplify the growing human impact on environmental systems. Nitrogen and phosphorus are among the primary nutrients of concern in both agricultural runoff and urban stormwater (USEPA, 2000). The nutrients in urban stormwater are mainly from urban landscape runoff (such as fertilizers, detergents, and plant debris), atmospheric deposition, and failing septic systems (Terrence Institute, 1994). The growing increase in the volume of stormwater runoff from urban catchments along with the concentration of pollutants within the runoff impairs urban watersheds physically, chemically, and biologically (Walsh et al., 2005). The degree and impacts to water bodies is significant; it affects water quality, water quantity, habitat and biological resources, public health, and the aesthetic appearance of waterways.

Unlike urban stormwater, animal waste is a major source of nutrient contamination in agricultural runoff. Agricultural runoff, particularly from concentrated animal feeding operations (CAFOs), is recognized as one of the largest sources of contamination of surface water in the United States (Puckett, 1994). CAFO is a legal designation for farms housing over a certain number of animals on site. Large

CAFOs, which can have thousands to hundreds of thousands of animals, can produce enough manure in a year to rival the sewage production of major US cities (GAO, 2008), and while all human waste is treated, there are no regulated treatment facilities for CAFOs. CAFO manure wastewater is high in nutrients like nitrogen and phosphorous, heavy metals, salts, hormones, antibiotics, and pathogens like e-coli. When this runoff reaches surface water, it can cause algal blooms leading to eutrophication and the death of aquatic species, and the other chemicals can also have devastating ecological and human health effects.

CAFO waste management in most countries follows "best management practices" which includes the storage of manure wastewater in treatment lagoons, holding ponds, and underground pits. It is typically disposed of via land application, and sprayed onto cropland as a fertilizer. Rain and flooding events can, and often do, result in overflow from the holding ponds as well as washout of the land-applied manure from fields. Leaks in holding ponds and storage tanks, as well as over application onto fields, can cause nutrients and other contaminants to leach into the groundwater as well. Rather than developing and enforcing stringent regulations, most countries require CAFOs to develop nutrient management plans (NMPs) for the land-applied wastewater by accounting for the nutrient uptake capacity of the crops as well as the nutrient- and water-holding characteristics of the soil. The NMPs tend not to prevent over application of the manure and surface and groundwater pollution (Bradford et al., 2008; GAO, 2008).

Due to the ongoing negative environmental and human impacts that result despite NMPs, there is an immediate need for CAFO treatment technologies. If the wastewater could undergo pretreatment to significantly lower contaminant concentrations, land application systems may be much easier to manage successfully. Further, if the wastewater could be treated within acceptable levels, then it could even be reused at the farm or discharged to local surface water without harm to the environment. To avoid significant increases in the cost of CAFO and the price of meat, the industry needs low-cost, low-maintenance treatment technologies for their wastewater.

## 1.2 Biochar application as sorption and plant growth media

Biochar has the potential to decrease the leaching of nutrients from agricultural systems to surface and groundwater. In a study conducted with cattle slurry and different amounts of spruce biochar added to the surface of a boreal grass field and analyzed for 3 years, the surface application of biochar found to inhibit nutrient loss (i.e., N and P) in runoff was reduced if an effective biochar was applied (Saarnio et al., 2018). In a review paper by Mohanty et al. (2018), they discussed contaminant removal mechanisms by biochar, summarized specific biochar properties that enhance targeted contaminants removal from stormwater, and identified challenges and opportunities to retrofit biochar in low-impact development (LID) to optimize stormwater treatment.

In practice, using biochar only as an adsorbent means that once all the adsorption sites are filled, the biochar is spent and is no longer useful to treat the wastewater. It then must be removed and replaced with fresh media. The useful life of biochar as a filtration media is dependent on the biochar's sorption capacity and the concentration of the contaminants of concern in the wastewater. Depending on the type of adsorbed contaminants, the biochar can be used for plant nutrients recovery and mixed into soil as a "slow-release fertilizer" (Ghezzehei et al., 2014; Taghizadeh-Toosi et al., 2012). For use as a soil fertilizer, biochar that has filtered only nutrients and no toxic chemicals may be used; otherwise, it may contain metal that could be toxic to the plants or could introduce toxic chemicals into food crops (Chaukura et al., 2016). Spent biochar can be disposed of by burning; however, if there are toxic volatile compounds within either the biochar structure or adsorbed by the biochar, combustion is not a desirable method of disposal (Lehmann et al., 2011).

Biochar can also be used as a media for microbial filtration from wastewater. In a study by Kätzl et al. (2014), biochar was found to be equal or more effective than sand for the treatment of municipal wastewater through the growth of biofilms. Biochar has been found to increase microbial activity in soils (Gregory et al., 2014; Lehmann et al., 2011), but the reasons for this are not well understood. It is likely that a combination of interconnected factors are responsible, potentially including increased energy supply due to the labile carbon content of biochar (Hamer et al., 2004), microbial sheltering in biochar pore structures (Lehmann et al., 2011; Thies and Rilling, 2009), pH increases and redox conditions brought on by biochar additions (Joseph et al., 2010), the removal of toxins from soil pore water through adsorption, and an increase in plant root exudates due to an increase in plant growth associated with biochar addition to soils (Gregory et al., 2014).

Microbial activity stimulation in biochar-amended soils has also been suggested as a potential mechanism for the increased plant growth found in many studies (Lehmann et al., 2011). Many studies which found increased microbial activity after biochar addition also noted that microbial community structure biodiversity increased (Khodadad et al., 2011), often resulting in an increase in rhizome-associated microbes and fungi which form symbiotic relationships with plant roots (Jin, 2010). The increased microbial activity can lead to an increase in plant resistance to stressors-like diseases (Matsubara et al., 2002) and drought (Herrmann et al., 2004).

Many studies have found biochar addition to soil to increase roots, above-ground plant biomass, plant germination, and growth rates. These effects have been attributed to a combination of microbial activity and direct effects from biochar physicochemical properties that lead to water and nutrient retention, toxin immobilization, changes in pH, and soil aeration (Lehmann et al., 2011).

Because of its contaminant adsorption ability and its positive impact on plant growth, biochar has been studied as a soil amendment for the phytostabilization and phytoextraction of heavy metal contaminated soils with some success (Paz-Ferreiro et al., 2014). Houben et al. (2013) observed that biochar reduced the bioavailability of cadmium (Cd) and zinc (Zn) and increased plant biomass. Fellet et al. (2014) used biochar to increase plant uptake of lead (Pb) by phytoextractors, although they saw no change in biomass, whereas Chirakkara and Reddy (2015) found increased Cd and Pb uptake along with increased plant biomass. These studies make use of both the direct effect of biochar on contaminants, and the indirect effect through the environmental and ecological effects of plants.

## 1.3 Constructed wetlands for the treatment of wastewater

CTWs offer a mature technology to treat wastewater from many sources. The use of CTWs as a low-cost, low-maintenance, passive or semipassive, solar-driven, and aesthetically pleasing technology to treat stormwater and agricultural runoff continues to grow. One disadvantage of CTWs is that they tend to take up large land areas and may require two or more stages to effectively remove all the contaminants of concern in a waste stream. For instance, vertical flow wetlands successfully target ammonia, typically through microbial oxidation, but are not effective denitrifiers. In contrast, horizontal flow wetlands tend to promote nitrification (Vymazal, 2008). Phosphate removal is similarly complex, requiring specialized steps that rely on chemical removal processes like sorption

(Kadlec and Wallace, 2008). Some other drawbacks of using CTWs include poor performance in winter months and the need for pretreatment or dilution of the wastewater to reduce any toxicity effects. In northern latitudes during winter months, reduced plant uptake is caused by dormancy or plant death. Thus, the plants used in CTWs need to be hardy, have good longevity, have high-nutrient uptake capacities, and have ability to withstand sudden changes in nutrient concentrations.

By combining biochar and plants in a wastewater treatment system, the biochar with its large specific surface area and CEC should function as a sort of sponge that slows the transport of contaminants through the wetland. This retardation provides additional residence time for the plant roots to take up nutrients bound to the biochar and the residuals dissolved in the wastewater. The removal of the sorbed nutrients by plants should keep the biochar sorption sites from becoming saturated. Additional benefits include the biochar enhancement of plant growth and promotion of microbial activity which should further enhance treatment in the system.

Research conducted by Rizzo (2015) confirmed the effectiveness of biochar in the reduction of thermotolerant coliforms and *Escherichia coli* throughout all the tests conducted in biochar-amended soils. Although uniform reduction in removal of total suspended solids (TSSs), total nitrogen (TN), and phosphorus (TP) in the stormwater was not observed, the addition of biochar did reduce the concentration levels of pollutants within the soil than the 100% soil treatments. De Rozari et al. (2016) studied the effectiveness of sand media amended with biochar and two plants species (*Melaleuca quinquenervia* and *Cymbopogon citratus*) in removing phosphorus from sewage effluent in a constructed wetland (CW). Although the sand-only CW performed better than the biochar-amended CW, after flushing due to major rain event, there was no significant difference between sand and sand augmented with 20% biochar.

There is increased interest in using more reactive media in CTWs to provide additional treatment mechanisms. The reactive media include natural minerals (like bauxite and dolomite), industrial by-products (slags), and man-made products like Filtralite (Vohla et al., 2011). Meanwhile, recently there has been growing interest in the application of biochar to increase the specific surface area of media in CTWs and the physical and chemical removal of contaminants (De Rozari et al., 2015; Ghezzehei et al., 2014). To enhance their pollutant removal capacity, CTW can be augmented with natural or engineered geomedia that meet the following criteria: They should be economical, readily available, and have capacity to remove a wide range of stormwater pollutants in conditions expected in wetlands. Abedi and Mojiri (2019) studied the removal of contaminants from synthetic waste water in constructed wetland bioreactors of two parallel cylinders (lysimeters) with one containing modified biochar and gravel and common reed plants while the control contained only gravel and common reed plants. The modified biochar, zeolite, gravel, and reed-constructed wetland performed better in treating synthetic wastewater; removing all of COD, ammonia, phenols, Pb and Mn at optimum pH (6.3) and retention time of 57.4 h.

Zhou et al. (2017) studied vertical-flow-constructed wetlands (VFCWs) with intermittent aeration as an enhancement for the removal efficiency of organics and nitrogen for wastewater treatment. Their results suggested that adding biochar to intermittent aerated VFCWs could be an effective and appropriate strategy for low C/N wastewater treatment. Lin et al. (2008) demonstrated in pilot tests that nitrate removal rates in both free water surface flow (FWS) and subsurface flow (SSF) wetlands increased with increasing hydraulic loading rate (HLR) until a maximum value was reached. Floating treatment wetlands (FTWs) with monoculture plantings and mixed plantings showed that nutrient assimilation within plant tissues did not correlate with overall remediation performance for monocultures or mixtures, as tissue accumulation varied by nutrient and mixture (Chance et al., 2020).

Biochar can adsorb pollutants, improve water-retention capacity of soil, retain and slowly release nutrients for plant uptake, and help sustain microbiota in soil and plants atop; all these attributes could help improve removal of contaminants in stormwater treatment systems (Mohanty et al., 2018). The biochar incorporated into constructed wetland is an effective alternative for low-cost wastewater treatment in treating BOD, COD, TDS, and turbidity. Biochar wetlands and normal wetlands have both shown significant reduction in pollutant removal. Biochar wetlands as compared to normal wetlands did not show much difference in reduction efficiency, pH, total nitrogen, and chromium (Vijay et al., 2019).

## 2 Objectives

This study aims to further quantify and contribute to the current understanding of environmentally green and sustainable treatment practices for wastewater impacted by animal manure, with a focus on CAFOs. Biochar as a phytoremediation enhancer has been studied more for soil than for water treatment, and very few studies have so far used biochar in a wetland environment. There is a great potential for a synergistic effect of biochar and plants to enhance the removal of contaminants by wetlands. De Rozari et al. (2015) found biochar-amended sand to be more effective for removing chemical oxygen demand (COD), biological oxygen demand (BOD), suspended solids (TSSs), and coliforms from septic tank sludge than sand alone in planted wetland mesocosms. Since farmers are more likely to apply low-cost treatment systems that provide them with added income, the biochar can be made on-site using their on-farm agricultural waste products like crop residues, solid manure, or chicken litter. In addition, a biochar-enhanced CTW for CAFO wastewater should produce treated wastewater reused or recycled on the farm and carbon sequestration credits that are bankable.

## 3 Methods

### 3.1 Biochar media characterization

The softwood biochar produced by pyrolysis at 550°C was obtained from a commercial supplier and subjected to grain size analysis prior to use in the batch sorption and CTW tests. The softwood biochar was selected for two reasons: (1) availability in large quantities for large-scale applications in SE USA, and (2) the fast growth rate of softwood trees means it is a renewable resource. The grain size distribution of the biochar was determined by sieving using #5, 10, 20, and 40 sieve sizes (method ASTM D422-63 (2007)).

The grain size distribution was 0% gravel, 18.2% coarse sand, 80% medium sand, and 1.8% fine sand or smaller (Fig. 14.1). The biochar particles finer than 0.425 mm (#40 mesh) were separated, and the >0.425 mm biochar fraction was used in both the batch sorption and CTW tests.

## 4 Batch sorption of nutrients in CAFO wastewater by biochar

The removal of each nutrient from aqueous solution was measured in separate batch tests. Stock solutions of the three nutrients of interest ($NH_4^+$-N, $NO_3^-$-N, and $PO_4^{3-}$-P) were prepared by dissolving $KNO_3$, $(NH_4)_2SO_4$, and $K_2HPO_4$, separately in distilled water to obtain the desired concentrations specified below. To determine the equilibration time for the approach to sorption

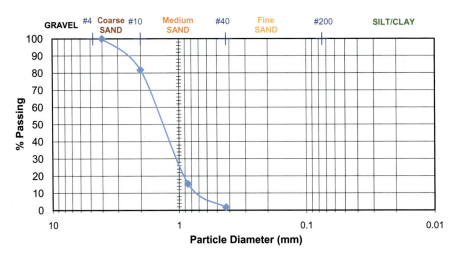

**Fig. 14.1** Biochar grain size distribution, chart provided by UM Lowell Geotechnical Engineering Research Laboratory.

equilibrium, a preliminary sorption kinetics experiment was performed using solutions of ~100 mg L$^{-1}$ NH$_3$-N, ~5 mg L$^{-1}$ NO$_3$-N, or ~100 mg L$^{-1}$ PO$_4^{3-}$, and mixing times of 5 min, 10 min, 20 min, 2 h, 10 h, and 24 h. Based on the results of the kinetics test, 24 h was determined as adequate to attain sorption equilibrium.

Sorption isotherms were not determined for NO$_3^-$ and PO$_4^{3-}$ because the kinetics experiment showed insignificant sorption of either of them by softwood biochar. To each 50-mL nominal volume crimp top vial was added 5 g of biochar and 45 mL of the prepared NH$_3$-N solution. The NH$_3$-N concentrations were ~20, 40, 60, 80, 100, 150, and 200 mg L$^{-1}$, respectively. The tests were set up in duplicate or in triplicate if the variance in the results was >5%. The control samples consisted of nutrient solutions alone and biochar with distilled water, respectively, and handled in parallel. The prepared samples and control vials were mixed continuously on a rotary shaker at a speed of 3.3 rpm for 24 h.

To determine if the nutrients were reversibly sorbed, the spent biochar samples from the sorption isotherm experiment were dried in an oven for 24 h at 40°C. Duplicate samples of the biochar (~5 g each) were each mixed with 45 mL of distilled water, equilibrated on the mixer for 48 h, filtered, and the solution was analyzed for NH$_3$-N. The solution from each sample was filtered through a Fisher Brand P8 qualitative filter prior to analysis. The filtered solution was analyzed with a HACH DR3900 spectrophotometer (Hach Method 10031 for high-range ammonia nitrogen, Hach Method 10206 for low-range nitrate nitrogen, and Hach Method 8048 for low-range reactive phosphorous).

## 4.1 Results of nutrient batch sorption tests

The preliminary batch sorption kinetics tests indicated that the approach to sorption equilibrium for $NH_4^+$-N was achieved in about 24 h. There was no measurable adsorption of $NO_3^-$-N or $PO_4^{3-}$-P, confirming that, like most inactivated softwood biochar, the biochar's surface was primarily negatively charged and suitable for adsorption of ammonium cations. Therefore, only the $NH_4^+$-N equilibrium sorption isotherms were measured.

The Langmuir isotherm model was a better fit to the data ($R^2 = 0.92$). The linearized form of the Langmuir model applied to the batch sorption data is $\frac{1}{S} = \frac{1}{K_L C S_m} + \frac{1}{S_m}$, where $S$ = sorbed solute concentration (mg g$^{-1}$), $C$ = solute solution concentration (mg L$^{-1}$), $K_L$ = Langmuir coefficient, and $S_m$ = sorption maximum (mg g$^{-1}$). The Langmuir coefficient ($K_L$) was estimated at 0.026, and the sorption maximum ($S_m$) was 280 mg kg$^{-1}$. This adsorption maximum for $NH_4^+$-N is quite low compared to literature values for many other types of wood biochars, whose $S_m$ range from 730 to 44,640 mg kg$^{-1}$ (Cui et al., 2016; Kizito et al., 2015; Sarkhot et al., 2013; Wang et al., 2015; Wang et al., 2015; Zeng et al., 2013). In the desorption test, 97% of the adsorbed $NH_4^+$-N was desorbed from the biochar into distilled water, indicating its potential bioavailability to plants in a wetland system.

Although the loading of $NH_4^+$-N to this biochar was relatively low, it was still used in the greenhouse CTW tests because sorption is not the only mechanism of nitrogen removal in a CTW system. In addition, there is evidence that the sorption effectiveness of biochar tends to increase with aging in soils as it undergoes biotic and abiotic oxidation changes that increase its CEC (Cheng et al., 2006, 2008). This combined with the biochar's potential stimulation of the growth of microbial communities within the CTWs is expected to contribute to enhanced pollutant removal over time (Fig. 14.2).

Most fresh biochars are primarily negatively charged with a low anion exchange capacity (AEC). This apparently explains the findings of many studies that have found biochar to have a low capacity for removing $NO_3^-$-N from water (Ghezzehei et al., 2014; Kameyama et al., 2016). The removal of nitrate nitrogen is potentially related to the biochar's pyrolysis temperature, with higher temperatures resulting in some $NO_3^-$-N adsorption (Mizuta et al., 2004; Kameyama et al., 2012). Some biochars have also been found to remove $PO_4^{3-}$ even with a low AEC, probably due to $PO_4^{3-}$ coprecipitation with other elements on the biochar surface, for example $Mg^{2+}$ (Yao et al., 2011; Zeng et al., 2013).

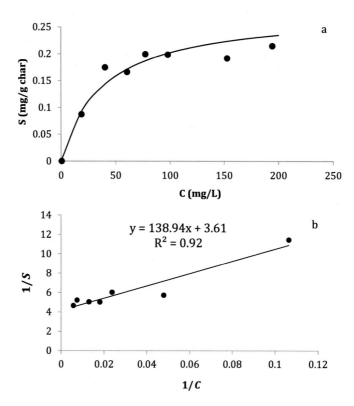

**Fig. 14.2** (A, B) Langmuir isotherm for adsorption of $NH_4^+$-N by softwood biochar.

## 5 Greenhouse study

### 5.1 Setup of constructed treatment wetland

The role of biochar in the enhancement of nutrient removal from wastewater was investigated using four simulated constructed treatment wetland (CTW) reactors in a greenhouse (Fig. 14.3). These reactors were 140 L rectangular planter tanks with dimensions $1 \times 0.3 \times 0.3$ m. At the bottom of each reactor, there was a drainage channel topped with a perforated (0.5 cm holes) false bottom to support the media in the reactor and allow effluent drainage into the channel. This channel spanned the entire length of the reactors and ended in a drainage port for treated effluent removal. A 2.5-cm layer of pea gravel was added to the bottom of each of the four.

Reactors, then landscape fabric, and a 2.5-cm layer of sand on top to keep any biochar from washing out of the reactors. Reactor 1 (Fig. 14.3, R1) was then filled with 20 cm of biochar (approx. 15.7 kg), sieved first over a #40 mesh sieve to remove fines and

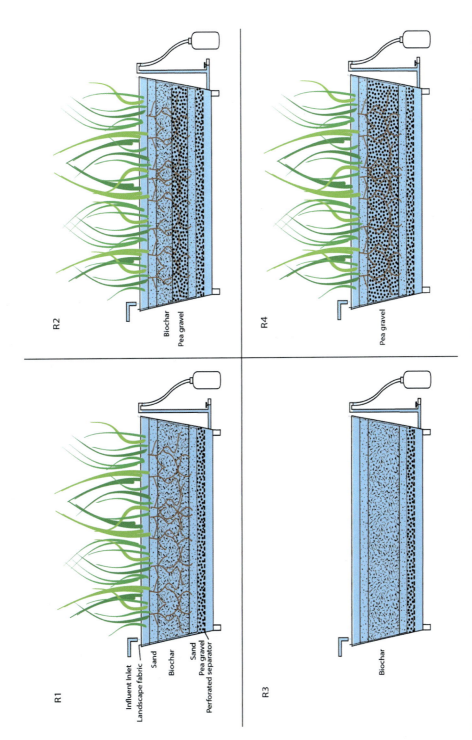

**Fig. 14.3** Layout of experimental wetland reactors. R1 is the higher biochar media (20 cm) planted reactor, R2 is the half biochar (10 cm) planted, R3 is the high biochar unplanted, and R4 is the no-biochar-planted reactor.

prevent clogging. This was topped by another 2.5-cm layer of sand to keep the biochar from floating while simultaneously functioning as a prefilter for suspended solids. This reactor was planted.

Reactor 2 (Fig. 14.3, R2) was similarly designed, except that it contained half the amount of biochar in CTW R1, so that the treatment media consisted of a 10-cm layer of pea gravel and a 10-cm layer of biochar, and was planted like R1. Reactor 3 (Fig. 14.3, R3) was packed with the same quantity of biochar as in R1 except that it was unplanted. Reactor 4 (Fig. 14.3, R4) contained only pea gravel with no biochar and was planted. R1, R2, and R4 provide a gradient from R1 "full biochar," R2 "half biochar," and R4 "no biochar." R3 provided the biochar control, representing the contribution of biochar only to the nutrient removal while R4 served as the plant control by providing the contribution of the plants only. Each reactor had 1 in. of wastewater above the packed treatment media.

An effluent hose was attached to the drainage port on each reactor as illustrated in Fig. 14.3. The hose rose vertically and then bent at the same level as the top of the planter tank so it matched up with the water level inside the tank. As wastewater was added to the reactors, the water level would rise and the pressure would cause the downward movement of the wastewater through the tanks into the drainage channel, out the port, and up the hose where it dripped out while equalizing the water level on each side. This kept the effluent flow rate consistent with the level of water in the reactor so that the effluent flow rate depended both on the influent flow rate and on the evapotranspiration rate from each reactor. Another hose was connected to a spigot at the bottom to allow for full, quick, and complete draining of the reactors. Each reactor had an influent hose on the opposite side of the reactor effluent drainage port through which water and CAFO wastewater were pumped into the reactor from a 200 L CAFO wastewater drum. The diluted CAFO wastewater was supplied to the reactors from a 200 L CAFO reservoir that contained enough wastewater to last for 2 days. The reservoir was refilled with a new "batch" of diluted wastewater every 48 h.

At the start of the greenhouse CTW tests in Summer (August 2015), reactors R1, R2, and R4 were initially planted with six common cattails (*Typha latifolia*), three soft rush (*Juncus effusus*), two willow (*Salix*) branch cuttings, and a handful of duckweed (Fig.14.4) each. The plants were collected from ponds and lakes around the University of Georgia (UGA), Athens, GA. Cattails and rush are both commonly used in CTW, and cattails especially are known for their hardiness and ability to withstand stressors-like sudden spikes in nutrient concentration as well as in seasonal

**Fig. 14.4** Photograph of reactors after initial planting (August 7, 2015).

temperature changes. The reactors were filled with an equal mixture (1:1 v/v) of clean water and dairy wastewater collected from the UGA dairy farm lagoon. The reactors were kept full for 3 months to allow the plants to acclimate, establish large roots, and spread.

Three different tests lasting several weeks each were conducted by applying wastewater to the reactors during different seasons. An initial test was conducted in November 2015, a second in April–June 2016, and a third in October 2016. For the first test, the influent flow rate was $\sim$2 L h$^{-1}$ for a residence time in each reactor of 33.5 h. Diluted swine wastewater was used for these tests instead of dairy wastewater as it was easier to collect from the UGA swine farm anaerobic digester. The swine wastewater, however, had a much higher N concentration (>1200 mg L$^{-1}$ $NH_4^+$-N, undiluted) than the dairy wastewater ($\sim$200 mg L$^{-1}$ $NH_4^+$-N, undiluted). The initial November 2015 test performed with the 2× diluted swine wastewater resulted in the death of plants in R4 (no biochar), so they were removed and healthy plants from R1 and R2 were transplanted, and all the reactors were allowed time for the plants to reestablish fully before the subsequent tests. None of the planted reactors had any rush, willow, or duckweed left; however, there were some other aquatic plants, parrot's feather (*Myriophyllum aquaticum*) and knotweed (*Persicaria*) that had volunteered along with the original plants when they were collected. These were divided and planted equally within all three planted reactors (R1, R2, and R4). Photographs of the reactors were taken at the start of the test and subsequently every few days for the duration to document the growth and health of the aquatic plants.

For the second test, the flow rate was lowered to $1\,L\,h^{-1}$ for a residence time of 67 h. The test ran for 53 days, and each reactor treated a total of 1272 L of the wastewater. For the first 2 weeks, the wastewater was diluted $>10\times$ to obtain an initial $NH_3$-N concentration in the range of $\sim 50$–113 mg $L^{-1}$. Thereafter, the influent concentration was doubled. Photographs were taken of the reactors at the start of the test as well as every other day for the duration of the test. These, along with yardsticks, were used to measure the growth of the cattails in each tank over the course of the test. The plants were cut down to $\sim 2.5$ cm above the water in all the reactors after the second test. The biomass was dried in an oven at 44°C and 24% humidity and weighed. The reactors were allowed to regrow until the third test when they were all cut to the same height of 3 ft.

For the third test run during October 2016, the wastewater was diluted to the nutrient concentration level of the wastewater used in second test. The test's duration was 11 days at a flow rate of $1\,L\,h^{-1}$ for a total of 264 L treated by each reactor. A total of three sample sets were taken during the third test, and photographs were also taken to document the relative plant growth in each of the three planted CTW reactors.

## 5.2 CTW sampling and data analysis

The baseline concentration consisted of 1-L influent samples taken from the 200-L reservoir prior to the treatment of each fresh 200 L batch of the CAFO wastewater. Effluent samples were taken at intervals of 45.5 h (for the $2\,L\,h^{-1}$ influent rate) or 91 h ($1\,L\,h^{-1}$) after treatment began, corresponding to half of the inflow volume. For each test, the effluent flow rate ($F_e$) of each reactor was measured at 6 a.m. and 6 p.m. during a 24-h day. The effluent flow rates were generally less than the influent flow rate ($F_i$), with the difference being attributed to the evapotranspiration rate ($E$) of each reactor ($L\,h^{-1}$):

$$E = F_i - F_e \tag{14.1}$$

The samples from the second test were analyzed for total suspended solids (TSSs), $PO_4^{3-}$, chemical oxygen demand (COD), total Kjeldahl nitrogen (TKN), $NH_4^+$-N, $NO_3^-$-N, and total minerals (Al, B, Ca, Cu, Fe, K, Mg, Mn, Na, P, S, Zn). The TSS was estimated gravimetrically by drying 20 mL of the wastewater sample in an oven at 104°C. $PO_4^{3-}$ was measured using the Hach spectrophotometer method. The other parameters were analyzed by the

UGA Extension Soil, Plant and Water Analysis Laboratory. During the third test, the samples were analyzed only for $NH_4^+$-N and $PO_4^{3-}$ using the Hach spectrophotometer methods.

The amount of each measured constituent removed in each reactor was calculated using the evapotranspiration rates ($E$) and influent and effluent concentrations ($C_i$ and $C_e$) by applying Eqs. (14.2)–(14.6). The mass loading ($M$) of each pollutant into each reactor over a given period ($T$) in mg is as follows:

$$M = C_i * F_i * T \quad (14.2)$$

where $C_i$ = influent concentration (mg L$^{-1}$) and $F_i$ = influent flow rate (L h$^{-1}$); the total volume ($V_E$) of wastewater removed from the reactor by evapotranspiration ($E$) over a given period ($T$) in L is as follows:

$$V_E = T * E \quad (14.3)$$

the total volume ($V_R$) of wastewater remaining in the reactor after evapotranspiration in L is as follows:

$$V_R = (F_i * T) - V_E \quad (14.4)$$

where $F_i$ = influent flow rate (L h$^{-1}$), $T$ = time (h), and $V_E$ = volume of water removed by evapotranspiration as defined in Eq. (14.2); the total mass ($M_R$) of pollutant removed from the wastewater in the reactor (mg) is as follows:

$$M_R = M - (C_e * V_R) \quad (14.5)$$

where $M$ = total mass of pollutant loaded into each reactor (mg, Eq. 14.4), $C_e$ = effluent concentration (mg L$^{-1}$), and $V_R$ = volume of water remaining after evapotranspiration (L, Eq. 14.3), and the concentration ($C_R$) removed by each reactor (mg L$^{-1}$) is as follows:

$$C_R = \frac{M_R}{T * F_i} \quad (14.6)$$

where $M_R$ = mass of pollutant removed (mg, Eq. 14.5), $T$ = time (h), and $F_i$ = influent flow rate (L h$^{-1}$).

Percent removal was calculated using Eq. (14.7) is as follows:

$$R(\%) = (M - M_R) * 100 \quad (14.7)$$

For the second test, a two-way (2 × 3) type 3 ANOVA was performed to understand the effect of influent concentration and amount of char on the contaminant removal capacity of the planted reactors. The low initial concentration loading had only two observations per reactor, while the high initial concentration loading had three, making the ANOVA unbalanced, so a type 3 was applied. R statistical software was used to do the ANOVA, using the "ANOVA" function from the package "car" which can be easily set to do a type 3 ANOVA. For each parameter that the reactors removed from the wastewater, the effects of influent concentration ("low" or "high"), amount of biochar ("0," "50," or "100"), and the interaction between them were tested for significance. If there was no significant ($P > .1$) interaction, this was removed from the ANOVA so that only each factor's individual effect was evaluated. If there was a significant ($P < .1$) effect, Tukey's honestly significant difference (Tukey's HSD) test was performed to see which levels were different from one another. For some of the parameters, the ANOVA had to be altered to account for influent concentrations not matching the wastewater dilution or effluent parameter concentrations that were higher than the influent. For the third greenhouse test, the CTW performance was compared using one-way ANOVA and Tukey's HSD.

## 6  Greenhouse CTW results and discussion

**CTW Test #1:** The initial treatment test carried out in November (Fall season) was performed with 2× diluted swine wastewater from the University of Georgia CAFO lagoon. The initial concentrations of nitrogen and phosphorous ions were 636.2 mg L$^{-1}$ for NH$_3$-N, 9.45 mg L$^{-1}$ for NO$_3^-$-N, 30.8 mg L$^{-1}$ for PO$_4^{3-}$-P, and pH = 8.01. Within 3 days of this initial test, all the plants in R4 (i.e., the no-biochar reactor) began yellowing and showed signs of stress, which was attributed to the high influent nutrient concentrations. The reactors were drained and refilled with the lower strength wastewater (diluted to NH$_3$-N concentration of ~200 mg L$^{-1}$) in order to mitigate the high-nitrogen loading that was apparently toxic to the plants. By the end of the month, all plants in R4 had died. Although the plants in R1 (high biochar loading—HBL) and R2 (50% < HBL) also wilted, not nearly as many of the plants died (Fig. 14.5). It was observed that half of the cattails and all the rush, willow, and duckweed in R2 died while the remaining cattails showed some yellowing. In R1, only a few (~4 of 30 cattails, 2 rush, all willow) died and many were yellowing. The latter observations confirmed the decrease in plant mortality in a CTW bioreactor with high biochar fraction in the growth media. Biochar appears to improve cattails' resistance to stressors-like sudden nutrient spikes and seasonal temperature changes. The biochar tended to buffer the stress on the plants in R1 and R2.

**Fig. 14.5** Photograph of treatment wetland reactors in November showing response to high nutrient shock from swine wastewater. The no-biochar reactor (R4) experienced 100% plant mortality, while the biochar reactors (R1 and R2) experience partial mortality.

The improved plant resistance to the nutrient shock in R1 and R2 relative to R4 can be attributed to adsorption of some fraction of the $NH_4^+$-N by biochar (Spokas et al., 2012), thereby decreasing its concentration in the water column and plant rhizosphere. Using the Langmuir isotherm results, the calculated maximum loading of $NH_4^+$-N onto the biochar in R1 would be ~15.7 kg and ~7.85 kg for biochar in R2. Secondly, biochar has been shown to significantly increase the number of bacteria and the microbial activity of denitrifiers in the bioreactor (Mohanty et al., 2018). It has also been shown that there are many species of microbes and fungi whose presence increases the ability of plants to withstand nutrient, salt, and heavy metal toxicity (Dimkpa et al., 2009). It is also likely that the onset of cold weather during this test (November) further decreased the plants' resilience to the nutrient shock.

**CTW Test #2:** At the start of the second test, all of the planted CTW reactors had approximately the same number of cattail stems (>30), a small clump of parrot feather and a small clump of knotweed (Fig. 14.6). The tallest cattails in each reactor were within 5 cm of each other at an average height of about 0.97 m (measured from the top of the reactor tank). Over the course of the 55 days of Test #2, all the plants in the reactor grew considerably; however, growth was greater in R1 and R2 than in R4 (Fig. 14.7). At the end of the test, the

**Fig. 14.6** Photograph of treatment wetland reactors at the beginning of second test with line indicating the average maximum height of the tallest plant in each reactor.

**Fig. 14.7** Photograph of treatment wetland reactors at the end of second test with lines indicating the maximum height of the tallest plant in each reactor. The two biochar-planted reactors (R1 and R2) had the tallest cattails, while R4 (no biochar) had the shortest.

tallest cattails in R1 were 0.4 m and R2 0.5 m taller than in R4. The knotweed and parrot feather clumps also grew larger in the biochar reactors than in the control. The dry weight of the harvested plant biomass from each reactor was as follows: R1 = 1.48 kg (32% greater than R4), R2 = 1.32 kg (18% greater than R4), and R4 = 1.12 kg.

The influent concentration (IC) of the monitored constituents in the wastewater and the amount of biochar both influenced the removal efficiency observed in the planted reactors, which explains why R1 and R2 outperformed the other two reactors. Table 14.1 shows ANOVA results for all the monitored pollutants where either IC or biochar amount made a significant difference in the planted reactor performance. For all constituents, the ANOVA performed showed that the IC and biochar amount had no significant interaction, so the interaction term was removed. The IC significantly ($P < .05$) affected TSS removal from the planted reactors, measured in $mg L^{-1}$ and percentage removal. Table 14.1 contains the Tukey's HSD test results and shows the significance of the differences between pollutant removal due to IC and biochar amount. For TSS, a significantly ($P < .05$) higher amount was removed from the higher concentration wastewater. In contrast, biochar significantly ($P < .05$) affected the percentage removal but *not* the effluent concentration expressed in $mg L^{-1}$ (Table 14.1). This is probably due to the variation in influent concentrations (ICs) causing extra variation in removal concentrations, if removal amounts by each reactor are affected by the IC. This extra variation may be minimized when the removal percentage (the percent of the IC removed by each reactor) is calculated.

Fig. 14.8 shows the influent and effluent TSS concentrations over time. The total suspended solid (TSS) concentrations in R1- and R2-treated effluents were consistently lower than for R3 and R4. There was no significant difference between the removal of TSS in R1 and R2 CTWs, suggesting that there is a threshold of biochar mass to realize optimum performance that cannot be inferred from the results of this study. Both did, however, remove a significantly higher ($P < .05$) percent of TSS from the influent than R4. Overall R1's mean removal was 51% greater than R3 and 16% greater than R4. R3, with biochar only and no plants, removed the least TSS, which is attributed mainly to adsorption and/or entrapment by the biochar and sand media.

For all four reactors, removal percentages of the TSS increased with time (Fig. 14.8) and the IC, and the actual concentration of solids removed more than doubled as the plant biomass increased (Fig. 14.9). The TSS removal from the wastewater tends to be dominated by physical removal processes such as adsorption to plant root and biochar surfaces. Since the removal increased over time

**Table 14.1** Results of influent and effluent $NH_4^+$-N, $PO_4^{3-}$-P, and pH during CAFO wastewater treatment Test #3.

| CTW | $NH_4^+$-N | $PO_4^{3-}$-P | pH |
|---|---|---|---|
| Influent | 107.2 | 27.8 | 8.43 |
| R1 Effluent | 36.3 | 4.2 | 7.72 |
| R2 Effluent | 55.2 | 4.0 | 7.44 |
| R3 Effluent | 61.8 | 11.3 | 8.02 |
| R4 Effluent | 69.3 | 8.3 | 7.68 |
| Influent | 116.1 | 34.2 | 8.48 |
| R1 Effluent | 41.6 | 5.6 | 6.97 |
| R2 Effluent | 56.8 | 8.2 | 7.71 |
| R3 Effluent | 51.0 | 10.8 | 7.91 |
| R4 Effluent | 68.2 | 7.6 | 7.54 |
| Influent | 98.1 | 76.1 | 8.75 |
| R1 Effluent | 35.5 | 8.5 | 8.02 |
| R2 Effluent | 51.3 | 8.0 | 7.35 |
| R3 Effluent | 51.9 | 15.8 | 7.69 |
| R4 Effluent | 57.9 | 7.8 | 7.24 |

Concentrations are in mg L$^{-1}$, and each sampling event occurred 4 days after a new influent stock of the CAFO wastewater was started.

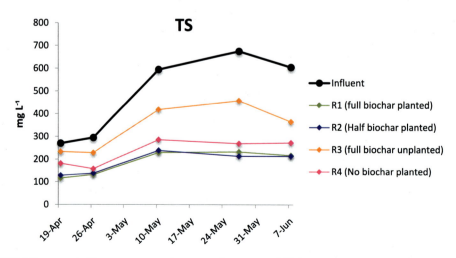

**Fig. 14.8** TSS influent and effluent concentrations over the course of 53 days.

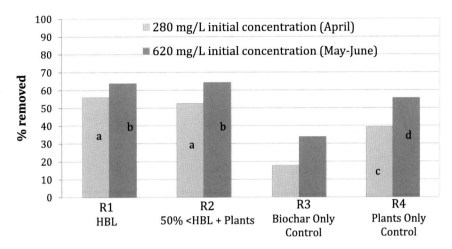

**Fig. 14.9** Percent (%) solids removal from each CTW reactor. *Light bars* represent average removal from lower (280 mg L$^{-1}$) TSS loading, while *dark bars* represent average removal from higher (620 mg L$^{-1}$) TSS loading. *Different letters* represent a statistical difference exist between the means of TSS removed in the planted reactors, while *same letter* means the results are not statistically different and vice versa.

and as such the adsorption surfaces were not saturated, the biodegradable fraction of TSS was likely metabolized by microorganisms. Other investigators have confirmed increases in microbial community abundance and composition due to biochar application to plant growth substrate (Lehmann et al., 2011). The enhanced microbial growth contributes to higher biotransformation leading to the greater TSS removal in R1 and R2 compared to R3 and R4.

The removal trends for COD were similar to TSS (Figs. 14.8 and 14.10). COD is a measure of the amount of oxygen required to degrade the organic material in the wastewater that can be oxidized by other chemicals as well as by microbes. In constructed wetlands, the primary COD removal mechanism is microbial oxidation, although plants can remove some COD as well. The influent concentration significantly ($P < .05$) affected COD removal in mg L$^{-1}$ but not the percentage suggesting that for each planted reactor, the overall COD removal capacity can be better described as a percent of the IC by the effluent concentration. The reactors removed a significantly ($P < .05$) higher concentration from the higher IC than the lower influent dose loading. Although the COD removal efficiencies of R1 and R2 were not significantly different from each other in terms of the percentages of COD, R1 and

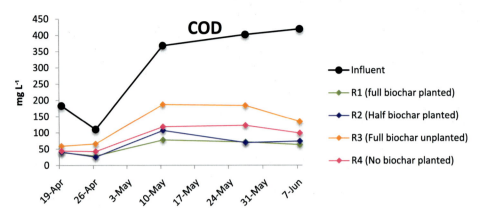

**Fig. 14.10** Chemical oxygen demand (COD) influent and effluent concentrations over the course of 53 days.

R2 did remove a significantly higher ($P < .05$ and $P < .1$, respectively) percent than R4 and R3. R1 removed 29.1% and 12.2% more than R3 and R4, respectively.

Fig. 14.11 shows that for each CTW reactor, the percentage removal of COD was statistically the same when the IC was doubled. Biochars have also been found to remove COD through adsorption (Berger, 2012). The COD removal in R3 is attributed to a combination of adsorption by biochar and microbial removal activity, meanwhile in R4, it is attributed to the plants and microbial activity. Since R1 and R2 removed the most COD, the increased plant biomass and biochar's large surface area available for adsorption and enhancement of microbial activity increased COD removal. This result is comparable to a similar study by

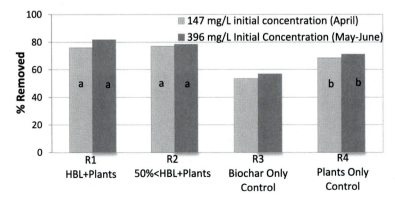

**Fig. 14.11** COD removal from each tank. *Light-colored bars* represent average removal from lower initial concentration influent period. *Dark-colored bars* represent average removal from higher concentration influent. *Different letters* represent a statistical difference exist between the means of COD removed in the planted reactors, while *same letter* means the results are not statistically different and vice versa.

De Rozari et al. (2015) which found planted wetlands with biochar-amended sand media removed more biological oxygen demand than those with sand and plants only.

Fig. 14.12 shows influent and effluent concentrations of TKN, $NH_4^+$-N, and organic N (calculated as TKN-$NH_4^+$-N) over time. The TKN is a measure of the total amount of $NH_4^+$-N and organic N. The primary source of TKN in most swine wastewater is $NH_4^+$-N. For the most part, all reactors reduced the amount of organic N in the wastewater, with one noticeable spike in organic N concentration in the effluent of R3. Organic nitrogen can be released into a wetland system by a variety of processes including microbial mineralization of solid organic material (e.g., organic TSS), desorption from wetland media, nitrogen fixation as well as release from plants and organisms (Hollister et al., 2013; Kadlec, 2009). It is unclear exactly which process caused this release in R3. Overall, the concentrations of organic N were significantly lower than $NH_4^+$-N.

The removal of $NH_4^+$-N in the three planted reactors increased with loading and time (Fig. 14.12). A comparison of mass removal at the start and end of the treatment tests in each planted CTW reactor as percent shows that $NH_4^+$-N removal in R1 increased by 43.3%, R2 by 30.0%, and R4 by 18.0%. These increases can be attributed to the observed increase in plant growth corresponding to increased plant uptake of $NH_4^+$-N over time. The removal of $NH_4^+$-N in R3 was dominated by adsorption onto the biochar surface and to a lesser extent to algae and microbial metabolism (Ghezzehei et al., 2014). If sorption alone contributed to $NH_4^+$-N removal in R3, the mass loading would have exceeded the sorption capacity of the biochar and caused complete breakthrough in the effluent concentration. Considering the influent concentration of $NH_4^+$-N, the total volume treated over 2 weeks, and the maximum loading of $NH_4^+$-N onto the softwood biochar, the total mass of $NH_4^+$-N of 17,472 mg added to R3 far exceeds the calculated theoretical adsorption maximum of 4396 mg. Therefore, other removal mechanisms such as microbial oxidation or transformation of $NH_4^+$-N to ammonia and/or nitrogen gases also contributed.

In a CTW reactor, the contribution of algae and microorganisms to the removal of $NH_4^+$-N is expected to be significant. The relatively higher removal of $NH_4^+$-N in R4 than in R3 confirmed the importance of plants in $NH_4^+$-N removal and use as a plant nutrient. The synergistic benefits of biochar and plants were demonstrated in the R1 and R2 CTWs that consistently removed TKN, $NH_4^+$-N, and organic N better than R3 and R4. The high rate of nutrient removal contributed to the measured high rate of plant growth and total plant biomass produced.

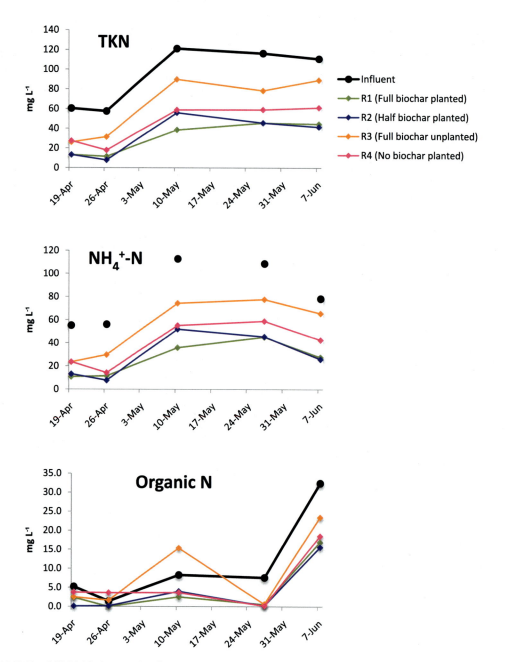

**Fig. 14.12** Total Kjeldahl nitrogen, $NH_4^+$-N, and Organic N (calculated as TKN-$NH_4^+$-N) concentrations over 53 days.

Unlike TSS and COD removal, the IC had a significant effect ($P <.05$) on the removal of $NH_4^+$-N. A higher IC of $NH_4^+$-N resulted in a significantly ($P <.05$) lower percent removal in the planted reactors (i.e., R1 and R2). The mass of biochar had a moderately significant ($P <.1$) effect on $NH_4^+$-N mg $L^{-1}$ removal and a significant ($P <.05$) effect on the percentage removal of the planted reactors. The removal of $NH_4^+$-N in R1 and R2 was not significantly different, but R1 removed a moderately significantly ($P <.1$) higher amount than R4 and both R1 and R2 removed a significantly ($P <.05$) higher percent $NH_4^+$-N than R4. The overall $NH_4^+$-N removal in R1 was 50% greater than in R3 and 23% greater than in R4 (Fig. 14.13).

The concentration of $NO_3^-$-N in the influent CAFO wastewater was low ($<1.0$ mg $L^{-1}$) and fluctuated in the effluent of all four reactors (Fig. 14.14). The increase in $NO_3^-$-N in the effluent tended to be relatively higher in the biochar-planted reactors. The increased levels of $NO_3^-$-N in reactor effluent (over that of the influent) probably indicate periods of nitrification that were greater than any $NO_3^-$-N removal by plant uptake and denitrification by microorganisms. The planted biochar CTWs, R1 and R2, had relatively higher nitrate concentrations in their effluent. Since $NO_3^-$-N samples were collected during the day and not at night, there is no data to determine how light and dark cycles influence nitrification and denitrification in the CTW. Contrarily, Lin et al. (2008) demonstrated in pilot tests that nitrate removal rates in both free water surface flow (FWS) and subsurface flow (SSF) wetlands increased with increasing hydraulic loading rate (HLR) until a maximum value was reached.

$PO_4^{3-}$ is the primary source of P in swine wastewater, and the difference between Total P and $PO_4^{3-}$ is probably organic P. Fig.14.15 shows a progressive increase in effluent TP over in the controls, before decreasing to below the influent concentration during the last 2 weeks of the test. The effluent TP and $PO_4^{3-}$-P concentrations measured in the effluents of R1 and R2 were consistently lower than controls (i.e., R3 and R4). The organic P (calculated as *Total P-$PO_4^{3-}$-P*) concentrations were higher than the influent for some monitoring events. The spikes in R3 and R4 are most likely due to changes in microbial and plant root zone activity as both utilize P for growth and can convert $PO_4^{3-}$-P to organic P. Plants can release organic P back into a wetland and can release large amounts in a short period of time when they are dying (Dunne and Reddy, 2005); however, there was no indication of dying plants in R4 during the time of its P spike. Microbial activity or die off and reduction of ferric to ferrous iron caused by changes in redox potential within these reactors are likely causes of P release into the system. There were corresponding

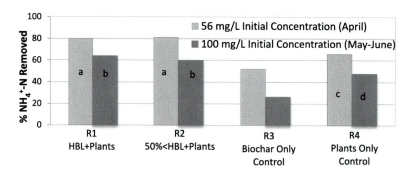

**Fig. 14.13** NH$_4^+$-N removal from each tank. *Light-colored bars* represent average removal from the lower initial concentration influent period. *Dark-colored bars* represent average removal from higher concentration influent. *Different letters* represent a statistical difference between means of removal by the planted reactors.

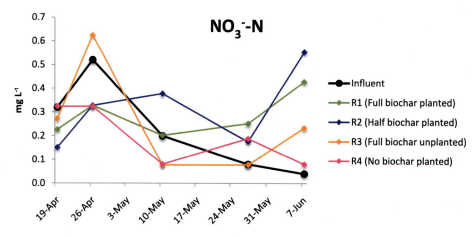

**Fig. 14.14** NO$_3^-$-N influent and effluent concentrations over the course of 53 days.

releases of Fe in the effluents of the two control reactors (R3 and R4) on the same dates as the P releases that may provide evidence of this (Fig. 14.17). The results of R1 and R2 suggest that the interaction between biochar and plants in the CTW minimizes the organic P release from the wetland reactors by stabilizing the redox potential.

Although the effluent PO$_4^{3-}$-P concentration was significantly ($P <.05$) influenced by doubling the PO$_4^{3-}$-P IC from 4 to 9.5 mg L$^{-1}$, the overall percent removal was not. The decrease in effluent PO$_4^{3-}$-P concentration in mg L$^{-1}$ at higher IC loading was significantly ($P <.05$) higher than from the lower IC loading. Fig. 14.16 shows that for all four reactors, the amount of inorganic P removed increased with time and the IC (i.e., April–May vs May–June). Since the

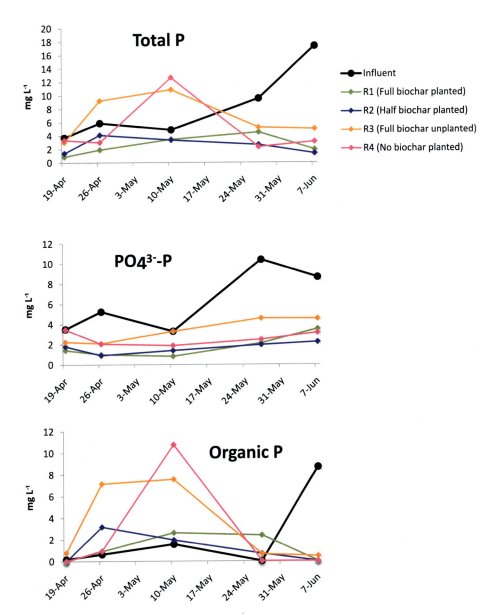

**Fig. 14.15** Total P, $PO_4^{3-}$-P, and organic P (Total P-$PO_4^{3-}$-P) influent and effluent concentrations over 53 days.

softwood biochar showed poor sorption of $PO_4^{3-}$-P in batch sorption isotherm tests, the removal of $PO_4^{3-}$-P from the CAFO wastewater in R3 was attributed to algae and other microbes living in the reactor's media and adsorption to the layers of sand and gravel. R4

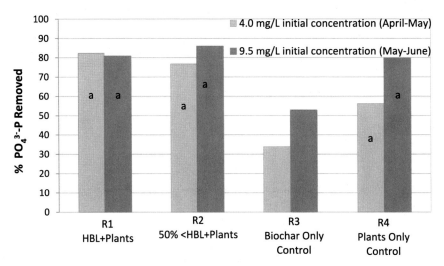

**Fig. 14.16** $PO_4^{3-}$-P removal from each tank. *Light-colored bars* represent average removal from lower initial concentration influent period. *Dark-colored bars* represent average removal from higher concentration influent. *Different letters* (i.e., a & b) represent a statistical difference between means of removal by the planted reactors. *Same letter* (i.e., a & a) represents no statistical difference between means of removal by the planted reactors.

removed $PO_4^{3-}$-P better than R3 indicating that plants and the many microbes growing in the sand and gravel within the rhizosphere were more efficient at $PO_4^{3-}$-P removal from the CAFO wastewater than just biochar alone. The biochar had no statistically significant effect on the planted reactors' (R1 and R2) $PO_4^{3-}$-P removal. Despite the biochar's ability to remove $PO_4^{3-}$-P on its own in R3, it does not appear to enhance plant uptake of $PO_4^{3-}$-P in R1 and R2. This is comparable to results in a similar study by De Rozari et al. (2016) where planted wetlands with only sand media removed more P than ones amended with biochar. This is likely because sand media have a buildup of surface minerals amendable to P adsorption compared to the negative-charged surfaces of the softwood biochar used in this test.

Ongoing work in our laboratory shows that aluminum-modified peanut shell biochar was effective in removing $PO_4^{3-}$-P from wastewater, while the magnesium-modified biochar was effective for the removal of both $PO_4^{3-}$-P and $NH_4^+$-N than the unmodified PSB. Thus, substituting AL- and Mg-modified biochars for softwood biochar could potentially yield significantly higher $PO_4^{3-}$-P removal in the biochar control (R3) and planted CTW (R1 and R2). It has also been observed that in a manure solution, other anions can compete with $PO_4^{3-}$ for coprecipitates or adsorption sites, decreasing the amount of $PO_4^{3-}$ adsorbed compared to removal from pure solutions (Ghezzehei et al., 2014). The latter study also found that the higher the manure $PO_4^{3-}$ concentration, the more $PO_4^{3-}$ was adsorbed potentially due to a higher concentration of coprecipitates.

The removal of Ca, Fe, K, Mg, Na, and S from the CAFO wastewater in the CTW reactors was monitored, and results of CTW Test #2 are shown in Fig. 14.17. The CTWs with biochar growth media

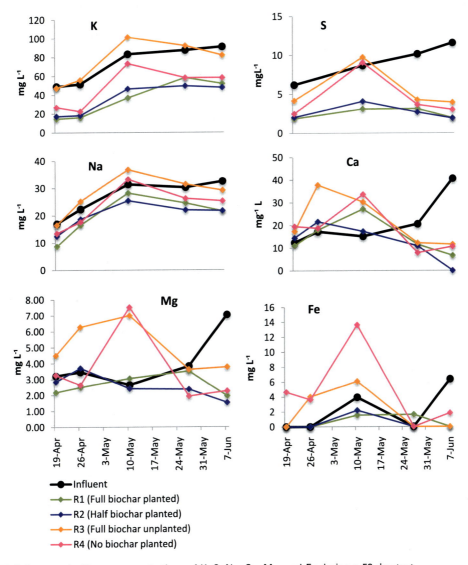

**Fig. 14.17** Influent and effluent concentrations of K, S, Na, Ca, Mg, and Fe during a 53-day test.

were generally more effective in reducing the latter metals and S in their effluent to below the influent concentrations, more so for K, S, and Na (Fig. 14.17). The success of biochars to adsorb heavy metals, salts, organics, other chemicals, and suspended solids from wastewater (Mohan et al., 2014; Niazi et al., 2016) makes biochar an important potential media for the filtration of various mixed contaminants in wastewaters. Thus, biochar has been applied to successfully filter multiple contaminants from storm (Reddy et al., 2014), dairy (Ghezzehei et al., 2014), industrial—brewery (Huggins et al., 2016), and municipal (Kätzl et al., 2014) wastewaters.

The biochar and plants together improved the removal of S in R1 and R2, but not in the controls (R3 and R4). The IC of most of these constituents varied in a way inconsistent with the wastewater dilution, although they all increased over time. Fig. 14.17 also shows spikes in K, S, and Na effluent concentrations from one or more of the reactors on May 10, the same date of the spike in Organic N and Organic P. The data for K in the IC were consistent enough with the dilution to perform the ANOVA test, but it was performed twice, once with all the data and once without the data from May 10, because Fig. 14.17 also shows a release of K in R3 and R4 (Fig. 14.18).

Among the planted CTW reactors, the average mass of K removed in R1 and R2 was significantly greater than in R4. The higher K concentration in the effluent of R3 than in the influent confirms the leaching of K from the biochar. The amount of biochar in the planted reactor significantly ($P < .05$) affected the amount of K removed in the planted reactors when the data from

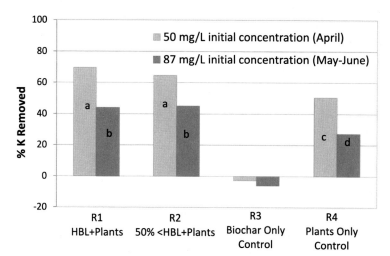

**Fig. 14.18** K removal from each tank. *Light-colored bars* represent average removal from lower initial concentration CAFO wastewater, while *dark-colored bars* represent average removal from higher influent concentration. *Same letters* indicate no statistical difference while *different letters* represent a statistical difference between means in the planted reactors.

May 10 were included; however, the significance was moderate ($P <.1$) when the data from May 10 were not included in the statistical analysis. CTW R1 and R2 did not remove significantly different percentages of K, but both removed significantly ($P <.05$) more than R4 when the data from May 10 were considered and moderately ($P <.1$ R1) to significantly ($P <.05$ R2) more when discarding May 10 suggesting plant uptake as the mechanism of K removal in the R1, R2, and R4. The relatively greater percent removal of K in R1 and R2 than in R4 corresponds to their greater plant biomass. Fig. 14.19 also showed no Na removal from R3, indicating that any Na removal is entirely due to plant uptake. The removal of K was greater than Na expected since K is a plant macronutrient. There was no statistically significant effect of biochar on the removal of Na in the planted tanks.

The release of nutrients from the CTW observed in the effluents was generally higher in the biochar-enhanced CTW (R1 and R2) than in the controls (R3 and R4). R3 released the most organic N (Fig. 14.13), P (Fig. 14.15), K, S, Na, Ca, Mg, and Fe (Fig.14.17). Meanwhile, R4 also released to a lesser extent P, K, S, Na, Ca, Mg, and Fe. Organic P was the only nutrient of concern released from CTW R1 and R2. Nutrient releases in wetlands can be due to redox and pH changes or decomposition due to microbe or plant death. If the releases were just due to a change in redox, the COD should also change significantly; however, Fig. 14.10 does not show any increase in the COD in the planted CTWs. Since most of the recorded releases were observed on May 10 when influent and effluent pH were at their highest (Fig. 14.20), it is unlikely that the spikes were due to a sudden dissolution of

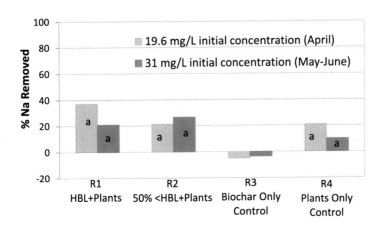

Fig. 14.19 Na removal from each tank. *Light-colored bars* represent average removal from lower initial concentration influent period. *Dark-colored bars* represent average removal from higher concentration influent. *Same letters* indicate no statistical difference while *different letters* represent a statistical difference between means in the planted reactors.

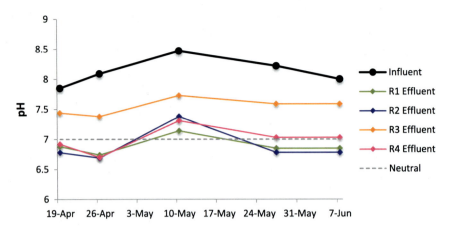

**Fig. 14.20** Influent and effluent pH over the course of Test #2.

adsorbed nutrients. More so, the pH in the effluent of R3 remained high during the week of May 10.

**CTW Test #3:** The enhanced removal of $NH_4^+$-N and $PO_4^{3-}$-P coupled with buffering of the high influent pH of the CAFO wastewater at the circumneutral range observed in the planted CTW with biochar media from CTW Test #2 was sustained (Table 14.1). The ANOVA performed with data from both CTW #2 and #3 showed the effect of time (aging of the CTWs) on performance. As a result, the CTWs removed significantly ($P <.05$) more $NH_4^+$-N and $PO_4^{3-}$-P during the third test when microbial growth and acclamation, as well as increased root mass, were highest. $NH_4^+$-N removal was moderately significantly ($P <.1$) with no significant difference between R1 and R2. Meanwhile, CTW R1 removed significantly ($P <.05$ for $mg L^{-1}$ and $P <.1$ for percent removal) more $NH_4^+$-N than R4. This is consistent with the results from the *sec*ond treatment test. When compared to CTW Test #2, the mass of $NH_4^+$-N removed increased by 34%, 31%, 91%, and 41% in R1, R2, R3, and R4, respectively. The remarkably high removal in R3 was unexpected and attributed to increase in algae and microbial biomass and aging of the biochar which is known to enhance the adsorption of $NH_4^+$-N as more negatively charged sites are formed. In between Tests #2 and #3, the reactors received tap water only, so some nutrients may have become limited in the planted reactors as the plants regrew. It is likely that if the nutrient-rich wastewater was provided, the planted reactors might have experienced even more rapid microbial and plant growth and increased their nutrient removal efficiency.

During CTW Test #3, the removal of $PO_4^{3-}$-P from the reactors was similar to the results of CTW #2. All the planted reactors each experienced similar increases in removal from the 2nd to the 3rd test as follows: 22%, 24%, 18%, and 23% in R1, R2, R3, and R4, respectively. R3 showed the lowest increase in $PO_4^{3-}$-P removal which again confirms the role of plants in the performance of biochar treatment systems for CAFO and similar wastewater.

Gupta et al. (2016) also evaluated the efficiency of constructed wetlands by using biochar as media. Their wetland beds with dimension (1 m × 0.33 m × 0.3 m) were prepared using gravels and biochar and cultivated with the Canna species. Synthetic wastewater, rather than actual CAFO wastewater as in this study, was treated at an average flow rate of $1.2 \times 10^{-7}$ m³/s achieving a retention time of 3 days. As in this study, they compared their CTW performance by comparing the controls and experimental wetland beds. Their wetlands with biochar were more efficient as compared to the wetland with gravels alone with average removal rate of 91.3% COD, 58.3% TN, 58.3% $NH_3$, 92% $NO_3$-N, 79.5% TP, and 67.7% $PO_4$. It should be noted that the residence time in the study by Gupta et al. (2016) was longer than for this study.

# 7 Conclusions

The constructed treatment wetland (CTW) bioreactors amended with softwood biochar in their growth media performed better than the CTW with plants without biochar, and the no-plant bioreactor containing only biochar. The enhanced removal of $NH_4^+$-N and $PO_4^{3-}$-P coupled with buffering of the high influent pH of the CAFO wastewater at the circumneutral pH range in the planted CTW with biochar media was sustained in the three separate treatment events. Plant growth in the biochar-amended CTWs was significantly faster than in the planted gravel reactor, and by the end of the tests had produced larger amounts of aboveground plant biomass. The biochar tended to increase plant tolerance of high nutrient concentrations and cold weather, which could potentially increase the capacity of a wetland system to treat wastewater during the winter months. The presence of biochar in a CTW media may also decrease the need to pretreat wastewater prior to its addition to a CTW. This effect, however, is dependent on the amount of biochar in the system as the plants in the reactor with the most biochar had the highest survival rate when the reactors were overloaded with nutrients.

The study results confirmed that the CTW with biochar growth media was most effective in removing nutrients from the CAFO wastewater due to the combined contributions of nutrient adsorption by the biochar, uptake by the plants, and potentially microbial metabolism. Interestingly, doubling the mass of biochar in the CTW growth media appears to have provided minimal additional benefit for most of the parameters (i.e., nutrients and metals) tested. This suggests that the enhancement of nutrient removal by the aquatic plants and microbes was more important than the adsorption contributed by the biochar. Therefore, the presence of biochar, at an undetermined threshold quantity, tends to retain the constituents and significantly enhance biological removal of the nutrients from the CAFO wastewater. The overall removal efficiency of all reactors increased over time for many of the constituents monitored.

Compared to many other biochars, the softwood biochar used in this study showed a relatively low adsorption capacity for $NH_4^+$-N, so it is possible that the use of a biochar with a much higher sorption capacity would have greatly improved the efficiency of the CTWs. There are also biochars with a high anion adsorption capacity, and there are also several feedstock modifications ("designer" biochars) and biochar activation methods (using steam or acid, for example) that have been found to increase cation as well as anion adsorption onto the biochar surface (Borchard et al., 2012; Liu et al., 2015). The use of such biochars in conjunction with plants for constructed wetland water treatment needs to be studied further.

Additional studies that provide a better understanding of the role of biochar fraction in the growth media of CTW for CAFO wastewater treatment are needed. Such studies should determine the optimum ratio of biochar to gravel or sand in CTW systems. It would also be useful to investigate the effects of using biochar in conjunction with other filter media used in wetlands, for example, calcium carbonates, peat, zeolite, etc. A greater understanding of biochar-microbe and biochar-microbe-plant interactions within a biochar wetland system could provide a better understanding of the specific mechanisms of biochar-enhanced wastewater treatment observed in this study. Although this study was performed at a relatively short, small scale in a greenhouse, the study results highlight the value of large-scale biochar-planted wetland pilot tests at the field scale.

Combining the well-established effectiveness of constructed treatment and natural wetlands to sustainably treat wastewater and biochar to enhance the filtration efficiency of treatment wetlands offers us simple and low-cost techniques to treat CAFO and other nutrient-rich wastewater and reduce our carbon footprint. Since biochar is quite recalcitrant and easily produced from different types of biomass, it offers a very promising media for treatment for long-term application in treatment wetlands where the improved plant growth will increase treatment capacity over many years.

## Questions and problems

1. What three parameters most influenced the removal of nutrients from constructed treatment wetlands with biochar growth media?
2. What mechanisms contributed to the effective treatment of the concentrated animal feeding operation (CAFO) wastewater?
3. What enhancement can be made to the constructed treatment wetland with biochar to increase the efficiency of phosphate ($PO_4^{3-}$-P) removal?
4. What enhancement can be made to the constructed treatment wetland with biochar to increase the efficiency of ammonium/ammonia nitrogen ($NH_4^+$-N) removal?
5. What parameters dominate the success of a constructed treatment wetland?
6. Besides $NH_4^+$-N and $PO_4^{3-}$-P, what other constituents were removed from the constructed treatment wetland with biochar growth media more efficiently than in the CTW with sand and gravel growth media without biochar?
7. Are constructed treatment wetlands with biochar growth media effective for nitrate removal? Why or why not?
8. How did the performance of the constructed treatment wetlands with biochar in the growth media perform with time?
9. How did the influent pH change during treatment?
10. Did increasing the amount of biochar by 50% in the constructed treatment wetland significantly improve performance?

## References

Abedi, T., Mojiri, A., 2019. Constructed wetland modified by biochar/zeolite addition for enhanced wastewater treatment. Environ. Technol. Innov. 16, 100472.

Berger, C., 2012. Biochar and Activated Carbon Filters for Greywater Treatment—Comparison of Organic Matter and Nutrients Removal (Ph.D. thesis). Swedish University of Agricultural Sciences.

Borchard, N., Wolf, A., Laabs, V., Aeckersberg, R., Scherer, H.W., Moeller, A., Amelung, W., 2012. Physical activation of biochar and its meaning for soil fertility and nutrient leaching—a greenhouse experiment. Soil Use Manag. 28, 177–184.

Bradford, S.A., Segal, E., Zheng, W., Wang, Q., Hutchins, S.R., 2008. Reuse of concentrated animal feeding operation wastewater on agricultural lands. J. Environ. Qual. 37 (5 suppl), S97–S115.

Chance, L.M.G., Majsztrik, J.C., Bridges, W.C., Willis, S.A., Albano, J.P., White, S.A., 2020. Comparative nutrient remediation by monoculture and mixed species plantings within floating treatment wetlands. Environ. Sci. Technol. 54, 8710–8718.

Chaukura, N., Gwenzi, W., Tavengwa, N., Manyuchi, M.M., 2016. Biosorbents for the removal of synthetic organics and emerging pollutants: opportunities and challenges for developing countries. Environ. Dev. 19, 84–89.

Cheng, C., Lehmann, J., Thies, J.E., Burton, S.D., Engelhard, M.H., 2006. Oxidation of black carbon by biotic and abiotic processes. Org. Geochem. 37, 1477–1488.

Cheng, C., Lehmann, J., Engelhard, M.H., 2008. Natural oxidation of black carbon in soils: changes in molecular form and surface charge along a climosequence. Geochim. Cosmochim. Acta 72, 1598–1610.

Chirakkara, R.A., Reddy, K.R., 2015. Biomass and chemical amendments for enhanced phytoremediation of mixed contaminated soils. Ecol. Eng. 85, 265–274. https://doi.org/10.1016/j.ecoleng.2015.09.029.

Cui, X., Hao, H., Zhang, C., He, Z., Yang, X., 2016. Capacity and mechanisms of ammonium and cadmium sorption on different wetland-plant derived biochars. Sci. Total Environ. 539, 566–575.

De Rozari, P., Greenway, M., El Hanandeh, A., 2015. An investigation into the effectiveness of sand media amended with biochar to remove BOD5, suspended solids and coliforms using wetland mesocosms. Water Sci. Technol. J. Int. Assoc. Water Pollut. Res. 71, 1536.

De Rozari, P., Greenway, M., El Hanandeh, A., 2016. Phosphorus removal from secondary sewage and septage using sand media amended with biochar in constructed wetland mesocosms. Sci. Total Environ. 569–570, 123–133.

Dimkpa, C., Weinand, T., Asch, F., 2009. Plant–rhizobacteria interactions alleviate abiotic stress conditions. Plant Cell Environ. 32, 1682–1694.

Dunne, E.J., Reddy, K.R., 2005. Phosphorus biogeochemistry of wetlands in agricultural watersheds. In: Nutrient Management in Agricultural Watersheds: A Wetland Solution. Wageningen Academic Publishers, Wageningen, The Netherlands, pp. 105–119.

Fellet, G., Marmiroli, M., Marchiol, L., 2014. Elements uptake by metal accumulator species grown on mine tailings amended with three types of biochar. Sci. Total Environ., 468–469. https://doi.org/10.1016/j.scitotenv.2013.08.072.

GAO, 2008. Concentrated Animal Feeding Operations: EPA Needs More Information and a Clearly Defined Strategy to Protect Air and Water Quality From Pollutants of Concern. Government Accountability Office.

Ghezzehei, T.A., Sarkhot, D.V., Berhe, A.A., 2014. Biochar can be used to capture essential nutrients from dairy wastewater and improve soil physico-chemical properties. Solid Earth 5, 953–962. https://doi.org/10.5194/se-5-953-2014.

Gregory, S.J., Anderson, C., Arbestain, M.C., McManus, M.T., 2014. Response of plant and soil microbes to biochar amendment of an arsenic-contaminated soil. Agric. Ecosyst. Environ. 191, 133–141.

Gupta, P., Ann, T.-w., Lee, S.-M., 2016. Use of biochar to enhance constructed wetland performance in wastewater reclamation. Environ. Eng. Res. 21 (1), 36–44.

Hamer, U., Marschner, B., Brodowski, S., Amelung, W., 2004. Interactive priming of black carbon and glucose mineralisation. Org. Geochem. 35, 823–830.

Herrmann, S., Oelmüller, R., Buscot, F., 2004. Manipulation of the onset of ectomycorrhiza formation by indole-3-acetic acid, activated charcoal or relative humidity in the association between oak microcuttings and *Piloderma croceum*: influence on plant development and photosynthesis. J. Plant Physiol. 161, 509–517.

Hollister, C.C., Bisogni, J.J., Lehmann, J., 2013. Ammonium, nitrate, and phosphate sorption to and solute leaching from biochars prepared from corn stover (*Zea mays* L.) and oak wood (*Quercus* spp.). J. Environ. Qual. 42, 137.

Houben, D., Evrard, L., Sonnet, P., 2013. Beneficial effects of biochar applicationto contaminated soils on the bioavailability of Cd, Pb and Zn and the biomassproduction of rapeseed (*Brassica napus* L.). Biomass Bioenergy 5, 196–204.

Huggins, T.M., Latorre, A., Biffinger, J.C., Ren, Z.J., 2016. Biochar based microbial fuel cell for enhanced wastewater treatment and nutrient recovery. Sustainability 8, 169. https://doi.org/10.3390/su8020169.

Jin, H., 2010. Characterization of Microbial Life Colonizing Biochar and Biochar-Amended Soils (Ph.D. thesis). Cornell University.

Joseph, S.D., Camps-Arbestain, M., Lin, Y., Munroe, P., Chia, C.H., Hook, J., van Zwieten, L., Kimber, S., Cowie, A., Singh, B.P., Lehmann, J., Foidl, N., Smernik, R.J., Amonette, J.E., 2010. An investigation into the reactions of biochar in soil. Aust. J. Soil Res. 48, 501–515.

Kameyama, K., Miyamoto, T., Shiono, T., Shinogi, Y., 2012. Influence of sugarcane bagasse-derived biochar application on nitrate leaching in calcaric dark red soil. J. Environ. Qual. 41, 1131–1137.

Kadlec, R., 2009. Comparison of free water and horizontal subsurface treatment wetlands. Ecol. Eng. 35 (2), 159–174.

Kadlec, R.H., Wallace, S., 2008. Treatment Wetlands, second ed. CRC Press. ISBN 9781566705264.

Kameyama, K., Miyamoto, T., Iwata, Y., Shiono, T., 2016. Influences of feedstock and pyrolysis temperature on the nitrate adsorption of biochar. Soil Sci. Plant Nutr. 62, 180–184.

Kätzl, K., Lübken, M., Alfes, K., Werner, S., Marschner, M., Wichern, M., 2014. Slow sand and slow biochar filtration of raw wastewater. In: Progress in Slow Sand and Alternative Biofiltration Processes. IWA Publishing, pp. 297–305.

Khodadad, C.L.M., Zimmerman, A.R., Green, S.J., Uthandi, S., Foster, J.S., 2011. Taxa-specific changes in soil microbial community composition induced by pyrogenic carbon amendments. Soil Biol. Biochem. 43, 385–392.

Kizito, S., Wu, S., Kipkemoi, K.W., Lei, M., Lu, Q., Bah, H., Dong, R., 2015. Evaluation of slow pyrolyzed wood and rice husks biochar for adsorption of ammonium nitrogen from piggery manure anaerobic digestate slurry. Sci. Total Environ. 505, 102–112.

Lehmann, J., Rillig, M.C., Thies, J., Masiello, C.A., Hockaday, W.C., Crowley, D., 2011. Biochar effects on soil biota—a review. Soil Biol. Biochem. 43, 1812–1836.

Lin, Y.-F., Shuh-Ren, J., Lee, D.-Y., Yih-Feng, C., 2008. Nitrate removal from groundwater using constructed wetlands under various hydraulic loading rates. Bioresour. Technol. 99 (16), 7504–7513.

Liu, F., Zuo, J., Chi, T., Wang, P., Yang, B., 2015. Removing phosphorus from aqueous solutions by using iron-modified corn straw biochar. Front. Environ. Sci. Eng. 9, 1066–1075.

Matsubara, Y., Hasegawa, N., Fukui, H., 2002. Incidence of Fusarium root rot in asparagus seedlings infected with arbuscular mycorrhizal fungus as affected by several soil amendments. J. Jpn. Soc. Hortic. Sci. 71, 370–374.

Mizuta, K., Matsumoto, T., Hatate, Y., Nishihara, K., Nakanishi, T., 2004. Removal of nitrate-nitrogen from drinking water using bamboo powder charcoal. Bioresour. Technol. 95, 255–257.

Mohan, D., Sarswat, A., Ok, Y.S., Pittman Jr., C.U., 2014. Organic and inorganic contaminants removal from water with biochar, a renewable, low cost and sustainable adsorbent—a critical review. Bioresour. Technol. 160, 191–202.

Mohanty, S.K., Valenca, R., Berger, A.W., Yu, I.K.M., Xiong, X., Saunders, T.M., Tsang, D.C.W., 2018. Plenty of room for carbon on the ground: potential applications of biochar for stormwater treatment (Review). Sci. Total Environ. 625, 1644–1658.

Niazi, N.K., Murtaza, B., Bibi, I., Shahid, M., White, J.C., Nawaz, M.F., Bashir, S., Shakoor, M.B., Choppala, G., Murtaza, G., Wang, H., 2016. Removal and recovery of metals by biosorbents and biochars derived from biowastes. In: Prasad, M.-N.V., Shih, K. (Eds.), Environmental Materials and Waste: Resource Recovery and Pollution Prevention. Academic Press, pp. 150–160.

Paz-Ferreiro, J., Lu, H., Fu, S., Méndez, A., Gascó, G., 2014. Use of phytoremediation and biochar to remediate heavy metal polluted soils: a review. Solid Earth 5, 65–75.

Puckett, L.J., 1994. Nonpoint and Point Sources of Nitrogen in Major Watersheds of the United States. U.S. Geological Survey.

Reddy, K.R., Dastgheibi, S., Xie, T., 2014. Evaluation of biochar as a potential filter media for the removal of mixed contaminants from urban storm water runoff. J. Environ. Eng. 140, 4014043.

Rizzo, G., 2015. Use of Biochar Geostructures for Urban Stormwater Water Cleanup (Thesis). University of Southern Queensland.

Saarnio, S., Raty, M., Hyrkas, M., Virkajarvi, P., 2018. Biochar addition changed the nutrient content and runoff water quality from the top layer of a grass field during a simulated snowmelt. Agric. Ecosyst. Environ. 265, 156–165.

Sarkhot, D.V., Ghezzehei, T.A., Berhe, A.A., 2013. Effectiveness of biochar for sorption of ammonium and phosphate from dairy effluent. J. Environ. Qual. 42, 1545–1554.

Spokas, K.A., Novak, J.M., Venterea, R.T., 2012. Biochar's role as an alternative N-fertilizer: ammonia capture. Plant Soil 350, 35–42.

Taghizadeh-Toosi, A., Clough, T.J., Sherlock, R.R., Condron, L.M., 2012. Biochar adsorbed ammonia is bioavailable. Plant Soil 350, 57–69.

Terrence Institute, 1994. Urbanization and Water Quality. A Guide to Protecting the Environment in Cooperation with EPA. Terrence Institute, Washington, DC.

Thies, E., Rilling, M.C., 2009. Characteristics of biochar: biological properties. In: Lehmann, J. (Ed.), Biochar for Environmental Management. Earthscan, London.

USEPA, 2000. Biosolid Technology Factsheet. Land Application of Biosolids. United States Environmental Protection Agency. https://www.epa.gov/sites/default/files/2018-11/documents/land-application-biosolids-factsheet.pdf.

Vijay, M.V., Sudarsan, J.S., Nithiyanantham, S., 2019. Sustainability of constructed wetlands using biochar as effective absorbent for treating wastewaters. Int. J. Energy Water Resour. https://doi.org/10.1007/s42108-019-00025-9.

Vohla, C., Koiv, M., Bavor, H.J., Chazarenc, F., Mander, U., 2011. Filter materials for phosphorus removal from wastewater in treatment wetlands—a review. J. Ecol. Eng. 37, 70–89. https://doi.org/10.1016/j.ecoleng.2009.08.003.

Vymazal, J., 2008. Constructed wetlands, surface flow. Encycl. Ecol. https://doi.org/10.1016/B978-008045405-4.00079-3.

Walsh, C.J., Roy, A.H., Feminella, J.W., Cottingham, P.D., Groffman, P.M., Morgan, R.P., 2005. The urban stream syndrome: current knowledge and the search for a cure. J. N. Am. Benthol. Soc. 24, 706–723.

Wang, B., Lehmann, J., Hanley, K., Hestrin, R., Enders, A., 2015. Adsorption and desorption of ammonium by maple wood biochar as a function of oxidation and pH. Chemosphere 138, 120–126.

Yao, Y., Gao, B., Inyang, M., Zimmerman, A.R., Cao, X., Pul-Lammanappallil, P., Yang, L., 2011. Removal of phosphate from aqueous solution by biochar derived from anaerobically di-gested sugar beet tailings. J. Hazard. Mater. 190, 501–507. https://doi.org/10.1016/j.jhazmat.2011.03.083.

Zeng, Z., Zhang, S.-d., Li, T.-q., Zhao, F.-l., He, Z.-l., Zhao, H.-p., Yang, X.-e., Wang, H.-l., Zhao, J., Rafiq, M.T., 2013. Sorption of ammonium and phosphate from aqueous solution by biochar derived from phytoremediation plants. J. Zhejiang Univ. Sci. B (Biomed. Biotechnol.) 14, 1152–1161.

Zhou, X., Wang, X., Zhang, H., Wu, H., 2017. Enhanced nitrogen removal of low C/N domestic wastewater using a biochar-amended aerated vertical flow constructed wetland. Bioresour. Technol. 241, 269–275.

# 15

# Biochar and biochar composites for oil sorption

Chanaka Navarathna[a], Prashan M. Rodrigo[a],
Vishmi S. Thrikawala[b], Arissa Ramirez[a], Todd E. Mlsna[a],
Charles U. Pittman, Jr.[a], and Dinesh Mohan[c]

[a]Department of Chemistry, Mississippi State University, Mississippi State, MS, United States. [b]College of Chemical Sciences, Institute of Chemistry Ceylon, Rajagiriya, Sri Lanka. [c]School of Environmental Sciences, Jawaharlal Nehru University, New Delhi, India

## 1 Introduction

Oil plays a vital role in day-to-day life for most of the world. It is incredibly significant in the world economy, used to fuel most means of transportation, and contributes to more than half of the U.S. energy consumption (Frieling and Madlener, 2017). Because oil has many advantageous qualities to those who have access to it, oil demand could increase, despite efforts to decrease $CO_2$ emissions, as economics continue to develop and the world population continues to grow. The continuous demand for oil mandates future oil spills will occur.

Oil spills are detrimental to the environment. Extracting and transporting oil lead to spills (Price et al., 2003). Most spills take place in the ocean or other water bodies, which puts wildlife in these areas at risk. For example, in 1989, the catastrophic Exxon Valdez oil spill occurred in Alaska. This was one of the worst spills in U.S. history, reaching more than 1100 miles of the Alaskan coast. Approximately $40.8 \times 10^6$ of crude oil were spilled in the Bligh Reef in Prince William Sound, Alaska (Peterson, 2001). The cleanup that followed resulted in more damage to the environment (Houghton et al., 1991). The population of fucoid algae and shellfish began to decrease, which, in turn, affected the population of animals that depended on them for food, including fish, whales, and sea otters (Loughlin et al., 1996). In addition to the effect on the environment, the spill also affected the economy by harming the area's tourism and the fishing industries (Goldberg, 1994). This devastating event led to a significant change in how the United States viewed oil, its

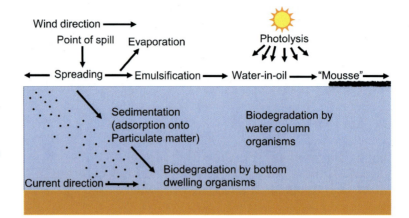

**Fig. 15.1** Pathways by which spilled oil may enter the marine ecosystem. Modified from Kingston, P.F., 2002. Long-term environmental impact of oil spills. Spill Sci. Technol. Bull. 7, 53–61.

transport, and research involving oil spill cleanup. On April 20, 2010, an explosion at the British Petroleum (BP) Deepwater Horizon oil rig eventually released more than 130 million gallons of crude oil into the Gulf of Mexico (Spencer and Fitzgerald, 2013). This was the largest oil spill in the U.S. waters. It continues to have an impact on fish and other marine species (Spencer and Fitzgerald, 2013).

When oil is spilled into a water body, most oil floats on the water surface. Some oil then evaporates into the atmosphere and some dissolves into the water. Features of an oil spill behavior in water are seen in Fig. 15.1 (Kingston, 2002). An ecosystem's recovery can be aided by biological, chemical, and physical remediation strategies enacted after the spill. Each method can aid in success under different circumstances. However, each presents its own issues, including introducing new pollution onto the environment, harming organisms residing in the area, significant manpower expenditures, and other costs. Various cleaning categories are compared in Table 15.1. The use of some nonenvironmentally friendly methods seems counterintuitive. However, cleanup methods will cause secondary environmental problems. The Deepwater Horizon event was countered by injecting huge amounts of surfactants into the oil and gas stream deep below the surface in an attempt to disperse the rising oil column before it reached the surface. This was done with the hope that microbial degradation of the oil would be enhanced. This was controversial and its effects have been intensively studied and these studies continue.

The purpose of cleaning oil spills is to reduce the harm they cause. Therefore, the optimal oil spill removal method would be

**Table 15.1 Comparison of various oil cleanup methods.**

| Classification | Methods | Examples | Advantages | Limitations | Environmental impact | Cost |
|---|---|---|---|---|---|---|
| Chemical | In situ burning | Combustion | Effective at quickly removing quantities of oil | Environment and safety concerns | Formation of large quantities of harmful smoke and viscous residues after combustion | Lowest |
| | Chemical treatment | Dispersion, solidifier emulsification | Simple operation, suitable to treat a large, polluted area | Little effect on viscous oil, ineffective in calm water, high initial and/or running costs | Being harmful to aquatic organisms | High |
| Biological | Bioremediation | Microorganism degradation | Good oil removal efficiency, low operating cost | Ineffective spill with large coherent mass, may affect aquatic life | Friendly | Low |
| Physical | Mechanical | Skimmers, booms | Efficient | Labor-intensive, time-consuming | Friendly | Highest |
| | Adsorption | Use of oil sorbents | Good oil removal efficiency, simple operation, practically feasible, less secondary pollution | Labor-intensive, recycling requires oil removal, high-density adsorbent can sink easily | Friendly, its biodegradation depends on the used sorbents | Low |

*Reproduced from Liu, H., Geng, B., Chen, Y., Wang, H., 2017. Review on the aerogel-type oil sorbents derived from nanocellulose. ACS Sustain. Chem. Eng. 5, 49–66, with the permission of the American Chemical Society.*

cheap and effective and cause as little harm to the environment as possible. Methods that might meet this requirement include bioremediation, mechanical removal, and the use of adsorbents. The mechanical removal of oil is most effective and should be used before wide dispersal when it becomes expensive. Bioremediation is not effective in all spills unless oil is highly dispersed and can have negative effects like lowering dissolved oxygen levels. Therefore, the use of adsorbents is an option for consideration.

Oil sorption involves the use of oleophilic sorbents that, when added to an oil spill, modifies the oil from a liquid to a semisolid state when the oil permeates the sorbent. Oil attached to the surfaces can give the solid sorbent a viscous surface coating. Oleophilic substances have an affinity for oil and can therefore absorb the spilled oil. Adsorption is a two-fold process where a solid adheres to oil or takes up the oil into its surface, where this layer becomes thicker. If sponge-like, the sorbent swells. Adsorption occurs through adhesive attraction where the oil clings onto the oleophilic solid adsorbent and progressively imbibes through pore filling and diffusion. This does not cause significant swelling for nonsponge-like solids. The oil can then be removed from the spill location by retrieving the sorbent, *as long as the adsorbent still floats*. A suitable oil sorbent can be characterized by having high oleophilic qualities, high hydrophobicity, low cost, ready availability, reusability or biodegradability, and easy retrieval.

Oil sorbents can be categorized into bio-based sorbents, inorganic mineral sorbents, and synthetic polymer sorbents. Bio-based sorbents are advantageous for oil spill cleanup because they have the lowest cost and efficiently biodegrade. These sorbents include peat moss, straw, wood pellets, and corncobs. Their deficiency is the tendency to absorb both oil and water, decreasing the amount of oil absorption (Kelle et al., 2013). Inorganic mineral sorbents (iron oxides (Wood et al., 2016), montmorillonite clay (Ugochukwu et al., 2014), silica ( Jada et al., 2002), and alumina (Franco et al., 2014)) are inexpensive and accessible in bulk (Adebajo et al., 2003). They can adsorb large amounts of oil compared to their size (Adebajo et al., 2003). However, they often sink in water because of their high densities, making both the adsorbents and adsorbed oil tough to retrieve and recycle. An example of this sorbent is volcanic ash or perlite (Bastani et al., 2006). Synthetic polymer sorbents typically adsorb more oil than bio-based sorbents. However, their poor biodegradability is not suitable for a large-scale cleanup. Synthetic polymer sorbent examples include polypropylene, polyurethane foam, and polyester (Oribayo et al., 2017). Each can be used for oil spill cleanup, but due to their lack of biodegradability, water retention, expensive landfill disposal of the exhausted sorbent, inefficient reclamation of low hydrocarbon concentrations, they are far less employed than bio-based adsorbents (Vocciante et al., 2019).

Biochar will adsorb heavy metals (Inyang et al., 2016), oxyanions (Li et al., 2018), organic dyes, hormones, and pharmaceutical residues (Park et al., 2016; Peiris et al., 2017). Biochar can absorb both oil and water quickly. When biochar is modified to become increasingly hydrophobic, it becomes an effective oil sorbent (Navarathna et al., 2020). Biochar composites modified with iron oxides are magnetic and can be retrieved by magnets from the water once oil adsorbs. After neat biochar adsorbs oil, the oil-laded char can be disposed of by combustion. This can produce useable heat and destroy this toxic waste.

## 1.1 Natural organic sorbents and biochar

While synthetic sorbents can exhibit high oil adsorptivity, drawbacks include their high cost, reduced biodegradability, and operating complexity. Therefore, organic sorbents from natural sources are preferred over synthetic sorbents due to their high abundance, cost, and biodegradability, and a reduced threat to nature (Navarathna et al., 2020; Asadpour et al., 2013). Rice straw, corncob, sawdust, and fibers from wood, cotton, kapok, wool, kenaf, and cattail, milkweed, rice husks, coconut husks, hay, feathers, and bagasse are examples of organic sorbents used for oil spill cleanup (Asadpour et al., 2013). Organic fibrous sorbents give higher adsorption capacities at lower densities than most other natural and synthetic sorbents. This allows them to float on water, assisting sorption (Adebajo and Frost, 2004). However, many can also exhibit high water uptake. The low hydrophobicity of biological microporous adsorption filters reduces their ability to adsorb oil (Pasila, 2004). The waxy coating surrounding kapok fiber sorbent filters increases hydrophobicity; hence, they have a higher oil sorption capacity. The oil sorption capacity of kapok, cotton, and milkweed floss is between 20 and 50 g of oil per g of sorbent (Choi et al., 1993; Deschamps et al., 2003).

Activated carbon (AC) and biochar show preferred activity versus conventional inorganic material-based sorbents for adsorbing many contaminants well. Indeed, they have been widely studied to remove many pollutants from wastewater and residues (Asadpour et al., 2013; Inyang et al., 2016; Peiris et al., 2017; Silvani et al., 2017; Dai et al., 2019; Navarathna et al., 2020). AC's and biochars' high availability and high efficacy stem from their large surface areas, porosities, and solid structures, making them candidates for oil removal. Recent research efforts have shifted from AC toward biochar (BC), which is produced through the pyrolysis of plant- and animal-based biomass at elevated temperatures and reduced oxygen concentrations. BC could have a greater scope than more expensive AC for treating oil spills (Navarathna et al., 2020), metal removal from wastewater (Inyang et al., 2016), soil amendment (Yu et al., 2019) and in air purifiers (Chen et al., 2017). BC's sorption capacity, like AC, can be further improved by modifying the surface area, surface functional groups, and microporosity (Moreira et al., 2017; Peiris et al., 2019). However, biochar will adsorb water along with contaminants unless prepared at a very high temperature because hydrophilic oxygen functions are present on its surface. When water imbibes BC's porosity replacing air, the particle density increases, and it sinks in water. However, by modifying BC's surfaces to be hydrophobic, this problem can be alleviated, and only less dense oil will penetrate into this porous volume.

## 1.2 Biochar

BC adsorbs species on its high specific surface area, surface functional groups, and high porosity remaining from plant morphology and the thermal decomposition outgassing. The internal surface area comes from longitudinal pores, residual structures, and pores varying from tiny ultra-micro to large-macro sizes. These vary with preparation temperature, time, and biomass source (Kalus et al., 2019).

The feedstock's ash, moisture content, fixed carbon fraction, % of volatiles, lignin, cellulose, hemicellulose content, and inorganic content all depend on the biomass type used (Angın, 2013; Shivaram et al., 2013). A higher feedstock lignin content increases the biochar porosity, and properties will also change depending on the amount of water, particle size, and biomass shape used in the pyrolysis (Nartey and Zhao, 2014). A few examples illustrate the tremendous compositional variability of BC versus the pyrolysis temperature used to prepare the char. When date seeds were pyrolyzed between 350°C (1 h) and 550°C (3 h), the carbon content rose from 64.4% to 82.2%. However, the BC yield was only ~24% (Mahdi et al., 2017) after raising the pyrolysis temperature of wastewater sludge from 300°C to 400°C. The carbon content decreased from 25.6% to 20.2% and then remained constant as the temperature was raised to 700°C. Pitch pine wood chips exhibited increasing carbon contents of 63.9%, 70.7%, and 90.5% at 300°C, 400°C, and 500°C pyrolysis temperatures, respectively, using fast pyrolysis (Kim et al., 2012).

Higher temperature pyrolysis removes surface polar functional groups on BC, making BC surfaces more hydrophobic, enhancing the adsorption compatibility with oils. Oil capacity depends on the surface area, pore size, pore volume, and surface shape of the adsorbent. Hydrophobic surface pores will repel water.

## 1.3 Tailoring biochar to enhance hydrophobicity

Modification of BC surfaces with the linear C-11 fatty acid, lauric acid, enhances oil adsorption capacities (Sidik et al., 2012), and increases the contact angle, buoyancy, and hydrophobicity of BC. Sidik et al., 2012 showed that hydrophobicity could be increased by lauric acid decoration. These modified biochars with increased buoyancy become better adsorbents than unmodified BC (Sidik et al., 2012). Fatty acid decoration, in general, increases BC's water contact angle and buoyancy in water. This latter factor is very important because when BC is spread on an oil slick in the ocean, the BC must float for a long time to allow the oil to penetrate into its porosity. Once this occurs, no water will imbibe. Hence, the internal volume and surface roughness volume will be filled with lower density oil instead of higher density water or salt water. Furthermore, more oil will now adhere to the surface layer of oil, expanding the adsorbent's particles' volume with oil. This increases buoyancy. Thus, any compound that raises surface hydrophobicity will increase buoyancy and raise oil uptake. Only a small amount of work has been carried out in the application of BC for oil spill cleanup.

Our group experimented on enhancing BC's hydrophobicity by modifying a BC with lauric acid (Navarathna et al., 2020) (Fig. 15.2). Commercial Douglas fir biochar (surface area 695 $m^2/g$, pore volume 0.264 $cm^3/g$, average pore size 13–19.5 Å, and particle density of ~0.1 $cm^3/g$) (BC) from Biochar Supreme Inc., was used. It was produced by a proprietary process in an updraft wood gasifier at 900°C (1–30 s). This BC is hydrophobic due to its high C/H ratio (C=81.5%, N=1.2%, H=2.1%, O=13.0%, and Ash=2.3%) (C/H (39.3) and O/C (0.15)). Nevertheless, it did not float on water. Instead, water uptake occurred, and it sank. After 2% lauric acid modification, the BC oil uptake increased from 4.2 to 11.0 g/g. Lauric acid-enhanced hydrophobicity increased the contact angle from 127.9 to 140.9 degrees and reduced water influx into the BC (Fig. 15.2). This 2% lauric acid/BC did not sink in water for 3 weeks. Magnetite iron oxide nanoparticle deposition onto BC coupled with lauric acid modification permitted recovery of the exhausted adsorbent by a magnetic field as an alternative to filtration. Simple recovery with a magnet can simplify recovery

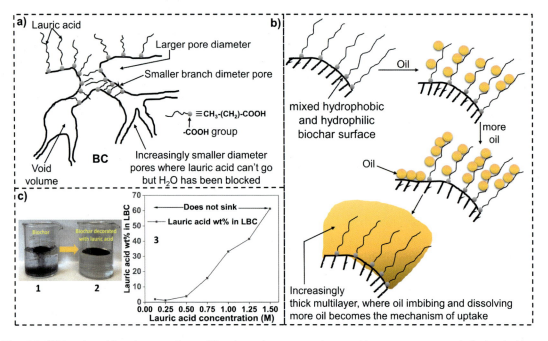

**Fig. 15.2** (A) Lauric acid surface uptake on BC, where the outer surface and large pores are made hydrophobic, blocking water access to substantial pore volume further below the surface where lauric acid cannot penetrate. (B) Oil molecules adhering to lauric acid adsorbed at the biochar surfaces. More oil adsorbs to lauric acid, to already adsorbed oil, and to hydrophobic biochar surface regions. Finally, oil takes up more oil, building a thick coating; (C1) neat biochar sinks in water, (C2) biochar after decoration with lauric acid floats, and (C3) effect of lauric acid dose on BC sinking (lauric acid weight % (wt/wt%); 0.125 M = 2.0%, 0.5 M = 1.2%, and 0.5 M = 3.9%). *Modified from Navarathna, C.M., Bombuwala Dewage, N., Keeton, C., Pennisson, J., Henderson, R., Lashley, B., Zhang, X., Hassan, E. B., Perez, F., Mohan, D., 2020. Biochar adsorbents with enhanced hydrophobicity for oil spill removal. ACS Appl. Mater. Interfaces 12, 9248–9260.*

because filtering oily particles is messy. Composites modified with superparamagnetic iron oxide nanoparticles have enhanced oil cleanup properties and fast sorbent recovery (Raj and Joy, 2015).

Five different biochar adsorbents were used for oil spill removal (Navarathna et al., 2020): Douglas fir BC, BC modified with lauric acid (LBC), iron oxide-modified BC (MBC), BC modified with lauric acid followed by iron oxide (LMBC), and BC modified with iron oxide followed by lauric acid (MLBC) (Navarathna et al., 2020). All five removed oil in ≤15 min uptake periods. The uptake rates followed pseudo-second-order kinetics. In addition, oil sorption capacities spanned the range of 3–11 g oil/g adsorbent. Oil uptake increased with temperature.

Four adsorbents were prepared by fast pyrolysis in a fluidized bed pyrolyzer using a 4–5 s residence time under an $N_2$ flow to test the crude oil from oil/water biphasic mixtures (Kandanelli et al., 2018). Three BCs were made using sawdust or rice husk powders.

Sawdust powder was treated at 450°C to generate SDBC-450 (B1) and rice husk powders were treated at 450°C and 550°C to generate RHBC-450 (B2) and RHBC-550 (B3), respectively. A fourth BC was made from rice husk powder at 400°C (RHBC-SP) (B4) in a batch autoclave by slow pyrolysis under an $N_2$ flow (Kandanelli et al., 2018). Fast-pyrolysis BC yields were in the 22%–25% range, while the slow-pyrolysis BC had a yield of 70%.

The TGA (Fig. 15.3A) of BCs from fast pyrolysis (at 450°C and 550°C) underwent 30%–50% decomposition in the 380–600°C temperature range, confirming the presence of residual cellulose, hemicellulose, and lignin. RHBC-SP BC (yellow) had the minimum weight loss (7%–8%). Broad FTIR 3299 $cm^{-1}$ peaks of these BCs confirmed the presence of –OH groups from residual hemicellulose/cellulose in RHBC-450 (red) and SDBC-450 (green) to a significant extent and a lesser extent in RHBC-550 (blue) (Fig. 15.3B). This –OH stretching peak was absent in RHBC-SP prepared by slow pyrolysis at 400°C, suggesting that an effective removal of hydroxyl groups occurs with time. This increases biochar hydrophobicity. The same trend was observed for $sp^3$ C–H stretching (2960–2820 $cm^{-1}$). The C–C stretching (at 1589 $cm^{-1}$) arose from the sugar backbone and was minimal in RHBC-SP. In addition, Si-O bands (1075 $cm^{-1}$) were observed from the presence of silica (Fig. 15.3B) derived from the high silica content in rice husks. $SiO_2$ reduces Si-OH surface groups, increasing biochar hydrophobicity and could aid oil adsorption. Adsorption of oil was instantaneous (<5 min) in most of the cases. RHBC-450 had the highest oil uptake (3.23 oil/g biochar) of the four biochars tested for all types of crude oils with different densities (API index), total acid number (TAN), and sulfur content.

The powder X-ray diffraction (PXRD) peak at 22.7 degrees for RHBCs in Fig. 15.3D. is for the cellulosic peaks in BC. These are for partly carbonized regions giving the aromatic $d$-spacing expected in stacked layers of aromatic systems. The original cellulose here has been dehydroxylated and decomposed partially in RHBC-450, RHBC-550, and SDBC-450 from fast pyrolysis to give XRD peaks at 15.2, 24.5, 35.5, and 37.5 degrees, which were absent in RHBC-SP, made by slow pyrolysis at 400°C. RHBC-SP (blue) graphitic peaks appeared at 22.7 and 45 degrees, further confirming that carbonized cellulosic residues are present in higher amounts in fast-pyrolysis BC. In temperature range employed, most of the slow-pyrolysis biochar is a polyaromatic carbon. Very little residual cellulose or hemicellulose remains in slow-pyrolysis biochar (Fig. 15.3C) (Kandanelli et al., 2018). $^{13}C$ NMR spectra of the RHBCs found aromatic carbons from lignin and from biomass thermal decomposition formed partially carbonized aromatic skeletons (Fig. 15.3D). Similar pyrolysis conditions tended to give

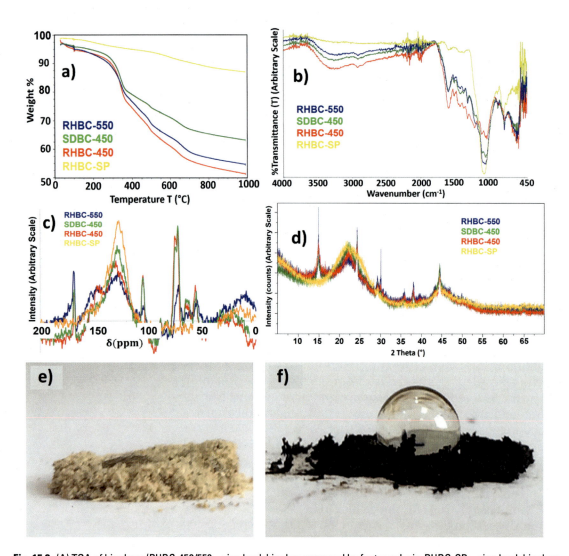

**Fig. 15.3** (A) TGA of biochars (RHBC-450/550—rice husk biochar prepared by fast pyrolysis, RHBC-SP—rice husk biochar prepared by slow pyrolysis and SDBC-450—sawdust biochar prepared by slow pyrolysis) under $N_2$ atmosphere, (B) FTIR spectra of biochar samples, (C) CPMAS $^{13}C$ NMR spectra of biochars ($\delta$: 0 to 200 ppm from right to left on X-axis), (D) P-XRD of biochar samples and water droplet over (E) rice husks (F) RHBC-450. *Modified from Kandanelli, R., Meesala, L., Kumar, J., Raju, C.S.K., Peddy, V.R., Gandham, S., Kumar, P., 2018. Cost effective and practically viable oil spillage mitigation: Comprehensive study with biochar. Mar. Pollut. Bull. 128, 32–40.*

similar extents of pyrolysis, regardless of the feed. No contact angle measurements were made, but these BCs were hydrophobic, in sharp visual contrast to the hydrophilic rice husk feedstocks (Fig. 15.3E and F). Water droplets on the feed biomass rice husks simply sank through, whereas water droplets on the biochar RHBC-450 remained as a high contact angle drop on top without passing through the husks, indicating intrinsic hydrophobicity. The same thing happened with all the biochar samples.

Washed rice husks were pyrolyzed in a fixed-bed stainless steel reactor at 480°C for 3h in a vacuum (~1.33 Pa) (Angelova et al., 2011). The resulting BC contained ~49% $SiO_2$ and had a ~240 $m^2$/g $N_2$ BET specific surface area (~1.053 $cm^3$/g pore volume) and PZC of 7. This BC had a variety of pore sizes; 0.04–1.00 μm (27.6%), 1.0–10.0 μm (10.7%), 10.0–40.0 μm (26.0%), and >40 μm (35.7%) and a bulk density of 102 kg/$m^3$. Low bulk density is beneficial to the formation of capillary spaces between its particles and ultimately leads to increased oil sorption capacity. GC-MS analysis of acetone extracts from the carbonized rice husk residues of this BC contained oleophilic high-molecular alkanes that had been present on the BC surface, promoting oil sorption (Angelova et al., 2011). The rice husk char has demonstrated high sorption capacities for gasoline (3.7 g/g), diesel (5.5 g/g), light crude oil (6.0 g/g), motor oil (7.5 g/g), and heavy crude oil (9.2 g/g). The neat rice husk BC and resulting conglomerate oil adsorbent (slime) had good buoyancy, floating for ten days. An indicator related to the sorption capacity is the C/$SiO_2$ weight ratio (~1.04) in the rice husk char (Angelova et al., 2011). Higher contents of $SiO_2$ give higher oil sorption capacities, perhaps due to the BC pore blockage and leading to low water uptake. Also, if silica's surface Si–OH groups were converted to Si–O–Si bridges, this would lower silica's hydrophilicity.

Biochar that is produced by gasification of pinewood pellets (DIN A1 quality) (6mm diameter and 3.1–40nm length) (at 850°C) could uptake a lot of toluene (an oil surrogate) (Silvani et al., 2017). The resulting biochar had a surface area of 343 ± 2 $m^2$/g (224 $m^2$/g as micropores and 119 $m^2$/g for *meso-* and macropores). Its average particle size was <0.1mm. SEM analysis displayed this BC's heterogeneous pore distribution, with varying shapes and dimensions ranging from 25 to 140 nm. Elemental analysis revealed ~78% C, indicating that this biochar is very hydrophobic. Toluene uptake capacities were 268 ± 35.81 and 197 ± 15.41 mg/g in deionized water and synthetic seawater, respectively. These uptakes were higher than the toluene uptake by Norit-activated carbon, type Darco (Sigma Aldrich, Catalog Number 242241) which is derived from lignite coal, 0.5–1 mm, 712 ± 2 $m^2$/g surface area, and 0.736 $cm^3$/g pore volume. For comparison, the Norit-activated carbon uptake capacities were 136 ± 8.16 mg/g in deionized water and 166 ± 10.4 mg/g in seawater.

Sohaimi et al. prepared four different textile sludge BCs (Sohaimi et al., 2017). The dried biomass was sorted into particles with an average size of 0.45–0.50mm and pyrolyzed at 105°C [BC105 (24.13% C)], 200°C [BC200 (29.02% C)], 400°C [BC400

(27.97%)], and 700°C [BC700 (39.32% C)], respectively. Carbon content increased with the pyrolysis temperature. In turn, the hydrophobicity increased, and BC700 (60% yield, PZC=7.4±2, 195 m$^2$/g surface area, and 0.25 cm$^3$/g pore volume) was the most hydrophobic of the four BCs prepared. The FTIR spectrum of BC700 did not have C–H stretching and bending peaks for –CH$_2$ and –CH$_3$ at (2919–2925 cm$^{-1}$) and (1350–1460 cm$^{-1}$). These peaks were exhibited by the other BCs made at lower temperatures, in further agreement with BC700's increased hydrophobicity. The authors only reported the adsorption capacities for BC700. Langmuir cooking oil uptake capacities on BC700 were 349 mg/L (pH 7.4 and 1 mg/mL dose), 500 mg/L (pH 3 and 1 mg/mL), and 323 mg/L (pH 7.4 and 2 mg/L dose).

Crab shells were carbonized to BC (CSB) by heating at 10°C/min from room temperature to 700°C and holding at 2h under 100 mL/min nitrogen after a pretreatment with 2M HCl (6h) (Cai et al., 2019). The KOH-activated analog of CSB (CSAB) was prepared by mixing 3:1 KOH:CSB (w/w) and then pyrolyzing in a tube furnace at 800°C for 1h under a nitrogen flow (100 mL/min). The Langmuir adsorption capacity of diesel oil on CSAB (57.74 mg/g) was higher than it was on CSB (30.71 mg/g). This was attributed to the surface area increase from 307 m$^2$/g (0.32 cm$^3$/g pore volume) on CSB to 2441 m$^2$/g (1.682 cm$^3$ pore volume) for CSAB.

## 2 Uptake performance

### 2.1 pH dependence

Sorption studies generally show no or little pH dependency on oil uptake (Table 15.2). For instance, the oil uptake on five different Douglas fir biochar variants (BC, LBC, MBC, LMBC, and MLBC) (see Section 1.3) for engine, transmission, and machine oils remained almost unchanging over the wide pH range (1–13) (Navarathna et al., 2020) (Fig. 15.4A). This lack of dependence on pH is a result of oil sorption occurring via hydrophobic interactions between oil hydrocarbons and alkyl chains covering the biochar surfaces (Fig. 15.2B).

Despite most research reporting no or negligible pH dependency for petroleum oils, cooking oils are derived from fats and fatty acids. The fatty acids are esterified by methanol or ethanol for the case of biodiesel fuels. Thus, ester hydrolysis can occur under high or very low pH conditions. Cooking oil uptake on textile sludge biochar (TSB) was increased from 97.4 to 194.7 mg/g when pH rose to 11 from 3 (Sohaimi et al., 2017). A similar trend has been observed on palm oil mill effluent (POME) uptake on

**Table 15.2 Comparison of oil uptake performance of different biochars.**

| Adsorbent | Oil type | pH | Stirring rate (rpm) | Temperature (°C) | Time (min) | Oil amount | Dose | Surface area (m²/g) | Sorption capacity (g/g) | Kinetics | Reference |
|---|---|---|---|---|---|---|---|---|---|---|---|
| Rice husk (RH) char | Black seawater Gasoline Diesel Light crude oil Motor oil Heavy crude oil | 7 | – | 20 | 10 | – | 1 g | 240 | 1.1 3.7 5.5 6.0 7.5 9.2 | – | Angelova et al. (2011) |
| Crab shell-activated biochar (CSAB) | Diesel oil | 7 | 150 | 30 | 240 | 100–500 mg/L | 0.2 g | 2441 | 0.058 (Langmuir, $R^2 = 0.99$) | $k_1 = 0.17\,\text{min}^{-1}$ $R^2 = 0.98$ $q_e = 79.8\,\text{mg/g}$ $k_2 = 0.023\,\text{g/mg·min}$ $R^2 = 0.99$ $q_e = 93.9\,\text{mg/g}$ | Cai et al. (2019) |
| Crab shell biochar (CSB) | | | | | | | | 307 | 0.031 (Langmuir, $R^2 = 0.99$) | $k_1 = 0.16\,\text{min}^{-1}$ $R^2 = 0.89$ $q_e = 65.0\,\text{mg/g}$ $k_2 = 0.089\,\text{g/mg·min}$ $R^2 = 0.99$ $q_e = 73.1\,\text{mg/g}$ | |
| Rice husk biochar (RHBC-450) | Crude oil | – | 350 | 25 | 10–15 s +5 min draining time | 500 g | 150–250 mg | – | 3.23 | – | Kandanelli et al. (2018) |
| Saw dust biochar (SDBC-450) | | | | | | | | | 2.4 | | |
| Douglas fir biochar (BC) | Transmission oils | 7 | Static | 25 | 15 | 2 g/25 mL | 0.25 g | 695.1 | 3.0–4.1 (Sips, $R^2 \sim 0.99$) | Pseudo 2nd order model $R^2 > 0.99$ | Navarathna et al. (2020) |
| | | | | | | | | 35.9 | 3.3–3.6 | | |

*Continued*

Table 15.2 Comparison of oil uptake performance of different biochars—cont'd

| Adsorbent | Oil type | pH | Stirring rate (rpm) | Temperature (°C) | Time (min) | Oil amount | Dose | Surface area ($m^2/g$) | Sorption capacity (g/g) | Kinetics | Reference |
|---|---|---|---|---|---|---|---|---|---|---|---|
| Lauric acid BC (LBC) | | | | | | | | | | | |
| Magnetic BC (MBC) | | | | | | | | 312.6 | 4.2–5.1 | | |
| Lauric-magnetic BC (LMBC) | | | | | | | | 37.3 | 3.2–4.9 | | |
| Magnetic-lauric BC (MLBC) | | | | | | | | 30.6 | 3.0–3.9 | | |
| BC | Machine oils | | | | | | | 695.1 | 2.3–3.2 | | |
| LBC | | | | | | | | 35.9 | 2.8–3.2 | | |
| MBC | | | | | | | | 312.6 | 1.5–1.9 | | |
| LMBC | | | | | | | | 37.3 | 3.5–4.8 | | |
| MLBC | | | | | | | | 30.6 | 3.8–5.0 | | |
| BC | Engine oils | | | | | | | 695.1 | 3.6–4.7 | | |
| LBC | | | | | | | | 35.9 | 3.1–4.2 | | |
| MBC | | | | | | | | 312.6 | 0.4–1.2 | | |
| LMBC | | | | | | | | 37.3 | 0.5–2.7 | | |
| MLBC | | | | | | | | 30.6 | 1.6–2.1 | | |
| BC | Crude oil in simulated seawater | | | | | | | 695.1 | 6.87 | — | |
| LBC | | | | | | | | 35.9 | 9.4 | | |
| MBC | | | | | | | | 312.6 | 3.31 | | |
| LMBC | | | | | | | | 37.3 | 5.7 | | |
| MLBC | | | | | | | | 30.6 | 6.18 | | |

| Biochar | Contaminant | | | | | | | | | | Reference |
|---|---|---|---|---|---|---|---|---|---|---|---|
| Pinewood biochar | Toluene in DI water | – | 15 | 25 | 1440 | 35–350 mg/mL | 0.02 g/0.02 L | 343 | 268±35.81 (Langmuir, $R^2=0.992$) | $k_1=1.93\pm0.353\,h^{-1}$, $R^2=0.99$, $q_e=178\pm6.67\,g/g$ | Silvani et al. (2017) |
|  | Toluene synthetic seawater |  |  |  |  |  |  |  | 197±15.4 (Langmuir, $R^2=0.995$) | $k_1=2.32\pm0.466\,h^{-1}$, $R^2=0.99$, $q_e=89.4\pm3.76\,g/g$ |  |
| Textile sludge biochar (BC700) | Simulated oil wastewater | 3 | 300 | 20 | 60 |  | 25 mg/25 mL | 195.32 | 500 mg/L (Langmuir, $R^2=0.83$) | $k_1=0.0246\,min^{-1}$, $R_2=0.95$, $q_e=105.3\,mg/L$, $k_2=0.003\,g/mg\,min$, $R^2=0.99$, $q_e=105.3\,mg/L$ | Sohaimi et al. (2017) |
| Maplewood biochar | Crude oil in seawater | – | 150 | 20 | 1020 |  | 10 mL | 0.5 g | 303–332 | 3.8–6.2 | – | Nguyen and Pignatello (2013) |

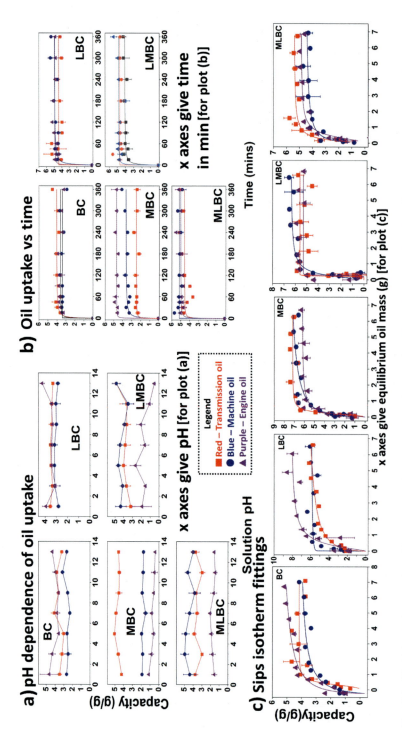

**Fig. 15.4** (A) pH dependences of machine, transmission, and engine oil uptake (at 25°C and 1 h of equilibration), (B) pseudo-second-order adsorption vs time model plots (at 25°C and pH 7), and (C) Sips isotherm fits of oil sorption (25°C, pH = 7, 1h of equilibration) for the BC, LBC, MBC, LMBC, and MLBC (see Section 1.3) adsorbents. Adapted from Navarathna, C.M., Bombuwala Dewage, N., Keeton, C., Pennisson, J., Henderson, R., Lashley, B., Zhang, X., Hassan, E. B., Perez, F., Mohan, D., 2020. Biochar adsorbents with enhanced hydrophobicity for oil spill removal. ACS Appl. Mater. Interfaces 12, 9248–9260.

synthetic rubber powder. Sodium hydroxide added during pH adjustments resulted in the formation of glycerol and fatty acid sodium salts (saponification), which are soluble in water. Therefore, this underestimates residual oil concentrations in solution determined by UV-Vis spectroscopy (Ahmad et al., 2005). The optimum pH for cooking oil adsorption from simulated wastewater was ~7 (Sohaimi et al., 2017). This benefits the treatment's final discharge after adsorption since it will have a neutral pH. Thus, no additional pH adjustment will be required before discharge.

Crab shell biochar (CSB) and activated crab shell biochar (CSAB) have parabolic sorption uptakes versus pH for petroleum diesel oil which is composed of 75% aliphatic hydrocarbons ($C_{10}H_{20}$–$C_{15}H_{28}$) and about 25% of aromatic hydrocarbons. The highest uptake occurred at pH 7 (82.2 mg/g for CSB and 62.4 mg/g for CSAB) (Cai et al., 2019). Most likely, strong acidity can cause oil coalescence and increase the size of oil droplets leading to low oil uptake. The stability of diesel oil emulsions increase in alkaline conditions, making droplets with negative surface charges in water harder to remove from water and adsorb on the negatively charged biochar surfaces (Sohaimi et al., 2017).

Deionized water and seawater (Nguyen and Pignatello, 2013) were used in simulated oil spills. The aqueous pH and influence of the pH on oil uptake were never reported in this study. This topic should be further studied in future petroleum spill research because highly acidic or basic pH may influence the degree of oil micelle formation. However, the vast majority of spills in seawater or fresh waters will not occur in strongly basic or acidic pHs.

## 2.2 Kinetics and stirring effects

Papers summarized in Table 15.2 report rate law data that could be fitted into pseudo-second-order models with high regression constants ($r^2 > 0.99$) (Fig. 15.4B) (Sohaimi et al., 2017; Cai et al., 2019; Navarathna et al., 2020). However, Silvani et al. fit their rate data into a pseudo-first-order model, also with a high regression coefficient ($r^2 > 0.99$) (Silvani et al., 2017). The data in the papers in Table 15.2 reporting rate laws were refit into pseudo-first-order models, and these gave high regression coefficients. However, none were above 0.979 (Cai et al., 2019). Thus, the Boyd diffusion model was employed to understand the uptake mechanism further, and the rate-limiting step of oil uptake was found to be controlled by film diffusion (Sohaimi et al., 2017). Cai et al. (2019), without evidence or further explanation, it was speculated that the oil uptake was controlled by chemisorption involving the exchange of valence electrons from the $\pi$ electron-rich regions on the basal planes of carbon. More studies are needed to understand what model fits oil uptake on biochar best. If a biochar has a really high uptake capacity, that means that multiple layers of oil have been layered on the surface. Thus, much of the "adsorbed" oil has become attracted to other oil molecules forming a thick multilayer coating covering the surface. Once one or two layers of oil cover the surface, subsequent oil layers do not interact with the biochar surface, itself. Only a small fraction of the captured oil has interacted with the adsorbent biochar surface.

Turbulent conditions may have an impact on petroleum or bioderived oil uptake by lowering the pore diffusion. Also, turbulence can cause shear forces at the sorbent's surface to shear away thicker regions of sorbed oil, thinning the adsorbed layers and reducing oil uptake. Crude oil sorption onto rice and sawdust biochars were studied under four different uptake conditions: (1) from deionized water under static conditions, (2) from salt water (5% wt NaCl) under static conditions, (3) deionized water under stirring conditions, and (4) from salt water (5% wt NaCl) stirring conditions (Fig. 15.5) (Kandanelli et al., 2018). Crude oil uptake

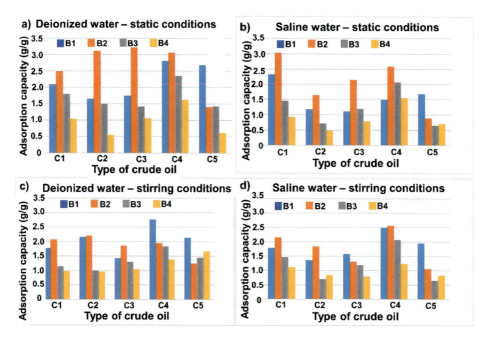

**Fig. 15.5** Crude oil (C1—30.26 API (American Petroleum Institute) density, C2—27.12, C3—28.12, C4—33, and C5—18.82) adsorption under both static and stirred conditions to SDBC-450 (B1), RHBC-450 (B2), RHBC-550 (B3), and RHBC-SP (B4) biochars (see Section 1.3) (A) in deionized water, (B) saline water, with stirring, (C) in deionized water, and (D) saline water. *Modified from Kandanelli, R., Meesala, L., Kumar, J., Raju, C.S.K., Peddy, V.R., Gandham, S., Kumar, P., 2018. Cost effective and practically viable oil spillage mitigation: Comprehensive study with biochar. Mar. Pollut. Bull. 128, 32–40.*

dropped 5%–15% in turbulent conditions compared to static conditions. The effect of different mixing rates on oil uptake was not reported for biochars. However, palm oil (mill effluent) uptake on synthetic rubber powder increased when the stirring rate was increased from 0 to 250 rpm, unlike the crude oil described above (Ahmad et al., 2005). The residual palm oil concentrations in water declined to 200 from 800 mg/L. The mass transfer rate increases as the stirring speed increases. Surface film resistance is lowered, and residual oil can more readily access the adsorbent surface. Creating turbulent conditions is more representative of real-world situations, and this aspect should be investigated further for biochar.

## 2.3 Isotherms and sorption thermodynamics

The use of isotherm models for oil uptake may be misleading due to the large amounts of oil uptake that occurs. Oil attached to the biochar extends far beyond monolayer or classic multilayer sorption. Oil uptake quickly becomes, oil associating with many oil layers, forming a continuum above the biochar surfaces. As a result, the adsorbent surface no longer plays a chemical role in adsorbing additional oil. The sorption capacity was calculated using a variety of isotherm models (Table 15.2). The Langmuir model was used to predict the petroleum-based diesel oil uptake capacity onto crab shell biochars (CSB and CSAB) (Table 15.2) (Cai et al., 2019). To predict oil uptake capacities on lauric acid/magnetite-modified Douglas fir biochar analogs, the Sips model (a hybrid of the Langmuir and Freundlich models) was used (Fig. 15.4c) (Navarathna et al., 2020). This model describes heterogeneous adsorption and overcomes the Freundlich isotherm model limitations associated with rising adsorbate concentrations (Sips, 1948). The Sips model reduces into the Freundlich isotherm (multilayer coverage) at low concentrations and Langmuir isotherm (monolayer coverage) at high concentrations. Sips parameters are influenced by process conditions, such as pH, temperature, and concentration.

Sorption can be exothermic ($-\Delta H$) or endothermic ($+\Delta H$). Endothermic petroleum oil adsorptions were found using crab shell biochar and Douglas fir biochar decorated with lauric acid and magnetite (Cai et al., 2019; Navarathna et al., 2020). Exothermic adsorption of cooking oil on textile sludge biochar (BC700) was demonstrated by a negative $\Delta H°$ ($-28.5$ kJ/mol) (Sohaimi et al., 2017). Thus, the total amount of cooking oil adsorbed on BC700 decreased as temperature increased. The magnitude of this enthalpy change, 28.5 kJ/mol, indicated that cooking oil adsorption occurred by physisorption. The adsorption of petroleum-type diesel oils onto crab shell-derived biochar was investigated at 0–50°C (Cai et al., 2019). As the temperature rose from 0°C to 30°C, crab shell biochar (CSB) and activated crab shell biochar (CSAB) capacities increased to 64.4 and 90.4 mg/g, from 46.2 and 57.5 mg/g, respectively. As the temperature further rose to 50°C, the adsorption capacities then shrank to ~55 mg/g for CSB and ~78 mg/g for CSAB. The $\Delta H$ remains endothermic. So, other factors must cause this drop in capacity between 30°C and 50°C. As temperature rises, the translational energy of the oil molecules increases, allowing it to more easily overcome surface attractions with the adsorbent. These $\Delta H$ values of oil/oil vs oil/char interaction forces stay approximately the same over this range. Oil associates with oil (in water media) vs interactions on oil/oil or oil/char uptake. The role of the entropy change rises as temperature increases. The absolute value of $T\Delta S$ will play a bigger role. It will decrease capacity from 30°C to 50°C by countering the endothermic influence of $\Delta H$. The oil-solvent interaction forces became relatively stronger than the oil-adsorbent interaction forces as the petroleum-derived diesel oil's solubility increased, lowering the oil:biochar partition coefficients ($K_d$). This makes uptake less favorable as the temperature rises from 30°C to 50°C (Cai et al., 2019). Crude oil uptake increased slightly (endothermic adsorption) as temperature increased from 10°C to 40°C for all sorbents derived from Douglas fir biochar (BC, LBC, MBC, LMBC, and MLBC) (see Section 1.3) (Navarathna et al., 2020).

## 2.4 Regeneration and recycling

Spilled oil can either be left in the environment to biodegrade or recovered. Recovery requires a variety of technical methods and materials. Using adsorbents to remove oil often requires removing oil from adsorbents and recycling the regenerated adsorbent.

Bioderived cooking oil, loaded onto the textile sludge biochar (BC700), was regenerated by extracting the oil with dilute isopropanol [0.1 M], 150 mg/mL (Fig. 15.6A) (Sohaimi et al., 2017). After the first three cycles, a 10%–30% (165–120 mg/g) decrease in oil uptake capacities occurs. A further oil uptake reduction to 45% (98.7 mg/g) was observed in the fourth cycle. In the fifth cycle, the uptake remained almost constant (95.6 mg/g). A better eluant would allow more cycles with higher oil uptake to be achieved (but maybe not more total cycles).

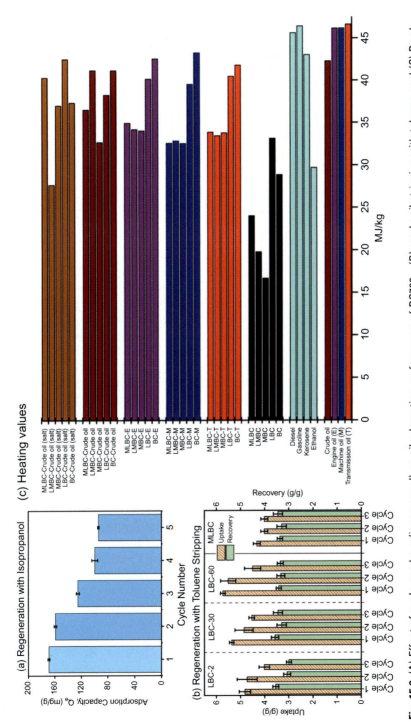

**Fig. 15.6** (A) Effect of cycle numbers (isopropanol) on oil adsorption performance of BC700. , (B) crude oil stripping with toluene, and (C) Bomb calorimeter heating values. (A) Modified from Sohaimi, K., Ngadi, N., Mat, H., Inuwa, I., Wong, S., 2017. Synthesis, characterization and application of textile sludge biochars for oil removal. J. Environ. Chem. Eng. 5, 1415–1422; (B and C)Adapted from Navarathna, C.M., Bombuwala Dewage, N., Keeton, C., Pennisson, J., Henderson, R., Lashley, B., Zhang, X., Hassan, E.B., Perez, F., Mohan, D., 2020. Biochar adsorbents with enhanced hydrophobicity for oil spill removal. ACS Appl. Mater. Interfaces 12, 9248–9260.

Lauric acid Douglas fir biochar (LBC) and magnetite-decorated lauric acid Douglas fir biochar (MLBC) (see Section 1.3) were used for petroleum-derived crude oil adsorption-desorption recycling after being stripped with pure toluene (Fig. 15.6B) (Navarathna et al., 2020). An LBC/crude oil mixture was filtered through filter paper and washed four times with toluene. An MLBC/oil mixture was separated from a simulated oil/water spill via a horseshoe magnet dipped and rotated within 50 mL of toluene 4–5 times. The magnetic sorbent with its sorbed oil was conveniently separated in this procedure, avoiding messy filtration. Then, this isolated oily sorbent was extracted with toluene. The toluene washing solutions were then evaporated to remove the toluene and isolate the recovered oils. Three adsorption-desorption cycles were repeated. LBC-30 (30% of lauric acid) oil uptakes for three cycles were 5.4, 4.9, and 4.5 g/g, respectively, and oil recoveries (stripped oil) of 3.6, 3.2, and 3.5 g/g over three cycles. LBCs with only 2% by weight (LBC-2) and higher lauric acid contents (LBC-60, 60% w/w lauric acid) exhibited similar adsorption-desorption trends. MLBC sorptions, as described above, had 4.3, 4.0, and 3.9 g/g crude oil uptakes, respectively, and recoveries of 3.4, 3.3, and 3.4 g/g. The amounts of oil adsorbed decreased progressively after each cycle. This could result from a loss of lauric from MLBC surfaces or some biochar losses upon after recovering it from a filter paper or magnet. Overall, LBC showed a more robust cyclic adsorption performance than MLBC (Navarathna et al., 2020).

Recovery via distillation or extraction is only one option. Biochar has a good fuel value (Douglas fir biochar—28.9 MJ/kg). The sorbed oil has an even higher fuel value per unit weight. Thus, the retrieved oil-laden adsorbent can be burned directly as a fuel. Fuel values between 32.6 and 41.2 MJ/kg were found for various oil-laden biochar sorbents (Fig. 15.6C) (Navarathna et al., 2020). This eliminates the need for an oil separation step from the adsorbent when recycling the adsorbent is not mandated. Alternatively, after recovering oil via recycling steps, the eventually spent biochar with its final oil load can be burned. Crude (from both deionized and seawater), machine, engine, and transmission oil-laden BC, LBC, MBC, MLBC, and LMBC (see Section 1.3) had fuel values in the 30–43 MJ/kg range. These heating values compared well with commercial fuels, such as ethanol, kerosene, gasoline, and diesel (30–45 MJ/kg). Thus, in furnaces or combustion equipment where biochar particles can be tolerated, these recovered oil-laden adsorbents can be used as fuels. It seems obvious that they can be added to coal-fed power plants. This route has a significant advantage in that it avoids the disposal of potentially toxic waste.

Alternatively, the oil-laden material can be left in the environment to allow natural oil attenuation, as biochar is environmentally friendly, and its presence is not regarded as undesirable. Biochar particles are buoyant. They will float in the photic zone of the seawater column, where photolysis and aerobic heterotrophic activity are most active. Maple wood biochar (BC500) and commercial biochar (Agrichar (BEST Energies, Australia)) were tested for biodegradation of Texas crude oil (Nguyen and Pignatello, 2013). Compared to seawater alone, biochar addition to seawater resulted in a slight increase in $CO_2$ release from 80 to 85 µmol for Agrichar. The oil-seawater-Agrichar gave the highest $CO_2$ release of 120 µmol. When going from seawater to oil-seawater-biochar, a similar increasing trend (from 175 to 300 mol) was observed for BC500. This increase was statistically significant. The same experiment was repeated in the presence of $PO_4^{3-}$, $NO_3^-$, $Fe^{3+}$, and $Cl^-$ which are inorganic microbial nutrients. $CO_2$ release increased to 450 mol from 300 mol, indicating increased oil degradation. Biochar particles may provide a favorable and stable environment for the colonization of oil-degrading microorganisms, that could access a hydrocarbon reservoir within the particle pore network. It is possible to reduce crude oil toxicity by reducing oil bioavailability to higher organisms, but this has yet to be tested (Nguyen and Pignatello, 2013).

## 2.5 Adsorption from seawater

Oil adsorption capacities on biochar can vary depending on the conditions in which the tests are conducted. Crude oil uptake on five Douglas fir biochar-based adsorbents (BC, LBC, MBC, LMBC, and MLBC) (see Section 1.3) revealed a 5% decrease in crude oil uptake capacities in seawater compared to deionized water (Navarathna et al., 2020). Toluene's adsorption capacities from deionized water onto wood pellet biochar were $268 \pm 35.8$ mg/g. These were higher than toluene's capacities from synthetic seawater (40 g/L of sea salts in deionized water) of $197 \pm 15.4$ mg/g (Silvani et al., 2017). Crude oil uptakes on rice husk biochars and sawdust biochar (see Section 1.3) in saltwater were 5%–10% lower versus uptakes from deionized water in both static and stirring conditions (Fig. 15.5) (Kandanelli et al., 2018). The dissolved ions in saltwater (high ionic strength) could cause a change in the thermodynamics ($\Delta G$) when an uptake reaction from oil on saltwater occurs to polar surface regions in the sorbents versus hydrophobic surfaces.

## 3 Uptake mechanisms

Biochar's high porosity, surface area, and hydrophobicity (high contact angle, high carbon content, and low O/C ratio) make it a highly effective substance for removing nonpolar/less-polar organic pollutants and oil. Van der Waal's interactions dominate in aliphatic oil uptake (Fig. 15.2B). Upon crude oil uptake, the contact angle of the lauric acid-decorated Douglas biochar (LBC-2) raised from 127.9 to 142.7 degrees. Thus, oil uptake makes the biochar surface more hydrophobic (Navarathna et al., 2020). As a result, more oil molecules will adsorb more firmly onto the surface, eventually resulting in oil blobs and a final thick layer of oil on the biochar (Fig. 15.2B). Thiacloprid sorption on Maze straw and pig manure biochar is also attributed to hydrophobic interactions (Zhang et al., 2018; Dai et al., 2019). Oils with more aromatic electron-rich or electron-deficient components might interact with the biochar surface via electron-donor acceptor interactions.

## 4 Conclusions, challenges, and gaps

As the need for effective oil spill cleanup measures becomes increasingly imperative, research is showing biochars to have promising potential use as an oil adsorbent. BC is generally cost-effective and environmentally friendly. With proper modifications and production conditions, it is a candidate for oil spill use. These production conditions depend mainly on the pyrolysis temperature and time and source material. Promising modifications include the addition of lauric acid, or other cheaper fatty acids from tallow or waste cooking oils, to increase BC's buoyancy. Adding a magnetic oxide, like $Fe_3O_4$, allows the BC to be magnetically removed from the water after adsorption. BC has the potential to be recycled by stripping oil from the spent adsorbent using toluene or other hydrocarbon solvent mixtures. Recycling further decreases its cost by decreasing the amount of BC needed. The biochar sorbents discussed in this chapter have three advantages for removing oil from spills: (1) a significant reduction in the spilled oil content, (2) the ability to use the oil/BC as a heat source, or (3) recovery of the sorbed oil. Thus, effective adsorbents can be prepared from biochar resources. One enticing possibility is the disposal of waste biochar/oil residues from oil spills by burning them for process heat recovery. More research is needed to better understand the kinetics, isotherms, and mechanisms of BC-crude oil uptake and determine what large-scale BC spill cleanup would be like.

## Questions and problems

1. List the conventional oil clean up methods and discuss their disadvantages.
2. Why is high temperature pyrolysis biochar considered to be an ideal candidate for oil recovery?
3. Give some important features of the lauric acid magnetic biochar (MLBC) adsorbent discussed in this chapter and give the importance of each feature for oil recovery.
4. Explain why MLBC biochar floats on water.
5. What are the different analytical techniques used in the oil uptake experiments to determine the remaining oil/water mixture?
6. Give two characterization techniques for biochar hydrophobicity.
7. In most studies, oil uptake was generally considered independent of the pH. What is the reason for this?
8. Why might plant-based or animal fat-based oil show a pH dependence?
9. Explain why the oil sorption results in multilayer uptake on biochar?
10. How can turbulent conditions influence oil uptake on biochar?
11. Give examples of different oil-laden biochar recycling or regeneration or disposal techniques discussed in this chapter.

## References

Adebajo, M.O., Frost, R.L., 2004. Acetylation of raw cotton for oil spill cleanup application: an FTIR and 13C MAS NMR spectroscopic investigation. Spectrochim. Acta A Mol. Biomol. Spectrosc. 60, 2315–2321.

Adebajo, M.O., Frost, R.L., Kloprogge, J.T., Carmody, O., Kokot, S., 2003. Porous materials for oil spill cleanup: a review of synthesis and absorbing properties. J. Porous. Mater. 10, 159–170.

Ahmad, A., Sumathi, S., Hameed, B., 2005. Adsorption of residue oil from palm oil mill effluent using powder and flake chitosan: equilibrium and kinetic studies. Water Res. 39, 2483–2494.

Angelova, D., Uzunov, I., Uzunova, S., Gigova, A., Minchev, L., 2011. Kinetics of oil and oil products adsorption by carbonized rice husks. Chem. Eng. J. 172, 306–311.

Angın, D., 2013. Effect of pyrolysis temperature and heating rate on biochar obtained from pyrolysis of safflower seed press cake. Bioresour. Technol. 128, 593–597.

Asadpour, R., Sapari, N.B., Tuan, Z.Z., Jusoh, H., Riahi, A., Uka, O.K., 2013. Application of sorbent materials in oil spill management: A review. Caspian J. Appl. Sci. Res. 2.

Bastani, D., Safekordi, A., Alihosseini, A., Taghikhani, V., 2006. Study of oil sorption by expanded perlite at 298.15 K. Sep. Purif. Technol. 52, 295–300.

Cai, L., Zhang, Y., Zhou, Y., Zhang, X., Ji, L., Song, W., Zhang, H., Liu, J., 2019. Effective adsorption of diesel oil by crab-shell-derived biochar nanomaterials. Materials 12, 236.

Chen, Y., Zhang, X., Chen, W., Yang, H., Chen, H., 2017. The structure evolution of biochar from biomass pyrolysis and its correlation with gas pollutant adsorption performance. Bioresour. Technol. 246, 101–109.

Choi, H.-M., Kwon, H.-J., Moreau, J.P., 1993. Cotton nonwovens as oil spill cleanup sorbents. Text. Res. J. 63, 211–218.

Dai, Y., Zhang, N., Xing, C., Cui, Q., Sun, Q., 2019. The adsorption, regeneration and engineering applications of biochar for removal organic pollutants: a review. Chemosphere 223, 12–27.

Deschamps, G., Caruel, H., Borredon, M.-E., Bonnin, C., Vignoles, C., 2003. Oil removal from water by selective sorption on hydrophobic cotton fibers. 1. Study of sorption properties and comparison with other cotton fiber-based sorbents. Environ. Sci. Technol. 37, 1013–1015.

Franco, C.A., Cortés, F.B., Nassar, N.N., 2014. Adsorptive removal of oil spill from oil-in-fresh water emulsions by hydrophobic alumina nanoparticles functionalized with petroleum vacuum residue. J. Colloid Interface Sci. 425, 168–177.

Frieling, J., Madlener, R., 2017. Fueling the US Economy: Energy as a Production Factor From the Great Depression Until Today.

Goldberg, V.P., 1994. Recovery for economic loss following the exxon "Valdez" oil spill. J. Leg. Stud. 23, 1–39.

Houghton, J.P., Lees, D.C., Driskell, W.B., Mearns, A.J., 1991. Impacts of the Exxon Valdez spill and subsequent cleanup on intertidal biota—1 year later. Int. Oil Spill Conf. 1991, 467–475. American Petroleum Institute.

Inyang, M.I., Gao, B., Yao, Y., Xue, Y., Zimmerman, A., Mosa, A., Pullammanappallil, P., Ok, Y.S., Cao, X., 2016. A review of biochar as a low-cost adsorbent for aqueous heavy metal removal. Crit. Rev. Environ. Sci. Technol. 46, 406–433.

Jada, A., Chaou, A.A., Bertrand, Y., Moreau, O., 2002. Adsorption and surface properties of silica with transformer insulating oils. Fuel 81, 1227–1232.

Kalus, K., Koziel, J.A., Opaliński, S., 2019. A review of biochar properties and their utilization in crop agriculture and livestock production. Appl. Sci. 9, 3494.

Kandanelli, R., Meesala, L., Kumar, J., Raju, C.S.K., Peddy, V.R., Gandham, S., Kumar, P., 2018. Cost effective and practically viable oil spillage mitigation: Comprehensive study with biochar. Mar. Pollut. Bull. 128, 32–40.

Kelle, H., Eboatu, A., Ofoegbu, O., Udeozo, I., 2013. Determination of the viability of an agricultural solid waste; corncob as an oil spill sorbent mop. IOSR J. Appl. Chem. 6, 30–57.

Kim, K.H., Kim, J.-Y., Cho, T.-S., Choi, J.W., 2012. Influence of pyrolysis temperature on physicochemical properties of biochar obtained from the fast pyrolysis of pitch pine (Pinus rigida). Bioresour. Technol. 118, 158–162.

Kingston, P.F., 2002. Long-term environmental impact of oil spills. Spill Sci. Technol. Bull. 7, 53–61.

Li, R., Wang, J.J., Gaston, L.A., Zhou, B., Li, M., Xiao, R., Wang, Q., Zhang, Z., Huang, H., Liang, W., 2018. An overview of carbothermal synthesis of metal–biochar composites for the removal of oxyanion contaminants from aqueous solution. Carbon 129, 674–687.

Loughlin, T.R., Ballachey, B.E., Wright, B., 1996. Overview of studies to determine injury caused by the Exxon Valdez oil spill to marine mammals. In: Exxon Valdez Oil Spill Symposium, American Fisheries Society Symposium 18, pp. 798–808.

Mahdi, Z., El Hanandeh, A., Yu, Q., 2017. Influence of pyrolysis conditions on surface characteristics and methylene blue adsorption of biochar derived from date seed biomass. Waste Biomass Valorization 8, 2061–2073.

Moreira, M., Noya, I., Feijoo, G., 2017. The prospective use of biochar as adsorption matrix–a review from a lifecycle perspective. Bioresour. Technol. 246, 135–141.

Nartey, O.D., Zhao, B., 2014. Biochar preparation, characterization, and adsorptive capacity and its effect on bioavailability of contaminants: an overview. Adv. Mater. Sci. Eng. 2014.

Navarathna, C.M., Bombuwala Dewage, N., Keeton, C., Pennisson, J., Henderson, R., Lashley, B., Zhang, X., Hassan, E.B., Perez, F., Mohan, D., 2020. Biochar adsorbents with enhanced hydrophobicity for oil spill removal. ACS Appl. Mater. Interfaces 12, 9248–9260.

Nguyen, H.N., Pignatello, J.J., 2013. Laboratory tests of biochars as absorbents for use in recovery or containment of marine crude oil spills. Environ. Eng. Sci. 30, 374–380.

Oribayo, O., Feng, X., Rempel, G.L., Pan, Q., 2017. Synthesis of lignin-based polyurethane/graphene oxide foam and its application as an absorbent for oil spill clean-ups and recovery. Chem. Eng. J. 323, 191–202.

Park, J.-H., Ok, Y.S., Kim, S.-H., Cho, J.-S., Heo, J.-S., Delaune, R.D., Seo, D.-C., 2016. Competitive adsorption of heavy metals onto sesame straw biochar in aqueous solutions. Chemosphere 142, 77–83.

Pasila, A., 2004. A biological oil adsorption filter. Mar. Pollut. Bull. 49, 1006–1012.

Peiris, C., Gunatilake, S.R., Mlsna, T.E., Mohan, D., Vithanage, M., 2017. Biochar based removal of antibiotic sulfonamides and tetracyclines in aquatic environments: a critical review. Bioresour. Technol. 246, 150–159.

Peiris, C., Nayanathara, O., Navarathna, C.M., Jayawardhana, Y., Nawalage, S., Burk, G., Karunanayake, A.G., Madduri, S.B., Vithanage, M., Kaumal, M., 2019. The influence of three acid modifications on the physicochemical characteristics of tea-waste biochar pyrolyzed at different temperatures: a comparative study. RSC Adv. 9, 17612–17622.

Peterson, C.H., 2001. The "Exxon Valdez" Oil Spill in Alaska: Acute, Indirect and Chronic Effects on the Ecosystem.

Price, J.M., Johnson, W.R., Marshall, C.F., Ji, Z.-G., Rainey, G.B., 2003. Overview of the oil spill risk analysis (OSRA) model for environmental impact assessment. Spill Sci. Technol. Bull. 8, 529–533.

Raj, K.G., Joy, P.A., 2015. Coconut shell based activated carbon–iron oxide magnetic nanocomposite for fast and efficient removal of oil spills. J. Environ. Chem. Eng. 3, 2068–2075.

Shivaram, P., Leong, Y.-K., Yang, H., Zhang, D., 2013. Flow and yield stress behaviour of ultrafine Mallee biochar slurry fuels: the effect of particle size distribution and additives. Fuel 104, 326–332.

Sidik, S., Jalil, A., Triwahyono, S., Adam, S., Satar, M., Hameed, B., 2012. Modified oil palm leaves adsorbent with enhanced hydrophobicity for crude oil removal. Chem. Eng. J. 203, 9–18.

Silvani, L., Vrchotova, B., Kastanek, P., Demnerova, K., Pettiti, I., Papini, M.P., 2017. Characterizing biochar as alternative sorbent for oil spill remediation. Sci. Rep. 7, 1–10.

Sips, R., 1948. On the structure of a catalyst surface. J. Chem. Phys. 16, 490–495.

Sohaimi, K., Ngadi, N., Mat, H., Inuwa, I., Wong, S., 2017. Synthesis, characterization and application of textile sludge biochars for oil removal. J. Environ. Chem. Eng. 5, 1415–1422.

Spencer, D.C., Fitzgerald, A., 2013. Three ecologies, transversality and victimization: the case of the British petroleum oil spill. Crime Law Soc. Chang. 59, 209–223.

Ugochukwu, U.C., Jones, M.D., Head, I.M., Manning, D.A., Fialips, C.I., 2014. Biodegradation and adsorption of crude oil hydrocarbons supported on "homoionic" montmorillonite clay minerals. Appl. Clay Sci. 87, 81–86.

Vocciante, M., Finocchi, A., De Folly, D., Auris, A., Conte, A., Tonziello, J., Pola, A., Reverberi, A.P., 2019. Enhanced oil spill remediation by adsorption with interlinked multilayered graphene. Materials 12, 2231.

Wood, M.H., Casford, M., Steitz, R., Zarbakhsh, A., Welbourn, R., Clarke, S.M., 2016. Comparative adsorption of saturated and unsaturated fatty acids at the iron oxide/oil interface. Langmuir 32, 534–540.

Yu, H., Zou, W., Chen, J., Chen, H., Yu, Z., Huang, J., Tang, H., Wei, X., Gao, B., 2019. Biochar amendment improves crop production in problem soils: a review. J. Environ. Manag. 232, 8–21.

Zhang, P., Sun, H., Min, L., Ren, C., 2018. Biochars change the sorption and degradation of thiacloprid in soil: insights into chemical and biological mechanisms. Environ. Pollut. 236, 158–167.

# Biochar and biochar composites for poly- and perfluoroalkyl substances (PFAS) sorption

Chanaka Navarathna[a], Michela Grace Keel[a],
Prashan M. Rodrigo[a], Catalina Carrasco[a], Arissa Ramirez[a],
Hailey Jamison[a], Dinesh Mohan[b], and Todd E. Mlsna[a]

[a]Department of Chemistry, Mississippi State University, Mississippi State, MS, United States. [b]School of Environmental Sciences, Jawaharlal Nehru University, New Delhi, India

## 1 Introduction

Per-/poly-fluoroalkyl substances (PFAS) are man-made organofluorine compounds introduced in 1940 and are classified as anthropogenic organic pollutants (Buck et al., 2011). They have a wide range of applications and can be found in fire retardants, textiles, leathers, paper, carpet, food packaging materials, nonstick cooking pans, surface-active agents in waterproof clothing, fabric, paints, surfactants, emulsifiers, waxes, polymer additives, and have many industrial applications [cleaning agents, photolithography, semiconductors, metal plating (Cr)]. Most firefighting foams have aqueous film-forming foams containing highly diverse organic moieties that bind variable length fluorocarbon moieties (Jovicic et al., 2018). These applications exploit PFAS extensive surface activity properties, high thermal and chemical stability, water repellent properties, and acid–base resistivity (Du et al., 2014; Banks et al., 2020).

PFAS are usually classified as long-chain (length of $C_8$ or higher) and short-chain (below $C_8$). Perfluorooctanoic acid, or its salt (PFOA), and perfluorooctane sulfonic acid, or its salt (PFOS), are widely used long-chain PFAS compounds. They show higher toxicity than short-chain perfluoro compounds (Gagliano et al., 2020). PFOA and PFOS can be products of the degradation of other long-chain PFAS or directly discharged into the environment from manufacturing processes and other industrial applications. Thus, PFAS are found in various environmental matrices worldwide.

PFAS were created, used, and disposed of without oversight for the last half-century before federal, state, and local governments began to act. As a result, food, air, and water have become polluted worldwide due to the release and use of PFAS-containing products. These compounds are widely found in former fluorochemical manufacturing facilities (DuPont, etc.), water treatment plants, landfills and sediment dredging, U.S. Department of Defense facilities, air force crash sites, aqueous film-forming foam (AFFF) training facilities, oil refineries and storage facilities, metal coating and plating facilities, wiring and semiconductors in the electronic and aerospace industry, and large rail yards (Cordner et al., 2019). At least 172 sites in 40 US states have been reported for the presence of PFAS. Some states, including Michigan, have recently declared an emergency due to the high levels of PFOS and PFOA found in water.

When PFAS are released into the environment, it can lead to groundwater and soil matrices pollution. Hence, it is a public health concern due to their high persistence, potential toxicity to biotic ecosystems, and environmentally bio-accumulative nature (Giesy and Kannan, 2002; Beesoon and Martin, 2015). It has been proposed that significant human exposure may occur via the contamination of drinking water or the consumption of seafood that previously ingested contaminated water. The exact health risks associated with exposure are unknown. However, PFOS earned classification as an animal carcinogen by the U.S. EPA in 2005. A lack of conclusive studies has prevented the complete understanding of how these chemicals affect humans. Epidemiological studies have shown an association between PFOA exposure and increased cholesterol, diabetes mortality, prostate cancer, and pancreatic cancer. Almost every U.S. citizen has traces of these substances in their body. Biomonitoring studies demonstrated their presence in blood, breast milk, umbilical cord blood, amniotic fluid, placenta, and other tissues. PFAS have high strength C—F bonds (ranging 485–585 kJ/mol); hence, these show high thermal and chemical stability and high persistence due to their long half-life (Lin et al., 2010; Espana et al., 2015).

PFAS have a hydrophobic region that contains C—F bonds at each carbon atom. They can be either branched or long-chain. The highly branched or substituted PFAS are more chemically inert than long-chain PFAS. Calculated and experimental $pK_a$ values indicate that most PFAS compounds are acidic (Table 16.1) (Gagliano et al., 2020). Therefore, at environmentally relevant pH values (near neutral), they exist in a negatively charged form. Perfluorocarboxylic acids and perfluorosulphonic acids show low vapor pressures with increasing carbon chain length, which suggests that they are not

Table 16.1 Physical and chemical properties of selected PFAS (Gagliano et al., 2020)

| PFAS acronym | PFAS name and molecular formula | Molecular weight (g/mol) | $pK_a$ | Log $K_{OW}$ (L/kg) | Solubility (mg/L) |
|---|---|---|---|---|---|
| **Perfluoroalkyl acids (PFAAs)—Perfluoroalkylcarboxylic acids** | | | | | |
| PFBA | Perfluorobutanoic acid [$CF_3(CF_2)_2COOH$] | 214.04 | 1.07 | 2.31 | Miscible |
| PFPeA | Perfluoropentanoic acid [$CF_3(CF_2)_3COOH$] | 264.05 | 0.34 | 3.01 | $112.6 \times 10^3$ |
| PFHxA | Perfluorohexanoic acid [$CF_3(CF_2)_4COOH$] | 314.05 | −0.13 | 3.48 | $21.7 \times 10^3$ |
| PFHpA | Perfluoroheptanoic acid [$CF_3(CF_2)_5COOH$] | 364.06 | −2.29 | 3.65 | $4.2 \times 10^3$ |
| PFOA | Perfluorooctanoic acid [$CF_3(CF_2)_6COOH$] | 414.07 | −0.16 to 3.8 | 4.81 | $3.4$–$9.5 \times 10^3$ |
| PFNA | Perfluorononanoic acid [$CF_3(CF_2)_7COOH$] | 464.08 | −0.17 | 5.48 | $9.5 \times 10^3$ |
| PFDA | Perfluorodecanoic acid [$CF_3(CF_2)_8COOH$] | 514.09 | −5.2 | 6.51 | $9.5 \times 10^3$ |
| PFUnDA | Perfluoroundecanoic acid [$CF_3(CF_2)_9COOH$] | 564.09 | −5.2 | 7.21 | 4 |
| PFDoDA | Perfluorododecanoic acid [$CF_3(CF_2)_{10}COOH$] | 614.10 | −5.2 | 7.92 | 0.7 |
| PFTrDA | Perfluorotridecanoic acid [$CF_3(CF_2)_{11}COOH$] | 664.11 | -5.2 | 8.62 | 0.2 |
| PFTeDA | Perfluorotetradecanoic acid [$CF_3(CF_2)_{12}COOH$] | 714.12 | −5.2 | 9.32 | 0.03 |
| PFPeDA | Perfluoropentadecanoic acid [$CF_3(CF_2)_{13}COOH$] | 764.12 | −5.2 | – | – |
| **Perfluoroalkyl acids (PFAAs)—Perfluoroalkylsulfonic acids** | | | | | |
| PFBS | Perfluorobutanesulfonic acid [$CF_3(CF_2)_3SO_3H$] | 300.10 | -3.31 | 1.82 | $(46.2–56.6) \times 10^3$ |
| PFHxS | Perfluorohexanesulfonic acid [$CF_3(CF_2)_5SO_3H$] | 400.11 | 0.14 | 3.16 | $2.3 \times 10^3$ |
| PFHpS | Perfluoroheptanesulfonic acid [$CF_3(CF_2)_6SO_3H$] | 450.12 | – | – | – |
| PFOS | Perfluorooctanesulfonic acid [$CF_3(CF_2)_7SO_3H$] | 500.13 | −6.0 to −2.6 | 4.49 | $(1.52–1.57) \times 10^3$ |
| PFDS | Perfluorodecanesulfonic acid [$CF_3(CF_2)_9SO_3H$] | 600.14 | −3.24 | 6.83 | 2 |
| **Perfluoroalkane sulfonamides (FASAs)** | | | | | |
| FOSA | Perfluorooctane sulfonamide [$CF_3(CF_2)_7SO_2NH_2$] | 499.14 | 3.37 | 5.8 | – |

*Continued*

Table 16.1 Physical and chemical properties of selected PFAS (Gagliano et al, 2020)—cont'd

| PFAS acronym | PFAS name and molecular formula | Molecular weight (g/mol) | $pK_a$ | Log $K_{OW}$ (L/kg) | Solubility (mg/L) |
|---|---|---|---|---|---|
| N-MeFOSA | N-Methyl-Perfluorooctane sulfonamide [$CF_3(CF_2)_7SO_2NHCH_3$] | 513.17 | – | – | 0.2 |
| | Perfluoroalkane sulfonamido substances (FASEs) | | | | |
| FOSE | Perfluorooctane sulfonamidoethanol [$CF_3(CF_2)_7SO_2NH(CH_2)_2OH$] | 543.19 | – | – | 0.9 |
| N-MeFOSE | N-Methyl-Perfluorooctane sulfonamidoethanol [$CF_3(CF_2)_7SO_2N(CH_3)(CH_2)_2OH$] | 557.22 | – | – | 0.3 |
| **Fluorotelomer substances** | | | | | |
| 4:2 FTOH | 4:2 Fluorotelomer alcohol [$CF_3(CF_2)_3(CH_2)_2OH$] | 264.10 | – | 0.93 | 974 |
| 6:2 FTOH | 6:2 Fluorotelomer alcohol [$CF_3(CF_2)_5(CH_2)_2OH$] | 364.10 | – | 2.43 | 18.8 |
| 8:2 FTOH | 8:2 Fluorotelomer alcohol [$CF_3(CF_2)_7(CH_2)_2OH$] | 464.10 | – | 3.84 | 0.194 |
| 10:2 FTOH | 10:2 Fluorotelomer alcohol [$CF_3(CF_2)_9(CH_2)_2OH$] | 564.10 | – | 6.20 | 0.011 |

easily volatilized when the carbon chain becomes larger. Hence, they cannot be easily removed from natural water bodies by air stripping.

Water solubility and hydrophilicity of PFAS can be predicted from the Log $K_{OW}$ value (Table 16.1) (Gagliano et al., 2020). Log-$K_{OW}$ can be defined as the ratio of the concentration of PFAS in the octanol phase to the concentration of PFAS in the aqueous phase. Log $K_{OW}$ decreases with increasing solubility in the aqueous phase. The PFAS solubility in the aqueous phase depends on the charge of the hydrophilic functional group and the carbon chain length. Perfluorosulphonic (PFSA) acids show higher water solubility than perfluoro carboxylic acid (PFCA) when the carbon chain length is the same because of the increased availability of forming H-bonds between PFAS hydrophilic heads and water molecules. Fluorotelomer substances show low water solubility since their hydrophilic heads are uncharged. Short-chain PFAS have lower Log $K_{OW}$ (Table 16.1) values than large chain PFAS

(Gagliano et al., 2020). Therefore, they can easily contaminate natural bodies of water. The sorption coefficient of PFAS can be interpreted from the Log $K_{OC}$ value. Log $K_{OC}$ can be defined as the ratio between the sorption coefficient $K_d$ and the organic carbon content of the sorbent material. The Log $K_{OC}$ value helps to predict the sorption of nonpolar organic compounds onto soils and sediments. Sorption studies reveal that log $K_{OC}$ values increase when increasing the PFAS carbon chain length (Gagliano et al., 2020).

Recent studies reveal PFOA and PFOS found in different bodies of water such as groundwater, seawater, rainwater, and atmosphere and biological samples (Prevedouros et al., 2006). PFAS-contaminated drinking water has become a critical issue over the past two decades. The United States Environmental Protection Agency (EPA) produced a striking statement on May 19th of 2016 on this issue. In this statement, drinking water is categorized as contaminated if the PFOA and PFOS concentration levels are above 70 ppt, either individually or combined (USEPA, 2016). In addition, some states have adopted their own maximum contaminant levels (MCLs) lower than the EPA limit [Vermont—20 ppt, New Jersey (13–14) ppt, California (13–14) ppt]. From 2002 to date, the MCL has been changed from 150,000 to 70 ppt (Cordner et al., 2019). Most industries have phased out PFAS and adapted substitutes with shorter-half-lives (GenX, ADONA, and F—53B) that are more prone to environmental degradation (Munoz et al., 2019).

## 1.1 Current PFAS cleanup methods

PFAS analysis and remediation became very challenging and tedious because of the low MCL. Strategies adopted vary from conservation practices and outreach activities to active treatment of these from point and nonpoint sources. Several strategies including adsorption (using activated carbon (McCleaf et al., 2017), graphene (Lath et al., 2018), metal oxides (Zhang et al., 2021b), metal organic frameworks (Barpaga et al., 2019), aerogels (Tian et al., 2021), polymeric adsorbents, ion-exchange (Liu et al., 2021a), degradation/destruction (advanced oxidation, ozone treatment, chlorination, photolysis, sonolysis, electrochemical oxidation, biodegradation (Ahmed et al., 2020)), coagulation (Xiao et al., 2013), and sedimentation (flocculation, electrocoagulation) have been studied or used to remove PFAS from both drinking and wastewaters (Du et al., 2014; Ross et al., 2018; Lee et al., 2019; Zhang et al., 2019). However, these methods are expensive, produce toxic byproducts, and have mixed success (Fig. 16.1) (Ross et al., 2018) removing PFAS from concentrations near or below the EPA limit of 70 ppt or state-

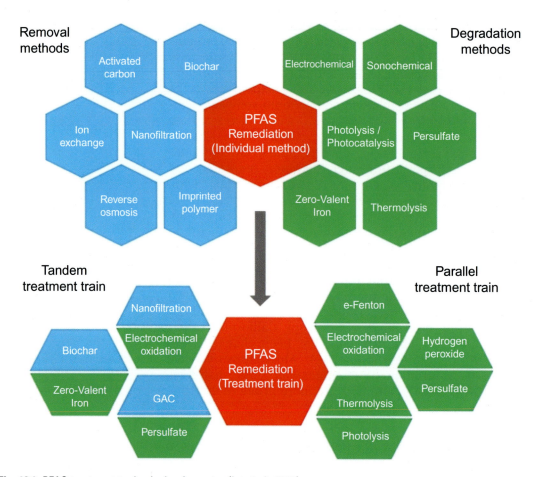

**Fig. 16.1** PFAS treatment technologies for water (Lu et al., 2020).

specific (<20 ppt) MCL. Additional drawbacks include practicality limitations, scale-up limitations, production of undesired toxic byproducts, and, more importantly, the cost (Ross et al., 2018). Therefore cheap, effective, and environmentally friendly practices are needed to remediate PFAS. While there are quite a few materials that function to adsorb PFAS, biochar and engineered biochar can be easily produced, eco-friendly, and have been shown to be very powerful adsorbents.

The use of biochar stands as a more effective technique as it is cheap, effective, readily available, and nontoxic (Lehmann and Joseph, 2015). Biochar is a byproduct that results from biomass pyrolysis under different anaerobic conditions, typically resulting in a high carbon yield. They have a higher affinity toward organic contaminants (pharmaceuticals (Patel et al., 2019), dyes (Navarathna et al., 2020), hormones (Dong et al., 2018), pesticides

(Yavari et al., 2015)) by forming hydrophilic, hydrophobic, or ionic interactions with the carbonaceous biochar surface (Ahmad et al., 2014). This chapter provides an up-to-date overview of PFAS remediation using biochar and biochar composites, including biochar tailoring for PFAS sorption, various adsorption benchmarks (pH optimization, dose, kinetics, isotherms), suitability in applied water treatment (column breakthrough studies, regeneration, competitive studies, pilot-scale plants), and mechanisms.

## 1.2 Biochar and biochar composites for PFAS remediation

Activated carbon (AC) has been widely employed in PFAS remediation in laboratory and field settings (Zhang et al., 2019). However, granular AC systems for the treatment of aqueous firefighting foam (AFFF) impacted (PFOA and PFOS) sources of water likely achieve poor removal compared to biochars (Fig. 16.2) (Guo et al., 2017). Calgon Filtersorb300 AC (F300, Calgon Carbon Co., Pittsburgh, PA, 997 $m^2$/g SA) and seven biochars were employed for remediation of the groundwater sample containing $10.7 \pm 0.9$ PFOA and $33.1 \pm 0.6$ μg/L PFOS (Fig. 16.2). The seven biochars included MCG (Mountain Crest Gardens, Gropro Inc., Etna, CA) biochar (351 $m^2$/g SA), and pine needle (PN)-derived biochars produced upon slow pyrolysis for 6h at (under $N_2$, heating rate 5°C/min)

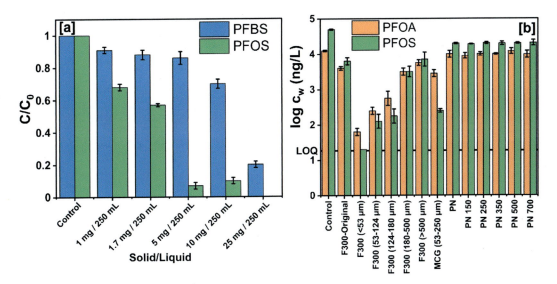

**Fig. 16.2** (A) The $C/C_0$ variation of PFBS and PFOS onto F300 (<53 μm) with different solid-to-liquid ratios after 5 days of sorption and (B) Logarithmic aqueous PFOA and PFOS concentrations (logCw) after 5 days of equilibration with F300 GAC (at varying particle size fractions), MCG-biochar, and PN-biochar (produced at various temperatures) (Xiao et al., 2017).

150 (42.3 m²/g), 250 (0.964 m²/g), 350 (34.3 m²/g), 500 (18.2 m²/g), and 700°C (37.3 m²/g).

Biochar pyrolytic temperature plays a vital role in PFAS adsorption. The amount of biochar surface functional groups depends on the pyrolysis temperatures and pyrolysis conditions (Ahmad et al., 2014; Guo et al., 2017). Low-temperature pyrolysis results in a large number of ionic groups on the biochar surface, which promotes electrostatic interactions with adsorbates. In contrast, high-temperature pyrolyzed biochars like graphene and carbon nanotubes interact mainly by pi-pi interactions and other types of hydrophobic interactions (Deng et al., 2015).

Four biochars were derived using corn-straw for PFOS removal pyrolyzing at 250°C (BC250), 400°C (BC400), 550°C (BC550), and 700°C (BC700) temperatures (Guo et al., 2017). Langmuir sorption capacities for PFOS (0.5–10 mg/L) increased from 135.53 mg/g to 169.30 mg/g with increased pyrolysis temperature from 250°C to 700°C. This was attributed to a sharp increase in surface area from 2.5 m²/g (0.013 cm³/g pore volume, 45.19% C) to 298.58 m²/g (0.199 cm³/g, 70.22% C). Decarboxylation and decarbonylation occur at high pyrolysis temperatures, resulting in the loss of biochar polar functional groups. At higher temperatures, aliphatic alkyl and ester C=O groups and phenolic groups destruct completely and deshield the aromatic core while increasing the aromaticity of the BC. SEM images (Figs. 16.3A–D) (Guo et al., 2017) show that the biochar surface became progressively "cleaner" from BC250

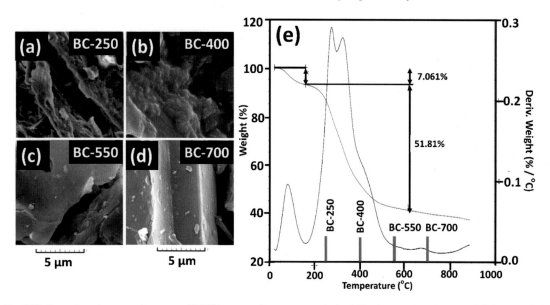

**Fig. 16.3** Scanning electron microscopy (SEM) images of corn-straw-derived biochar produced at 250°C (A), 400°C (B), 550°C (C), 700°C (D), and (E) TG and DTG of raw material under an ultrahigh-purity nitrogen atmosphere (Guo et al., 2017).

to BC700 as hemicellulose and cellulose decomposed and fully carbonized. Simultaneous thermogravimetric and differential thermogravimetric (TG–DTG) analysis for corn straw showed weight losses at 200–550°C (Fig. 16.3E). The weight loss at 85°C was attributed to free water loss, and the drying process was completed before 160°C. The decomposition peaks at 265°C and 362°C were due to hemicelluloses and cellulose losses, respectively. BC250 can be considered precursory biomass (natural organic matter) with partial hemicellulose decomposition. BC400 underwent complete hemicellulose decomposition as well as partial cellulose decomposition. Hemicellulose and cellulose decomposition is complete in both BC550 and BC700 (Fig. 16.3E).

Crushed corn straw (0.5–0.9 mm particle size) biochar (BC700) was prepared under $N_2$ at 700°C (heating rate = 10°C/min) and holding for 1.5 h (Guo et al., 2019). BC700 had 65.6% organic C, 298 $m^2$/g surface area, and 0.201 $cm^3$/g pore volume and removed ~1100 µg/g of PFOS/g in 48 h (150 rpm, batch conditions). The addition of BC700 to a sediment sample with 1.38% organic C at 2% and 5% w/w doses resulted in a marked PFOS sorption improvement from 5.10 to 7.5 and 10.6, respectively. The distribution coefficient ($K_d$) value for PFOS was 2.36 for the sediment and 3045 for BC. The Log $K_{oc}$ values for PFOS varied from 0.23 (sediment) to 1.60 (BC), while those for PFOS varied from 0.23 (sediment) to 1.67 (BC). When BC700 was added to the sediment, PFOS mobility in the water-sediment system decreased. This was further enhanced by increasing the biochar dose (Guo et al., 2019). Chars were prepared by pyrolysis (400°C for 2 h.) of maize (*Zea mays*) straw (M400) (C% 73.1, 7.21 $m^2$/g surface area), and willow (*Salix babylonica*) sawdust (W400) (C % 66.6, 11.6 $m^2$/g surface area) (Chen et al., 2011). Sorption rate of PFOS (100 mg/L) to W400 (91.6 mg/g) and M400 (164 mg/g) was low, and the equilibrium was reached in 384 h.

The adsorption behavior of PFAS ($C_0 = 0.1$ µg/L and pH 7.5 ± 0.5) [perfluoroalkyl carboxylates (PFCAs) (PFBA, PFPeA, PFHxA, PFHpA, PFOA, PFNA, PFDA, PFUnDA, PFDoA, PFTeDA, PFHxDA, PFOcDA), $C_4$, $C_6$, $C_8$ perfluoroalkyl sulfonates (PFSAs) (PFBS, PFHxS, PFOS), perfluorooctane sulfonamide (FOSA), and 6:2 and 8:2 fluorotelomer sulfonic acids (FTSAs)] to 44 inorganic and organic sorbents including several biochars is displayed in Fig. 16.4 (Sörengård et al., 2020). Biochar 1 (<4 mm particle size, 520 $m^2$/g surface area, 0.014 $cm^3$/g pore volume, and 53.2% C) originated from mixed wood at 700°C and a residence time of 20 min with a Pyreg 500W unit at Swiss Biochar, Switzerland. The biochar 2 (1.5 mm, 48 $m^2$/g surface area, 0.010 $cm^3$/g pore volume, and 19.2% C) was produced via slow pyrolysis (30 min) of

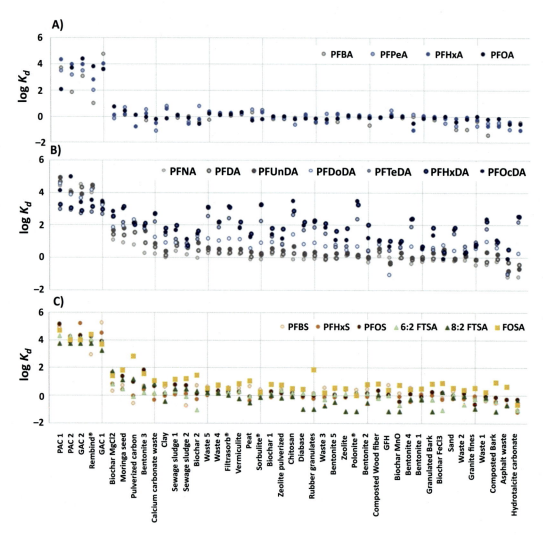

**Fig. 16.4** Mean log $K_d$ values [L/kg] ($n = 2$) for sorption of PFAS to different adsorbent materials ($n = 44$). (A) Short-chain $C_3$-$C_7$ PFCAs, (B) long-chain $C_8$-$C_{17}$ PFCAs, and (C) PFSAs, FTSAs, and FOSA. The adsorbent materials are listed from highest to lowest mean log $K_d$ of all PFAS (left to right on the x-axis) (Sörengård et al., 2020).

paper mill waste at 500°C using a BEST energies daisy reactor (Cashton, WI, USA). Three other biochars modified with $MgCl_2$ (66.1 m²/g surface area, 60% C, derived by slurring agricultural crop residues feedstock in $MgCl_2$, slow pyrolysis at 600°C, "Ecoera," Sweden), MnO [463.1 m²/g surface area, 0.022 cm³/g pore volume, 78.95% C, Loblolly pine (*Pinus taeda*) wood slurried in $MnCl_2 \cdot 4H_2O$ and pyrolyzed in a tube furnace (10°C/min, 600°C, holding for 1 h) under $N_2$] and $FeCl_3$ [99.9 m²/g surface area, 48.4%

C, commercial biochar manufactured in Sweden, "Skogens Kol" (80% birchwood (*Betula* sp.), and 20% Norway spruce (*Picea abies*), by slow pyrolysis (380–430°C) and subsequent chemical co-precipitation of magnetite] were also tested. Among the materials tested, $MgCl_2$-treated biochar was the best non-AC sorbent for the sorption of PFAS. This can be attributed to the $MgCl_2$-treated biochar's high organic carbon content. Also, residual $Mg^{2+}$ on the biochar surface can coordinate with the PFAS carboxylic and sulfonic groups and further increase the PFAS uptake. Biochars' origin and their specific surface area are critical for PFAS sorption efficiency to biochar (Guo et al., 2017).

Sawdust derived biochar (SDN600) (395.51 $m^2$/g surface area, 87.08C%) and red mud modified biochar (RMSDN600) (120.65 $m^2$/g surface area, 11.85C%) were synthesized by (1) pyrolysis of sawdust (~2 mm) at 600°C for 2 h (under 100 mL/min $N_2$ flow), (2) ball-milling sawdust with red mud was pyrolyzed under same conditions (Hassan et al., 2020). Red mud particles clog the biochar pores, resulting in a smaller surface area. Because of paramagnetic/diamagnetic $FeOH \cdot 4H_2O$, $TiO_2$, and $Al_2O_3$ mineral phases, RMSDN600 does not perform well as a magnetic sorbent, despite the tiny amounts of magnetic $Fe_3O_4$ detected in XANES analysis. However, these mineral phases have oxyanion and heavy metal removal capabilities that enable the simultaneous sorption of PFOS from complex water systems. SDN600 and RMSDN600 displayed PFOS ($C_0 = 0.5$–325 mg/L, pH 3.1, and 24 h) adsorption capacities of 178.1 mg/g and 194.6 mg/g, respectively. They were comparable to PFOS sorption capacities reported for GAC, PAC, and clay kaolinite which has similar sorption phases to SDN600 and RMSDN600 (Hassan et al., 2020).

Magnetic materials are appealing in water remediation because the exhausted adsorbent can be retrieved using an external magnet in batch sorption (Yi et al., 2020). *Calotropis gigantea* fibers (CGFs) were charred (under Ar) for 4 h at 400°C, 500°C, 600°C, 700°C, and 800°C (5°C/min) in a pipe furnace (Niu et al., 2020). Biochar prepared at 600°C had the highest PFOA (~50 mg/g) and PFOS (~77 mg/g) adsorption capacities. CGF and $Fe_3O_4$ nanoparticles (NPs) (prepared using $FeSO_4 \cdot 7H_2O$ and NaOH treatment) were mixed at 1:3 (w/w) ratio and heated at 400°C (5°C/min) for 4 h (under Ar) to obtain magnetite/magnetic carbonized CGF (MC-CGF). On MC-CGF, PFOA and PFOS ($C_0 = 25$–250 mg/L and pH 3) had sorption capacities of 204.7 and 195.5 mg/g, respectively. In addition, PFBA, PFBA, PFHxS, and PFDA were sorbed under the same conditions, and their capacities were 25, 27.5, 37.5, and 52.5 mg/g, respectively (Niu et al., 2020).

Alkali treatment to biochar can manipulate its surface structure and porosity. A three-dimensional (3D) hierarchically microporous biochars (HMBs) were prepared using discarded coconut shells (Zhou et al., 2021). Briefly, coconut shells were treated with 0, 0.6, 1.2, and 2.4 M KOH and pyrolyzed at 900°C to obtain HMB900-0, HMB900-0.6, HMB900-1.2, and HMB900-2.4, respectively. The HMB surface area increased from 462 (HMB900-0) to 1322 (HMB900-2.4) m$^2$/g, and the water contact angles increased from 30.3° to 129.1°. High KOH concentrations increase nucleophilic and decomposition reactions of biochar ether groups, and other impurities result in decreasing the oxygen-containing groups and improving HMB hydrophobicity (Zhou et al., 2021). HMB900-2.4 displayed high PFOA ($C_0 = 100$ mg/L) uptake (423 mg/g) with considerably fast (30 min) kinetics, while HMB900-0, HMB900-0.6, and HMB900-1.2 capacities were below 100 mg/g. HMB900-2.4 had an expectational Langmuir sorption capacity of 1269 mg/g for PFOA (0.04–300 mg/L), attributed to their high hydrophobicity and surface area (Zhou et al., 2021).

Pinyon Pine Juniper (PJ) wood char (73% C) and its KOH-activated analogue (67.5–68.5% C) were tested to treat PFOA effluent to obtain irrigation water (Yanala and Pagilla, 2020). Biochar was produced in a transportable metal kiln with mixtures of pinyon pine (*Pinus monophylla*) and juniper (*Juniperus osteosperma*) upon slow pyrolysis. PFOA ($C_0 \sim 3.7$ µg/L) Thomas model capacity from the continuous flow experiments for native biochar was 31.5 µg/g. However, both KOH-activated analogue and commercially available GAC outperformed native biochar (both individually and together), and their breakthroughs were not observed.

## 1.3 Short-chain PFAS removal

Recent regulations and restrictions on using long-chain PFAS have resulted in a significant shift in the industry toward short-chain alternatives (Ateia et al., 2019). However, short-chain PFAS removal is equally important and not thoroughly studied. Reed straw-derived biochar (RESCA) exhibited exceptional removal efficiencies (>92%) toward short-chain PFBA and PFBS at a low environment-relevant concentration of 1 µg/L, which was over six times greater than Calgon GAC (SA 1120 m$^2$/g) (Fig. 16.5A) (Liu et al., 2021b). Common reeds (*Phragmites australis*) were charred in an electric dual-zone split tube furnace at 500°C, 700°C, and 900°C to obtain RESCA-500 (SA 280 m$^2$/g), RESCA-700 (SA 387 m$^2$/g), and RESCA-900 (SA 731 m$^2$/g). The rapid short-chain PFAS sorption was associated with the scattered mesopore (2–10 nm in diameter) distribution in RESCA-900.

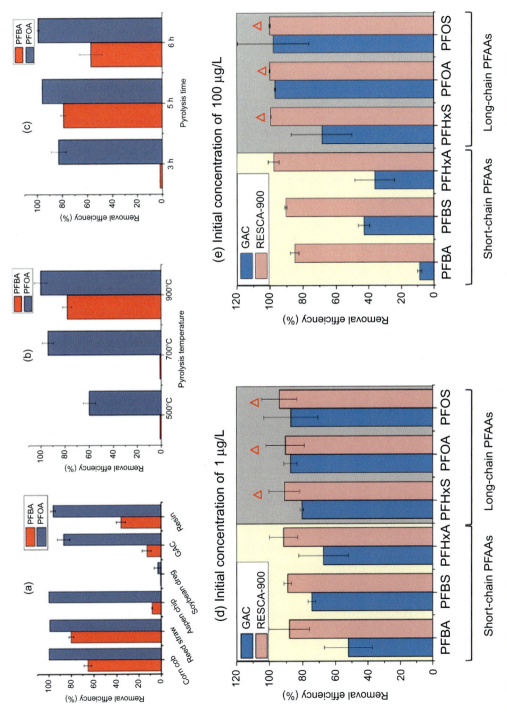

**Fig. 16.5** (A) PFBA and PFOA removal efficiencies exhibited by biochars synthesized with different feedstock materials, (B) PFBA and PFOA removal efficiencies exhibited by RESCAs, (C) PFBA and PFOA removal efficiencies exhibited by RESCA-900 for different pyrolysis durations (PFAS $C_0 = 100\,\mu g/L$, 0.2 g/L doses, and 24 h), removal efficiencies of short-chain and long-chain PFAAs by RESCA-900 and GAC (D) $C_0 = 1\,\mu g/L$ and (E) $C_0 = 100\,\mu g/L$. Red halo triangles on the top of columns represent that concentrations of PFAAs were reduced to below MDLs (i.e., 30–60 ng/L) after adsorption (Liu et al., 2021b).

GAC removed 87±6% of PFOA but was poor in adsorbing PFBA (13±4%) from 100 μg/L solutions. Biochars synthesized with carbon-rich feedstocks (i.e., corn cobs, reed straws, or aspen chips) showed nearly 100% PFOA removal (Fig. 16.5A) (Liu et al., 2021b), implying the vital contribution of surface hydrophobicity. Notably, RESCA-900 showed an exceptionally high PFBA removal efficiency of 80±2%.

An increase in pyrolysis temperature is critical to enhancing the PFAS adsorption by RESCAs. Adsorption of PFBA was minimal (<5%) for RESCA-500 and RESCA-700 (Fig. 16.5B) (Liu et al., 2021b). Notable PFBA removal was only observed for RESCA-900. In contrast, PFOA adsorption increased from 60% to 94% when going from RESCA-500 to RESCA-700. In addition, PFOA removal was complete in RESCA-900. Thus, pyrolysis at 700°C may be sufficient to generate biochars effective for removing long-chain PFAS. In comparison, higher pyrolysis temperatures (900°C) are required for producing biochars for short-chain PFAS removal.

Pyrolysis duration (from 3 to 6 h) greatly affected RESCAs' PFBA adsorption (Fig. 16.5C) (Liu et al., 2021b). RESCA900 showed a parabolic PFBA removal trend maximizing at 5 h. However, adsorption of PFOA was enhanced in longer time pyrolyzed chars, probably due to the increased biochar hydrophobicity at high pyrolytic temperatures. RESCA900 was superior versus GAC for PFBA, PFBS, PFHxA, PFHxS, PFOA, PFOS at two initial concentrations (1 and 100 μg/L) used in batch sorption (Fig. 16.5D and E) (Liu et al., 2021b). RESCA900 removed 92%–96% of three short-chain PFAAs (i.e., PFBA, PFBS, and PFHxA) significantly greater than GAC (71%–79%). In addition, RESCA-900 reduced three long-chain PFAS below the 30–60 ng/L method detection limit (MDL). This suggests RESCA-900's potential as an adsorbent alternative that can effectively remove PFOS and PFOA at environment-relevant concentrations to meet the stringent 70 ng/L EPA health advisory limit.

At $C_0 = 100$ μg/L, short-chain PFAS removal was 80%–89%, and long-chain PFAS were removed entirely (Fig. 16.5E). GAC had poor performance with the following trend PFOS (89%) > PFHxS (76%) ≥ PFOA (75%) > PFBS (43%) ≥ PFHxA (38%) > PFBA (18%). The PFBA removal by hardwood sawdust biochar was 87% at 1 μg/L with slow kinetics (3 h) (Inyang and Dickenson, 2017). In contrast, RESCA-900 can remove 92±1% of PFBA within 24 h. Overall, RESCA represents a green adsorbent alternative for the feasible and scalable treatment of a broad spectrum of PFAAs of different chain lengths and functional moieties.

Waste wood timber commercial Douglas fir biochar (459 m$^2$/g surface area) produced in a proprietary process (900°C, 1–30 s resident time, Biochar Supreme, Everson, WA) removed PFAS

($C_0 = 1000\,\mu g/L$) in 12–48 h (Zhang et al., 2021a). PFAS sorption capacity was chain-length dependent, with the following order: PFOS (70.42 ± 21.5 µmol/g) > PFOA (52.08 ± 14.8 µmol/g) > PFBA (48.31 ± 12.2 µmol/g) > PFBS (23.36 ± 7.4 µmol/g).

## 1.4 Sorption benchmarks

### 1.4.1 pH dependence on PFAS uptake

PFAS sorption is known to be primarily governed by electrostatic effects, ion-exchange, and hydrophobic effects (Zhang et al., 2019). It is critical to investigate the effect of solution pH on PFAS adsorption because electrostatic and ion exchange effects can change as the surface charges of biochar change with pH. Most PFAS compounds are acidic due to the strong electron-withdrawing effects of the fluorine atom in the C—F backbone. For example, PFOS and PFOA are strong acids with $pK_a$ values of −3.27 and −0.2, respectively (Table 16.1). They remain negatively charged in water from pH 1 to 14.

Corn straw biochar (BC700) ($pH_{PZC} = 10.3$) displayed a slight drop in PFOS sorption capacity from 63 to 55 mg/g when the solution pH increased from 3 to 10 (Guo et al., 2019). BC700 is positively charged below pH 10.3. Thus, at lower pH, a robust electrostatic attraction existed between the positively charged biochar surface and the negatively charged PFOS, resulting in a relatively high adsorption affinity by forming PFOS aggregates. At pH values higher than the $pH_{PZC}$ surface, functional groups get deprotonated leading to electrostatic repulsion between the surface and negatively charged PFOS. However, the protonation and deprotonation not only depend on the PZC of the materials but also depend on the individual $pK_a$ values of each existing adsorbent functional group (Guo et al., 2017).

Sawdust-derived biochar (SDN600) ($pH_{PZC} = 6.51$), ball-milled SDN600, and red mud (RMSDN600) ($pH_{PZC} = 10.04$) have shown a similar PFOS sorption behavior with increasing solution pH (Hassan et al., 2020). At pH 3.1, the PFOS adsorption values on SDN600 and RMSDN600 were 178.1 mg/g and 194.6 mg/g, respectively. At pH 10.2, they have dropped to 124 mg/g and 132 mg/g for SND600 and RMSDN600, respectively. RMSDN600 has a higher PFOS adsorption capacity versus SDN600 due to the high degree of electrostatic attractions posed by positively charged -$FeOH_2^+$, -$SiOH_2^+$, -$TiOH_2^+$, and -$AlOH_2^+$ surface functionalities (Hassan et al., 2020). Robust PFOS adsorption in both SDN600 and RMSDN600 over a wide pH window has attributed to hydrophobic interactions between the PFOS -CF backbone and biochar surface insignificantly related to solution pH.

PFOA, PFOS, PFBA, and PFBS sorption on both GAC ($pH_{PZC} = 7.2$) and Douglas fir biochar ($pH_{PZC} = 6.9$) decreased by 27.5%, 32.0%, 21.1%, and 25.5%, respectively, when pH increased from 3 to 9 (Zhang et al., 2021a). At pH 7, the surface of GAC would be positively charged and the biochar would exhibit a negative charge. With decreasing solution pH, the electrostatic attraction between anionic PFAS and positively charged adsorbents increased, resulting in a significant increase in PFAS sorption and vice versa. Hydrophobic interactions should have played an important role, especially when PFAS were sorbed onto negatively charged GAC and biochar at basic pH (Zhang et al., 2021a).

Hierarchically, microporous biochar's (HMB900-2.4) ($pH_{PZC} = 6.9$) PFOA sorption capacities and rates dropped with increasing pH (Zhou et al., 2021). About 80% PFOA was adsorbed in the pH range of 2–3. The removal percentage (20%–30%) and rates dropped when the pH increased to 10, due to the weakened electrostatic interactions with the surface charge changes.

The PFOA/PFOS adsorption capacities on carbonized magnetite/magnetic *Calotropis gigantea* fiber (MC-CGF) ($pH_{PZC} = 3.0$) decreased monotonically with increasing pH (Niu et al., 2020). MC-CGF is negatively charged over pH 3–10. The electrostatic repulsions would prevent PFOA and PFOS from diffusing onto the MC-CGF surface (Niu et al., 2020). However, MC-CGF still had considerable adsorption capacities for PFOA and PFOS at pH 3–7, suggesting that electrostatic interactions were not the key PFAS uptake mechanism.

Adsorption at environmentally relevant pH (i.e., near neutral) is essential. It can be accomplished by leveraging the PFAS and sorbent hydrophobic interactions. However, electrostatic effects may still have an impact on PFAS diffusion to the biochar surface. Many of the studies discussed in this book chapter have not investigated the effect of pH on sorption. Those experiments either buffered the pH at 7 or did not change the pH at all. For example, PFAS containing effluent at pH 7.9 was treated using Pinyon Pine Juniper (PJ) biochar to produce irrigation water (Yanala and Pagilla, 2020). Batch experiments for sorbent screening and kinetics were conducted using AFFF-impacted site water with a pH of 7.5 for biochars and GAC (Guo et al., 2017). Nevertheless, pH optimization is critical for achieving the material's maximum sorption capacities.

### 1.4.2 Adsorption kinetics

Adsorption kinetic data fittings are important in determining whether adsorption occurs due to chemisorption, physisorption, or other phenomena like intra-particle diffusion (Wu et al., 2020).

The pseudo-first-order (PFO) (Corbett, 1972) and second-order (PSO) (Brian et al., 1961) models frequently described PFAS sorption (Zhang et al., 2019). For instance, the PSO model fit well ($R^2 > 0.82$) for the sorption of PFBA, PFBS, PFHxA, PFHxS, PFOA, and PFOS on reed straw-derived biochar (RESCA) (Liu et al., 2021b). PSO constants ($k_2$) spanned from 1.37 to 8.38 (L/mgh). The equilibrium is achieved faster (8–12 h) for short-chain ($C_0 = 100$ µg/L) PFAS versus long-chains PFAS (20–24 h). This was an intriguing observation because removing short-chain PFAS is difficult because their hydrophobicity is lower than that of longer-chain PFAS, reducing the degree of hydrophobic interactions with the biochar surface.

PFOA, PFOS, PFBA, PFBS sorption onto Douglas fir biochar followed the PSO kinetics with $R^2 > 0.98$. Conversely, the PFO model fit well ($R^2 = 0.92$–0.93) versus the PSO model ($R^2 = 0.82$–0.87) for PFOA and PFOS sorption on magnetic carbonized *Calotropis gigantea* fiber (MC-CGF). The PFOS sorption rate ($k_1 = 0.01753$ min$^{-1}$) was lower than that of PFOA ($k_1 = 0.04347$ min$^{-1}$), and this was attributed to the longer C—F chain length of PFOS which required more time to diffuse to the same sorption site (Niu et al., 2020). PFOS adsorption on corn straw (Guo et al., 2017), sawdust (Hassan et al., 2020), and sawdust/red mud (Hassan et al., 2020) biochars followed PSO kinetics with $R^2 > 0.99$. Gasified (at 900°C) steam-activated pinewood (PWC) and hardwood sawdust (HWC) gave high $R^2$ (0.8–0.9) PSO fits for PFOA sorption (Inyang and Dickenson, 2017). In addition to PFO and PSO models, the Elovich (Aharoni and Tompkins, 1970) and intra-particle diffusion (IPD) (Boyd et al., 1947) models have been applied in PFAS kinetic data treatment. The Elovich model assumes that the sorbent surface is energetically heterogeneous, while the IPD model assumes that the external diffusion and boundary layer diffusion are negligible; intraparticle diffusion is the only rate-controlling step, which is usually valid for the well-mixed solution. Corn straw (Guo et al., 2019), willow (W400) (Chen et al., 2011), and maize straw (M400) (Chen et al., 2011) biochars gave good Elovich fits ($R^2 = 0.88$–0.98) for PFOS sorption. PFOA sorption on both gasified hardwood sawdust (HWC) and steam-activated pinewood (PWC) gave the best fits for both Elovich ($R^2 = 0.90$–0.97) and IPD models ($R^2 = 0.98$–0.99) (Inyang and Dickenson, 2017). Moreover, PFOS uptake kinetic data on maize straw and willow chars fitted to IPD (Chen et al., 2011). However, the IPD model plots did not go through the origin, suggesting that IPD is not the only rate limiting step (Chen et al., 2011).

Table 16.2 summarizes PFAS adsorption kinetic data obtained for various biochars and their composites.

### 1.4.3 Sorption isotherms and thermodynamics

The use of isotherm models may be misleading since more than one layer of PFAS can be adsorbed. If this occurs, the adsorbent surface is no longer playing the key chemical role in adsorbing the additional layer(s) of PFAS. The possibility of a PFAS multilayer sorption has been suggested in previous studies to occur through hemimicelle or micelle formation on carbonaceous surfaces (Zhang et al., 2019). This is expected because PFAS are surfactants since they exist as deprotonated anions in water with exceedingly hydrophobic tails. PFAS are also known to have unusual partitioning characteristics compared to hydrocarbons. For example, attempts to measure the octanol–water partitioning coefficient for PFOS have been restricted by forming the third "fluoro"-phase (Zhao et al., 2011). In addition, PFAS sorption during sonication onto granular activated carbon showed multilayer BET sorption behavior in a previous study (Zhao et al., 2011).

Despite the inaccuracy of isotherm models to predict sorption capacities, many studies determined maximum PFAS uptake capacities using Langmuir (Langmuir, 1918) and Freundlich (Freundlich, 1907) models (Table 16.3). PFOS sorption on corn straw chars (BC250, BC400, BC550, and BC700) was thought to follow monolayer Langmuir ($R^2 = 0.95$–0.96) adsorption (Guo et al., 2017). The Freundlich model also gave reasonable fits ($R^2 = 0.93$–0.97). Maize and willow chars fitted well to both Langmuir and Freundlich models with high $R^2$ values (0.98–0.99) (Chen et al., 2011). Gasified hardwood sawdust (HWC), steam activated gasified pine wood (PWC) displayed $R^2$ values >0.95 for both Langmuir and Freundlich models for both PFOS and PFOA (Inyang and Dickenson, 2017). The Freundlich sorption intensity ($n$) spanned from 0.29 to 0.33 (Inyang and Dickenson, 2017). When $n \sim 1$, partitioning is sought to be the dominant sorption mechanism. The low $n$ values suggest that other interactions, including hydrophobic interactions, may have existed. The Redlich-Peterson (RP) isotherm model was used to assess PFOS and PFOA uptake capacities on magnetic carbonized *Calotropis gigantea* fiber (MC-CGF) as the Langmuir and Freundlich $R^2$ values were considered to be low at 0.85–0.95. The RP model fits had a high $R^2$ (0.96–0.98) value. The sorption process becomes more heterogeneous with each subsequent layer of PFAS adsorption, and PFAS micelle formation may influence uptake, casting doubt on the use of these empirical isotherms.

Table 16.2 PFAS sorption kinetic data for biochars and engineered biochar

| Biochar adsorbent | PFAS type | Concentration ($C_0$) | pH | Equilibrium time | Pseudo-first order (PFO) | Pseudo-second order (PSO) | Elovich | Reference |
|---|---|---|---|---|---|---|---|---|
| Maize straw char (M400) | PFOS | 100 mg/L | 7.0 (phosphate buffer) | 16 days | — | $q_e = 41.1$ (mg/g) $k_2 = 7.74 \times 10^{-4}$ (mg/h g) $R^2 = 0.94$ | $a = 25.2$ (mg/hg) $b = 0.212$ (g/mg) $R^2 = 0.92$ | Chen et al. (2011) |
| Willow sawdust char (W400) | | | | | | $q_e = 70.4$ (mg/g) $k_2 = 6.04 \times 10^{-4}$ (mg/hg) $R^2 = 0.94$ | $a = 53.2$ (mg/hg) $b = 0.119$ (g/mg) $R^2 = 0.95$ | |
| Gasified hardwood sawdust (HWC) | PFBA | 1 μg/L | 7.2 (phosphate buffer) | 24 h | $q_e = 0.58$ (μg/g) $k_1 = 2.1$ (d$^{-1}$) $R^2 = 0.50$ | $q_e = 0.6$ (μg/g) $k_2 = 3.5$ (μg/dg) $R^2 = 0.48$ | $a = 6.6$ (μg/dg) $b = 10.8$ (μg/g) $R^2 = 0.42$ | Inyang and Dickenson (2017) |
| | PFOA | | | 7 days | $q_e = 2.2$ (μg/g) $k_1 = 0.27$ (d$^{-1}$) $R^2 = 0.99$ | $q_e = 2.4$ (μg/g) $k_2 = 0.1$ (μg/dg) $R^2 = 0.98$ | $a = 1.7$ (μg/dg) $b = 2.0$ (μg/g) $R^2 = 0.97$ | |
| Gasified steam-activated pinewood (PWC) | PFBA | | | 24 h | $q_e = 0.36$ (μg/g) $k_1 = 4.5$ (d$^{-1}$) $R^2 = 0.82$ | $q_e = 2.9$ (μg/g) $k_2 = 0.009$ (μg/dg) $R^2 = $ N/A | $a = 237$ (μg/dg) $b = 51.2$ (μg/g) $R^2 = $ N/A | |
| | PFOA | | | 7 days | $q_e = 2.0$ (μg/g) $k_1 = 1.6$ (d$^{-1}$) $R^2 = 0.57$ | $q_e = 2.1$ (μg/g) $k_2 = 1.2$ (μg/dg) $R^2 = 0.8$ | $a = 59$ (μg/dg) $b = 3.8$ (μg/g) $R^2 = 0.90$ | |
| Corn straw-derived biochar BC250 | PFOS | 10 mg/L | 7 (0.01 mol/L CaCl$_2$, 200 mg/L NaN$_3$) | 15 h | $q_e = 49.0$ (mg/g) $k_1 = 2.09$ (h$^{-1}$) $R^2 = 0.95$ | $q_e = 50.96$ (mg/g) $k_2 = 0.07$ (g/mg h) $R^2 = 0.99$ | $a = 2.78 \times 10^5$ (mg/hg) $b = 3.53$ (mg/g) $R^2 = 0.98$ | Guo et al. (2017) |

*Continued*

Table 16.2 PFAS sorption kinetic data for biochars and engineered biochar—cont'd

| Biochar adsorbent | PFAS type | Concentration ($C_0$) | pH | Equilibrium time | Pseudo-first order (PFO) | Pseudo-second order (PSO) | Elovich | Reference |
|---|---|---|---|---|---|---|---|---|
| Corn straw-derived biochar BC400 | | | | | $q_e = 51.27$ (mg/g) $k_1 = 2.07$ (h$^{-1}$) $R^2 = 0.91$ | $q_e = 53.66$ (mg/g) $k_2 = 0.06$ (g/mg h) $R^2 = 0.96$ | $a = 9.44 \times 10^4$ (mg/h g) $b = 4.07$ (mg/g) $R^2 = 0.98$ | Zhang et al. (2021a) |
| Corn straw-derived biochar BC550 | | | | | $q_e = 54.24$ (mg/g) $k_1 = 1.70$ (h$^{-1}$) $R^2 = 0.88$ | $q_e = 57.10$ (mg/g) $k_2 = 0.05$ (g/mg h) $R^2 = 0.95$ | $a = 1.78 \times 10^4$ (mg/h g) $b = 5.04$ (mg/g) $R^2 = 0.97$ | |
| Corn straw-derived biochar BC700 | | | | | $q_e = 57.42$ (mg/g) $k_1 = 1.54$ (h$^{-1}$) $R^2 = 0.91$ | $q_e = 60.25$ (mg/g) $k_2 = 0.04$ (g/mg h) $R^2 = 0.97$ | $a = 1.72 \times 10^4$ (mg/h g) $b = 5.33$ (mg/g) $R^2 = 0.97$ | |
| Douglas fir biochar | PFOA | 1000 µg/L | 7 | 48 h | – | $q_e = 29.33$ (µmol/g) $k_2 = 0.0099$ (g/µmol h)[73] $R^2 = 0.99$ | | |
| | PFOS | | | | | $q_e = 36.36$ (µmol/g) $k_2 = 0.0116$ (g/µmol h)[73] $R^2 = 0.99$ | | |
| | PFBA | | | 24 h | | $q_e = 13.52$ (µmol/g) $k_2 = 0.0340$ (g/µmol h)[73] $R^2 = 0.99$ | | |

| Adsorbent | PFAS | Concentration | Conditions | Time | Parameters | | References |
|---|---|---|---|---|---|---|---|
| | PFBS | | | 12 h | $q_e = 18.66$ (µmol/g) $k_2 = 0.0164$ (g/µmol h)[73] $R^2 = 0.99$ | | |
| Reed straw-derived biochar (RESCA) | PFBA | 100 µg/L | — | 20–24 h | $k_2 = 1.13$ (L/mg h) $R^2 = 0.98$ | — | Liu et al. (2021b) |
| | PFBS | | | | $k_2 = 1.23$ (L/mg h) $R^2 = 0.85$ | | |
| | PFHxA | | | | $k_2 = 2.40$ (L/mg h) $R^2 = 0.89$ | | |
| | PFHxS | | | 8–12 h | $k_2 = 6.59$ (L/mg h) $R^2 = 0.97$ | | |
| | PFOA | | | | $k_2 = 9.61$ (L/mg h) $R^2 = 0.95$ | | |
| | PFOS | | | | $k_2 = 8.38$ (L/mg h) $R^2 = 0.97$ | | |
| Corn straw biochar | PFOS | 2 mg/L | 7 (0.01 mol/L $CaCl_2$, 200 mg/L $NaN_3$) | 24 h | $q_e = 1040$ (µg/g) $k_1 = 2.28$ (min$^{-1}$) $R^2 = 0.79$ | $q_e = 1148$ (µg/g) $k_2 = 0.017$ (g/mg min) $R^2 = 0.99$ | Guo et al. (2019) |
| Sediment +2% Corn straw biochar | | | | 36 h | $q_e = 7.50$ (µg/g) $k_1 = 0.26$ (min$^{-1}$) $R^2 = 0.96$ | $q_e = 8.43$ (µg/g) $k_2 = 0.566$ (g/mg h) $R^2 = 0.99$ | |
| Sediment +5% Corn straw biochar | | | | | $q_e = 10.6$ (µg/g) $k_1 = 0.34$ (min$^{-1}$) $R^2 = 0.94$ | $q_e = 11.4$ (µg/g) $k_2 = 0.412$ (g/mg h) $R^2 = 0.99$ | |

*Continued*

Table 16.2 PFAS sorption kinetic data for biochars and engineered biochar—cont'd

| Biochar adsorbent | PFAS type | Concentration ($C_0$) | pH | Equilibrium time | Pseudo-first order (PFO) | Pseudo-second order (PSO) | Elovich | Reference |
|---|---|---|---|---|---|---|---|---|
| Coconut shell microporous biochar (HMB900-2.4) | PFOA | 100 mg/L | 3.8 | 30 min | — | $q_e = 423$<br>$R^2 = 0.99$ | — | Zhou et al. (2021) |
| Magnetic carbonized *Calotropis gigantea* fiber (MC-CGF) | PFOA | 50 mg/L | 3.0 | 1 h | $q_e = 41.85$ (mg/g)<br>$k_1 = 0.04347$ (min$^{-1}$)<br>$R^2 = 0.94$ | $q_e = 45.60$ (mg/g)<br>$k_2 = 1.440 \times 10^{-3}$ (g/mg h)<br>$R^2 = 0.83$ | | Niu et al. (2020) |
| | PFOS | | | 2 h | $q_e = 52.05$ (mg/g)<br>$k_1 = 0.01753$ (min$^{-1}$)<br>$R^2 = 0.92$ | $q_e = 63.89$ (mg/g)<br>$k_2 = 2.699 \times 10^{-4}$ (g/mg h)<br>$R^2 = 0.87$ | | |
| Sawdust derived biochar (SDN600) | PFOS | 248.48 mg/L | 5 | 12 h | $q_e = 5.20$ (mg/g)<br>$k_1 = 0.001$ (h$^{-1}$)<br>$R^2 = 0.26$ | $q_e = 185.19$ (mg/g)<br>$k_2 = 0.07$ (g/mg h)<br>$R^2 = 0.99$ | — | Hassan et al. (2020) |
| red mud modified Biochar (RMSDN600) | | | | | $q_e = 5.10$ (mg/g)<br>$k_1 = 0.001$ (h$^{-1}$)<br>$R^2 = 0.09$ | $q_e = 188.67$ (mg/g)<br>$k_2 = 0.56$ (g/mg h)<br>$R^2 = 1.00$ | | |

Table 16.3 PFAS sorption isotherm data for biochars and engineered biochar.

| Biochar adsorbent | PFAS type | Concentration range ($C_0$) | Temperature (°C) | Isotherm parameters | | Reference |
|---|---|---|---|---|---|---|
| | | | | Langmuir | Freundlich | |
| Maize straw char (M400) | PFOS | 1–500 mg/L | 25 | $q_m = 164$ (mg/g) $b = 0.011$ (L/mg) $R^2 = 0.99$ | $K = 7.27$ $(mg^{(1-n)}L^n g^{-1})$ $n = 0.459$ $R^2 = 0.99$ | Chen et al. (2011) |
| Willow sawdust char (W400) | | | | $q_m = 91.6$ (mg/g) $b = 0.010$ (L/mg) $R^2 = 0.99$ | $K = 5.23$ $(mg^{(1-n)}L^n g^{-1})$ $n = 0.492$ $R^2 = 0.99$ | |
| Gasified hardwood sawdust (HWC) | PFOA | 0.01–100,000 µg/L (wastewater) | 22 | $K = 0.31$ (L/mg) $S_{max} = 31.7$ (mg/g) $R^2 = 0.99$ | $K_F = 8.5$ $(mg^{(1-n)}L^n g^{-1})$ $n = 0.32$ $R^2 = 0.98$ | Inyang and Dickenson (2017) |
| | | 0.01–100,000 µg/L (Lake Mead) | | $K = 0.35$ (L/mg) $S_{max} = 41.2$ (mg/g) $R^2 = 0.99$ | $K_F = 11.5$ $(mg^{(1-n)}L^n g^{-1})$ $n = 0.32$ $R^2 = 0.99$ | |
| Gasified steam-activated pinewood (PWC) | | 0.01–100,000 µg/L (wastewater) | | $K = 0.52$ (L/mg) $S_{max} = 27.7$ (mg/g) $R^2 = 0.99$ | $K_F = 8.7$ $(mg^{(1-n)}L^n g^{-1})$ $n = 0.29$ $R^2 = 0.95$ | |
| | | 0.01–100,000 µg/L (Lake Mead) | | $K = 0.35$ (L/mg) $S_{max} = 41.3$ (mg/g) $R^2 = 0.99$ | $K_F = 11.2$ $(mg^{(1-n)}L^n g^{-1})$ $n = 0.33$ $R^2 = 0.98$ | |
| Corn straw-derived biochar BC250 | PFOS | 0.5–10.0 mg/L | 35 | $q_m = 135.53$ (mg/g) $K_L = 0.67$ (L/mg) $R^2 = 0.95$ | $K_F = 58.61$ $(mg^{(1-n)}L^n g^{-1})$ $1/n = 0.83$ $R^2 = 0.93$ | Guo et al. (2017) |
| Corn straw-derived biochar BC400 | | | | $q_m = 146.52$ (mg/g) $K_L = 0.75$ (L/mg) $R^2 = 0.96$ | $K_F = 68.99$ $(mg^{(1-n)}L^n g^{-1})$ $1/n = 0.82$ $R^2 = 0.94$ | |

Continued

Table 16.3 PFAS sorption isotherm data for biochars and engineered biochar—cont'd

| Biochar adsorbent | PFAS type | Concentration range ($C_0$) | Temperature (°C) | Isotherm parameters | | Reference |
|---|---|---|---|---|---|---|
| | | | | Langmuir | Freundlich | |
| Corn straw-derived biochar BC550 | | | | $q_m = 166.42$ (mg/g) $K_L = 0.82$ (L/mg) $R^2 = 0.98$ | $K_F = 83.14$ (mg$^{(1-n)}$L$^n$g$^{-1}$) $1/n = 0.81$ $R^2 = 0.97$ | |
| Corn straw-derived biochar BC700 | | | | $q_m = 169.30$ (mg/g) $K_L = 1.02$ (L/mg) $R^2 = 0.97$ | $K_F = 97.82$ (mg$^{(1-n)}$L$^n$g$^{-1}$) $1/n = 0.79$ $R^2 = 0.95$ | |
| Douglas fir biochar | PFOA | 100–1000 µg/L | 25 | $q_m = 52.08 \pm 14.8$ (µmol/g) $b = 0.99 \pm 0.31$ (L/µmol) $R^2 = 0.92$ | $K_F = 26.22 \pm 6.7$ (µmol$^{(1-n)}$L$^n$g$^{-1}$) $1/n = 0.68 \pm 0.14$ $R^2 = 0.97$ | Zhang et al. (2021a) |
| | PFOS | | | $q_m = 70.42 \pm 21.5$ (µmol/g) $b = 0.01 \pm 0.01$ (L/µmol) $R^2 = 0.94$ | $K_F = 52.13 \pm 13.9$ (µmol$^{(1-n)}$L$^n$g$^{-1}$) $1/n = 0.67 \pm 0.05$ $R^2 = 0.94$ | |
| | PFBA | | | $q_m = 48.31 \pm 12.2$ (µmol/g) $b = 0.19 \pm 0.06$ (L/µmol) $R^2 = 0.95$ | $K_F = 6.61 \pm 1.2$ (µmol$^{(1-n)}$L$^n$g$^{-1}$) $1/n = 0.82 \pm 0.19$ $R^2 = 0.94$ | |
| | PFBS | | | $q_m = 23.36 \pm 7.4$ (µmol/g) $b = 0.76 \pm 0.23$ (L/µmol) $R^2 = 0.91$ | $K_F = 9.61 \pm 1.4$ (µmol$^{(1-n)}$L$^n$g$^{-1}$) $1/n = 0.69 \pm 0.10$ $R^2 = 0.94$ | |

| | | | | | | |
|---|---|---|---|---|---|---|
| Granular Activated Carbon (GAC) | PFOA | | | $q_m = 86.21 \pm 12.7$ (μmol/g)<br>$b = 1.48 \pm 0.65$ (L/μmol)<br>$R^2 = 0.99$ | $K_F = 82.14 \pm 9.8$ ($\mu mol^{(1-n)} L^n g^{-1}$)<br>$1/n = 0.71 \pm 0.16$<br>$R^2 = 0.97$ | |
| | PFOS | | | $q_m = 123.45 \pm 45.8$ (μmol/g)<br>$b = 4.5 \pm 0.78$ (L/μmol)<br>$R^2 = 0.96$ | $K_F = 121.03 \pm 14.1$ ($\mu mol^{(1-n)} L^n g^{-1}$)<br>$1/n = 0.65 \pm 0.08$<br>$R^2 = 0.88$ | |
| | PFBA | | | $q_m = 23.98 \pm 9.6$ (μmol/g)<br>$b = 0.30 \pm 0.06$ (L/μmol)<br>$R^2 = 0.94$ | $K_F = 5.12 \pm 0.9$ ($\mu mol^{(1-n)} L^n g^{-1}$)<br>$1/n = 0.77 \pm 0.21$<br>$R^2 = 0.94$ | |
| | PFBS | | | $q_m = 31.44 \pm 6.1$ (μmol/g)<br>$b = 0.43 \pm 0.12$ (L/μmol)<br>$R^2 = 0.89$ | $K_F = 8.991 \pm 1.1$ ($\mu mol^{(1-n)} L^n g^{-1}$)<br>$1/n = 0.83 \pm 0.32$<br>$R^2 = 0.91$ | |
| Reed straw-derived biochar (RESCA) | PFBA | 100–250,000 μg/L | $24 \pm 2$ | $q_m = 6.88$ (mg/g)<br>$K_L = 3.25$ (mg/L)<br>$R^2 = 0.96$ | $K_F = 2.05$ ($\mu mol^{(1-n)} L^n g^{-1}$)<br>$1/n = 0.35$<br>$R^2 = 0.98$ | Liu et al. (2021b) |
| | PFBS | | | $q_m = 11.11$ (mg/g)<br>$K_L = 0.3$ (mg/L)<br>$R^2 = 0.87$ | $K_F = 6.02$ ($\mu mol^{(1-n)} L^n g^{-1}$)<br>$1/n = 0.22$<br>$R^2 = 0.96$ | |
| | PFHxA | | | $q_m = 13.52$ (mg/g)<br>$K_L = 3.19$ (mg/L)<br>$R^2 = 0.99$ | $K_F = 3.75$ ($\mu mol^{(1-n)} L^n g^{-1}$)<br>$1/n = 0.42$<br>$R^2 = 0.99$ | |
| | PFHxS | | | $q_m = 15.83$ (mg/g)<br>$K_L = 1.9$ (mg/L)<br>$R^2 = 0.81$ | $K_F = 7.36$ ($\mu mol^{(1-n)} L^n g^{-1}$)<br>$1/n = 0.21$<br>$R^2 = 0.92$ | |

*Continued*

Table 16.3 PFAS sorption isotherm data for biochars and engineered biochar—cont'd

| Biochar adsorbent | PFAS type | Concentration range ($C_0$) | Temperature (°C) | Isotherm parameters Langmuir | Isotherm parameters Freundlich | Reference |
|---|---|---|---|---|---|---|
| | PFOA | | | $q_m = 17.91$ (mg/g) $K_L = 1.23$ (mg/L) $R^2 = 0.94$ | $K_F = 8.41$ ($\mu mol^{(1-n)} L^n g^{-1}$) $1/n = 0.27$ $R^2 = 0.95$ | |
| | PFOS | | | $q_m = 34.62$ (mg/g) $K_L = 1.74$ (mg/L) $R^2 = 0.96$ | $K_F = 11.43$ ($\mu mol^{(1-n)} L^n g^{-1}$) $1/n = 0.3$ $R^2 = 0.92$ | |
| Corn straw biochar | PFOS | 0.2–4 mg/L | 25 ± 0.1 | — | $K_F = 2205$ ($\mu g\,g^{-1}$) ($\mu g\,mL^{-1}$)$^{-n}$ $1/n = 0.476$ $R^2 = 0.87$ | Guo et al. (2019) |
| Sediment +2% Corn straw biochar | | | | | $K_F = 7.29$ ($\mu g\,g^{-1}$) ($\mu g\,mL^{-1}$)$^{-n}$ $1/n = 0.399$ $R^2 = 0.87$ | Zhou et al. (2021) |
| Sediment +5% Corn straw biochar | | | | | $K_F = 11.2$ ($\mu g\,g^{-1}$) ($\mu g\,mL^{-1}$)$^{-n}$ $1/n = 0.381$ $R^2 = 0.78$ | |
| Coconut shell microporous biochar (HMB900-2.4) | PFOA | 0.05 to 100 mg/L | 25 | $q_m = 1269$ mg/g | — | Zhou et al. (2021) |
| Magnetic carbonized *Calotropis gigantea* fiber (MC-CGF) | PFOA | 25–250 mg/L | 30 | $q_m = 204.7$ (mg/g) $b = 0.0118$ (mg/L) $R^2 = 0.93$ | $K = 8.499$ (mg/g) (mg/L)$^n$ $1/n = 0.5453$ $R^2 = 0.85$ | Niu et al. (2020) |
| | PFOS | | | $q_m = 195.5$ (mg/g) $b = 0.0155$ (mg/L) $R^2 = 0.95$ | $K = 11.64$ (mg/g) (mg/L)$^n$ $1/n = 0.4925$ $R^2 = 0.87$ | |

| Sawdust derived biochar (SDN600) | PFOS | 0.5–325 mg/L | 25 ± 1 | $q_m = 178.92 \pm 15.27$ (mg/g) $K_L = 0.006 \pm 0.001$ (mg/L) $R^2 = 0.99$ | Niu et al. (2020) |
|---|---|---|---|---|---|
| | | | | $K_F = 4.242 \pm 0.58$ (mg/g)(mg/L)$^n$ $1/n = 0.59 \pm 0.08$ $R^2 = 0.87$ | Hassan et al. (2020) |
| Red mud modified Biochar (RMSDN600) | | | | $q_m = 168.63 \pm 16.50$ (mg/g) $K_L = 0.009 \pm 0.002$ (mg/L) $R^2 = 0.98$ $K_F = 0.5984 \pm 1.15$ (mg/g)(mg/L)$^n$ $1/n = 0.55 \pm 0.13$ $R^2 = 0.98$ | Hassan et al. (2020) |

Other isotherm models, such as the Sips model, need to be considered when assessing isotherm capacities. The Sips model (Sips, 1948) describes heterogeneous adsorption and overcomes the Freundlich isotherm model limitations associated with rising adsorbate concentrations. The Sips model reduces to the Freundlich isotherm (multilayer coverage) at low concentrations and to the Langmuir isotherm (monolayer coverage) at high concentrations, respectively. Sips parameters are influenced by process conditions such as pH, temperature, and concentration (Sips, 1948).

Temperature may influence PFAS micelle formation and aggregation. Only one study has reported PFAS sorption thermodynamics. Corn straw BC (BC700) was tested for PFOS uptake, varying the sorption temperatures from 15°C to 35°C at pH 7 (Guo et al., 2017). Uptake capacities increased for all initial concentrations ($C_0 = 0.5$, 1.0, 2.0, 6.0, and 10.0 mg/L) showing endothermic sorption ($\Delta H° = 92.26$ kJ/mol). At higher temperatures, less PFAS aggregation may occur. Also, adsorption sites become kinetically more accessible, and monolayer formation could be high. The magnitude of $\Delta H°$ indicated chemisorptive type adsorption by a charge-assisted hydrogen bond (oxygen-containing functional groups). The entropy change was minimal (0.34 kJ/mol/K), but its positivity indicates an increase in randomness at the PFOS-BC700 interface.

### 1.4.4 Regeneration

PFAS were reported to be effectively desorbed using organic solvents (ethanol, methanol, acetone, etc.) (Gagliano et al., 2020). Their hydrophilic anionic head responds to inorganic salt-containing aqueous solutions ($NH_4Cl$, NaOH, and NaCl). PFOA sorbed on hierarchically microporous coconut shell biochar (HMB900-2.4) was stripped with methanol (Zhou et al., 2021). The PFOA removal efficiency was maintained at 65% even after five cycles. After reuse, the surface area remained unchanged, which is evidence of the sorbent stability. PFOS- and PFOA-loaded Magnetite/magnetic carbonized *Calotropis gigantea* fibers (MC-CGF) were also regenerated using methanol after rapidly separating the adsorbent using a NdFeB magnet from the batch sorption solution (Niu et al., 2020). The PFOS and PFOA sorption performance remained above 50% after six cycles. The decrease in adsorption was attributed to MC-CGF mass losses, leaching of $Fe_3O_4$ NPs, etc. The use of inorganic salts for the regeneration of short-chain PFAS has not been reported for biochar, but it is worth investigating.

Adsorption or regeneration will only move PFAS from one matrix to another and do not neutralize toxicity (Liu et al., 2020). Granular $Fe^0$ (ZVI, 0.25–1.19 mm) [Connelly-GPM Inc. (Chicago, IL, USA)] and hardwood (oak)-based biochar (pyrolyzed at 700°C, 99.99% C), Cowboy Charcoal Co. (Brentwood, TN, USA) were tested for PFOA, PFHpA, PFOS, PFHpS, PFHxS, and PFBS adsorption and simultaneous degradation.

The solution $F^-$ contents were tested, and mass balance was used to calculate the degradation percentage. ZVI+BC (at 1:2 BCE: ZVI weight ratio) reactive media removed 60% and 94% of the input PFOA ($C_0 = 18{,}500$ μg/L), and PFOS respectively. Using ZVI alone, less input PFOA (20%) and PFOS (90%) were removed compared to (ZVI+BC). When compared to PFOA and PFOS, short-chain PFCAs and PFSAs were less effectively removed. Both ZVI alone and (ZVI+BC) removed approximately 17% PFHpA ($C_0 = 26$ μg/L). The combination of the two reactive media removed approximately 60%–70% of PFHpS (330 μg/L), 30%–40% of the PFHxS (13 μg/L), and 20% of the input PFBS (6 μg/L). The removal of short-chain PFCAs and PFSAs by ZVI and (ZVI+BC) decreased as chain length decreased. This proposed method could be an environmentally sustainable alternative versus advanced oxidation, reduction, and incineration (Kucharzyk et al., 2017). It requires less energy, is relatively efficient, and is co-effective. However, the production of undesired gaseous products such as $CHF_3$ and ZVI corrosion in $H_2O$ to iron oxides ($Fe_3O_4$ and $Fe_2O_3$) and iron hydroxides (FeOOH, $Fe(OH)_2$, and $Fe(OH)_3$) has to be considered (Liu et al., 2020).

In a semi-pilot fluebed pyrolysis-combustion unit (Kundu et al., 2021), PFAS-laden biosolids were combusted to (1) destruct PFAS and (2) convert it to biochar for reuse in PFAS recovery from wastewaters (Kundu et al., 2021). This process (at 500–600°C) produced 36%–45% biochar with 55.29 to 79.87 m$^2$/g surface area and 27.2%–28.3% C. Both PFOS ($C_0 = 480$ μg/L) and PFOA ($C_0 = 24$ μg/L) were excellently removed (~90%) from PFAS containing natural water samples (locations were not revealed). Overall, biosolid-biochar demonstrated >80% adsorption of long-chain PFAS and 19%–27% adsorption of short-chain PFAS from PFAS-contaminated water. The lower production cost of the biosolids biochar may still make it appealing for commercial use (Kundu et al., 2021). The PFAS combustion gaseous products $CF_4$ and $C_2F_6$, on the other hand, must be addressed because they cause secondary environmental issues.

### 1.4.5 Competitive ion effects

The presence of inorganic anions, cations, and dissolved organic matter/carbon (DOM/DOC) in the solution may influence PFAS removal. Competitive adsorption effects are likely to emerge and exacerbate during long-term operation in the field as the adsorbents are laden with PFAS and other chemicals if absorbents are not properly regenerated.

When sorption is carried out in 0.1 mM NaCl media, the PFOA and PFOS sorption capacities on magnetite/magnetic carbonized CGF (MC-CGF) decreased from 43 mg/g to 39 mg/g (Niu et al., 2020). Further increases to 1 and 10 mM resulted in increased capacities of 48 mg/g and 55 mg/g, respectively. This was due to the low PFOA/PFOS solubility in saline solutions versus deionized water, which resulted in PFAS salting-out on the biochar surface (Niu et al., 2020). When $MgCl_2$ and $FeCl_3$ were dissolved, the capacities increased, even at low concentrations of 0.1 mM. Further molarity increments increased the capacities even more, which was more significant than the effects of NaCl. This was thought to be due to surface-charge neutralization, in which $Mg^{2+}$ can neutralize the negative charge on MC-CG, divalent cation bridging effect, in which $Mg^{2+}/Fe^{3+}$ ions promote the formation of a salt-bridge between MC-CGF and PFOA/PFOS and PFAS anions can complex with $Fe^{3+}$ (see Section 3).

Background DOC in water is known to reduce the sorption affinity of sorbents for PFAS either through direct competition for adsorption sites, pore blockage, or altering the surface charge of biochar (Vo et al., 2022). Four water samples that contain relatively high DOCs (chloroethane, 1,4-dioxane, benzenes, toluene, etc., $C_{total}$ = 1.5–8.4 mg/L) were spiked with PFBA, PFBS, PFHxA, PFHxS, PFOA, and PFOS ($C_0$ = 50 µg/L for each PFAS) and sorbed through mini-filters packed with reed straw-derived biochar (RESCA-900). Results were compared with a parallel treatment with deionized water spiked with identical PFAS doses to determine the impacts of DOCs and co-contaminants (Liu et al., 2021b). DOCs in all samples were efficiently removed to below the detection limit of 1 mg/L. Among all four tested samples, the averaged removal efficiencies for RESCA-900 filters were 56% for PFBA, 61% for PFBS, 63% for PFHxA, 73% for PFHxS, 73% for PFOA, and 83% for PFOS. RESCA-900 displayed >65% removal of long-chain PFAS, while the short-chain removal was 50%–80%. Compared to the DI water, short-chain PFAS removal efficiencies were reduced by 18% (PFBA), 9% (PFBS), and 9% (PFHxA) in groundwater samples of low DOCs and 29% (PFBA), 19% (PFBS), and 26% (PFHxA) in groundwater samples of high

DOCs. Compared to short-chain PFAS, the effect of DOC in groundwater on long-chain PFAS adsorption was less pronounced. Thus, when DOC levels in groundwater are high (e.g., 8 mg/L), pretreatments to reduce DOCs are recommended to improve and extend the effective adsorption of short-chain PFAS.

Hardwood sawdust biochar (HWC) and pinewood biochar (PWC) treated PFOA (0.01–100,000 µg/L) in DOC (4.9 mg/L) spiked lake water (LW) and effluent water (EW) (2.0 mg/L) (Inyang and Dickenson, 2017). Lower PFOA Langmuir capacities ($q_{max}$ = 27.7 mg/g for PWC and 31.7 mg/g for HWC) were found in EW versus LM ($q_{max}$ = 41.3 mg/g for PWC and $S_{max}$ = 41.2 mg/g for HWC). EW with a higher DOC content likely led to higher DOM occupancy on biochar pores, ultimately resulting in the blockage of sorption sites to PFOA (Inyang and Dickenson, 2017).

The removal of individual PFAS from a mixture of PFOA, PFOS, PFBA, and PFBS was 32.8%–57.9% lower than in single component systems for Douglas fir biochar (Zhang et al., 2021a). Strong competition between co-existing compounds and PFAS on the sorbent surface is expected, and slow kinetics for PFAS adsorption can be caused by pore blockage and diffusion-controlled transport into the sorbent pores. Long-chain PFAS were preferentially adsorbed on adsorbents, owing to a stronger hydrophobic interaction than short-chain PFAS. The removal efficiencies of PFAS were in the following order: PFOS > PFOA > PFBC > PFBA, indicating that long-chain PFOA and PFOS were less affected by competitive effects (Zhang et al., 2021a).

The effect of inorganic anions on PFAS sorption was not reported in the articles that were reviewed. Inorganic anions that typically co-exist with PFAS in wastewater streams, such as $CO_3^{2-}$, $PO_4^{3-}$, $SO_4^{2-}$, $AsO_4^{3-}$, and $Cr_2O_7^{2-}/CrO_4^{2-}$, may compete for PFAS uptake by forming electrostatic interactions with the biochar. Future research should investigate this topic.

## 2 Breakthrough column studies and pilot scale testing

Breakthrough column studies and pilot-scale testing are essential to assess a material's suitability for applied PFAS treatment and to understand their limitations for specific applications such as achieving drinking water advisory limits. Mathematical models: Adams-Bohart, Yoon-Nelson, and Thomas (both linear and nonlinear) (Chu, 2010) are commonly applied to predict and understand PFAS column sorption breakthrough capacities.

Charred slashed materials from PJ wood (SBC) and its KOH-activated variant treated effluent water to obtain irrigation water (Yanala and Pagilla, 2020). Biochar between 0.5 and 0.6 mm (25 g) and 1–1.16 mm (15 g) particle sizes were filled in a polyvinyl chloride (PVC) column (5 cm dia. and 50 cm length). The larger size column used in this experiment allowed for the use of larger grain size biochar, which aids in addressing pressure loss that may occur along the length of the biochar in the column (Yanala and Pagilla, 2020). This column treated PFOA 3.7 ng/L ($C_{max}$=871 ng/g) containing effluent and had a 31,625 ng/g capacity using the Thomas model ($R^2$=0.734). The slight data misfit might be related to a range of $C_e/C_o$ (from 0 to 0.58), which is narrow for PFOA. This biochar can effectively treat 8467 L of PFOA (3.7 ng/L) containing water. The harvest typically requires 3,048,000 L of irrigation water, which was achieved with 0.360 kg of biochar. The KOH-activated variant and GAC used in the study did not reach a breakthrough during the experiment period (Yanala and Pagilla, 2020).

Flow rate is an important operational factor in field-deployed adsorption devices, influencing the contact time for adsorbents with pollutants of concern (e.g., PFAS) (Liu et al., 2021b). High flow rates are desirable because they allow adsorption devices to be smaller. However, they run the risk of reducing removal efficiency due to the short contact time.

Reed straw biochar (RESCA) (0.1 g) packed columns (1 cm dia.) were used to filter 10 mL Milli-Q water containing trichloroethylene, benzene, and 1,4-dioxane etc. collected from the sites located in California and New Jersey, US spiked with a mixture of all six PFAS (50 μg/L each) (Liu et al., 2021b). When the flow rate increased from 15 to 30 mL/min, no significant change of PFAS removal efficiencies was observed. However, further increase of the flow rate to 45 mL/min surprisingly exhibited significantly higher removal of all PFAS (>80%) except PFBA (~65%) compared to two other slower flow rates. The flow rate increase may create external pressure that drives the liquid penetration into the internal pores of RESCA-900 packed in the mini-filters. However, for PFBA, which is the most hydrophilic PFAS tested, the increased pore penetration was insufficient to offset the hindrance caused by the shortened contact time, resulting in a decrease in the overall PFBA removal efficiency. Adsorption of short-chain PFAAs in RESCA-900 filters was less kinetically restricted because adsorption sites are primarily surface mesopores (Liu et al., 2021b).

Bench and pilot-scale sorption tests were used to compare the performance of pyrolyzed peach pit (PPC) [6.4 m$^2$/g $S_{BET}$, 69.1% C] wood and poultry litter (PLC) [530°C, 14.8 m²/g $S_{BET}$, 0.03 cm³/g $V_{BET}$, 16.6% C], spruce pine wood (SWC) [650°C, 98.8 m²/g $S_{BET}$, 0.01 cm³/g $V_{BET}$, 85.7% C], and hardwood sawdust pellets (HWC) [900°C, 453 m²/g, $S_{BET}$, 0.11 cm³/g $V_{BET}$, 92.8% C] in removing PFAS from field waters (Inyang and Dickenson, 2017). Screening tests with organic matter-free water revealed that HWC ($K_d = 41$ L/g) and PWC ($K_d = 49$ L/g) biochars had the best PFOA removal performance, comparable to bituminous coal GAC ($K_d = 41$ L/g). HWC or SWC biochars have comparable PFOA removal performance to GAC in organic matter-free water. Because of stronger hydrophobic interactions, HWC and SWC biochars could remove 3–4 times the amount of PFOA removed from wastewater, albeit at a slower sorption rate than PFBA. Breakthrough of nine PFAS was investigated (Inyang and Dickenson, 2017), but only four PFAS (PFPnA, PFHxA, PFOA, and PFOS) were consistently detected in the filter effluents at concentrations above their method detection limit (0.5–5 ng/L). Average influent concentrations of these four PFAAs in the pilot influents were: $17.6 \pm 2.4$ ng/L (PFPnA), $9.5 \pm 1.4$ ng/L (PFHxA), $7.3 \pm 0.8$ ng/L (PFOA), and $4.1 \pm 5.6$ ng/L (PFOS). Ambient filter influent and effluent concentrations of the remaining five target PFAS (PFBA, PFHpA, PFHxS, PFNA, and PFDA) were consistently below or close to the method detection limit during the pilot study (11 months, 35,663 BV). Treatment performance of HWC for PFPnA and PFHxA was comparable to anthracite. However, partial removal of PFOA (10%) and PFOS (20%) at 2963 BV was observed. It is recommended that biochar(s) be produced from more rigid precursor materials, such as coconut or palm kernel shells, that can withstand physical abrasion without disintegrating for higher throughput. PFOS was removed to a greater extent (22% at 10,351 BV) than PFOA (7% at 10,351 BV). Biochar would need to be replaced at <19,100 BV, well before the breakdown of biochar is noticeable, for PFOA removal by HWC.

Rapid small-scale column tests (RSSCT) are a scaled-down version of a pilot- or full-scale system. RSSCT experiments were conducted using a 5 mL syringe (column dia./particle dia. >50, 2.5 cm length) filled with 500 mg sugarcane biochar (Fig. 16.6) (Vo et al., 2022) to treat PFAS containing groundwater samples. Shorter chain PFAS demonstrated shallower breakthrough curves and smaller bed volumes (BV). The BV order was PFOS > PFHxS > PFHxA. The operation of RSSCT and BV depended on various engineering factors such as initial PFAS concentrations (10 μg/L in this work), numbers of PFAS, depth of the packed bed, and characteristics of the sorbent.

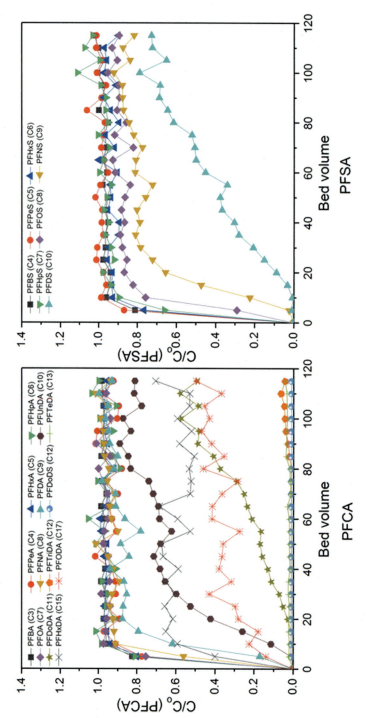

**Fig. 16.6** Full breakthrough profiles of 19 PFAS (PFCA and PFSA groups) versus sorption time in AW using biochar in RSSCT packed-bed column (Vo et al., 2022).

Biochar can be a suitable filter medium for the removal of PFAS in on-site wastewater treatment systems (OWTSs), especially for longer-chain ($\geq C_6$) PFAS (Dalahmeh et al., 2019). Pine-spruce wood biochar (Vindelkol AB, Vindeln, Sweden) filters were used to replace or complement sand filters for PFAS ($C_0 = 1500$–$4900$ ng/L) removal from wastewaters. Four treatments with column filters (1) biochar (BC) without biofilm (BC-no-biofilm), (2) biochar with active biofilm (BC-active-biofilm), (3) biochar with inactive biofilm (BC-inactive-biofilm), and (4) sand with active biofilm (Sand-active-biofilm) were separately performed for 22 weeks at 50 L/m² day hydraulic loading. BC-no-biofilm removed 20%–60% short-chain PFAS with 0–88 ng/g with sorption capacity. The longer chain removal was 90%–99% ($q_{max} = 73$–$168$ ng/g) (Dalahmeh et al., 2019). The presence of biofilm and solids in BC-active-biofilm and wastewater solids in BC-inactive-biofilm reduced the availability and number of PFAS adsorption sites compared to the BC-no-biofilm. Inactivation of the biofilm, on the other hand, resulted in lower PFAS removal efficiencies for $C_{5-11}$ PFCAs, $C_4$, $C_6$, $C_8$ PFSAs, and FOSA, most likely because the biofilm degraded organic matter, increasing the availability and number of adsorption sites compared to BC-inactive-biofilm. Sand-active-biofilm had poor PFAS removal (0–70%) except for FOSA (90%) and a lower adsorption capacity (0.0–7.5 ng/g). In general, shorter-chain PFAS were more difficult to remove than longer-chain PFAS in all biochar treatments. Furthermore, $C_4$, $C_6$, and $C_8$ PFAS showed 10%–30%, 10%–50%, and 20%–30% higher average removal efficiency versus long-chain PFAS (Dalahmeh et al., 2019).

## 3 Sorption interactions

Electrostatic attractions (Punyapalakul et al., 2013; Niu et al., 2020; Zhang et al., 2021a), hydrophobic effects (Zareitalabad et al., 2013; Zhang et al., 2021a), π-π bonding (Du et al., 2014), hydrogen bonding (Deng et al., 2009), ion-exchange (Wang et al., 2012), and Van der Waals forces (Deng et al., 2012), have been claimed to be involved in the adsorption of PFAS on various adsorbents including biochar (Fig. 16.7). However, π-π bonding is not very common due to the absence of π-electrons in the PFAS C—F backbone. Van der Waal forces might seem unimportant in adsorption because of the low fluorine polarizabilities and small molecular sizes of PFAS (Du et al., 2014). PFAS molecules phase separate in both hydrocarbons and water and have low water solubilities (when neutral) (PFOS: 680 mg/L and PFOA: 9500 mg/L)

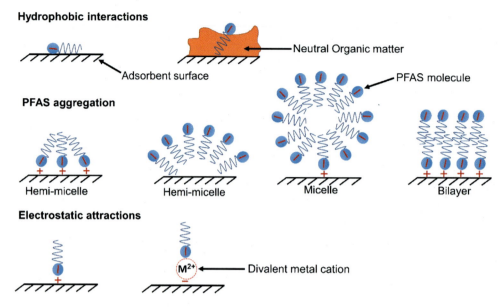

Fig. 16.7 Summary of PFAS sorption mechanisms on the biochar surface (Du et al., 2014).

(Adam, n.d.). However, adsorption onto extended aromatic (graphene-like) carbonaceous surfaces may be thermodynamically preferable to remaining in water phase, even if the adsorbent's hydrophobic surfaces give fairly large contact angles with PFAS droplets. Sorption onto carbonaceous surfaces can occur through hydrophobic interactions, aggregation, and electrostatic attractions (Fig. 16.7) (Du et al., 2014). The functional groups on the biochar surface can act as hydrogen bond acceptors or donors, allowing them to bond to adsorbate (PFAS carboxylic/sulfonic groups) via hydrogen bonding. Sorption onto the metal oxide phases may occur through electrostatic (Eq. 16.1) or hydrogen bonding.

$$\equiv M - OH_2^+ + CF_3(CF_2)_n COO^- \text{(electrostic attraction)} \rightarrow$$
$$\equiv M - OOC(CF_2)_n CF_3 \text{ (inner sphere complex)} + H_2O \quad (16.1)$$

Additionally, some adsorbed $(RCOO)_2M$, $(RSO_3)_2M$, $(RCOO)_3M$, $(RSO_3)_3M$ (R=perfluorocarbon) occurring from formation from PFAS anions complexing with soluble $M^{3+}$ or $M^{2+}$ could take place. Thus, esterification of M-OH sites on the metal oxide surface is a possible pathway of sorption onto metal oxide surfaces.

## 4 Conclusions, gaps, and future studies

Low-cost biochar and engineered biochar adsorbents could serve in large scale water treatment to adsorb PFAS where the cost of activated carbon makes their use impossible. The operation cost can be reduced significantly if an alternative adsorbent can be found which is readily available, easily recovered from solution, and is cheap. Wood, agricultural byproducts, and energy crops provide an abundant and renewable supply of lignocellulosic feeds for biorefineries. A solid char byproduct from biorefinery liquid fuel production could provide low-cost adsorbents to purify PFAS contaminated waters. Renewable biofuels are closer to carbon neutral vs. fossil fuels. Thus, developing a large-scale biofuel and bio-derived chemicals industry would result in large quantities of biochar becoming available for soil conditioning, carbon storage, and adsorbent used.

Although some studies have shown that the feedstock, high pyrolysis temperature, and activation procedure all play a role in biochar's PFAS uptake ability, isotherm studies use unrealistically high concentrations. Thus, even though sufficient adsorption capacities have been observed, they are difficult to replicate in real-world conditions at environmentally relevant concentrations (ng/L to sub μg/L). Results from fixed bed experiments, on the other hand, can be scaled up to full-scale plants. Hence, fixed bed experiments are needed to establish biochar as a potential adsorbent to remove emerging contaminants from wastewater. Furthermore, no studies on the biochar-based sorption of the recently introduced PFAS substitute such as GenX, ADONA, and F—53B have been reported. These topics should be researched further in future studies.

## Questions and problems

1. What are the current PFAS cleaning methods?
2. Why is PFAS removal considered difficult versus other pollutants?
3. Why do PFAS molecules generally have a high acidity?
4. Give two biochar requirements to be an ideal candidate for PFAS removal.
5. What are the advantages of biochar magnetization?
6. This chapter has discussed a zero valent iron (ZVI)/biochar material that can be used in simultaneous adsorption and degradation of PFAS. What PFAS degradation products were observed/speculated in that study?
7. Why shorter chain PFAS compounds are difficult to remove using biochar?

8. Why isotherm model fittings may not be applicable in estimating the PFAS sorption capacities?
9. List the PFAS-loaded biochar regeneration methods described in this chapter.
10. List the main PFAS sorption interactions with the biochar surface.

# References

Adam, M., n.d. Per- and polyfluoroalkyl substances (PFAS). Chemistry and Behavior, Technology Integration and Information Branch. https://clu-in.org/contaminantfocus/default.focus/sec/Per-_and_Polyfluoroalkyl_Substances_(PFASs)/cat/Chemistry_and_Behavior/.

Aharoni, C., Tompkins, F., 1970. Kinetics of adsorption and desorption and the Elovich equation. Adv. Catal. 21, 1–49. Elsevier.

Ahmad, M., Rajapaksha, A.U., Lim, J.E., Zhang, M., Bolan, N., Mohan, D., Vithanage, M., Lee, S.S., Ok, Y.S., 2014. Biochar as a sorbent for contaminant management in soil and water: a review. Chemosphere 99, 19–33.

Ahmed, M.B., Alam, M.M., Zhou, J.L., Xu, B., Johir, M.A.H., Karmakar, A.K., Rahman, M.S., Hossen, J., Hasan, A.K., Moni, M.A., 2020. Advanced treatment technologies efficacies and mechanism of per-and poly-fluoroalkyl substances removal from water. Process Saf. Environ. Prot. 136, 1–14.

Ateia, M., Maroli, A., Tharayil, N., Karanfil, T., 2019. The overlooked short-and ultrashort-chain poly-and perfluorinated substances: a review. Chemosphere 220, 866–882.

Banks, D., Jun, B.-M., Heo, J., Her, N., Park, C.M., Yoon, Y., 2020. Selected advanced water treatment technologies for perfluoroalkyl and polyfluoroalkyl substances: a review. Sep. Purif. Technol. 231, 115929.

Barpaga, D., Zheng, J., Han, K.S., Soltis, J.A., Shutthanandan, V., Basuray, S., McGrail, B.P., Chatterjee, S., Motkuri, R.K., 2019. Probing the sorption of perfluorooctanesulfonate using mesoporous metal–organic frameworks from aqueous solutions. Inorg. Chem. 58, 8339–8346.

Beesoon, S., Martin, J.W., 2015. Isomer-specific binding affinity of perfluorooctanesulfonate (PFOS) and perfluorooctanoate (PFOA) to serum proteins. Environ. Sci. Technol. 49, 5722–5731.

Boyd, G., Adamson, A., Myers Jr., L., 1947. The exchange adsorption of ions from aqueous solutions by organic zeolites. II. Kinetics1. J. Am. Chem. Soc. 69, 2836–2848.

Brian, P., Hurley, J., Hasseltine, E., 1961. Penetration theory for gas absorption accompanied by a second order chemical reaction. AICHE J. 7, 226–231.

Buck, R.C., Franklin, J., Berger, U., Conder, J.M., Cousins, I.T., De Voogt, P., Jensen, A.A., Kannan, K., Mabury, S.A., van Leeuwen, S.P., 2011. Perfluoroalkyl and polyfluoroalkyl substances in the environment: terminology, classification, and origins. Integr. Environ. Assess. Manag. 7, 513–541.

Chen, X., Xia, X., Wang, X., Qiao, J., Chen, H., 2011. A comparative study on sorption of perfluorooctane sulfonate (PFOS) by chars, ash and carbon nanotubes. Chemosphere 83, 1313–1319.

Chu, K.H., 2010. Fixed bed sorption: setting the record straight on the Bohart–Adams and Thomas models. J. Hazard. Mater. 177, 1006–1012.

Corbett, J.F., 1972. Pseudo first-order kinetics. J. Chem. Educ. 49, 663.

Cordner, A., Vanessa, Y., Schaider, L.A., Rudel, R.A., Richter, L., Brown, P., 2019. Guideline levels for PFOA and PFOS in drinking water: the role of scientific uncertainty, risk assessment decisions, and social factors. J. Expo. Sci. Environ. Epidemiol. 29, 157.

Dalahmeh, S.S., Alziq, N., Ahrens, L., 2019. Potential of biochar filters for onsite wastewater treatment: effects of active and inactive biofilms on adsorption of per-and polyfluoroalkyl substances in laboratory column experiments. Environ. Pollut. 247, 155–164.

Deng, S., Shuai, D., Yu, Q., Huang, J., Yu, G., 2009. Selective sorption of perfluorooctane sulfonate on molecularly imprinted polymer adsorbents. Front. Environ. Sci. Eng. China 3, 171–177.

Deng, S., Zhang, Q., Nie, Y., Wei, H., Wang, B., Huang, J., Yu, G., Xing, B., 2012. Sorption mechanisms of perfluorinated compounds on carbon nanotubes. Environ. Pollut. 168, 138–144.

Deng, S., Nie, Y., Du, Z., Huang, Q., Meng, P., Wang, B., Huang, J., Yu, G., 2015. Enhanced adsorption of perfluorooctane sulfonate and perfluorooctanoate by bamboo-derived granular activated carbon. J. Hazard. Mater. 282, 150–157.

Dong, X., He, L., Hu, H., Liu, N., Gao, S., Piao, Y., 2018. Removal of 17β-estradiol by using highly adsorptive magnetic biochar nanoparticles from aqueous solution. Chem. Eng. J. 352, 371–379.

Du, Z., Deng, S., Bei, Y., Huang, Q., Wang, B., Huang, J., Yu, G., 2014. Adsorption behavior and mechanism of perfluorinated compounds on various adsorbents—a review. J. Hazard. Mater. 274, 443–454.

Espana, V.A.A., Mallavarapu, M., Naidu, R., 2015. Treatment technologies for aqueous perfluorooctanesulfonate (PFOS) and perfluorooctanoate (PFOA): a critical review with an emphasis on field testing. Environ. Technol. Innov. 4, 168–181.

Freundlich, H., 1907. Über die adsorption in lösungen. Z. Phys. Chem. 57, 385–470.

Gagliano, E., Sgroi, M., Falciglia, P.P., Vagliasindi, F.G., Roccaro, P., 2020. Removal of poly-and perfluoroalkyl substances (PFAS) from water by adsorption: role of PFAS chain length, effect of organic matter and challenges in adsorbent regeneration. Water Res. 171, 115381.

Giesy, J.P., Kannan, K., 2002. Perfluorochemical surfactants in the environment. Environ. Sci. Technol. 36, 146A–152A.

Guo, W., Huo, S., Feng, J., Lu, X., 2017. Adsorption of perfluorooctane sulfonate (PFOS) on corn straw-derived biochar prepared at different pyrolytic temperatures. J. Taiwan Inst. Chem. Eng. 78, 265–271.

Guo, W., Lu, S., Shi, J., Zhao, X., 2019. Effect of corn straw biochar application to sediments on the adsorption of 17α-ethinyl estradiol and perfluorooctane sulfonate at sediment-water interface. Ecotoxicol. Environ. Saf. 174, 363–369.

Hassan, M., Liu, Y., Naidu, R., Du, J., Qi, F., 2020. Adsorption of Perfluorooctane sulfonate (PFOS) onto metal oxides modified biochar. Environ. Technol. Innov. 19, 100816.

Inyang, M., Dickenson, E.R., 2017. The use of carbon adsorbents for the removal of perfluoroalkyl acids from potable reuse systems. Chemosphere 184, 168–175.

Jovicic, V., Khan, M.J., Zbogar-Rasic, A., Fedorova, N., Poser, A., Swoboda, P., Delgado, A., 2018. Degradation of low concentrated perfluorinated compounds (PFCS) from water samples using non-thermal atmospheric plasma (NTAP). Energies 11, 1290.

Kucharzyk, K.H., Darlington, R., Benotti, M., Deeb, R., Hawley, E., 2017. Novel treatment technologies for PFAS compounds: a critical review. J. Environ. Manage. 204, 757–764.

Kundu, S., Patel, S., Halder, P., Patel, T., Marzbali, M.H., Pramanik, B.K., Paz-Ferreiro, J., de Figueiredo, C.C., Bergmann, D., Surapaneni, A., 2021. Removal of PFAS from biosolids using a semi-pilot scale pyrolysis reactor and the application of biosolids derived biochar for the removal of PFAS from contaminated water. Environ. Sci.: Water Res. Technol. 7, 638–649.

Langmuir, I., 1918. The adsorption of gases on plane surfaces of glass, mica and platinum. J. Am. Chem. Soc. 40, 1361–1403.

Lath, S., Navarro, D.A., Losic, D., Kumar, A., McLaughlin, M.J., 2018. Sorptive remediation of perfluorooctanoic acid (PFOA) using mixed mineral and graphene/carbon-based materials. Environ. Chem. 15, 472–480.

Lee, S.-H., Cho, Y.-J., Lee, M., Lee, B.-D., 2019. Detection and treatment methods for perfluorinated compounds in wastewater treatment plants. Appl. Sci. 9, 2500.

Lehmann, J., Joseph, S., 2015. Biochar for Environmental Management: Science, Technology and Implementation. Routledge.

Lin, C.-Y., Lin, L.-Y., Chiang, C.-K., Wang, W.-J., Su, Y.-N., Hung, K.-Y., Chen, P.-C., 2010. Investigation of the associations between low-dose serum perfluorinated chemicals and liver enzymes in US adults. Am. J. Gastroenterol. 105, 1354–1363.

Liu, Y., Ptacek, C.J., Baldwin, R.J., Cooper, J.M., Blowes, D.W., 2020. Application of zero-valent iron coupled with biochar for removal of perfluoroalkyl carboxylic and sulfonic acids from water under ambient environmental conditions. Sci. Total Environ. 719, 137372.

Liu, C., Chu, J., Cápiro, N.L., Fortner, J.D., Pennell, K.D., 2021a. In-situ sequestration of perfluoroalkyl substances using polymer-stabilized ion exchange resin. J. Hazard. Mater. 422, 126960.

Liu, N., Wu, C., Lyu, G., Li, M., 2021b. Efficient adsorptive removal of short-chain perfluoroalkyl acids using reed straw-derived biochar (RESCA). Sci. Total Environ. 798, 149191.

Lu, D., Sha, S., Luo, J., Huang, Z., Jackie, X.Z., 2020. Treatment train approaches for the remediation of per-and polyfluoroalkyl substances (PFAS): a critical review. J. Hazard. Mater. 386, 121963.

McCleaf, P., Englund, S., Östlund, A., Lindegren, K., Wiberg, K., Ahrens, L., 2017. Removal efficiency of multiple poly-and perfluoroalkyl substances (PFAS) in drinking water using granular activated carbon (GAC) and anion exchange (AE) column tests. Water Res. 120, 77–87.

Munoz, G., Liu, J., Duy, S.V., Sauvé, S., 2019. Analysis of F-53B, Gen-X, ADONA, and emerging fluoroalkylether substances in environmental and biomonitoring samples: a review. Trends Environ. Anal. Chem. 23, e00066.

Navarathna, C.M., Dewage, N.B., Karunanayake, A.G., Farmer, E.L., Perez, F., Mlsna, T.E., Pittman, C.U., 2020. Rhodamine B adsorptive removal and photocatalytic degradation on MIL-53-Fe MOF/magnetic magnetite/biochar composites. J. Inorg. Organomet. Polym. Mater. 30, 214–229.

Niu, B., Yang, S., Li, Y., Zang, K., Sun, C., Yu, M., Zhou, L., Zheng, Y., 2020. Regenerable magnetic carbonized Calotropis gigantea fiber for hydrophobic-driven fast removal of perfluoroalkyl pollutants. Cellul. 27, 5893–5905.

Patel, M., Kumar, R., Kishor, K., Mlsna, T., Pittman Jr., C.U., Mohan, D., 2019. Pharmaceuticals of emerging concern in aquatic systems: chemistry, occurrence, effects, and removal methods. Chem. Rev. 119, 3510–3673.

Prevedouros, K., Cousins, I.T., Buck, R.C., Korzeniowski, S.H., 2006. Sources, fate and transport of perfluorocarboxylates. Environ. Sci. Technol. 40, 32–44.

Punyapalakul, P., Suksomboon, K., Prarat, P., Khaodhiar, S., 2013. Effects of surface functional groups and porous structures on adsorption and recovery of perfluorinated compounds by inorganic porous silicas. Sep. Sci. Technol. 48, 775–788.

Ross, I., McDonough, J., Miles, J., Storch, P., Thelakkat Kochunarayanan, P., Kalve, E., Hurst, J., Dasgupta, S.S., Burdick, J., 2018. A review of emerging technologies for remediation of PFAS. Remediat. J. 28, 101–126.

Sips, R., 1948. On the structure of a catalyst surface. J. Chem. Phys. 16, 490–495.

Sörengård, M., et al., 2020. Adsorption behavior of per-and polyfluoroalkyl substances (PFASs) to 44 inorganic and organic sorbents and use of dyes as proxies for PFAS sorption. J. Environ. Chem. Eng. 8 (3), 103744.

Tian, D., Geng, D., Mehler, W.T., Goss, G., Wang, T., Yang, S., Niu, Y., Zheng, Y., Zhang, Y., 2021. Removal of perfluorooctanoic acid (PFOA) from aqueous solution by amino-functionalized graphene oxide (AGO) aerogels: influencing factors, kinetics, isotherms, and thermodynamic studies. Sci. Total Environ. 783, 147041.

USEPA, 2016. Lifetime health advisories and health effects support documents for perfluorooctanoic acid and perfluorooctane sulfonate. Fed. Regist. 81, 33250–33251.

Vo, H.N.P., Nguyen, T.M.H., Ngo, H.H., Guo, W., Shukla, P., 2022. Biochar sorption of perfluoroalkyl substances (PFAS) in aqueous film-forming foams-impacted groundwater: effects of PFAS properties and groundwater chemistry. Chemosphere 286, 131622.

Wang, F., Liu, C., Shih, K., 2012. Adsorption behavior of perfluorooctanesulfonate (PFOS) and perfluorooctanoate (PFOA) on boehmite. Chemosphere 89, 1009–1014.

Wu, C., Klemes, M.J., Trang, B., Dichtel, W.R., Helbling, D.E., 2020. Exploring the factors that influence the adsorption of anionic PFAS on conventional and emerging adsorbents in aquatic matrices. Water Res. 182, 115950.

Xiao, F., Simcik, M.F., Gulliver, J.S., 2013. Mechanisms for removal of perfluorooctane sulfonate (PFOS) and perfluorooctanoate (PFOA) from drinking water by conventional and enhanced coagulation. Water Res. 47, 49–56.

Xiao, X., Ulrich, B.A., Chen, B., Higgins, C.P., 2017. Sorption of poly-and perfluoroalkyl substances (PFAS) relevant to aqueous film-forming foam (AFFF)-impacted groundwater by biochars and activated carbon. Environ. Sci. Technol. 51, 6342–6351.

Yanala, S.R., Pagilla, K.R., 2020. Use of biochar to produce reclaimed water for irrigation use. Chemosphere 251, 126403.

Yavari, S., Malakahmad, A., Sapari, N.B., 2015. Biochar efficiency in pesticides sorption as a function of production variables—a review. Environ. Sci. Pollut. Res. 22, 13824–13841.

Yi, Y., Huang, Z., Lu, B., Xian, J., Tsang, E.P., Cheng, W., Fang, J., Fang, Z., 2020. Magnetic biochar for environmental remediation: a review. Bioresour. Technol. 298, 122468.

Zareitalabad, P., Siemens, J., Hamer, M., Amelung, W., 2013. Perfluorooctanoic acid (PFOA) and perfluorooctanesulfonic acid (PFOS) in surface waters, sediments, soils and wastewater—a review on concentrations and distribution coefficients. Chemosphere 91, 725–732.

Zhang, D., Zhang, W., Liang, Y., 2019. Adsorption of perfluoroalkyl and polyfluoroalkyl substances (PFAS) from aqueous solution—a review. Sci. Total Environ. 694, 133606.

Zhang, D., He, Q., Wang, M., Zhang, W., Liang, Y., 2021a. Sorption of perfluoroalkylated substances (PFAS) onto granular activated carbon and biochar. Environ. Technol. 42, 1798–1809.

Zhang, J., Pang, H., Gray, S., Ma, S., Xie, Z., Gao, L., 2021b. PFAS removal from wastewater by in-situ formed ferric nanoparticles: solid phase loading and removal efficiency. J. Environ. Chem. Eng. 9, 105452.

Zhao, D., Cheng, J., Vecitis, C.D., Hoffmann, M.R., 2011. Sorption of perfluorochemicals to granular activated carbon in the presence of ultrasound. J. Phys. Chem. A 115, 2250–2257.

Zhou, Y., Xu, M., Huang, D., Xu, L., Yu, M., Zhu, Y., Niu, J., 2021. Modulating hierarchically microporous biochar via molten alkali treatment for efficient adsorption removal of perfluorinated carboxylic acids from wastewater. Sci. Total Environ. 757, 143719.

# Simple modeling approaches to monitor and predict organic contaminant adsorption by biochar

Kyle K. Shimabuku[a] and R. Scott Summers[b]
[a]Department of Civil Engineering, Gonzaga University, Spokane, WA, United States. [b]Department of Civil, Environmental, and Architectural Engineering, University of Colorado Boulder, Boulder, CO, United States

## 1 Introduction

Biochar can efficiently remove many organic contaminants (OCs) from water (Shimabuku et al., 2016; Kearns et al., 2014, 2019). It has distinct advantages over more established technologies used to control OCs in drinking water (e.g., advanced oxidation processes, activated carbon) because it is less expensive and more environmentally friendly (Thompson et al., 2016). For example, biochar production also generates energy, whereas activated carbon generation consumes substantial amounts of energy (Thompson et al., 2016). In low-income and remote regions, biochar is one of the few technologies that can enable communities to protect themselves from OC contamination of drinking water sources. It can be produced in simple, low-cost production systems, such as cookstoves, using locally available and abundant feedstocks, including food waste (e.g., nutshells, corn cobs) and fast-growing biomass (e.g., bamboo) (Kearns et al., 2014, 2015, 2019). However, the physical and chemical processes that govern OC adsorption by biochar are complex, which can pose a barrier to its adoption in drinking water treatment applications.

One complexity surrounds the heterogeneity of biochar properties because it can be produced under a wide range of conditions from nearly any carbon-rich precursor material (e.g., wood, food waste, or wastewater sludge). These production parameters govern its physicochemical properties and, as a result, its OC removal efficiency (Lattao et al., 2014). For instance, several studies have shown that OC adsorption capacities of biochar made from the same feedstock but at different temperatures can vary by over an order of magnitude (Shimabuku et al., 2016; Kearns et al., 2019; Lattao et al., 2014). The same range in performance can be found at a set production condition but with different feedstock materials (Shimabuku et al., 2016; Kearns et al., 2014, 2019; Bentley et al., 2021). Thus, it is important to be able to identify efficient biochar adsorbents.

Another challenge is determining how water quality conditions might impact biochar performance for OC removal. One important water quality parameter is the initial concentration ($C_0$) of the target OC, as increasing initial concentrations can result in a nonlinear decrease in removal (Shimabuku et al., 2017). The presence of other known OCs can also reduce biochar removal efficiency for a target OC by competing for adsorption sites (Shimabuku et al., 2017; Sander and Pignatello, 2005). Background dissolved organic matter (DOM), which constitutes the complex mixture of mostly unknown organic compounds of natural and anthropogenic origin, is a key water quality parameter that influences biochar performance because DOM can directly compete for and block adsorption sites, which taken together is termed "fouling" (Kearns et al., 2019, 2021). Thus, models that can predict biochar performance at different OC initial concentrations in the presence of DOM are needed.

A common method of treating water with biochar is with flow-through fixed-bed biochar columns that are often referred to as "adsorbers." Since various water quality parameters can influence OC removal by biochar, pilot testing with the water source of interest may be necessary to design and evaluate the anticipated performance of biochar adsorbers. However, such testing can incur substantial costs and require long run times (greater than several months). Also, adsorption under flow-through conditions is a nonsteady-state process and OCs eventually breakthrough adsorbers requiring media to be replaced that can require frequent OC analysis of the effluent to ensure adequate OC removal. Such OC monitoring can be expensive and even infeasible in communities that do not have access to OC analytical instruments (e.g., liquid chromatography-mass spectrometry).

To overcome some of the hurdles for utilizing biochar to remove OCs from drinking water, this chapter focusses on simple modeling approaches that can be used to:
- identify efficient biochar adsorbents for OCs based on surface area and hydrogen to carbon (H:C) ratios
- simulate OC removal performance in the presence of DOM at any target OC concentration and known OC competitor

concentration as long as OC concentrations remain below some concentration threshold
- predict adsorber performance based on batch test results
- monitor OC breakthrough in biochar adsorbers by measuring relatively inexpensive, simple, and rapid DOM optical properties

## 2 Relationships between OC adsorption capacity with surface area and H:C ratios

Surface area and elemental analyses are two analytical techniques often used to characterize biochar that are included in the group of standard analyses recommended in the International Biochar Initiative's *Standardized Product Definition and Product Testing Guidelines for Biochar That Is Used in Soil-Version 2.1*. There are different approaches to characterizing surface area involving different gasses (e.g., $CO_2$) and theoretical assumptions surrounding pore geometry (e.g., density functional theory) needed to estimate surface area. Nitrogen ($N_2$) gas adsorption modeled with the Brunauer–Emmett–Teller (BET) theory is the most common approach. Though $N_2$-BET surface area characterization has some limitations in its ability to quantify OC accessible surface area (as $N_2$ gas requires freezing temperatures that can block $N_2$ penetration when tarry/rubbery substances are present) (Lattao et al., 2014), general trends often emerge between $N_2$-BET surface area and OC adsorption capacity. For example, Fig. 17.1A is adapted from Shimabuku et al. (2016) and estimates biochar performance observed in nonequilibrium batch adsorption tests using a contact time of 30 min. The biochar doses needed to remove 75% ($Dose_{75}$) of sulfamethoxazole (SMX), which is an antibiotic that often enters the environment from wastewater treatment plants, that was initially present at 100 ng/L are plotted against biochar surface areas in Fig. 17.1A. As surface area increases, the dose required to remove SMX decreases, indicating greater adsorption efficiency, regardless of if experiments were performed in deionized water (DI) or surface water (Fig. 17.1A). Similarly, Fig. 17.1B, adapted from Kearns et al. (2014), also shows that as surface area increases, adsorption capacity ($q_{100}$) for the pesticide 2,4-D in DI water increases when the equilibrium concentration was 100 μg/L.

In both studies, the differences in surface area are associated with different production conditions in kilns and gasifiers covering a range of highest treatment temperatures (350–900°C) and biochar feedstock materials, including wood, food waste (e.g., corncobs), and wastewater sludge (e.g., biosolids). Thus, it

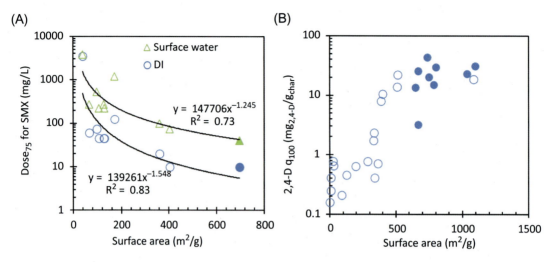

Fig. 17.1 (A) Biochar (hollow symbols) and activated carbon (filled symbols) doses needed to achieve 75% SMX removal ($C_0$ 100 ng/L) versus BET surface area for experiments performed in deionized water (DI) and surface water. (B) 2,4-D adsorption capacity of biochar (hollow circles) and activated carbon (filled circles) when the equilibrium concentration was 100 µg/L ($q_{100}$) versus BET surface area.

appears surface area can be used to assess the performance of a variety of biochar materials and that higher surface area materials exhibit greater adsorption efficiencies for OCs. However, there are diminishing returns when continuing to increase surface area as Fig. 17.1 shows there is a plateau in performance that emerges around 400–500 m²/g. Since surface area accessible N₂ gas (MW=28 g/mol) may not be accessible to OCs that are larger (MW>100 g/mol), OC removal also depends on biochar pore size distribution and geometry (Lattao et al., 2014), which partly explains the scatter in the data from the general trends. While surface area may not be able to predict the performance of biochar with a high degree of certainty, it can be an effective tool to screen for high-performing biochars.

Ash-free matter (AFM) doses to achieve 75% SMX removal are plotted versus hydrogen to carbon ratios (H:C) in Fig. 17.2, which was adapted from Shimabuku et al. (2016). It was determined that the adsorption capacity of ash for SMX was negligible. Also, ash does not contribute to biochar hydrogen and carbon content measurements. Therefore, AFM biochar doses to 75% removal are most appropriate to plot versus H:C ratios, especially since the high-ash biochars had ash contents in excess of 60%, whereas the low-ash biochars had ash contents <6%.

For the low-ash biochar, a strong correlation was observed between the AFM-Dose$_{75}$ values and H:C ratios. One reason is that

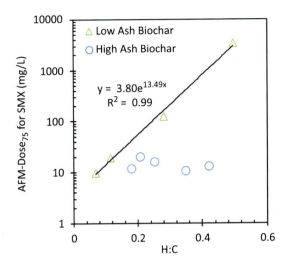

**Fig. 17.2** Biochar ash-free matter (AFM)-Dose$_{75}$ versus H:C ratios for high-ash and low-ash biochars (C$_0$ 100 ng/L and DI).

H:C ratios are a proxy for aromaticity. More aliphatic, hydrogen-rich biochar structures possess high H:C ratios, whereas more aromatic, graphene-like structures contain less hydrogen and have lower H:C ratios. Biochar aromaticity can enhance its affinity for aromatic OCs, such as SMX, through pi–pi electron donor–accepter interactions (Zhu and Pignatello, 2005).

Increasing the highest treatment temperature during biochar production has been associated with an increase in aromaticity, a loss of polar moieties like hydroxyl and carboxyl functional groups that have been shown to inhibit OC adsorption (Zhu et al., 2005), and an increase in surface area (Keiluweit et al., 2010). Therefore, when producing biochar from a given feedstock but at different temperatures, H:C ratios also covary with these other physicochemical properties. Such phenomena likely explain why H:C ratios correlated with AFM-Dose$_{75}$ values for the low-ash biochars but not for the high-ash biochars (Fig. 17.2). Low-ash biochars were all produced from wood but at the highest treatment temperatures varying between 350°C and 850°C. Conversely, the high-ash biochar was produced at a set temperature of 850°C but from different wastewater sludge precursor materials. Differences in H:C ratios resulted from changes in wastewater sludge content and not the intensity of pyrolysis reactions that expel polar functional groups and develop pore structure and surface area (Keiluweit et al., 2010). Thus, H:C ratios may only correlate with biochar performance when H:C ratios capture the transformation of biochar properties associated with changes in the highest treatment temperature.

These findings that were collected for anionic compounds, SMX and 2,4-D, are consistent with a large meta-analysis from 29 biochar studies that looked at the adsorption of neutral OCs (Hale et al., 2016). While surface area measurements appear to relate to biochar performance regardless of production conditions and feedstock materials, the utility of H:C ratios may be largely limited to a given feedstock material produced at different temperatures. Both parameters could be useful in QA/QC applications to ensure sufficient pyrolysis intensity and duration has occurred, especially when the highest treatment temperatures are uncertain. Many commercial biochar production systems that rely on gasification have variable production temperatures through the reactor making it difficult to determine the highest treatment temperatures.

## 3 Modeling the influence of OC concentration on adsorption by biochar

Under single-solute conditions (i.e., when an OC is the only adsorbate present) biochar (as well as other carbonaceous adsorbents, like activated carbon) batch adsorption data fall on a single isotherm plot (i.e., solid-phase versus liquid-phase concentration at equilibrium) irrespective of if the experiment is run at a constant OC $C_0$ and varied biochar dose or constant biochar dose and varied OC $C_0$. Isotherm models, such as the Freundlich model, can simulate biochar performance at different OC $C_0$ under these conditions. However, when background DOM is present, which is ubiquitous in drinking water sources, this is no longer the case as depicted in Fig. 17.3, which is adapted from Shimabuku et al. (2017). This figure shows cyclohexanol adsorption by wood biochar produced in a furnace at 850°C with a 2 h contact time in surface water with a dissolved organic carbon concentration of 5 mg/L. The variable concentration dataset was generated using cyclohexanol $C_0$s between 340 and 9,600,000 ng/L with a biochar dose of 750 mg/L. The variable dose data were generated by holding the cyclohexanol $C_0$ constant at 1000 ng/L and changing the biochar dose between 400 and 1400 mg/L.

The different isotherm adsorption behavior is displayed in Fig. 17.3 because as the equilibrium concentration, $C_e$, increases as a result of an increasing $C_0$ for the variable concentration plot, cyclohexanol is better able to outcompete background DOM for adsorption sites. Alternatively, as the equilibrium concentration,

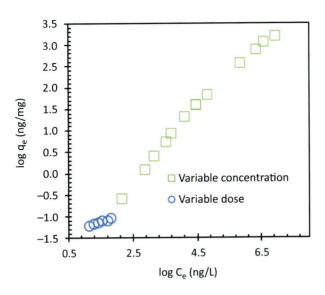

**Fig. 17.3** Biochar adsorption of cyclohexanol in surface water DOM, data generated by varying the biochar dose but holding the cyclohexanol concentration constant (variable dose) and keeping the biochar dose constant but varying the cyclohexanol concentration (variable concentration).

$C_e$, increases as a result of a decreasing biochar dose, DOM can better outcompete cyclohexanol for adsorption sites because DOM is a mixture of adsorbates that vary in their competitiveness for adsorption sites. At low biochar doses, only the most competitive DOM constituents are present and they can outcompete target adsorbates for the limited number of biochar adsorption sites.

One approach to model the effect of target adsorbate $C_0$ is using the ideal adsorbed solution theory (IAST) model that utilizes a fictitious compound representing the mixture of competitive DOM adsorbates called the equivalent background compound (EBC) (Najm et al., 1991). This model that was initially developed for activated carbon is fairly complex and requires adsorption experiments to be performed in deionized water and the DOM-containing water of interest. Observing the variable concentration isotherm in Fig. 17.3 reveals that at lower $C_e$s, associated with lower $C_0$s, the isotherm is linear. When an isotherm is linear, an increase in $C_0$ results in a proportional increase in the fraction of the OC that is adsorbed. This phenomenon permits a simple modeling approach where it can be assumed that the fractional removal of an OC (i.e., the initial concentration of the target adsorbate minus the final concentration normalized by the initial concentration of the target adsorbate) in the presence

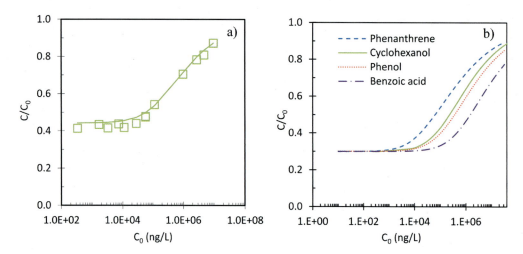

**Fig. 17.4** (A) Cyclohexanol fraction remaining versus its $C_0$ fit with the IAST-EBC model. (B) IAST-EBC model simulations of phenanthrene, cyclohexanol, phenol, and benzoic acid fraction remaining versus their respective $C_0$.

of DOM remains constant regardless of the $C_0$ of the OC until some concentration threshold is exceeded and the isotherm no longer remains linear.

To demonstrate the aforementioned simple modeling approach, the data from the variable concentration isotherm in Fig. 17.3 are replotted in Fig. 17.4A to show the fraction of cyclohexanol remaining versus the cyclohexanol $C_0$. When the cyclohexanol $C_0$ is less than approximately 100,000 ng/L in Fig. 17.4A, the fraction remaining stays constant at around 0.43, demonstrating that the fraction remaining (and fractional removal) of cyclohexanol is independent of $C_0$ so long as it stays below 100,000 ng/L.

In Shimabuku et al. (2017), it was found that the IAST-EBC model adequately fit experimental data as shown in Fig. 17.4A. IAST-EBC model simulations are provided in Fig. 17.4B that can be used to determine where the concentration threshold occurs (i.e., the $C_0$ of the OC at which a further increase in $C_0$ corresponds to an increase in the fraction remaining). Fig. 17.4B reveals that, according to the IAST-EBC model, the concentration threshold is dependent on the OC of interest. It was also determined in Shimabuku et al. (2017) that the concentration threshold also depends on the concentration and character of background DOM and biochar dose. The threshold may also differ depending on the biochar, as appeared to be the case for SMX adsorption to wood and wastewater sludge-based biochars in Shimabuku et al.

(2016). For most OCs and waters containing DOM, the concentration threshold is typically at a $C_0$ greater than 10,000 ng/L (10 µg/L). Since OCs tend to be present at the low to sub-microgram per liter range in most drinking water sources, it usually can be assumed that the fractional removal of an OC is independent of its $C_0$. It should be noted that the presence of other specific OCs in drinking water sources typically will not impact the adsorption of a target OC as long as the sum of their concentrations is also below some threshold that can be assumed to be around 10 µg/L, which was validated in Kennedy and Summers (2015) with an activated carbon column adsorber.

## 4 Estimating biochar adsorber performance from batch adsorption data

One way to model the performance of a full-scale biochar column adsorber is to perform pilot testing where full-sized media are used in a more narrow pilot column, which requires less materials and water to be used to assess biochar performance, but long operational times (e.g., several months or longer) can be required. The rapid small-scale column test (RSSCT) has been developed that can be operated in a fraction of the time of a pilot test, but it requires a more complex scaling approach with varying degrees of success as such tests often overestimate the performance of a full-scale adsorber (Kearns et al., 2020). Alternatively, batch equilibrium testing can be used to determine the batch use rate that subsequently can be used to estimate the column use rate. The batch use rate is the dose (mass/volume) of biochar adsorbent needed to achieve a certain level of OC removal. The column use rate is mass of GAC in the adsorber column relative to the volume of water treated at the time the target OC breakthrough (GAC mass/volume treated).

The natural log of batch use rates to achieve 90% removal of various OCs (e.g., SMX, 2,4-D, and *per-* and polyfluoroalkyl substances) versus the natural log of use rates to 10% breakthrough (or 90% removal) for a given OC and biochar is shown in Fig. 17.5, which is adapted from Kearns et al. (2021). Although the batch versus column use rates do not fall on a 1:1 plot, there is a good correlation between the batch and column use rate data.

The greater use rates observed in columns (i.e., more adsorbent is needed in columns to achieve the same level of removal) can be related to the greater level of DOM fouling that occurs under flow-through conditions. Since DOM concentrations are typically a few orders of magnitude greater than that of OCs in

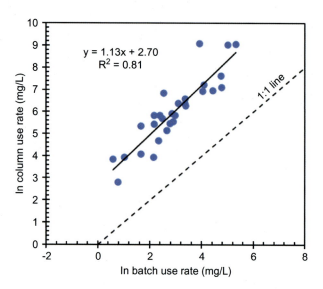

**Fig. 17.5** Column use rates versus batch use rates for 90% OC removal.

drinking water sources, DOM can outcompete OCs for many adsorption sites. In addition, DOM tends to break through column adsorbers faster than OCs (as described in the following section), which enables DOM to preload adsorption sites deeper in the biochar bed and prior to the arrival of the more strongly adsorbing OCs. Contrastingly, in a batch mode, OCs and DOM are simultaneously exposed to biochar adsorbents, which avoids DOM preloading, and the diffusion coefficient of the OCs is higher than that of many DOM components, which allows the OCs to reach internal adsorption sites more quickly. Despite these differences, the extent to which DOM fouls biochar adsorbents in batch versus column tests is similar for various OC–biochar combinations. These results show it is possible to estimate column use rates from batch test results, permitting a simple approach to predicting biochar adsorber performance for many more OC–biochar combinations in a given amount of time.

## 5 Using ultraviolet absorbance to monitor OC breakthrough in biochar column adsorbers

Optical properties such as ultraviolet absorbance at 254 nm ($UVA_{254}$), which primarily indicates the presence of aromatic DOM, can be used to monitor DOM adsorption and breakthrough in a biochar adsorber. $UVA_{254}$ and SMX breakthrough curves in

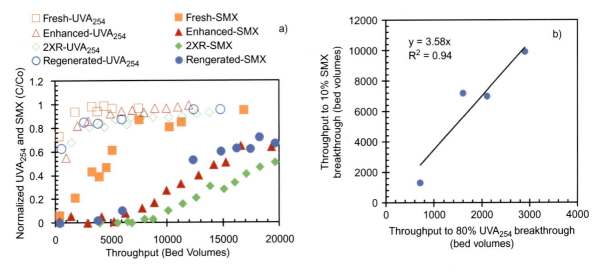

**Fig. 17.6** Adsorption of $UVA_{254}$ and SMX in surface water DOM for four biochars: (A) $UVA_{254}$ and SMX breakthrough curves. (B) Bed volumes to 10% SMX breakthrough versus bed volumes to 80% $UVA_{254}$ breakthrough.

surface water DOM are shown in Fig. 17.6, modified from Greiner et al. (2018), for four different biochar adsorbents. Fresh biochar was made at 850°C in a furnace for 2 h. Enhanced biochar is the fresh biochar that was reheated at 600°C for 2 h. Regenerated biochar is the fresh biochar that was exhausted in a column adsorber for SMX and reheated at 600°C for 2 h. 2XR biochar is the regenerated biochar that was exhausted for SMX and reheated at 600°C for 2 h (for a discussion of biochar regeneration, see Bentley et al. (2022) in this book).

In general, biochars that showed later $UVA_{254}$ breakthroughs also exhibited delayed SMX breakthroughs. This effect can be clearly seen in Fig. 17.6B where the bed volumes to 80% $UVA_{254}$ breakthrough and 10% SMX breakthrough determined from the breakthrough curves in Fig. 17.6A show a strong correlation. These results suggest that biochars that have a greater adsorption strength for a target OC (e.g., SMX) will also more strongly adsorb DOM that produces a $UVA_{254}$ signal. Such a phenomenon can be leveraged to facilitate the monitoring of biochar adsorbers and determine when adsorbent media needs to be changed out. $UVA_{254}$ can be measured rapidly, in the field, and at a fraction of the cost of quantifying OC concentrations that typically require chromatographic coupled with mass spectrometry instrumentation. In remote and low-resource settings, $UVA_{254}$ analysis may be one of the only appropriate analytic techniques.

Since SMX is a more poorly adsorbed compound relative to many other OCs that could be detected in drinking water sources, more strongly adsorbed compounds may still be adequately removed even if SMX breaks through and requires biochar media to be changed out. Thus, using 80% $UVA_{254}$ breakthrough as the threshold for when biochar needs to be changed out with the surface water studied in Greiner et al. (2018) could be an appropriate setpoint to conservatively replace exhausted biochar. The $UVA_{254}$ threshold that should be used to determine when to change out biochar adsorbents likely depends on the affinity of $UVA_{254}$ absorbing constituents and on the strength of the DOM fouling effect as discussed in Kearns et al. (2021). Thus, the $UVA_{254}$ threshold should be determined by developing $UVA_{254}$ and target OCs in the water of interest.

## 6 Summary

Biochar exhibits heterogeneous physicochemical properties, and its performance as an OC removal technology can vary widely depending on those properties. Coupled with complex OC adsorption phenomena, such as DOM fouling, that govern its performance in water treatment applications, the implementation of biochar adsorbents can be challenging. This chapter demonstrated that results from commercially available surface area and elemental analyses can be used to screen for efficient biochar adsorbents. A simple modeling approach outlined that if the $C_0$ of an OC is less than some concentration threshold, it can be assumed that the percent removal of the OC (or its fractional removal) is independent of its $C_0$. For most OCs and waters, the concentration threshold is greater than $10\,\mu g/L$ if DOM, which is ubiquitous in drinking water sources, is present. A linear relationship was developed between the natural logarithms of batch and column use rates at set OC percent removals, which could be used to estimate biochar column adsorber performance using batch tests that are simpler and less resource and time intensive relative to many other methods, such pilot or RSSCT studies. A low-cost and rapid monitoring approach was presented by measuring $UVA_{254}$ breakthrough to indicate target OC breakthrough. Taken together, this chapter outlines methods that can simplify each phase of biochar adsorbent implementation for OC control by (i) informing the selection of efficient biochar adsorbents, (ii) predicting biochar adsorber performance from batch tests and at different OC $C_0$s, and (iii) identifying when biochar media become exhausted for conservative OCs and need to be changed out.

## Questions and problems

1. Why consider biochar to remove organic contaminants from water instead of more established technologies?
2. Provide two reasons why implementing biochar to treat organic contaminants in water can be challenging.
3. What is a shortcoming of using nitrogen ($N_2$) gas analysis to characterize biochar surface area?
4. If only surface area and elemental analyses are available to identify an efficient biochar adsorbent for organic contaminants, which analysis is preferable? Why?
5. When dissolved organic matter is present, how does this impact adsorption isotherms?
6. What factors impact the concentration threshold denoted by when it can no longer be assumed that organic contaminant fractional removal by biochar is not impacted by the organic contaminant initial concentration?
7. Why might an alternative approach to pilot testing be considered when modeling the performance of a full-scale biochar adsorber for organic contaminant control?
8. Provide one reason why batch and column adsorber use rates can be different.
9. Why monitor a dissolved organic matter optical property instead of the organic contaminant concentration when determining when biochar filter media need to be replaced?
10. Why might the relationship between UV absorbance at 254 and organic contaminant breakthrough change for different waters?

## References

Bentley, M.J., Solomon, M.E., Marten, B.M., Shimabuku, K.K., Cook, S.M., 2021. Evaluating landfill leachate treatment by organic municipal solid waste-derived biochar. Environ. Sci.: Water Res. Technol. 7 (11), 2064–2074.

Bentley, M., Kennedy, A., Summers, R.S., 2022. Optimizing biochar adsorption relative to activated carbon in water treatment, Chapter in this book.

Greiner, B.G., Shimabuku, K.K., Summers, R.S., 2018. Influence of biochar thermal regeneration on sulfamethoxazole and dissolved organic matter adsorption. Environ. Sci.: Water Res. Technol. 4, 169–174.

Hale, S.E., Arp, H.P.H., Kupryianchyk, D., Cornelissen, G., 2016. A synthesis of parameters related to the binding of neutral organic compounds to charcoal. Chemosphere 144, 65–74.

Kearns, J.P., Wellborn, L.S., Summers, R.S., Knappe, D.R.U., 2014. 2,4-D adsorption to biochars: effect of preparation conditions on equilibrium adsorption capacity and comparison with commercial activated carbon literature data. Water Res. 62, 20–28.

Kearns, J.P., Shimabuku, K.K., Mahoney, R.B., Knappe, D.R.U., Scott Summers, R., 2015. Meeting multiple water quality objectives through treatment using locally generated char: improving organoleptic properties and removing synthetic organic contaminants and disinfection by-products. J. Water, Sanit. Hyg. Develop. 5, 359–372.

Kearns, J.P., Shimabuku, K.K., Knappe, D.R., Summers, R.S., 2019. High temperature co-pyrolysis thermal air activation enhances biochar adsorption of herbicides from surface water. Environ. Eng. Sci. 36, 710–723.

Kearns, J., Dickenson, E., Knappe, D., 2020. Enabling organic micropollutant removal from water by full-scale biochar and activated carbon adsorbers using predictions from bench-scale column data. Environ. Eng. Sci. 37, 459–471.

Kearns, J., Dickenson, E., Aung, M.T., Joseph, S.M., Summers, S.R., Knappe, D., 2021. Biochar water treatment for control of organic micropollutants with UVA surrogate monitoring. Environ. Eng. Sci. 38 (5), 298–309.

Keiluweit, M., Nico, P.S., Johnson, M.G., Kleber, M., 2010. Dynamic molecular structure of plant biomass-derived black carbon (biochar). Environ. Sci. Technol. 44, 1247–1253.

Kennedy, A.M., Summers, R.S., 2015. Effect of DOM size on organic micropollutant adsorption by GAC. Environ. Sci. Technol. 49, 6617–6624.

Lattao, C., Cao, X., Mao, J., Schmidt-Rohr, K., Pignatello, J.J., 2014. Influence of molecular structure and adsorbent properties on sorption of organic compounds to a temperature series of wood chars. Environ. Sci. Technol. 48, 4790–4798.

Najm, I.N., Snoeyink, V.L., Richard, Y., 1991. Effect of initial concentration of a SOC in natural water on its adsorption by activated carbon. J. Am. Water Works Ass. 83, 57–63.

Sander, M., Pignatello, J.J., 2005. Characterization of charcoal adsorption sites for aromatic compounds: insights drawn from single-solute and bi-solute competitive experiments. Environ. Sci. Technol. 39, 1606–1615.

Shimabuku, K.K., Kearns, J.P., Martinez, J.E., Mahoney, R.B., Moreno-Vasquez, L., Summers, R.S., 2016. Biochar sorbents for sulfamethoxazole removal from surface water, stormwater, and wastewater effluent. Water Res. 96, 236–245.

Shimabuku, K.K., Paige, J.M., Luna-Aguero, M., Summers, R.S., 2017. Simplified modeling of organic contaminant adsorption by activated carbon and biochar in the presence of dissolved organic matter and other competing adsorbates. Environ. Sci. Technol. 51, 10031–10040.

Thompson, K.A., Shimabuku, K.K., Kearns, J.P., Knappe, D.R., Summers, R.S., Cook, S.M., 2016. Environmental comparison of biochar and activated carbon for tertiary wastewater treatment. Environ. Sci. Technol. 50 (20), 11253–11262.

Zhu, D., Pignatello, J.J., 2005. Characterization of aromatic compound sorptive interactions with black carbon (charcoal) assisted by graphite as a model. Environ. Sci. Technol. 39, 2033–2041.

Zhu, D., Kwon, S., Pignatello, J.J., 2005. Adsorption of single-ring organic compounds to wood charcoals prepared under different thermochemical conditions. Environ. Sci. Technol. 39, 3990–3998.

# Nanoscale zero-valent iron-decorated biochar for aqueous contaminant removal

Xuefeng Zhang[a], Tharindu Karunaratne[a], Chanaka Navarathna[b], Jilei Zhang[a], and Charles U. Pittman Jr.[b]

[a]Department of Sustainable Bioproducts, Mississippi State University, Starkville, MS, United States. [b]Department of Chemistry, Mississippi State University, Mississippi State, MS, United States

## 1 Introduction

Recently, nanoscale zero-valent iron (nZVI) has become one of the most widely used environmental nanomaterials for water remediation because of its low cost, environmental friendliness, and high reactivity toward contaminant removal (Bae et al., 2018). nZVI provides higher surface area than bulk iron. It can adsorb contaminants from solutions through physisorption or chemisorption. Moreover, nZVI can also transform toxic contaminants into nontoxic or less toxic compounds through redox reactions because of the low redox potential ($E^0$, $-0.44$ V) (Zhang et al., 2021a). For instance, nZVI can sequester heavy metal cations (e.g., $Cu^{2+}$, $Ag^+$, $Au^{2+}$, $Pb^{2+}$, etc.) by reduction (Liu et al., 2018a; Tang et al., 2020; Gil et al., 2018; Li et al., 2020a; Ling et al., 2018). Moreover, nZVI can also uptake toxic cadmium and arsenic ions chemisorption and Fenton oxidation (Liu et al., 2021a; Yang et al., 2020). Furthermore, nZVI can remove toxic organics, halogenated organic compounds, and nonmetal inorganic species through combined mechanisms including reduction, dechlorination, surface complexation, precipitation, and Fenton degradation (Zhang et al., 2021a; Liu et al., 2018a).

Despite the high efficiency and wide applicability of nZVI particles in water remediation, several limiting issues were reported. First, nZVI particles suffer severe agglomeration after production because of van der Waals forces, electrostatic attraction, and magnetic forces, greatly reducing their surface area and reactivity toward contaminants. Moreover, this aggregation significantly decreased nZVI particle mobility in water, limiting its application for underground water remediation (Tosco et al., 2014). Secondly, nZVI particles suffer rapid passivation during transportation, storage, and usage because of high reactivity for forming iron oxides/hydroxides surface layers that greatly decrease their water remediation efficiency through redox reactions (Bae et al., 2018; Wu et al., 2020). Thirdly, nZVI particles have low selectivity to the target pollutants in aqueous solutions when passivated by water, dissolved oxygen, and other solutes (Jie et al., 2020). To help overcome these issues, various carrier materials such as clay, silicon dioxide, zeolite, metal-organic framework (MOF), graphene, porous carbon, and biochar have been explored to disperse and stabilize nZVI particles (Zou et al., 2016; Ezzatahmadi et al., 2017; Alotaibi et al., 2017; Su et al., 2016; Li et al., 2018, 2017; Zhu et al., 2009; Das and Bezbaruah, 2021; Liu et al., 2014). Biochar supports have been widely used because of their low cost, high specific surface area, and rich surface functional groups. The high specific surface area aids the dispersion of nZVI particles and alleviates the aggregation issue (Zhou et al., 2021). The surface functional groups can serve as the active sites for nucleation and growth of nZVI particles on biochar surface, and also as a stabilization agent for nZVI particles during water remediation through forming stable interact with leached Fe ions (Qian et al., 2017; Yang et al., 2018a). Moreover, biochar is conductive with a short-range ordered graphene structure that serves as an electron shuttle to transfer electrons between nZVI and contaminants (Zhao et al., 2020; Liu et al., 2020a). Furthermore, surface functional groups (e.g., phenol, quinone) in biochar can accept or donate electrons during the contaminant removal, thereby enhancing the removal efficiency (Liu et al., 2021a; Klüpfel et al., 2014; Oh et al., 2017). In this chapter, the recent advances for the synthesis and characterization of biochar-supported nZVI (nZVI@BC) are reviewed. The removal of heavy metal(loid)s and organic contaminants from water using nZVI@BC is summarized and the removal mechanisms are discussed.

## 2 Synthesis of nZVI and biochar-supported nZVI

nZVI has been synthesized through both top-down and bottom-up approaches. The top-down synthesis refers to the disintegration or breakdown of bulk iron materials into iron nanoparticles through physical or chemical methods such as optics, etching, and grinding (Li et al., 2009). The high-speed ball milling method has been commercialized for large-scale nZVI production because of the low energy consumption and short processing time (Mu et al., 2017). On the other hand, the bottom-up approach refers to the assembly of iron atoms or ions together to form iron nanoparticles. This approach includes borohydride reduction, carbothermal reduction, ultrasound-assisted synthesis, and electrochemical methods (Wu et al., 2020; Zou et al., 2016; Crane and Scott, 2012; Huber, 2005; Yan et al., 2012). Besides the above-mentioned strategies, nZVI has also been produced by the thermal reduction of iron oxides in a reducing atmosphere and the green synthesis strategy using plant-derived materials (Mittal et al., 2013). A high-quality review paper on the synthesis of nZVI has been written by Stefaniuk et al. (2016). Fig. 18.1 shows the morphology of nZVI synthesized from different methods under transmission electron microscopy (TEM).

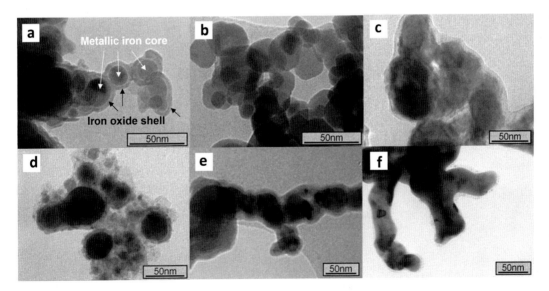

**Fig. 18.1** TEM images show the morphology of iron-based nanoparticles prepared by different methods. (A) nZVI synthesized by the reduction of aqueous $Fe^{2+}$ using sodium borohydride; (B) nanoscale magnetite ($Fe_3O_4$), purchased from Sigma-Aldrich (PubChem Substance ID:24882829); (C) NANOFER STAR, purchased from NANO IRON, s.r.o.; (D) nZVI synthesized by the carbothermal reduction of aqueous $Fe^{2+}$; (E) nZVI synthesized by the reduction of aqueous $Fe^{2+}$ using green tea polyphenols; and (F) nZVI synthesized by the reduction of aqueous $Fe^{2+}$ using sodium borohydride and then annealed under vacuum (<10mbar) at 500°C for 24 h. Modified from Crane, R.A., Scott, T.B., 2012. Nanoscale zero-valent iron: future prospects for an emerging water treatment technology. J. Hazard. Mater. 211–212, 112–125. https://doi.org/10.1016/j.jhazmat.2011.11.073. Copyright 2012, Elsevier.

## 2.1 Synthesis of nZVI/BC by borohydride reduction

Borohydride ($BH_4^-$) reduction of ferrous salt is the commonly used method for the preparation of nZVI@BC because of its ease of synthesis. nZVI particles obtained from borohydride reduction usually have a spherical shape with a diameter between 30 and 100 nm. The spherical nZVI particles have a core-shell structure that composes of a crystallized metallic iron core and a thin-layered amorphous iron oxide shell (Fig. 18.1A). In a typical nZVI@BC preparation process, biochar samples are dispersed in an aqueous solution containing $Fe^{2+}$ and/or $Fe^{3+}$ ions, followed by the addition of sodium borohydride ($NaBH_4$) for $Fe^{2+/3+}$ reduction (Eqs. 18.1 and 18.2). The overall synthesis process is usually conducted in an oxygen-free environment with the purging gas of nitrogen or argon to prevent the oxidation of nZVI (Yan et al., 2012; Stefaniuk et al., 2016).

$$2Fe^{2+} + BH_4^- + 3H_2O \rightarrow 2Fe^0 \downarrow + H_2BO_3^- + 4H^+ + 2H_2 \uparrow \quad (18.1)$$

$$4Fe^{3+} + 3BH_4^- + 9H_2O \rightarrow 4Fe^0 \downarrow + 3H_2BO_3^- + 12H^+ + 6H_2 \uparrow \quad (18.2)$$

Biochar contains negatively charged moieties such as hydroxyls, quinones, and carbonyls. These groups serve to anchor $Fe^{2+/3+}$ complexation and nucleation, as well as growth of nZVI particles (Qian et al., 2017; Xu et al., 2019). Biochar surfaces also attach nZVI particles which have formed in suspension. Moreover, these functional groups could form stable interactions with oxide/hydroxide shells of nZVI particles, lower their mobility, and alleviate the agglomeration of nZVI particles (Wang et al., 2019a; Yuan et al., 2017). The properties of nZVI@BC such as particle size of nZVI, surface oxidation degree, loading of nZVI particles, surface area, and reactivity of the composite can be tailored by manipulating the synthesis condition including biochar type, $Fe^{2+/3+}$ concentration, solution pH, $BH_4^-$ volume, delivery rate, mixing speed, and precursor concentrations (e.g., biochar to $Fe^{2+/3+}$ and $BH_4^-$ to $Fe^{2+/3+}$ ratios) (Liu et al., 2021a; Zhou et al., 2021, 2020; Mortazavian et al., 2018). Table 18.1 summarized nZVI@BC synthesized from various sources and their application in water remediation.

## 2.2 Synthesis of nZVI@BC by carbothermal reduction

Although borohydride reductive nZVI synthesis is simple and convenient on a lab scale, this approach is not applicable for large-scale nZVI production because of the toxicity and high cost of borohydrides. The low-cost conversion of iron oxides or iron

Table 18.1 Summary of biochar-supported nZVI particles synthesized from various sources through borohydride reduction and their application in water remediation.

| Biochar origin | Reduction condition | Size of nZVI | SSA[a] of nZVI@BC (m$^2$/g) | PZC[b] of nZVI@BC | Contaminants studied | References |
|---|---|---|---|---|---|---|
| Coffee ground | Oxic condition, 25°C, 1.50 h | <100 nm | 15.23 ± 0.27 | N.A. | Pb(II), Cd(II), As(III), As(V) | Park et al. (2019) |
| Iron sludge | N$_2$ purging 1 h, 25°C, 1.50 h | 30 nm | 43.97 | N.A. | Sb(III) | Wei et al. (2020) |
| Rice hull | N$_2$ purging 1 h, 25°C, 1.50 h, pH 5.0 | 30 nm | 205.35 | N.A. | Trichloroethylene | Yan et al. (2015) |
| Rice husk | N$_2$ purging 1 h, 25°C, 1 h 20 min | 200–950 nm | 61.03 | N.A. | Nitrobenzene | Zhang et al. (2018a) |
| Oak saw dust | Oxic condition, 25°C, 1.50 h | 200 nm | 264.9 | N.A. | Nitrobenzene Aniline | Wei et al. (2019) |
| Wheat straw/ carboxymethyl cellulose | N$_2$ purging 30 min, 25°C for 1.5 h | 100 nm | 11.6 | 3.0–4.0 | Cr(VI) | Zhang et al. (2019a) |
| Maize straw | N$_2$ purging 30 min, 20–25°C for 24.5 h | 40–100 nm | 30.5 | N.A. | Cr(VI) | Wang et al. (2019b) |
| Animal bone | Oxic condition, 25°C for 1 h | 0.5 μm | 509.21 | N.A. | Cr(VI) | Liu et al. (2021b) |
| Longan shells | N$_2$ purging 30 min, 45°C for 6 h | N.A. | 785.81 | N.A. | U(VI) | Zhang et al. (2021b) |
| Corn stalks | N$_2$ purging 30 min, 25°C for 3.5 h | 100–230 nm | N.A. | N.A. | Pb(II) | Li et al. (2020a) |
| Waste date palm | N$_2$ purging 30 min, 25°C for 1.5 h, pH 5.0 | N.A. | 220.92 | 5.2 | As(V) | Ahmad et al. (2020) |
| Sludge derived | N$_2$ purging 15 min, 25°C for 4 h | N.A. | 81.5914 | 4.32 | Cr(III), Cr(VI) | Qiu et al. (2020) |
| Corncob waste | N$_2$ purging 20 min, 25°C for 25 h | 91 nm | 50.0–90.2 | 7.6 | V(V) | Fan et al. (2020) |
| Palm fiber powder | N$_2$ purging 1 h min, 25°C for 3 h | 40–60 nm | 60.08 | N.A. | Cd(II), As(III) | Liu et al. (2020a) |

[a]SSA: specific surface area, determined by N$_2$ adsorption at −196°C.
[b]PZC: point of zero charges.

salts to metallic iron particles can also be achieved by the thermal reduction route. High-carbon content coke has been used for metallic iron and steel production from iron core since the 18th century (Babich and Senk, 2019). Reductive gases such as hydrogen and carbon monoxide reduce iron oxides to zero-valent iron at temperatures above 600°C (Zhang et al., 2017a). Because reductive gases and solid carbon are generated during biomass thermal degradation, copyrolysis of biomass with iron salts/oxides has been used for the synthesis of nZVI and nZVI@BC (Zhang et al., 2021a; Wu et al., 2015). Unlike borohydride reduction, carbothermal reduction produces a variety of iron phases including iron oxide ($Fe_2O_3$), metallic iron (ferrite and austenite), and iron carbide nanoparticles (Zhang et al., 2017a, 2018b; Yan et al., 2018a; Neeli and Ramsurn, 2018; Mun et al., 2015). Generally, the properties of the generating nZVI particles depend on the temperature, carbon/biomass origin, ramping rate, purging gas, and iron source (Zhang et al., 2017a, 2018b; Neeli and Ramsurn, 2018; Mun et al., 2015; Yan et al., 2018b,c).

Our group investigated the cocarbonization of lignin with iron nitrate and found that the iron phase transformation undergoes the following steps: (1) decomposition of iron nitrate to form iron oxides ($Fe_2O_3$) at 200–300°C; and (2) reduction of $Fe_2O_3$ to FeO at 550°C and subsequently to metallic iron nanoparticles at 600°C (Zhang et al., 2017a; Zhang, 2016). We also demonstrated that iron particles interact with solid amorphous carbon at high temperatures. nZVI particles produced through copyrolysis of lignin and iron nitrate at 600°C showed a body-centered cubic structure, namely α-Fe (Zhang et al., 2017a). At 700°C, face-centered cubic iron nanoparticles (γ-Fe) were formed because of the dissolution of amorphous carbon into α-Fe (Zhang et al., 2018b). γ-Fe nanoparticles predominate at 900°C. At 1000°C, further dissolution of amorphous carbon into γ-Fe generates iron carbides ($Fe_3C$).

The pyrolysis of lignocellulosic materials generates lots of carbonaceous gases, which can react with iron at temperatures above 600°C to form iron carbides and graphitic carbons (Neeli and Ramsurn, 2018; Mun et al., 2015; Yan et al., 2018c; Zhang et al., 2018c). However, the formation of iron carbides from solid amorphous carbon usually requires a high temperature of 1000°C because of the slow kinetics for solid carbon diffusion to iron nanoparticles (Zhang et al., 2018b). The properties of nZVI@BC such as SSA, nZVI particle size, iron phase, iron crystal size, and graphitic carbon content can be manipulated by adjusting the synthesis conditions (e.g., biomass precursor, temperature, iron precursor, etc.) (Zhang et al., 2018b; Yan et al., 2018a,b; Wang et al., 2017a; Meng et al., 2020). Table 18.2 summarizes the

Table 18.2 Summary of nZVI@BC synthesized from various sources through carbothermal reduction and their application in water remediation.

| Biochar origin | Carbonization condition | Size of nZVI | SSA[a] of nZVI@BC ($m^2/g$) | PZC[b] of nZVI@BC | Contaminant studied | References |
|---|---|---|---|---|---|---|
| Bamboo | 5°C/min to 800–1200°C, 1 h, Ar gas of 300 mL/min | 5 μm | N.A. | N.A. | N.A. | Wu et al. (2015) |
| Starch | 5°C/min to 700–900°C, 2 h, $N_2$ gas | 200 nm | 25–1017 | N.A. | U | Zhang et al. (2019b) |
| Corn stover | 5°C/min to 800°C, 2 h, $N_2$ gas of 80 mL/min | 50 nm | 224.8 | N.A. | Cr(III) Cr(VI) | Li et al. (2019a) |
| Lignocellulose waste | 5°C/min to 850°C, 2 h, $N_2$ gas of 80 mL/min | ~4 nm | 140–220 | N.A. | Organic pollutants (PMS, DMPO, TEMP, BPA) | Dai et al. (2019) |
| Coir pith | 600°C, 1 h, $N_2$ gas of 35 mL/min | 100–400 nm | 315 | N.A. | Trichloroethylene | Su et al. (2013) |
| Bamboo | 45°C/min to 600°C, 5 h, $N_2$ gas of 200 mL/min | N.A. | N.A. | N.A. | Chlorobenzene | Zhang and Wu (2016) |
| Sewage sludge | 5°C/min to 500°C, 1 h, $N_2$ gas of 200 mL/min | N.A. | N.A. | N.A. | Cr(III), Cr(VI) | Liu et al. (2020b) |
| Bamboo | 600–1000°C, $N_2$ gas | 3–10 μm | 484.6 | N.A. | Methylene blue | Xing et al. (2019) |
| Starch | 5°C/min to 700–900°C, 1 h, $N_2$ gas | N.A. | 378.7 | N.A. | Cr(VI) | Zhuang et al. (2014) |
| Lignin | 10°C/min to 1000°C, 1 h, $N_2$ gas of 1 L/min | 5–15 nm | 10 | 7.8 | Pb(II), As(III), phosphate, nitrate | Zhang et al. (2021a) |

[a]SSA: specific surface area, determined by $N_2$ adsorption at −196°C.
[b]PZC: point of zero charge.

properties of nZVI@BC synthesized from various sources through carbothermal reduction and their application in water remediation.

## 3 Advantages of biochar as the nZVI support

nZVI particles suffer several issues for water remediation applications, such as aggregation (Liu et al., 2015), passivation (Bae et al., 2018), low mobility (Tosco et al., 2014), and reduced electron transfer (Wang et al., 2019a). Applying biochar as a nZVI particle carrier to achieve immobilization is beneficial from the following four aspects. First, biochar immobilization can reduce the nZVI aggregation and enable the tailoring of its particle size (Wang et al., 2019a). Biochar contains abundant pores and with large specific surface area, which allows the good dispersion of nZVI (Zhou et al., 2021). Biochar surface functionalities such as hydroxyls can serve as the fast nucleation sites during nZVI synthesis through the $NaBH_4$ process (Wang et al., 2019a), thus leading to a smaller crystal size (Wang et al., 2017b). Moreover, the surface functional groups (e.g., hydroxyls) can also interact with nZVI to form stable bonds, thereby reducing the intraparticle interactions among nZVI particles and their aggregation (Ruiz-Torres et al., 2018; Jiang et al., 2019). The particle size of nZVI can be manipulated by the selection of biomass origins and the production conditions. For instance, cellulose-based biochar favors the formation of small nZVI particles (20–100 nm) using $NaBH_4$ reduction, while hemicellulose produces nZVI particles with a medium size (50–150 nm), and lignin-based biochar produces large nZVI particles (100–200 nm) (Zhou et al., 2021).

Biochar supports also been known to decrease surface passivation and iron ion leaching from nZVI (Wang et al., 2019a; Park et al., 2019; Lei et al., 2018). It is believed that surface functional groups in biochars could interact with nZVI and react with the leached Fe ions to form stable structures, thereby reducing the nZVI corrosion and the iron ion leaching (Qian et al., 2017; Jiang et al., 2019). Moreover, biochars pose abundant micro-, meso-, and macropores, which not only provide the room for nZVI to form and accommodate but also lower the diffusion rate of oxidative species for nZVI oxidation (Li et al., 2018).

Another advantage of utilizing biochar for nZVI supports is the enhanced reactivity of nZVI toward contaminant removal through reduction and catalytic degradation. Biochar is a type of pyrogenic carbon that containing short-range ordered graphene crystallites (Zhang et al., 2017b). These endow biochar with a high electron transfer rate. This facilitates the reduction and catalytic degradation of contaminants (Mortazavian et al., 2018; Jiang et al., 2019). Moreover, biochar surface functionalities assist in electron shuttle, accepting or donating electrons among the biochar, nZVI particle, and contaminant solution phases (Xu et al., 2019; Yu et al., 2015). For instance, quinone moieties severe as the electron acceptor to accelerate the arsenic removal through Fenton oxidation (Liu et al., 2021a), and phenolic moieties, and particularly ortho- and para-dihydroxy aromatic rings in biochar donate electrons that reduce contaminants as they are countered to quinones (Klüpfel et al., 2014).

## 4 Characterization of nZVI@BC

nZVI particles usually are spherical with diameters between 50 and 100 nm, which are composed of metallic iron cores and iron oxide shells (Fig. 18.1A). The shell with a thickness of 2–4 nm is formed rapidly after nZVI production due to the surface oxidation of the $Fe^0$ core. Moreover, the shell has an amorphous structure, consisting of both Fe(II) and Fe(III) oxides (Wang et al., 2009a). nZVI particles tend to aggregate into chains or cluster-like structures (Figs. 18.1 and 18.2) because of the electrostatic attraction and magnetic forces among particles (Crane et al., 2011). The size of nZVI particles, the grain size of $Fe^0$ core, thickness and chemical composition of nZVI particles' oxidation layer, and degree of aggregation have key influences on nZVI particle performances

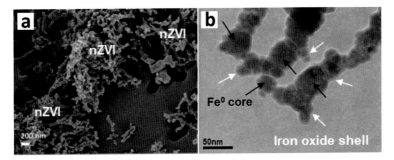

**Fig. 18.2** SEM and TEM characterizations of nZVI/BC composites prepared from the coffee ground through borohydride reduction approach. Reproduced with permission from Park, M.H., Jeong, S., Lee, G., Park, H., Kim, J.Y., 2019. Removal of aqueous-phase Pb(II), Cd(II), As(III), and As(V) by nanoscale zero-valent iron supported on exhausted coffee grounds. Waste Manag. 92, 49–58. https://doi.org/10.1016/j.wasman.2019.05.017 © 2019 Elsevier Inc.

in water remediation. When nZVI particles are supported on biochar, the biochar's porosity, surface area, elemental composition, surface functionalities also affect their water remediation performance.

## 4.1 Electron microscopies

The morphology of nZVI@BC is usually investigated by scanning electron microscopy (SEM), where the average particle size and particle size distribution of nZVI particles can be obtained by statistical analysis. When equipped with energy-dispersive X-ray spectroscopy (EDS), SEM-EDS analysis provides elemental composition information of nZVI particles and biochars, as well as the distribution of nZVI in biochars. Less common, atomic force microscopy (AFM) is also be used for particle size analyses of nZVI (Khajouei et al., 2017; Arthy and Phanikumar, 2016).

TEM is also useful to study morphology, particle size, and distribution of nZVI particles. Similar to SEM, TEM-EDS can also provide elemental composition information of nZVI particles and biochars with a better resolution. Moreover, high-resolution TEM allows visualization of the nZVI crystal structure, that is, the crystal lattices and orientation of the $Fe^0$ core and iron oxide shell. Wang et al. investigated the oxidation of nZVI under ambient conditions using TEM (Wang et al., 2005). A critical size of 8 nm exists for nZVI oxidation. nZVI particles with diameters less than 8 nm were fully oxidized under ambient conditions. The Caberra-Mott model simulation results indicated that forming a 1 nm thick iron oxides layer only takes 0.2 fs, and forming a 2 nm oxides layer needs 40 s, while it takes 600 weeks for the growth of 2 nm oxides layer to 4 nm. This was because the oxidation of nZVI is driven by the diffusion of iron atoms from the core to the shell, while the increasing oxide layer thickness reduces the diffusion rate exponentially. This study also answered the question of why the thickness of nZVI's oxidation layer is mainly between 2 and 4 nm.

TEM is also very useful to visualize the interactions between nZVI and contaminants during their uptake. This provides unique perspectives to interpret removal mechanisms (Tuček et al., 2017; Ling et al., 2015). Most recently, a spherical aberration-corrected scanning transmission electron microscope (Cs-STEM) has been used to map the surface of nZVI at the atomic level. Zhang et al. studied the reactions of heavy metals with nZVI particles using Cs-STEM equipped with an EDS (Fig. 18.3) (Ling et al., 2018, 2017; Liu et al., 2018b; Ling and Zhang, 2017). Because of the high resolution (0.1 nm) and low detection limit (0.02 wt%.), the

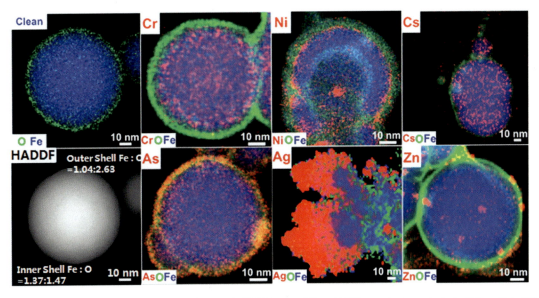

**Fig. 18.3** (A, B) The high-angle dark-field (HAADF) STEM image (A) and EDS mapping spectral of nZVI particle (B); (C–H) EDS mapping spectral of nZVI particles after the removal of Cr (C), As (D), Ni (E), Ag (F), Cs (G), and Zn (H). Data from Ling, L., Huang, X., Li, M., Zhang, W., 2017. Mapping the reactions in a single zero-valent Iron nanoparticle. Environ. Sci. Technol. 51, 14293–14300. https://doi.org/10.1021/acs.est.7b02233 © 2017 American Chemical Society.

STEM-EDS system is capable of generating atomic-scale details on the elemental distributions and migration of atoms in the nZVI. It can also map the mass transfer, reactions, and deposition of contaminants onto nZVI quantitatively.

### 4.2 X-ray spectroscopies

X-ray diffraction (XRD) analysis can provide iron phase and crystal information of nZVI particles (Fig. 18.4). The XRD peaks centered at 44.8 degrees and 43.7 degrees are usually the indications of the existence of metallic body-centered cubic (BCC) iron (α-Fe) and face-centered cubic (FCC) iron (γ-Fe), respectively. The NaBH$_4$ reduction approach produces nZVI particles contain singe phase α-Fe, while the carbothermal reduction approach produces nZVI particles that usually contain multiple iron phases (e.g., α-Fe and γ-Fe) (Zhang et al., 2021a, 2018b). Additionally, iron carbides (e.g., Fe$_3$C, Fe$_5$C$_2$, etc.) are commonly seen in nZVI@BC samples synthesized through the carbothermal reduction approach because of the carbon diffusion into metallic iron nanoparticles (Neeli and Ramsurn, 2018; Mun et al., 2015; Zhang et al., 2018c). The crystal size of iron nanoparticles can be determined using the Scherrer equation (Zhang et al., 2017a). However, XRD could not provide the crystal information from the iron oxide shell

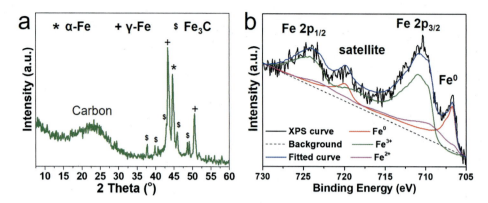

**Fig. 18.4** XRD (A) and XPS (B) pattern of nZVI@BC through carbothermal reduction of iron nitrate with the presence of lignin at 1000°C. Reproduced with permission from Zhang, X., Navarathna, C.M., Leng, W., Karunaratne, T., Thirumalai, R.V.K.G., Kim, Y., Pittman Jr., C.U., Mlsna, T., Cai, Z., Zhang, J., 2021. Lignin-based few-layered graphene-encapsulated iron nanoparticles for water remediation. Chem. Eng. J. 417, 129199. https://doi.org/10.1016/j.cej.2021.129199 © 2021 Elsevier Inc.

of nZVI particles. This is because either the crystal size is too low or because of a low crystallinity of the iron oxide shell.

The surface characteristics of nZVI critically influence the physical and chemical properties of nZVI during contaminant removal. Iron corrosion, particle stabilization, pollutant adsorption, and reoxidation, and reduction transformations can be modified. XPS is a surface characterization technique that has been widely used to investigate nZVI surface characteristics, as well as the interactions between nZVI and contaminants (Zhang et al., 2021a). The thin nZVI surface iron oxide forms spontaneously during synthesis and undergoes continuous and slow evolution throughout the life of the nanoparticles. XPS analysis confirmed a surface oxide layer composed mainly of FeOOH, FeO, $\alpha/\gamma$-$Fe_2O_3$, and $Fe_3O_4$ (Zhang et al., 2021a; Signorini et al., 2003; Wang et al., 2009b). The thickness of the iron oxide layer can also be determined by XPS (Martin et al., 2008), as well as the distribution of $Fe^{2+}$ and $Fe^{3+}$ in the layer (Mu et al., 2017).

## 5 nZVI@BC for water remediation

### 5.1 Removal of heavy metal(loid)s using nZVI@BC

Heavy metals are defined as metals with a density higher than 5000 kg/m³, and there are about 45 examples in nature. Some heavy metals such as nickel, lead, chromium, mercury, and cadmium are poisonous to organisms. Arsenic is a highly toxic metalloid usually seen in natural spring water. Because the chemical properties and migration behavior in the environment of arsenic are similar to heavy metals, arsenic has been also classified as a heavy metal (Tchounwou et al., 2012). Recently, increasing water pollution by heavy metals has resulted in severe global public health and ecological issues worldwide. The chemical forms of heavy metal ions are complex and pH dependent. Copper ($Cu^{2+}$), lead ($Pb^{2+}$), and cadmium ($Cd^{2+}$) exist in the water as cations, while arsenic ($AsO_3^{3-}$ and $AsO_4^{3-}$), chromium ($CrO_4^{2-}$ and $CrO_7^{2-}$), and selenium ($SeO_3^{2-}$ and $SeO_4^{2-}$) exist in the water as oxyanions. Moreover, heavy metals exist in water and soil usually at very low concentration levels, which makes decontamination challenges. Furthermore, the metallic oxyanions usually have a polyvalent state, and their valent states and anion forms are very sensitive to environmental conditions such as solution pH and redox potential. These characteristics make oxyanion removal from water more difficult.

nZVI can remove various heavy metals from water via diverse mechanisms including reduction, physisorption, coprecipitation, chemisorption, and Fenton oxidation (Fig. 18.5). Metals having a redox potential ($E_h^0$) higher than that of $Fe^0$, for example, $Cu^{2+}$, $Pb^{2+}$, $Cr^{6+}$, $Se^{2+}$, and $UiO^{2+}$, undergo the reduction and coprecipitation both on nZVI and biochar surfaces (Zhang et al., 2021a,b; Liu et al., 2018a; Yang et al., 2018a; Zhao et al., 2020). Metals that

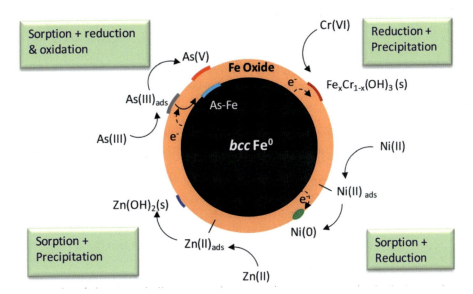

**Fig. 18.5** Role of the core-shell structure in contaminant sequestration. Reproduced with permission from Yan, W., Lien, H.-L., Koel, B.E., Zhang, W., 2012. Iron nanoparticles for environmental clean-up: recent developments and future outlook. Environ. Sci. Process. Impacts 15, 63–77. https://doi.org/10.1039/C2EM30691C © 2013 Royal Society of Chemistry.

have lower redox potentials than $Fe^0$, for example, Zn, the adsorb and precipitate on nZVI@BC as the main removal mechanisms (Yang et al., 2018a). Additionally, As(III)/As(V) is also removed by the Fenton oxidation by nZVI@BC (Zhang et al., 2021a; Yang et al., 2020; Liu et al., 2020a; Wang et al., 2017b).

The nZVI core-shell structure exhibits different functions in heavy metal removal. The iron oxide shell adsorbs pollutants from water by coordination and electrostatic attraction. Conversely, the $Fe^0$ core uptakes contaminants through a series of redox reactions. Moreover, the corrosion of the $Fe^0$ core by water generates $HO^-$ ions that could raise the local solution pH and facilitate heavy metal removal through precipitation. Furthermore, the $Fe^{2+}$ and $Fe^{3+}$ ions generated from $Fe^0$ corrosion could facilitate heavy metal removal (e.g., $Cr^{6+}$, $As^{5+}$) through coprecipitation.

Using biochars as a supporting material enhanced heavy metal removal efficiency by nZVI (Zhao et al., 2020; Zhang et al., 2021b; Liu et al., 2019a; Diao et al., 2018). The recent heavy metal removal studies by nZVI@BC are listed in Table 18.3. The role of biochar in heavy metal removal is enhanced by two factors. First, biochar's surface functional groups attract metal cations and anions on its surface through chemisorption or surface complexation. Second, biochar with a high electron transfer rate allows electron transfer from nZVI to its surface, promoting reduction of adsorbed ions. The performance of nZVI/BC varies with the types of metal ions, the biochar surface characteristics, as well as the biochar production conditions. Generally, a negatively charged biochar surface binds with cations while a positively charged surface attracts anions. Meanwhile, surface functional groups such as C=O and OH of biochar can form coordination complexes with ions such as $Pb^{2+}$ and $Cu^{2+}$ (Liu et al., 2018a; Yang et al., 2018a; Diao et al., 2018). Biochar produced at higher pyrolysis temperature exhibits higher electric conductivity (more graphitization) but has fewer surface functional groups for metal ion complexation (Qian et al., 2017; Zhao et al., 2020; Zhang et al., 2017b, 2019c). During heavy metal removal, metal ions are adsorbed on biochar surface, followed by electron transfer from $Fe^0$ core to biochar through an iron oxide shell. The conductive graphene and surface functions (Zhao et al., 2020; Qin et al., 2020) shuttle electrons into biochar and to the adsorbed heavy metals. This is followed by the heavy metal sequestration by mechanisms such as reduction, precipitation, Fenton oxidation, etc. (Zhang et al., 2021a).

Dong et al. found that the pretreatment of biochar improved the removal efficiency of heavy metals by nZVI@BC (Rajapaksha et al., 2016). Specific surface area (SSA) of biochar was found to increase by 1.5–7 times after pretreatment with KOH, $H_2O_2$, and

Table 18.3 Summary of removal performance of nZVI@BC for different heavy metals.

| Biochar origin | Removal condition | Temp. (°C) | pH | $q_e$ (mg/g) | References |
|---|---|---|---|---|---|
| Sewage sludge | Oxic | 25 | 2.0 | 60.61 ($As^{5+}$) | Liu et al. (2021c) |
| Corn stalk | Oxic | 25 | 6.0 ($Cd^{2+}$) | 33.81 ($Cd^{2+}$) | Yang et al. (2020) |
| | | | 4.0 ($As^{3+}$) | 148.5 ($As^{3+}$) | |
| Biochar | Oxic | 25 | 4.0 | 31.53 ($Cr^{6+}$) | Fan et al. (2019) |
| Herbal residue | Oxic | 25 | 2.5 | 126.12 ($Cr^{6+}$) | Gao et al. (2018) |
| Biochar | Oxic | 30 | 6.0 | 55.14 (U) | Zhang et al. (2019b) |
| Bamboo | Oxic | 22 ± 0.5 | NA | 135.61 ($Ag^+$) | Wang et al. (2017a) |
| Amino-modified biochar | Oxic | 25 | 5.0–6.0 | 12.4 ($Cd^{2+}$) | Yang et al. (2018b) |
| | | | | 8.89 ($Ni^{2+}$) | |
| | | | | 7.87 ($Cu^{2+}$) | |
| | | | | 20.1 ($Cr^{6+}$) | |
| | | | | 28.7 ($As^{5+}$) | |
| Corn stalk | Oxic | 25 ± 2 | 5.0 | 17.8 ($Cr^{6+}$) | Dong et al. (2017a) |
| Rice straw | Oxic | 25 | 4.0 | 40.0 ($Cr^{6+}$) | Qian et al. (2017) |
| Corn stalk | Oxic | 25 ± 1 | 7.0 | 195.1 ($Pb^{2+}$) | Yang et al. (2018a) |
| | | | | 161.9 ($Cu^{2+}$) | |
| | | | | 109.7 ($Zn^{2+}$) | |
| Shell biomass | Oxic | 30 | 6.0 | 93.01 ($Pb^{2+}$) | Wang et al. (2020) |
| | | | | 46.07 ($Cd^{2+}$) | |
| Corn stalk | Oxic | 25 ± 1 | 6.0 | 480.9 ($Pb^{2+}$) | Li et al. (2020a) |
| Lignin | Oxic | 25 | 6.0 | 127.8 ($Pb^{2+}$) | Zhang et al. (2021a) |
| | | | 7.0 | 107.2 ($As^{3+}$) | |

HCl, which ensured a better dispersion of nZVI onto biochars. The uptake of Cr(VI) through reduction and precipitation was greatly enhanced. nZVI@BC obtained from acid-treated biochar exhibited the best Cr(VI) oxyanion removal performance, because acid treatment protonated the biochar, creating a more positively charged surface that attracted Cr(VI) oxyanions onto the nZVI@BC for reduction and precipitation.

Qian et al. investigated the effect of biochar production temperature on the Cr(VI) removal performance of nZVI@BC (Qian et al., 2017). Biochars were produced from the pyrolysis of rice straw at 100–700°C under oxygen-limited conditions (Qian and Chen, 2013). They found that biochar produced at moderate pyrolysis temperature (i.e., 400–500°C) exhibited better Cr(VI) removal performance when loaded with nZVI. Carboxyl groups and silicon minerals on biochars were believed to serve as dual support sites for nZVI reduction and precipitation during nZVI@BC preparation. Biochars produced the moderate pyrolysis temperature seemed to contain more carboxyl groups and silicon minerals, which led to a higher nZVI loading and thereby increased the Cr(VI) removal capacity. However, the authors haven't provided the quantitative comparison of carboxyl and silicon mineral contents, as well as the nZVI loading in biochars. Therefore, the differences in Cr(VI) removal capacity among different temperature biochars could also attribute to other characteristics of biochars such as electron transfer capacity (Klüpfel et al., 2014).

Later, Yang et al. decorated nZVI onto corn stalk biochar and used the nZVI@BC for simultaneous removal of Cd(II) and As(III) from aqueous solution (Yang et al., 2020). The production temperature of biochars affected the removal performance of Cd(II) and As(III). Corn stalk-based biochar produced at moderate pyrolysis temperature (e.g., 500°C) exhibited higher Cd(II) and As(III) uptake capacities. The uptake capacities of Cd(II) and As(III) were found to increase with nZVI loading. This demonstrated that nZVI caused Cd(II) and As(III) removal. At the optimized nZVI loading (i.e., 50wt%), the maximum Langmuir capacities of nZVI@BC for Cd(II) and As(III) were 33.8 and 148.5 mg/g, respectively. Synergistic effects were found in the mixed Cd(II) and As(III) solution that considerably enhanced the uptake capacity of Cd(II) (179.9 mg/g) and As(III) (158.5 mg/g) by nZVI@BC. Although the cause of the synergistic effect was unclear, this study demonstrates nZVI@BC could be a good candidate for the treatment of wastewater containing both Cd(II) and As(III).

Li et al. investigated As(III) removal under aerobic conditions. The oxidation and removal of As(III) were found to occur simultaneously (Liu et al., 2021a). Specifically, the corrosion of nZVI in the aerobic aqueous solution forms $HO^-$, $O_2^{\bullet-}$, $H_2O_2$, and iron hydroxides. $O_2^{\bullet-}$ was responsible for the oxidation of As(III) to As(V), while iron hydroxides were responsible for the removal of As(III) to As(V) through chemisorption. The high electron transfer capacity of biochar facilitates nZVI corrosion and further enhances the generation of $O_2^{\bullet-}$. This increases the As(III) oxidation rate to As(V) and removal. Moreover, the surface quinone groups in biochar also play important role in As(III) oxidation. They were acting as the electron acceptor and forming semiquinone moieties that activate oxygen to generate $O_2^{\bullet-}$.

## 5.2 Removal of organics using nZVI@BC

Organic pollutants such as phenols, chlorinated organic compounds (COCs), endocrine-disrupting chemicals, dyes, pesticides, and polyaromatic compounds present in contaminated water and soils have induced severe environmental problems (Saha et al., 2017). nZVI can remove oxidative organic contaminants through reduction reactions because of its low standard reduction potential (−0.44 eV) (Mu et al., 2017). nZVI@BC degrade and remove COCs from aqueous solutions has been widely reported (Meng et al., 2020; Liu et al., 2019b; Zhang et al., 2021c; Wu et al., 2019; Shan et al., 2021). In these studies, $Fe^0$ and $Fe^{2+}$ were served as the main reductants for COC dehalogenation through the following reaction paths (Eqs. 18.3 and 18.4) (Mu et al., 2017). nZVI@BC outperformed nZVI for the removal of COCs. This is because biochar can adsorptively capture COCs from an aqueous solution for nZVI reduction and also enhance the electron transfer between COCs and nZVI (Tang et al., 2011; Dong et al., 2017b; Li et al., 2019b).

$$RX + Fe^0 + H^+ \rightarrow RH + X^- + Fe^{2+} \quad (18.3)$$

$$2Fe^{2+} + RX + H^+ \rightarrow 2Fe^{3+} + RH + X^- \quad (18.4)$$

In anoxic aqueous conditions, nZVI is mainly acted as a reductant for organic contaminant removal and degradation. When exposing to aerobic aqueous solutions, nZVI can produce reactive oxygen species (ROS), hydroxyl radical ($^\bullet OH$), and superoxide radical ($^\bullet O_2^-$) (Eqs. 18.5–18.9) (Mu et al., 2017; Wang et al., 2019a). These radicals can degrade organic contaminants including bisphenol (Liu et al., 2018a), malachite green (Diao et al., 2018), and 4-chlorophenol (Li et al., 2020b).

$$O_2 + Fe^0 + 2H^+ \rightarrow Fe^{2+} + H_2O_2 \quad (18.5)$$

$$H_2O_2 + Fe^0 + 2H^+ \rightarrow Fe^{2+} + 2H_2O \quad (18.6)$$

$$H_2O_2 + Fe^{2+} \rightarrow Fe^{3+} + {}^\bullet OH + OH^- \quad (18.7)$$

$$H_2O_2 + Fe^{2+} \rightarrow Fe_{(IV)} = O^{2+} + H_2O \ (pH > 5) \quad (18.8)$$

$$Fe^{2+} + O_2 \rightarrow Fe^{3+} + O^{\bullet 2-} \ (pH \sim 7) \quad (18.9)$$

Meng et al. prepared nZVI@BC from copyrolysis of lignin and iron salt at 900°C (Meng et al., 2020). The nZVI@BC was composed of iron carbide ($Fe_5C_2$ and $Fe_3C$) cores shelled with graphitic carbon layers (Fig. 18.6A–D). The presence of iron carbides played a critical effect on the dichlorination of trichloroethene from anaerobic water. The iron carbides reacted with water forming atomic

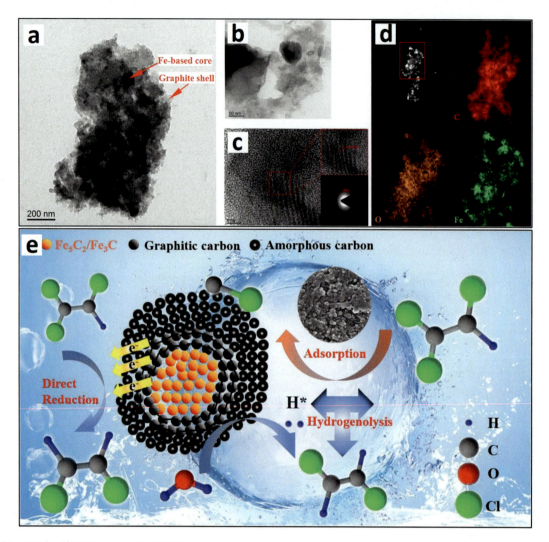

**Fig. 18.6** (A–C) TEM images of nZVI@BC show iron nanoparticles are encapsulated in carbon shell (A, B), the selected area diffraction pattern of TEM image (C) shows the diffraction pattern of less crystalline $Fe_5C_2$, and the TEM-EDS maps (C) show the element distribution of C, O, and Fe in nZVI@BC; (D) proposed TCE removal mechanism by Fe-LBC30. Modified from Meng, F., Li, Z., Lei, C., Yang, K., Lin, D., 2020. Removal of trichloroethene by iron-based biochar from anaerobic water: key roles of Fe/C ratio and iron carbides. Chem. Eng. J. 413, 127391. https://doi.org/10.1016/j.cej.2020.127391 © 2020 Elsevier Inc.

hydrogen (H•), which degrades trichloroethene through hydrolysis. The authors also investigated the effect of Fe/C on trichloroethene dichlorination efficiency. A lower or higher Fe/C ratio deteriorated the overall dichlorination performance of nZVI@BC. At the optimal iron loading (Fe/C ratio of $1:10^4$ to 1:30), high removal efficiencies of >97% were achieved when the initial trichloroethene concentration was 10.4 mg/L.

Most recently, nZVI@BC was used as a peroxydisulfate (PDS) activator to generate sulfate radicals ($SO_4^{\bullet-}$) for oxidative degradation of organic contaminants (Liu et al., 2018a; Jiang et al., 2019). For example, nZVI@BC can remove bisphenol through both adsorption and oxidative degradation with the presence of PDS (Liu et al., 2018a). In a typical removal process, bisphenol A was adsorbed on the biochar surface, then $Fe^0$ passivated to $Fe^{2+}$, which activated persulfate to form oxidative $SO_4^{\bullet-}$. Meantime, the passivation of $Fe^0$ with the presence of dissolved oxygen generated $OH^{\bullet}$ radicals through the Fenton reaction. Both $SO_4^{\bullet-}$ and $OH^{\bullet}$ were responsible for bisphenol A degradation to small molecules such as *p*-isopropenyl phenol, 4-isopropyl phenol, 4-hydroxy acetophenone, *p*-hydroquinone, fumaric acid, and 2-hydroxypropionic acid.

nZVI@BC was also used to activate persulfate for atrazine degradation (Zhang et al., 2021c). The atrazine degradation activation energy of nZVI@BC was 83 kJ/mol, which was significantly lower than that of heat activation (141 kJ/mol). A high atrazine removal rate of 93.8% was achieved for nZVI@BC. Various reactions including alkyl oxidation, dealkylation, dehydrogenation, dechlorination hydroxylation, and olefination occurred during atrazine degradation (Fig. 18.7). Persistent free radicals (PFRs) including $^1O_2$, $SO_4^{\bullet-}$, and $^{\bullet}OH$ were responsible for the degradation of atrazine.

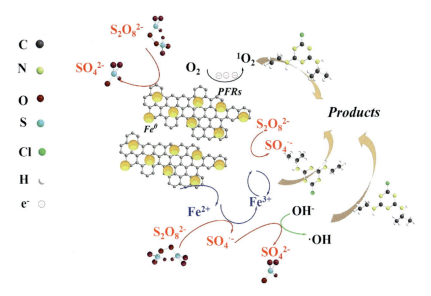

**Fig. 18.7** Multimechanisms of nZVI@BC for highly efficient activating persulfate to oxidatively degrade atrazine.
Modified from Zhang, Y., Jiang, Q., Jiang, S., Li, H., Zhang, R., Qu, J., Zhang, S., Han, W., 2021. One-step synthesis of biochar supported nZVI composites for highly efficient activating persulfate to oxidatively degrade atrazine. Chem. Eng. J. 420, 129868. https://doi.org/10.1016/j.cej.2021.129868 © 2021 Elsevier Inc.

Li et al. prepared highly stable nZVI@BC composites from the cocarbonization of glucose and iron oxides for 4-chlorophenol (4-CP) removal (Li et al., 2020b). nZVI particles were uniformly distributed into the biochar and stable in the air for more than 120 days. nZVI@BC reacted with dissolved oxygen and generated hydroxyl radicals (HO$^\bullet$, Eqs. 18.5 and 18.7) in aqueous solution. HO$^\bullet$ played the dominant role for 4-CP degradation. With the presence of persulfate, the formation of HO$^\bullet$ was enhanced (Eq. 18.10), which increased 4-CP degradation rate from 50% to 82% after 15 min of reaction. The removal of 4-CP underwent two steps (Fig. 18.8): (1) dichlorination of 4-CP to phenols through Fe$^0$ reduction and (2) degradation of phenol hydroquinone and small molecules through HO$^\bullet$ attacking.

$$SO_4^{-\bullet} + H_2O \rightarrow HO^\bullet + SO_4^{2-} + H^+ \qquad (18.10)$$

Wu et al. prepared nZVI shelled with graphitic layers and embedded mesoporous carbon, designated as MCFe, by a chelation-assisted coassembly and carbothermal reduction strategy using

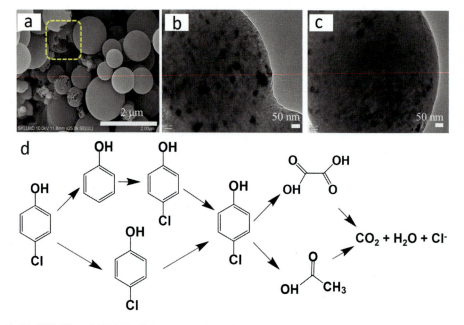

Fig. 18.8 (A–C) SEM (A) and TEM (B, C) images of nZVI/BC prepared from glucose and iron oxides through cocarbonization at 700°C, (D) the proposed degradation pathway of 4-CP with the presence of nZVI@BC and persulfate. Modified from Li, S., Tang, J., Liu, Q., Liu, X., Gao, B., 2020. A novel stabilized carbon-coated nZVI as heterogeneous persulfate catalyst for enhanced degradation of 4-chlorophenol. Environ. Int. 138, 105639. https://doi.org/10.1016/j.envint.2020.105639. © 2016 Elsevier Inc.

citrate, phenolic resin, and iron oxide (Wu et al., 2019). The MCFe was an effective PDS activator for the degradation of 2,4,6-tricholorophenol (TCP) in aqueous solutions. When the initial TCP concentration was 0.5 mmol/L, more than 95% of TCP was degraded within 3 h with the presence of MCFe (482 mg/L) and PDS (2 mmol/L). The turnover frequency of MCFe for TCP degradation is approximately 2 times higher than that of commercial nZVI. The high efficiency of MCFe was attributed to the continuous release of Fe(II) that generated reactive oxygen species (ROS) including $^{\bullet}SO_4^-$, $HO^{\bullet}$, $^{\bullet}O_2^-$, and $^1O_2$ for TCP degradation. The TCP was first attacked by ROS for dichlorination, followed by the formation of quinones and cracking of aromatic rings to form small molecules including $CO_2$ and $H_2O$ (Fig. 18.9). The graphitic layered shell encapsulated $Fe^0$ is beneficial because it enhanced the electron transfer on the interfaces among $Fe^0$ core, mesoporous carbon, and TCP. The porous structure of mesoporous carbon also assists the sorption and transfer of TCP during its degradation.

## 6 Conclusions and future perspectives

The use of biochar as a support material not only reduces the aggregation and rapid passivation of nZVI, but also improved the nZVI removal efficiency for aqueous contaminants by serving as an electron shuttle and donor/acceptor. The removal mechanisms vary with the type of contaminants which mainly include reduction, precipitation, Fenton oxidation, physisorption, and chemisorption. To date, the majority of studies used the borohydride reduction method to prepare nZVI@BC. Despite its easy synthesis, the use of expensive and hazardous borohydride salt is a concern. On the other hand, studies on nZVI@BC preparation by carbothermal reduction route via cocarbonization of biomass and iron compound are still limited. Consider the complex structure of biomass and complicated pyrolysis reactions, more studies are needed to understand the iron reduction, crystal growth, and migration processes during cocarbonization. Moreover, previous research has shown that cocarbonization of biomass and iron compounds generated graphene-shelled nZVI. However, the role of graphene shells on the stabilization, migration, and reaction process of nZVI during contaminant removal is still unclear. Thereby, further investigation on this topic is needed. Furthermore, limited studies were conducted on the application of nZVI@BC for the remediation of real polluted water and soil conditions. New phenomena and mechanisms may be discovered

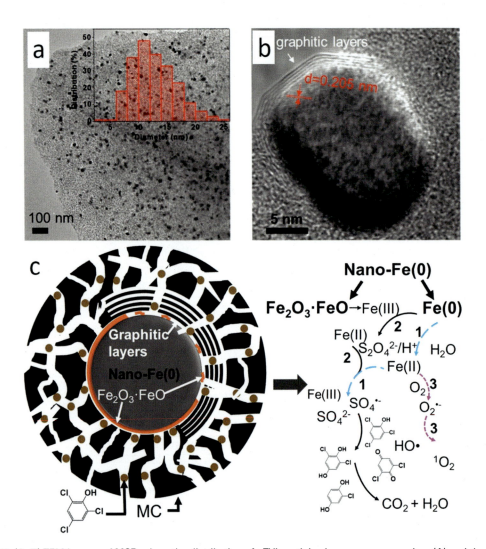

**Fig. 18.9** (A, B) TEM images of MCFe show the distribution of nZVI particles in mesoporous carbon (A) and the single nZVI particle was shelled with graphitic layer (B); (C) degradation mechanisms of TCP by MCFe. Revised from Wu, Y., Chen, X., Han, Y., Yue, D., Cao, X., Zhao, Y., Qian, X., 2019. Highly efficient utilization of nano-Fe(0) embedded in mesoporous carbon for activation of peroxydisulfate. Environ. Sci. Technol. 53, 9081–9090. https://doi.org/10.1021/acs.est.9b02170. © 2019 American Chemical Society.

when applying nZVI@BC in real environmental conditions. Therefore, the treatment of real water and soil using nZVI@BC should be a part of future studies. The driving force is to make nZVI@BC and perhaps graphene layer encapsulated nZVI@BC on a large scale at a low cost.

## Questions and problems

1. What are the advantages of using nZVI for water remediation?
2. What are the issues with using nZVI for water remediation?
3. What's the redox potential ($E^0$) of nZVI?
4. What are the main mechanisms of nZVI toward Ni(II), Cu(II), and Pb(II) removal?
5. What are the main mechanisms of nZVI toward Zn(II) removal?
6. What are the main mechanisms of nZVI toward Cr(VI) removal?
7. What are the main mechanisms of nZVI toward As(VI) removal?
8. Please list the functions of nZVI's core-shell structure for contaminant removal.
9. What are typical methods for nZVI@BC preparation?
10. What are the advantages of using biochar as nZVI support?
11. What are the roles of biochar functionalities during nZVI@BC preparation through borohydride reduction?
12. List the characterization techniques for nZVI@BC.
13. List the mechanisms of nZVI for organic contaminant removal.
14. What factors affect the structure and properties of nZVI@BC prepared from borohydride reduction?
15. What factors affect the structure and properties of nZVI@BC prepared from carbothermal reduction?

## Acknowledgments

This work is supported by the USDA National Institute of Food and Agriculture (NIFA), Agriculture and Food Research Initiative (AFRI) program, under grant No. 2020-65210-30763. This manuscript is publication #SB1047 of the Sustainable Bioproducts, Mississippi State University (MSU). This publication is a contribution of the Forest and Wildlife Research Center, MSU.

## References

Ahmad, M., Usman, A.R.A., Hussain, Q., Al-Farraj, A.S.F., Tsang, Y.F., Bundschuh, J., Al-Wabel, M.I., 2020. Fabrication and evaluation of silica embedded and zero-valent iron composited biochars for arsenate removal from water. Environ. Pollut. 266, 115256. https://doi.org/10.1016/j.envpol.2020.115256.

Alotaibi, K.M., Shiels, L., Lacaze, L., Peshkur, T.A., Anderson, P., Machala, L., Critchley, K., Patwardhan, S.V., Gibson, L.T., 2017. Iron supported on bioinspired green silica for water remediation. Chem. Sci. 8, 567–576. https://doi.org/10.1039/C6SC02937J.

Arthy, M., Phanikumar, B.R., 2016. Efficacy of iron-based nanoparticles and nano-biocomposites in removal of Cr3+. J. Hazard. Toxic Radioact. Waste 20, 04016005. https://doi.org/10.1061/(ASCE)HZ.2153-5515.0000317.

Babich, A., Senk, D., 2019. 13—Coke in the iron and steel industry. In: Suárez-Ruiz, I., Diez, M.A., Rubiera, F. (Eds.), New Trends Coal Conversion. Woodhead Publishing, pp. 367–404, https://doi.org/10.1016/B978-0-08-102201-6.00013-3.

Bae, S., Collins, R.N., Waite, T.D., Hanna, K., 2018. Advances in surface passivation of nanoscale zerovalent iron: a critical review. Environ. Sci. Technol. 52, 12010–12025. https://doi.org/10.1021/acs.est.8b01734.

Crane, R.A., Scott, T.B., 2012. Nanoscale zero-valent iron: future prospects for an emerging water treatment technology. J. Hazard. Mater. 211–212, 112–125. https://doi.org/10.1016/j.jhazmat.2011.11.073.

Crane, R.A., Dickinson, M., Popescu, I.C., Scott, T.B., 2011. Magnetite and zero-valent iron nanoparticles for the remediation of uranium contaminated environmental water. Water Res. 45, 2931–2942. https://doi.org/10.1016/j.watres.2011.03.012.

Dai, X.-H., Fan, H.-X., Yi, C.-Y., Dong, B., Yuan, S.-J., 2019. Solvent-free synthesis of a 2D biochar stabilized nanoscale zerovalent iron composite for the oxidative degradation of organic pollutants. J. Mater. Chem. A 7, 6849–6858. https://doi.org/10.1039/C8TA11661J.

Das, T.K., Bezbaruah, A.N., 2021. Comparative study of arsenic removal by iron-based nanomaterials: potential candidates for field applications. Sci. Total Environ. 764, 142914. https://doi.org/10.1016/j.scitotenv.2020.142914.

Diao, Z.-H., Du, J.-J., Jiang, D., Kong, L.-J., Huo, W.-Y., Liu, C.-M., Wu, Q.-H., Xu, X.-R., 2018. Insights into the simultaneous removal of Cr6+ and Pb2+ by a novel sewage sludge-derived biochar immobilized nanoscale zero valent iron: coexistence effect and mechanism. Sci. Total Environ. 642, 505–515. https://doi.org/10.1016/j.scitotenv.2018.06.093.

Dong, H., Deng, J., Xie, Y., Zhang, C., Jiang, Z., Cheng, Y., Hou, K., Zeng, G., 2017a. Stabilization of nanoscale zero-valent iron (nZVI) with modified biochar for Cr(VI) removal from aqueous solution. J. Hazard. Mater. 332, 79–86. https://doi.org/10.1016/j.jhazmat.2017.03.002.

Dong, H., Zhang, C., Hou, K., Cheng, Y., Deng, J., Jiang, Z., Tang, L., Zeng, G., 2017b. Removal of trichloroethylene by biochar supported nanoscale zero-valent iron in aqueous solution. Sep. Purif. Technol. 188, 188–196. https://doi.org/10.1016/j.seppur.2017.07.033.

Ezzatahmadi, N., Ayoko, G.A., Millar, G.J., Speight, R., Yan, C., Li, J., Li, S., Zhu, J., Xi, Y., 2017. Clay-supported nanoscale zero-valent iron composite materials for the remediation of contaminated aqueous solutions: a review. Chem. Eng. J. 312, 336–350. https://doi.org/10.1016/j.cej.2016.11.154.

Fan, Z., Zhang, Q., Gao, B., Li, M., Liu, C., Qiu, Y., 2019. Removal of hexavalent chromium by biochar supported nZVI composite: batch and fixed-bed column evaluations, mechanisms, and secondary contamination prevention. Chemosphere 217, 85–94. https://doi.org/10.1016/j.chemosphere.2018.11.009.

Fan, C., Chen, N., Qin, J., Yang, Y., Feng, C., Li, M., Gao, Y., 2020. Biochar stabilized nano zero-valent iron and its removal performance and mechanism of pentavalent vanadium(V(V)). Colloids Surf. A Physicochem. Eng. Asp. 599, 124882. https://doi.org/10.1016/j.colsurfa.2020.124882.

Gao, J., Yang, L., Liu, Y., Shao, F., Liao, Q., Shang, J., 2018. Scavenging of Cr(VI) from aqueous solutions by sulfide-modified nanoscale zero-valent iron supported by biochar. J. Taiwan Inst. Chem. Eng. 91, 449–456. https://doi.org/10.1016/j.jtice.2018.06.033.

Gil, A., Amiri, M.J., Abedi-Koupai, J., Eslamian, S., 2018. Adsorption/reduction of Hg(II) and Pb(II) from aqueous solutions by using bone ash/nZVI composite: effects of aging time, Fe loading quantity and co-existing ions. Environ. Sci. Pollut. Res. 25, 2814–2829. https://doi.org/10.1007/s11356-017-0508-y.

Huber, D.L., 2005. Synthesis, properties, and applications of iron nanoparticles. Small 1, 482–501. https://doi.org/10.1002/smll.200500006.

Jiang, S.-F., Ling, L.-L., Chen, W.-J., Liu, W.-J., Li, D.-C., Jiang, H., 2019. High efficient removal of bisphenol A in a peroxymonosulfate/iron functionalized biochar system: mechanistic elucidation and quantification of the contributors. Chem. Eng. J. 359, 572–583. https://doi.org/10.1016/j.cej.2018.11.124.

Jie, S., Haoyang, F.U., Wei, W., Weixian, Z., Lan, L., 2020. Research progress on the characterization and modification of nanoscale zero-valent iron for applications. Environ. Chem., 2959–2978. https://doi.org/10.7524/j.issn.0254-6108.2020070803.

Khajouei, M., Jahanshahi, M., Peyravi, M., Hoseinpour, H., Shokuhi Rad, A., 2017. Anti-bacterial assay of doped membrane by zero valent Fe nanoparticle via in-situ and ex-situ aspect. Chem. Eng. Res. Des. 117, 287–300. https://doi.org/10.1016/j.cherd.2016.10.042.

Klüpfel, L., Keiluweit, M., Kleber, M., Sander, M., 2014. Redox properties of plant biomass-derived black carbon (biochar). Environ. Sci. Technol. 48, 5601–5611. https://doi.org/10.1021/es500906d.

Lei, C., Sun, Y., Tsang, D.C.W., Lin, D., 2018. Environmental transformations and ecological effects of iron-based nanoparticles. Environ. Pollut. 232, 10–30. https://doi.org/10.1016/j.envpol.2017.09.052.

Li, S., Yan, W., Zhang, W., 2009. Solvent-free production of nanoscale zero-valent iron (nZVI) with precision milling. Green Chem. 11, 1618–1626. https://doi.org/10.1039/B913056J.

Li, X., Zhao, Y., Xi, B., Meng, X., Gong, B., Li, R., Peng, X., Liu, H., 2017. Decolorization of methyl orange by a new clay-supported nanoscale zero-valent iron: synergetic effect, efficiency optimization and mechanism. J. Environ. Sci. 52, 8–17. https://doi.org/10.1016/j.jes.2016.03.022.

Li, Z., Wang, L., Meng, J., Liu, X., Xu, J., Wang, F., Brookes, P., 2018. Zeolite-supported nanoscale zero-valent iron: new findings on simultaneous adsorption of Cd(II), Pb(II), and As(III) in aqueous solution and soil. J. Hazard. Mater. 344, 1–11. https://doi.org/10.1016/j.jhazmat.2017.09.036.

Li, S., You, T., Guo, Y., Yao, S., Zang, S., Xiao, M., Zhang, Z., Shen, Y., 2019a. High dispersions of nano zero valent iron supported on biochar by one-step carbothermal synthesis and its application in chromate removal. RSC Adv. 9, 12428–12435. https://doi.org/10.1039/C9RA00304E.

Li, H., Chen, S., Ren, L.Y., Zhou, L.Y., Tan, X.J., Zhu, Y., Belver, C., Bedia, J., Yang, J., 2019b. Biochar mediates activation of aged nanoscale ZVI by *Shewanella putrefaciens* CN32 to enhance the degradation of pentachlorophenol. Chem. Eng. J. 368, 148–156. https://doi.org/10.1016/j.cej.2019.02.099.

Li, S., Yang, F., Li, J., Cheng, K., 2020a. Porous biochar-nanoscale zero-valent iron composites: synthesis, characterization and application for lead ion removal. Sci. Total Environ. 746, 141037. https://doi.org/10.1016/j.scitotenv.2020.141037.

Li, S., Tang, J., Liu, Q., Liu, X., Gao, B., 2020b. A novel stabilized carbon-coated nZVI as heterogeneous persulfate catalyst for enhanced degradation of 4-chlorophenol. Environ. Int. 138, 105639. https://doi.org/10.1016/j.envint.2020.105639.

Ling, L., Zhang, W., 2017. Visualizing arsenate reactions and encapsulation in a single zero-valent iron nanoparticle. Environ. Sci. Technol. 51, 2288–2294. https://doi.org/10.1021/acs.est.6b04315.

Ling, L., Pan, B., Zhang, W., 2015. Removal of selenium from water with nanoscale zero-valent iron: mechanisms of intraparticle reduction of Se(IV). Water Res. 71, 274–281. https://doi.org/10.1016/j.watres.2015.01.002.

Ling, L., Huang, X., Li, M., Zhang, W., 2017. Mapping the reactions in a single zero-valent iron nanoparticle. Environ. Sci. Technol. 51, 14293–14300. https://doi.org/10.1021/acs.est.7b02233.

Ling, L., Huang, X.-Y., Zhang, W.-X., 2018. Enrichment of precious metals from wastewater with core–shell nanoparticles of iron. Adv. Mater. 30, 1705703. https://doi.org/10.1002/adma.201705703.

Liu, F., Yang, J., Zuo, J., Ma, D., Gan, L., Xie, B., Wang, P., Yang, B., 2014. Graphene-supported nanoscale zero-valent iron: removal of phosphorus from aqueous solution and mechanistic study. J. Environ. Sci. 26, 1751–1762. https://doi.org/10.1016/j.jes.2014.06.016.

Liu, M., Wang, Y., Chen, L., Zhang, Y., Lin, Z., 2015. Mg(OH)2 supported nanoscale zero valent Iron enhancing the removal of Pb(II) from aqueous solution. ACS Appl. Mater. Interfaces 7, 7961–7969. https://doi.org/10.1021/am509184e.

Liu, C.-M., Diao, Z.-H., Huo, W.-Y., Kong, L.-J., Du, J.-J., 2018a. Simultaneous removal of Cu2+ and bisphenol A by a novel biochar-supported zero valent iron from aqueous solution: synthesis, reactivity and mechanism. Environ. Pollut. 239, 698–705. https://doi.org/10.1016/j.envpol.2018.04.084.

Liu, A., Wang, W., Liu, J., Fu, R., Zhang, W., 2018b. Nanoencapsulation of arsenate with nanoscale zero-valent iron (nZVI): a 3D perspective. Sci. Bull. 63, 1641–1648. https://doi.org/10.1016/j.scib.2018.12.002.

Liu, X., Lai, D., Wang, Y., 2019a. Performance of Pb(II) removal by an activated carbon supported nanoscale zero-valent iron composite at ultralow iron content. J. Hazard. Mater. 361, 37–48. https://doi.org/10.1016/j.jhazmat.2018.08.082.

Liu, Y., Sohi, S.P., Liu, S., Guan, J., Zhou, J., Chen, J., 2019b. Adsorption and reductive degradation of Cr(VI) and TCE by a simply synthesized zero valent iron magnetic biochar. J. Environ. Manag. 235, 276–281. https://doi.org/10.1016/j.jenvman.2019.01.045.

Liu, K., Li, F., Cui, J., Yang, S., Fang, L., 2020a. Simultaneous removal of Cd(II) and As(III) by graphene-like biochar-supported zero-valent iron from irrigation waters under aerobic conditions: synergistic effects and mechanisms. J. Hazard. Mater. 395, 122623. https://doi.org/10.1016/j.jhazmat.2020.122623.

Liu, L., Liu, X., Wang, D., Lin, H., Huang, L., 2020b. Removal and reduction of Cr(VI) in simulated wastewater using magnetic biochar prepared by co-pyrolysis of nano-zero-valent iron and sewage sludge. J. Clean. Prod. 257, 120562. https://doi.org/10.1016/j.jclepro.2020.120562.

Liu, K., Li, F., Zhao, X., Wang, G., Fang, L., 2021a. The overlooked role of carbonaceous supports in enhancing arsenite oxidation and removal by nZVI: surface area versus electrochemical property. Chem. Eng. J. 406, 126851. https://doi.org/10.1016/j.cej.2020.126851.

Liu, K., Li, F., Tian, Q., Nie, C., Ma, Y., Zhu, Z., Fang, L., Huang, Y., Liu, S., 2021b. A highly porous animal bone-derived char with a superiority of promoting nZVI for Cr(VI) sequestration in agricultural soils. J. Environ. Sci. 104, 27–39. https://doi.org/10.1016/j.jes.2020.11.031.

Liu, L., Zhao, J., Liu, X., Bai, S., Lin, H., Wang, D., 2021c. Reduction and removal of As(V) in aqueous solution by biochar derived from nano zero-valent-iron (nZVI) and sewage sludge. Chemosphere 277, 130273. https://doi.org/10.1016/j.chemosphere.2021.130273.

Martin, J.E., Herzing, A.A., Yan, W., Li, X., Koel, B.E., Kiely, C.J., Zhang, W., 2008. Determination of the oxide layer thickness in core – shell zerovalent iron nanoparticles. Langmuir 24, 4329–4334. https://doi.org/10.1021/la703689k.

Meng, F., Li, Z., Lei, C., Yang, K., Lin, D., 2020. Removal of trichloroethene by iron-based biochar from anaerobic water: key roles of Fe/C ratio and iron carbides. Chem. Eng. J. 413, 127391. https://doi.org/10.1016/j.cej.2020.127391.

Mittal, A.K., Chisti, Y., Banerjee, U.C., 2013. Synthesis of metallic nanoparticles using plant extracts. Biotechnol. Adv. 31, 346–356. https://doi.org/10.1016/j.biotechadv.2013.01.003.

Mortazavian, S., An, H., Chun, D., Moon, J., 2018. Activated carbon impregnated by zero-valent iron nanoparticles (AC/nZVI) optimized for simultaneous adsorption and reduction of aqueous hexavalent chromium: material characterizations and kinetic studies. Chem. Eng. J. 353, 781–795. https://doi.org/10.1016/j.cej.2018.07.170.

Mu, Y., Jia, F., Ai, Z., Zhang, L., 2017. Iron oxide shell mediated environmental remediation properties of nano zero-valent iron. Environ. Sci. Nano 4, 27–45. https://doi.org/10.1039/C6EN00398B.

Mun, S.P., Cai, Z., Zhang, J., 2015. Preparation of Fe-cored carbon nanomaterials from mountain pine beetle-killed pine wood. Mater. Lett. 142, 45–48. https://doi.org/10.1016/j.matlet.2014.11.053.

Neeli, S.T., Ramsurn, H., 2018. Synthesis and formation mechanism of iron nanoparticles in graphitized carbon matrices using biochar from biomass model compounds as a support. Carbon 134, 480–490. https://doi.org/10.1016/j.carbon.2018.03.079.

Oh, S.-Y., Seo, Y.-D., Ryu, K.-S., Park, D.-J., Lee, S.-H., 2017. Redox and catalytic properties of biochar-coated zero-valent iron for the removal of nitro explosives and halogenated phenols. Environ. Sci. Process. Impacts 19, 711–719. https://doi.org/10.1039/C7EM00035A.

Park, M.H., Jeong, S., Lee, G., Park, H., Kim, J.Y., 2019. Removal of aqueous-phase Pb(II), Cd(II), As(III), and As(V) by nanoscale zero-valent iron supported on exhausted coffee grounds. Waste Manag. 92, 49–58. https://doi.org/10.1016/j.wasman.2019.05.017.

Qian, L., Chen, B., 2013. Dual role of biochars as adsorbents for aluminum: the effects of oxygen-containing organic components and the scattering of silicate particles. Environ. Sci. Technol. 47, 8759–8768. https://doi.org/10.1021/es401756h.

Qian, L., Zhang, W., Yan, J., Han, L., Chen, Y., Ouyang, D., Chen, M., 2017. Nanoscale zero-valent iron supported by biochars produced at different temperatures: synthesis mechanism and effect on Cr(VI) removal. Environ. Pollut. 223, 153–160. https://doi.org/10.1016/j.envpol.2016.12.077.

Qin, C., Wang, H., Yuan, X., Xiong, T., Zhang, J., Zhang, J., 2020. Understanding structure-performance correlation of biochar materials in environmental remediation and electrochemical devices. Chem. Eng. J. 382, 122977. https://doi.org/10.1016/j.cej.2019.122977.

Qiu, Y., Zhang, Q., Gao, B., Li, M., Fan, Z., Sang, W., Hao, H., Wei, X., 2020. Removal mechanisms of Cr(VI) and Cr(III) by biochar supported nanosized zero-valent iron: synergy of adsorption, reduction and transformation. Environ. Pollut. 265, 115018. https://doi.org/10.1016/j.envpol.2020.115018.

Rajapaksha, A.U., Chen, S.S., Tsang, D.C.W., Zhang, M., Vithanage, M., Mandal, S., Gao, B., Bolan, N.S., Ok, Y.S., 2016. Engineered/designer biochar for contaminant removal/immobilization from soil and water: potential and implication of biochar modification. Chemosphere 148, 276–291. https://doi.org/10.1016/j.chemosphere.2016.01.043.

Ruiz-Torres, C.A., Araujo-Martínez, R.F., Martínez-Castañón, G.A., Morales-Sánchez, J.E., Guajardo-Pacheco, J.M., González-Hernández, J., Lee, T.-J., Shin, H.-S., Hwang, Y., Ruiz, F., 2018. Preparation of air stable nanoscale zero valent iron functionalized by ethylene glycol without inert condition. Chem. Eng. J. 336, 112–122. https://doi.org/10.1016/j.cej.2017.11.047.

Saha, J.K., Selladurai, R., Coumar, M.V., Dotaniya, M.L., Kundu, S., Patra, A.K., 2017. Organic pollutants. In: Saha, J.K., Selladurai, R., Coumar, M.V., Dotaniya, M.L., Kundu, S., Patra, A.K. (Eds.), Soil Pollution—An Emerging Threat to Agriculture. Springer, Singapore, pp. 105–135, https://doi.org/10.1007/978-981-10-4274-4_5.

Shan, A., Idrees, A., Zaman, W.Q., Abbas, Z., Ali, M., Rehman, M.S.U., Hussain, S., Danish, M., Gu, X., Lyu, S., 2021. Synthesis of nZVI-Ni@BC composite as a stable catalyst to activate persulfate: trichloroethylene degradation and insight mechanism. J. Environ. Chem. Eng. 9, 104808. https://doi.org/10.1016/j.jece.2020.104808.

Signorini, L., Pasquini, L., Savini, L., Carboni, R., Boscherini, F., Bonetti, E., Giglia, A., Pedio, M., Mahne, N., Nannarone, S., 2003. Size-dependent oxidation in iron/iron oxide core-shell nanoparticles. Phys. Rev. B 68, 195423. https://doi.org/10.1103/PhysRevB.68.195423.

Stefaniuk, M., Oleszczuk, P., Ok, Y.S., 2016. Review on nano zerovalent iron (nZVI): from synthesis to environmental applications. Chem. Eng. J. 287, 618–632. https://doi.org/10.1016/j.cej.2015.11.046.

Su, Y., Cheng, Y., Shih, Y., 2013. Removal of trichloroethylene by zerovalent iron/activated carbon derived from agricultural wastes. J. Environ. Manag. 129, 361–366. https://doi.org/10.1016/j.jenvman.2013.08.003.

Su, F., Zhou, H., Zhang, Y., Wang, G., 2016. Three-dimensional honeycomb-like structured zero-valent iron/chitosan composite foams for effective removal of inorganic arsenic in water. J. Colloid Interface Sci. 478, 421–429. https://doi.org/10.1016/j.jcis.2016.06.035.

Tang, H., Zhu, D., Li, T., Kong, H., Chen, W., 2011. Reductive dechlorination of activated carbon-adsorbed trichloroethylene by zero-valent iron: carbon as electron shuttle. J. Environ. Qual. 40, 1878–1885. https://doi.org/10.2134/jeq2011.0185.

Tang, C., Ling, L., Zhang, W., 2020. Pb(II) deposition-reduction-growth onto iron nanoparticles induced by graphitic carbon nitride. Chem. Eng. J. 387, 124088. https://doi.org/10.1016/j.cej.2020.124088.

Tchounwou, P.B., Yedjou, C.G., Patlolla, A.K., Sutton, D.J., 2012. Heavy metal toxicity and the environment. In: Luch, A. (Ed.), Molecular, Clinical and Environmental Toxicology. Environmental Toxicology, vol. 3. Springer, Basel, pp. 133–164, https://doi.org/10.1007/978-3-7643-8340-4_6.

Tosco, T., Petrangeli Papini, M., Cruz Viggi, C., Sethi, R., 2014. Nanoscale zerovalent iron particles for groundwater remediation: a review. J. Clean. Prod. 77, 10–21. https://doi.org/10.1016/j.jclepro.2013.12.026.

Tuček, J., Prucek, R., Kolařík, J., Zoppellaro, G., Petr, M., Filip, J., Sharma, V.K., Zbořil, R., 2017. Zero-valent iron nanoparticles reduce arsenites and arsenates to As(0) firmly embedded in core–shell superstructure: challenging strategy of arsenic treatment under anoxic conditions. ACS Sustain. Chem. Eng. 5, 3027–3038. https://doi.org/10.1021/acssuschemeng.6b02698.

Wang, C.M., Baer, D.R., Thomas, L.E., Amonette, J.E., Antony, J., Qiang, Y., Duscher, G., 2005. Void formation during early stages of passivation: initial oxidation of iron nanoparticles at room temperature. J. Appl. Phys. 98, 094308. https://doi.org/10.1063/1.2130890.

Wang, C., Baer, D.R., Amonette, J.E., Engelhard, M.H., Antony, J., Qiang, Y., 2009a. Morphology and electronic structure of the oxide shell on the surface of iron nanoparticles. J. Am. Chem. Soc. 131, 8824–8832. https://doi.org/10.1021/ja900353f.

Wang, Q., Snyder, S., Kim, J., Choi, H., 2009b. Aqueous ethanol modified nanoscale zerovalent iron in bromate reduction: synthesis, characterization, and reactivity. Environ. Sci. Technol. 43, 3292–3299. https://doi.org/10.1021/es803540b.

Wang, S., Zhao, M., Zhao, Y., Wang, N., Bai, J., Feng, K., Zhou, Y., Chen, W., Wen, F., Wang, S., Wang, X., Wang, J., 2017a. Pyrogenic temperature affects the particle size of biochar-supported nanoscaled zero valent iron (nZVI) and its silver removal capacity. Chem. Speciat. Bioavailab. 29, 179–185. https://doi.org/10.1080/09542299.2017.1395712.

Wang, S., Gao, B., Li, Y., Creamer, A.E., He, F., 2017b. Adsorptive removal of arsenate from aqueous solutions by biochar supported zero-valent iron nanocomposite: batch and continuous flow tests. J. Hazard. Mater. 322, 172–181. https://doi.org/10.1016/j.jhazmat.2016.01.052.

Wang, S., Zhao, M., Zhou, M., Li, Y.C., Wang, J., Gao, B., Sato, S., Feng, K., Yin, W., Igalavithana, A.D., Oleszczuk, P., Wang, X., Ok, Y.S., 2019a. Biochar-supported nZVI (nZVI/BC) for contaminant removal from soil and water: a critical review. J. Hazard. Mater. 373, 820–834. https://doi.org/10.1016/j.jhazmat.2019.03.080.

Wang, H., Zhang, M., Li, H., 2019b. Synthesis of nanoscale zerovalent iron (nZVI) supported on biochar for chromium remediation from aqueous solution and soil. Int. J. Environ. Res. Public Health 16, 4430. https://doi.org/10.3390/ijerph16224430.

Wang, Z., Wu, X., Luo, S., Wang, Y., Tong, Z., Deng, Q., 2020. Shell biomass material supported nano-zero valent iron to remove $Pb^{2+}$ and $Cd^{2+}$ in water. R. Soc. Open Sci. 7, 201192. https://doi.org/10.1098/rsos.201192.

Wei, G., Zhang, J., Luo, J., Xue, H., Huang, D., Cheng, Z., Jiang, X., 2019. Nanoscale zero-valent iron supported on biochar for the highly efficient removal of nitrobenzene. Front. Environ. Sci. Eng. 13, 61. https://doi.org/10.1007/s11783-019-1142-3.

Wei, D., Li, B., Luo, L., Zheng, Y., Huang, L., Zhang, J., Yang, Y., Huang, H., 2020. Simultaneous adsorption and oxidation of antimonite onto nano zero-valent iron sludge-based biochar: indispensable role of reactive oxygen species and redox-active moieties. J. Hazard. Mater. 391, 122057. https://doi.org/10.1016/j.jhazmat.2020.122057.

Wu, M., Ma, J., Cai, Z., Tian, G., Yang, S., Wang, Y., Liu, X., 2015. Rational synthesis of zerovalent iron/bamboo charcoal composites with high saturation magnetization. RSC Adv. 5, 88703–88709. https://doi.org/10.1039/C5RA13236C.

Wu, Y., Chen, X., Han, Y., Yue, D., Cao, X., Zhao, Y., Qian, X., 2019. Highly efficient utilization of nano-Fe(0) embedded in mesoporous carbon for activation of peroxydisulfate. Environ. Sci. Technol. 53, 9081–9090. https://doi.org/10.1021/acs.est.9b02170.

Wu, Y., Guan, C.-Y., Griswold, N., Hou, L., Fang, X., Hu, A., Hu, Z., Yu, C.-P., 2020. Zero-valent iron-based technologies for removal of heavy metal(loid)s and organic pollutants from the aquatic environment: recent advances and perspectives. J. Clean. Prod. 277, 123478. https://doi.org/10.1016/j.jclepro.2020.123478.

Xing, J., Liu, Q., Zheng, K., Ma, J., Liu, X., Yang, H., Peng, X., Nie, S., Wang, K., 2019. Synergistic effect of Fenton-like treatment on the adsorption of organic dye on bamboo magnetic biochar. BioResources 14, 714–724.

Xu, X., Huang, H., Zhang, Y., Xu, Z., Cao, X., 2019. Biochar as both electron donor and electron shuttle for the reduction transformation of Cr(VI) during its sorption. Environ. Pollut. 244, 423–430. https://doi.org/10.1016/j.envpol.2018.10.068.

Yan, W., Lien, H.-L., Koel, B.E., Zhang, W., 2012. Iron nanoparticles for environmental clean-up: recent developments and future outlook. Environ. Sci. Process. Impacts 15, 63–77. https://doi.org/10.1039/C2EM30691C.

Yan, J., Han, L., Gao, W., Xue, S., Chen, M., 2015. Biochar supported nanoscale zerovalent iron composite used as persulfate activator for removing trichloroethylene. Bioresour. Technol. 175, 269–274. https://doi.org/10.1016/j.biortech.2014.10.103.

Yan, Q., Li, J., Zhang, X., Zhang, J., Cai, Z., 2018a. In situ formation of graphene-encapsulated iron nanoparticles in carbon frames through catalytic graphitization of kraft lignin. Nanomater. Nanotechnol. 8, 1–12. https://doi.org/10.1177/1847980418818955.

Zhang, X., Navarathna, C.M., Leng, W., Karunaratne, T., Thirumalai, R.V.K.G., Kim, Y., Pittman Jr., C.U., Mlsna, T., Cai, Z., Zhang, J., 2021a. Lignin-based few-layered graphene-encapsulated iron nanoparticles for water remediation. Chem. Eng. J. 417, 129199. https://doi.org/10.1016/j.cej.2021.129199.

Yang, D., Wang, L., Li, Z., Tang, X., He, M., Yang, S., Liu, X., Xu, J., 2020. Simultaneous adsorption of Cd(II) and As(III)by a novel biochar-supported nanoscale zero-valent iron in aqueous systems. Sci. Total Environ. 708, 134823. https://doi.org/10.1016/j.scitotenv.2019.134823.

Zou, Y., Wang, X., Khan, A., Wang, P., Liu, Y., Alsaedi, A., Hayat, T., Wang, X., 2016. Environmental remediation and application of nanoscale zero-valent Iron and its composites for the removal of heavy metal ions: a review. Environ. Sci. Technol. 50, 7290–7304. https://doi.org/10.1021/acs.est.6b01897.

Zhu, H., Jia, Y., Wu, X., Wang, H., 2009. Removal of arsenic from water by supported nano zero-valent iron on activated carbon. J. Hazard. Mater. 172, 1591–1596. https://doi.org/10.1016/j.jhazmat.2009.08.031.

Zhou, M., Yang, X., Sun, R., Wang, X., Yin, W., Wang, S., Wang, J., 2021. The contribution of lignocellulosic constituents to Cr(VI) reduction capacity of biochar-supported zerovalent iron. Chemosphere 263, 127871. https://doi.org/10.1016/j.chemosphere.2020.127871.

Yang, F., Zhang, S., Sun, Y., Cheng, K., Li, J., Tsang, D.C.W., 2018a. Fabrication and characterization of hydrophilic corn stalk biochar-supported nanoscale zero-valent iron composites for efficient metal removal. Bioresour. Technol. 265, 490–497. https://doi.org/10.1016/j.biortech.2018.06.029.

Zhao, M., Zhang, C., Yang, X., Liu, L., Wang, X., Yin, W., Li, Y.C., Wang, S., Fu, W., 2020. Preparation of highly-conductive pyrogenic carbon-supported zero-valent iron for enhanced Cr(VI) reduction. J. Hazard. Mater. 396, 122712. https://doi.org/10.1016/j.jhazmat.2020.122712.

Yuan, Y., Bolan, N., Prévoteau, A., Vithanage, M., Biswas, J.K., Ok, Y.S., Wang, H., 2017. Applications of biochar in redox-mediated reactions. Bioresour. Technol. 246, 271–281. https://doi.org/10.1016/j.biortech.2017.06.154.

Zhou, M., Zhang, C., Yuan, Y., Mao, X., Li, Y., Wang, N., Wang, S., Wang, X., 2020. Pinewood outperformed bamboo as feedstock to prepare biochar-supported zero-valent iron for Cr6+ reduction. Environ. Res. 187, 109695. https://doi.org/10.1016/j.envres.2020.109695.

Zhang, D., Li, Y., Tong, S., Jiang, X., Wang, L., Sun, X., Li, J., Liu, X., Shen, J., 2018a. Biochar supported sulfide-modified nanoscale zero-valent iron for the reduction of nitrobenzene. RSC Adv. 8, 22161–22168. https://doi.org/10.1039/C8RA04314K.

Zhang, S., Lyu, H., Tang, J., Song, B., Zhen, M., Liu, X., 2019a. A novel biochar supported CMC stabilized nano zero-valent iron composite for hexavalent chromium removal from water. Chemosphere 217, 686–694. https://doi.org/10.1016/j.chemosphere.2018.11.040.

Zhang, Q., Wang, Y., Wang, Z., Zhang, Z., Wang, X., Yang, Z., 2021b. Active biochar support nano zero-valent iron for efficient removal of U(VI) from sewage water. J. Alloys Compd. 852, 156993. https://doi.org/10.1016/j.jallcom.2020.156993.

Zhang, X., Yan, Q., Hassan, E.B., Li, J., Cai, Z., Zhang, J., 2017a. Temperature effects on formation of carbon-based nanomaterials from kraft lignin. Mater. Lett. 203, 42–45. https://doi.org/10.1016/j.matlet.2017.05.125.

Zhang, X., Yan, Q., Li, J., Chu, I.-W., Toghiani, H., Cai, Z., Zhang, J., 2018b. Carbon-based nanomaterials from biopolymer lignin via catalytic thermal treatment at 700 to 1000 °C. Polymers 10, 183. https://doi.org/10.3390/polym10020183.

Yan, Q., Li, J., Zhang, X., Zhang, J., Cai, Z., 2018b. Synthetic bio-graphene based nanomaterials through different iron catalysts. Nanomaterials 8, 840. https://doi.org/10.3390/nano8100840.

Yan, Q., Zhang, X., Li, J., Hassan, E.B., Wang, C., Zhang, J., Cai, Z., 2018c. Catalytic conversion of Kraft lignin to bio-multilayer graphene materials under different atmospheres. J. Mater. Sci. 53, 8020–8029. https://doi.org/10.1007/s10853-018-2172-0.

Zhang, X., 2016. Synthesis and Characterization of Carbon-Based Nanomaterials From Lignin (Ph.D.). Mississippi State University. https://search.proquest.com/docview/1858816366/abstract/9BE7891B9F3543B6PQ/1.

Zhang, X., Yan, Q., Li, J., Zhang, J., Cai, Z., 2018c. Effects of physical and chemical states of iron-based catalysts on formation of carbon-encapsulated iron nanoparticles from kraft lignin. Materials 11, 139. https://doi.org/10.3390/ma11010139.

Zhang, H., Ruan, Y., Liang, A., Shih, K., Diao, Z., Su, M., Hou, L., Chen, D., Lu, H., Kong, L., 2019b. Carbothermal reduction for preparing nZVI/BC to extract uranium: insight into the iron species dependent uranium adsorption behavior. J. Clean. Prod. 239, 117873. https://doi.org/10.1016/j.jclepro.2019.117873.

Zhang, X., Wu, Y., 2016. Application of coupled zero-valent iron/biochar system for degradation of chlorobenzene-contaminated groundwater. Water Sci. Technol. 75, 571–580. https://doi.org/10.2166/wst.2016.503.

Zhuang, L., Li, Q., Chen, J., Ma, B., Chen, S., 2014. Carbothermal preparation of porous carbon-encapsulated iron composite for the removal of trace hexavalent chromium. Chem. Eng. J. 253, 24–33. https://doi.org/10.1016/j.cej.2014.05.038.

Zhang, X., Yan, Q., Leng, W., Li, J., Zhang, J., Cai, Z., Hassan, E.B., 2017b. Carbon nanostructure of kraft lignin thermally treated at 500 to 1000 °C. Materials 10, 975. https://doi.org/10.3390/ma10080975.

Yu, L., Yuan, Y., Tang, J., Wang, Y., Zhou, S., 2015. Biochar as an electron shuttle for reductive dechlorination of pentachlorophenol by *Geobacter sulfurreducens*. Sci. Rep. 5, 16221. https://doi.org/10.1038/srep16221.

Zhang, Y., Xu, X., Zhang, P., Zhao, L., Qiu, H., Cao, X., 2019c. Pyrolysis-temperature depended quinone and carbonyl groups as the electron accepting sites in barley grass derived biochar. Chemosphere 232, 273–280. https://doi.org/10.1016/j.chemosphere.2019.05.225.

Yang, J., Ma, T., Li, X., Tu, J., Dang, Z., Yang, C., 2018b. Removal of heavy metals and metalloids by amino-modified biochar supporting nanoscale zero-valent iron. J. Environ. Qual. 47, 1196–1204. https://doi.org/10.2134/jeq2017.08.0320.

Zhang, Y., Jiang, Q., Jiang, S., Li, H., Zhang, R., Qu, J., Zhang, S., Han, W., 2021c. One-step synthesis of biochar supported nZVI composites for highly efficient activating persulfate to oxidatively degrade atrazine. Chem. Eng. J. 420, 129868. https://doi.org/10.1016/j.cej.2021.129868.

# 19

# An insight into the sorptive interactions between aqueous contaminants and biochar

**Chathuri Peiris[a], Janeshta C. Fernando[a], Y. Vindula Alwis[a], Namal Priyantha[a,b], and Sameera R. Gunatilake[a]**
[a]College of Chemical Sciences, Institute of Chemistry Ceylon, Rajagiriya, Sri Lanka. [b]Department of Chemistry, University of Peradeniya, Peradeniya, Sri Lanka

## 1 Introduction

Biochar (BC) is a low-cost, porous, carbonaceous material derived from the thermal decomposition of lignocellulose-rich biomass under anaerobic conditions (Fernando et al., 2021). It is primarily utilized for immobilizing a wide range of inorganic and organic contaminants, and is a promising soil amendment and carbon sequestration technique (Ahmad et al., 2014; Peiris et al., 2022).

A distinctive characteristic of BC is the adaptability of its surface chemistry. Surface properties of BC, such as its functionality, porosity, and polarity, directly correlate with its sorptive interactions with contaminants, associated with remediating numerous organic and inorganic pollutants (Peiris et al., 2020). According to literature reported thus far, the surface interactions of BC have been defined as adsorption mechanisms. However, by definition, a mechanism is considered a step-by-step sequence of reactions, hence, the term "interactions" has been employed instead of "mechanism," in presenting the surface phenomena involving BC and pollutants.

Sorption properties of BC are affected by the precursor, pyrolysis temperature, pyrolysis method, type of modification performed, and experimental conditions employed for sorption (Hassan et al., 2020). Therefore, a thorough understanding of BC chemistry is vital when selecting a suitable type to remove a specific analyte of interest during remediation.

Biochar has gained widespread attention as a multidisciplinary subject with its popularity among chemists, environmentalists, engineers, and agriculturists. Moreover, the process of removal of contaminants by BC has gained equal research interest. The study of BC-adsorbate interactions requires a comprehensive understanding of the surface chemistry and adsorption properties of BC. Although available literature provides an individualized idea regarding these interactions that take place, these data remain scattered. In the presented work, the authors have attempted to provide a generalized idea of how sorptive interactions vary in relation to surface chemistry and the types of BC-adsorbate interactions that take place.

## 2 Structural changes that occur in biomass during biochar production

Common sources of feedstock utilized in BC production are lignocellulose-rich biomass consisting of agricultural and forestry residues. Biomass with trace amounts of lignocellulose such as sewage sludge, manure, algae, bone char, and animal residue have also been used. Both forms of biomass contribute in various ways to the biomass-BC conversion process (Rangabhashiyam and Balasubramanian, 2019; Li and Jiang, 2017).

The main chemical components present in lignocellulosic biomass are cellulose, hemicellulose, and lignin with percentage dry weights of approximately 20–60, 15–60, and 5–40 wt%, respectively (Rangabhashiyam and Balasubramanian, 2019). Cellulose is a polysaccharide with $\beta$-D-glucopyranose sugar units, whereas hemicellulose is a complex, branched polysaccharide composed of pentose and hexose sugars with a low degree of polymerization in comparison to cellulose. On the other hand, lignin is a complicated, highly branched molecular structure with cross-linked 3D phenolic polymers (Cheng and Li, 2018; Anca-Couce, 2016). Temperature ranges at which hemicellulose, cellulose, and lignin approximately decompose are 180–240°C, 230–310°C, and 160–900°C, respectively (Peiris et al., 2019).

During pyrolysis, cellulose undergoes depolymerization to produce oligosaccharides. Further cleavage and intramolecular rearrangement lead to the formation of levoglucosan intermediate (Liu et al., 2015). Pyrolysis of hemicellulose follows a similar process as cellulose, forming 1,4-anhydro-D-xylopyranose as an intermediate (Liu et al., 2015). The formation of a highly stable aromatic network of BC involves multiple reactions including de-carbonylation, dehydration, ring opening, aromatization, dehydrogenation, and deoxygenation of the aforementioned intermediates (Fig. 19.1). Additionally, radical reactions can occur

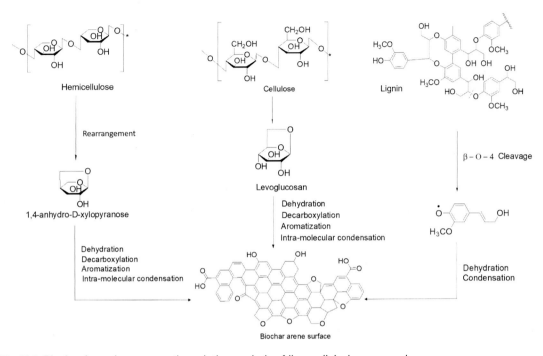

**Fig. 19.1** Biochar formation process through the pyrolysis of lignocellulosic compounds.

at high temperatures. In contrast to cellulose and hemicellulose, pyrolysis of lignin follows a different mechanism that involves different free radical reaction pathways. Free radicals formed by breaking the β-O-4 lignin linkage initiate such a reaction, which may further react through rearrangement, electron abstraction, or radical-radical interactions, to form a stable BC network (Fig. 19.1) (Liu et al., 2015; Brebu and Vasile, 2010).

Inorganic species, such as K, Fe, Pb, and Cd, and hetero atoms, such as N, P, and S, are abundantly found in nonlignocellulosic biomass. Such elements can act as catalysts and increase the BC yield. However, it can have a negative effect on the thermochemical stability of the BC and can even delay the formation of stable polycyclic aromatic carbons (Li and Jiang, 2017; Cheng and Li, 2018; Lehmann and Joseph, 2015). This provides reasons for nonlignocellulosic biomass being less common than lignocellulosic biomass in BC production.

Generally, a variety of oxygen-containing functional groups can be obtained from feedstock which are rich in cellulose or hemicellulose. On the other hand, increasing the proportion of lignin contributes to an increased BC yield due to its larger carbon content and thermal resilience of aromatic monomers as compared to the high oxygen and aliphatic carbon content of cellulose and hemicellulose. Moreover, BC produced at low temperatures has an abundance of carboxylic, hydroxyl, lactones, carbonyl, and phenolic moieties, whereas BC produced at high temperature is richer in orderly arranged aromatic domains (Peiris et al., 2020).

## 3 Common functional moieties in biochar and adsorbates

Interactions between functional groups in the BC surface and adsorbate molecules play a major role in the adsorption process (Peiris et al., 2020). Knowledge on the chemical environment around the moieties in BC and how they interact with the functional groups in different adsorbates are vital when predicting the sorptive interactions that take place. In this section, common functional groups that are available in BC, as well as in adsorbates, and their chemical interactions are summarized.

The hydroxy group (–OH) is one of the most common functional groups available in both adsorbents and adsorbates. The high electronegativity of oxygen and the lone pairs available on the oxygen atom make this an excellent electron donor group. It is also a good site for the formation of hydrogen bonds, as the highly polarized O—H bond makes it a hydrogen bond donor. Oxygen atom in the hydroxyl group is considered as a hard base, which can easily complex with hard metal ions (Pearson, 1963). These characteristics are influenced by the molecular structure to which the hydroxy moiety is attached. For example, the presence of neighboring intramolecular H-bond acceptor groups can reduce the ability of an –OH to form intermolecular H-bonds. Nevertheless, if the –OH group is attached to a highly conjugated or an aromatic system, the electron donor properties of the –OH group are suppressed due to the involvement of lone pairs in resonance. Furthermore, conjugated –OH groups often have improved acidic properties forming anionic groups, owing to the resonance stabilization of the conjugate base formed after deprotonation. As a result, additional electrostatic attractions (EAs) can be observed in phenolate moieties.

Similar to hydroxyl groups, ethers act as potential electron donors. The positive inductive effect of the alkyl groups attached to the ether oxygen can improve the electron donor properties; however, if it is a bulky group, interacting with other chemical species can be hindered due to steric reasons. If aromatic groups are present in the ether moiety, the electron donor properties are reduced not only due to steric effects, but also due to the resonance effect with the lone pairs of the oxygen atom (Bent, 1968). Furthermore, alkyl or aromatic chains connected to the ether oxygen create a suitable environment to interact with nonpolar structures via hydrophobic interactions.

Carbonyl groups behave as both electron donors and acceptors; thus, they can be considered as important binding sites due to their electron donor-acceptor (EDA) interactions. Electron-rich carbonyl oxygen is a good electron donor, while the electron-deficit carbonyl carbon is a good acceptor. When conjugated systems like arenes are connected to carbonyl groups, the arene surface can act as a $\pi$-acceptor, owing to the negative mesomeric effect shown by the carbonyl carbon.

Like the –OH group, amines can also show electron donor and H-bond forming properties, which are affected by the resonance stabilization and intermolecular H-bonds, respectively. As discussed under ethers, the strength of the interactions in secondary or tertiary amines may be altered by steric, resonance, or inductive effects offered by the substituent groups (Pearson, 1963). A unique characteristic of amines is its ability to get easily protonated owing to the high Lewis basicity of the nitrogen atom. The positive charge received by protonation allows amines to form EA with negatively charged species. Unlike protonated amines, which can be discharged by deprotonating, quaternary ammonium salts bear a permanent positive charge. Secondary and tertiary amines together with quaternary ammonium salts form hydrophobic interactions of varying strengths depending on the chain length and the polarity of the substituent groups. Lone pairs available on amines make them excellent ligands, which can complex well either with hard or hard-soft borderline metals (Pearson, 1963). The nitro group ($-NO_2$) is another frequently found nitrogen containing functionality, especially in nitric-acid-modified BC. Nitro group interacts strongly as a highly polarized electron donor group.

Carboxylic acids and their derivatives (such as amides and esters) are another type of functional group, commonly found on the BC surface as well as in adsorbates. Interactions of carboxylic acids/amides/esters can be considered as a combination of those of the carbonyl groups and alcohols/amines/ethers, respectively. The deprotonated form of the acid-carboxylate anion forms EA with the positively charged species. However, because the lone pair on nitrogen is delocalized by resonance, the nitrogen atom of amides acts as weak ligands and cannot be protonated easily and thus hardly tends to form EA. Lactones and lactams, which are cyclic esters and amides, respectively, show similar interaction patterns as their noncyclic analogues.

In natural and synthetic chemical environments, sulfur is considered an isosteric substitution for oxygen. Thus, thiols, sulfides, and thioesters can be considered to have quite similar properties as their oxygen containing analogues – alcohols, ethers, and esters, respectively. However, due to the low electronegativity of sulfur, they are not considered H-bond donors, and due to the large atomic radius of sulfur, they are considered to be soft bases. Sulfoxides, sulfones, sulfinic acids, and sulfonic acids are the oxidized, and sulfur-containing functionalities that are commonly found in adsorbates. In these groups, similar to the carbonyl carbon, sulfur exists as an electrophilic center, while oxygen atoms are electron donating in nature. Therefore, these moieties can form interactions as both electron donors and electron acceptors. Additionally, sulfinic and sulfonic acids are good H-bond donors and form strong EA once deprotonated. (Bissantz et al., 2010)

Highly conjugated scaffolds, such as arenes, should not be neglected when studying the adsorbent-adsorbate interactions, as it constitutes the BC surface and many adsorbates, such as polyaromatic hydrocarbons (Peiris et al., 2017). Arenes, being planar structures sandwiched by two delocalized electron clouds, can be considered as strong $\pi$ donors. The donating nature of the arene is highly sensitive to the inductive and mesomeric properties of the functional moieties attached to it. The positive inductive and mesomeric effects exerted by different functional groups elevate the $\pi$ donating ability of the arene system. On the other hand, the negative inductive and mesomeric effects of the attached moieties lessen the $\pi$ donation by the arenes. Typically alkyl groups exert positive inductive effects, while negative inductive effects are caused by halo groups. Positive mesomeric effects are brought about by hydroxy, alkoxy, amino and halo groups, whereas negative mesomeric effects result due to nitro groups.

## 4 Biochar-based sorptive interactions

Depending on the nature of the contaminant (adsorbate) and the adsorbent, BC-based interactions would be either chemical (ionic, covalent, or coordinate) or physical (intermolecular forces) (Tong et al., 2019). Electron donor-acceptor interactions, cation-$\pi$ bonding, H-bonding, electrostatic interactions, ion exchange, surface complexation, and inter- and intraparticle diffusion followed by transfer to meso-, macro-, and micropores are the common processes involved in remediating contaminants (Rajapaksha et al., 2016; Xu et al., 2020). Dipole–dipole interactions are a type of intermolecular forces which occur as a result of an EA between the permanent dipoles of different molecules. These interactions are attractions rather than bonds due to the uneven distribution of electrons between the two dipoles, in which the partial positive charge on one molecule is electrostatically attracted to the partial negative charge on a neighboring molecule (Singh, 2014).

## 4.1 Electron donor-acceptor interactions

The interactions that occur between electron-rich and electron-deficient moieties are defined as EDA interactions. The donor or acceptor properties of a specific molecule arises due to molecular polarization (Peiris et al., 2017).

The contribution by a functional moiety either as a donor or an acceptor is elaborated in Section 3. Generally, BC is considered as an electron-rich donor due to its aromatic nature. For example, in the removal of sulfonamide, electron donor-acceptor interactions occur between the electron-rich arene network of BC and the protonated (acceptor) aniline ring of the sulfonamide (Fig. 19.2A) (Peiris et al., 2017).

On the other hand, EDA interactions are also considered as Lewis acid base interactions when a lone pair of electrons is transferred to an acceptor moiety. For example, the interaction between the lone pair of the amine group of the sulfonamide antibiotic and the arene surface of BC is defined as a Lewis acid–base interaction (Peiris et al., 2017; Kah et al., 2017).

## 4.2 Cation-π interactions

Cation-π interactions exist as noncovalent interactions between cations and the π electrons of aliphatic or aromatic compounds (Fig. 19.2B). The interaction is said to be favorable when the cationic moieties are within the range of 6.0 Å of the face of an aromatic ring (Cheng et al., 2008). Moreover, the cation must align above or below the π electron cloud, for the attractive forces to occur, while the cations are repelled when they are oriented along the plane of the ring (Mitchell and Means, 2018). According to the study conducted by Cheng et al., the electron-donating groups could strengthen the cation-π interactions as opposed to electron withdrawing groups (Cheng et al., 2008).

Cation-π interactions can be classified into three types depending on the properties of the cation: (1) interactions between the inorganic metal cations such as $Na^+$, $K^+$, $Mg^{2+}$, and $Ca^{2+}$, and aromatic systems; (2) interaction between organic

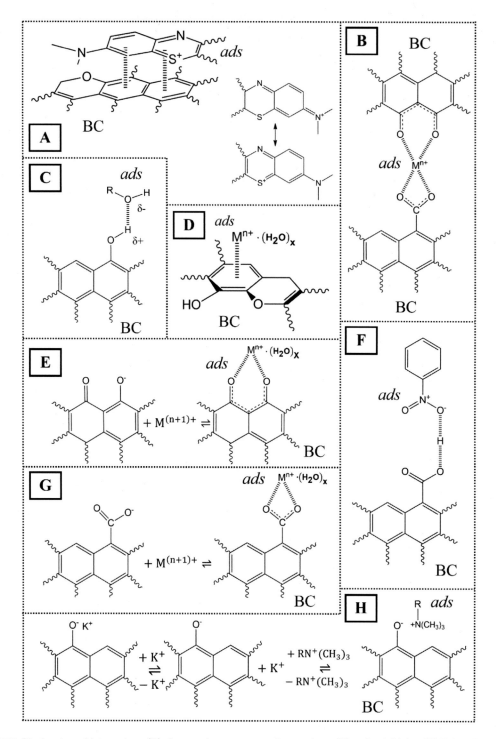

**Fig. 19.2** Biochar-based interactions: (A) electron donor-acceptor interactions, (B) cation bridging, (C) H-bonding, (D) cation-π interactions, (E) ion exchange, (F) charge-assisted H-bonding, (G) electrostatic interactions, and (H) ion exchange. Note: The ion represented as $M^{n+}$ can be a metal cation such as $Pb^{2+}$ and $Cd^{2+}$. The interactions are not possible for metals which form oxyanions. Metal cations are subjected to solvation, and cations with high valency may not undergo such interactions due to extensive solvation. "BC" = biochar surface and "*ads*" = adsorbate.

cations and aromatic systems; and (3) interaction between the atoms possessing partial positive charges (e.g., H in an N—H bond) and aromatic systems. (Cheng et al., 2008) Several studies have reported the existence of cation-π interactions in BC-based sorption (Teixidó et al., 2011; Ji et al., 2009; Yu et al., 2018).

## 4.3 Hydrogen bonding

Even though a specific group of secondary interactions is named as H-bonds, it is a type of strong dipole–dipole interactions. The specialty of these interactions is that the positive pole, known as an H-bond donor or a proton donor, is always a hydrogen atom bound to a highly electronegative atom, namely, nitrogen, oxygen, and fluorine (Fig. 19.2C) (Nekoei and Vatanparast, 2019). A species bearing a lone pair, preferably possessing a total or partial negative charge, can act as an H-bond acceptor. Once the hydrogen atom starts interacting with the H-bond acceptor, the hydrogen-electronegative atom bond length increases, thus sharing the hydrogen almost equally between the H-bond acceptor and the electronegative atom to which hydrogen is originally attached. (Nekoei and Vatanparast, 2019) The excellent combination of a small interacting distance and a high partial positivity developed on hydrogen results in H-bonds having an outstanding strength (10–40 kJ mol$^{-1}$), as opposed to other dipole–dipole interactions (3–4 kJ mol$^{-1}$) (McMurry, 2010). Molecules or adsorbent surfaces with –NH– and –OH functional groups can form strong H-bonds by functioning both as H-bond acceptors as well as donors. There are three main processes through which the strength of a H-bond is enhanced, as described below:

First, when the acceptor moiety is negatively charged, the electron-donor property is strong, therefore, the H-bond acceptor property is high. Such enhancement of H-bond strength is known as "negative charge assisted hydrogen bonding" or (−) CAHB (Gilli and Gilli, 2009). These H-bonds are as strong as ion-dipole interactions. Secondly, anionic functional groups such as carboxylate, sulfonate, and phenolate or simple anions such as nitrates and phosphates can efficiently form such strong H-bonds (Fig. 19.2D). Even though some aprotic functional groups, such as carbonyl groups, are naturally incapable of acting as H-bond donors, they can do so after protonation. Thirdly, when solvated with polar-protic solvents, such as water, especially under acidic conditions, such functional groups are protonated. Strong H-bonding of a protonated functional group with another H-bond acceptor group is known as "positive charge-assisted hydrogen bonding" or (+) CAHB (Gilli and Gilli, 2009). The strength of H-bonds formed by neutral species can be enhanced by the charges created through their resonance structures. This is known as "resonance-assisted hydrogen bonding (RAHB)," which is commonly found in resonance-stabilized functional groups, such as amides, acids, and quinoids (Gilli and Gilli, 2009).

In addition to the aforementioned conventional H-bonding, another common type of H-bonding is π–hydrogen bonding, in which systems rich in π electrons (e.g., aromatic systems) act as H-bond acceptors. BC surface consists of an aromatic system, which is a potential π-hydrogen bond acceptor system. Therefore, these interactions play a significant role in the adsorption of molecules having H-bond donor functional groups on to BC. Electron-donating groups attached to an aromatic system would improve the π-acceptor properties of the BC surface, thus leading to stronger π–hydrogen bonds (Nekoei and Vatanparast, 2019).

## 4.4 Electrostatic interactions, precipitation interaction, and cation exchange

Electrostatic attractions are noncovalent, long-range interactions occurring between permanently charged groups (Fig. 19.2E). These interactions could be attractive or repulsive and can be quantified using the zeta potential. The zeta potential is the electrical potential at the interface which separates mobile fluid from fluid that remains attached to the surface. The extent of EA is affected by the solution pH and ionic strength of the medium due to the screening effect (Peiris et al., 2020). For instance, if the adsorbate is a cation and the medium is rich with other cations, a positively charged shield can be created on the BC surface. As a result, EA is hindered due to the repulsive forces occurring between the induced positively charged BC surface and the cation of interest.

Precipitation is another common BC-based interaction mainly involved in removing metal ions. It is often associated with neutral to alkaline aqueous conditions, relatively high concentrations of the heavy metal concerned, low solubility of the metal compound, or low specific sorption sites (Echeverria et al., 1998). Lead is a common analyte which forms precipitates if the BC surface contains phosphates or carbonates (Cao et al., 2009). The formation of the phosphate precipitate is shown below:

$$6\,HPO_4^{2-}(aq) + 9\,Pb^{2+}(aq) + 6\,OH^-(aq) \rightarrow Pb_9(PO_4)_6(s) + 6\,H_2O(l)$$

Ion exchange is another process associated with the remediation of cations and anions, such as $Pb^{2+}$, $Cd^{2+}$, $Cr^{3+}$, $PO_4^{3-}$, and $AsO_4^{3-}$, and involves a displacement of the cations and anions of the BC surface with the analyte ions of interest. Ion exchange is more likely to occur as an exchange with the labile ions or electrostatically bound ions of the BC surface, thereby enabling the retention of cations and anions (Fig. 19.2F). Solution pH is a vital factor in determining the ion exchange phenomenon. It influences the surface charge of the adsorbent and ionization of the adsorbate, despite the competition caused by $H^+$ or $OH^-$ (Mukherjee et al., 2011). Thus, cation exchange is favored when the $pK_a$ of the adsorbate is greater than the PZC of the adsorbent, whereas anion exchange is favored when the $pK_a$ of the adsorbate is lower than the PZC of the adsorbent. Cation and anion exchange selectivity is reported as follows according to the lyotropic series (Singh et al., 2017).

Cation series: $Al^{3+} > Ca^{2+} > Mg^{2+} > K^+ \approx NH_4^+ > Na^+$.
Anion series: $SiO_4^{4-} > PO_4^{3-} >> SO_4^{2-} > NO_3^- = Cl^-$.

## 4.5 Surface complexation and cation bridging

Surface complexation and cation bridging are quite similar to each other. Surface complexation is when cations on the surface of the BC interact with anionic moieties of adsorbates to facilitate adsorption (Fig. 19.2G). Contrarily, cation bridging involves the bridging of a negatively charged adsorbate with a negatively charged site on BC surface through a metal cation (Fig. 19.2H) (MacKay and Vasudevan, 2012). Both these interactions can occur through either inner or outer sphere complexation (water bridging) (Kah et al., 2017; Goldberg, 2014). Generally, outer sphere bridging is favored when the cation is monovalent, whereas divalent cations favor both outer and inner sphere bridging (Kah et al., 2017; Ji et al., 2010; Han et al., 2013). Although numerous studies have shown increased sorption capacities in the presence of divalent cations, Jia et al. has reported a decreased capacity in oxytetracycline removal (Jia et al., 2013).

## 4.6 Transfer to pores (pore filling)

Sorption of organic and inorganic contaminants through the micro-, macro-, or mesopores of BC is known as pore filling, as the overlapping of pore wall potentials leads to stronger binding. However, it is important to note that pore filling is achievable only when the pore size of the adsorbate is 1.7 times larger than the second widest dimension of the solute of interest. Consequently, retaining of bulky molecules can be hindered in the presence of smaller pores due to the size exclusion effect (Peiris et al., 2017; Pelekani and Snoeyink, 2000).

### 4.7 Hydrophobic interaction

In contrast to most other interactions, which are controlled by favorable enthalpies of attraction, hydrophobic interactions are governed by entropy. Usually, polar compounds get dissolved in water by forming strong intermolecular attractions with the solvent molecules. However, hydrophobic solutes, such as nonpolar organic compounds, are distributed and stabilized in water, by forming clathrate hydrate cages surrounding the hydrophobic solutes (Englezos, 1993). When hydrophobic solutes, distributed in water interact with another such solute or with the hydrophobic surface of BC, these clathrate cages will be broken and the hydrophobic entities will interact with each other by forming weak dispersion forces. Even though the hydrophobic interactions are not energetically strong, their formation is thermodynamically feasible, because of the increase in entropy that results from breaking the highly structured clathrate cages (Bissantz et al., 2010). Well-carbonized BC is less hydrophilic, as most of the polar functionalities are removed. (Chun et al., 2004) Such BC is better at adsorbing hydrophobic molecules (e.g., trichloroethylene) owing to their ability to form hydrophobic interactions (Ahmad et al., 2013).

### 4.8 Catalytic degradation by radicals

Recent literature reports the potential of BC to act as a catalytic degradation material due to the presence of Environmentally Persistent Free Radicals (EPFRs) (Peiris et al., 2020). The persulfates and peroxides in the media can be activated by these EPFRs in BC, thus producing reactive oxygen species (ROS) which aids the efficacious degradation of many inorganic and organic contaminants (Peiris et al., 2020). For instance, Yu et al. has reported the degradation of estrogen with the aid of $H_2O_2$ (Yu et al., 2019). EPFRs produced during pyrolysis converts a fraction of $H_2O_2$ to $O_2^{-\bullet}$ which act as electron donors, whereas the other fraction was converted to $OH^{\bullet}$. The $OH^{\bullet}$ and the $O_2^{-\bullet}$ generated during the process aids the degradation of estrogens through a radical reaction mechanism (Yu et al., 2019).

# 5 Factors effecting the BC-adsorbate interactions

Sorption can be affected by a range of factors such as feedstock type, pyrolysis method, pyrolysis temperature, modification methods, and the experimental conditions. Detailed explanations of the effect of each parameter on the adsorption phenomena associated with BC are provided in the following sections.

## 5.1 Effect of feedstock

The most common feedstocks utilized in BC production can be categorized into four main groups: green waste (GW), animal waste (AW), municipal waste (MW), and sludge waste (SW) (López-Cano et al., 2018; Ippolito et al., 2020). Green waste can be further classified into agricultural waste (AgW), crop waste (CW), and wood waste (WW), whereas pig manure, cattle manure, and other feces are grouped under animal waste. (Hassan et al., 2020; Ippolito et al., 2020)

It has been reported that GW is high in organic matter as such waste is composed of large amounts of cellulose, hemicellulose, and lignin, in comparison to other types of feedstock (López-Cano et al., 2018). However, the organic matter content of BC would vary depending on the type of GW used, due to the composition variation of cellulose, hemicellulose, and lignin. Literature reports that the lignin and cellulose fraction is high when the GW is rich in branches and grasses, whereas in the presence of leaves and grass, the hemicellulose and cellulose fraction is high (López-Cano et al., 2018). On the other hand, the oxygen content of GW and manure waste is higher than that of WW and MW (Hassan et al., 2020). The feedstock type is one of the determining factors for the content of surface functional groups (SFGs) on the BC surface. As previously stated, with the increase in the organic matter fraction of the biomass, the oxygen content of BC also increases, indicating that BC produced from GW and manure waste is rich in oxygenated SFGs (O-SFGs) in comparison to WW (Hassan et al., 2020). Thus, BC derived from GW and manure waste are more polar, enabling interactions, such as H-bonding and EA, with organic contaminants and cations.

However, a notable difference between BC produced from GW and manure-derived BC is that the latter is rich in inorganic minerals, with both macro- and micronutrients, including phosphates, nitrates, Ca, Mg, Fe, and Mn (Hassan et al., 2020; Ippolito et al., 2020). Consequently, adsorbates, anionic compounds in particular, can be retained by manure-derived BC via surface complexation, cation bridging, ion exchange, and precipitation processes. However, it should be stressed that high amounts of inorganic constituents would also have a negative effect on the sorptive interactions. Leaching of inorganic ions creates a competitive effect for the binding sites, thereby hindering the BC-adsorbate interactions resulting in a decreased removal rate. On the other hand, BC derived from WW has a higher carbon content and aromatic structure which is capable of promoting EDA and hydrophobic interactions owing to its electron-donating property and hydrophobicity (Hassan et al., 2020; Ippolito et al., 2020).

Cation exchange capacity (CEC), surface area (SA), pore volume (PV), and pH are significant characteristics which influence the surface interactions associated with BC. It has been reported that BC derived from CW leads to a higher CEC as opposed to that of other types of feedstock (Ippolito et al., 2020). Thus, cations, such as $Pb^{2+}$, $Cd^{2+}$, and $K^+$, can be retained through the cation exchange process. Moreover, crop-based and manure-based BC exhibit a higher pH value than that of wood-based BC suggesting that the former two types would probably be rich in basic SFGs and ash consisting of hydroxides/carbonates of alkaline metals (Ippolito et al., 2020). These hydroxides and carbonates would thereby facilitate EA and precipitations with cations depending on the pH employed for sorption. For example, $Pb^{2+}$ can form $Pb(OH)_2$ precipitate after interacting with the surface -OH groups of BC.

Considering the SA variation with respect to feedstock choice, it is clear that wood-based BC, when compared to other feedstock types, usually has a greater SA and PV (Hassan et al., 2020; Ippolito et al., 2020). Hassan et al. have also stated that wood-derived BC exhibits the highest SA as compared to BC produced by MW and GW. This study also states that BC derived from hardwood, such as eucalyptus, cottonwood, oak, mahogany, walnut, teak, birch, and beech, possesses higher SA than softwood BC (Hassan et al., 2020). Therefore, one of the predominant interactions associated with the hardwood BC is considered to be the trapping of contaminants into the pores.

## 5.2 Effect of pyrolysis method and pyrolysis temperature

Biochar can be mainly produced through slow pyrolysis, fast pyrolysis, hydrothermal carbonization, and gasification processes (Lehmann and Joseph, 2015). Usually, BC prepared via fast pyrolysis exhibits greater SA along with increased porosity, while BC produced via slow pyrolysis is rich in ash content, Fe, and nitrate ions (Ippolito et al., 2020). Thus, the predominant interaction associated with the BC produced from fast pyrolysis is likely to be pore filling, while slow-pyrolyzed BC can facilitate SC, EA, and EDA due to its increased mineral and hydroxide content.

Biochar produced under absolute $N_2$ environment is generally rich in O-SFGs as compared to BC produced under oxygen-deficient conditions (Hassan et al., 2020; Luo et al., 2015). Hydrothermally produced BC is also reported to have polar functionalities as the production is carried out in the presence of water (Peiris et al., 2020). Thus, the polar moieties of such BC are capable of forming EA and H-bonding interactions with the adsorbate. Additionally, it is reported that fast-pyrolyzed BC is abundant in S, K, Ca, and Mg as compared to slow-pyrolyzed BC. However, the pyrolysis method is reported to have only a little effect on the total macroelement content of the BC relative to feedstock selection and temperature (Ippolito et al., 2020).

Depending on the duration of pyrolysis and the type of thermal treatment, many of the functional groups would change from one state to another. Biochar produced within the operational temperature range of 300–600°C contains an abundance of various aliphatic functional groups, such as hydroxyls and ketones. Further increase in temperature up to 900°C leads to the formation of an arene surface (Hassan et al., 2020; Lehmann and Joseph, 2015). Generally, low-temperature BC (LTBC) is rich in O-SFGs and hydrophilic groups, while high-temperature BC (HTBC) is more hydrophobic. Capability in forming EA and H-bonding is more feasible with LTBC, whereas EDA, cation-$\pi$ bonding, and hydrophobic interactions are the predominant interactions associated with HTBC (Hassan et al., 2020; Peiris et al., 2017).

Moreover, pyrolysis temperature influences the nutrient content of BC as well. Increase in pyrolysis temperature would increase the ash content and C, P, Ca, Mg, K, and Fe content. However, the availability of K, Mg, and Fe content might decrease upon increasing the temperature beyond 800°C (Ippolito et al., 2020).

It is important to consider that, at low pyrolysis temperatures and short pyrolysis times, oxygen-centered EPFRs are formed, whereas at high pyrolysis temperatures and longer pyrolysis times, carbon-centered EPFRs are formed. Thus, both LTBC and HTBC can promote catalytic degradation by BC.

## 5.3 Effect of modification

Numerous innovative methodologies have been adopted for modifying BC to enhance adsorption properties. Modifications are performed either pre- or post-pyrolysis of the BC. The two most common methods employed in the modification of BC are chemical activation and physical activation. Chemical activation includes acid modification, magnetization, base modification, oxidizing agent modification, and metal impregnation (Rajapaksha et al., 2016; Wang and Wang, 2019). All these modifications lead to enhanced capacity by increasing the surface functionality, SA, porosity, or the mineral content. Modification incorporates specific properties into the BC surface; thus, a predominant sorption process can be interpreted in the removal of a particular adsorbate. Therefore, a proper understanding of how a modification would influence the sorptive interactions is vital for effective removal.

Treatment of BC with acidic substances, such as $HNO_3$, $H_2O_2$, $H_3PO_4$, $H_2SO_4$, and HCl, leads to an increase in the O-SFG content on the BC surface. Generally, nitric and peroxide acid treatments are reported to increase the carboxylic and phenolic fraction of BC due to their strong oxidizing ability (Lin et al., 2012). Furthermore, fixation of O-SFGs is also possible when modified with $HNO_3$ (Yakout et al., 2015). Similarly, base treatment using KOH or NaOH increases the oxygen content and surface basicity of the BC surface (Rajapaksha et al., 2016; Fan et al., 2010). Thus, these O-SFGs aid the adsorption of metal ions through EA, whereas adsorption of organic contaminants occur mainly via H-bonding. Moreover, acid modification could introduce carbonyl groups, making the BC surface favorable toward π-π interactions. Additionally, phosphoric acid treatment has the ability to introduce phosphate moieties onto the BC surface which is advantageous in remediating potentially toxic metal ions through EA and precipitation.

Considering the surface morphological changes upon acid treatment, $H_2SO_4$/$HNO_3$ treatment is said to increase the meso- and macropore fraction of BC due to the conversion of micropores into meso- and macropores (Yakout et al., 2015). Furthermore, formation of chemical species, such as $K_2O$ and $K_2CO_3$, during KOH/NaOH modification results in widening of pores and formation of new pores due to diffusion of such species into the internal structure of the BC matrix (Mao et al., 2015). Thus, the presence of macropores would probably favor entering bulky molecules, eliminating the size exclusion effect caused by micropores (Peiris et al., 2017).

N-containing SFGs (N-SFGs) can be introduced onto the BC through $HNO_3$ treatment and surface amination as well as through modification using chitosan (Rajapaksha et al., 2016). Introduction of N-SFGs, such as amide, imide, lactame, and pyrrolic and pyridinic acids facilitate the sorption of metal cations, namely, $Cu^{2+}$, $Zn^{2+}$, and $Cd^{2+}$, through surface complexation (Rajapaksha et al., 2016; Shafeeyan et al., 2010).

New innovative methodologies have been used to engineer the production and modification of BC by impregnating with minerals/metals and by magnetization, which would lead to extended applications (Yao et al., 2014; Zhang et al., 2013). Most commonly used clay minerals in such methodologies are montmorillonite, gibbsite, and kaolinite as well as metal oxides (Rajapaksha et al., 2016). Mineral impregnation and magnetization enhance the sorption of anions through surface complexation and EA due to the introduction of metals onto the BC surface. For example, BC modified with montmorillonite and kaolinite increases the Al and Fe content of BC, while magnetic modification increases the Fe content (Yao et al., 2014; Zhang et al., 2013). Physical modification, which does not involve chemicals, mainly uses steam activation to remove trapped constituents of incomplete combustion resulting in high SA of BC (Rajapaksha et al., 2016).

However, these modifications can also negatively affect the interactions surrounding sorptive removal. The decreased mineral content of BC upon demineralization during acid treatment might hinder the capability in forming SC and EA interactions, due to the limited mineral content. Moreover, reduction of the pore size due to group fixation results in decreased SA (Yakout et al., 2015). Metal impregnation at higher concentrations could also reduce the accessibility of pores due to pore blockage.

## 5.4 Effect of pH

Adjustment of solution pH is capable of altering the surface charge of the adsorbent and the degree of ionization of the adsorbate resulting in changes in sorption capacities (Peiris et al., 2020). Therefore, optimization of solution pH is crucial to enhance the sorption capacity, with respect to the intended adsorbate, to desired values. Accurate values of the surface charge of BC are usually determined through point of zero charge ($pH_{PZC}$) measurements. The BC surface is positively charged when the solution pH is lower than the $pH_{PZC}$. Therefore, positively charged BC surfaces can form EA with negatively charged adsorbates (Peiris et al., 2017). Furthermore, H-bonding and EDA interactions with organic contaminants are also possible at lower pH (pH < 4), as the SFGs are in the protonated form. However, when the pH is increased up to 8, a greater fraction of dissociated carboxylic and lactonic groups are present on the BC surface resulting in a net negative surface charge (Singh et al., 2017) ($pK_a$ of carboxylic group: 2–4; $pK_a$ of lactonic group: 4–7). Thus, toxic metal removal through EA is feasible at this pH. It is also important to note that H-bonding can occur even at this pH to a certain extent with phenolic moieties due to its high $pK_a$ value (Singh et al., 2017).

## 6 Prediction of contaminant-biochar interactions

Understanding the sorption of contaminants on BC is an important issue, yet a challenging task, mainly due to not being able to generalize it as the interaction between the adsorbate and adsorbent. The BC-adsorbate interactions depend on several factors, including the nature of BC (e.g., type of BC, source of BC), process variables (e.g., type of modifier, nature of modifier, conditions of modification), and experimental variables (e.g., solution temperature, concentration of adsorbate, and type of adsorbate). Moreover, experimental findings are usually fitted to various models, and conclusions in many instances have been made merely based on regression coefficient values, rather than considering a microscopic approach.

The holistic approach including investigations on pH experiments, kinetics, isotherm models, thermodynamics, and surface property characterization accompanied by sufficient mathematical analysis should thus be considered although interpretations are performed in many ways focusing on different aspects. Correlation between the solution pH and adsorption capacity could be considered as valuable evidence to be used in predicting the sorptive interactions between BC and contaminants. Interactions such as EA, H-bonding, CAHB, and EDA are possible when the adsorption capacity is pH-dependent. Further, the relative contribution of each type of intermolecular interaction could be quantified to some extent based on the correlation between adsorption capacity and solution pH. Nevertheless, such approaches have not been much attended to. On the other hand, when the adsorption capacity is unchanged over a pH range, the dominant interactions are considered to be pore filling and hydrophobic interactions with less involvement of the functional moieties (Peiris et al., 2020).

Application of different kinetics models and thermodynamics relationships would also assist in predicting the nature of the sorptive interactions through mathematical calculations, followed by modeling. The frequently used pseudo-first-order (PFO) and pseudo-second-order (PSO) models confirm the physisorptive and chemisorptive behavior of the process, respectively (Kołodyńska et al., 2012). Additionally, the reaction enthalpy

($\Delta H$) values of thermodynamics analysis confirm chemisorption when $\Delta H$ values are greater than $40 \text{ kJ mol}^{-1}$, whereas values less than $20 \text{ kJ mol}^{-1}$ are indicative of physisorption, based on bonding considerations (Navarathna et al., 2019; Monárrez-Cordero et al., 2018). The above interpretation clearly indicates that fitting experimental data to the PFO model should not lead to $\Delta H$ values more than $40 \text{ kJ mol}^{-1}$, and vice versa. However, this is not the case in some experimental findings, and in such instances, mechanistic interpretation should carefully be performed, considering the interactions at a molecular level.

Based on $\Delta H$ values of adsorbate-BC interactions reported in literature, it can be stated that both physisorptive interactions (e.g., EA, H-bonding, EDA interactions, pore filling, hydrophobic, EAA) and chemisorptive interactions (e.g., surface complexation, cation bridging, precipitation, and ion exchange) are possible (Tong et al., 2019). However, literature reports in numerous occasions that the process is chemisorptive if the adsorption is governed by EDA, EA, etc. Even though kinetic and thermodynamic studies confirm chemisorptive behavior, it is erroneous to assume such interpretations, as EDA is classified under physisorptive interactions. It can however be argued that the predominant type of interactions depends on the microscopic environment of the system.

It is important to note that, even though H-bonding is considered as a physisorptive interaction, the formation of CAHB is considered to be chemisorptive due to the enthalpy values above $37 \text{ kJ mol}^{-1}$ (Tong et al., 2019; Pignatello et al., 2017). For instance, Ni et al. has reported CAHB interactions with an enthalpy value of $100 \text{ kJ mol}^{-1}$ during the adsorption of allelopathic aromatic acid ions on BC proving a covalent like character (Tong et al., 2019; Ni et al., 2011).

Another misinterpretation often encountered in literature is fitting the adsorption data to a particular empirical isotherm based on linear regression analysis used for mechanistic predictions. The fact that the extent of sorption, whether it is monolayer or multilayer formation, should ideally be determined by monitoring the BC surface after being exposed to adsorbate solutions must be emphasized. However, the surface coverage is, most of the time, determined through solution analysis. The amount of adsorbate transferred from the solution phase to the solid adsorbent phase is used in many publications to decide whether the adsorption is monolayer or multilayer, without paying much attention to microscopic characteristics of the system. In certain instances, transfer of adsorbate in excess could lead to monolayer formation followed by other modes of mass transfer, such as transfer to pores, interparticle diffusion and intraparticle diffusion, without forming a multilayer.

As the BC surface is inhomogeneous having different pore sizes and surface characteristics, the BC-adsorbate interactions can be so complex that the possibility of having both chemisorption and physisorption under a given set of experimental conditions cannot be ruled out.

Thus, subjecting adsorption results to kinetic model fitting and thermodynamic analysis is required to avoid possible misinterpretations caused when predicting the sorptive interactions solely based on batch sorption experiments. Results of the above approaches, together with microscopic characteristics of the adsorbate-BC system and energy requirements associated with different bonding types, should thus be considered together in predicting a possible interaction for individual systems.

## 7 Conclusions

Surface characteristics of BC such as surface acidity and basicity, morphological characteristics, surface charge, mineral composition, and the physicochemical properties of the adsorbate are the factors that determine the predominant BC-adsorbate interactions. Generally, the number of dissociable functional groups present in BC primarily influences H-bond formation and EA. The extent of EA is highly dependent on the acidic functional groups with a lower pKa such as carboxylic acids as opposed to stronger acids such as phenolic acids which significantly contribute to H-bond formation. Moreover, BCs which are high in phenolic moieties can facilitate EDA interaction due to its electron-donating properties. Furthermore, cation pi-bonding and EDA were observed to predominate in BC produced at higher temperatures owing to its rich arene surface. The prevalent acid and base modifications were observed to cause enhanced or suppressed sorption capacities. Introduction of oxygenated and nitrogenated SFGs upon modifications facilitate interactions which are predominantly electrostatic and H-bonding. However, group fixation at the pores by such treatments might hinder the entry of bulky molecules restricting the extent of pore filling, resulting in the conversion of macropores to micropores. A clear understanding of the sorptive interactions and factors that affect the adsorption process is important as it aids the production of effective BC for the removal of any analyte of interest.

## Questions and problems

1. What main interactions are most likely to occur between green manure BC and adsorbates?
2. Name several instances of common chemisorptive and physisorptive interactions.
3. What are the modifications that assist in improving the oxygen content of the biochar surface?
4. What are the main processes occurring during the biomass-biochar conversion?
5. Why does hydrothermally produced BC have polar functionalities?
6. Why are hydrophobic interactions thermodynamically feasible?
7. Why are bulky molecules hindered in the presence of smaller pores?
8. What factors affect the properties of biochar?
9. What factors affect the extent of electrostatic interactions?
10. How does the experimental pH affect the surface charge of the biochar?

## References

Ahmad, M., Lee, S.S., Rajapaksha, A.U., Vithanage, M., Zhang, M., Cho, J.S., Lee, S.-E., Ok, Y.S., 2013. Trichloroethylene adsorption by pine needle biochars produced at various pyrolysis temperatures. Bioresour. Technol. 143, 615–622.

Ahmad, M., Rajapaksha, A.U., Lim, J.E., Zhang, M., Bolan, N., Mohan, D., Vithanage, M., Lee, S.S., Ok, Y.S., 2014. Biochar as a sorbent for contaminant management in soil and water: a review. Chemosphere 99, 19–33.

Anca-Couce, A., 2016. Reaction mechanisms and multi-scale modelling of lignocellulosic biomass pyrolysis. Prog. Energy Combust. Sci. 53, 41–79.

Bent, H.A., 1968. Structural chemistry of donor-acceptor interactions. Chem. Rev. 68 (5), 587–648.

Bissantz, C., Kuhn, B., Stahl, M., 2010. A medicinal chemist's guide to molecular interactions. J. Med. Chem. 53 (14), 5061–5084.

Brebu, M., Vasile, C., 2010. Thermal degradation of lignin—a review. Cellul. Chem. Technol. 44 (9), 353.

Cao, X., Ma, L., Gao, B., Harris, W., 2009. Dairy-manure derived biochar effectively sorbs lead and atrazine. Environ. Sci. Technol. 43 (9), 3285–3291.

Cheng, F., Li, X., 2018. Preparation and application of biochar-based catalysts for biofuel production. Catalysts 8 (9), 346.

Cheng, J., Luo, X., Yan, X., Li, Z., Tang, Y., Jiang, H., Zhu, W., 2008. Research progress in cation-π interactions. Sci. China, Ser. B: Chem. 51 (8), 709–717.

Chun, Y., Sheng, G., Chiou, C.T., Xing, B., 2004. Compositions and sorptive properties of crop residue-derived chars. Environ. Sci. Technol. 38 (17), 4649–4655.

Echeverria, J., Morera, M., Mazkiaran, C., Garrido, J., 1998. Competitive sorption of heavy metal by soils. Isotherms and fractional factorial experiments. Environ. Pollut. 101 (2), 275–284.

Englezos, P., 1993. Clathrate hydrates. Ind. Eng. Chem. Res. 32 (7), 1251–1274.

Fan, Y., Wang, B., Yuan, S., Wu, X., Chen, J., Wang, L., 2010. Adsorptive removal of chloramphenicol from wastewater by NaOH modified bamboo charcoal. Bioresour. Technol. 101 (19), 7661–7664.

Fernando, J.C., Peiris, C., Navarathna, C.M., Gunatilake, S.R., Welikala, U., Wanasinghe, S.T., Madduri, S.B., Jayasinghe, S., Mlsna, T.E., Ferez, F., 2021. Nitric acid surface pre-modification of novel Lasia spinosa biochar for enhanced methylene blue remediation. Groundw. Sustain. Dev. 14, 100603.

Gilli, G., Gilli, P., 2009. The Nature of the Hydrogen Bond: Outline of a Comprehensive Hydrogen Bond Theory. vol. 23 Oxford University Press.

Goldberg, S., 2014. Application of surface complexation models to anion adsorption by natural materials. Environ. Toxicol. Chem. 33 (10), 2172–2180.

Han, X., Liang, C.-F., Li, T.-Q., Wang, K., Huang, H.-G., Yang, X.-E., 2013. Simultaneous removal of cadmium and sulfamethoxazole from aqueous solution by rice straw biochar. J. Zhejiang Univ. Sci. B 14 (7), 640–649.

Hassan, M., Liu, Y., Naidu, R., Parikh, S.J., Du, J., Qi, F., Willett, I.R., 2020. Influences of feedstock sources and pyrolysis temperature on the properties of biochar and functionality as adsorbents: a meta-analysis. Sci. Total Environ. 744, 140714.

Ippolito, J.A., Cui, L., Kammann, C., Wrage-Mönnig, N., Estavillo, J.M., Fuertes-Mendizabal, T., Cayuela, M.L., Sigua, G., Novak, J., Spokas, K., 2020. Feedstock choice, pyrolysis temperature and type influence biochar characteristics: a comprehensive meta-data analysis review. Biochar, 1–18.

Ji, L., Chen, W., Duan, L., Zhu, D., 2009. Mechanisms for strong adsorption of tetracycline to carbon nanotubes: a comparative study using activated carbon and graphite as adsorbents. Environ. Sci. Technol. 43 (7), 2322–2327.

Ji, L., Chen, W., Bi, J., Zheng, S., Xu, Z., Zhu, D., Alvarez, P.J., 2010. Adsorption of tetracycline on single-walled and multi-walled carbon nanotubes as affected by aqueous solution chemistry. Environ. Toxicol. Chem. 29 (12), 2713–2719.

Jia, M., Wang, F., Bian, Y., Jin, X., Song, Y., Kengara, F.O., Xu, R., Jiang, X., 2013. Effects of pH and metal ions on oxytetracycline sorption to maize-straw-derived biochar. Bioresour. Technol. 136, 87–93.

Kah, M., Sigmund, G., Xiao, F., Hofmann, T., 2017. Sorption of ionizable and ionic organic compounds to biochar, activated carbon and other carbonaceous materials. Water Res. 124, 673–692.

Kołodyńska, D., Wnętrzak, R., Leahy, J., Hayes, M., Kwapiński, W., Hubicki, Z., 2012. Kinetic and adsorptive characterization of biochar in metal ions removal. Chem. Eng. J. 197, 295–305.

Lehmann, J., Joseph, S., 2015. Biochar for Environmental Management: Science, Technology and Implementation. Routledge.

Li, D.-C., Jiang, H., 2017. The thermochemical conversion of non-lignocellulosic biomass to form biochar: a review on characterizations and mechanism elucidation. Bioresour. Technol. 246, 57–68.

Lin, Y., Munroe, P., Joseph, S., Henderson, R., Ziolkowski, A., 2012. Water extractable organic carbon in untreated and chemical treated biochars. Chemosphere 87 (2), 151–157.

Liu, W.-J., Jiang, H., Yu, H.-Q., 2015. Development of biochar-based functional materials: toward a sustainable platform carbon material. Chem. Rev. 115 (22), 12251–12285.

López-Cano, I., Cayuela, M.L., Mondini, C., Takaya, C.A., Ross, A.B., Sánchez-Monedero, M.A., 2018. Suitability of different agricultural and urban organic wastes as feedstocks for the production of biochar—part 1: physicochemical characterisation. Sustainability 10 (7), 2265.

Luo, L., Xu, C., Chen, Z., Zhang, S., 2015. Properties of biomass-derived biochars: combined effects of operating conditions and biomass types. Bioresour. Technol. 192, 83–89.

MacKay, A.A., Vasudevan, D., 2012. Polyfunctional ionogenic compound sorption: challenges and new approaches to advance predictive models. Environ. Sci. Technol. 46 (17), 9209–9223.

Mao, H., Zhou, D., Hashisho, Z., Wang, S., Chen, H., Wang, H.H., 2015. Preparation of pinewood-and wheat straw-based activated carbon via a microwave-assisted potassium hydroxide treatment and an analysis of the effects of the microwave activation conditions. Bioresources 10 (1), 809–821.

McMurry, J.E., 2010. Fundamentals of Organic Chemistry. Cengage Learning.

Mitchell, M.O., Means, J., 2018. Cation – Π interactions in biochemistry: a primer. J. Chem. Educ. 95 (12), 2284–2288.

Monárrez-Cordero, B., Sáenz-Trevizo, A., Bautista-Carrillo, L., Silva-Vidaurri, L., Miki-Yoshida, M., Amézaga-Madrid, P., 2018. Simultaneous and fast removal of $As^{3+}$, $As^{5+}$, $Cd^{2+}$, $Cu^{2+}$, $Pb^{2+}$ and $F-$ from water with composite Fe-Ti oxides nanoparticles. J. Alloys Compd. 757, 150–160.

Mukherjee, A., Zimmerman, A., Harris, W., 2011. Surface chemistry variations among a series of laboratory-produced biochars. Geoderma 163 (3–4), 247–255.

Navarathna, C.M., Karunanayake, A.G., Gunatilake, S.R., Pittman Jr., C.U., Perez, F., Mohan, D., Mlsna, T., 2019. Removal of arsenic (III) from water using magnetite precipitated onto Douglas fir biochar. J. Environ. Manage. 250, 109429.

Nekoei, A.-R., Vatanparast, M., 2019. π-Hydrogen bonding and aromaticity: a systematic interplay study. Phys. Chem. Chem. Phys. 21 (2), 623–630.

Ni, J., Pignatello, J.J., Xing, B., 2011. Adsorption of aromatic carboxylate ions to black carbon (biochar) is accompanied by proton exchange with water. Environ. Sci. Technol. 45 (21), 9240–9248.

Pearson, R.G., 1963. Hard and soft acids and bases. J. Am. Chem. Soc. 85 (22), 3533–3539.

Peiris, C., Gunatilake, S.R., Mlsna, T.E., Mohan, D., Vithanage, M., 2017. Biochar based removal of antibiotic sulfonamides and tetracyclines in aquatic environments: a critical review. Bioresour. Technol. 246, 150–159.

Peiris, C., Nayanathara, O., Navarathna, C.M., Jayawardhana, Y., Nawalage, S., Burk, G., Karunanayake, A.G., Madduri, S.B., Vithanage, M., Kaumal, M., 2019. The influence of three acid modifications on the physicochemical characteristics of tea-waste biochar pyrolyzed at different temperatures: a comparative study. RSC Adv. 9 (31), 17612–17622.

Peiris, C., Nawalage, S., Wewalwela, J.J., Gunatilake, S.R., Vithanage, M., 2020. Biochar based sorptive remediation of steroidal estrogen contaminated aqueous systems: a critical review. Environ. Res. 191, 110183.

Peiris, C, Wathudura, P.D., Gunatilake, S.R., Gajanayake, B., Wewalwela, J.J., Abeysundara, S., Vithanage, M., 2022. Effect of acid modified tea-waste biochar on crop productivity of red onion (*Allium cepa* L.). Chemosphere 288, 132551. https://doi.org/10.1016/j.chemosphere.2021.132551.

Pelekani, C., Snoeyink, V.L., 2000. Competitive adsorption between atrazine and methylene blue on activated carbon: the importance of pore size distribution. Carbon 38 (10), 1423–1436.

Pignatello, J., Mitch, W.A., Xu, W., 2017. Activity and reactivity of pyrogenic carbonaceous matter toward organic compounds. Environ. Sci. Technol. 51 (16), 8893–8908.

Rajapaksha, A.U., Chen, S.S., Tsang, D.C., Zhang, M., Vithanage, M., Mandal, S., Gao, B., Bolan, N.S., Ok, Y.S., 2016. Engineered/designer biochar for contaminant removal/immobilization from soil and water: potential and implication of biochar modification. Chemosphere 148, 276–291.

Rangabhashiyam, S., Balasubramanian, P., 2019. The potential of lignocellulosic biomass precursors for biochar production: performance, mechanism and wastewater application—a review. Ind. Crop Prod. 128, 405–423.

Shafeeyan, M.S., Daud, W.M.A.W., Houshmand, A., Shamiri, A., 2010. A review on surface modification of activated carbon for carbon dioxide adsorption. J. Anal. Appl. Pyrolysis 89 (2), 143–151.

Singh, M.K., 2014. Industrial Practices in Weaving Preparatory. Woodhead Publishing India PVT. Limited.

Singh, B., Camps-Arbestain, M., Lehmann, J., 2017. Biochar: A Guide to Analytical Methods. Csiro Publishing.

Teixidó, M., Pignatello, J.J., Beltrán, J.L., Granados, M., Peccia, J., 2011. Speciation of the ionizable antibiotic sulfamethazine on black carbon (biochar). Environ. Sci. Technol. 45 (23), 10020–10027.

Tong, Y., McNamara, P.J., Mayer, B.K., 2019. Adsorption of organic micropollutants onto biochar: a review of relevant kinetics, mechanisms and equilibrium. Environ. Sci.: Water Res. Technol. 5 (5), 821–838.

Wang, J., Wang, S., 2019. Preparation, modification and environmental application of biochar: a review. J. Clean. Prod. 227, 1002–1022.

Xu, Y., Yu, X., Xu, B., Peng, D., Guo, X., 2020. Sorption of pharmaceuticals and personal care products on soil and soil components: influencing factors and mechanisms. Sci. Total Environ. 753, 141891.

Yakout, S.M., Daifullah, A.E.H.M., El-Reefy, S.A., 2015. Pore structure characterization of chemically modified biochar derived from rice straw. Environ. Eng. Manag. J. 14 (2), 473–480.

Yao, Y., Gao, B., Fang, J., Zhang, M., Chen, H., Zhou, Y., Creamer, A.E., Sun, Y., Yang, L., 2014. Characterization and environmental applications of clay–biochar composites. Chem. Eng. J. 242, 136–143.

Yu, W., Lian, F., Cui, G., Liu, Z., 2018. N-doping effectively enhances the adsorption capacity of biochar for heavy metal ions from aqueous solution. Chemosphere 193, 8–16.

Yu, J., Zhu, Z., Zhang, H., Chen, T., Qiu, Y., Xu, Z., Yin, D., 2019. Efficient removal of several estrogens in water by Fe-hydrochar composite and related interactive effect mechanism of $H_2O_2$ and iron with persistent free radicals from hydrochar of pinewood. Sci. Total Environ. 658, 1013–1022.

Zhang, M., Gao, B., Varnoosfaderani, S., Hebard, A., Yao, Y., Inyang, M., 2013. Preparation and characterization of a novel magnetic biochar for arsenic removal. Bioresour. Technol. 130, 457–462.

# 20

# Nanobiochar for aqueous contaminant removal

Tej Pratap[a], Abhishek Kumar Chaubey[a], Manvendra Patel[a], Todd E. Mlsna[b], Charles U. Pittman, Jr.[b], and Dinesh Mohan[a]

[a]School of Environmental Sciences, Jawaharlal Nehru University, New Delhi, India. [b]Department of Chemistry, Mississippi State University, Mississippi State, MS, United States

## 1 Introduction

Water consumption is steadily increasing due to population growth, industrialization, and urbanization (Abd El Hameed et al., 2015; Gafoor et al., 2021; Patel et al., 2019; Xiao et al., 2020; Xu et al., 2019). Such exponential consumption led to the growth of many contaminants in aquatic systems (Abd El Hameed et al., 2015; Gafoor et al., 2021). Both organic and inorganic aqueous contaminants have been extensively reported (Agrafioti et al., 2014a; Chen et al., 2017; Fosso-Kankeu et al., 2016; Kumar et al., 2019a,b; Patel et al., 2019; Premarathna et al., 2019; Tang et al., 2015; Wang et al., 2015; Xiao et al., 2020; Zhou et al., 2014). Adsorptive removal of these contaminants provides a low-capital-cost, easy-to-handle, and highly acceptable technology (Xu et al., 2019). The nanoadsorbents including graphene, carbon nanotubes, nanosilica, metal nanoparticles, metal oxide nanocrystals, nanospheres, and nanobiochar are effective for laboratory-scale aqueous contaminant sorption (Okitsu et al., 2007; Patil and Joshi, 2007; Ramanayaka et al., 2020a,b; Sharma et al., 2009; Xu et al., 2019). Of these, nanobiochar emerged as the more sustainable adsorbent for inorganic and organic contaminants sorption based on its lower cost, excellent adsorption capacity, and superior regeneration capability

(Ramanayaka et al., 2020a,b; Taheran et al., 2018; Zhang et al., 2019a,b). Nanobiochar is suitable in agricultural applications and as catalysts, biomolecule carriers, an alternative to carbon black and sensing material (Dong et al., 2018; Dong et al., 2017; Naghdi et al., 2017; Nath et al., 2019; Qin et al., 2020; Ramanayaka et al., 2020a,b; Xiao et al., 2020).

Biochar has been classified into macro- (>1 μm), colloidal (1 μm–100 nm), and nanobiochar (<100 nm) (Song et al., 2019). Reducing biochar to the nanosized range enhanced its contaminant adsorption capabilities (Naghdi et al., 2017). Reducing its size to "nanobiochar" enhanced specific surface area, the number of surface functional groups per unit weight, and modified pore structures (Kumar et al., 2020; Xiao et al., 2020). Nanobiochar can overcome pristine biochar problems such as limited surface area, less diverse functional groups, and a few number of pores per unit weight of char (Song et al., 2019; Zhang et al., 2019a,b). Various methods have been used to develop nanobiochar economically (Kumar et al., 2020; Lyu et al., 2016; Ramanayaka et al., 2020a,b). Nanobiochars efficiently removed inorganic and organic aqueous contaminants (Lyu et al., 2016; Naghdi et al., 2017; Peng et al., 2018; Ramanayaka et al., 2020a,b; Xiao et al., 2020; Xu et al., 2019; Zhang et al., 2019a,b). Since nanobiochar is prepared from agricultural and municipal wastes, it provides an alternative for solid waste management (Ramanayaka et al., 2020b). This chapter provides an overview of nanobiochar development, characterization, and applications for aqueous contaminant (organic and inorganic) removal.

## 2  Nanobiochar preparation methods

Nanobiochar fabrication techniques include ball milling, centrifugation, microwave pyrolysis, and thermal heat flash treatment (Fig. 20.1) (Genovese et al., 2015; Ramanayaka et al., 2020a,b; Wallace et al., 2019). Top-down and bottom-up approaches are used for fabricating nanostructures (Kumar et al., 2020). In the top-down approach, macrobiochar size is reduced to the nanolevel, while in the bottom-up approach, nanobiochar is built up from the atomic scales (Naghdi et al., 2017). Ball milling, centrifugation, arc discharge, and laser ablation are top-down methods. Chemical precipitation and microwave pyrolysis are bottom-up techniques (Kumar et al., 2020; Ullah et al., 2014). However, bottom-up approaches are not applicable for nanobiochar preparation because biochar preparation requires biomass pyrolysis (Pratap et al., 2020). Ball milling is the most widely employed fabrication

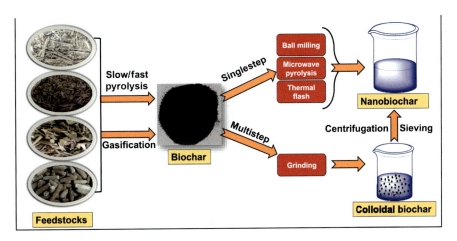

**Fig. 20.1** Nanobiochar preparation methods.

technique for nanobiochar preparation (Kumar et al., 2020; Ramanayaka et al., 2020a,b). Approximately 98% of the engineered nanobiochars have been fabricated by mechanical grinding (Ramanayaka et al., 2020b).

## 2.1 Ball milling

Ball milling is a green method as it does not require harmful chemicals (Naghdi et al., 2017). Ball milling has an advantage of achieving nanosized biochar without causing biochar microcrystalline structure damage (Ramanayaka et al., 2020b). Ball milling is a top-down method. Mechanical forces reduce biochar size to nanobiochar dimensions (Naghdi et al., 2017). Ball milling starts with the lignocellulosic biomass conversion into biochar using thermo-chemical processes, including gasification, hydrothermal carbonization, and slow and fast pyrolysis (Cho et al., 2019; Shaheen et al., 2019). This is followed by biochar particle size reduction in a ball mill. Ball milling is a nonequilibrium process which mechanically decreases the size of the macromolecules to the ultrafine or nanoscale (Ullah et al., 2014). During ball milling, rapid extrusion and high energy shear forces greatly reduced the particle size and improved the homogeneity and surface biochar functionality (Kumar et al., 2020). Ball milling is capable of breaking chemical bonds, increasing the specific surface area, and creating new surfaces using high energy (Kumar et al., 2020; Lyu et al., 2020; Xu et al., 2019). To date, wood, agricultural residues, forest residues, animal manure, municipal solid wastes, and sewage sludge have been used for biochar fabrication (Oleszczuk et al., 2016). Similarly, nanobiochars have also been prepared from a variety of feedstocks, including rice husk, peanut shell, sugarcane bagasse, rice straw, grass oil palm empty fruit bunch, bamboo, wood, and soybean stover (Ramanayaka et al., 2020a,b). Woody biomass is rich in lignin while agricultural residue contains high levels of hemicelluloses (Ramanayaka et al., 2020b). Agricultural waste provides better grindability and broad particle size distributions due to its high hemicellulose concentration (Weber and Quicker, 2018). Hemicellulose provides better grindability versus lignin and is suitable for nanobiochar preparation (Ramanayaka et al., 2020b).

Surface area, porous structure, pore size distribution, charge density, zero-point charge (ZPC), surface functional groups, heteroatom content and speciation, aromatic, and graphitic structures are distinctive characteristics that depend upon the feedstock, pyrolysis conditions, and pre- or postprocessing (Kumar et al., 2020; Naghdi et al., 2017; Ramanayaka et al., 2020a,b) (Fig. 20.2). Attritor mills, planetary ball mills, tumbler ball mills, and vibrator ball mills are commonly used to reduce biochar size (Ullah et al., 2014). The planetary ball mill is most widely used (Kumar et al., 2020). Ball-mill selection depends on the requirement (amount, size, and uniformity of nanobiochar), techno-economic analysis, and operational convenience (Kumar et al., 2020). For instance, vibrator and planetary ball mills are suitable for handling a small amount of biochar (<100 g) (Karinkanta et al., 2018). For larger biochar amounts (100 g–10 kg), attritor ball mills (running time in hours) are very useful

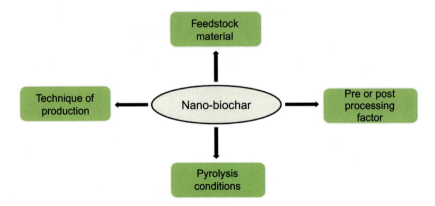

**Fig. 20.2** Factors affecting nanobiochar properties.

for commercial nanobiochar productions. Vibrator disc milling is suitable for nanobiochar production with uniform particle sizes and shape (Ramanayaka et al., 2020b). Disc milling produces low-surface-area nanobiochars (5–370 $m^2/g$), which may be ascribed to the presence of nanopores and graphitic nature (Lyu et al., 2018a,b; Naghdi et al., 2017; Naghdi et al., 2019; Ramanayaka et al., 2020a,b; Weber and Quicker, 2018). Both stainless steel and agate balls have been used in the ball milling process (Naghdi et al., 2017).

Ball mill type, milling media, milling time, container, milling speed, temperature, ball-to-powder ratio, extent of drum filling, and milling atmosphere are important factors in nanobiochar preparation (Kumar et al., 2020; Ullah et al., 2014). During milling, larger biochar particles get broken down into smaller ones due to internal strain manufactured by high pressure during impacts (Kumar et al., 2020; Ullah et al., 2014). The kinetic energy transfer intensity from the balls to the carbonaceous material varies in different ball mill types (Kumar et al., 2020). Chemical (oxidation) and physical destruction during milling can be avoided by filling the ball-milling chamber with gases like argon, ammonia, and hydrocarbon gases (Xing et al., 2016).

Longer milling time can result in agglomeration of the small primary particles (Ullah et al., 2014). Strong agglomeration is favored by high surface area, but can be prevented by using surfactants during milling (Ullah et al., 2014). Surfactants played a critical role in preventing the close contact of the fine particles by reducing surface tension and providing steric barriers (Ullah et al., 2014). Ball-to-powder ratio is an important parameter, determining the time required to achieve the nanosized biochar (Ullah et al., 2014). Its value can range from as low as 1:1 to as high as 220:1 (Ullah et al., 2014). Therefore, optimization of ball-to-powder ratio is very important (Ullah et al., 2014). Higher ball-to-powder ratios result into greater energy transfer during ball collisions with the grinding media. Furthermore, this evolved energy may lead to the phase transformation of particles (Kumar et al., 2020; Ullah et al., 2014).

The milling atmosphere can contaminate or produce specific functionalities in nanobiochar (Suryanarayana, 2001). Generally, the milling medium is filled with inert gases to prevent cross-contamination. The presence of a specific gas as the milling atmosphere can also introduce specific functionalities on the nanobiochar. For example, hydrogen, nitrogen, and oxygen can produce hydrides, nitrides, and oxides, respectively, on nanobiochar surfaces (Ullah et al., 2014; Xu et al., 2019).

Ball milling is categorized into wet and dry ball milling (Kumar et al., 2020; Ullah et al., 2014; Yuan et al., 2020). Dry milling is preferred for finely ground concentrates, which are difficult to wet filter (Kumar et al., 2020). Dry milling is a cost-effective to increase micropores and total surface area of biochar, but this method has limited efficiency (Kumar et al., 2020). Wet ball milling helps to achieve a narrow size distribution, smooth morphology, and less agglomeration (Jung et al., 2015). Wet biochar ball milling in the presence of organic solvents (heptane, hexane, and ethanol) results into biochar BET surface area enhancement up to 194 $m^2/g$ versus $3 m^2/g$ of unmilled biochar (Peterson et al., 2012). In wet milling, milling media-to-biochar mass ratio is a critical parameter. Best results were achieved at a ratio of 100:1 (Peterson et al., 2012). Wet ball milling gives improved efficiency over dry ball milling because the solvent adsorption on the freshly formed surface lowers its surface energy preventing agglomeration (Dolgin et al., 1986; Kumar et al., 2020; Yuan et al., 2020). A comparative evaluation of dry vs wet ball milling (in presence of water) operated for 12h produced into better particle dispersion and diverse functional groups in wet milling (Yuan et al., 2020). Wet ball milling also resulted in smaller and better particle distributions along with an increased O/C ratio (0.096 over 0.080) versus dry milling (Yuan et al., 2020).

## 2.2 Centrifugation

Centrifugation involves macrobiochar grinding using a smashing machine and sieving using a fine pore-size mesh. The sieved biochar containing nanobiochar was formed into a suspension (Song et al., 2019). The suspension is further sonicated and centrifuged to obtain nanobiochar (Song et al., 2019). Despite the advantages of ball milling, centrifugation was also used for nanobiochar preparation (Ramanayaka et al., 2020a; Song et al., 2019; Zhang et al., 2013). Table 20.2 summarizes the nanobiochar properties obtained by centrifugation. This table will help in determining the optimum parameters required for centrifugation method.

For instance, woody biochar (remaining after gasification of *Gliricidia sepium*) obtained from a thermal power plant was crushed into the finer particles and sieved (<3mm). The sieved biochar was washed with distilled water and oven-dried at 60°C for 24h (Ramanayaka et al., 2020a). The biochar obtained was preconditioned at a temperature of −80°C for a period of 3 days and ground using disc mill (Ramanayaka et al., 2020a). Ethanol was added with dry biochar and disk milled at 1000rpm for 2min. The colloidal biochar obtained was dispersed in ethanol and centrifuged at 2000 rpm for 2 h. A supernatant containing nanobiochar was obtained. The supernatant was collected using sonication for 30 min at 50 kHz and dried in petri dish at 50°C in a vacuum oven (Ramanayaka et al., 2020a). Dried nanobiochar was scraped off with a spatula (Ramanayaka et al., 2020a,b). This nanobiochar had a 28 m$^2$/g surface area. The nanobiochar has plenty of aldehyde/ketone, carbonyl, Si—O, and phenolic –OH functional groups. These enhance nanobiochar sorption capabilities (Ramanayaka et al., 2020a). Similarly, nanobiochar was prepared from premade plant biochar by dispersing this sieved biochar in deionized water followed by stirring. Ultrasonication was then carried out for 30 min to obtain a suspension that was permitted to settle for 24 h. The liquid and solid phases were separated using the siphon. Suspensions were centrifuged for 30 min at 3500 × g to remove the nanobiochar supernatant (Song et al., 2019).

Similar methodology was also used to prepare peanut shell nanobiochar (Liu et al., 2018). Peanut shell biochar was sieved (75–150 μm), and this colloidal biochar was dispersed into water and settled for 24 h to separate the biochar particles that were > 100 nm. The unsettled suspension was centrifuged, and the portion of suspension having nanobiochar was taken out of the liquid (Liu et al., 2018). A similar method was adopted to prepare the wheat straw, grass, and rice husk nanobiochars (Lian et al., 2018; Oleszczuk et al., 2016).

## 2.3 Other methods

Apart from ball milling and centrifugation, microwave pyrolysis and thermal flash heating approaches were used to prepare nanobiochar (Genovese et al., 2015; Wallace et al., 2019). Flash heating uses a different approach to synthesize biochar nanosheets (Genovese et al., 2015). Corncob feedstock was also used for nanobiochar preparation. Feedstock pretreatment soaked corncob chunks in 1 M $HNO_3$ at 75°C for 4 h. Then, flash heating (950°C for 45 s) was performed in a muffle furnace (Genovese et al., 2015). Exfoliated nanosheet biochar was obtained without further mechanical treatment (Genovese et al., 2015).

Microwave pyrolysis is an another nanobiochar preparation method (Wallace et al., 2019). This is a single-step nanobiochar synthesis process. Softwood microwave pyrolysis at 2100, 2400, and 2700 W (for 15–20 min) generated nanobiochars with BET surface areas of 14.44, 28.65, and 9.96 m$^2$/g, respectively. Increase in power level (from 2100 to 2400 W) also increased the surface area and micropore area due to fast release of residual volatiles (Wallace et al., 2019). However, a further increase in power (2700W) resulted in pore shrinkage and closing, reducing the surface area and the number of micropores (Wallace et al., 2019).

Low nanobiochar yields were obtained from other methods than ball milling. Low nanobiochar yield is the major limitation in commercializing these preparation methods (Ramanayaka et al., 2020b). Nanobiochar centrifugation yields depend upon feedstock type, ash content of the bulk biochar, and biochar suspension color (Song et al., 2019). Generally, dark-colored suspensions gave high yields, whereas light-colored suspensions produced low yields (Song et al., 2019).

## 2.4 Preparation of modified nanobiochar

Nanobiochar can be modified to enhance their adsorption capacities (Li et al., 2020; Lyu et al., 2020; Nath et al., 2019; Xu et al., 2019). Nitrogen doping, thiol modification, and adding magnetic material phases can modify nanobiochar capacities (Li et al., 2020; Lyu et al., 2020; Xu et al., 2019). Adding nitrogen-containing groups ($-NH_2$ and $C \equiv N$) to the surface can enhance capacity (Xu et al., 2019). Nitrogen incorporation methods to the nanobiochar surface include pretreatment with the nitrogen-containing precursors (ionic liquids, organic polymers) (Paraknowitsch et al., 2010), posttreatment with ammonia at high temperature (Kundu et al., 2010), chemical bonding via nitration and reduction (Yang and Jiang, 2014), and N-doping during ball milling.

Nitrogen doping increases surface sorption sites, basicity, and can bring-in positive charge on the biochar surface. These changes favor contaminant adsorption (Xu et al., 2019). Nitrogen-doped hickory wood ball-milled biochar was prepared by mixing 18 mL of 29% ammonium hydroxide with 1.8 g biochar (prepared at 450°C) in a planetary ball mill (300 rpm, 12 h). This biochar was washed with deionized water, collected by filtration, and dried at 80°C for 48 h (Xu et al., 2019). The N-doped ball-milled biochar had a higher surface area (497 $m^2$/g) versus pristine (7.36 $m^2$/g) and ball-milled biochars (342 $m^2$/g) (Xu et al., 2019). This increase was due to pore generation in the biochar via ammonium hydroxide's activating properties during milling (Xu et al., 2019). N-doped biochar prepared from bagasse at different pyrolysis temperatures exhibited similar trends (Xu et al., 2019). Nitrile and amine functions were formed during N-doping (Xu et al., 2019). Biochar prepared at 450°C gave more nitrogen doping (2.41%–2.65%) than those prepared at 600°C (1.18%–1.82%) (Xu et al., 2019). Nitrogen doping during ball milling enhanced the reactive red adsorption capacities from 3.66 mg/g to 37.4 mg/g (Xu et al., 2019). Nitrogen doping increases the sorption capacity of ball-milled biochar toward reactive red dye and acidic $CO_2$ (Xu et al., 2019). The sorption capacity of nitrogen-doped biochar for the reactive red and $CO_2$ increased from 3.66%, 10 16.2%, 31.6% to 55.2%, respectively (Xu et al., 2019).

Thiol modification of nanobiochar produces enhanced capacities for Hg (II) and $CH_3Hg^+$ via complexation and ligand exchange mechanisms (Lyu et al., 2020). Thiol modification increases surface negative charge (by introducing –SH) on nanobiochars. This can enhance electrostatic interactions with the positively charged contaminants including heavy metals (Ji et al., 2019; Lyu et al., 2020). Chemical impregnation is widely used to incorporate thiol on biochar surfaces (Huang et al., 2019).

Thiol-modified ball-milled biochar was developed by mixing 3 g of poplar wood chips biochar with 120-mL solution [114 mL ethanol +3.6 mL water +2.4 mL 3-MPTS (3-mercaptopropyltrimethoxysilane] (Lyu et al., 2020). 3-MPTS was added dropwise to the solution (Lyu et al., 2020). Balls with diameter 15 mm, 5 mm, and 3 mm with ball mass ratio 2:5:3 were used (total ball mass 300 g). An operational speed of 300 rpm for 12 h was maintained, while altering the milling direction every 6 h. Finally, the milled product was washed with deionized water and ethanol, and freeze-dried for 48 h (Lyu et al., 2020). Thiol modification improved the physicochemical properties leading to enhanced adsorption capacities. Ball milling significantly increased functional groups, facilitating the immobilization of 3-MPTS onto the ball-milled biochar surface through —C—O—Si⟨⟩SH. These changes demonstrated the positive impacts that thiol group introduction has on mercury adsorption (Lyu et al., 2020). Higher adsorption capacities of 105 for $CH_3Hg^+$ and 320 mg/g for $Hg^+$ with thiol-modified biochar versus pristine biochar gave capacities of 8.21 for $CH_3Hg^+$ and 105.7 mg/g for $Hg^+$ (Lyu et al., 2020).

Nanobiochar loaded with $Fe_3O_4$ can be prepared by treating the precursor biomass with iron solution (1 L solution containing 8.5 g $FeCl_3$ and 3.4 g $FeCl_2$ was mixed with 100 g wheat straw for 24 h) and pyrolyzed to form magnetized nanobiochar (Li et al., 2020). The suspension was vacuum-filtered and oven-dried at 80°C for 24 h. Fe-impregnated wheat straw biomass was pyrolyzed (400–700°C) in tubular furnace under $N_2$ environment (Li et al., 2020). Magnetic nanobiochar was then prepared by ball-milling the pristine magnetic biochar (Li et al., 2020). The product was dispersed in deionized water, and nanobiochar/$Fe_3O_4$ hybrid was separated using a magnet (Li et al., 2020). This nanobiochar successfully enhanced tetracycline and mercury adsorption (Li et al., 2020).

A magnetized nanobiochar/$Fe_3O_4$ was also prepared by the coprecipitation method carried out in $N_2$ environment. First, the biochar nanoparticles were prepared by grinding and centrifugation from the bagasse biochar developed at 400–800°C under $N_2$ environment. Biochar nanoparticles (3.29 g) were suspended in 300-ml ultrapure water with constant stirring at 70°C. Then, 1.37 g $FeSO_4 \cdot 7H_2O$ and 2.67 g $FeCl_3 \cdot 6H_2O$ were dissolved in 50-mL ultrapure water and mixed with the nanobiochar suspension. The resulting mixture was held for 10 min at 70°C, and about 10.97 mL $NH_3 \cdot H_2O$ was added with incubation for 90 min at 300 rpm (Dong et al., 2018). The resultant magnetic biochar/$Fe_3O_4$ nanoparticles were washed and used for 17β-estradiol adsorption (Dong et al., 2018). The reactions involved in the formation of $Fe_3O_4$ on the surface of nanobiochar composite are given in Eqs. (20.1)–(20.4).

$$2Fe(OH)_2 + H_2O + \frac{1}{2}O_2 \rightarrow 2Fe(OH)_3 \quad (20.1)$$

$$Fe(OH)_3 \rightarrow FeOOH + 2H_2O \quad (20.2)$$

$$Fe(OH)_2 + 2FeOOH \rightarrow Fe_3O_4 + 2H_2O \quad (20.3)$$

$$2Fe(OH)_2 + \frac{1}{2}O_2 \rightarrow Fe_2O_3 + 2H_2O \quad (20.4)$$

This coprecipitation with iron salts for the nanobiochar not only improves the physicochemical properties like H/C and O/C ratio, pore volume and pore size, surface functional groups, and specific surface area, but also significantly increases the adsorption capacities for the organic as well as inorganic contaminants of nanobiochar (Dong et al., 2018; Li et al., 2020; Xu et al., 2019).

## 3 Comparative evaluation of pristine and nanobiochar properties

Nanobiochar inherits most of its precursor bulk biochar characteristics (Xu et al., 2017). Significant increase in physicochemical properties (surface area, functional groups, mineral content, and sorption capacity) of nanobiochar vs bulk biochar was reported (Fig. 20.3A) (Lyu et al., 2020; Naghdi et al., 2017; Ramanayaka et al., 2020a,b; Song et al., 2019; Xu et al., 2019). An increase in colloidal stability further increases the favorability toward large-scale production reducing aggregation during

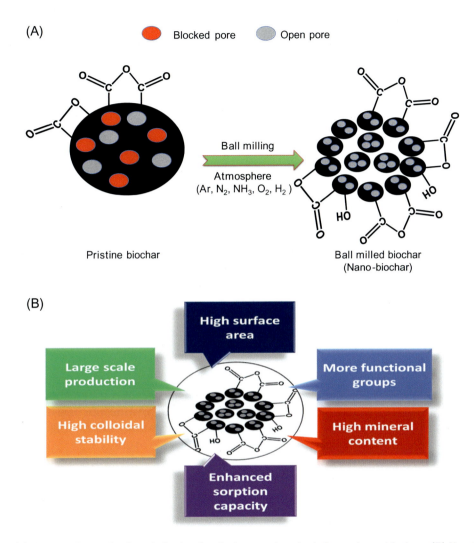

**Fig. 20.3** (A) Comparative evaluation of physicochemical properties of pristine and nanobiochars. (B) Nanobiochar characteristics. (A) *Adapted with permission from Lyu, H., Gao, B., He, F., Zimmerman, A.R., Ding, C., Huang, H., Tang, J., 2018a. Effects of ball milling on the physicochemical and sorptive properties of biochar: experimental observations and governing mechanisms. Environ. Pollut. 233, 54–63. doi:10.1016/j.envpol.2017.10.037. Copyright @ Elsevier.*

grinding or milling (Fig. 20.3B) (Liu et al., 2018; Lyu et al., 2017a,b; Oleszczuk et al., 2016; Song et al., 2019; Xiao et al., 2020). Nanobiochar applicability for toxic contaminant sorption is more favorable due to the change in aromaticity, specific surface area, and incorporation of specific functional groups (especially acidic functional groups) onto nanobiochar surface (Zhang et al., 2020; Zhang et al., 2019a, b).

Nanobiochar properties depend upon feedstock and the conditions during preparation (Ramanayaka et al., 2020a,b). Physicochemical property (specific surface area, total pore volume, pore size, O/C ratio, and H/C ratio) differences between pristine and nanobiochars are compared and summarized in Tables 20.1 and 20.2. Table 20.1 provides nanobiochar preparation details through ball milling, whereas Table 20.2 summarizes the information for nanobiochar prepared by other methods. In general, nanobiochar is characterized by high specific surface area (SSA) which is 1- to 97-fold greater than pristine biochar (Xiang et al., 2020a,b; Xiao et al., 2020; Yuan et al., 2020). For instance, the pristine bone biochars had SSAs of 2.76, 22.90, and 52.78 $m^2/g$ prepared at 300°C, 450°C, and 600°C, respectively. These were lower than those made after ball milling where the SSAs increased to 35.49, 199.5, and 313.09 $m^2/g$, respectively (Xiao et al., 2020). High adsorption capacities were obtained for Cd(II), Cu(II), and Pd(II) using ball-milled high SSA biochar. A decline was reported in average pore diameter of ball-milled biochars from 14.48 to 11.74 nm, 10.15 to 8.67 nm, and 8.22 to 6.46 nm when prepared at 300°C, 450°C, and 600°C, respectively (Xiao et al., 2020). Slight increases in O/C and H/C ratios were observed for ball-milled biochar versus pristine biochar (Table 20.1).

Table 20.2 summarizes the specific surface area, total pore volume, molar O/C, and H/C ratios of pristine biochars and nanobiochars prepared from other methods than ball-milling. Again, large increase in specific surface area (from 1.1 to ~48 fold) was reported for nanobiochar vs its pristine biochar (Table 20.2). Pore volume and surface area of rice hull biochar prepared at 600°C were 0.036 $cm^3/g$ and 27 $m^2/g$, respectively. These increased to 0.062 $cm^3/g$ and 123.3 $m^2/g$ for nanorice hull biochar (Yue et al., 2019). The surface area of nanobiochars prepared from peanut shell biochars increased from 3.67 to 63.6 $m^2/g$ (for biochar produced at 300°C), 4.75 to 78.6 $m^2/g$ (for biochar produced at 400°C), 34 to 230 $m^2/g$ (for biochar prepared at 500°C), and 8.82 to 264 $m^2/g$ (for biochar prepared at 600°C), respectively (Liu et al., 2018).

The pore volume, surface area, molar O/C and H/C ratios of miscanthus, wicker, and wheat straw biochars and nanobiochars were compared (Oleszczuk et al., 2016). Nanobiochars had smaller pore sizes and a larger surface area than their macrobiochar analogs. The surface area of miscanthus nanobiochar increased from 0.76 to 36.39 $m^2/g$. That for wicker rose from 11.38 to 18.25 $m^2/g$, while wheat straw biochar increased from 26.27 to 29.56 $m^2/g$. Zeta potential of miscanthus (−83.6), wicker wood

**Table 20.1** Pristine and nanobiochar (prepared by ball-milling method) properties.

| Precursor biomass | Pyrolysis conditions | Ball-milling conditions | Specific surface area (m²/g) [pristine/ nanobiochar] | Increase (fold) in specific surface area | Total pore volume (cm³/g) [pristine/ nanobiochar] | Pore size (nm) [pristine/ biochar/ nanobiochar] | O/C ratios [pristine/ biochar/ nanobiochar] | H/C ratio [pristine/ biochar/ nanobiochar] | References |
|---|---|---|---|---|---|---|---|---|---|
| Cow bone meal | Pyrolyzed in a tube furnace under continuous N₂ flow (200 mL min⁻¹) at 300°C | Wet milling, ball-to-powder ratio of 100:1, ball machine speed of 300 rpm for 12 h. | 2.76/35.49 | 12.86 | 0.017/0.163 | 14.48/11.74 | 0.32/0.36 | 1.13/1.15 | Xiao et al. (2020) |
| | 450°C | | 22.90/199.5 | 8.71 | 0.074/0.367 | 10.15/8.67 | 0.39/0.43 | 0.65/0.68 | |
| | 600°C | | 52.78/313 | 5.94 | 0.097/0.453 | 8.22/6.46 | 0.46/0.46 | 0.42/0.51 | |
| Sawdust | Pyrolysis at 600°C for 2 h, ramp rate = 10°C min⁻¹, N₂ atmosphere | Biochar was ball-milled in planetary ball mill (ball diameter of 6–10 mm for 2 h) | 154/328 | 2.13 | | | 0.069/0.069 | 0.341/0.340 | Yuan et al. (2020) |
| | | 12 h | 154/360 | 2.34 | | | 0.069/0.080 | 0.341/0.338 | |
| | | Biochar was mixed with water in a ratio 1:3 (g mL⁻¹), then grounded with planetary ball mill for 2 h | 154/325 | 2.11 | | | 0.069/0.069 | 0.341/0.351 | |
| | | 12 h | 154/334 | 2.17 | | | 0.069/0.096 | 0.341/0.354 | |
| Wheat straw | Pyrolysis at 600°C, lab-scale tube furnace, heating rate = 2°C min⁻¹ in a 100 mL min⁻¹ N₂ flow rate for 1 h | | 6.89/130.14 | 18.89 | 7.04/22.49 | | 0.04/0.06 | 0.19/0.22 | Cao et al. (2019) |

*Continued*

**Table 20.1** Pristine and nanobiochar (prepared by ball-milling method) properties—cont'd

| Precursor biomass | Pyrolysis conditions | Ball-milling conditions | Specific surface area ($m^2/g$) [pristine/nanobiochar] | Increase (fold) in specific surface area | Total pore volume ($cm^3/g$) [pristine/nanobiochar] | Pore size (nm) [pristine biochar/nanobiochar] | O/C ratios [pristine biochar/nanobiochar] | H/C ratio [pristine biochar/nanobiochar] | References |
|---|---|---|---|---|---|---|---|---|---|
| Poplar woodchips | Feedstock was heated in muffle furnace in ambient atmosphere at 300°C | | 1.21/7.92 | | 0.003/0.034 | | 0.43/0.54 | 1.06/1.20 | Xiao et al. (2020) |
| | 500°C | | 167/151 | | 0.013/0.063 | | 0.45/0.45 | 0.67/0.79 | |
| | 700°C | | 415/156 | | 0.081/0.045 | | 0.24/0.51 | 0.27/0.47 | |
| Wheat stalk | Oven-dried wheat stalk was chopped into 1 cm length, pyrolyzed in presence of $N_2$ (flow rate = 50 mL min$^{-1}$; heating rate = 15°C min$^{-1}$ at 300°C) | Planetary ball milling, ball-to-powder ratio of 100:1 with 300 rpm for 12 h, and direction was altered every 3 h | 3.90/13.50 | 3.46 | 0.0174/0.076 | 4.84/4.03 | | | Xiang et al. (2020a) |
| | 450°C | | 33.69/139.89 | 4.15 | 0.0544/0.3366 | 3.0/2.27 | | | |
| | 600°C | | 211.56/257.50 | 1.22 | 0.1370/0.2678 | 2.28/2.41 | | | |
| Poplar wood chips | Feedstock was oven-dried at 80°C for 12 h and pyrolyzed at 300°C for 3 h | Biochar (3 g) was mixed with 120-ml mixture (3.6 mL $H_2O$ + 114 mL $C_2H_5OH$ + 2.4 mL 3-MPTS), 300-g agate balls were used (speed 300 rpm for 12 h with direction alteration after every 6 h) | 1.83/61.34 | 33.52 | 0.003/0.291 | 7.76/19.05 | 0.27/0.43 | | Lyu et al. (2020) |

| Material | Condition | Milling parameters | | | | | | | Reference |
|---|---|---|---|---|---|---|---|---|---|
| Wheat straw | 300°C for 2h | Planetary ball milling, ball-to-powder ratio of 1:100, stainless steel balls size = 5mm, operational speed = 300 rpm (24h with change in rotation direction after every 3h). | 1.25/10.8 | 8.64 | 0.002/0.043 | 21.3/24.6 | 0.42/0.45 | 0.91/1.40 | Zhang et al. (2019b) |
| | 500°C for 2h | | 2.98/289 | 96.98 | 0.005/0.050 | 23/48.1 | 0.39/0.40 | 0.57/0.65 | |
| | 700°C for 2h | | 328/401 | 1.22 | 0.031/0.076 | 3.42/15.1 | 0.32/0.34 | 0.11/0.27 | |
| Rice husk | 300°C for 2h | | 1.80/9.29 | 5.16 | 0.004/0.040 | 33.1/21.0 | 0.55/0.57 | 1.22/1.36 | |
| | 500°C for 2h | | 10.8/190 | 17.59 | 0.024/0.036 | 12.7/46.3 | 0.48/0.53 | 0.30/0.64 | |
| | 700°C for 2h | | 294/329 | 1.12 | 0.035/0.080 | 4.46/23.5 | 0.44/0.49 | 0.16/0.22 | |
| Sugarcane bagasse | 300°C for 2h | Planetary ball milling, ball diameter = 6mm, operational speed = 300 rpm with rotation and direction altered every 0.5h. | —/10.8 | 10.8 | 0.007/0.052 | 31.7 | — | — | Lyu et al. (2018a) |
| | 450°C for 2h | | 51/331 | 6.49 | 0.008/0.099 | 3.6 | — | — | |
| | 600°C for 2h | | 359/364 | 1.01 | 0.009/0.125 | 3.6/3.4 | — | — | |
| Bamboo | 300°C for 2h | | 2.0/8.3 | 4.15 | 0.005/0.038 | — | — | — | |
| | At 450°C for 2h | | 4.7/299 | 63.62 | 0.003/0.083 | 21.6/14.5 | — | — | |
| | At 600°C for 2h | | 59/276 | 4.68 | 0.002/0.097 | 3.5 | — | — | |
| Hickory wood | 300°C for 2h in nitrogen atmosphere | Planetary ball milling, ball-to-mass ratio of 1:100 operational speed = 300 rpm with alteration of direction after every 3h. | 0.8/5.6 | 7.00 | 0.096/0.079 (ml/g) | 14.59/13.34 | 0.28/0.24 | 0.80/0.80 | Xiang et al. (2020b) |
| | 450°C for 2h in nitrogen atmosphere | | 9.8/284.74 | 29.05 | 0.436/0.304 | 2.45/2.13 | 0.14/0.17 | 0.55/0.60 | |
| | 600°C for 2h in nitrogen atmosphere | | 221.5/304.84 | 1.37 | 0.471/0.307 | 2.71/2.02 | 0.05/0.11 | 0.36/0.42 | |

Table 20.2 Pristine biochar and nanobiochar (prepared by other than ball milling) properties.

| Precursor | Pyrolysis conditions | Nanobiochar preparation | Specific surface area (m³/g) [pristine biochar/nanobiochar] | Increase (fold) in specific surface area | Total pore volume (cm³/g) [pristine/nanobiochar] | Molar O/C ratio [pristine biochar/nanobiochar] | H/C ratio [pristine biochar/nanobiochar] | References |
|---|---|---|---|---|---|---|---|---|
| Rice hull | Carbonized at 600°C in oxygen-limited conditions. Rice hulls were washed with tap water and passed with 2-mm mesh. | Nanobiochar was obtained from the biochar powder by mixing with deionized water and followed by quiescent for 2 h, the upper suspension layer was siphoned off and centrifuged at 10,000 rpm for 30 min, the suspension was freeze-dried, and nanobiochar was obtained. | 27.1/123.3 | 4.55 | 0.036/0.062 | | | Yue et al. (2019) |
| Corncob | Flash heat treatment at 900°C for 45 s | Corncob was pretreated with 1 M $HNO_3$ at 75°C for a period of 4 h. The flash heat treatment was given to the pretreated biomass at 950°C for 45 s in muffle furnace. | 7.91/543.68 | 68.7 | 0.0018/0.25 | 0.0146/0.0706 | | Genovese et al. (2015) |

| Feedstock | Preparation | | | | | | Reference |
|---|---|---|---|---|---|---|---|
| Peanut shell | Biomass was carbonized at 300–600°C for 2h under 500mL/N₂ flow in a tube furnace. | | 3.67/63.6 | 17.33 | 0.28/0.40 | 0.86/0.88 | Liu et al. (2018) |
| | 0.7-g bulk biochar was dispersed in 40-ml vial with 35-ml deionized water and dispersed for 15min at 25°C under sonication at 120W. The suspension was set quiescently for 24h to settle the particles >1000nm. Nanobiochar (size ≤1000nm) was retained by centrifugation | 300°C | | | | | |
| | | 400°C | 4.75/78.6 | 16.55 | 0.21/0.28 | 0.71/0.81 | |
| | | 500°C | 7.34/230 | 31.34 | 0.15/0.19 | 0.53/0.70 | |
| | | 600°C | 8.82/264 | 29.93 | 0.11/0.15 | 0.38/0.38 | |
| Wheat straw | The biochar was produced by slow pyrolysis of bagasse under nitrogen at 400–800°C | | 26.27/29.56 | 1.13 | 0.02566/00.0002 | 0.392/0.667 | Oleszczuk et al. (2016) |
| Wicker | | 400°C | 11.38/18.25 | 1.60 | 0.00614/0.00014 | 0.559/0.491 | |
| Miscanthus | Feedstock was thermo-pyrolyzed (350–700°C) in an oxygen-poor atmosphere (<1% O₂). | | 0.76/36.39 | 47.88 | 0.00104/0.00068 | 0.537/0.521 | |

(−88.0), and wheat straw (−88.4) nanobiochars was higher than their pristine precursors (−36.1, −34.9, and −56.6, respectively). Quartz (SiO$_2$) and calcite (CaCO$_3$) crystalline forms dominated macrobiochars while other crystalline forms KHCO$_3$, K(AlSiO$_4$), K$_2$SO$_4$, K$_2$Ca(CO$_3$)$_2$, CaCO$_3$, and Fe$_2$O$_3$ were also reported in nanobiochars (Oleszczuk et al., 2016).

## 4 Nanobiochars for aqueous contaminant removal

Biochars have been successfully applied for contaminant removal, agricultural applications, substitutes for carbon black, and sensing materials (Ardebili et al., 2020; Mohan et al., 2014; Patel et al., 2019; Vimal et al., 2019). Reduction in biochar size has a positive contribution on the surface area, surface functional groups, and adsorption capacity enhancement (Kumar et al., 2020; Liu et al., 2020a; Naghdi et al., 2017; Pratap et al., 2020). Nanobiochar has been successfully utilized for contaminants removal (Ardebili et al., 2020; Chen et al., 2017; Nath et al., 2019; Qin et al., 2020; Ramadan and Abd-Elsalam, 2020; Ramanayaka et al., 2020a,b; Pratap et al., 2020). Nanobiochar, like biochar, also has applications as catalysts, adsorbents, biomolecule carriers, sensing materials, electrodes, battery materials, plant growth enhancers, and as an alternative for carbon black (Banek et al., 2018; Jiang et al., 2020; Lyu et al., 2018b; Naghdi et al., 2019; Saxena et al., 2014).

However, a majority of the published research has focused on its use as an adsorbent (Liu et al., 2020b; Lyu et al., 2018a; Naghdi et al., 2017; Ramanayaka et al., 2020b; Xiang et al., 2020a). The sorption of organic and inorganic contaminants from aqueous media by nanobiochar is discussed in this section.

### 4.1 Organic contaminant removal

#### 4.1.1 Removal of conventional organic contaminants

Organic contaminants including dyes (Lyu et al., 2018b), CH$_3$Hg$^+$ (Lyu et al., 2018a), and volatile organic compounds (Xiang et al., 2020a,b) have been successfully removed using nanobiochar. Bagasse biochar prepared in a tubular furnace at 300–600°C for 1.5h under N$_2$ was converted into nanobiochar through planetary ball milling with a ball-to-biochar ratio of 100:1 at 300 rpm for 12h where an alteration of direction was used every 3h (Lyu et al., 2018a,b). This nanobiochar provided excellent methylene blue sorption capacities (213 mg/g and 354 mg/g at pH 4.5 and 7.5, respectively) (Lyu et al., 2018b). These capacities were much higher than their pristine (unmilled) precursors (5 mg/g and 3.5 mg/g at 4.5 and 7 pH, respectively) (Lyu et al., 2018b). At pH 4.5, π–π interactions of attractive electron-rich graphitic biochar surfaces and the electron-poor aromatic methylene blue were the major adsorptive interactions on the ball-milled bagasse nanobiochar (Lyu et al., 2018b; Zhang et al., 2012). This increased MB sorption capacity suggested that ball milling exposes the graphitic structure on the biochar surface (Lyu et al., 2018b). This is clearly evident as the ball milling increased the surface area by 2.3–33 times in nanobiochars as compared to unmilled biochars. At pH 7.5, both π–π interactions between methylene blue's aromatic portion and milled biochar's graphitic surface and more electrostatic interactions were involved leading to very high sorption capacity. At this pH, electrostatic attractions take place between the positively charged biochar surface ($P_{ZPC}=2.7$) and negatively charged hydroxyl and carboxyl groups of methylene blue (pKa=7.0 and 4.8 for hydroxyl and carboxyl groups, respectively). However, with the increase in pH, a slight reduction in the π–π interaction-driven organic compounds adsorption on the carbonaceous adsorbents was reported (Chen et al., 2014). The pseudo-second-order equation fit the kinetic data better than the pseudo-first-order and Elovich equations (Lyu et al., 2017b, 2018b). Sorption equilibrium was achieved in 8 and 16 h for ball-milled and unmilled biochars, respectively (Lyu et al., 2018b). Thus, ball-milled nanobiochar provided higher capacity and faster kinetics than its pristine analog. Biochar pore networks were altered during ball milling, enlarging pore openings, and further facilitating intraparticle diffusion (Lyu et al., 2018b). This enhanced adsorption capacity also reveals the importance of internal biochar surface area for adsorption (Lyu et al., 2018b). The methylene blue adsorption capacity (354 mg/g) was higher than biochar/AlOOH composite (85 mg/g), graphene-coated biochar (174 mg/g), activated carbon (270 mg/g), and graphite oxide (351 mg/g) (Lyu et al., 2018a,b). Low-surface-area (10.8 mg/g)-milled nanobiochar had high methylene blue sorption (354 mg/g) than activated carbon (surface area 823 mg/g, sorption capacity 270 mg/g) because of high surface functionalities. Activated carbon adsorbed methylene blue mostly via van der Waal forces, while large functionalities enhanced sorption on nanobiochar through π–π and electrostatic interactions (Lyu et al., 2018b).

Nitrogen-containing functionalities are active sites for enhanced sorption performance of the acidic/negatively charged contaminants (Nguyen and Lee, 2016; Shao et al., 2018). N-doping adds basic N-functions which protonate over a wide pH range producing positive surface charge. This promotes electrostatic reactive red dye adsorption (He and Hu, 2012). N-doped ball-milled biochar had a 19 times higher reactive red sorption capacity (31.3–37.4 mg/g) than pristine biochar (1.66–3.66 mg/g) (Xu et al., 2019). N-doped ball-milled biochar generated a two-fold increase in sorption capacity (31.3 mg/g) over ball-milled biochar without N-doping (15.57 mg/g) (Xu et al., 2019). N-doping also significantly increased the zeta potential and PZC of the ball-milled biochar (Xu et al., 2019). Positive charge formation allows strong electrostatic interaction leading to the adsorption of anionic contaminants (Tian et al., 2018).

Thiol-modified nanobiochar was developed from poplar wood chips biochar by using ball milling. Thiol-modified nanobiochar helped in enhancing surface area by opening the internal pores and increasing external surface area. Ball milling helped driving the reaction of 3-mercaptopropyltrimethoxysilane (3-MPTS) that bonded the 3- mercaptopropylsilyl group to the surface (Lyu et al., 2020). 3-MPTS modification increased the surface area from 1.83 to 5.21 m2/g by eliminating tar particles formed on biochar pores during pyrolysis. Increased net negative charge on thiol-modified ball-milled biochar surface resulted from the acidic -SH functions ($-SH + H_2O \leftrightarrows -S^- + H_3O^+$). The $S^-$ reacts with $CH_3Hg^+$ binding it ($-S - HgCH_3$) to the surface (Lyu et al., 2020). Both Freundlich and Langmuir sorption isotherm models fit the $CH_3Hg^+$ adsorption data ($R^2 > 0.904$). The Freundlich linearity constant ($n$) was <1, indicating $CH_3Hg^+$ chemisorption occurred (Chen and Chen, 2009; Lyu et al., 2020). Fig. 20.4 provides the possible

| Organic contaminants | Heavy metals |
|---|---|
| Surface adsorption | Surface complexation |
| Pore filling | Electrostatic attraction |
| Hydrogen bonding | Intraparticle diffusion |
| π-π interaction | Ligand exchange |
| Hydrophobic interaction | Cation-π interaction |
| Electrostatic attraction | Surface adsorption |

**Fig. 20.4** Possible mechanisms for aqueous inorganic and organic contaminant remediation using ball-milled nanobiochar (Li et al., 2020; Lian and Xing, 2017; Lyu et al., 2020; Lyu et al., 2018a,b; Xiang et al., 2020a,b; Zhang et al., 2019a,b).

mechanisms responsible for the adsorption of inorganic and organic contaminants on nanobiochar.

Ball-milled biochar also effectively eliminated VOCs (Xiang et al., 2020b). Ethanol, acetone, cyclohexane, chloroform, and toluene were effectively removed from vapor steam using ball-milled hickory wood nanobiochar (Xiang et al., 2020b). Ball milling enhanced the VOCs adsorption capacities from 6.5–55.1 to 23.4–103.4 mg/g at 20°C and 50 mL min$^{-1}$ vapor flow (Xiang et al., 2020b). High specific surface area and oxygen-containing surface functional groups interacting with the VOCs controlled adsorption (Xiang et al., 2020b). Oxygen functionalities, as discussed earlier, are added to the biochar surface during ball milling. They promoted polar VOCs adsorption through H-bonding and dipole–dipole interactions (Zhang et al., 2019a,b). Rapid VOCs adsorption was observed during the initial 10–15 min, and equilibrium was achieved within 1 h at 20°C and 50 mL min$^{-1}$ vapor flow. Adsorption equilibrium for ball-milled biochar was rapidly attained versus the pristine biochar (Xiang et al., 2020b). VOC boiling points and dipole moments influenced the adsorption onto the ball-milled nanobiochar (Xiang et al., 2020b). High-boiling-point VOCs were easily adsorbed due to strong intermolecular forces and low vapor pressures (Zhang et al., 2019a,b). As a result, VOCs like cyclohexane, ethanol, and toluene with higher boiling points (80.7°C, 78.2°C, and 110.6°C, respectively) better adsorb onto the ball-milled biochar (Xiang et al., 2020b). Ball-milled biochar has excellent reusability and recyclability (Xiang et al., 2020b). Removal efficiencies between 81.2 and 91.4% were maintained for the VOCs on the ball-milled biochar after five successive sorption/desorption cycles. Ball-milled nanobiochar has emerged as a sustainable and environment-friendly adsorbent for VOCs removal (Xiang et al., 2020b).

### 4.1.2 Adsorption of emerging contaminants

Emerging contaminants include pharmaceuticals, pesticides, personal care products, new industrial chemicals, synthetic hormones, and endocrine disruptors (Daughton, 2004; Daughton and Ternes, 1999; Patel et al., 2019; Richardson, 2009). Nanobiochar has now been studied for carbamazepine (Naghdi et al., 2019), oxytetracycline (Ramanayaka et al., 2020a), glyphosate (Ramanayaka et al., 2020a), 17β-estradiol (Dong et al., 2018), sulfamethoxazole, sulfapyridine (Huang et al., 2020), and tetracycline sorption (Li et al., 2020) (Table 20.3). Woody nanobiochar developed through crushing, sieving, and centrifugation removed glyphosate and oxytetracycline with maximum sorption

Table 20.3 Sorption of organic and inorganic contaminants from water using nanobiochar.

| Nanobiochar | Contaminants | Experimental conditions | | | | Maximum adsorption capacity (mg/g) | References |
|---|---|---|---|---|---|---|---|
| | | Temperature (K) | Initial concentration (mg/L) | pH | Adsorbent dose (g/L) | | |
| Cow bone meal ball-milled biochar prepared at 300°C | Cd (II) | 298 | 50 | 5 | | 66.3 | Xiao et al. (2020) |
| | Cu (II) | | 50 | | | 159.3 | |
| | Pb (II) | | 120 | | | 339.3 | |
| 450°C | Cd (II) | | 50 | | | 122.1 | |
| | Cu (II) | | 50 | | | 184.3 | |
| | Pb (II) | | 120 | | | 428.8 | |
| 600°C | Cd (II) | | 50 | | | 165.8 | |
| | Cu (II) | | 50 | | | 217.6 | |
| | Pb (II) | | 120 | | | 558.9 | |
| Magnetic nanobiochar prepared at 400°C | 17β-estradiol | 283 | 0.6–3.0 | 4 | 200 | 46.9 | Dong et al. (2018) |
| | | 298 | | | | 50.2 | |
| | | 313 | | | | 31.4 | |
| | | 283 | | | | 48.8 | |
| | | 298 | | | | 41.7 | |
| | | 313 | | | | 33.7 | |
| Magnetic nanobiochar prepared at 600°C | | 283 | | | | 42.1 | |
| Magnetic nanobiochar prepared at 800°C | | 298 | | | | 34.1 | |
| | | 313 | | | | 24.8 | |
| Ball-milled bamboo biochar | Ammonium | 293 | 0–100 | 6 | 0.25 | 22.9 | Qin et al. (2020) |
| Ball-milled wheat stalk biochar | Tetracycline hydrochloride | 298 | 25 | 4–11 | 0.2 | 84.6 | Xiang et al. (2020a) |

| Material | Adsorbate | T (K) | Conc. range | pH | Dose | Capacity | Reference |
|---|---|---|---|---|---|---|---|
| Ball-milled biochar prepared at 450°C from Hickory chips | Sulfamethoxazole | 298 | 10 | 3.5–8.5 | | 100.3 | Huang et al. (2020) |
| Ball-milled biochar prepared at 450°C from bamboo | Sulfapyridine | | | | | 57.9 | |
| Pine wood-derived nanobiochar | Carbamazepine | 298 | 0.0005–0.02 | 6 | 0.1 | 0.04 | Naghdi et al. (2019) |
| Ball-milled biochar | Galaxolide | 293 | 0–5 | | 0.75 | 0.61–2.1 | Zhang et al. (2019b) |
| Ball-milled magnetic biochar | Tetracycline | 297 | 40 | 5 | 0.4 | 268.3 | Li et al. (2020) |
| | Hg | | 50 | | | 127.4 | |
| Ball-milled bagasse biochar | Methylene blue | 293 | 10–200 | 4.5 | 0.16 | 213.0 | Lyu et al. (2018a) |
| | | | | 7.5 | | 354.0 | |
| Iron oxide permeated mesoporous rice husk nanobiochar | As | 298 | 0.01–1 | 7–12 | 10 | 5987 | Nath et al. (2019) |
| Poplar wood chips ball-milled biochar | Hg (II) | 298 | 5 | 7 | 0.04 | 320.1 | Lyu et al. (2020) |
| | CH$_3$Hg$^+$ | | 1 | | | 104.9 | |
| Ball-milled biochar | Ni (II) | 293 | 10–200 | 6 | 0.40 | 114.0 | Lyu et al. (2018a,b) |
| Wheat straw | Pb (II) | 298 | 100 | 5 | 1 | 134.7 | Cao et al. (2019) |
| Hickory wood chips-vermiculites nanobiochar | As (V) | 298 | 20 | 6 | 1 | 20.1 | Li et al. (2020) |
| Dendro nanobiochar | Glyphosate | 303 | 5–250 | 7 | 1 | 83.0 | Ramanayaka et al. (2020a) |
| | Oxytetracycline | | 10–500 | 9 | | 520 | |
| | Cd (II) | | 5–300 | 9 | | 922 | |
| | Cr (VI) | | 1–25 | 4 | | 7.46 | |

capacities of 83 and 520 mg/g, respectively (Ramanayaka et al., 2020a). Freundlich, Langmuir, Hills, and Redlich-Peterson models fit the glyphosate and oxytetracycline adsorption data. Hills equation fit the sorption data best (Ramanayaka et al., 2020a). Polar surface functions like –OH, C=O, -NH contributed to high removal capacities (Ramanayaka et al., 2020a). Physisorption was the main mode of adsorption (Ramanayaka et al., 2020a). Pine wood nanobiochar's carbamazepine adsorption capacity was higher than that of carbon nanotubes, activated carbon, and graphene oxide (Naghdi et al., 2019). This nanobiochar has a carbamazepine removal efficiency of up to 95% and a 74-μg capacity (Naghdi et al., 2019). Carbamazepine adsorption was initially fast (71% within 30 min), due to the nanosized high-surface-area biochar with plenty of adsorption sites (Naghdi et al., 2019). The presence of the Tween 80 surfactant as matrix in the adsorption media enhanced carbamazepine sorption efficiency from 42 to 66% as the Tween 80/carbamazepine mole ratio of Tween 80 to carbamazepine increased from 0 to 1 (Naghdi et al., 2019).

Hickory chips and bamboo ball-milled nanobiochar removed sulfapyridine and sulfamethoxazole with sorption capacities of 58 and 100 mg/g, respectively (Table 20.3) (Huang et al., 2020). Increasing surface amounts of $–CH_2$, C=O/C=C and -CO functions through ball milling enhanced the affinity toward polar groups of sulfamethoxazole and sulfapyridine which form H-bonds (Huang et al., 2020). Hydrophobic, electrostatic, and π-π interactions were also suggested to aid sorption (Huang et al., 2020). Biochar made from pyrolyzed wheat stalk and then milled in a planetary ball mill at a ball-to-powder ratio of 100:1 at 300 rpm for a period of 12 h effectively adsorbed tetracycline hydrochloride from aqueous solution (Xiang et al., 2020a). Ball milling enhanced the tetracycline hydrochloride adsorption capacity of 3-fold (84.55 mg/g) versus pristine biochar (28.64 mg/g) (Xiang et al., 2020a). BET surface area, t-plot external surface area, mesoporous volume, and total pore volume were linearly correlated to the tetracycline hydrochloride adsorption capacity (Xiang et al., 2020a). Both the pseudo-first- and pseudo-second-order equations fit the kinetic data ($R^2 \leq 0.99$) well (Xiang et al., 2020a).

Nanobiochar might be a candidate-sustainable adsorbent for 17β-estradiol sorption based on its adsorption capacity of 50.24 mg/g (Dong et al., 2018). $Fe_3O_4$/nanobiochar composites had 14.5 times more surface area versus pristine biochar (Dong et al., 2018). 17β-estradiol adsorption depends on the pyrolysis temperature used to make the pristine biochar precursor. The major sorption interactions changed from hydrogen bonding to π–π interactions using biochars made at higher pyrolysis temperature (Dong et al., 2018). $Fe_3O_4$/nanobiochar composites were readily regenerated by treating the adsorbed 17β-estradiol with ozone, rendering it available for next adsorption cycle (Dong et al., 2018). Biochar particle size reduction by ball milling promoted both adsorption kinetics and mass transfer (Kumar et al., 2020). The kinetics of methylene blue on ball-milled nanobiochar was enhanced. Equilibrium was reached in 8h versus 16h for the precursor to ball-milled biochar (Lyu et al., 2018b). Sorption capacity of the ball-milled nanobiochar developed from hickory chips biochar was also studied in the real wastewater. This nanobiochar successfully removed sulfapyridine and sulfamethoxazole from the real wastewater (Huang et al., 2020). Adsorption capacities were 58.6mg/g and 25.7mg/g for sulfapyridine and sulfamethoxazole, respectively (Huang et al., 2020). Thus, this ball-milled nanobiochar should be further evaluated to remove emerging contaminants from wastewater under environmental conditions (Huang et al., 2020).

Similar findings have appeared of removing emerging contaminants from aqueous media using nanobiochar (Ramanayaka et al., 2020a,b). Adsorption capacities of 81.87mg/g and 519.95mg/g were obtained for glyphosate and tetracycline, respectively, under laboratory conditions (Table 20.3). Feedstock selection, optimization of milling, and other preparation parameters were varied to maximize adsorption capacities.

## 4.2 Inorganic contaminants removal

Nanobiochar sorption capacities for inorganic contaminants were similar or higher than other carbonaceous adsorbents, including carbon nanotubes, activated carbon, and graphene oxide (Lyu et al., 2018a,b; Xiao et al., 2020). More surface functional groups, graphitic nature, enhanced surface area, and the presence of humic acid-like components are responsible for this enhanced sorption (Lyu et al., 2020). Few studies are available on inorganic contaminant (anions and heavy metals) removal using nanobiochar (Cao et al., 2019; Li et al., 2020; Lyu et al., 2018a,b; Lyu et al., 2020; Nath et al., 2019; Ramanayaka et al., 2020a,b; Xiao et al., 2020). Nanobiochar proved effective for the removal of cations ($Hg^{2+}$) and anions ($F^-$) from the wastewater (Lyu et al., 2020). Inorganic contaminant sorption on different nanobiochars is discussed in subsequent paragraphs.

*4.2.1 Heavy metals*

Cu, Pb, Hg, As, Ni, Cd, and Cr were successfully removed from aqueous solution using nanobiochar (Table 20.3) (Cao et al., 2019; Li et al., 2020; Nath et al., 2019; Ramanayaka et al., 2020a; Xiao et al., 2020). A high Cd(II) adsorption capacity of 922.3 mg/g was achieved at pH = 7.0 using graphitic nanobiochar prepared by ball milling (Ramanayaka et al., 2020a). The sorption equilibrium data were well-fitted to the Hill ($R^2 = 0.997$) and Freundlich ($R^2 = 0.98$) models (Ramanayaka et al., 2020a). This capacity confirms the excellent cation removal by the negatively charged nanobiochar surface (Ramanayaka et al., 2020a). Nanobiochar ($pH_{pzc} = 7.4$) surface is negatively charged in the solution pH range of 3 to 7 because of the surface functional groups like aldehyde/ketone (C=O), phenolic (-OH), and Si—O. In aqueous media, cadmium mainly exists as $Cd^{2+}$, and easily binds to the negatively charged nanobiochar surface at $pH > 7.4$ ($pH_{pzc}$) (Ramanayaka et al., 2020a). The negatively charged dichromate ions ($Cr_2O_7^{2-}$) were repelled by the negatively charged nanobiochar surface at $pH > 7.4$ (Ramanayaka et al., 2020a). However, $Ca^{2+}$ available on the nanobiochar surface provides electrostatic binding sites for the $Cr_2O_7^{2-}$ ions (Ramanayaka et al., 2020a). $Cr_2O_7^{2-}$ ions reduced to $Cr^{3+}$ were attached to the negatively charged functional groups (aldehyde/ketone, phenolic) via surface complexation (Ramanayaka et al., 2020a).

Enhanced Ni(II) adsorption occurred on ball-milled nanobiochars prepared from different feedstocks and at varying pyrolysis temperatures (Lyu et al., 2018a). These nanobiochars were prepared using bamboo, bagasse, and hickory wood chips by pyrolysis at 300°C, 450°C, and 600°C followed by ball milling (Lyu et al., 2018a). Ni(II) adsorption capacities increased for all the nanobiochars vs their unmilled precursors, due to an increase in the surface oxygen functions (lactonic, carboxyl, and hydroxyl) during ball milling (Lyu et al., 2018a). Maximum adsorption capacities of 230–650 mmol/kg were achieved for the ball-milled bagasse biochar prepared at 600°C versus 26–110 mmol/kg obtained for its unmilled biochar precursor (Lyu et al., 2018a). Ni(II) adsorption onto ball-milled biochar was pH and dose-dependent (Lyu et al., 2018a). Increasing pH from 1.6 to 6.0 raised the Ni(II) uptake from 134 to 696 mmol/kg (Lyu et al., 2018a). Rapid adsorption kinetics was displayed onto the ball-milled bagasse biochar prepared at 600°C due to faster intraparticle diffusion rates (Liang et al., 2012; Lyu et al., 2018a). The Elovich model best fits the kinetic data ($R^2 = 0.987$). The pseudo-first-, pseudo-second-, and Ritchie $n$th-order rate equations were also applied to fit the kinetic data ($R^2 > 0.93$ for all) (Lyu et al., 2018a).

A maximum Pb(II) adsorption capacity of 134.7 mg/g was achieved using ball-milled wheat straw biochar (Cao et al., 2019). Ion exchange, acid functional group complexation, and precipitation with minerals were claimed to be possible Pb(II) adsorption pathways (Cao et al., 2019). Ball milling followed by carbonization enhances the ion exchange mechanism of Pb(II) onto low wheat straw biochar doses (Cao et al., 2019). Pb(II) exchanged with $K^+$, $Na^+$, $Ca^{2+}$, and $Mg^{2+}$ (Cao et al., 2019). Increase in the number of nanobiochar particles in solution led to their aggregation, thus reducing the number of exposed surface active sites per unit nanobiochar. This results in the decline in Pb(II) sorption capacity per gram nanobiochar (Cao et al., 2019). EDX analysis revealed a decline in Ca, K, and Mg concentrations in the lead-loaded ball-milled straw biochar. This is consistent with ion exchange occurring during Pb adsorption. It is worth noting that ball milling and carbonization significantly enhanced the amount of ion exchange occurring at low adsorbent dose (Cao et al., 2019).

Iron oxide permeated mesoporous rice husk nanobiochar (IPMN) removed As(V) efficiently with a huge adsorption capacity of 5987 mg/g (Nath et al., 2019). The enhanced capacity is due to the formation of mono- and bidentate surface complexes between the active surface sites located on IPMN and arsenic [As(V)] ions in alkaline pH range (pH=7–12) (Nath et al., 2019).

IPMN also exhibited better recyclability. It retained removal efficiencies of ~61.81%, ~65.76%, and ~67.20% after five cycles of sorption–desorption. Desorption was carried out using 0.5 M NaCl, 0.5 M NaOH, and a phosphate saline buffer. Maximum As(V) removal percentages of ~94, ~91, and ~86 were obtained under standard laboratory, simulated field conditions, and actual field conditions, respectively (Nath et al., 2019). Arsenic adsorption capacity of IPMN was higher than many other reported Fe-modified biochars (Agrafioti et al., 2014b; Cope et al., 2014; Hu et al., 2015; Kumar et al., 2019a). The huge adsorption capacity (5987 mg/g) was mainly due to exceptionally high surface area (1736.81 $m^2$/g) (Nath et al., 2019). The IPMN matrix was mesoporous. Iron oxide was impregnated on the surface, leading to an increase of active FeO sorption sites. The presence of FeO on the surface contributed to overall As(V) adsorption enhancement. This is evident from the reduction in FeO bending vibrations at 611 $cm^{-1}$ in FTIR after As(V) sorption (Nath et al., 2019).

Hg(II) sorption was studied onto pristine biochar prepared by pyrolyzing poplar wood chips at 300°C for 3 h, thiol-modified biochar produced by chemical impregnation (CIS-biochar), and thiol-modified biochar through ball milling (BMS-biochar) (Lyu et al., 2020). 3-Mercaptopropyltrimethoxysilane (3-MPTS) was used to incorporate thiol functional groups (–SH) onto these biochar surfaces (Ji et al., 2019; Lyu et al., 2020). Thiol modification increased the number of Hg adsorption sites (Huang et al., 2019). Furthermore, thiols created more negative surface charges onto nanobiochar surfaces, resulting in more electrostatic adsorption for positively charged heavy metals and methyl mercury (Huang et al., 2019). The Hg(II) sorption capacity of BMS biochar was the highest (320 mg/g), followed by CIS biochar (176 mg/g) and pristine biochar (106 mg/g). The large Hg(II) capacity was due to surface adsorption on a large surface area, combined with ligand exchange, surface complexation, and electrostatic attractions (Lyu et al., 2020).

Table 20.4 illustrates that chemisorption (including covalent bonding), pore filling, π-π attractive interactions, hydrogen bonding, hydrophobic interactions, electrostatic attractions, and ligand exchange are contributing factors in monolayer and multilayer sorption of organic and inorganic contaminants on nanobiochar (Dong et al., 2018; Huang et al., 2020; Lyu et al., 2020; Naghdi et al., 2019; Nath et al., 2019; Qin et al., 2020; Ramanayaka et al., 2020a,b; Xiang et al., 2020a,b). The use of Langmuir or Freundlich isotherm models and either pseudo-first- and second-order rate equations well fitted the majority of equilibrium and kinetic data obtained for the organic and inorganic contaminant adsorption on nanobiochar (Table 20.4). Municipal and agricultural waste conversion to nanobiochar provides one way to reduce this waste while making adsorbents. If biochars are used to sequester carbon long term in soils, while providing carbon tax credits, this may become more attractive.

## 5 Conclusions and future recommendations

Reducing the size of pristine biochar to the nanolevel enhances specific surface areas, porosity, mineral exposure, and the amount of surface functional groups versus its precursor biochar. These factors lead to improved sorption of organic and inorganic contaminants. Ball milling is the most used for preparation, but other methods such as centrifugation advanced for the nanobiochar preparation. Positive and negative nanobiochar impacts on flora and fauna must be studied. The use of biochars in agriculture was not covered in this chapter, but this work is heating up. The subclass of nanobiochars may have a role in this future direction. The following recommendations are made for future research:

Table 20.4 Equilibrium and kinetic parameters obtained from adsorption of organic and inorganic contaminants from water.

| Adsorbate | Adsorbent | Surface area ($m^2/g$) | pH | Temp. (K) | Conc. | Isotherm study | Langmuir capacity | Freundlich capacity | Temp. (K) | Conc. | Pseudo-first order | Pseudo-second order | Mechanism | References |
|---|---|---|---|---|---|---|---|---|---|---|---|---|---|---|
| Carbamazepine | Pine wood nanobiochar | 47.25 | 6 | 298 | 0.5–20 ng/mL | Partition–adsorption | 40 (ng/mg) | 0.082 (ng/mg)(L/ng)1/n | 298 | 5 ng/mL | 1.202 1/h | 95.21 (g/mg/min) | Chemisorption | Naghdi et al. (2019) |
| Tetracycline hydrochloride | Wheat stalk ball milled biochar | 139.89 | 4–11 | 298 | 5–30 mg/L | | 96.69 mg/g | 13.264 $(mg^{(1-n)} l^n/g)$ | 298 | 25 mg/L | 0.4017 1/min | 0.0074 (g/(mg.min) | Surface adsorption Pore filling | Xiang et al. (2020a) |
| Arsenic | Iron oxide permeated mesoporous rice husk nanobiochar | 1736.81 | 2–12 | 298 | 10–10,0000 ppb | Langmuir–Freundlich Sips model | – | – | 298 | – | 0.440 L/min | 0.022 $(gmg^{-1} min^{-1})$ | Strong covalent interaction | Nath et al. (2019) |
| Ammonium | Ball-milled biochar | 298.6 | 6 | 298 | 0–100 mg/L | Hill | 22.9 mg/g | 1.13 $(mg^{(1-n)} l^n/g)$ | 298 | 50 mg/L | 2.783 1/h | 0.357 $(gmg^{-1} min^{-1})$ | Monolayer adsorption | Qin et al. (2020) |
| Glyphosate | Dendro nanobiochar | 28 | 3–8 | 303 | 5–250 mg/L | Langmuir–Freundlich | – | 0.039 (mg/g)/(mg/L)n | | | | | Chemosorptive mechanism | Ramanayaka et al., (2020a) |
| Oxytetracycline | | | | | 10–500 mg/L | | – | | | | | | | |
| Cd (II) | | | | | 5–300 mg/L | Temkin | – | 0.45 (mg/g)/(mg/L)n | | | | | | |
| Cr (VI) | | | | | 1–25 mg/L | Redlich–Peterson | – | | | | | | | |
| Sulfamethoxazole | Ball-milled biochar | | 3.5–8.5 | 298 | 10 mg/L | | 5.683 (L/mg) | 18.700 $(mg^{(1-n)} l^n/g)$ | 298 | 10 mg/L | 1.947 1/h | 0.127 (g/mg/h) | π–π interaction, hydrogen bonding, hydrophobic interaction, and electrostatic interaction | Huang et al. (2020) |
| Sulfapyridine | | | | | | | 6.207 (L/mg) | 40.920 $(mg^{(1-n)} l^n/g)$ | | | 3.414 1/h | 0.125 (g/mg/h) | | |
| Hg (II) | Thiol-modified nanobiochar | 61.34 | 7 | 298 | 1.0–25 mg/L | – | 175.6 mg/g | – | 298 | 5 mg/L | | | Electrostatic attraction, ligand exchange, surface adsorption, and surface complexation | Lyu et al. (2020) |
| $CH_3Hg^+$ | | | | | 0.1–8 mg/L | | 58.0 mg/g | – | | 1 mg/L | | | | |

Continued

**Table 20.4** Equilibrium and kinetic parameters obtained from adsorption of organic and inorganic contaminants from water—cont'd

| | | | | Isotherm study | | | | | Kinetics | | | |
|---|---|---|---|---|---|---|---|---|---|---|---|---|
| Adsorbate | Adsorbent | Surface area ($m^2$/g) | pH | Temp. (K) | Conc. | Langmuir capacity | Freundlich capacity | Temp. (K) | Conc. | Pseudo-first order | Pseudo-second order | Mechanism | References |
| Tetracycline | Ball-milled magnetic nanobiochar | 296.3 | 2–9 | – | 1–150 mg/L | 268.3 mg/g | 87.12 $(mg^{1-1/n} l^{1/n} g^{-1})$ | 297 | 40 mg/L | 0.379 $h^{-1}$ | $1.48 \times 10^{-2}$ (mg/g/h) | Hg-C π bond formation, electrostatic attraction, and surface complexation | Li et al. (2020) |
| Hg (II) | | | | | | 127.4 mg/g | 58.78 $(mg^{1-1/n} l^{1/n} g^{-1})$ | | | 0.326 $h^{-1}$ | $9.75 \times 10^{-3}$ (mg/g/h) | | |
| 17β-estradiol | Magnetic nanobiochar prepared at 400°C | 166.87 | 3–9 | 298 | 0.6–3.0 mg/L | 50.24 mg/g | 18.84 (mg/g).(mg/l) | 298 | – | 1.26 $min^{-1}$ | 0.30 (g/(mg min)) | Hydrogen bonding, hydrophobic interaction, and π-π interaction | Dong et al. (2018) |
| | Magnetic nanobiochar prepared at 600°C | | | | | 41.71 mg/g | 18.75 (mg/g).(mg/l) | | | 1.70 $min^{-1}$ | 0.47 (g/(mg min)) | | |
| | Magnetic nanobiochar prepared at 800°C | | | | | 34.06 mg/g | 16.06 (mg/g).(mg/l) | | | 1.17 $min^{-1}$ | 0.52 (g/(mg min)) | | |

- Specific commercially useful adsorption processes need to be demonstrated and employed.
- Research on the impact on ecology and human health is lacking, so studies focusing on the ecotoxicological effects of nanobiochars (positive and negative) must be carried out.
- Cost–benefit analyses and life cycle assessments must be carried out to demonstrate nanobiochar suitability for aqueous contaminant removal.
- Field-scale investigations are mostly lacking because of small nanobiochar yields. Thus, production enhancement research is needed.
- Physical and chemical properties of nanobiochar depend on biomass types, pyrolysis conditions, and modes of preparation. As more research becomes available, meta-analyses covering multiple factors should be conducted to increase nanobiochar applications.
- Continue chemical and physical nanobiochar modifications while alerting for new applications.
- A key future use of nanobiochars is as a soil modifier and of agricultural importance (in future). This will play into carbon sequestration issues (and perhaps help agricultural industries with carbon tax credits).
- Can increasingly smaller nanobiochar particles be a source of organic carbon enhancement in soils under the influence of microorganisms (fungi and bacteria)?

## Questions and problems

Question 1. What are the major emerging aquatic contaminants?

Question 2. Name the common nanomaterials used for aqueous contaminant removal?

Question 3. What is nanobiochar?

Question 4. How is nanobiochar different from its pristine biochar precursor?

Question 5. What major pathways are involved in the removal of aqueous contaminants using nanobiochar?

Question 6. Which method is best for the preparation of nanobiochar?

Question 7. What are the advantages of the nanobiochar vs other nanomaterials?

Question 8. Name the common feedstocks used for nanobiochar preparation.

Question 9. Identify the major properties of pristine biochar and nanobiochar?
Question 10. Mention different methods used for nanobiochar modification?
Question 11. Which method is most suitable for nanobiochar preparation at large scale?
Question 12. What are the other applications of nanobiochar other than as an adsorbent?

# References

Abd El Hameed, A.H., Eweda, W.E., Abou-Taleb, K.A., Mira, H., 2015. Biosorption of uranium and heavy metals using some local fungi isolated from phosphatic fertilizers. Ann. Agric. Sci. 60 (2), 345–351.

Agrafioti, E., Kalderis, D., Diamadopoulos, E., 2014a. Arsenic and chromium removal from water using biochars derived from rice husk, organic solid wastes and sewage sludge. J. Environ. Manage. 133, 309–314. https://doi.org/10.1016/j.jenvman.2013.12.007.

Agrafioti, E., Kalderis, D., Diamadopoulos, E., 2014b. Ca and Fe modified biochars as adsorbents of arsenic and chromium in aqueous solutions. J. Environ. Manage. 146, 444–450. https://doi.org/10.1016/j.jenvman.2014.07.029.

Ardebili, S.M.S., Taghipoor, A., Solmaz, H., Mostafaei, M., 2020. The effect of nanobiochar on the performance and emissions of a diesel engine fueled with fusel oil-diesel fuel. Fuel 268. https://doi.org/10.1016/j.fuel.2020.117356, 117356.

Banek, N.A., Abele, D.T., McKenzie, K.R., Wagner, M.J., 2018. Sustainable conversion of lignocellulose to high-purity, highly crystalline flake potato graphite. ACS Sustain. Chem. Eng. 6 (10), 13199–13207. https://doi.org/10.1021/acssuschemeng.8b02799.

Cao, Y., Xiao, W., Shen, G., Ji, G., Zhang, Y., Gao, C., Han, L., 2019. Carbonization and ball milling on the enhancement of Pb(II) adsorption by wheat straw: competitive effects of ion exchange and precipitation. Bioresour. Technol. 273, 70–76. https://doi.org/10.1016/j.biortech.2018.10.065.

Chen, B., Chen, Z., 2009. Sorption of naphthalene and 1-naphthol by biochars of orange peels with different pyrolytic temperatures. Chemosphere 76 (1), 127–133. https://doi.org/10.1016/j.chemosphere.2009.02.004.

Chen, H., Gao, B., Li, H., 2014. Functionalization, pH, and ionic strength influenced sorption of sulfamethoxazole on graphene. J. Environ. Chem. Eng. 2 (1), 310–315.

Chen, L., Chen, X.L., Zhou, C.H., Yang, H.M., Ji, S.F., Tong, D.S., Zhong, Z.K., Yu, W.-H., Chu, M.Q., 2017. Environmental-friendly montmorillonite-biochar composites: facile production and tunable adsorption-release of ammonium and phosphate. J. Clean. Prod. 156, 648–659. https://doi.org/10.1016/j.jclepro.2017.04.050.

Cho, D.W., Yoon, K., Ahn, Y., Sun, Y., Tsang, D.C.W., Hou, D., Ok, Y.S., Song, H., 2019. Fabrication and environmental applications of multifunctional mixed metal-biochar composites (MMBC) from red mud and lignin wastes. J. Hazard. Mater. 374, 412–419. https://doi.org/10.1016/j.jhazmat.2019.04.071.

Cope, C.O., Webster, D.S., Sabatini, D.A., 2014. Arsenate adsorption onto iron oxide amended rice husk char. Sci. Total Environ. 488–489, 554–561. https://doi.org/10.1016/j.scitotenv.2013.12.120.

Daughton, C.G., 2004. PPCPs in the environment: future research—beginning with the end always in mind. In: Pharmaceuticals in the Environment. Springer, pp. 463–495.

Daughton, C.G., Ternes, T.A., 1999. Pharmaceuticals and personal care products in the environment: agents of subtle change? Environ. Health Perspect. 107 (Suppl. 6), 907–938. https://doi.org/10.1289/ehp.99107s6907.

Dolgin, B., Vanek, M., McGory, T., Ham, D., 1986. Mechanical alloying of Ni, CO, and Fe with Ti. Formation of an amorphous phase. J. Non Cryst. Solids 87 (3), 281–289.

Dong, X., Qi, X., Liu, N., Yang, Y., Piao, Y., 2017. Direct electrochemical detection of bisphenol a using a highly conductive graphite nanoparticle film electrode. Sensors (Basel) 17 (4), 836. https://doi.org/10.3390/s17040836.

Dong, X., He, L., Hu, H., Liu, N., Gao, S., Piao, Y., 2018. Removal of 17β-estradiol by using highly adsorptive magnetic biochar nanoparticles from aqueous solution. Chem. Eng. J. 352, 371–379.

Fosso-Kankeu, E., Waanders, F., Potgieter, J., 2016. Enhanced adsorption capacity of sweet sorghum derived biochar towards malachite green dye using bentonite clay. In: International Conference on Advances in Science, Engineering, Technology and Natural Resources (ICASETNR-16) Nov.

Gafoor, A., Kumar, S., Begum, S., Rahman, Z., s., 2021. Elimination of nickel (II) ions using various natural/modified clay minerals: a review. Mater. Today Proc. 37, 2033–2040.

Genovese, M., Jiang, J.H., Lian, K., Holm, N., 2015. High capacitive performance of exfoliated biochar nanosheets from biomass waste corn cob. J. Mater. Chem. A 3 (6), 2903–2913. https://doi.org/10.1039/c4ta06110a.

He, C., Hu, X.J., 2012. Functionalized ordered mesoporous carbon for the adsorption of reactive dyes. Adsorpt. J. Int. Adsorpt. Soc. 18 (5–6), 337–348. https://doi.org/10.1007/s10450-012-9410-6.

Hu, X., Ding, Z., Zimmerman, A.R., Wang, S., Gao, B., 2015. Batch and column sorption of arsenic onto iron-impregnated biochar synthesized through hydrolysis. Water Res. 68, 206–216.

Huang, Y., Gong, Y.Y., Tang, J.C., Xia, S.Y., 2019. Effective removal of inorganic mercury and methylmercury from aqueous solution using novel thiol-functionalized graphene oxide/Fe-Mn composite. J. Hazard. Mater. 366, 130–139. https://doi.org/10.1016/j.jhazmat.2018.11.074.

Huang, J., Zimmerman, A.R., Chen, H., Gao, B., 2020. Ball milled biochar effectively removes sulfamethoxazole and sulfapyridine antibiotics from water and wastewater. Environ. Pollut. 258. https://doi.org/10.1016/j.envpol.2019.113809, 113809.

Ji, J., Chen, G., Zhao, J., 2019. Preparation and characterization of amino/thiol bifunctionalized magnetic nanoadsorbent and its application in rapid removal of Pb (II) from aqueous system. J. Hazard. Mater. 368, 255–263. https://doi.org/10.1016/j.jhazmat.2019.01.035.

Jiang, C., Bo, J.Y., Xiao, X.F., Zhang, S.M., Wang, Z.H., Yan, G.P., Wu, Y.G., Wong, C.P., He, H., 2020. Converting waste lignin into nano-biochar as a renewable substitute of carbon black for reinforcing styrene-butadiene rubber. Waste Manag. 102, 732–742. https://doi.org/10.1016/j.wasman.2019.11.019.

Jung, H.J., Sohn, Y., Sung, H.G., Hyun, H.S., Shin, W.G., 2015. Physicochemical properties of ball milled boron particles: dry vs. wet ball milling process. Powder Technol. 269, 548–553. https://doi.org/10.1016/j.powtec.2014.03.058.

Karinkanta, P., Ämmälä, A., Illikainen, M., Niinimäki, J.J.B., Bioenergy., 2018. Fine grinding of wood–Overview from wood breakage to applications. Biomass Bioenergy 113, 31–44.

Kumar, H., Patel, M., Mohan, D., 2019a. Simplified batch and fixed-bed design system for efficient and sustainable fluoride removal from water using slow Pyrolyzed okra stem and black gram straw biochars. ACS Omega 4 (22), 19513–19525. https://doi.org/10.1021/acsomega.9b00877.

Kumar, R., Patel, M., Singh, P., Bundschuh, J., Pittman Jr., C.U., Trakal, L., Mohan, D., 2019b. Emerging technologies for arsenic removal from drinking water in rural and peri-urban areas: methods, experience from, and options for Latin America. Sci. Total Environ. 694. https://doi.org/10.1016/j.scitotenv.2019.07.233, 133427.

Kumar, M., Xiong, X.N., Wan, Z.H., Sun, Y.Q., Tsang, D.C.W., Gupta, J., Gao, B., Cao, X.D., Tang, J.C., Ok, Y.S., 2020. Ball milling as a mechanochemical technology for fabrication of novel biochar nanomaterials. Bioresour. Technol. 312. https://doi.org/10.1016/j.biortech.2020.123613, 123613.

Kundu, S., Xia, W., Busser, W., Becker, M., Schmidt, D.A., Havenith, M., Muhler, M., 2010. The formation of nitrogen-containing functional groups on carbon nanotube surfaces: a quantitative XPS and TPD study. Phys. Chem. Chem. Phys. 12 (17), 4351–4359. https://doi.org/10.1039/b923651a.

Li, R., Zhang, Y., Deng, H., Zhang, Z., Wang, J.J., Shaheen, S.M., Xiao, R., Rinklebe, J., Xi, B., He, X., Du, J., 2020. Removing tetracycline and Hg(II) with ball-milled magnetic nanobiochar and its potential on polluted irrigation water reclamation. J. Hazard. Mater. 384. https://doi.org/10.1016/j.jhazmat.2019.121095, 121095.

Lian, F., Xing, B., 2017. Black carbon (biochar) in water/soil environments: molecular structure, sorption, stability, and potential risk. Environ. Sci. Technol. 51 (23), 13517–13532. https://doi.org/10.1021/acs.est.7b02528.

Lian, F., Yu, W., Wang, Z., Xing, B., 2018. New insights into black carbon nanoparticle-induced dispersibility of goethite colloids and configuration-dependent sorption for phenanthrene. Environ. Sci. Technol. 53 (2), 661–670.

Liang, Q.Q., Zhao, D.Y., Qian, T.W., Freeland, K., Feng, Y.C., 2012. Effects of stabilizers and water chemistry on arsenate sorption by polysaccharide-stabilized magnetite nanoparticles. Ind. Eng. Chem. Res. 51 (5), 2407–2418. https://doi.org/10.1021/ie201801d.

Liu, G.C., Zheng, H., Jiang, Z.X., Zhao, J., Wang, Z.Y., Pan, B., Xing, B.S., 2018. Formation and physicochemical characteristics of Nano biochar: insight into chemical and colloidal stability. Environ. Sci. Technol. 52 (18), 10369–10379. https://doi.org/10.1021/acs.est.8b01481.

Liu, W., Li, Y., Feng, Y., Qiao, J., Zhao, H., Xie, J., Fang, Y., Shen, S., Liang, S., 2020a. The effectiveness of nanobiochar for reducing phytotoxicity and improving soil remediation in cadmium-contaminated soil. Sci. Rep. 10 (1), 858. https://doi.org/10.1038/s41598-020-57954-3.

Liu, P., Zhang, Y.H., Feng, N.C., Zhu, M.L., Tian, J.C., 2020b. Potentially toxic element (PTE) levels in maize, soil, and irrigation water and health risks through maize consumption in northern Ningxia, China. BMC Public Health 20 (1), 1–13. https://doi.org/10.1186/s12889-020-09845-5.

Lyu, H., Gong, Y., Gurav, R., Tang, J., 2016. Potential application of biochar for bioremediation of contaminated systems. In: Biochar Application. Elsevier, pp. 221–246.

Lyu, H., Gao, B., He, F., Ding, C., Tang, J.C., Crittenden, J.C., 2017a. Ball-milled carbon nanomaterials for energy and environmental applications. ACS Sustain. Chem. Eng. 5 (11), 9568–9585. https://doi.org/10.1021/acssuschemeng.7b02170.

Lyu, H., Tang, J., Huang, Y., Gai, L., Zeng, E.Y., Liber, K., Gong, Y., 2017b. Removal of hexavalent chromium from aqueous solutions by a novel biochar supported nanoscale iron sulfide composite. Chem. Eng. J. 322, 516–524.

Lyu, H., Gao, B., He, F., Zimmerman, A.R., Ding, C., Huang, H., Tang, J., 2018a. Effects of ball milling on the physicochemical and sorptive properties of biochar: experimental observations and governing mechanisms. Environ. Pollut. 233, 54–63. https://doi.org/10.1016/j.envpol.2017.10.037.

Lyu, H.H., Gao, B., He, F., Zimmerman, A.R., Ding, C., Tang, J.C., Crittenden, J.C., 2018b. Experimental and modeling investigations of ball-milled biochar for the removal of aqueous methylene blue. Chem. Eng. J. 335, 110–119. https://doi.org/10.1016/j.cej.2017.10.130.

Lyu, H., Xia, S., Tang, J., Zhang, Y., Gao, B., Shen, B., 2020. Thiol-modified biochar synthesized by a facile ball-milling method for enhanced sorption of inorganic Hg(2+) and organic CH3Hg(.). J. Hazard. Mater. 384. https://doi.org/10.1016/j.jhazmat.2019.121357, 121357.

Mohan, D., Sarswat, A., Ok, Y.S., Pittman Jr., C.U., 2014. Organic and inorganic contaminants removal from water with biochar, a renewable, low cost and sustainable adsorbent–a critical review. Bioresour. Technol. 160, 191–202.

Naghdi, M., Taheran, M., Brar, S.K., Rouissi, T., Verma, M., Surampalli, R.Y., Valero, J.R., 2017. A green method for production of nanobiochar by ball milling-optimization and characterization. J. Clean. Prod. 164, 1394–1405. https://doi.org/10.1016/j.jclepro.2017.07.084.

Naghdi, M., Taheran, M., Pulicharla, R., Rouissi, T., Brar, S.K., Verma, M., Surampalli, R.Y., 2019. Pine-wood derived nanobiochar for removal of carbamazepine from aqueous media: adsorption behavior and influential parameters. Arab. J. Chem. 12 (8), 5292–5301. https://doi.org/10.1016/j.arabjc.2016.12.025.

Nath, B.K., Chaliha, C., Kalita, E., 2019. Iron oxide permeated mesoporous rice-husk nanobiochar (IPMN) mediated removal of dissolved arsenic (As): Chemometric modelling and adsorption dynamics. J. Environ. Manage. 246, 397–409. https://doi.org/10.1016/j.jenvman.2019.06.008.

Nguyen, M.V., Lee, B.K., 2016. A novel removal of $CO_2$ using nitrogen doped biochar beads as a green adsorbent. Process Saf. Environ. Prot. 104, 490–498. https://doi.org/10.1016/j.psep.7016.04.007.

Okitsu, K., Mizukoshi, Y., Yamamoto, T.A., Maeda, Y., Nagata, Y., 2007. Sonochemical synthesis of gold nanoparticles on chitosan. Mater. Lett. 61 (16), 3429–3431. https://doi.org/10.1016/j.matlet.2006.11.090.

Oleszczuk, P., Cwikla-Bundyra, W., Bogusz, A., Skwarek, E., Ok, Y.S., 2016. Characterization of nanoparticles of biochars from different biomass. J. Anal. Appl. Pyrolysis 121, 165–172. https://doi.org/10.1016/j.jaap.2016.07.017.

Paraknowitsch, J.P., Zhang, J., Su, D., Thomas, A., Antonietti, M., 2010. Ionic liquids as precursors for nitrogen-doped graphitic carbon. Adv. Mater. 22 (1), 87–92. https://doi.org/10.1002/adma.200900965.

Patel, M., Kumar, R., Kishor, K., Mlsna, T., Pittman Jr., C.U., Mohan, D., 2019. Pharmaceuticals of emerging concern in aquatic systems: chemistry, occurrence, effects, and removal methods. Chem. Rev. 119 (6), 3510–3673. https://doi.org/10.1021/acs.chemrev.8b00299.

Patil, P.R., Joshi, S.S., 2007. Polymerized organic–inorganic synthesis of nanocrystalline zinc oxide. Mater. Chem. Phys. 105 (2–3), 354–361.

Peng, Z., Zhao, H., Lyu, H., Wang, L., Huang, H., Nan, Q., Tang, J., 2018. UV modification of biochar for enhanced hexavalent chromium removal from aqueous solution. Environ. Sci. Pollut. Res. 25 (11), 10808–10819. https://doi.org/10.1007/s11356-018-1353-3.

Peterson, S.C., Jackson, M.A., Kim, S., Palmquist, D.E., 2012. Increasing biochar surface area: optimization of ball milling parameters. Powder Technol. 228, 115–120. https://doi.org/10.1016/j.powtec.2012.05.005.

Pratap, M.P., Pittman, C.U., Nguyen, T.A., Mohan, D., 2020. Nanobiochar: A Sustainable Solution for Agricultural and Environmental Applications. Elsevier.

Premarathna, K.S.D., Rajapaksha, A.U., Sarkar, B., Kwon, E.E., Bhatnagar, A., Ok, Y.-S., Vithanage, M., 2019. Biochar-based engineered composites for sorptive decontamination of water: a review. Chem. Eng. J. 372, 536–550. https://doi.org/10.1016/j.cej.2019.04.097.

Qin, Y., Zhu, X., Su, Q., Anumah, A., Gao, B., Lyu, W., Zhou, X., Xing, Y., Wang, B., 2020. Enhanced removal of ammonium from water by ball-milled biochar. Environ. Geochem. Health 42 (6), 1579–1587. https://doi.org/10.1007/s10653-019-00474-5.

Ramadan, M.M., Abd-Elsalam, K.A., 2020. Micro/Nano biochar for sustainable plant health: Present status and future prospects. In: Carbon Nanomaterials for Agri-Food and Environmental Applications. Elsevier, pp. 323–357.

Ramanayaka, S., Tsang, D.C.W., Hou, D., Ok, Y.S., Vithanage, M., 2020a. Green synthesis of graphitic nanobiochar for the removal of emerging contaminants in aqueous media. Sci. Total Environ. 706. https://doi.org/10.1016/j.scitotenv.2019.135725, 135725.

Ramanayaka, S., Vithanage, M., Alessi, D.S., Liu, W.J., Jayasundera, A.C.A., Ok, Y.S., 2020b. Nanobiochar: production, properties, and multifunctional applications. Environ. Sci.-Nano 7 (11), 3279–3302. https://doi.org/10.1039/d0en00486c.

Richardson, S.D., 2009. Water analysis: emerging contaminants and current issues. Anal. Chem. 81 (12), 4645–4677. https://doi.org/10.1021/ac9008012.

Saxena, M., Maity, S., Sarkar, S., 2014. Carbon nanoparticles in 'biochar' boost wheat (Triticum aestivum) plant growth. RSC Adv. 4 (75), 39948–39954.

Shaheen, S.M., Niazi, N.K., Hassan, N.E.E., Bibi, I., Wang, H.L., Tsang, D.C.W., Ok, Y.-S., Bolan, N., Rinklebe, J., 2019. Wood-based biochar for the removal of potentially toxic elements in water and wastewater: a critical review. Int. Mater. Rev. 64 (4), 216–247. https://doi.org/10.1080/09506608.2018.1473096.

Shao, J.G., Zhang, J.J., Zhang, X., Feng, Y., Zhang, H., Zhang, S.H., Chen, H.P., 2018. Enhance $SO_2$ adsorption performance of biochar modified by $CO_2$ activation and amine impregnation. Fuel 224, 138–146. https://doi.org/10.1016/j.fuel.2018.03.064.

Sharma, Y.C., Srivastava, V., Singh, V.K., Kaul, S.N., Weng, C.H., 2009. Nano-adsorbents for the removal of metallic pollutants from water and wastewater. Environ. Technol. 30 (6), 583–609. https://doi.org/10.1080/09593330902838080.

Song, B., Chen, M., Zhao, L., Qiu, H., Cao, X., 2019. Physicochemical property and colloidal stability of micron- and nano-particle biochar derived from a variety of feedstock sources. Sci. Total Environ. 661, 685–695. https://doi.org/10.1016/j.scitotenv.2019.01.193.

Suryanarayana, C., 2001. Mechanical alloying and milling. Prog. Mater. Sci. 46 (1–2), 1–184.

Taheran, M., Naghdi, M., Brar, S.K., Verma, M., Surampalli, R.Y., 2018. Emerging contaminants: here today, there tomorrow! Environ. Nanotechnol. Monit. Manage. 10, 122–126.

Tang, J., Lv, H., Gong, Y., Huang, Y., 2015. Preparation and characterization of a novel graphene/biochar composite for aqueous phenanthrene and mercury removal. Bioresour. Technol. 196, 355–363. https://doi.org/10.1016/j.biortech.2015.07.047.

Tian, C., Feng, C., Wei, M., Wu, Y., 2018. Enhanced adsorption of anionic toxic contaminant Congo red by activated carbon with electropositive amine modification. Chemosphere 208, 476–483. https://doi.org/10.1016/j.chemosphere.2018.06.005.

Ullah, M., Ali, M.E., Abd Hamid, S.B., 2014. Surfactant-assisted ball milling: a novel route to novel materials with controlled nanostructure—a review. Rev. Adv. Mater. Sci. 37 (1–2), 1–14.

Vimal, V., Patel, M., Mohan, D., 2019. Aqueous carbofuran removal using slow pyrolyzed sugarcane bagasse biochar: equilibrium and fixed-bed studies. RSC Adv. 9 (45), 26338–26350. https://doi.org/10.1039/c9ra01628g.

Wallace, C.A., Afzal, M.T., Saha, G.C., 2019. Effect of feedstock and microwave pyrolysis temperature on physio-chemical and nano-scale mechanical properties of biochar. Bioresour. Bioprocess. 6 (1). https://doi.org/10.1186/s40643-019-0268-2.

Wang, S., Gao, B., Zimmerman, A.R., Li, Y., Ma, L., Harris, W.G., Migliaccio, K.W., 2015. Removal of arsenic by magnetic biochar prepared from pinewood and natural hematite. Bioresour. Technol. 175, 391–395. https://doi.org/10.1016/j.biortech.2014.10.104.

Weber, K., Quicker, P., 2018. Properties of biochar. Fuel 217, 240–261. https://doi.org/10.1016/j.fuel.2017.12.054.

Xiang, W., Wan, Y., Zhang, X., Tan, Z., Xia, T., Zheng, Y., Gao, B., 2020a. Adsorption of tetracycline hydrochloride onto ball-milled biochar: governing factors and mechanisms. Chemosphere 255. https://doi.org/10.1016/j.chemosphere.2020.127057, 127057.

Xiang, W., Zhang, X.Y., Chen, K.Q., Fang, J.N., He, F., Hu, X., Tsang, D.C.W., Ok, Y.S., Gao, B., 2020b. Enhanced adsorption performance and governing mechanisms of ball-milled biochar for the removal of volatile organic compounds (VOCs). Chem. Eng. J. 385, 123842. https://doi.org/10.1016/j.cej.2019.123842.

Xiao, J., Hu, R., Chen, G., 2020. Micro-nano-engineered nitrogenous bone biochar developed with a ball-milling technique for high-efficiency removal of aquatic Cd(II), Cu(II) and Pb(II). J. Hazard. Mater. 387. https://doi.org/10.1016/j.jhazmat.2019.121980, 121980.

Xing, T., Mateti, S., Li, L.H., Ma, F., Du, A., Gogotsi, Y., Chen, Y., 2016. Gas protection of two-dimensional nanomaterials from high-energy impacts. Sci. Rep. 6 (1), 35532. https://doi.org/10.1038/srep35532.

Xu, F., Wei, C., Zeng, Q., Li, X., Alvarez, P.J.J., Li, Q., Qu, X., Zhu, D., 2017. Aggregation behavior of dissolved black carbon: implications for vertical mass flux and fractionation in aquatic systems. Environ. Sci. Technol. 51 (23), 13723–13732. https://doi.org/10.1021/acs.est.7b04232.

Xu, X.Y., Zheng, Y.L., Gao, B., Cao, X.D., 2019. N-doped biochar synthesized by a facile ball-milling method for enhanced sorption of $CO_2$ and reactive red. Chem. Eng. J. 368, 564–572. https://doi.org/10.1016/j.cej.2019.02.165.

Yang, G.X., Jiang, H., 2014. Amino modification of biochar for enhanced adsorption of copper ions from synthetic wastewater. Water Res. 48, 396–405. https://doi.org/10.1016/j.watres.2013.09.050.

Yuan, Y., Zhang, N., Hu, X., 2020. Effects of wet and dry ball milling on the physicochemical properties of sawdust derived-biochar. Instrum. Sci. Technol. 48 (3), 287–300.

Yue, L., Lian, F., Han, Y., Bao, Q., Wang, Z., Xing, B., 2019. The effect of biochar nanoparticles on rice plant growth and the uptake of heavy metals: implications for agronomic benefits and potential risk. Sci. Total Environ. 656, 9–18. https://doi.org/10.1016/j.scitotenv.2018.11.364.

Zhang, M., Gao, B., Yao, Y., Xue, Y., Inyang, M., 2012. Synthesis, characterization, and environmental implications of graphene-coated biochar. Sci. Total Environ. 435-436, 567–572. https://doi.org/10.1016/j.scitotenv.2012.07.038.

Zhang, H., Luo, Y., Makino, T., Wu, L., Nanzyo, M., 2013. The heavy metal partition in size-fractions of the fine particles in agricultural soils contaminated by waste water and smelter dust. J. Hazard. Mater. 248-249, 303–312. https://doi.org/10.1016/j.jhazmat.2013.01.019.

Zhang, G., Liu, Y., Zheng, S., Hashisho, Z., 2019a. Adsorption of volatile organic compounds onto natural porous minerals. J. Hazard. Mater. 364, 317–324. https://doi.org/10.1016/j.jhazmat.2018.10.031.

Zhang, Q., Wang, J., Lyu, H., Zhao, Q., Jiang, L., Liu, L., 2019b. Ball-milled biochar for galaxolide removal: sorption performance and governing mechanisms. Sci. Total Environ. 659, 1537–1545. https://doi.org/10.1016/j.scitotenv.2019.01.005.

Zhang, K., Wang, Y., Mao, J., Chen, B., 2020. Effects of biochar nanoparticles on seed germination and seedling growth. Environ. Pollut. 256. https://doi.org/10.1016/j.envpol.2019.113409, 113409.

Zhou, Y.M., Gao, B., Zimmerman, A.R., Chen, H., Zhang, M., Cao, X.D., 2014. Biochar-supported zerovalent iron for removal of various contaminants from aqueous solutions. Bioresour. Technol. 152, 538–542. https://doi.org/10.1016/j.biortech.2013.11.021.

# Life cycle analysis of biochar use in water treatment plants

**Md Mosleh Uddin and Mark Mba Wright**
*Department of Mechanical Engineering, Iowa State University, Ames, IA, United States*

## 1 Introduction

Biochar production, an ancient technology, was mainly used for producing cooking fuel in many regions of the world. The use of biochar for agricultural activities is also historically well-known, albeit, with different names. Climate change concerns have rejuvenated the interest on biochar technologies due to their positive impacts on carbon sequestration, waste management, and energy production. Biochar is considered a means for long-term carbon sequestration because of its stable carbon content that can be stored in the soil. Under the soil, biochar's carbon content can remain buried for hundreds of years. Soil amendment and remediation, and agricultural soil application are among the primary uses of biochar. Moreover, the spectrum of biochar usage is rapidly expanding. Some applications are more environmentally advantageous than others.

Nevertheless, a common challenge for developing a solution to environmental problems is shifting environmental impacts from one area to another. For instance, the environmental benefits of biochar use can be offset by negative impacts during biochar production stages, which depend on various factors, such as geographical locations, source of biochar, conversion technologies, and traditional waste management practices. Therefore, a holistic analysis is necessary to account for the total environmental impacts of any product system or service at every stage of the life cycle.

Life cycle assessment (LCA) or life cycle analysis is *the systematic evaluation of direct and indirect environmental impacts through the life cycle of a product system or service*, including raw material extraction, transportation, production, distribution, usage, and disposal or recycle. It is a widely recognized methodology of quantifying net environmental impact of any product or process. All input and output materials and energy flow of a product or process is systematically collected to quantify the environmental impacts, based on scientific evidence. Some LCA categories assess the potential socioeconomic effect of any product or process on society. The results are very useful for sustainable product development and informed decision making for the policy makers.

LCA has been widely used to evaluate the net environmental impacts of different biochar production pathways and applications. From an LCA perspective, biochar as a soil amendment sequester carbon and reduce greenhouse gas (GHG) emissions. However, biochar production incurs various undesirable impacts on other environmental categories. LCA provides tools for systematically assessing all types of environmental impacts of biochar production and usage.

In this chapter, first, we will describe the fundamentals of LCA and the standard structure set by the International Standard Organization (ISO). Second, the environmental impacts of biochar production will be presented in the LCA framework. Lastly, the application of LCA for biochar production will be shown based on a case study.

## 2 Fundamentals of life cycle analysis (LCA)

In the 1960s, LCA began as a tool for accounting of cumulative energy demand (CED) for chemical intermediates production. The LCA methodology has been standardized by the International Standards Organization (ISO) in the 14000 series through a series of evolution phases. According to ISO 14040 standard (ISO/IEC, n.d.) LCA should:

1. compile an inventory of all relevant input materials, energy, and environmental emissions;
2. evaluate all potential environmental impacts associated with the inputs and outputs; and
3. interpret the impact assessment results to make objective decisions.

The structure of LCA, as defined by ISO, consists of four stages: (i) goal and scope definition, (ii) life cycle inventory analysis, (iii) life cycle impact assessment, and (iv) interpretation. Fig. 21.1 illustrates the structure of the LCA methodology. This standard structure aligns with many existing frameworks that use similar terminologies to maintain standard and transparent communication of the information (Klöpffer and Grahl, 2014).

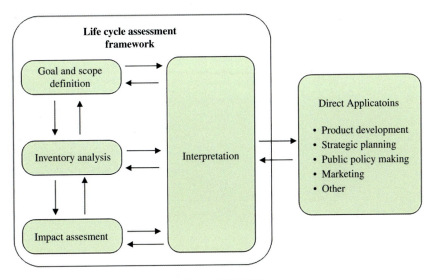

**Fig. 21.1** Structure of life cycle assessment according to ISO 14040.

## 2.1 Goal and scope definition

LCA starts with a clear goal statement of the analysis being performed. The overarching goal of every LCA is to minimize resource consumptions and environmental impacts to ensure sustainability. Typical objectives may include selecting products with the least environmental impacts, identifying process "hotspot," optimizing production processes, supporting product certification, developing benchmark processes, etc.

After the goal statement is identified, the next step is to define the scope. Scope definition means what is considered in the analysis and what is not. Due to complex interdependency among different product life cycle stages, accounting for every process might be prohibitive. Judicious selection and omission of life cycle stages are acceptable with a clear justification based on the scope definition.

**System boundary:** A system boundary shows all unit processes that are included in the analysis. A graphical representation of the system boundary is preferred along with textual descriptions. A complete LCA should include all stages starting from raw material and energy extraction to the product's disposal (cradle-to-grave). Nevertheless, the system boundary can exclude downstream processes after manufacturing (e.g., distribution, usage, disposal) to compare different products performing similar functions. This approach is called a cradle-to-gate system boundary. Fig. 21.2 illustrates the system boundary of a fictitious product's

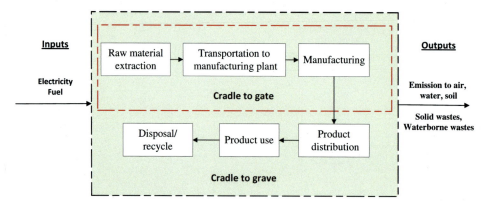

**Fig. 21.2** System boundary of a product life cycle.

life cycle. The red dashed line shows the cradle-to-gate system boundary, and the black dashed line shows the cradle-to-grave system boundary.

However, arbitrary stages can be excluded from the analysis aligning with the defined goal. System boundaries can also define the geographical boundary and time period within which the LCA results are valid.

**Functional unit:** A functional unit (FU) can either be the function performed by a product (e.g., miles driven by 1 gal of diesel) or the performance of any process (e.g., time required to produce 1 unit of product). LCA often compares the environmental impacts of different products and processes. FU defines the common basis for the comparative analysis among different products or processes. The material-energy consumptions, environmental emissions, waste productions of the analyzed product, or process are normalized based on the FU.

## 2.2 Life cycle inventory (LCI)

The life cycle inventory (LCI) is the systematic compilation and quantification of all inputs and outputs of any product's life cycle stages for environmental impact evaluation (ISO 14040:2006). The LCI contains materials and energy flow information, measurements of environmental emissions of each life cycle stage. Unit processes are the assembly of several subsystems that transform one type of product or energy into another form while incurring environmental impacts. A systematic flow chart is helpful to visualize the total life cycle of a product, where each unit process within the system boundary is represented with its associated inputs and outputs.

LCI inputs typically include material consumption, energy consumption, and resource transportation. On the other hand, the outputs include environmental emissions to air, water, and soil, solid/liquid waste, etc. LCI is typically compiled by summing the input-output flows of the constituting unit processes. Fig. 21.3 illustrates the LCI flows of a hypothetical product life cycle.

LCI flows are collected from various sources such as laboratory test results, pilot or commercial plants, mathematical models, simulations, published scientific journals, or other databases. Estimated data are often used if the required data unavailable or very difficult to collect. Nevertheless, the estimations should be based on sound scientific or engineering analysis. The influence of any estimated data on the final results can be analyzed by sensitivity analysis to quantify the range of possible deviation.

There are both commercial and free LCA databases that are good sources of LCI data. Commercial LCA software packages such as SimaPro and GaBi provide access to multiple LCI databases but typically require a license to unlock most of their information and functionality. Note that commercial databases often lack transparency as the process of data evaluation and underlying assumptions might not be well documented. The US Life cycle inventory (USLCI) is an example of a free, government-funded, publicly available LCI database, collaboratively developed by different government research organizations and industrial partners (https://www.nrel.gov/lci/). Completeness, accuracy, and data consistency are crucial for the later LCA stages and have direct consequences in decision making. However, a trade-off between

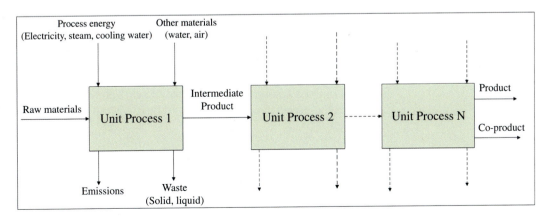

**Fig. 21.3** Schematic LCI flows of a product life cycle.

data quality and data collection effort always exists to justify the goal of the analysis.

LCI data for all unit processes are accumulated and normalized to the defined functional unit. In another word, the required inputs/outputs of each unit process to produce one functional unit are determined. Total emissions or the emission inventory from any process (or combination of multiple processes) is calculated using Eq. (21.1):

$$Emission\ inventory = \sum\nolimits_{e \in E} Activity\ or\ production\ rate\ (e) \times Emission\ factor\ (e) \qquad (21.1)$$

Here, $E$ represents the set of all unit processes in the product life cycle. The emission factor reflects the quantity of released pollutants (for example, $CO_2$, $NO_x$, $SO_x$, etc.) or environmental impact associated with the rate of activity or production. Emission factors are experimentally determined that can be collected from published literature or LCA databases.

**Impact allocation methods:** A common challenge for any production system is the allocation of LCI flows if multiple products are produced from a single process. Assigning materials-energy consumptions, and environmental emissions, to each product is not possible if they cannot be produced independently. Instead, LCI flows can be allocated either based on the physical relationship of the co-products (e.g., mass, energy, etc.) or based on the economic value of the co-products (e.g., market price) where physical relationships are hard to quantify.

However, ISO guidelines 14044 recommend adopting the system expansion method instead of the allocation method whenever possible. In the system expansion method, the co-products are assumed to displace other products with similar functions outside the system boundary. Therefore, the emissions associated with the displaced product are avoided and subtracted from the total impact of the considered product system. For example, nitrogen-rich digestate is produced along with biogas during the anaerobic digestion of animal manure. Application of digestate in an agricultural field can replace an equivalent amount of synthetic nitrogen fertilizer and avoid the emissions that would have been produced for its production.

## 2.3 Life cycle impact assessment (LCIA)

LCI data provide a quantitative measure of different types of environmental emissions throughout the product life cycle. However, the real impact on human health and environment is not defined yet. Life cycle impact assessment (LCIA) evaluates the environmental impact of any product system based on the compiled LCI data. It explains the potential impacts based on scientific methodologies. For instance, LCIA can predict the impact of releasing 1-ton $N_2O$ during the production of nitrogen fertilizer. Although these predictions are not necessarily absolute, they provide a meaningful basis for comparative analysis among different products. ISO 14042 defines the standard process of performing LCIA categorized as mandatory and optional steps.

*Mandatory steps* include selection and definition of impact categories, assignment of LCI flows to selected impact categories, transformation to a common unit, and weighting of different flows within an impact category.

First, environmental emissions throughout the product life cycle are assigned to different impact categories concerning human health, ecosystem, or resource depletion (area of protection). Some commonly used impact categories are global warming potential (GWP), ozone depletion, acidification, eutrophication, smog formation, ecotoxicity, human health toxicity, fossil fuel depletion, etc. These impact categories explain specific issues related to protected areas. For example, climate change is a key environmental concern caused by GHG emissions, such as $CO_2$, $CH_4$, and $N_2O$. Eutrophication is a major concern that can damage marine habitats caused by an oversupply of nutrients (such as phosphorus, nitrogen). The selection of impact categories depends on the goal of the study. Fig. 21.4 shows the impact categories commonly used by LCA practitioners.

*Second*, LCI flows of life cycle stages are *assigned* to the selected impact categories. For example, $CO_2$ is assigned to the global warming category; $SO_2$ emission is assigned to the acidification category. These assignments are based on scientific evidences. One flow can be assigned to multiple impact categories based on its impacts.

*Third*, all flows in any individual impact category are transformed into a common unit, called the *category indicator*. It helps to express the category impact as a single unit. For example, the category indicator for global warming is kg of $CO_{2\ eq}$/FU.

*Fourth*, the relative weight of different flows within an impact category, compared to the category indicator, is called the *characterization factor*. Characterization factors convert different quantities of emissions into a single category indicator. These factors are determined based on scientific models, considering various factors, such as the lifetime of a specific gas in the atmosphere. For example, the impact of 1 kg of $N_2O$ on global warming is 298 times that of 1 kg $CO_2$ over a 100-year period. Therefore, the characterization factor for $N_2O$ is 298 kg $CO_{2\ eq}$.

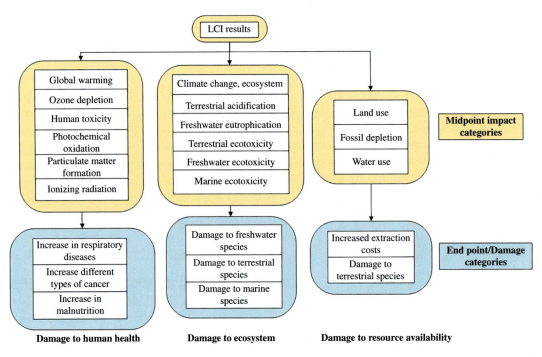

**Fig. 21.4** Midpoint and endpoint life cycle impact categories are grouped based on their impacts to human health, the ecosystem, and resource availability.

Impact indicators of LCI flows on a specific impact category are calculated using Eq. (21.2),

$$Indicator = \sum_{e \in E} Emission\ inventory\ in\ terms\ of\ category\ indicator\ (e) \times Characterization\ factor(e)$$

(21.2)

*Optional steps* include normalization, grouping, weighting, etc., which can vary based on the goal and scope of the study. For a detailed discussion of optional steps, readers are referred to Klöpffer and Grahl (2014).

**Midpoint versus endpoint analysis:** Environmental emissions have very complex cause-effect relationships with different impact categories. *Endpoint impact categories* measure the effect on several areas of concern, such as human health, ecosystem, and resource availability. *Midpoint impact categories* estimate an intermediate measurement between the emission and the environmental impacts just before their impacts on the areas of concerns. Midpoint impact categories are easier to measure because they deal with a single environmental problem and do not require making many assumptions. On the other hand, endpoint impact modeling becomes more complex and suffers from significant uncertainty. For example, eutrophication, a midpoint-level impact category, quantifies the potential impact of nitrogen and phosphate concentration on marine water. In comparison, endpoint impact category would quantify the impact of eutrophication on the ecosystem, such as the extinction potential of marine species. Fig. 21.4 represents different midpoint impact categories along with their potential endpoint effects or damage to human health, ecosystem, and resource availability.

## 2.4 LCIA methods

A good number of LCIA methods are available differing on the types and number of categories included. For evaluating only GWP, IPCC (2013) characterization factors are mostly used. For evaluating multiple impact categories, ReCiPe (Huijbregts et al., 2017), CML (Guinee, 2002), and IMPACT 2002+ (Jolliet et al., 2003) are some of the widely used methods. TRACI (Tools for the Reduction and Assessment of Chemical and other environmental Impacts) is a LCIA method developed by US EPA using US-based input parameters (Bare et al., 2003). Due to the variations and complexity of different LCIA methods, practitioners use LCA software. Most LCA software packages offer flexibility to use any of the above-mentioned LCIA methods.

## 2.5 Interpretation

The interpretation stage combines and summarizes the LCI and LCIA results to provide recommendations and facilitate decision making, according to the goal and scope. ISO 14044 describes the three steps of interpretation: (a) identification of the significant issues; (b) completeness, sensitivity, and consistency check; and (c) conclusions, recommendations, and limitations.

Identification of "hotspots" (cause of the most significant environmental impacts) is crucial to guide future research or process optimization. The engineering estimates and assumptions that bridge the data gap in the LCI phase can have considerable influence on the results. Explaining these limitations, along with the results, is equally essential for a comprehensive analysis. Techniques, like sensitivity analysis and uncertainty analysis, provide a way to analyze the influence of estimated data on the environmental impact. Sometimes, the results may not be sufficient to reach a conclusion due to the underlying uncertainty.

Nevertheless, this still serves as a valuable aid for the decision maker. The clear, easy to understand, and unbiased communication of LCA results is important for the creditability of any study.

## 3 LCA of biochar production

Scientists have published various studies describing the environmental benefits of biochar use in different applications, including wastewater treatment. However, biochar's environmental benefits depend on how it is produced and utilized. Several LCA studies have explored different scenarios for biochar use and arrived at varying conclusions. This section describes the key aspects of conducting a biochar LCA.

### 3.1 Goal and scope

The goal of a biochar LCA primarily includes a comparative evaluation of different biochar feedstocks, conversion technologies, and applications. The analysis often compares biochar performance in different applications with conventionally used materials. For example, using biochar for water filtration can be compared to chemical filtration using chlorine. In most of the published LCA, biochar is not the main product of the process but is considered a byproduct.

The functional unit can widely vary based on the goal of the production process. If the aim of biochar production process is waste management, upstream functional units are used, such as 1 ton of corn stover used. On the other hand, if the aim of biochar production is to produce energy from bio-oil and syngas, the functional unit would be 1 MJ of energy produced. For the process with the sole purpose of biochar production, 1 ton of biochar produced can be the most suitable functional unit (Llorach-Massana et al., 2017; Muñoz et al., 2017). Based on the intended use of biochar, a variety of functional units can be defined, such as "impact from an average village household utilizing available cocoa waste" (Sparrevik et al., 2014), "impact per produced ton of maize per year" (Sparrevik et al., 2013), etc.

The system boundary of biochar production processes depends on various aspects, including feedstock source and availability, existing practices with the feedstocks, utilization scope of co-products, geographical location, etc. Typically, feedstock production is not considered in the system boundary if the feedstock is a waste resource such as wastewater sludge, corn stover, or discarded waste going to the soil, water, or landfills (Oldfield et al., 2018). However, if the energy crops are purposely grown for biochar production, agricultural processes are included in the system boundary (Matuštík et al., 2020). The system boundary typically includes the end-use of biochar to fully quantify the environmental benefits. Conventional biochar end-uses include soil amendment, combustion, water treatment, etc.

## 3.2 Life cycle stages of biochar production

Ideally, a complete LCA combines emissions from each life cycle stage of a product system to evaluate the net environmental impact. Life cycle stages can widely vary for biochar production depending on the wide range of feedstock options, conversion technologies, and end-uses. These variations further increase with different supply chain context depending on the geographical locations. However, a typical biochar production scenario includes life cycle stages such as (i) feedstock production and collection, (ii) feedstock handling and preprocessing, (iii) thermochemical conversion, (iv) biochar distribution, and (v) end-use, as shown in Fig. 21.5. All variations in the biochar production supply chain can be included within these major stages.

**Feedstock production and collection:** Cellulosic biomass residues are common feedstocks for biochar production. They include crop residues (either field residue from agricultural land or processing residues such as nut shells, bagasse, etc.), food waste, yard trimmings, forestry waste, and animal manures (International Biochar Initiative, n.d.). If the feedstock is produced as the residue or byproduct of another process, the feedstock production-related emissions are distributed among the main products, or assigned entirely to the primary product. For example, the emissions of corn stover are a fraction of the emissions required to grow corn. In that case, the primary product does not need to be included in the system boundary because it would be produced regardless of whether the byproduct is utilized or not. However, for the purposely grown feedstock, such as switchgrass, emissions related to the production and harvesting should be included. In that case, material inputs for the feedstock production, such as fertilizer, pesticides, diesel for agricultural equipment, can contribute significantly to the total life cycle emissions. Other activities, such as the removal of forest

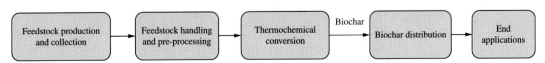

**Fig. 21.5** Major life cycle stages of biochar production.

residues from the field, might cause nutrient losses in the soil, leading to the degradation of soil quality. Therefore, additional fertilizer inputs will be required in subsequent years.

Collecting and transporting feedstock from the source to the process plant produces emissions via fossil fuel combustion, such as diesel combustion in a tractor. For distributed feedstock sources, transport-related emissions might significantly increase the total emissions. Feedstock collection may require intermediate or on-site storage facilities to create a buffer for the supply chain. This can produce emissions from construction materials production and other emissions like volatile organic compounds (VOC) and particulate matters. If convenient for the type of feedstock source, on-site conversion facilities could significantly reduce the transportation-related emissions and provide more flexibility in feedstock collection logistics.

**Feedstock processing and handling:** Biochar production technologies require specific feedstock properties (e.g., moisture, ash content, size, etc.) requiring preprocessing, and handling mechanism for efficient feedstock conversion. Generally, drying is required to meet the technology-specific moisture content for biochar production. Dryers operate with either hot gas or steam generated from waste process heat or fuel combustion. Combustion of fossil fuel for drying causes resource depletion and environmental emissions. Other preprocessing might include chipping, hammering, grinding, and screening, etc.

Most of the equipment pieces are operated by grid electricity or diesel combustion. The complete life cycle emissions for the grid electricity or diesel combustion should be accounted for as indirect emissions. Using bio-oil or syngas (from the biochar production process) for the feedstock preprocessing can reduce the net direct (via combustion of fuel) and indirect emissions (e.g., off-site emissions for natural gas extraction, processing, and transmission).

**Thermochemical conversion:** Conversion of biomass to biochar is the most influential life cycle stage of biochar production due to its highest atmospheric emissions (Dutta and Raghavan, 2014; Roberts et al., 2010). A wide spectrum of technologies are available for biochar production based on two primary conversion processes: (i) pyrolysis and (ii) gasification. From ancient biochar production technology, such as pit kilns, to highly advanced pyrolysis technology, most emissions occur during the conversion process. Modern technologies employ emission controls to prevent the release of GHGs and toxic compounds to the environment. Bio-oil and syngas are two byproducts produced along the biochar production process. The percentage yield of biochar, bio-oil, and syngas depends on the feedstock type, conversion technology, and process parameters (e.g., residence time, operating temperature, heating rate, etc.). For instance, a slow pyrolysis yields 35% biochar, 35% gas, and 30% liquid, while fast pyrolysis yields 10% biochar (Bridgwater et al., 1999). Depending on the end-use of these products, additional technologies are added to the plant such as oil/gas handling system, storage, distribution, safety, etc., resulting in additional environmental impacts.

Allocation of environmental impacts among the products depends on the intent of the conversion process. Biochar is usually a byproduct of biofuel production via pyrolysis in a biorefinery. On the other hand, bio-oil, heat, and syngas are considered as byproducts when biochar production is the primary intent. System expansion methods are widely used for biochar LCA to assign environmental burden to each product. For the biochar production scenario, heat produced by the syngas combustion can be considered equivalent to burning natural gas for producing equal amount of heat. The life cycle emissions of that amount of natural gas are subtracted as "avoided burden" credit.

**Biochar transportation to demand point:**

Postprocessing of biochar depends on the end-use and the mode of transportation to the demand sites. Biochar pelletization increases product density and reduces handling and transportation costs. Pelletized biochar is suitable for residential and commercial energy use. Electricity or fuel use in pelleting mills adds environmental emissions to the biochar life cycle. On the other hand, for some applications such as pharmaceutical use, powdered biochar is necessary, which increases the environmental risk due to dust particle emission.

Bulk transportation of biochar using specialized rail or truck is often preferred for large-scale applications and wholesale markets (Anderson et al., 2016). The distance from the plant to the demand site determines the environmental loads from transportation. Various packaging mediums are being used for biochar transportation, for example, large polyethylene bags, metal and plastic drums, buckets, etc. (Anderson et al., 2016).

**Biochar application:**

A common definition of biochar is *the charcoal used for soil application*. The primary environmental benefit of biochar application unanimously comes from the C sequestration in the soil. Biochar application to soil is attributed to manifold benefits such as improved soil condition, high water-holding capacity, improved nutrients availability, pollutants removal, high crop yield, etc. These benefits bring positive environmental impacts by reducing fertilizer requirement and associated emissions. Various preprocessing techniques before soil application, such as grinding, screening, and mixing with compost or mulch, have environmental impacts from fossil fuel use. Agricultural machineries like planters, tillers, seeders, or spreader for biochar application add additional environmental burden.

Biochar is a potential fuel for energy production with high energy content. It can substitute the combustion of fossil coal, and natural gas for heat and electricity production. Both fossil coal and natural gas emit significant environmental emissions in the extraction, processing, and distribution stages. Biochar for energy purposes is advantageous for GWP reduction; however, it can negatively impact other LCIA parameters, for instance, cumulative energy demand (Peters et al., 2015). The reason is that biomass pyrolysis is more energy-intensive than the direct combustion of other fossil fuels.

Beyond the age-old use for soil and energy application, multifarious usages of biochar are prevalent in the modern applications. Nearly 55 potential usages in different categories have been reported by Schmidt and Wilson (2014). Some of the current applications include building insulation, air filtration, humidity regulation, wastewater treatment, pharmaceutical ingredients, fabric additive, food conservations, etc. All these applications have positive environmental and human health benefits either by replacing conventional products or by minimizing atmospheric emissions. Often, biochar used in other applications can be further used for the soil application, if it does not contain any hazardous compound for soil. Recently, biochar is used in animal husbandry for animal feed additive, and manure odor control, which can be ultimately used for soil application. Researchers find a reduction in enteric methane emission by using biochar as animal feed additive (Schmidt et al., 2019). This would bring cascading positive environmental impacts (Azzi et al., 2019).

## 3.3 Life cycle impact assessment

LCI flows throughout the life cycle stages of biochar production contribute to different human health and environmental impact categories. The net impacts vary significantly based on different biochar production and utilization pathways. Feedstock types, conversion technologies, process parameters, and end-uses determine the magnitude of different environmental emissions.

In all reported LCA, biochar production reduces GWP mostly by removing atmospheric $CO_2$ stored in the biomass as carbon (Muñoz et al., 2017; Peters et al., 2015; Rajabi Hamedani et al.,

2019). Biochar application as soil amendment provides C (stable C of biochar) sinks for a longer period. Biochar can reduce soil $N_2O$ emission by reducing nitrogen volatilization and leaching. $N_2O$ is an important GHG having 298 times more GWP than $CO_2$. Biochar is reported to increase fertilizer efficiency, water retention capacity, and product yield that can reduce nitrogen fertilizer requirement, leading to reduced eutrophication potential (Kong et al., 2014; Mašek et al., 2013). Positive environmental impacts on fossil fuel depletion, eutrophication, and human toxicity categories are mainly due to the avoided urea and natural gas production. Because higher amount of urea can cause toxicity and eutrophication in water streams.

Increasing pyrolysis temperature increases process energy consumptions and decreases biochar yield, leading to lower GWP reduction (Hammond et al., 2011). Produced syngas can contribute to GWP reduction by replacing natural gas production. Electricity required for the agricultural activities (during feedstock production stage) and conversion plant increases GWP, fossil fuel depletion, and human toxicity via off-site emissions. This indirect emissions can get even higher for region where electricity is mostly produced from fossil fuel. Transportation is also responsible for adding impact to fossil fuel depletion, human toxicity, and eutrophication.

Generally, pyrolysis process is the most contributing stage followed by preprocessing and handling (Dutta and Raghavan, 2014; Matuštík et al., 2020). In most cases, contribution of transportation is insignificant compared to other life cycle stages. GWP results are robust against uncertainty of process parameters and allocation method, whereas other impact categories, such as eutrophication, are highly sensitive to the allocation method. Carbon sequestration via soil application is sensitive to various factors like climate, soil conditions, and biochar properties, etc. (Galgani et al., 2014).

## 4 Case study: Biochar production from corn stover

This section provides a case study of an LCA of biochar production using slow pyrolysis. For the case study, a hypothetical industrial-scale, slow pyrolysis facility is assumed to produce biochar from corn stover and the biochar is used in the cornfield as a soil amendment. Here, biochar is the primary product and produced syngas is the byproduct. All inventory data for the life cycle stages are collected from published literature. USLCI database is the primary source of the emission factors (Table A1), and some emission factors are obtained from ecoinvent database. The following sections will explain the stepwise approach of performing LCA for biochar production for the case study.

## 4.1 Goal and scope

The goal of the study is to evaluate cumulative energy demand and climate change impacts of biochar production from corn stover at a slow pyrolysis facility in the United States. The impact assessment is performed for a time horizon of 100 years.

**Functional unit:** The main purpose of the process is to manage agricultural wastes (corn stover) and turn those into value-added products. Therefore, the functional unit is defined as the management of 1 tonne of dry corn stover.

**System boundary:** Fig. 21.6 illustrates the system boundary (with dashed line) of the biochar production process. The system boundary includes unit processes such as corn stover collection, preprocessing, pyrolysis, syngas combustion for heat production, and biochar application into soil. Corn stover production is not within the system boundary as corn stover is a waste product from corn farming. Nevertheless, harvesting corn stover from the field

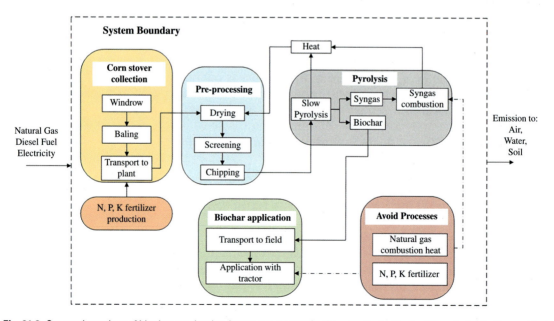

**Fig. 21.6** System boundary of biochar production from corn stover via slow pyrolysis. *Dashed lines* indicate the avoided processes.

removes a portion of soil nutrients that would come with the decomposition of the stover. Therefore, the application of additional fertilizer is considered inside the system boundary. Pyrolysis process heat is supplied by the combustion of produced syngas and bio-oil. However, a small amount of grid natural gas is used for the process start-up. The application of biochar as a soil amendment can replace fertilizer production. This avoided burden is treated as credit for the total impact calculation following the system expansion method. Additionally, excess heat produced by syngas and bio-oil combustion replaces natural gas production and combustion.

According to several studies, biochar application can replace the nitrogen fertilizer requirement due to its N content. It is not clear how much N would be bioavailable for the crop uptake. Nevertheless, biochar increases fertilizer application efficiency by several mechanisms, like ammonia adsorption, runoff/leaching prevention, and lowering soil acidity as discussed in other chapters. For this analysis, a 7.2% increase in fertilizer-use efficiency is assumed based on average N, $P_2O_5$, and $K_2O$ application rates of 154, 65, and 94 kg/ha for a corn field (Roberts et al., 2010).

Environmental impacts of equipment manufacturing (harvesting equipment, pyrolysis facility, transportation vehicle) are not included in this analysis, but they are typically small compared to the lifetime emissions of the unit. Indirect land-use change and water consumption are also beyond the system boundary for this analysis.

## 4.2 Life cycle inventory

**Corn Stover Collection:** Decomposition of corn stover provides nutrients like N, P, and K to the field if left after the corn harvesting. Studies show that removing less than 33% of the available stover would not negatively affect soil quality. Nevertheless, supplementary synthetic fertilizer is applied to compensate for the nutrient loss.

In corn stover harvesting, the residuals (e.g., stalks, cobs, and husks) are gathered, windrowed for drying, baled, and moved to a roadside location. Different equipment pieces like windrow, baler, and tractor are used to perform individual tasks. Farm equipment pieces commonly use conventional diesel as the energy source. The total energy consumption for the corn stover collection stage is calculated as 188,000 BTU per ton of dry matter (Han et al., 2011). The chosen fertilizer replacement rates are 7.7 kg N, 2 kg P, and 12 kg K per dry ton stover harvested. The bale moisture content after field drying is 15%. Corn stover is transported to the conversion plant using a diesel-powered heavy-duty truck for 15 km.

**Corn Stover handling:** Stored corn stover at the plant goes through several preprocessing stages such as size reduction, screening, and drying. Size reduction processes using a knife mill require 32.3 MJ/tonne of diesel (Roberts et al., 2010). The feedstock is dried to 5% moisture content in a rotary steam drier with a fuel efficiency of 3.48 MJ/kg of water removed. A portion of the drying heat is supplied by pyrolysis heat, and the rest is met by downstream syngas combustion.

**Pyrolysis technology:** All pyrolysis process parameters are taken from Roberts et al. (2010) Corn stover is pyrolyzed at 450°C in a swept drum kiln through a continuous operation. The kiln is a highly airtight, horizontal cylindrical shell that is heated externally. Natural gas is used to provide the start-up heat estimated as 58 MJ/tonne feedstock. The feed stays in the kiln for several minutes and comes out as biochar. Heat produced during the pyrolysis reaction (an exothermic reaction) is used for feedstock drying. The volatile components of the feed convert into noncondensable gas called syngas, mixed with a small amount of tar, during the residence time. A portion of the syngas is burnt in a firebox to provide heat for the kiln. The syngas is a mixture of CO, $H_2$, $CH_4$, and other higher molecular weight hydrocarbon.

The remaining syngas is combusted with atmospheric air in a thermal oxidizer and produces high-temperature flue gas with a temperature ranging from 1000°C to 1100°C. The heat of the flue gas is transferred to the on-site application using an air duct heat exchanger with an efficiency of 75%. The heat for feedstock drying is also met by the flue gas heat. The characteristics of input corn stover and output biochar are shown in Table 21.1. Mass and energy balance for the pyrolysis unit is presented in Fig. 21.7.

**Biochar transportation to field and application:** A 15-km distance from plant to the corn field for biochar application is considered. A heavy-duty diesel truck is assumed as the means of transportation. Biochar is applied to the field using a conventional agricultural tractor. The diesel energy required for the tractor is 506 MJ/ha, and the biochar application rate is 5 ton/ha (Roberts et al., 2010).

**Avoided fertilizer production:** Biochar application decreases N, $P_2O_5$, and $K_2O$ fertilizer use by improving the fertilizer-use efficiency. This is mainly due to the reduction in N volatilization and leaching. For this study, a 7.2% increase in fertilizer-use efficiency is considered. It means that 7.2% of the required fertilizer production can be avoided. Table 21.2 describes all life cycle inventory

**Table 21.1 Corn stover and biochar properties.**

| Properties | Input<br>Corn stover | Output<br>Biochar |
|---|---|---|
| Mass (kg) | 1000 | 296 |
| Moisture content | 15% | – |
| C content (wt%) | 45[a] | 67.68 |
| Higher heating value (MJ/kg) | 16 | 30 |
| Stable C | | 80% |

[a] wt% of dry matter.

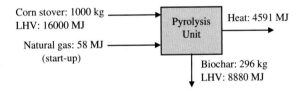

**Fig. 21.7** Mass and energy balance for the pyrolysis unit (heat losses, process heat for drying and pyrolysis are subtracted).

data of the considered process. Table 21.3 illustrates the energy input for different life cycle stages.

## 4.3 Life cycle impact assessment

LCIA shows the cumulative energy demand for the biochar production and application stages. Fig. 21.8 illustrates that corn stover collection consumes the highest (59%) portion of the total energy demand. The pyrolysis process is the second-highest energy-consuming stage from natural gas combustion. Preprocessing stage consumes approximately 9% of the total energy consumption, mainly for the size reduction process. Feedstock drying heat is provided by the process exhaust heat, requiring no additional energy supply. Transportation of feedstock to the facility and biochar to the field has the least energy demand calculated as 4% and 1%, respectively.

Environmental impacts are assessed at several midpoint impact categories: GWP, smog formation, acidification, and eutrophication. Emission inventories are calculated using Eq. (21.1).

Table 21.2 Inventory data for life cycle stages of 1 tonne of dry corn stover management.

| Life cycle stages | Amount | Unit | Purpose |
|---|---|---|---|
| Corn stover collection | | | |
| N fertilizer | 7.7 | kg | Compensation of N removal with stover |
| P fertilizer | 2 | kg | Compensation of P removal with stover |
| K fertilizer | 12 | kg | Compensation of K removal with stover |
| Diesel | 188,000 | BTU | Harvesting equipment |
| Transport | 15 | t-km | Corn stover transportation to plant |
| Preprocessing | | | |
| Diesel | 32.3 | MJ | Corn stover size reduction |
| Pyrolysis | | | |
| Natural gas combustion | 58 | MJ | Process start-up |
| Biochar application | | | |
| Transport | 4.4 | t-km | Biochar transportation to field |
| Diesel | 101.2 | MJ | Biochar soil application |
| Avoided processes | | | |
| Natural gas combustion | 4591 | MJ | Syngas heat production |
| N fertilizer production | 0.66 | kg | Increased fertilizer-use efficiency |
| P fertilizer production | 0.28 | kg | Increased fertilizer-use efficiency |
| K fertilizer production | 0.40 | kg | Increased fertilizer-use efficiency |

Table 21.3 Energy input for different processes.

| Processes | Amount (MJ) | Descriptions | References |
|---|---|---|---|
| Corn stover collection | 198.34 | Windrow, baling | Han et al. (2011) |
| Corn stover transportation | 14.61 | 15 km transport by diesel-powered truck 1 t-km requires 0.027 L diesel[a] | EcoInvent v2.0 |
| Preprocessing (Chipping) | 32.3 | Size reduction at 32.3 MJ/tonne feedstock. | Roberts et al. (2010) |
| Pyrolysis | 58 | Natural gas combustion for start-up | Roberts et al. (2010) |
| Biochar transportation | 4.33 | 15 km transport by diesel-powered truck 1 t-km requires 0.027 L diesel[a] | EcoInvent v2.0 |
| Biochar application | 29.95 | Diesel for tractor at 506 MJ/ha | Roberts et al. (2010) |

[a]LHV of diesel: 35.8 MJ.

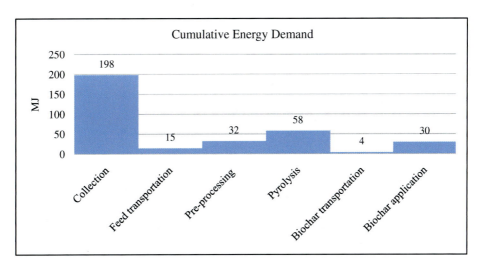

**Fig. 21.8** Cumulative energy demand of corn stover to biochar production and application.

Emission factors for various processes are obtained from the USLCI database, as given in Appendix. Impact indicators for the GWP impact category are calculated based on Eq. (21.2). However, the results of other impact categories are obtained using SimaPro due to the complexity in LCI flow assignments to different impact categories.

Fig. 21.9 shows the contribution of different biochar production stages in each category. Global warming is the most affected impact category followed by smog formation. Corn stover collection produces 57.85 kg $CO_{2\ eq}$/dry tonne of feedstock, resulting in 81% of the global warming. GHG emissions from fertilizer production and fuel used for harvesting are the main sources for these emissions. Pyrolysis is the second largest contributor (7%) for global warming followed by feedstock preprocessing (5%) and biochar application. Upstream extraction and on-site combustion of natural gas and diesel increase GWP. Transportation has the least significant influence calculated as 2% and 1%, for corn stover and biochar transportation, respectively.

The influence of corn stover collection is also dominant in all other impact categories. Preprocessing and pyrolysis equally contribute to acidification mainly due to the natural gas and diesel use. Eutrophication is not highly influenced by any production stages except for corn stover collection.

Effect of avoided processes: Table 21.4 shows the net environmental impacts considering the avoided emissions. Net GWP becomes negative mainly due to the avoidance of $CO_2$ emissions with the sequestered C. Avoided natural gas combustion and

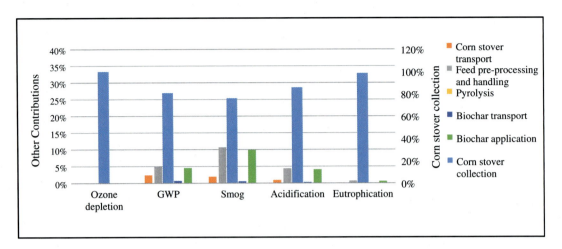

**Fig. 21.9** Contributions of different life cycle stages on different impact categories with corn stover collection (right axis) and other categories (left axis).

**Table 21.4** Net LCIA results for 1 tonne of corn stover management.

| Impact category | Unit | Value |
|---|---|---|
| Ozone depletion | kg CFC eq | 1.95431E-06 |
| Global warming | kg $CO_2$ eq | −834.32010 |
| Smog | kg $O_3$ eq | 5.08492 |
| Acidification | kg $SO_2$ eq | −1.86589 |
| Eutrophication | kg $SO_2$ eq | 0.24490 |

fertilizer production also add credits to acidification along with GWP, resulting in net negative emissions. However, other impact categories such as ozone depletion, smog, and eutrophication show net increases in emissions.

Alternative applications of biochar and syngas would show different environmental impacts. For example, biochar can be used as an adsorbent for tertiary wastewater treatment to remove nitrates and phosphates replacing fossil-based powdered activated carbon (PAC). Although biochar reduces negative environmental impacts associated with PAC and minimize wastewater treatment carbon footprint, the net benefit compared to soil application would be lower (Thompson et al., 2016). If syngas combustion replaces the heat produced by fossil coal combustion instead of natural gas, the environmental credit would be higher because coal extraction and combustion produces more environmental emissions than its natural gas counterpart.

## 4.4 Interpretation

LCIA provides substantial information to investigate the environmental impacts of a product or service. Based on the above results, we can identify several key factors to maximize the process performance while minimizing the environmental burden.

It is evident that corn stover collection is the most significant life cycle stage in terms of the environmental burden. For this case study, corn stover production is not included as the stover is considered as a waste byproduct of corn production. However, if there is any existing use of corn stover with significant economic or environmental benefit, diverting supply from that application could result in negative impacts. Generally, waste biomass derived biochar generates maximum environmental benefits.

For the case study, feedstock transportation has an insignificant contribution based on the assumption that the distance between the field and the plant is 15 km. For large conversion facilities, the collection of feedstocks from distributed sources would increase transportation emissions. Plant location should be optimized to minimize the feedstock transportation and storage costs and emissions.

The pyrolysis process contributes only 7% of the total GWP. At 450°C, the pyrolysis reaction becomes exothermic and self-sustaining. This process parameter determines the yield ratio of biochar, syngas, and bio-oil. When biochar production is the main target, a lower temperature is preferred. Below 350°C, the pyrolysis process becomes endothermic and requires additional energy supply. In that case, the net environmental impact would increase if the energy is supplied from nonrenewable sources. Therefore, changing the pyrolysis process parameters might change the process energy demand and their associated impacts.

Biochar soil application provides most of the environmental benefits due to carbon sequestration. If the biochar end-use changes, which is again dependent on various factors, the LCA results would be changed. For this analysis, we have assumed the use of synthetic fertilizer for the cornfield. If the conventional agricultural practice is to use organic fertilizers from other sources, the credit for this avoided emissions would change.

Due to the lack of data availability, other influences of biochar use, such as increase in crop yield, reduction in soil $N_2O$ emissions and nitrate leaching to streams, and long-term soil quality change, are not considered for this study. Other societal and economic sustainability metrics are not explored for this study that would provide a more comprehensive impact of biochar production.

## 5 LCA of biochar use in wastewater treatment

As an adsorbent of heavy metal and harmful micropollutants, biochar can play a significant role in wastewater treatment. The environmental benefits of using biochar in place of activated carbon are well documented in Alhashimi and Aktas (2017). Coal is the most common feedstock for activated carbon manufacturing. Therefore, replacing coal-based activated carbon with biochar reduces the environmental footprint of wastewater treatment plants. However, LCAs of biochar use for wastewater treatment are still scarce.

A comparison study shows that wood biochar offers significant environmental benefits relative to PAC in multiple LCI categories (Thompson et al., 2016). Fig. 21.10 shows a side-by-side comparison of the system boundaries for using PAC and wood biochar (from pine wood) for wastewater treatment. As shown, the two processes are similar and differ primarily on the starting feedstock. The life cycle of PAC includes PAC generation (coal extraction, hauling, energy inputs for conversion process), hauling, storage before dosing, removal, and landfilling. The biochar life cycle includes wood chip generation (forest harvesting, hauling,

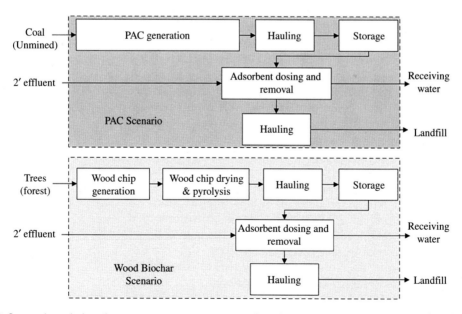

**Fig. 21.10** System boundaries of wastewater treatment process by using powdered activated carbon (PAC) and wood biochar adsorbent. Based on from Thompson, K.A., Shimabuku, K.K., Kearns, J.P., Knappe, D.R.U., Summers, R.S., Cook, S.M., 2016. Environmental comparison of biochar and activated carbon for tertiary wastewater treatment. Environ. Sci. Technol. 50, 11253–11262. https://doi.org/10.1021/acs.est.6b03239.

and chipping), conversion to biochar, hauling, storage before dosing, and landfilling. The end-of-life impacts of the adsorbents (e.g., emissions after disposal to landfill) are beyond the system boundary, and the common stages in both processes are excluded from the analysis.

> **Low- vs. moderate-capacity biochar:** The study categorizes wood biochar into two categories based on their adsorption capacity: low capacity and moderate capacity. Adsorption capacity can vary based on different pyrolysis parameters.
>
> **Low- vs. high-impact PAC:** High- and low-impact PAC is categorized based on the net environmental impacts for PAC generation. The net impact of PAC generation depends on several location-specific factors, such as hauling distance between coal mines and PAC production facilities, local electricity production mixes, etc.

Fig. 21.11 illustrates the comparative environmental impacts of using wood biochar and PAC for wastewater treatment in different impact categories. For two categories such as eutrophication and carcinogenics, wood biochar generates more environmental impacts than PAC. However, in four categories (ecotoxicity, acidification, ozone depletion, and fossil fuel depletion), biochar

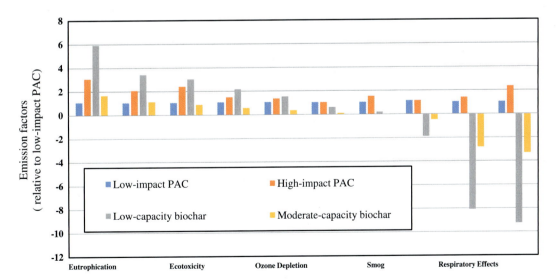

**Fig. 21.11** Comparative environmental impacts between powdered activated carbon (PAC) and wood biochar for wastewater treatment. (Descriptions of the legends are explained in the *shaded box*.) Data from Thompson, K.A., Shimabuku, K.K., Kearns, J.P., Knappe, D.R.U., Summers, R.S., Cook, S.M., 2016. Environmental comparison of biochar and activated carbon for tertiary wastewater treatment. Environ. Sci. Technol. 50, 11253–11262. https://doi.org/10.1021/acs.est.6b03239.

shows lower impact than PAC. The environmental impacts for wood biochar mainly comes from pyrolysis gas combustion (direct emission) and electricity consumption (indirect emission). The required amount of biochar is much higher than PAC for removing equal amount of pollutants, resulting in higher storage and disposal impacts. This impact increases proportionally with the decrease of biochar adsorption capacity.

Wood biochar shows negative emissions in four categories: smog, global warming, respiratory effects, and noncarcinogenics. The comparative benefits may differ for the biochar with lower adsorption capacity (low-capacity biochar). Nevertheless, the environmental benefits in global warming, respiratory effects, and noncarcinogenics remain positive (negative emissions). The reported higher environmental benefits of wood biochar come from the energy production during pyrolysis and the carbon sequestration with the spent biochar. Wood pyrolysis produces heat energy that replaces wood chips combustion for producing equivalent amount of energy. As mentioned earlier, the comparative benefit decreases for low-capacity wood biochar in all impact categories.

Another important point to note is that the impact of PAC is highly influenced by distance between the wastewater plant and coal resources, sources of electricity for the coal processing, and activation (high-impact PAC vs low-impact PAC). Typically, the hauling distance of wood biochar is significantly lower than PAC, resulting in lower impact in smog category because the forest residues are available near the wastewater treatment plant unlike coal mines.

## 6 Critical discussion

The environmental benefits of biochar production and utilization have been established, and the number of pilot and industrial-scale biochar facilities is growing globally. LCA is the only tool that can holistically evaluate the net benefit of this rapidly growing technology. Nevertheless, the wide variation in LCA methodologies makes this evaluation difficult for biochar production.

As discussed, biochar can be produced from nearly any carbon-containing feedstocks ranging from biomass to plastic materials. The sources of these feedstocks play a significant role in quantifying the net benefits for the biochar production supply chain. Feedstocks from waste sources such as agricultural residues, food waste, or the byproduct of crude oil refineries provide the maximum benefit in terms of LCA emissions. As the waste is an unavoidable output of the primary process, feedstock production emissions are not considered or partially accounted for. Nevertheless, if the waste is diverted from any conventional use that produces economic or environmental benefits, that should be considered. Converting waste into biochar avoids conventional waste management practices (such as landfilling, incineration) with low economic value and high environmental burden. Biochar production from purposely grown energy crops, on the other hand, is highly debated among researchers due to its possible impacts on food security or indirect land use change.

Geographic location can significantly affect the LCA results of biochar production. Feedstock availability and its current use impact the net environmental benefits. It can influence the environmental emissions related to feedstock collection and processing. The same product can be treated as waste in some places, and as a valuable product in other places. The end-use of biochar is also subject to regional differences. For example, biochar can be used for soil amendment in rural areas, whereas this use is limited in urban areas. Biochar application for energy use might be the best choice for nonagricultural locations, but this activity comes with lower environmental benefits.

The selection of avoided products or processes directly influences the LCA results, which are often controlled by the facility locations. The net benefit of replacing natural gas combustion with syngas is lower than replacing coal combustion because coal extraction and combustion generate more environmental emissions than natural gas. If the syngas is used for electricity production, the additional credits are higher in locations where electricity production is highly fossil fuel-based.

Most of the published biochar LCA focuses on the benefit of GWP reduction, mainly due to carbon sequestration. However, other environmental impacts should be considered. Some LCA studies caution about the negative impacts of biochar use on other impact categories like acidification, smog, and eutrophication. For agriculturally based feedstock (residues or energy crops), fertilizer application negatively impacts eutrophication and acidification. Electricity used for pyrolysis also has indirect emissions affecting these categories (Peters et al., 2015). Feedstock and biochar transportation add loads to fossil fuel depletion, eutrophication, and human toxicity, which are highly dependent on feedstock source to plant and plant to end-use site distances (Muñoz et al., 2017).

Most of the LCA studies did not consider socioeconomic impact of biochar production and applications (Homagain et al., 2016). The intervention of biochar application to the conventional fossil fuel-based product could create or displace existing jobs. The economics of biochar production depends on location-specific supply chains, conversion technologies, existing energy infrastructure, etc. A complete LCA should consider the volatility of socioeconomic impacts in the future for a specific project.

## 7 Conclusions

Biochar is a promising material for replacing coal-based activated carbon in wastewater treatment plants. One of its major advantages is a significant reduction in the environmental footprint of wastewater treatment compared to the conventional facilities. This chapter described how to use LCA to investigate and estimate the environmental impacts biochar use in wastewater treatment facilities.

The environmental impacts of biochar use in wastewater treatment plants depend on feedstock sources, biochar production technology, and biochar properties. Waste biomass achieves lower emissions than dedicated biomass in general. Biomass conversion technologies with high biochar yield and low energy inputs tend to have lower emissions per unit of biochar. Biochar with a high carbon content benefits from carbon sequestration credits. A comprehensive LCA can provide guidelines for reducing the environmental impacts of biochar production. Compared to coal-based activated carbon materials, biochar products almost always achieve a lower environmental footprint. However, the differences are minor in some cases, and more studies are needed to establish reliable metrics for biochar-use scenarios. There are limitations in published LCA studies that should be addressed in the future, including accounting for socioeconomic impacts, reducing variability in impact factors, and conducting long-term studies.

## Questions and problems

1. Briefly describe the four stages of life cycle analysis.
1. What is functional unit?
2. Briefly explain the system expansion method.
3. Describe cradle-to-grave and cradle-to-gate system boundaries using examples from biochar life cycle stages. Illustrate the example with diagrams.
4. Explain why the allocation method depends on the intent of the conversion process.

5. How is avoided burden calculated using the system expansion method?
6. What are the advantages of using midpoint impact categories over endpoint impact categories? List five midpoint and endpoint impact categories.
7. What are the possible off-site emissions for biochar production?
8. How does geographic location influence the environmental impacts of biochar?

# Appendix

Table A1 Emission factors for various unit processes from USLCI database.

| Unit processes | Impact categories | | | | |
| --- | --- | --- | --- | --- | --- |
| | Total kg $CFC_{eq}$ | Total kg $CO_{2\,eq}$ | Total kg $O_{3\,eq}$ | Total kg $SO_{2\,eq}$ | Total kg $N_{eq}$ |
| 1 m³ of natural gas combusted in industrial boiler | 1.73E-12 | 2.40E+00 | 5.20E-02 | 2.03E-02 | 2.02E-04 |
| 1 L of diesel combusted in industrial equipment | 1.30E-10 | 3.15E+00 | 1.39E+00 | 3.58E-02 | 2.61E-03 |
| 1 kWh of electricity from US grid mix | 1.41E-11 | 7.79E-01 | 5.23E-02 | 6.27E-03 | 9.50E-05 |
| 1 kg of nitrogen fertilizer produced at US plant | 7.73E-10 | 1.96E+00 | 2.62E-02 | 2.01E-02 | 1.98E-04 |
| 1 kg of phosphorus fertilizer produced at US plant | 4.73E-12 | 3.93E-01 | 4.35E-02 | 1.85E-02 | 1.60E-01 |
| 1 kg of potassium oxide fertilizer produced at plant | 1.74E-07 | 1.12E+00 | 7.83E-02 | 1.98E-02 | 1.58E-03 |
| 1 t-km transport, combination truck, diesel powered/US | 3.54E-12 | 9.27E-02 | 1.52E-02 | 4.68E-04 | 3.09E-05 |

# References

Alhashimi, H.A., Aktas, C.B., 2017. Life cycle environmental and economic performance of biochar compared with activated carbon: a meta-analysis. Resour. Conserv. Recycl. 118, 13–26. https://doi.org/10.1016/j.resconrec.2016.11.016.

Anderson, N.M., Bergman, R.D., Page-Dumroese, D.S., 2016. A supply chain approach to biochar systems. In: Biochar. Cambridge University Press, pp. 25–45, https://doi.org/10.1017/9781316337974.003.

Azzi, E.S., Karltun, E., Sundberg, C., 2019. Prospective life cycle assessment of large-scale biochar production and use for negative emissions in Stockholm. Environ. Sci. Technol. 53, 8466–8476. https://doi.org/10.1021/acs.est.9b01615.

Bare, J.C., Norris, G.A., Pennington, D.W., McKone, T., 2003. TRACI: the tool for the reduction and assessment of chemical and other environmental impacts. J. Ind. Ecol. 6, 49–78. https://doi.org/10.1162/108819802766269539.

Bridgwater, A., Czernik, S., Diebold, J., Meier, D., 1999. Fast Pyrolysis of Biomass: A Handbook. CPL Press.

Dutta, B., Raghavan, V., 2014. A life cycle assessment of environmental and economic balance of biochar systems in Quebec. Int. J. Energy Environ. Eng. 5, 1–11. https://doi.org/10.1007/s40095-014-0106-4.

Galgani, P., van der Voet, E., Korevaar, G., 2014. Composting, anaerobic digestion and biochar production in Ghana. Environmental-economic assessment in the context of voluntary carbon markets. Waste Manag. 34, 2454–2465. https://doi.org/10.1016/j.wasman.2014.07.027.

Guinee, J.B., 2002. Handbook on life cycle assessment operational guide to the ISO standards. Int. J. Life Cycle Assess. 7, 311–313. https://doi.org/10.1007/bf02978897.

Hammond, J., Shackley, S., Sohi, S., Brownsort, P., 2011. Prospective life cycle carbon abatement for pyrolysis biochar systems in the UK. Energy Policy 39, 2646–2655. https://doi.org/10.1016/j.enpol.2011.02.033.

Han, J., Elgowainy, A., Palou-Rivera, I., Dunn, J.B., Wang, M.Q., 2011. Well-to-Wheels Analysis of Fast Pyrolysis Pathways With GREET. Argonne Natl Lab, https://doi.org/10.2172/1036090.

Homagain, K., Shahi, C., Luckai, N., Sharma, M., 2016. Life cycle cost and economic assessment of biochar-based bioenergy production and biochar land application in Northwestern Ontario, Canada. For. Ecosyst. 3, 1–10. https://doi.org/10.1186/s40663-016-0081-8.

Huijbregts, M.A.J., Steinmann, Z.J.N., Elshout, P.M.F., Stam, G., Verones, F., Vieira, M., et al., 2017. ReCiPe2016: a harmonised life cycle impact assessment method at midpoint and endpoint level. Int. J. Life Cycle Assess. 22, 138–147. https://doi.org/10.1007/s11367-016-1246-y.

International Biochar Initiative, n.d. https://biochar-international.org/.

IPCC, 2013. Climate Change 2013: The Physical Science basis. Contribution of Working Group I to the Fifth Assessment Report of the Intergovernmental Panel on Climate Change.

ISO/IEC, n.d. Environmental Management-Life Cycle Assessment-Principles and Framework. ISO 14040:2006.

Jolliet, O., Margni, M., Charles, R., Humbert, S., Payet, J., Rebitzer, G., et al., 2003. IMPACT 2002+: a new life cycle impact assessment methodology. Int. J. Life Cycle Assess. 8, 324–330. https://doi.org/10.1007/BF02978505.

Klöpffer, W., Grahl, B., 2014. Life Cycle Assessment (LCA): A Guide to Best Practice., https://doi.org/10.1002/9783527655625.

Kong, S.H., Loh, S.K., Bachmann, R.T., Rahim, S.A., Salimon, J., 2014. Biochar from oil palm biomass: a review of its potential and challenges. Renew. Sustain. Energy Rev. 39, 729–739. https://doi.org/10.1016/j.rser.2014.07.107.

Llorach-Massana, P., Lopez-Capel, E., Peña, J., Rieradevall, J., Montero, J.I., Puy, N., 2017. Technical feasibility and carbon footprint of biochar co-production with tomato plant residue. Waste Manag. 67, 121–130. https://doi.org/10.1016/j.wasman.2017.05.021.

Mašek, O., Brownsort, P., Cross, A., Sohi, S., 2013. Influence of production conditions on the yield and environmental stability of biochar. Fuel 103, 151–155. https://doi.org/10.1016/j.fuel.2011.08.044. Elsevier.

Matuštík, J., Hnátková, T., Kočí, V., 2020. Life cycle assessment of biochar-to-soil systems: a review. J. Clean. Prod. 259, 120998. https://doi.org/10.1016/j.jclepro.2020.120998.

Muñoz, E., Curaqueo, G., Cea, M., Vera, L., Navia, R., 2017. Environmental hotspots in the life cycle of a biochar-soil system. J. Clean. Prod. 158, 1–7. https://doi.org/10.1016/j.jclepro.2017.04.163.

Oldfield, T.L., Sikirica, N., Mondini, C., López, G., Kuikman, P.J., Holden, N.M., 2018. Biochar, compost and biochar-compost blend as options to recover nutrients and sequester carbon. J. Environ. Manag. 218, 465–476. https://doi.org/10.1016/j.jenvman.2018.04.061.

Peters, J.F., Iribarren, D., Dufour, J., 2015. Biomass pyrolysis for biochar or energy applications? A life cycle assessment. Environ. Sci. Technol. 49, 5195–5202. https://doi.org/10.1021/es5060786.

Rajabi Hamedani, S., Kuppens, T., Malina, R., Bocci, E., Colantoni, A., Villarini, M., 2019. Life cycle assessment and environmental valuation of biochar production: two case studies in Belgium. Energies 12, 2166. https://doi.org/10.3390/en12112166.

Roberts, K.G., Gloy, B.A., Joseph, S., Scott, N.R., Lehmann, J., 2010. Life cycle assessment of biochar systems: estimating the energetic, economic, and climate change potential. Environ. Sci. Technol. 44, 827–833. https://doi.org/10.1021/es902266r.

Schmidt, H.-P., Wilson, K., 2014. The 55 uses of biochar. Biochar J. 2297-1114. www.biochar-journal.org/en/ct/2. (Accessed 12 August 2020).

Schmidt, H.P., Hagemann, N., Draper, K., Kammann, C., 2019. The use of biochar in animal feeding. PeerJ 2019. https://doi.org/10.7717/peerj.7373.

Sparrevik, M., Field, J.L., Martinsen, V., Breedveld, G.D., Cornelissen, G., 2013. Life cycle assessment to evaluate the environmental impact of biochar implementation in conservation agriculture in Zambia. Environ. Sci. Technol. 47, 1206–1215. https://doi.org/10.1021/es302720k.

Sparrevik, M., Lindhjem, H., Andria, V., Fet, A.M., Cornelissen, G., 2014. Environmental and socioeconomic impacts of utilizing waste for biochar in rural areas in Indonesia—a systems perspective. Environ. Sci. Technol. 48, 4664–4671. https://doi.org/10.1021/es405190q.

Thompson, K.A., Shimabuku, K.K., Kearns, J.P., Knappe, D.R.U., Summers, R.S., Cook, S.M., 2016. Environmental comparison of biochar and activated carbon for tertiary wastewater treatment. Environ. Sci. Technol. 50, 11253–11262. https://doi.org/10.1021/acs.est.6b03239.

# 22

# Optimizing biochar adsorption relative to activated carbon in water treatment

Matthew J. Bentley[a,*], Anthony M. Kennedy[b], and R. Scott Summers[a]

[a]Environmental Engineering, University of Colorado Boulder, Boulder, CO, United States. [b]Corona Environmental Consulting, LLC, Louisville, CO, United States
*Corresponding author: matthew.bentley@colorado.edu

## 1 Introduction

The evaluation of biochar as a more sustainable adsorbent in water treatment applications is a relatively new field. Fewer than 10 peer-reviewed publications were published annually on biochar water treatment prior to 2006, but the field has seen exponential growth with ~18,000 peer-reviewed publications in 2020, as shown in Fig. 22.1. While the field of biochar water treatment is fairly new, a much broader body of knowledge is available on carbonaceous sorbents, including activated carbon (AC). AC has a much larger and more established body of published, peer-reviewed literature, with nearly 100,000 peer-reviewed publications on water treatment in 2020, as shown in Fig. 22.1. AC studies can serve as a resource for biochar researchers due to the physicochemical and sorptive similarities with biochar. AC results can also be used as a baseline and help biochar researchers avoid common pitfalls and misunderstandings surrounding sorption and water treatment using carbonaceous materials.

Biochar for water treatment should be viewed as a research category within the broader body of work on carbonaceous materials: AC, charcoal, carbon black, graphene, carbon nanotubes, etc. (Pignatello et al., 2017). Among this group, AC has the greatest similarity to biochar due to its extensive internal porosity and comparable physicochemical properties that govern sorption processes in water treatment scenarios (Kah et al., 2017). Historically, many biochar studies have not directly compared novel laboratory-produced biochar results to those of commercially

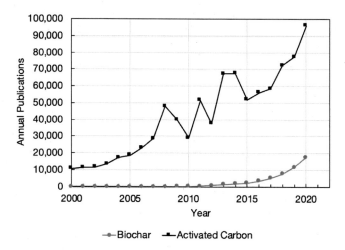

**Fig. 22.1** Number of annual publications with keywords "biochar water treatment" and "activated carbon water treatment" from 2000 through 2020 using Dimensions.ai database.

available ACs. This oversight has led to a widespread inability to evaluate or compare the relative performance of many biochars between studies and with commercially available alternatives (Gwenzi et al., 2017). In addition, the specific experimental conditions (e.g., sorbate, contact time, reactor configuration) of each study may further limit direct comparisons (Gwenzi et al., 2017), and even standard analytical techniques (e.g., BET surface area analysis, batch isotherms) exhibit inconsistencies across published studies (Tran et al., 2017).

AC is widely available throughout high-income countries (HICs) and is therefore important as a benchmark for sorption in water and wastewater treatment scenarios (Sontheimer et al., 1988; Summers et al., 2011). Direct performance comparisons between biochar and AC under experimental conditions representative of water and wastewater treatment are essential for researchers and engineers to evaluate the potential benefits of biochar over AC (Gwenzi et al., 2017). In those studies that have directly compared biochar and AC sorption performance under experimental conditions representative of water and wastewater treatment, conventional biochars tend to dramatically underperform commercial ACs, often by several orders of magnitude (Greiner et al., 2018; Kearns et al., 2014b; Kennedy et al., 2021; Shimabuku et al., 2016). The significant gap in sorption capacity between biochars and ACs could undermine the implementation of biochar in water treatment scenarios depending on the other factors that separate the two (e.g., environmental sustainability, cost, availability). In order to address the gap between biochar and AC performance in adsorption of organic compounds (i.e., dissolved organic matter (DOM), organic micropollutants (OMPs)), many enhancement and modification methods have been developed to improve biochar performance in water treatment scenarios (Ahmed et al., 2016; Rajapaksha et al., 2016).

Enhancing the performance of biochar has become a popular topic of research in recent years; however, many of the proposed improvement or modification methods can undermine the very advantages of biochar over AC (e.g., carbon negative, low cost, decentralized production, waste-to-resource) (Ahmed et al., 2016; Rajapaksha et al., 2016). Commonly, biochar has been viewed as an alternative to AC that does not undergo additional treatment (activation) steps, remaining a sort of precursor material with somewhat inferior sorbent properties that could be offset by its low cost and environmental impacts (Inyang and Dickenson, 2015; Keiluweit et al., 2010). Recent biochar improvement and modification methods blur the line between biochar and AC, and the benefits of such enhancements should be evaluated on a performance basis and against the impacts to the historical advantages of biochar over AC (Inyang and Dickenson, 2015). Some enhancement methods have been developed that show promise for producing biochars that can compete with commercial AC without significantly impacting biochar's favorable properties (Bentley and Summers, 2020; Greiner et al., 2018; Kearns et al., 2019).

Biochar should always be directly compared to commercial ACs under experimental conditions representative of water and wastewater treatment, and methods for improving biochar performance relative to AC should be developed that retain its advantageous attributes (e.g., low cost, carbon sequestration, decentralized production, waste-to-resource).

## 2 Non-adsorptive qualities

Before evaluating the relative performance of biochar and AC in water treatment, it is important to first outline and quantify the non-adsorptive qualities that might advantage biochar over AC in certain water and wastewater treatment applications.

### 2.1 Environmental impacts and sustainability

Biochar can be a more sustainable adsorbent than AC for water and wastewater treatment, but such a designation deserves further attention. The environmental impacts and sustainability of biochar and AC have been directly compared in wastewater treatment applications by Thompson et al. (2016). Using the 10 impact categories from EPA's Tool for the Reduction and Assessment of Chemical and Other Environmental Impacts (TRACI)—eutrophication, carcinogenics, ecotoxicity, acidification, ozone depletion, fossil fuel depletion, smog, global warming, respiratory effects, and non-carcinogenics—powdered activated carbon (PAC) and biochar were directly compared for sorption of organic micropollutants (OMPs) in a representative wastewater treatment scenario. Two different pine biochars were evaluated: a "low-capacity" biochar produced at full scale in a pyrolysis facility between 400 and 1200°C and a "moderate-capacity" biochar produced at 850 °C using a top-lit updraft gasifier, which allows for limited oxygen in the pyrolysis chamber (see Section 4.1 for more details). The use of wastewater (WW) biosolids biochar was also evaluated, and these biochars were all compared to a commercial PAC produced under a high-impact case and low-impact case.

Compared to the high-impact PAC case, moderate-capacity pine biochar resulted in lower environmental impacts, shown in Fig. 22.2, in every category and resulted in environmental benefits in four categories: smog, global warming, respiratory effects, and non-carcinogenics. Low-impact PAC had lower environmental impacts in only two categories—eutrophication and carcinogenics—compared to moderate-capacity biochar, highlighting the general environmental sustainability of biochars over even the lowest impact ACs across most environmental impact categories. Low-capacity wood biochar required much greater adsorbent doses to achieve equivalent OMP removal in WW treatment, and therefore exhibited higher environmental impacts than both PAC cases in five categories. Preparation of biochar feedstocks is an important step to consider in environmental impact analysis, as the biosolids biochar case exhibited higher environmental impacts due to drying the biosolids prior to pyrolysis (Thompson et al., 2016). These results point to the importance of producing high-capacity biochars that can compete with AC in order to retain the environmental benefits that are often assumed to be inherent in biochars. Environmental sustainability cannot be assumed for all biochars, but high-capacity biochars provide a significant opportunity to reduce environmental impacts associated with water and wastewater treatment. Environmental sustainability advantages have also been observed for biochar over AC in similar life cycle assessments focused on adsorption (Alhashimi and Aktas, 2017; Moreira et al., 2017).

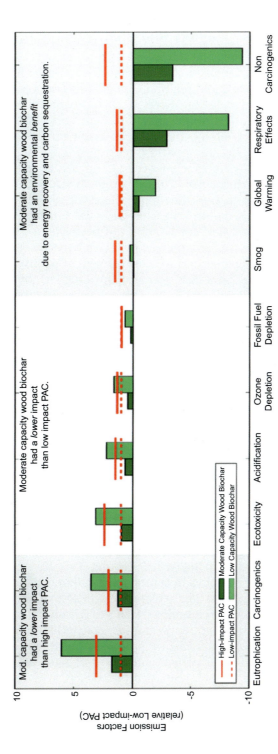

**Fig. 22.2** Comparison of relative environmental impact across 11 EPA TRACI categories for low-impact PAC *(red dashed lines)*, high-impact PAC *(red solid lines)*, moderate-capacity wood biochar *(dark green bars)*, and low-capacity wood biochar *(light green bars)*. Reproduced with permission from Thompson, K.A., Shimabuku, K.K., Kearns, J.P., Knappe, D.R.U., Summers, R.S., Cook, S.M., 2016. Environmental comparison of biochar and activated carbon for tertiary wastewater treatment. Environ. Sci. Technol. 50, 11253–11262. https://doi.org/10.1021/acs.est.6b03239. Copyright 2016 American Chemical Society.

## 2.2 Availability and cost

Another chief advantage of biochar over AC can be its widespread availability and lower cost. As a newer technology in water and wastewater treatment, biochar has not yet experienced economies of scale that lowers AC cost. Despite the limited scale of biochar production compared to commercial AC, commonly cited estimates of biochar and AC costs range from $50–1200 and $2900–3300 per metric ton, respectively, highlighting the cost advantage of biochar over AC on a mass basis (Kearns et al., 2014b; Meyer et al., 2011; Thompson et al., 2016). This wide range of costs creates uncertainty, however, about the cost benefits of using biochar in water and wastewater treatment scenarios, particularly when considering the performance gap between most biochars and commercial ACs. Using the lower and upper bounds of the biochar and AC cost ranges, respectively, provided above (i.e., the "best case scenario" for biochar), biochar doses or use rates would need to be within a factor of 66 compared to AC to be economically advantageous. Using values from the middle of the biochar and AC ranges, biochar doses or use rates would need to be within a factor of 5 compared to AC to be economically viable. This is certainly achievable, though the feedstock and production methods (primarily pyrolysis temperature) of biochars must be optimized to achieve this relative level of performance. When evaluating the use of biochar in water and wastewater treatment, direct comparisons with commercial AC are essential to gauge its economic viability.

Besides cost, another key advantage to biochar over AC is the global availability of biochar feedstocks and production technologies. Particularly in low- and middle-income countries (LMICs), AC can be unavailable or prohibitively expensive. Biochar can be inexpensively produced in LMICs using locally available, low-cost materials such as 200-L steel drums and waste agricultural biomass through methods similar to the Top-Lit UpDraft (TLUD) gasifier shown in Fig. 22.3 (Kearns et al., 2014a). Biochar can easily be produced in a decentralized fashion, while AC requires industrial processes (e.g., steam activation) that are only cost-effective at large-scale centralized facilities (Gwenzi et al., 2017). In alternative water treatment scenarios, such as landfill leachate (LL) treatment, biochars can be produced onsite using readily available waste sources, which could further motivate the application of biochar in such scenarios (Bentley et al., 2021). Even if biochar may be more expensive than AC in certain scenarios, other factors, such as local availability and lower environmental impacts, may favor its use over commercial AC products.

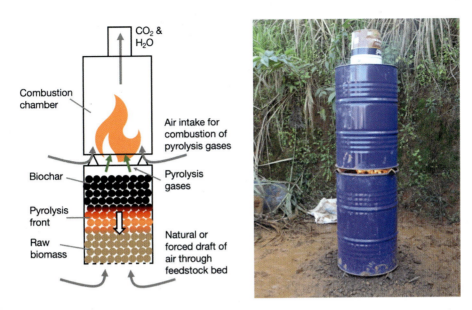

**Fig. 22.3** Schematic (left) and photo (right) of a Top-Lit UpDraft (TLUD) biochar gasifier feedstock is placed in the bottom steel drum, and a fire is started at the top, pyrolyzing biochar in a downward trending pyrolysis front as pyrolysis gases are released from the heated feedstock and combusted in the upper steel drum. Holes are cut or drilled in the bottom drum to allow for limited air flow through the feedstock bed, allowing for co-pyrolysis thermal air activation (CPTA—see Section 4.1), which induces a draft that allows for complete combustion of pyrolysis gases with little to no smoke. Additional air is provided to the combustion chamber through the gap between the two barrels. Once the pyrolysis front has moved through the entire feedstock bed, converting all of the biomass into biochar, the bed is quenched with water to prevent combustion of the biochar (Kearns et al., 2014a).

## 2.3 Production techniques

AC is produced through, first, pyrolysis/carbonization of an organic material (commonly coal, coconut, or wood) followed by a physical or chemical activation step (Marsh and Rodríguez-Reinoso, 2006). Biochar is typically produced from waste biomass (e.g., beetle-kill pine, WW biosolids, grasses, sugarcane bagasse) through a single-step pyrolysis process, though recent advances have led to the development of pre-, mid-, and postpyrolysis enhancement methods to optimize biochar sorption of OMPs (Ahmed et al., 2016; Bentley and Summers, 2020; Greiner et al., 2018; Kearns et al., 2019; Rajapaksha et al., 2016).

The most important factor in the pyrolysis process of biochars is the highest treatment temperature (HTT), which greatly influences biochar yield, porosity, and sorption of organic compounds (i.e., OMPs, DOM). Generally, higher HTTs result in lower biochar yields and more porous biochars that perform better as OMP adsorbents in water treatment (Ahmad et al., 2012; Chen et al., 2019; Shimabuku et al., 2016). Some studies have shown high sorption from low HTT biochars of certain OMPs, but these studies are often limited to sorption in deionized (DI) water and OMP concentrations orders of magnitude above typical environmental concentrations (i.e., tens to hundreds of ng/L) (Chen and Chen, 2009; Tan et al., 2015). As HTT increases in pyrolysis processes, porosity increases as cellulose and hemicellulose volatilize, and carbon present in raw biomass is mobilized, forming disordered/defective graphene sheets that form the carbon skeletal structure and pore walls in the resulting material (Keiluweit et al., 2010; Marsh and Rodríguez-Reinoso, 2006; Suhas et al., 2007). Amorphous carbon decreases with increasing HTT, forming turbostratic crystallites and increasing porosity up to and above 700°C (Keiluweit et al., 2010). Biochar produced at high HTTs (e.g., ~850°C) tend to exhibit higher surface area and OMP sorption capacities in water treatment (Kearns et al., 2014b; Lian and Xing, 2017; Shimabuku et al., 2016). While high HTTs are essential for producing high-performing biochars for water and wastewater treatment applications, increased time spent at HTT has been shown to further enhance OMP adsorption capacity and surface area (Kearns et al., 2014b; Shimabuku et al., 2016). The internal pore structure, particularly the surface area and pore size distribution, of a biochar tends to govern its OMP sorption capacity (Li et al., 2003a; Pelekani and Snoeyink, 1999), and enhancement methods that alter or increase the internal pore structure and surface area of biochar could improve their performance relative to AC (Rajapaksha et al., 2016; Sizmur et al., 2017).

Feedstock-inherent effects are also very important in biochar production, as biomass materials exhibit a wide range of inherent pore structures and chemical compositions (Liu et al., 2015; Wang et al., 2017). Feedstock-inherent mineral content, or ash, has been viewed as deleterious to pore development and OMP sorption capacity in biochar, likely due to (i) catalytic effects on carbon structures during pyrolysis and (ii) pore blockage by precipitated metal oxides and silicates postpyrolysis (Rodriguez Correa et al., 2017; Sun et al., 2013; Xu et al., 2017), though some studies indicate possible enhancement of biochar properties as a result of inherent ash content (Shimabuku et al., 2016). Biochars produced from materials with a highly developed inherent pore structure, such as pine and other woody biomass, tend to exhibit higher porosity and OMP sorption capacity than those produced from materials with a lower inherent pore structure, such as WW biosolids, essentially comprised of microbes and inert solids from WW treatment (Bentley and Summers, 2020; Shimabuku et al.,

2016). Some evidence is emerging that certain low-cost biochar enhancement methods may somewhat reduce the importance of feedstock selection by universally improving biochars across a broad range of feedstocks (Bentley et al., 2021). Feedstock and pyrolysis condition selection are important factors that govern the performance and sustainability of biochars in water and wastewater treatment scenarios, but more research is needed to identify optimal production conditions that maximize performance without negatively impacting biochar's potential environmental, availability, or economic advantages over AC.

## 3 Adsorptive performance comparison in water and wastewater treatment

Before comparing biochar and AC performance in water treatment, a few topics must be introduced to provide necessary background on adsorption processes that inform the interpretation of the results shared in the following sections. The performance of adsorbents could be, and often are throughout the literature, compared under many diverse experimental conditions. Reactor (e.g., contacting system) design, competing adsorbates (e.g., DOM), contact time, and treatment targets all significantly impact the relative performance of biochars and AC and must be considered when comparing results within and across studies.

Reactor design—that is, the use of completely mixed batch reactors (CMBR), or continuous-flow stirred tank reactors (CFSTR), that exhibit simultaneous adsorption of target OMPs and competing adsorbates, or flow-through column adsorbers that exhibit the adsorption of competing adsorbates deeper in the bed, termed preloading—significantly impacts the performance of carbonaceous adsorbents. The comparison of biochar and AC performance varies dramatically between these two common contacting systems due to fouling effects of ubiquitous DOM and other competing adsorbates with target OMPs (Greiner et al., 2018; Shimabuku et al., 2016; Ulrich et al., 2015). Fig. 22.4 depicts the simplified schematics of these two contacting system types.

Competing adsorbates such as DOM or other organic constituents can cause direct site competition and pore blockage/constriction effects—depicted in Fig. 22.5—that impact OMP removal by biochar and AC (Kearns et al., 2020; Kennedy and Summers, 2015; Li et al., 2003a, b). The degree to which an adsorbent experiences competition and pore blockage depends on the internal pore structure and pore size distribution of the adsorbent and can vary greatly between and within biochars and ACs (Li

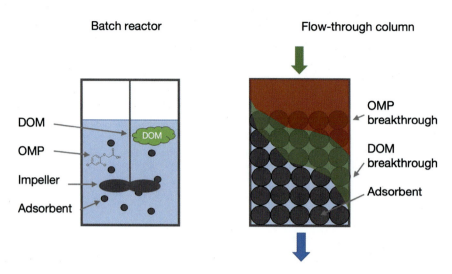

**Fig. 22.4** Simplified schematic of the two most common adsorbent reactor designs, (A) batch reactor and (B) flow-through column adsorber. Not to scale.

et al., 2003a, b; Pelekani and Snoeyink, 1999). Adsorption studies that do not include competing adsorbates, namely, background DOM, are not representative of water or wastewater treatment processes, and the results of such studies should be used with great caution.

Contact time, which is indicative of how close to equilibrium or steady state a treatment system reaches, can affect the relative performance of biochar and AC due to differences in internal pore structure and diffusion effects—as is shown in Fig. 22.6 for the uptake of 2,4-dichlorophenoxyacetic acid (2,4-D) (Bentley and Summers, 2020) and also shown by others for a variety of OMPs (Chen et al., 2012; Kennedy et al., 2021; Xiao and Pignatello, 2015). Equilibrium sorption experiments can be very helpful for modeling, identifying adsorption mechanisms, and evaluating the total sorption capacity of adsorbents, but OMP removal at equilibrium may not be representative of performance in water and wastewater treatment scenarios with shorter contact times (Shimabuku et al., 2016; Sontheimer et al., 1988; Tran et al., 2017). Adsorption studies should use contact times that are representative of full-scale water treatment processes to evaluate performance at relevant time scales as well as evaluating total sorption capacity.

Finally, there are many different water quality measures by which one could compare biochar and AC. Removal of DOM as measured by dissolved organic carbon (DOC) or ultraviolet

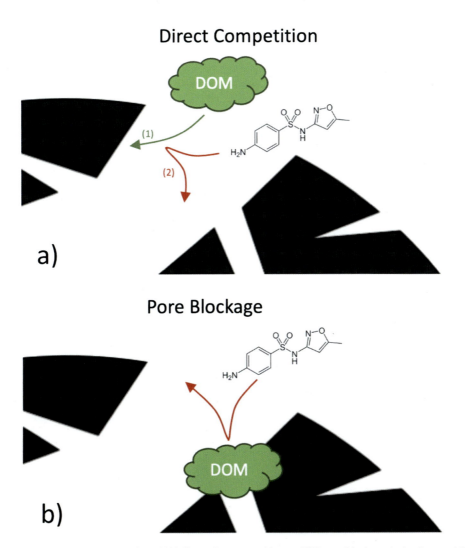

**Fig. 22.5** Simplified visual representation of (A) direct site competition and (B) pore blockage by background DOM with an OMP, such as sulfamethoxazole, an antibiotic. Numbers in (A) denote the order of adsorption, with DOM adsorbing first and taking an available adsorption site away from the SMX molecule.

absorbance at 254 nm (UVA$_{254}$) could be relevant for controlling disinfection byproduct (DBP) formation in water treatment (Bond et al., 2011; Shimabuku et al., 2017; Summers et al., 2020), or OMP removal could be targeted in water or wastewater treatment (Inyang and Dickenson, 2015; Kearns et al., 2015; Shimabuku et al., 2016). The chemical structure and properties of each targeted OMP will govern its adsorbability and removal in water treatment processes, which complicates the comparison

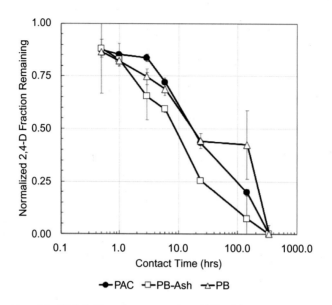

**Fig. 22.6** Relative removal (y-axis) of 2,4-dichlorophenoxyacetic acid (2,4-D) by pine biochar (PB), ash-pretreated pine biochar (PB-Ash), and PAC varies by adsorbent characteristics; therefore, contact time (x-axis) must be carefully selected in order to ensure that the results of experimental studies are representative of full-scale treatment processes. Reprinted from Bentley, M.J., Summers, R.S., 2020. Ash pretreatment of pine and biosolids produces biochars with enhanced capacity for organic micropollutant removal from surface water, wastewater, and stormwater. Environ. Sci. Water Res. Technol. 6, 635–644. https://doi.org/10.1039/C5EW00120J with permission from the Royal Society of Chemistry.

of biochar and AC (Kearns et al., 2020; Kennedy et al., 2015). An example of OMP-specific breakthrough, or increased effluent concentrations with increased water throughput from gradual exhaustion of adsorption capacity, in a flow-through granular activated carbon (GAC) adsorber is shown in Fig. 22.7, taken from Kennedy et al. (2015). In that study, OMP breakthrough was found to vary based on OMP chemical properties (e.g., log D, Abraham parameters polarizability ($S$), and McGowan molecular volume ($V$)) in addition to background DOM levels as measured by DOC (Kennedy et al., 2015). Typically, researchers should select OMPs that are conservative, meaning that they are more challenging to remove, to evaluate the performance of biochar and AC for the type of compounds that will control the adsorbent dose or filter lifetimes in water treatment processes (Kennedy et al., 2015). The impact of OMP initial concentration on biochar performance has been neglected in many biochar studies, and future studies should utilize initial OMP concentrations that are as close as possible to those found in environmental matrices (i.e., <100 µg/L) to ensure that the observed behavior accurately represents results at full scale (Shimabuku et al., 2016).

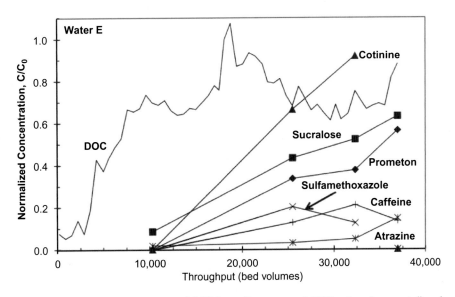

**Fig. 22.7** Full-scale breakthrough performance of GAC for a diverse set of OMPs at environmentally relevant concentrations relative to background DOM as measured by DOC. Reproduced with permission from Kennedy, A.M., Reinert, A.M., Knappe, D.R.U., Ferrer, I., Summers, R.S., 2015. Full- and pilot-scale GAC adsorption of organic micropollutants. Water Res. 68, 238–248. https://doi.org/10.1016/j.watres.2014.10.010. https://www.sciencedirect.com/science/article/abs/pii/S0043135414007052.

## 3.1 Adsorption in batch reactors

In CMBRs or CFSTRs, adsorbents are added to the water with the target adsorbate and mixed for a set time, or residence/contact time, before being removed from water by, most commonly, sedimentation or filtration. The contact time is typically only 30 min to a couple of hours; as such, the adsorbent particle size is small, <0.1 mm, so that the kinetics of uptake are faster and a large part of the adsorption capacity can be utilized in that time frame. In these reactors, OMPs and DOM are simultaneously exposed to the adsorbent and removal of OMPs, DOM (DOC and $UVA_{254}$) increases with contact time, although removal rate will decrease over time. Adsorbents are often applied in this way for seasonal (i.e., not year-round) contamination events, in response to a contaminant spill, or where fixed-bed infrastructure does not exist (Cook et al., 2001; Sontheimer et al., 1988). Because adsorbents are simultaneously exposed to DOM and OMPs, smaller OMPs are able to diffuse more quickly than larger DOM into the internal pore structure. Faster OMP diffusion rates during simultaneous loading result in higher OMP removal rates by biochars and AC relative to DOM-preloaded adsorbents, which replicate the deleterious effect of early DOM breakthrough in flow-through columns discussed in the next section (Li et al., 2003b; Quinlivan et al., 2005).

### 3.1.1 Removal of DOM in batch reactors

Biochar and AC removal of DOM in graywater were compared by Thompson et al. (2020) in batch reactors at 2-h contact times. Laboratory-produced biochars were compared with biochars produced at full scale using industrial pyrolysis processes and with a low-cost 200-L drum gasifier. Laboratory biochar DOM removal increased slightly with increasing pyrolysis temperature (HTT) from 800 to 900 °C, and full-scale bone biochar outperformed full-scale wood biochar. The gasifier-produced wood biochar performed significantly better than all the other evaluated biochars and removed 50% of DOM in 2 h compared to roughly 20%–30% for the other biochars. Similar results were found for $UVA_{254}$. When comparing the gasifier-produced biochar to a commercial AC, biochar removed about half as much DOM as AC at each respective dose. Pretreatments (e.g., coagulation, biodegradation) were evaluated to reduce the adsorbent doses required to achieve graywater reuse treatment targets, and the gasifier-produced biochar presented an attractive alternative to AC in graywater treatment due to its potentially lower cost.

Kennedy et al. (2021) compared an ash-pretreated pine biochar and a bituminous coal-based powdered AC (PAC) for the removal of DOC from a surface water (SW) and wastewater (WW) using CMBRs. While PAC removed ~76% of DOC in SW and WW at 200 mg/L, ash-pretreated pine biochar only removed ~26% of DOC at 200 mg/L after 60 min of contact time. At a dose of 50 mg/L, increasing contact time from 10 min to 2 h continuously increased DOC removal by PAC, but DOC removal by biochar plateaued after 10 min. Similar trends were observed for $UVA_{254}$. As such, this study highlighted the limitations of a typical pine biochar for removing DOM (DOC and $UVA_{254}$) in CMBRs.

### 3.1.2 Removal of OMPs in batch reactors

Kearns et al. (2014b) evaluated the use of biochar to remove a conservative OMP, 2,4-D, in equilibrium batch experiments in deionized (DI) water. Biochars produced at high HTT and longer pyrolysis times exhibited increased 2,4-D sorption capacity, but feedstock-specific effects were also notable, with wood chips performing the best among the tested materials. Performance of biochars produced at field and laboratory scales increased with HTT, and the high-temperature laboratory chars (wood and bamboo) and field-collected char (basudha) exhibited sorption capacities in DI water that are within the range of commercial AC. This study was limited to equilibrium sorption capacity without competing adsorbates, but the results point to the potential viability of biochar as an AC alternative for OMP removal in low-cost water treatment.

In a follow-up to this study, wood biochars and AC were compared in equilibrium batch tests in a representative surface water for the removal of trihalomethanes (THMs), warfarin (WFN), and methylisoborneol (MIB) (Kearns et al., 2015). THMs are a class of regulated disinfection byproducts (DBPs), while WFN and MIB are commonly studied OMPs in the categories of pharmaceuticals and taste and odor compounds, respectively. Biochars were produced at various HTTs using various field-scale methodologies—retort, cookstove, and 200 L drum oven—and evaluated under equilibrium batch conditions with a competing background matrix of surface water DOM. THM removal by high HTT (~900 °C) biochar produced using the drum oven gasifier was comparable to AC, while intermediate and low HTT biochars produced using the retort and cookstove exhibited capacities 3–25 times lower than the gasifier biochar. The 900°C cookstove biochar achieved equivalent sorption capacities of WFN and MIB, but lower-temperature chars required adsorbent doses 10–100 times higher than AC to achieve equivalent WFN/MIB removal. DOM and chlorine reduced the sorption capacity of biochars and AC, highlighting the importance of using representative treatment conditions in biochar studies. While low HTT biochars dramatically underperform commercial AC, high HTT biochars produced using cookstoves or drum oven gasifiers are very competitive with AC for the removal of DBPs and OMPs.

Shimabuku et al. (2016) evaluated a wide range of biochars for the removal of a conservative OMP—sulfamethoxazole (SMX)—from DI water, SW, stormwater (StW), and WW effluent in batch reactors. OMP sorption for a wide range of biochars was compared to AC at 60-min and 7-day contact times, representing kinetic and apparent-equilibrium time scales, respectively. The performance of gasifier-produced biochar, produced using the same 200-L drum oven gasifier described above, matched performance of AC in DI water and performed within a factor of 2–3 in SW, WW, and StW. Other higher-performing biochars in this study performed within a factor of 5, well within the range of potential economic viability (see Section 4.1). Biochar and AC rankings were the same across the diverse background matrices in this study, but the relative performance varied by background matrix, with WW and StW exerting a larger negative impact on OMP sorption than SW. SMX sorption by "low-ash" biochars in this study was negatively correlated with H:C ratio, indicating that more aromatic biochars exhibited higher SMX uptake. Lower H:C ratios are typically observed in higher HTT biochars, indicating, again, the importance of producing biochars at high pyrolysis temperatures for water treatment applications. Sorption kinetics of biochars and AC were found to be comparable within a 4-h contact time, despite significant differences in physicochemical properties and pore structure across adsorbents.

Bentley and Summers (2020) evaluated the performance of biochars produced from pine pellets and wastewater biosolids for the removal of SMX and 2,4-D in SW, WW, and StW at 3-h and 7-day contact times. Untreated biochars produced in a laboratory muffle furnace at 800 °C underperformed AC by over an order of magnitude, and WW biosolids biochar underperformed AC by around a factor of 100 across SW, WW, and StW background matrices. OMP sorption was linked to biochar surface area, particularly non-micropore surface area, due to the fouling effects of background DOM. DOM type impacted biochar performance more significantly than DOM concentration, which is further discussed in Section 3.3. AC exhibited slower kinetics than biochars in this study, indicating a potential advantage for the use of biochars in treatment processes with lower contact times. $UVA_{254}$ removal was found to be a conservative surrogate measure for OMP removal.

Biochars from a diverse set of organic waste feedstocks comprising the organic fraction of municipal solid waste were evaluated for the removal of nitrobenzene (NB) and 2,4-D in LL (Bentley et al., 2021). Biochar performance varied dramatically with feedstock type, and even the high HTT (850°C) biochars had much lower performance relative to that of AC sorption in 3-h batch tests. LL is a complex and challenging background matrix with high concentrations of competing DOM fractions that lower OMP removal rates of adsorbents. Of the feedstocks evaluated, the highest performing were office paper, pine needles, and peanut shells, which all performed within a factor of 10 of AC. Pine wood, grass, orange peels, and coffee were the more poorly performing adsorbents in LL, requiring doses over 10 times higher than AC to achieve equivalent levels of NB removal. Office paper biochar performed within a factor of 7 of AC in 3-h batch tests, and it, as well as three other representative feedstocks, was further improved by two emerging biochar enhancement methods (Sections 4.2 and 4.3). Despite the wide range of feedstocks evaluated in this study, no feedstock characteristics were found to correlate well with the resulting biochar's performance. The most promising feedstock characteristic was cellulose content, which showed a weak positive trend with micropore surface area and NB sorption in resulting biochars.

In certain cases, biochar and AC may be complementary adsorbents that can be used together to achieve high levels of DOM, $UVA_{254}$, and OMP removal across diverse background matrices. As previously discussed in relation to DOM removal, Kennedy et al. (2021) also evaluated the ash-pretreated biochar from Bentley and Summers (2020) (discussed further in Section 4.3) for OMP removal across a suite of common OMPs in CMBRs at representative contact times (<2 h). This enhanced biochar was able to match PAC performance for a number of OMPs, but vastly underperformed PAC for typically weakly adsorbing OMPs. Biochar adsorption of OMPs was also more negatively impacted by DOM than by PAC. Interestingly, the relative affinity for the suite of tested OMPs was opposite for biochar and PAC, indicating possible synergistic qualities of this biochar and PAC. Added simultaneously, DOM and OMP removal increased relative to the use of each adsorbent alone, indicating a potential to lower costs associated with commercial PAC by using both adsorbents in combination.

Overall, biochars can be effectively applied in batch scenarios to remove OMPs. Typically, laboratory- and full-scale biochars exhibit far lower performance than AC for the removal of the above-mentioned treatment targets in the presence of background DOM, but some production methods—such as the 200-L drum oven gasifier further discussed in Section 4.1—can produce biochars with comparable performance in batch sorption scenarios. While typical laboratory- and full-scale biochars tend to underperform AC in water treatment, many are still within the range of potential economic viability and present significant opportunity to reduce environmental impacts associated with water treatment processes. Low-cost enhancement processes could improve the viability and competitiveness of biochars with commercial AC, and a selection of such enhancements is presented in Section 4.

## 3.2 Adsorption in flow-through column adsorbers

When moving from the use of batch reactors (CMBRs or CFSTRs) (Fig. 22.6) to flow-through column adsorbers (Fig. 22.7) in biochar and AC sorption processes, it is essential to understand the different impacts that background DOM can have on OMP sorption capacity and adsorbent performance. While batch reactors are characterized by simultaneous adsorption of DOM and OMPs, leading primarily to the direct competition DOM fouling effect, flow-through columns allow for DOM to preload the adsorbent deeper in the bed because DOM mass transfer zones move through adsorption columns faster than OMP mass transfer zones (Fig. 22.4). The spatial and temporal effects of DOM in column adsorbers lead to both direct competition and pore blockage/constriction in biochars and AC (Kearns et al., 2020; Kennedy and Summers, 2015; Li et al., 2003b). Biochars often lack a broad pore size distribution, exhibiting most of their porosity in small micropores; therefore, they are often more affected by pore blockage fouling than ACs (Kearns et al., 2020; Li et al., 2003a; Pelekani and Snoeyink, 1999; Shimabuku et al., 2016; Ulrich et al., 2015). This increased susceptibility to DOM fouling effects often results in biochars being less competitive with commercial AC in column adsorbers compared to batch reactors. Modeling of biochar column breakthrough based on batch results was outside of the scope of this chapter, but it has been the focus of multiple studies and ongoing works (Bentley, 2020; Kearns et al., 2020; Kennedy et al., 2021). Generally, flow-through column adsorber data for biochars are lacking in the literature and should be prioritized in future work.

### 3.2.1 Removal of DOM in flow-through columns

Breakthrough of DOM, measured as DOC and $UVA_{254}$, is typically faster for biochar than AC in flow-through columns. AC typically has a more developed pore structure, particularly in mesopores, that enhances DOM uptake and reduces the effects of DOM fouling relative to more microporous biochars (Ding et al., 2008; Li et al., 2003a, b; Pelekani and Snoeyink, 1999). Greiner et al. (2018) observed $UVA_{254}$ breakthrough 34 times faster for conventional anaerobic pyrolysis (CAP) pine biochar produced at 850°C compared to GAC. Kearns et al. (2021) observed $UVA_{254}$ breakthrough in co-pyrolysis thermal air (CPTA) activation hardwood pellet biochar columns 8.5 and 6.5 times faster in SW and WW, respectively, compared to GAC. Overall, data on DOM breakthrough in biochar columns are severely limited and should be evaluated in future studies across DOM types using biochars produced with a variety of feedstocks and pyrolysis conditions.

### 3.2.2 Removal of OMPs in flow-through columns

Greiner et al. (2018) evaluated fresh and regenerated biochar in flow-through column tests for the removal of SMX and $UVA_{254}$ in a representative SW. Thermal regeneration was used to restore adsorption capacity by heating the exhausted biochar to 600°C with a two-hour hold time in oxygen-limited covered crucibles. SMX broke through the fresh biochar column 40 times faster than GAC, which is typical for biochars in columns due to the significant pore blockage fouling effects of background DOM along with biochars' typically limited internal pore structure. This study also evaluated regenerated and double-heated biochars, which is further discussed in Section 4.2.

The previously mentioned study (Kennedy et al., 2021) found that a synergistic effect between biochar and PAC also modeled OMP breakthrough for each adsorbent in hypothetical flow-through GAC columns. Even the ash-pretreated biochar used in the study underperformed GAC—as measured by relative adsorbent-use rate—by a factor of up to 38 and 264 in SW and WW, respectively. WW DOM more significantly impacted OMP breakthrough using biochar than GAC, and the performance of biochar was much lower than that in batch tests, indicating its increased susceptibility to DOM fouling. The OMPs selected for breakthrough modeling were the weakest adsorbing compounds for biochar and the strongest adsorbing compounds for GAC in batch tests, so the predicted difference in performance between biochar and GAC should be less than the values reported above for the other OMPs evaluated in this study. Nevertheless, this study highlights the large gap between biochar and GAC performance in column sorption applications due to the impact of background DOM. By theoretically combining biochar and GAC in column adsorbers, biochar-use rates were decreased significantly compared to using biochar alone.

Two companion studies by Kearns et al. (2020, 2021) evaluated the breakthrough of OMPs in SW, WW, and LL for a range of biochars produced under CAP and CPTA conditions. Biochars produced, regardless of temperature, using CAP drastically underperformed GAC for the removal of OMPs, with all CAP biochars seeing near-immediate breakthrough (<100 bed volumes) of 2,4-D, SMX, and simazine (SZN). CPTA biochars performed better than CAP biochars, with higher temperature and forced-draft biochars exhibiting the highest OMP sorption capacities in flow-through columns. The best biochar in this study performed within a factor of 16 and 34 compared to GAC for SMX *and 2,4-D removal, respectively, while the worst chars* experienced OMP breakthrough >500 times faster than GAC. This data is in conflict with batch data utilizing the same biochars (Kearns et al., 2019), which showed that CPTA biochars had comparable adsorption capacity with PAC in the presence of background DOM. The causes for biochars' lack of performance parity with AC in flow-through column adsorbers are further discussed in the following section on the impacts of background DOM on OMP sorption.

## 3.3 Impact of the background DOM character across reactor types

As previously discussed, DOM readily adsorbs to carbonaceous adsorbents and reduces OMP adsorption capacity through two primary, simplified mechanisms: direct site competition and pore blockage (Fig. 22.5). In batch reactors (CMBRs or CFSTRs) where DOM and OMPs are simultaneously exposed to adsorbents, direct site competition dominates (Li et al., 2003b; Pelekani and Snoeyink, 1999; Quinlivan et al., 2005; Shimabuku et al., 2016). In flow-through column adsorbers, time-dependent DOM preloading occurs in addition to direct site competition, further reducing sorption capacity (Corwin and Summers, 2010; Ding et al., 2008; Kennedy and Summers, 2015; Li et al., 2003b). Biochars have been produced using single-step pyrolysis processes (e.g., CPTA) that are competitive with GAC in batch reactors, but these biochars are unable to achieve parity with GAC in flow-through column adsorbers due to DOM fouling (Kearns et al., 2021). The prevailing explanation for the increased effect of pore blockage in columns is a less-developed pore structure, particularly large micropores and small mesopores (1.5–5 nm pore widths) that are responsible for accommodating DOM and reducing pore blockage effects (Ding et al., 2008; Li et al., 2003b; Pelekani and Snoeyink, 1999). If biochars are to eventually achieve parity or competitiveness with commercial ACs in flow-through column adsorbers, their pore structure must be modified/enhanced to increase DOM-accommodating pore sizes without sacrificing OMP-specific micropores. Any enhancements or modifications should focus on improving biochar sorption without undermining the key advantages of biochar over AC: cost, availability, and environmental sustainability.

To further complicate matters, background DOM does not exert a consistent impact on biochar performance across background matrices (e.g., SW, WW, StW, LL). Different background matrices have different DOM compositions (e.g., molecular weight, humic content, hydrophobicity, etc.), and these varied compositions affect adsorbability in different, and sometimes predictable, ways. Across studies, WW DOM is consistently found to reduce OMP sorption capacity to a greater extent than SW DOM in batch reactors (Bentley and Summers, 2020; Shimabuku et al., 2014, 2016) and column adsorbers (Kearns et al., 2021; Kennedy et al., 2015; Zietzschmann et al., 2016). This effect is attributed to the larger percentage of the small molecular weight $UVA_{254}$-absorbing DOM fraction, which has been found to be most competitive with OMPs in adsorption applications, along with the presence of other competing OMPs (Bentley and Summers,

2020; Shimabuku et al., 2016; Zietzschmann et al., 2016). The highly competitive nature of WW DOM can lead to lower concentrations of WW DOM exerting stronger competitive impacts on biochar than that of SW DOM at higher concentrations (Bentley and Summers, 2020). StW DOM is also highly competitive relative to SW DOM, although StW can have a high degree of variability based on its type and environmental conditions surrounding its origins (Bentley and Summers, 2020; Shimabuku et al., 2016).

LL is an incredibly complex and competitive background matrix that can vary greatly with landfill type, age, and environmental conditions. The primary DOM component in LL by concentration are volatile fatty acids (VFAs), which exert very little competitive effect in adsorption scenarios; yet, other lower-concentration DOM constituents exert incredible competitive effects on biochar compared to SW, WW, and StW (Bentley et al., 2021). While it is impossible to generalize the relative competitive effects of these background matrices due to site- and adsorbent-specific considerations, data compiled across studies suggesting that SW, WW, StW, and leachate DOM increased pine biochar doses by around 1.5, 4, 9, and 13 times, respectively, compared to DI water for the removal of 2,4-D in 3-h batch tests as shown in Fig. 22.8. These values represent a relative measure of the competitiveness of the background matrix but suffer from certain complexities based on comparison across studies and non-normalized DOM concentrations.

If biochars are to compete with ACs, they must be generated with sufficient pore structure to both (i) accommodate and

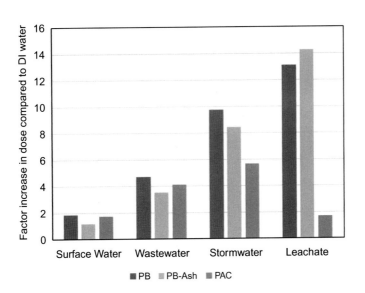

**Fig. 22.8** Relative impact of the background matrix on adsorbent dose required for pine biochar (PB), ash-pretreated pine biochar (PB-Ash), and powdered activated carbon (PAC). Surface water, wastewater, and stormwater data originate from Bentley and Summers (2020), while landfill leachate data originate from Bentley et al. (2021).

minimize the impact of background DOM and (ii) allow for many favorable sorption sites within pores suited for target OMPs (typically 1.5–2 × the OMP's kinetic diameter) (Quinlivan et al., 2005). DOM-inherent characteristics (e.g., $UVA_{254}$, fluorescence index) are being increasingly used to model and predict the performance of ACs and biochar across water treatment scenarios (Bentley and Summers, 2020; Kearns et al., 2021; Shimabuku et al., 2017). Future work should focus on applying biochars under representative treatment conditions that account for and quantify background DOM effects, and the impacts of various DOM types and fractions should be further evaluated to better understand the effects of background DOM on OMP sorption.

## 4 Emerging biochar enhancement methods

Because biochars often lack sufficient OMP and/or DOM sorption capacity to compete with AC, particularly in flow-through column adsorbers, researchers and practitioners have begun to develop enhancement methods to minimize the gap between biochar and AC performance. Many of the enhancement methods suggested throughout the literature have not been validated beyond laboratory scale and may undermine biochar's main advantages, blurring the lines between biochar and AC. It is beyond the scope of this chapter to provide an extensive review of all biochar enhancement methods and their products' performance relative to AC; however, comprehensive reviews of biochar enhancement/modification/improvement methods have been published in peer-reviewed journals (Ahmed et al., 2016; Cheng et al., 2021; Kazemi Shariat Panahi et al., 2020; Rajapaksha et al., 2016). The focus of this work is to highlight a selection of emerging improvement methods that show promise for producing high-performing biochars that can compete with AC in batch and/or flow-through column reactor scenarios.

Biochar enhancement methods can be quite diverse, so it is helpful to define certain categories and groups for these enhancement methods. The most obvious division between enhancement methods is where they fall in the production process: pre-, mid-, or postpyrolysis. Any feedstock treatment would be classified as prepyrolysis—such as ash pretreatment in Section 4.3. Enhancements occurring during the pyrolysis process would be classified as mid-pyrolysis—such as CPTA—as discussed in Section 4.1. Additional treatment steps that occur to the biochar after pyrolysis would be classified as postpyrolysis—such as double heating, Section 4.2. It is very important to understand the impacts of enhancements on the non-adsorptive qualities of biochar (i.e., cost, availability, and environmental impact), and the enhancement method summaries below will include these considerations wherever possible. The following methods are only a small selection of available biochar enhancement methods and are included in order of most studied to most emerging.

## 4.1 Co-pyrolysis thermal air activation

CPTA, also known as co-pyrolysis air oxidation (CPAO), is a simple and effective method for dramatically increasing biochar performance without sacrificing the simplicity, availability, and cost advantages of biochar over AC. CPTA uses a modified (i.e., non-traditional) pyrolysis chamber to allow some small amounts of oxygen (i.e., air) into contact with biochar during pyrolysis. The addition of air during pyrolysis allows for a greater degree of pore development by reaming pore walls and removing tarry substances that block the internal pore structure in biochars (Kearns et al., 2019; Xiao and Pignatello, 2016). CPTA biochars produced at higher HTTs exhibited higher mesopore surface area than conventional anaerobic pyrolysis (CAP) biochars, supporting the hypothesized mechanisms of pore reaming, clearing of tarry substances, and pore widening. High-temperature forced draft CPTA biochars improved OMP sorption capacity compared to CAP biochars by over an order of magnitude, which was linked to the increased mesopore surface area responsible for OMP sorption capacity in the presence of background DOM. CPTA biochars exhibited around 50% of the sorption capacity of ACs on a mass basis in a representative SW for 2,4-D adsorption, placing them well within the range of competitiveness in batch sorption scenarios (Kearns et al., 2019). Feedstock, pyrolysis temperature, and degree of forced air draft were found to be the key parameters governing the performance of the enhanced adsorbents in this study.

One major advantage of the CPTA process over other biochar enhancement methods is its universality and simplicity. CPTA can be used at any scale, from industrial pyrolysis facilities to 200-L steel drum ovens in remote villages in LMICs. Because this process requires no additional materials and is conducted mid-pyrolysis, there are no additional costs beyond construction and modification of the pyrolysis chamber to allow for air flow. Additionally, the exothermic nature of pyrolysis allows for this process to occur without additional energy inputs for heat generation, a major contributor to energy, cost, and environmental impact for ACs. Forced draft may require limited additional costs and energy associated with operation of a blower or fan, but these are minor compared to multistep enhancement processes that require additional materials and energy, such as physical or chemical activations. 200-L drum oven gasifiers have been used to produce CPTA biochars from local agricultural biomass in LMICs. Such simple designs make the production of high-performing biochar at field scale attainable for even remote, resource-limited communities that are impacted by chemical pollution (Kearns et al., 2015).

## 4.2 Regeneration and double heating

Once an adsorbent such as biochar or AC has been exhausted in water treatment, particularly when applied granularly in column filter adsorbers, it is beneficial to regenerate the adsorbent so it can be reused. Regeneration extends the lifetime of biochar and AC by removing DOM and OMPs previously sorbed and reopening pore structure and sorption sites to allow for its reuse in water treatment. Ultimately, regeneration can reduce cost and environmental impacts associated with adsorption processes by reducing demand for new adsorbent production. While many regeneration methods exist, thermal regeneration is the most promising for biochar because it does not require additional chemical inputs and can be implemented simply and cheaply across a range of applications (Greiner et al., 2018).

Greiner et al. (2018) evaluated the performance of pine biochars that had been produced under CAP and compared to a commercial AC for sorption of SMX in a representative SW. GAC exhibited ~40 times more sorption capacity for SMX than the fresh biochar due to the lower surface area and increased DOM fouling effects previously discussed. The exhausted biochar was then thermally regenerated in a covered crucible (limiting oxygen/air) at 600°C for 2h before being re-evaluated in the same SW. After the first regeneration, biochar sorption capacity increased by a factor of ~3.5, which contradicts the expected results based on AC regeneration/reactivation. In AC research, reactivation often leads to a slight reduction in performance upon each cycle, ultimately leading to a point of diminishing returns where AC is disposed of instead of being reactivated (Cannon et al., 1994; Marsh and Rodríguez-Reinoso, 2006; Sontheimer et al., 1988). Biochar exhibited the opposite trend upon regeneration, whereby its SMX adsorption capacity and surface area increased dramatically, likely due to mechanisms similar to CPTA and extended times spent at HTT. A second regeneration further increased surface area and sorption capacity, and further regenerations were not performed. The double-regenerated pine biochar performed within a factor of 5 of AC and could perhaps be further enhanced by combining regeneration with other enhancement methods. While fresh biochar may significantly underperform AC in flow-through columns, regeneration has potential to increase biochar bed life and overall lifetimes in water treatment scenarios that bring biochar within the range of economic viability for OMP sorption applications.

Double heating is an enhancement method that has emerged from positive regeneration effects on biochar performance. Essentially, biochars can undergo a regeneration process before they are ever applied as an adsorbent, and a similar positive impact on adsorbent performance can be observed. This effect was observed by Greiner et al. (Greiner et al., 2018) and has been further evaluated by Bentley et al. (2021) for a wider range of adsorbents in batch sorption with LL background DOM. While the previous study observed a performance increase around a factor of five for SMX in SW, the enhancement was a bit lower for 2,4-D and NB in LL batch sorption experiments. A variety of feedstocks were used to represent the most common organic fractions present in municipal solid waste and included pine wood, grass, orange peels, and office paper. For wood, grass, and orange peels, double heating improved the sorption of 2,4-D and NB in LL by a factor of around two for NB and around three for 2,4-D (Fig. 22.9). Office paper saw no change in 2,4-D sorption upon double heating but did see a doubling of NB sorption. Ultimately, the best double-heated biochar in this study only performed within a factor of 10 of AC, highlighting the particular challenges associated

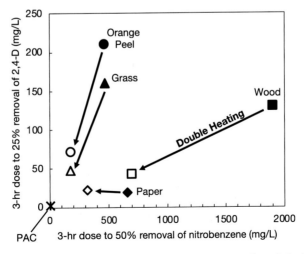

**Fig. 22.9** Level of enhancement observed through double heating for the sorption of nitrobenzene (NB) and 2,4-dichlorophenoxyacetic acid (2,4-D) in LL at 3-h contact times for activated carbon and biochars produced from pine wood, grass, orange peels, and office paper. Arrows depict the change in adsorbent dose required to achieve 50% and 25% removal of NB and 2,4-D, respectively, after double heating enhancement was performed Bentley et al. (2021).

with treating LL, which likely affected biochars more than AC for reasons previously discussed.

As is true of the CPTA process, regeneration and double heating could be simply and inexpensively applied in low-resource settings using a modified 200-L drum oven with a second chamber intended for reheating previously produced biochar. This would allow for the double heating process to use process heat from pyrolysis of new biochar rather than requiring additional energy input, thereby maintaining biochar's low cost and environmental impact.

### 4.3 Ash pretreatment

One emerging enhancement method, ash pretreatment, holds promise for inexpensively producing high-quality biochars that can compete with AC. Ash pretreatment uses a high pH, metal-laden solution derived from dissolved biomass ash to pretreat biomass feedstocks prior to pyrolysis (Bentley and Summers, 2020). The mechanisms of this enhancement method are still under investigation, but ash pretreatment has been shown to significantly enhance OMP sorption capacity in biochars. Ash pretreatment enhanced pine biochar sorption of 2,4-D by 8 to 19 times in SW, WW, and StW, performing within a factor of two of PAC—as shown in Fig. 22.10. Ash pretreatment slightly increased

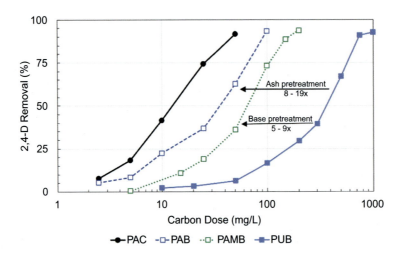

**Fig. 22.10** Adsorbent dose-response curves for powdered activated carbon (PAC), ash-pretreated pine biochar (PAB), base-pretreated pine biochar (PAMB), and untreated pine biochar (PUB). PAMB pretreatment solution pH matched the pH of the ash pretreatment solution. Enhancement level of ash and base pretreatment for pine biochar is denoted by black arrows, and ranges below arrows summarize the range of improvement observed through each method across surface water, wastewater, and stormwater background matrices from Bentley and Summers (2020).

micropore surface area and doubled non-micropore surface area, but these pore structure changes did not result in a change to biochar's susceptibility to fouling from SW, WW, or StW DOM. In this study, the non-micropore surface area was the best correlated factor with adsorbent performance in the presence of background DOM, likely due to the DOM-accommodation effect mentioned previously (Section 3.3). Across all background matrices in this study, ash-pretreated pine biochar was competitive with AC at 3-h contact times, though it was somewhat less competitive at 7-day contact times due to a lower equilibrium sorption capacity compared to AC. Ash pretreatment had small, mixed effects on WW biosolids biochar, likely due to the high amount of ash already present in biosolids (Bentley and Summers, 2020).

A follow-up study evaluated the impact of ash pretreatment across a range of biomass feedstocks present in the organic fraction of municipal solid waste for removing OMPs from LL in 3-h batch tests (Bentley et al., 2021). Ash pretreatment showed a stronger positive effect than double heating on wood and paper for the sorption of 2,4-D and NB and a stronger positive effect on grass for the sorption of NB (Fig. 22.11). Ash pretreatment enhanced biochar sorption of OMPs by up to a factor of 15, though the level of improvement was both feedstock- and OMP-specific. Ash pretreatment showed very little impact on orange peels for NB

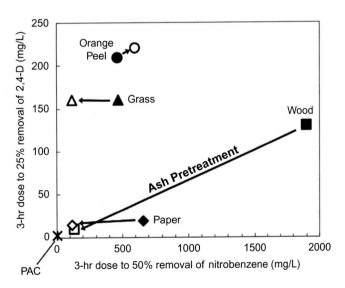

**Fig. 22.11** Level of enhancement observed through ash pretreatment for the sorption of nitrobenzene (NB) and 2,4-dichlorophenoxyacetic acid (2,4-D) in LL at 3-h contact times for activated carbon and biochars produced from pine wood, grass, orange peels, and office paper. Arrows depict the change in adsorbent dose required to achieve 50% and 25% removal of NB and 2,4-D, respectively, after ash pretreatment (Bentley et al., 2021)s.

or 2,4-D sorption. Overall, ash pretreatment enhanced micropore surface area in all feedstocks, and slightly reduced non-micropore surface area for most feedstocks—grass excluded. The mechanisms dictating the level and mechanisms of enhancement observed through ash pretreatment require further evaluation, but the results of this study show that it can be an effective enhancement method across diverse feedstock categories. Some feedstocks may not respond positively to ash pretreatment, so care must be taken during the feedstock selection process. Ash-pretreated pine wood and paper were highly competitive with AC in LL, indicating the utility of this pretreatment method for producing biochars with high sorption capacity for OMPs in the presence of challenging background matrices (Bentley et al., 2021).

As previously discussed, Kennedy et al. (2021) evaluated the performance of ash-pretreated pine biochar in CMBRs and predicted flow-through column performance based on the adsorption models, comparing results to PAC and GAC. The modeled sucralose breakthrough for biochar was 140 times faster for granular ash-pretreated biochar than for GAC, and the gap widened in WW compared to SW, which caused breakthrough to occur 570 times faster. Ash-pretreated pine biochar performance is promising in batch reactors but may lag significantly behind that of AC in flow-through column adsorbers.

Ash pretreatment presents both benefits and drawbacks compared to traditional biochar production processes. First, ash pretreatment is a widely applicable technique that requires materials that can be easily found or generated anywhere in the world (e.g., biomass ash, soaking basin), making it advantageous particularly in resource-limited communities and LMICs. Ash pretreatment, ideally, does not require additional energy input when a local ash source is readily available; however, the feedstock drying process could require the use of pyrolysis process heat or other energy input depending on site-specific considerations. If biomass ash is not readily available in a community from cooking, home heating, or a nearby industrial process, its generation could lead to deleterious environmental impacts (e.g., deforestation, carbon emissions). The additional step of presoaking the biomass prior to pyrolysis complicates biochar production and could interfere with the pyrolysis process if the feedstock is not sufficiently dried; therefore, education and operator training is essential to successful implementation of this enhancement method. Finally, more information is needed to generalize the impacts of ash pretreatment across feedstock types to simplify the feedstock selection process.

## 4.4 Base and alkali- and alkali-earth metal pretreatment

The final enhancement method covered in this chapter is base and alkali- and alkali-earth metal (AAEM) pretreatment. This method is essentially a simplification or standardization of the ash pretreatment process, as it isolates the primary hypothesized mechanisms of biochar structural and sorptive changes as a result of ash pretreatment. Bentley et al. (2022) evaluated the impact of base and AAEM addition to a feedstock pretreatment solution, similar to the method developed for ash pretreatment. As shown in Fig. 22.12, base pretreatment increased primarily micropore surface area, while AAEMs increased non-micropore surface area relative to base-only pretreatment. The catalytic activity of

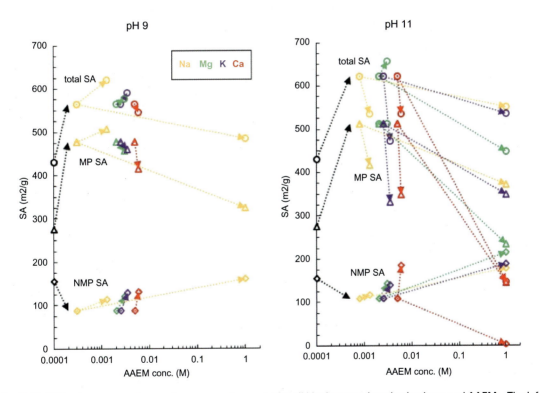

**Fig. 22.12** BET, micropore, and non-micropore surface area for all biochars produced using base and AAEMs. The left panel is for biochars produced at pH 9 + AAEMs, while the right panel is for biochars produced at pH 11 + AAEMs. The black points on the left denote untreated pine biochar surface area values, while the specific AAEMs are depicted by their respective flame colors. The x-axis denoted the concentration of each AAEM added into the base pretreatment solution. Arrows depict the change in biochar pore structure and surface area as a result of AAEM addition and increasing concentration. AAEMs were added at concentrations of $10^{-3}$ and 1 M Bentley et al., 2022.

AAEMs, defined in the study as the degree to which each AAEM shifted surface area from micropores to non-micropores, decreased from Ca > K > Na > Mg. Base and AAEM-pretreated biochars all outperformed untreated pine biochar, and base-only pretreated biochars performed the best—increasing OMP sorption by over an order of magnitude. Pretreatment with a pH 11 solution produced a biochar that nearly matched AC performance in the presence of background SW DOM. The increase in micropore and non-micropore was anticorrelated, with an increase in one pore size being offset by a decrease in the other. Higher micropore surface areas were correlated with higher OMP removal, and higher non-micropore surface areas led to higher degrees of DOM competition. AAEMs overall presented little advantage for the particular OMPs in this study, but they provide potential for pore size tailoring due to their unique catalysis mechanisms (e.g., intercalation, gasification) (Bentley et al., 2022).

Base and AAEM pretreatment are emerging techniques that present opportunities for high-specificity tailoring of biochar surface area. More work is certainly needed to evaluate the impacts of these methods across feedstock types and target OMPs in the presence of diverse background matrices, but the results are promising. Base and AAEM pretreatment presents the most challenging of the enhancement methods described in this section for resource-limited communities, as special materials and processes are required that are much more easily implemented at larger scales. Base (e.g., lye, slaked lime) is nearly universally available worldwide, but AAEMs are often laboratory grade, hard to find cheaply, and may not be widely available. This method suffers from the same drawbacks of ash pretreatment from a process complexity and energy efficiency standpoint, and may induce additional environmental impacts associated with the production, shipping, and use of the base and AAEMs.

## 5  Summary

- The production and use of biochar for water treatment applications is still a relatively new field of research, but it should be viewed as a subsection within the broader class of carbonaceous materials and adsorbents. Biochar researchers could learn much from the broad body of knowledge surrounding activated carbon, a chemically and physically similar material.

- Activated carbon is the industry standard adsorbent in water treatment applications for organics removal and therefore should be used to benchmark biochars' performance as an adsorbent. Biochars often exhibit lower sorption capacity for dissolved organic matter and organic micropollutants compared to commercial activated carbon.
- Biochar has several non-adsorptive advantages over activated carbon that motivate its use in water treatment applications, including, primarily, (i) lower environmental impact, (ii) lower cost, and (iii) widespread availability.
- Biochar performance as an aqueous adsorbent is significantly impacted by the production methods used to generate it, namely, highest treatment temperature and feedstock-inherent effects. Higher highest treatment temperatures and low-ash woody biomass tend to produce the highest quality biochars for water treatment.
- Four factors that significantly impact the performance of biochars relative to activated carbon include (i) reactor type, (ii) competing adsorbates, (iii) contact time, and (iv) treatment targets.
- Biochars have been produced that are competitive with commercial activated carbon in batch reactor scenarios in the presence of background organic matter for the removal of organic micropollutants. Unfortunately, biochars often dramatically underperform activated carbon in flow-through column adsorbers due to their higher susceptibility to preloading and pore blockage fouling by background dissolved organic matter.
- The characteristics and composition of the background dissolved organic matter significantly impact biochar and activated carbon performance and can lead to more significant reductions in biochar performance relative to activated carbon. Stormwater, wastewater, and landfill leachates tend to have increasingly competitive dissolved organic matter fractions compared to typical surface waters, which may lead to reduced biochar performance relative to activated carbon in the order of surface water, stormwater, wastewater, and landfill leachate.
- Biochar enhancement methods are increasingly available that provide opportunities to improve biochar sorption characteristics without removing the non-adsorptive advantages of biochar over commercial activated carbon. Examples of promising enhancement methods include co-pyrolysis thermal air

activation, regeneration (also referred to as double heating), ash pretreatment, and base and alkali and alkaline earth metal catalysis.

## Questions and problems
### Conceptual
1. What are three key advantages of biochar over AC?
2. What is the key pyrolysis variable governing the performance of biochars as adsorbents of DOM and OMPs?
3. Name the four factors that most impact relative biochar and AC performance as aqueous adsorbents.
4. Which type of reactor exhibits higher levels of DOM fouling (and therefore lower biochar performance relative to AC), and why?
5. Would you expect surface water or wastewater to have a larger impact on biochar adsorption capacity, and why?
6. What would be the primary purpose of a biochar enhancement method and why might it negate key advantages from Question 1?

### Computational
7. Granular biochar is being used to remove a pesticide, 2,4-dichlorophenoxyacetic acid (2,4-D), from a contaminated surface water in a flow-through column reactor. This biochar exhibits an initial adsorbent use rate (mass of adsorbent required per volume of treated water) of 0.42 kg/m3 before 2,4-D starts to break through the column. Once 10% breakthrough of 2,4-D is observed in the column, this biochar is regenerated and added back into the filter. This process continues until the adsorbent use rate deteriorates such that fresh biochar is needed to replace the regenerated biochar. Hypothetical adsorbent use rates for this biochar after each subsequent regeneration are shown in the table below.
    a. What is the total amount of water that can be treated by this biochar if 20 kg of biochar is used, assuming no mass losses upon regeneration?
    b. What is the overall adsorbent use rate of this biochar?
    c. If this biochar costs $1 per kg to produce and $0.2 per kg to regenerate, what cost would GAC need to be to achieve cost-parity based on the total water treated, assuming an AC-use rate of 0.02 kg/m3?

| # of regenerations | Adsorbent-use rate to 10% 2,4-D breakthrough (kg/m³) |
|---|---|
| 0 (fresh biochar) | 0.42 |
| 1 | 0.10 |
| 2 | 0.09 |
| 3 | 0.22 |
| 4 | 0.35 |
| 5 | 0.50 |

8. PAC ($3.30/kg) is added at a dose of 10 mg/L for 100 days each year to remove the taste and odor compound 2-methylisoborneol (MIB) at an initial concentration of 100 ng/L to below the odor threshold concentration of 10 ng/L. the water treatment plant's average flow for the 100 days is $20 \times 10^6$ L/d. ash-pretreated pine biochar ($0.20/kg) is considered as an alternative. What is the maximum ash-pretreated pine dose (in mg/L) that can be considered to compete with PAC based solely on adsorbent cost (e.g., not including sludge processing costs)?

# References

Ahmad, M., Lee, S.S., Dou, X., Mohan, D., Sung, J.K., Yang, J.E., Ok, Y.S., 2012. Effects of pyrolysis temperature on soybean stover- and peanut shell-derived biochar properties and TCE adsorption in water. Bioresour. Technol. 118, 536–544. https://doi.org/10.1016/j.biortech.2012.05.042.

Ahmed, M.B., Zhou, J.L., Ngo, H.H., Guo, W., Chen, M., 2016. Progress in the preparation and application of modified biochar for improved contaminant removal from water and wastewater. Bioresour. Technol. 214, 836–851. https://doi.org/10.1016/j.biortech.2016.05.057.

Alhashimi, H.A., Aktas, C.B., 2017. Life cycle environmental and economic performance of biochar compared with activated carbon: a meta-analysis. Resour. Conserv. Recycl. 118, 13–26. https://doi.org/10.1016/j.resconrec.2016.11.016.

Bentley, M.J., 2020. Enhancing Biochar Sorption of Organic Micropollutants in Water Treatment: Impacts of Ash Content and Background Dissolved Organic Matter. University of Colorado Boulder.

Bentley, M.J., Kearns, J.P., Murphy, B., Summers, R.S., 2022. Pre-pyrolysis metal and base addition catalyzes pore development and improves organic micropollutant adsorption to pine biochar. Chemosphere 286 (Rev.).

Bentley, M.J., Solomon, M.E., Marten, B., Shimabuku, K.K., Cook, S.M., 2021. Evaluating landfill leachate treatment by organic municipal solid waste-derived biochar. Environ. Sci. Water Res. Technol. 7 (11), 2064–2074 (Rev.).

Bentley, M.J., Summers, R.S., 2020. Ash pretreatment of pine and biosolids produces biochars with enhanced capacity for organic micropollutant removal from surface water, wastewater, and stormwater. Environ. Sci. Water Res. Technol. 6, 635–644. https://doi.org/10.1039/C5EW00120J.

Bond, T., Goslan, E.H., Parsons, S.A., Jefferson, B., 2011. Treatment of disinfection by-product precursors. Environ. Technol. 32, 1–25. https://doi.org/10.1080/09593330.2010.495138.

Cannon, F.S., Snoeyink, V.L., Lee, R.G., Dagois, G., 1994. Reaction mechanism of calcium-catalyzed thermal regeneration of spent granular activated carbon. Carbon N. Y. 32, 1285–1301. https://doi.org/10.1016/0008-6223(94)90114-7.

Chen, B., Chen, Z., 2009. Sorption of naphthalene and 1-naphthol by biochars of orange peels with different pyrolytic temperatures. Chemosphere 76, 127–133. https://doi.org/10.1016/j.chemosphere.2009.02.004.

Chen, W., Wei, R., Yang, L., Yang, Y., Li, G., Ni, J., 2019. Characteristics of wood-derived biochars produced at different temperatures before and after deashing: their different potential advantages in environmental applications. Sci. Total Environ. 651, 2762–2771. https://doi.org/10.1016/j.scitotenv.2018.10.141.

Chen, Z., Chen, B., Chiou, C.T., 2012. Fast and slow rates of naphthalene sorption to biochars produced at different temperatures. Environ. Sci. Technol. 46, 11104–11111. https://doi.org/10.1021/es302345e.

Cheng, N., Wang, B., Wu, P., Lee, X., Xing, Y., Chen, M., Gao, B., 2021. Adsorption of emerging contaminants from water and wastewater by modified biochar: a review. Environ. Pollut. 273, 116448. https://doi.org/10.1016/j.envpol.2021.116448.

Cook, D., Newcombe, G., Sztajnbok, P., 2001. The application of powdered activated carbon for MIB and geosmin removal: predicting PAC doses in four raw waters. Water Res. 35, 1325–1333. https://doi.org/10.1016/S0043-1354(00)00363-8.

Corwin, C.J., Summers, R.S., 2010. Scaling trace organic contaminant adsorption capacity by granular activated carbon. Environ. Sci. Technol. 44, 5403–5408. https://doi.org/10.1021/es9037462.

Ding, L., Snoeyink, V.L., Mariñas, B.J., Yue, Z., Economy, J., 2008. Effects of powdered activated carbon pore size distribution on the competitive adsorption of aqueous atrazine and natural organic matter. Environ. Sci. Technol. 42, 1227–1231. https://doi.org/10.1021/es0710555.

Greiner, B.G., Shimabuku, K.K., Summers, R.S., 2018. Influence of biochar thermal regeneration on sulfamethoxazole and dissolved organic matter adsorption. Environ. Sci. Water Res. Technol. 4, 169–174. https://doi.org/10.1039/c7ew00379j.

Gwenzi, W., Chaukura, N., Noubactep, C., Mukome, F.N.D., 2017. Biochar-based water treatment systems as a potential low-cost and sustainable technology for clean water provision. J. Environ. Manage. 197, 732–749. https://doi.org/10.1016/j.jenvman.2017.03.087.

Inyang, M., Dickenson, E., 2015. The potential role of biochar in the removal of organic and microbial contaminants from potable and reuse water: a review. Chemosphere 134, 232–240. https://doi.org/10.1016/j.chemosphere.2015.03.072.

Kah, M., Sigmund, G., Xiao, F., Hofmann, T., 2017. Sorption of ionizable and ionic organic compounds to biochar, activated carbon and other carbonaceous materials. Water Res. 124, 673–692. https://doi.org/10.1016/j.watres.2017.07.070.

Kazemi Shariat Panahi, H., Dehhaghi, M., Ok, Y.S., Nizami, A.S., Khoshnevisan, B., Mussatto, S.I., Aghbashlo, M., Tabatabaei, M., Lam, S.S., 2020. A comprehensive review of engineered biochar: production, characteristics, and environmental applications. J. Clean. Prod. 270, 122462. https://doi.org/10.1016/j.jclepro.2020.122462.

Kearns, J., Dickenson, E., Aung, M.T., Joseph, S.M., Summers, S.R., Knappe, D., 2021. Biochar water treatment for control of organic micropollutants with UVA surrogate monitoring. Environ. Eng. Sci. 38, 298–309. https://doi.org/10.1089/ees.2020.0173.

Kearns, J., Dickenson, E., Knappe, D., 2020. Enabling organic micropollutant removal from water by full-scale biochar and activated carbon adsorbers using predictions from bench-scale column data. Environ. Eng. Sci. 37, 459–471. https://doi.org/10.1089/ees.2019.0471.

Kearns, J.P., Knappe, D.R.U., Summers, R.S., 2014a. Synthetic organic water contaminants in developing communities: an overlooked challenge addressed by adsorption with locally generated char. J. Water Sanit. Hyg. Dev. 4, 422–436. https://doi.org/10.2166/washdev.2014.073.

Kearns, J.P., Shimabuku, K.K., Knappe, D.R.U., Summers, R.S., 2019. High temperature co-pyrolysis thermal air activation enhances biochar adsorption of herbicides from surface water. Environ. Eng. Sci. 36, 710–723. https://doi.org/10.1089/ees.2018.0476.

Kearns, J.P., Shimabuku, K.K., Mahoney, R.B., Knappe, D.R.U., Summers, R.S., 2015. Meeting multiple water quality objectives through treatment using locally generated char: improving organoleptic properties and removing synthetic organic contaminants and disinfection by-products. J. Water Sanit. Hyg. Dev. 5, 359–372. https://doi.org/10.2166/washdev.2015.172.

Kearns, J.P., Wellborn, L.S., Summers, R.S., Knappe, D.R.U., 2014b. 2,4-D adsorption to biochars: effect of preparation conditions on equilibrium adsorption capacity and comparison with commercial activated carbon literature data. Water Res. 62, 20–28. https://doi.org/10.1016/j.watres.2014.05.023.

Keiluweit, M., Nico, P.S., Johnson, M., Kleber, M., 2010. Dynamic molecular structure of plant biomass-derived black carbon (biochar). Environ. Sci. Technol. 44, 1247–1253. https://doi.org/10.1021/es9031419.

Kennedy, A.M., Arias-Paic, M., Bentley, M.J., Summers, R.S., 2021. Experimental and modeling comparisons of ash-treated pine biochar and activated carbon for the adsorption of dissolved organic matter and organic micropollutants. J. Environ. Eng. 147. https://doi.org/10.1061/(ASCE)EE.1943-7870.0001895.

Kennedy, A.M., Reinert, A.M., Knappe, D.R.U., Ferrer, I., Summers, R.S., 2015. Full- and pilot-scale GAC adsorption of organic micropollutants. Water Res. 68, 238–248. https://doi.org/10.1016/j.watres.2014.10.010.

Kennedy, A.M., Summers, R.S., 2015. Effect of DOM size on organic micropollutant adsorption by GAC. Environ. Sci. Technol. 49, 6617–6624. https://doi.org/10.1021/acs.est.5b00411.

Li, Q., Snoeyink, V.L., Mariãas, B.J., Campos, C., 2003a. Elucidating competitive adsorption mechanisms of atrazine and NOM using model compounds. Water Res. 37, 773–784. https://doi.org/10.1016/S0043-1354(02)00390-1.

Li, Q., Snoeyink, V.L., Mariñas, B.J., Campos, C., 2003b. Pore blockage effect of NOM on atrazine adsorption kinetics of PAC: the roles of PAC pore size distribution and NOM molecular weight. Water Res. 37, 4863–4872. https://doi.org/10.1016/j.watres.2003.08.018.

Lian, F., Xing, B., 2017. Black carbon (biochar) in water/soil environments: molecular structure, sorption, stability, and potential risk. Environ. Sci. Technol. https://doi.org/10.1021/acs.est.7b02528.

Liu, W.-J., Jiang, H., Yu, H.-Q., 2015. Development of biochar-based functional materials: toward a sustainable platform carbon material. Chem. Rev. 115, 12251–12285. https://doi.org/10.1021/acs.chemrev.5b00195.

Marsh, H., Rodríguez-Reinoso, F., 2006. Activated Carbon. Elsevier Ltd., https://doi.org/10.1016/b978-008044463-5/50020-0.

Meyer, S., Glaser, B., Quicker, P., 2011. Technical, economical, and climate-related aspects of biochar production technologies: a literature review. Environ. Sci. Technol. 45, 9473–9483. https://doi.org/10.1021/es201792c.

Moreira, M.T., Noya, I., Feijoo, G., 2017. The prospective use of biochar as adsorption matrix—a review from a lifecycle perspective. Bioresour. Technol. 246, 135–141. https://doi.org/10.1016/j.biortech.2017.08.041.

Pelekani, C., Snoeyink, V.L., 1999. Competitive adsorption in natural water: role of activated carbon pore size. Water Res. 33, 1209–1219. https://doi.org/10.1016/S0043-1354(98)00329-7.

Pignatello, J.J., Mitch, W.A., Xu, W., 2017. Activity and reactivity of pyrogenic carbonaceous matter toward organic compounds. Environ. Sci. Technol. 51, 8893–8908. https://doi.org/10.1021/acs.est.7b01088.

Quinlivan, P.A., Li, L., Knappe, D.R.U., 2005. Effects of activated carbon characteristics on the simultaneous adsorption of aqueous organic micropollutants and natural organic matter. Water Res. 39, 1663–1673. https://doi.org/10.1016/j.watres.2005.01.029.

Rajapaksha, A.U., Chen, S.S., Tsang, D.C.W., Zhang, M., Vithanage, M., Mandal, S., Gao, B., Bolan, N.S., Ok, Y.S., 2016. Engineered/designer biochar for contaminant removal/immobilization from soil and water: potential and implication of biochar modification. Chemosphere 148, 276–291. https://doi.org/10.1016/j.chemosphere.2016.01.043.

Rodriguez Correa, C., Otto, T., Kruse, A., 2017. Influence of the biomass components on the pore formation of activated carbon. Biomass Bioenergy 97, 53–64. https://doi.org/10.1016/j.biombioe.2016.12.017.

Shimabuku, K.K., Cho, H., Townsend, E.B., Rosario-Ortiz, F.L., Summers, R.S., 2014. Modeling nonequilibrium adsorption of MIB and sulfamethoxazole by powdered activated carbon and the role of dissolved organic matter competition. Environ. Sci. Technol. 48, 13735–13742. https://doi.org/10.1021/es503512v.

Shimabuku, K.K., Kearns, J.P., Martinez, J.E., Mahoney, R.B., Moreno-Vasquez, L., Summers, R.S., 2016. Biochar sorbents for sulfamethoxazole removal from surface water, stormwater, and wastewater effluent. Water Res. 96, 236–245. https://doi.org/10.1016/j.watres.2016.03.049.

Shimabuku, K.K., Kennedy, A.M., Mulhern, R.E., Summers, R.S., 2017. Evaluating activated carbon adsorption of dissolved organic matter and micropollutants using fluorescence spectroscopy. Environ. Sci. Technol. 51, 2676–2684. https://doi.org/10.1021/acs.est.6b04911.

Sizmur, T., Fresno, T., Akgül, G., Frost, H., Moreno-Jiménez, E., 2017. Biochar modification to enhance sorption of inorganics from water. Bioresour. Technol. 246, 34–47. https://doi.org/10.1016/j.biortech.2017.07.082.

Sontheimer, H., Crittenden, J.C., Summers, R.S., 1988. Activated Carbon for Water Treatment, second ed. DVGW-Forschungsstelle, Engler-Bunte-Institut, Universitat Karlsruhe (TH), Karlsruhe, Germany.

Suhas, Carrott, P.J.M., Ribeiro Carrott, M.M.L., 2007. Lignin—from natural adsorbent to activated carbon: a review. Bioresour. Technol. 98, 2301–2312. https://doi.org/10.1016/j.biortech.2006.08.008.

Summers, R.S., Knappe, D.R.U., Snoeyink, V.L., 2011. Adsorption of organic compounds by activated carbon (Chapter 14). In: Edzwald, J.K. (Ed.), Water Quality and Treatment, sixth ed. McGraw-Hill, New York, NY.

Summers, R.S., Shiokari, S.T., Johnson, S., Peterson, E., Yu, Y., Cook, S., 2020. Reuse treatment with ozonation, biofiltration, and activated carbon adsorption for total organic carbon control and disinfection byproduct regulation compliance. AWWA Water Sci. 2, 1–14. https://doi.org/10.1002/aws2.1190.

Sun, K., Kang, M., Zhang, Z., Jin, J., Wang, Z., Pan, Z., Xu, D., Wu, F., Xing, B., 2013. Impact of deashing treatment on biochar structural properties and potential sorption mechanisms of phenanthrene. Environ. Sci. Technol. 47, 11473–11481. https://doi.org/10.1021/es4026744.

Tan, X., Liu, Y., Zeng, G., Wang, X., Hu, X., Gu, Y., Yang, Z., 2015. Application of biochar for the removal of pollutants from aqueous solutions. Chemosphere. https://doi.org/10.1016/j.chemosphere.2014.12.058.

Thompson, K.A., Shimabuku, K.K., Kearns, J.P., Knappe, D.R.U., Summers, R.S., Cook, S.M., 2016. Environmental comparison of biochar and activated carbon for tertiary wastewater treatment. Environ. Sci. Technol. 50, 11253–11262. https://doi.org/10.1021/acs.est.6b03239.

Thompson, K.A., Valencia, E.W., Scott Summers, R., Cook, S.M., 2020. Sorption, coagulation, and biodegradation for graywater treatment. Water Sci. Technol. 81, 2152–2162. https://doi.org/10.2166/wst.2020.273.

Tran, H.N., You, S.J., Hosseini-Bandegharaei, A., Chao, H.P., 2017. Mistakes and inconsistencies regarding adsorption of contaminants from aqueous solutions: a critical review. Water Res. https://doi.org/10.1016/j.watres.2017.04.014.

Ulrich, B.A., Im, E.A., Werner, D., Higgins, C.P., 2015. Biochar and activated carbon for enhanced trace organic contaminant retention in stormwater infiltration systems. Environ. Sci. Technol. 49, 6222–6230. https://doi.org/10.1021/acs.est.5b00376.

Wang, S., Dai, G., Yang, H., Luo, Z., 2017. Lignocellulosic biomass pyrolysis mechanism: a state-of-the-art review. Prog. Energy Combust. Sci. https://doi.org/10.1016/j.pecs.2017.05.004.

Xiao, F., Pignatello, J.J., 2016. Effects of post-pyrolysis air oxidation of biomass chars on adsorption of neutral and ionizable compounds. Environ. Sci. Technol. 50, 6276–6283. https://doi.org/10.1021/acs.est.6b00362.

Xiao, F., Pignatello, J.J., 2015. Interactions of triazine herbicides with biochar: steric and electronic effects. Water Res. 80, 179–188. https://doi.org/10.1016/j.watres.2015.04.040.

Xu, X., Zhao, Y., Sima, J., Zhao, L., Mašek, O., Cao, X., 2017. Indispensable role of biochar-inherent mineral constituents in its environmental applications: a review. Bioresour. Technol. 241, 887–899. https://doi.org/10.1016/j.biortech.2017.06.023.

Zietzschmann, F., Stützer, C., Jekel, M., 2016. Granular activated carbon adsorption of organic micro-pollutants in drinking water and treated wastewater—aligning breakthrough curves and capacities. Water Res. 92, 180–187. https://doi.org/10.1016/j.watres.2016.01.056.

# 23

# Biochar-assisted advanced oxidation processes for wastewater treatment

Rahul Kumar Dhaka[a], Karuna Jain[a], Charles U. Pittman, Jr.[b], Todd E. Mlsna[b], Dinesh Mohan[c], Krishna Pal Singh[d,e], Pooja Rani[a], Sarita Dhaka[g], and Lukáš Trakal[f]

[a]Department of Chemistry and Centre for Bio-Nanotechnology (COBS&H), Chaudhary Charan Singh Haryana Agricultural University, Hisar, Haryana, India. [b]Department of Chemistry, Mississippi State University, Mississippi State, MS, United States. [c]School of Environmental Sciences, Jawaharlal Nehru University, New Delhi, India. [d]Biophysics Unit, College of Basic Sciences & Humanities, G.B. Pant University of Agriculture & Technology, Pantnagar, Uttarakhand, India. [e]Mahatma Jyotiba Phule University, Bareilly, Uttar Pradesh, India. [f]Department of Environmental Geosciences, Faculty of Environmental Sciences Czech University of Life Sciences Prague, Praha, Suchdol, Czech Republic. [g]Department of Chemistry, Sanatan Dharm (PG) College, Muzaffarnagar, Affiliated to Chaudhary Charan Singh University, Meerut, Uttar Pradesh, India

## 1 Introduction

Water is a fundamental need for society, but a limited amount of drinking water is accessible for humans. Freshwater availability is limited and declining at a global level due to population growth and climate change. Water is being polluted at a fast pace due to anthropogenic and natural activities. Hence, sustainable wastewater treatments must be developed on a priority basis (Kurian, 2021). Generally, wastewater is treated via physical, biological, and/or chemical methods. Physicochemical methods like adsorption, membrane separation, and sedimentation alone cannot efficiently remove contaminants completely. Biological treatments (e.g., activated sludge and biofilm) can solve these problems but their major drawbacks include high investment, operating costs, and long prepreparation cycles which hamper the acceptability of the process (Kurian, 2021).

Among the chemical treatment practices, AOPs are considered the best oxidative degradation methods for organic pollutants. AOPs can effectively remediate wastewater via in situ-generated reactive oxygen species (ROSs), including hydroxyl radicals ($^{\bullet}OH$), superoxide radical anions ($O_2^{\bullet-}$), hydroperoxyl radicals ($HO_2^{\bullet}$), and sulfate radical anions ($SO_4^{\bullet-}$). The mechanism of radical formation depends upon process-specific parameters, system design, and water quality. AOPs based on ozone, UV irradiation, and Fenton/Fenton-like processes are used at a large scale for wastewater treatment. AOPs based on electrochemistry, plasma, ultrasound, microwave, and electron beam are emerging technologies in wastewater treatment (Fig. 23.1). AOPs, in comparison to conventional technologies, are beneficial, as refractory pollutants are degraded without secondary waste generation (Miklos et al., 2018).

The homogeneous catalysis involving AOPs takes place via activation of persulfate and Fenton reactions. In the case of heterogeneous catalysis transition, metal oxides, metal-exchanged clays, zeolites, and carbonaceous materials act as catalysts and provide support to the catalyst (Kurian, 2021; Miklos et al., 2018). Moreover, a carbonaceous material like biochar is frequently being utilized in the area of environmental catalysis due to its eco-friendly and stable nature, cost-effectiveness, and wide availability. Biochar has been exploited in different AOPs as the catalyst due to its high surface area, porosity, conductive nature, and graphitic content. Moreover, biochar physicochemical properties can be engineered by optimizing process parameters (nature of precursor biomass, pretreatment, pyrolysis temperature, and time) for intended applications (Wang et al., 2019a). Biochar can be employed alone or as a support for metal/metal oxides and heteroatoms to install upgraded

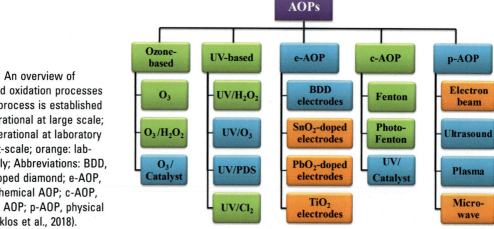

**Fig. 23.1** An overview of advanced oxidation processes (Green: process is established and operational at large scale; blue: operational at laboratory and pilot-scale; orange: lab-scale only; Abbreviations: BDD, boron-doped diamond; e-AOP, electrochemical AOP; c-AOP, catalytic AOP; p-AOP, physical AOP (Miklos et al., 2018).

properties. Biochar's large surface area, high porosity, and acidic and basic media stability contribute to the dispersion and stabilization of loaded materials (Liu et al., 2015). Moreover, oxygen-containing functional groups (OCFGs), persistent free radicals (PFRs), doped metals, and nonmetals on biochar surfaces play an important role in catalysis.

Biochar application in catalysis for environmental remediation is a sustainable approach. Recently, AOPs namely photocatalysis (Zhou et al., 2021), Fenton-like (Zhou et al., 2021), and persulfate activation (Zhou et al., 2021) have used biochar as a catalyst and catalyst support in aqueous phase organic pollutant degradations. However, the structural and compositional complexity of biochar in comparison to conventional carbonaceous materials make it difficult to relate its catalytic degradation mechanism to its functional structure (Cha et al., 2016). Herein, the potential roles of neat, doped, and metal-loaded biochars as catalysts in AOPs are presented.

## 2 Biochar-based catalysts

Biochar-based catalysts have displayed a remarkable performance in wastewater treatment (Zhao et al., 2021; Pan et al., 2021; Zhou et al., 2021). The following section discusses bare biochar, metal-loaded biochar, heteroatom-doped biochar, and co-doped biochar as catalysts used in AOPs.

### 2.1 Bare biochar

The catalytic activity of biochar toward organic pollutant removal is attributed to the presence of oxygen-containing functional groups (OCFGs), inorganic component, and persistent free radicals (PFRs) at its surface (Zhao et al., 2021; Zhou et al., 2021). The active functional groups on the biochar surface activate hydrogen peroxide ($H_2O_2$), peroxydisulfate ($S_2O_8^{2-}$), and peroxymonosulfate ($HSO_5^-$) to generate reactive oxygen species (ROSs) (Zhao et al., 2021; Zhou et al., 2021). Recently, researchers have reported pollutant abatement using the biochar/$H_2O_2$ system. For example, cornstalks, bamboo, and pig manure biochars activated $H_2O_2$ and degraded tetracycline through ·OH radicals generated by PFRs present on the biochar surface (Huang et al., 2019). Pig manure biochar most efficiently activated $H_2O_2$ due to its metal and phenolic content (Huang et al., 2019). Wheat straw biochar activated $H_2O_2$ and degraded the antibacterial drug sulfamethazine (Huang et al., 2016); the role of higher pyrolytic temperature used to prepare the biochar was significant in $H_2O_2$ activation and ROS generation (Huang et al., 2016). The $^{\bullet}$OH production was influenced by the surface chemistry and porosity of biochar (Huang et al., 2016). Pine needle, wheat straw, and maize biochars stimulated $H_2O_2$ to remove 2-chlorobiphenol (Fang et al., 2014). The individual efficiencies of $H_2O_2$ and biochar were low, but significantly improved when used in combination to remove 2-chlorobiphenol (~95%) (Fang et al., 2014).

Biochar has also been used as a persulfate activator. For example, sawdust biochar activated $S_2O_8^{2-}$ to degrade acid orange 7 dye (>90% removal) (He et al., 2019). Biochar transferred electrons to $S_2O_8^{2-}$ by an electronic shuttle followed by its decomposition into $^{\bullet}$OH and $SO_4^{\bullet-}$ (He et al., 2019). Similar to $H_2O_2$ activation, pyrolytic temperature and conditions also influence persulfate activation. For example, both biochar and activated carbon stimulated $S_2O_8^{2-}$ and degraded sulfamethoxazole (Liang et al., 2019). Biochar synthesized at a higher temperature (700°C) produced 88.7% of degradation as compared to the biochar generated at a low temperature (400°C, 30.4%). The solution phase degradation of sulfamethoxazole by the biochar/$S_2O_8^{2-}$ system shows good catalytic capability at generating $^{\bullet}$OH and $SO_4^{\bullet-}$ oxidizing radicals (Liang et al., 2019). Bare biochar has also been used as a catalyst in other AOPs like photocatalytic removal of diethyl phthalate (Fang et al., 2017) and sonocatalytic degradation of acetaminophen and naproxen (Im et al., 2014).

## 2.2 Modified biochar

Biochar can be modified using activation, surface functionalization, loading of metal/metal oxide, heteroatom-doping, and co-doping methods to enhance its catalytic performance in AOPs (Fig. 23.2) (Liu et al., 2015).

Biochar activation includes enhancement of specific surface area and porosity by inducing additional internal pores via physical or chemical routes. Precursor biomass, activating agent, and pyrolysis conditions decide the physicochemical nature of biochars (Kumar et al., 2020). Biochar can be physically activated by treating it with steam and/or $CO_2$ at high temperatures (>700°C) under controlled gas flow. The $CO_2$ environment enhances the catalytic efficiency of biochar-supported materials, creating surface defects and oxygen-containing functional groups (Liu et al., 2015; Kumar et al., 2020). Biochar can be chemically activated by impregnating with acid/alkaline/metal salt solutions ($H_2SO_4$, $H_3PO_4$, KOH, $ZnCl_2$, and $K_2CO_3$) followed by dry solid pyrolysis. These chemical agents develop new pores in biochar, suppress tar synthesis, and facilitate

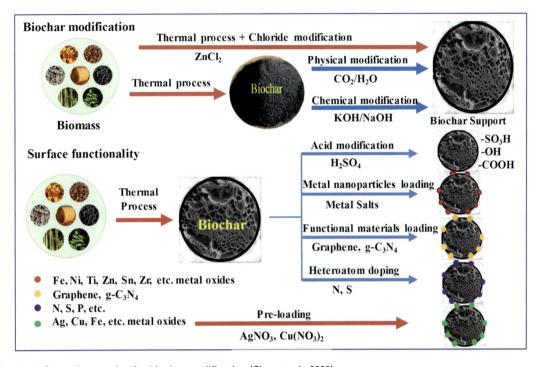

**Fig. 23.2** General strategies for biochar modification (Shan et al., 2020).

volatile organic compound formation. After activation, extra impregnating agents and salts must be cleaned. Pyrolysis temperature, rate, residence time, amount of chemical agents, and biomass type influence specific surface area and porosity and pore size distribution in the activated biochar (Liu et al., 2015; Kumar et al., 2020). Chemical activation is relatively more effective.

The properties of biochar as catalysts and/or catalyst supports may also be improved by surface functionalization or active material deposition. Metals and metal oxides can be preloaded onto the biomass matrix and/or impregnated to improve their catalytic performance (Shan et al., 2020; Pan et al., 2021). Acidic functionalization is a common approach for biochar surface modification (Shan et al., 2020). Sulfonation is performed via treatment with concentrated sulfuric acid and gaseous sulfur trioxide. Sulfonated biochars are useful catalysts for dehydration, biomass hydrolysis, and biodiesel production. Weak acids may also be incorporated into the biochar matrix (Shan et al., 2020; Pan et al., 2021).

Biochar can be loaded with graphene (Shan et al., 2020), graphene oxide (Shan et al., 2020), graphitic carbon nitride (g-$C_3N_4$), carbon nanotubes (CNTs) (Liu et al., 2015), and zerovalent iron (ZVI)

(Deng et al., 2018) to improve oxygen-containing functional groups, specific surface area, porosity, thermal steadiness, and catalytic property. The absorption, visible light response, and photocatalytic efficiency of the polymeric semiconductor, g-$C_3N_4$, were improved after immobilization onto biochar. Biochar effectively aids the separation of photogenerated charge carriers (Liu et al., 2015; Shan et al., 2020). Metal/metal oxides can be loaded onto biochar by soaking it into metal salt solutions followed by the thermal treatment which precipitates metal oxides on the biochar surface. Biochars loaded with Fe, Ni, or Fe-Ni nanoparticles can be used in cracking, photocatalysis, and pyrolysis reactions. The iron oxide ($Fe_2O_3$ or $Fe_3O_4$) loadings by chemical co-precipitation make biochar magnetic (Liu et al., 2015; Shan et al., 2020; Pan et al., 2021). Nonmetal-doped biochar is a sustainable metal-free catalyst for AOPs. N, S, B, and P doping will alter biochar's electronic structure and improve its catalytic activity. N-doping, for example, promotes electron transfer during catalysis, improving $HSO_5^-$, and catalyst surface interactions (Shan et al., 2020; Pan et al., 2021).

### 2.2.1 Biochar loaded with metals

Transition metals and their compounds are often used as catalysts. The catalytic potential of biochar may be improved via metal loading for application in AOPs. Metals are evenly distributed over biochar surfaces, minimizing metal leaching. Biochar prevents the agglomeration of metal nanoparticles and offers more available active sites, eventually enhancing the catalytic efficacy of these composites. Biochar has been widely explored as a support for both metals and metal oxides for use in wastewater treatment. A few representative reports of metal loadings on biochar are summarized in Table 23.1. Manganese oxide-loaded biochar activated $S_2O_8^{2-}$ to rapidly degrade dye Orange G. Fast degradation kinetics resulted from the generation of $HO^{\cdot}$ and $SO_4^{\cdot -}$ (Fan et al., 2019). Zero-valent iron-loaded biochar efficiently activated $S_2O_8^{2-}$ and degraded aromatic contaminants in alkaline wastewater. ZVI particles on the biochar enhanced persulfate surface activation sites (Luo et al., 2019).

In addition, the loading of binary metal oxides (Fe-Mn) on biochar improves its catalytic activity. The composite binary metal oxide/biochar enhanced the naphthalene degradation (44.9%–82.8% within 148 min) due to an increase in PFRs (Li et al., 2019a). Moreover, a $Co_3O_4$-loaded wheat straw biochar composite activated $HSO_5^-$ to catalyze chloramphenicol degradation (~95% removal within 10 min) (Xu et al., 2020a). A copper-doped graphitic biochar activated $S_2O_8^{2-}$ to better degrade

**Table 23.1 Application of biochar-based catalysts in different AOPs.**

| Process | Contaminant | Biochar-based catalysts/oxidants | Degradation efficiency | Responsible active species in degradation mechanism | References |
|---|---|---|---|---|---|
| Fenton-like | 2,4-Dichlorophenoxy-acetic acid | Pyrite biochar mixture/$H_2O_2$ | >95% | ·OH | (Zhu et al., 2020a) |
| | 2,4-Dichlorophenoxy-acetic acid | Liquor waste biochar (700°C)+Fe(III)/$H_2O_2$ | 94.8% | ·OH | (FeMethylene blue was degraded byng et al., 2021) |
| | Ciprofloxacin | FeS-$Fe_3O_4$@biochar (500°C)/$H_2O_2$ | ~100% | ·OH | (Wang et al., 2019c) |
| | Ciprofloxacin | $MnO_x$-$Fe_3O_4$-biochar/$H_2O_2$ | 92.8% | $O_2^{\cdot -}$ and ·OH | (Li et al., 2021a) |
| | Ciprofloxacin | Sewage sludge biochar/$H_2O_2$ | >80% | ·OH | (Li et al., 2019b) |
| | Phenol | FeAl-layered double hydroxide@biochar (0.25)/$H_2O_2$ | 85.3% | ·OH | (Fan et al., 2021) |
| | 4-Chlorophenol | Fe-rich biochar/$H_2O_2$ | 100% | ·OH | (Gan et al., 2020) |
| | Metronidazole | Magnetic biochar (400°C)/$H_2O_2$ | ~100% | ·OH | (Yi et al., 2020) |
| | Sulfamethazine | Swine manure biochar/$H_2O_2$ | 85% | PFRs and ·OH | (Deng et al., 2020) |
| | Sulfamethazine | Nanoscale zero-valent iron@biochar/$H_2O_2$ | 74% | ·OH | (Deng et al., 2018) |
| | Sulfamethoxazole | $CaO_2$-biochar-Fe(III)/$H_2O_2$ | 96% | $O_2^{\cdot -}$ and ·OH | (Zhang et al., 2022) |
| | Sulfamethoxazole | Schwertmannite@biochar/$H_2O_2$ | 100% | ·OH and $O_2^{\cdot -}$ | (Yang et al., 2021) |
| | Sulfamethoxazole | $Fe_xP$-biochar/$H_2O_2$ | 99% | ·OH and $O_2^{\cdot -}$ and $^1O_2$ | (Xie et al., 2020) |
| | Tetracycline | Pig manure biochar (500°C)/$H_2O_2$ | 100% | ·OH | (Huang et al., 2019) |

*Continued*

**Table 23.1 Application of biochar-based catalysts in different AOPs—cont'd**

| Process | Contaminant | Biochar-based catalysts/oxidants | Degradation efficiency | Responsible active species in degradation mechanism | References |
|---|---|---|---|---|---|
| | Trichloroethylene | Sewage sludge biochar/$H_2O_2$ | 83% | $\cdot OH$ | (Huang et al., 2020) |
| | Pyrene | $Fe_3O_4$-biochar/$H_2O_2$ | 99.5% | $\cdot OH$ and iron-oxo species | (Jian et al., 2021) |
| | Methylene blue | $Fe_2O_3$-$TiO_2$-biochar/$H_2O_2$ | 100% | $\cdot OH$ | (Chen et al., 2020) |
| | 1-Methyl-1-cyclohexane carboxylic acid | Wheat straw biochar/$H_2O_2$ | 100% | $\cdot OH$ | (Devi et al., 2020) |
| | Diethyl phthalate | Biochar - pine needles, wheat straws, and maize/$H_2O_2$ | 100% | $\cdot OH$ | (Fang et al., 2015) |
| | Norfloxacin | Sewage sludge biochar/$H_2O_2$ | 98.8% | $\cdot OH$ and $O_2^{\cdot -}$ | (Liu et al., 2018) |
| Photo-Fenton like | Tetracycline | $CuFeO_2$-biochar/$H_2O_2$ | 93% | $\cdot OH$, $e^-$ and $O_2^{\cdot -}$ | (Xin et al., 2021) |
| | Tetracycline | $MnFe_2O_4$-biochar/$H_2O_2$ | 95% | $\cdot OH$ | (Lai et al., 2019) |
| | Metronidazole | $CuFe_2O_4$-coated pretreated biochars/$H_2O_2$ | 96.3% | $O_2^{\cdot -}$, $\cdot OH$ and $h^+$ | (Cai et al., 2021) |
| | Acid Orange 7 | Biochar from sawdust and rice husk/$S_2O_8^{2-}$ | ~100% | $\cdot OH$ and $SO_4^{\cdot -}$ | (Li et al., 2020) |
| Persulfate activation | Malachite green | $CuFe_2O_4$@biochar/$S_2O_8^{2-}$ | 98.9% | $O_2^{\cdot -}$ and $SO_4^{\cdot -}$ | (Huang et al., 2021) |
| | Phenol, bisphenol A, 2,4-dichlorophenol | 13%-Mn-Fe biochar/$HSO_5^-$ | ~100% | $^1O_2$, $O_2^{\cdot -}$, $SO_4^{\cdot -}$ and $\cdot OH$ | (Kou et al., 2021) |
| | Phenol | Bagasse biochar calcined with $CaCl_2$ and KOH/$S_2O_8^{2-}$ | 90% | $^1O_2$, $SO_4^{\cdot -}$ and $\cdot OH$ | (Zhang et al., 2020) |
| | Bisphenol A | Fe-N co-doped biochar/$HSO_5^-$ | 97% | $^1O_2$, $SO_4^{\cdot -}$ and $\cdot OH$ | (Xu et al., 2020b) |
| | Bisphenol A | Polymeric carbon nitride-biochar hybrid-visible/$S_2O_8^{2-}$ | 97.23% | $^1O_2$, $SO_4^{\cdot -}$, $\cdot OH$ and $h^+$ | (Wang et al., 2021a) |

| | Pollutant | Catalyst/Oxidant | Efficiency | ROS | Reference |
|---|---|---|---|---|---|
| | p-Hydroxybenzoic acid | Fe$_3$O$_4$-biochar from *Myriophyllum aquaticum*/HSO$_5^-$ | 100% | $^1$O$_2$, O$_2^{\bullet-}$, SO$_4^{\bullet-}$ and ·OH | (Fu et al., 2019) |
| | 1,4-Dioxane | Pine needle biochar/HSO$_5^-$ | 84.2% | SO$_4^{\bullet-}$ and ·OH | (Ouyang et al., 2019) |
| | Atrazine | Nanoscale zero-valent iron@biochar/S$_2$O$_8^{2-}$ | 93.8% | SO$_4^{\bullet-}$, ·OH, PFRs and $^1$O$_2$ | (Zhang et al., 2021a) |
| | 17β-Estradiol | Nanoscale zero-valent iron-loaded porous graphitized biochar/S$_2$O$_8^{2-}$ | 100% | SO$_4^{\bullet-}$ and ·OH | (Ding et al., 2021) |
| | Norfloxacin | Nanoscale zero-valent iron-Ni@biochar/S$_2$O$_8^{2-}$ | 99% | SO$_4^{\bullet-}$ and ·OH | (Zhu et al., 2020b) |
| | Benzene | Fe$_x$S$_y$@biochar/S$_2$O$_8^{2-}$ | 99.8% | SO$_4^{\bullet-}$ and ·OH | (Zhu et al., 2022) |
| | Tetracycline | KOH-activated biochar/S$_2$O$_8^{2-}$ | 84.2% | $^1$O$_2$ | (Li et al., 2021b) |
| | Tetracycline | Magnetic nitrogen-doped sludge biochar/S$_2$O$_8^{2-}$ | 82.2% | SO$_4^{\bullet-}$ and ·OH | (Yu et al., 2019) |
| | Sulfamethazine | Fe-Mg oxide-biochar/S$_2$O$_8^{2-}$ | 99% | SO$_4^{\bullet-}$ | (Qin et al., 2020) |
| | Sulfadiazine | Zerovalent iron biochar/S$_2$O$_8^{2-}$ | 100% | $^1$O$_2$, O$_2^{\bullet-}$, SO$_4^{\bullet-}$ and ·OH | (Ma et al., 2021) |
| | Sulfamethoxazole | Wood chip graphitized biochar/S$_2$O$_8^{2-}$ | 98% | SO$_4^{\bullet-}$, ·OH, $^1$O$_2$ and O$_2^{\bullet-}$ | (Du et al., 2020) |
| | Triclosan | Sludge derived biochar/HSO$_5^-$ | 100% | SO$_4^{\bullet-}$, ·OH, $^1$O$_2$ | (Wang and Wang, 2019) |
| Photocatalysis | Methylene blue | Biochar-ZnO | 98% | h$^+$, ·OH, and O$_2^{\bullet-}$ | (He et al., 2021) |
| | Rhodamine B | Ag nanoparticles@biochar/S$_2$O$_8^{2-}$ | 99.9% | SO$_4^{\bullet-}$, ·OH, $^1$O$_2$ and O$_2^{\bullet-}$ | (Yu et al., 2021) |
| | Bisphenol A | Ag-Fe@Bamboo-derived activated biochar/S$_2$O$_8^{2-}$ | 95.6% | SO$_4^{\bullet-}$, ·OH and O$_2^{\bullet-}$ | (Talukdar et al., 2020) |
| | Estrone | Bi-Bi$_2$O$_3$ biochar | 94.9% | ·OH | (Zhu et al., 2020c) |
| | Carbamazepine | MgIn$_2$S$_4$-BiOCl-biochar | 84.9% | ·OH and O$_2^{\bullet-}$ | (Qi et al., 2022) |
| | Ibuprofen | Biochar-ZnAl$_2$O$_4$ | 100% | ·OH and O$_2^{\bullet-}$ | (Siara et al., 2022) |
| | Phenol | Iodine-doped biochar | 99.7% | O$_2^{\bullet-}$, H$_2$O$_2$, $^1$O$_2$, ·OH, e$^-$, h$^+$ | (Wang et al., 2021b) |

*Continued*

Table 23.1 Application of biochar-based catalysts in different AOPs—cont'd

| Process | Contaminant | Biochar-based catalysts/oxidants | Degradation efficiency | Responsible active species in degradation mechanism | References |
|---|---|---|---|---|---|
| | Tetracycline | Enteromorpha biochar modified with carbon nitride/$HSO_5^-$ | 90% | $O_2^{\bullet-}$, $^1O_2$ | (Tang et al., 2022) |
| | Ofloxacin & ciprofloxacin | $Bi_2WO_6$-$Fe_3O_4$-biochar | ~100% | $O_2^{\bullet-}$, $h^+$ and $\cdot OH$ | (Wang et al., 2021c) |
| | Iohexol | Biochar loaded with perylene diimide/$S_2O_8^{2-}$ | 100% | $h^+$, $\cdot OH$, $^1O_2$, $O_2^{\bullet-}$, and $SO_4^{\bullet-}$ | (Ji et al., 2021) |
| Sonocatalysis | Cefazolin sodium | Fe-Cu-layered double hydroxide-biochar | 97.6% | $\cdot OH$ | (Gholami et al., 2020) |
| | Bisphenol A | Sludge biochar-ultrasound/$S_2O_8^{2-}$ | 98% | $SO_4^{\bullet-}$ and $\cdot OH$ | (Diao et al., 2020) |
| | Bisphenol A | $MnO_2$-biochar/$H_2O_2$ | 100% | $\cdot OH$ and $O_2^{\bullet-}$ | (Jung et al., 2019) |
| | Gemifloxacin | ZnO-biochar | 96% | $\cdot OH$ | (Gholami et al., 2019) |
| Electrochemical | Nitrobenzene | Zn-Fe-modified biochar | ~94% | $\cdot OH$ | (Liu et al., 2021) |
| | 2,4-Dichlorophenol | Hydrochar | ~76% | $\cdot OH$ | (Cao et al., 2020) |
| | 4-Chlorophenol | Biochar electrode | 99.9% | $\cdot OH$ | (Xie et al., 2020) |
| Microwave assisted | 2,4-Dichlorophenoxy-acetic acid | Biochar from oak (OBC-900°C) | 82% | $\cdot OH$, $^1O_2$, $O_2^{\bullet-}$ | (Sun et al., 2020) |
| | Bisphenol S | Magnetic sludge-derived biochar/$H_2O_2$ | — | $\cdot OH$ and $h^+$ | (Lv et al., 2019) |

$\cdot OH$, hydroxyl radical; $^1O_2$, singlet oxygen; $e-$, electron; $h^+$, hole; $HSO_5^-$, peroxymonosulfate; $O_2^{\bullet-}$, superoxide radical; PFRs, persistent free radicals; $S_2O_8^{2-}$, peroxydisulfate; $SO_4^{\bullet-}$, sulfate radical anion.

rhodamine B, bisphenol A, phenol, and 4-chlorophenol than the pristine undoped biochar (Wan et al., 2019). The recent reviews contain more reports on metal-loaded biochar employed in wastewater remediations (Zhao et al., 2021; Zhou et al., 2021).

### 2.2.2 Biochar doped with heteroatoms

Metal as doping agents may create a leaching problem causing secondary pollution. Therefore, heteroatom doping is preferred to improve the catalytic performance of the biochar. Commonly used heteroatoms such as B, N, P, and S regulate the chemical and electrical properties of carbon and reactive sites of the biochar matrix. Nitrogen is a preferable doping candidate as it has a size similar to carbon; N can simply accommodate into a graphene sheet in the biochar. The difference in N and C electronegativity creates a structural defect, enhancing biochar's electron transfer capability. For example, corncob-derived N-doped biochars and urea in different ratios exhibit admirable catalytic activity toward $S_2O_8^{2-}$ activation for sulfadiazine degradation. The edge nitrogen configurations (like pyridinic N and pyrrolic N) generate more active sites than graphitic N in $S_2O_8^{2-}$ activation (Wang et al., 2019b). N-doped spirulina residue biochar activated $S_2O_8^{2-}$ to degrade sulfamethoxazole via a nonradical electron mechanism (Ho et al., 2019). In addition to N, S can also change the structural properties of biochar. Rice straw biochar activated $HSO_5^-$ for metolachlor herbicide removal which was further influenced by N doping (positive effect) and S doping (negative effect). The positive impact with N may be attributed to the displacement of C ($sp^2$) by N. The S on the other hand displaces oxygen-containing functional groups and disturbs the charge reallocation on the biochar matrix, resulting in poor $HSO_5^-$ adsorption and less degradation (Ding et al., 2020). Doping with other heteroatoms and their effect on different catalytic systems must be explored.

### 2.2.3 Co-doped biochar

Recently, the synthesis of co-doped biochar using elemental combinations (i.e., Fe-Cu, Fe-S, and N-S) is at the center of research interest. Metal-to-nonmetal co-doping is common. Rice straw N-doped magnetic biochar-supported $CoFe_2O_4$ catalyst, for example, activated $HSO_5^-$ and degraded metolachlor in river and groundwater (Liu et al., 2019). The $^•OH$ and $SO_4^{•-}$ were generated by the redox reaction between $Co^{2+}$ and $HSO_5^-$, followed by singlet oxygen production due to $HSO_5^-$ decomposition. Similarly, sludge-derived co-doped (MnOx-N) biochar activated $HSO_5^-$ and removed AO7 dye completely (within 40 min). The Mn-oxide, oxygen defects, porous structure, and the presence of N influenced this biochar's catalytic efficiency. The Mn-oxide acted as redox center and produced ·OH and $SO_4^{·-}$. The oxygen defects adsorbed dissolved oxygen and acted as catalytic reactive sites to produce singlet oxygen. Nitrogen regulated the charge distribution on the catalyst surface (Mian et al., 2019). Lately, a novel biochar composite (CoO/$Co_9S_8$@N-S-BC) prepared by pyrolyzing sludge biochar, urea, sulfur, and $CoCl_2·6H_2O$ effectively activated $HSO_5^-$ produced ·OH and $SO_4^{·-}$ and completely degraded sulfamethoxazole (0.08 mM within 10 min) (Wang and Wang, 2020). Likewise, N and Cu co-doped biochar activated $S_2O_8^{2-}$ and degraded tetracycline completely within 90 min (Zhong et al., 2020).

## 3 Biochar application in AOPs

Biochar finds application in adsorption, catalysis, soil conditioning, improving water holding capacity of the soil, carbon sequestration, and energy storage among the other areas. This section deals with the role of biochar's catalytic properties in AOPs and the mechanisms involved therein (Tables 23.1 and 23.2, Fig. 23.3).

### 3.1 Fenton-like AOP

The Fenton reaction is a traditional, cost-effective, and rapid AOP approach for wastewater treatment. It involves $Fe^{2+}$ and $H_2O_2$ (pH: 2.5–3.0) to generate reactive hydroxyl radicals which degrade many organic contaminants in water. The Fenton system has certain limitations like nonreusability of $Fe^{2+}$, requirement of acidic conditions, and production of Fe-based sludge.

These limitations can be dealt using Fenton-like systems (generally $Fe^{3+}$ and $H_2O_2$). These systems use several heterogeneous catalytic materials (natural minerals, metallic oxides, and carbonaceous materials) (Pan et al., 2021; Zhou et al., 2021). The pure Fenton reaction takes place by the generation of hydroxyl radical (·OH) as per the following Eq. (23.1).

$$Fe^{2+} + H_2O_2 \rightarrow Fe^{3+} + OH^- + ·OH \tag{23.1}$$

The reaction of $H_2O_2$ with $Fe^{3+}$ first gives $HO_2·$ radical. Moreover, Fenton-like systems involve the following reactions (Eqs. 23.2–23.7) (Hashemian, 2013).

$$Fe^{3+} + H_2O_2 \rightarrow Fe^{2+} + HO_2· + H^+ \tag{23.2}$$

$$Fe^{3+} + HO_2· \rightarrow Fe^{2+} + H^+ + O_2 \tag{23.3}$$

**Table 23.2** Functionalities and mechanism/active species in advanced oxidation systems (Zhou et al., 2021).

| System | Functional structure/component | Mechanism/important active species |
|---|---|---|
| Fenton-like system | Hydroxyl | $\cdot OH$ |
| | Quinone | $O_2^{\cdot -}$ |
| | Metal | $\cdot OH$, $O_2^{\cdot -}$ |
| | Oxygen-centered PFRs | $\cdot OH$ |
| | Carbon-centered PFRs | $O_2^{\cdot -}$ |
| | Pore structure | Concentration of pollutants |
| | Specific surface area | Degradation of pollutants |
| | Doped heteroatom | |
| Persulfate system | C=O | $^1O_2$, electron transfer pathway |
| | —OH/—COOH | $\cdot OH$, $SO_4^{\cdot -}$ |
| | Defect | $O_2^{\cdot -}$, $^1O_2$ |
| | Persistent free radicals | $\cdot OH$, $SO_4^{\cdot -}$, $O_2^{\cdot -}$ |
| | Doped heteroatom | Electron transfer pathway |
| | Pore structure | Concentration of pollutants |
| | Specific surface area | Degradation of pollutants |
| | Degree of graphitization | Favoring electron transfer |
| Photocatalytic system | Dissolved organic matter | Generating radicals |
| | Doped heteroatom | Reducing the bandgap |
| | Persistent free radicals | Generating radicals |
| | Oxygen-containing functional group | Reducing bandgap |
| | | Radical generation |
| | | PFR formation |
| | Pore structure | Favoring electron transfer |
| | Degree of graphitization | |

$$Fe^{2+} + \cdot OH \rightarrow Fe^{3+} + {}^-OH \quad (23.4)$$

$$\cdot OH + H_2O_2 \rightarrow HO_2^{\cdot} + H_2O \quad (23.5)$$

$$Fe^{2+} + HO_2^{\cdot} \rightarrow Fe^{3+} + HO_2^{-} \quad (23.6)$$

$$\cdot OH + \cdot OH \rightarrow H_2O_2 \quad (23.7)$$

Biochar and its modifications have frequently been employed as catalysts in Fenton-like systems for aqueous phase organic pollutant degradation (Pan et al., 2021; Zhou et al., 2021). A number of mechanisms (Table 23.1) have been reported in the literature for biochar-mediated Fenton-like systems. The reaction between

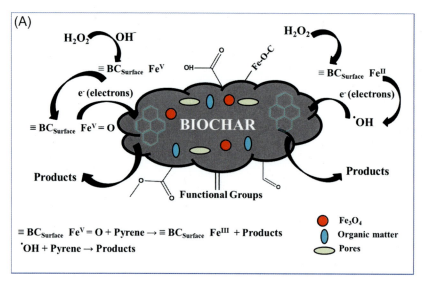

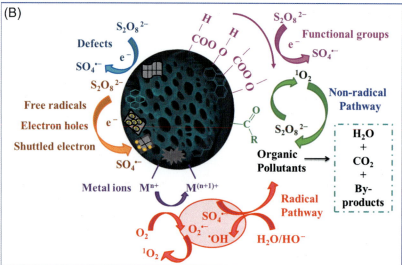

**Fig. 23.3** Reaction mechanisms followed in AOPs. (A) Fenton-like process (Jian et al., 2021), (B) Persulfate activation (Zhao et al., 2021),

*(Continued)*

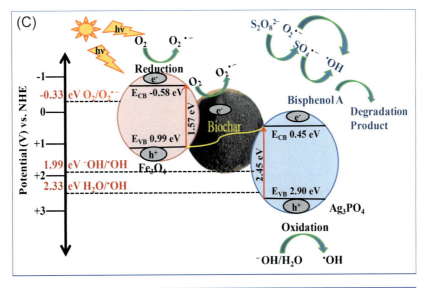

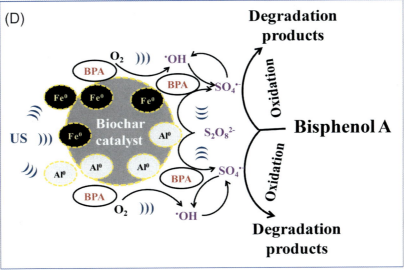

**Fig. 23.3, cont'd** (C) Photocatalytic (Talukdar et al., 2020), (D) Sonocatalytic (Diao et al., 2020).

biochar's functional group and $H_2O_2$ can be presented as follows (Eq. 23.8) (Zhou et al., 2021):

$$Biochar_{surface}\text{-}OH + H_2O_2 \rightarrow {}^{\bullet}CO + {}^{\bullet}OH + H_2O \qquad (23.8)$$

A natural pyrite-mixed biochar (BC) accelerated the catalytic oxidative degradation of herbicide 2,4-dicholorophenoxy acetic acid

(2,4-D) via a HO·-mediated Fenton-like system (Zhu et al., 2020a). The biochar addition improved the rate constant 1.98–2.39 times (with 0.1 g BC/L, and 1.5 g pyrite/L) as it increased ·OH production (2.72 times) as compared to the pyrite system. The transformation of $O_2^{·-}$ to ·OH and a quinone-like structure in biochar acts as an electron shuttle (Zhu et al., 2020a). Similarly, the addition of liquor waste biochars made by pyrolysis at 500 and 700°C (LB700 and LB500) into the Fenton process (using trace Fe content ranging from 0.2 to 1.8 mg/L) enhanced the herbicide, 2,4-D, degradation kinetics and simultaneously reduced the Fe-rich sludge. The rate of reaction was fast in the case of the biochar-assisted Fenton reaction (rate constant: 4.41–5.80 times greater versus using the Fenton process only or 3.09–4.97 faster than biochar/$H_2O_2$). Further, biochar addition specifically enhanced ·OH production and 2,4-D degradation in the case of the biochar prepared at 700°C (Feng et al., 2021).

Biochar-supported iron sulfide and oxide catalytically degraded ciprofloxacin in Fenton-like reactions. Biochar-dispersed metal species improve the catalytic efficiency toward ·OH (as major species)-mediated ciprofloxacin degradation. Biochar improved ·OH production and suppressed $H_2O_2$ consumption in the reaction. Biochar with hydroquinone/quinone moieties acted as an electron regulator and promoted the $Fe^{2+}/Fe^{3+}$ cycle (Wang et al., 2019c). In a Fenton-like system ($MnO_x$-$Fe_3O_4$/biochar), ciprofloxacin was degraded in the presence of hydroxylamine and oxalic acid (Li et al., 2021a). Hydroxylamine and oxalic acid acted as reducing agents which influenced the FeMn/biochar-$H_2O_2$ system performance by the mechanisms namely, changing pH, electrostatic interactions, and accelerating metal redox cycles (Li et al., 2021a). The metal redox cycle accelerated $H_2O_2$ decomposition to $O_2^{·-}$ to ·OH. Ciprofloxacin degradation efficiency was improved in the presence of oxalic acid and hydroxylamine (38.2–92.8%). The system (FeMn/biochar-$H_2O_2$-$HONH_2$) was also efficient in the case of real wastewater samples containing pharmaceuticals, dyes, and polycyclic aromatic hydrocarbons (Li et al., 2021a, b). Sewage sludge biochar degraded ciprofloxacin via heterogeneous Fenton-like catalysis through synergistic oxidation and adsorption. The HO· was the dominant species in this degradation (Li et al., 2019b).

Biochar-supported iron-aluminum layered double hydroxide (FeAl-LDH) activated $H_2O_2$ to degrade phenol (85.28% at pH 3) through both homogeneous and heterogeneous processes in the same system. Electron paramagnetic resonance (EPR) and radical quenching studies indicated hydroxyl radical as a major species responsible for phenol degradation. The catalyst was reusable with a small amount of iron leaching (Fan et al., 2021). Sludge cake biochar enriched with multivalent iron ($Fe_0$, $Fe_{0.95}C_{0.05}$, $Fe_3O_4$, and $FeAl_2O_4$) catalytically removed 4-chlorophenol in homogeneous and heterogeneous Fenton reactions. Homogeneous Fenton reaction (where $Fe^{2+}$ was contributed by $Fe_0$ and $Fe_{0.95}C_{0.05}$) predominated in the first step, while heterogeneous Fenton reaction initiated through $FeAl_2O_4$ and $Fe_3O_4$, was dominant in the degradation step. The complete removal of 4-chlorophenol was achieved with Fe-rich biochar with low Fe leaching (Gan et al., 2020).

A magnetic biochar-assisted system was used to degrade metronidazole in a Fenton-like system. Biochar prepared at 400°C when coupled with $H_2O_2$ completely decomposed metronidazole through surface-bound hydroxyl radicals. EPR studies and free radical quenching experiments were used to compare biochars (prepared at 300°C, 400°C, and 500°C) capacities to activate $H_2O_2$ to generate ˙OH (Yi et al., 2020). The differences in biochar catalytic activities were due to $Fe^{2+}$ in FeO (Yi et al., 2020). Sulfamethazine was degraded over swine manure biochar in heterogeneous Fenton reactions (~85% degradation within 30 min) (Deng et al., 2020). This biochar possessed a high catalytic efficiency due to electron transfer pathways but weak sulfamethazine adsorption affinity. The environmentally persistent free radicals were observed in manure biochar through electron paramagnetic resonance and X-ray photoelectron spectroscopy studies (Deng et al., 2020). Oxygen-centered PFRs had a key role which converted them into carbon-centered PFRs. The contribution of ˙OH was dominant. The catalytic ability of swine manure biochar was promoted through an acidic or alkaline medium. The biochar made at 600°C/$H_2O_2$ system was efficient in the real wastewater application, giving improved catalytic ability and excellent adaptability (Deng et al., 2020). A nano zerovalent iron-biochar (nZVI/BC) activated $H_2O_2$ for sulfamethazine removal via ˙OH radicals. BC prevented nZVI aggregation, adsorbed sulfamethazine, activated $H_2O_2$, and alleviated nZVI passivation. The catalyst nZVI/BC was reused three times giving progressively lower removal efficiencies (74.04% to 53.28%-II run and 38.02%-III run) due to the gradual loss of $Fe^0$ (Deng et al., 2018).

The Fenton-like system, calcium peroxide/biochar, and tartaric acid-chelated $Fe^{3+}$ catalyst increased $O_2^{˙-}$ radical-based $Fe^{2+}$ regeneration while degrading sulfamethoxazole (Zhang et al., 2020). $CaO_2$/BC reduced $Fe^{3+}$ efficiently, increased the rate of sulfamethoxazole degradation, and raised peroxide utilization more than $CaO_2$ or $H_2O_2$. $CaO_2$/BC Fenton-like reactions were correlated with PFRs in biochar. The PFRs acted as a shuttle. PFRs in $CaO_2$/BC enhanced the regeneration of $Fe^{2+}$ through superoxide radicals and promoted ˙OH formation in Fenton-like reactions even at acidic to neutral pH (Zhang et al., 2022). Schwertmannite-loaded biochar (Scg@BC) removed sulfamethoxazole, sulfadiazine, and sulfisoxazole 100, 91, and 93%, respectively, under the same conditions. Sulfamethoxazole degradation (in an Sch@BC/$H_2O_2$ system) through ˙OH and $O_2^{˙-}$ was confirmed by EPR and radical scavenging studies (Yang et al., 2021). Similarly, the FexP/biochar composite (FexP/BC-5) adsorbed and degraded 99% of sulfamethoxazole in 30 min via activation of dissolved oxygen (DO). The species ˙OH, $O_2^{˙-}$, and $^1O_2$ were produced in FexP/BC-5/DO by charge transfer and Fenton reaction. This system operated in a wide pH range (3.0–9.0) and showed recyclability up to 5 cycles (Xie et al., 2020).

Biochars derived from cornstalks, bamboo, and pig manure all activated $H_2O_2$, produced ˙OH, and effectively degraded tetracycline. A large number of PFRs at the biochar surface transfer the electrons to $H_2O_2$ and produce ˙OH. This ultimately increased the biochars' degradation efficiency. Additionally, the concentration of PFRs in the biochar mainly depends upon the phenolic and metal content of the feedstock (Huang et al., 2019). The Fe content and high specific surface area of sewage sludge biochar, produced using high-power microwaves, had a better catalytic effect while degrading trichloroethylene in the biochar catalyzed a heterogeneous Fenton reaction (pH 3.1, 300 W) (Huang et al., 2020). Magnetite-doped biochars ($Fe_3O_4$/BC) were used for pyrene oxidation in wastewater. The composite $Fe_3O_4$/BC5 removed 99.5% of pyrene within 720 min due to ˙OH and iron-oxo species which were confirmed by changes in $Fe_3O_4$/BC5 surface chemistry. Overall, radical and electron transfer oxidation pathways were proposed for pyrene degradation Fig. 23.3A (Jian et al., 2021).

Methylene blue was degraded by $Fe_2O_3$/$TiO_2$-functionalized biochar through adsorption and Fenton-like oxidation. The removal efficiency of $Fe_2O_3$/$TiO_2$-BC for total organic carbon exceeded 65% and ˙OH dominance was confirmed by radical quenching experiments (Chen et al., 2020). Many other studies have also examined different biochars serving as activators for $H_2O_2$ to generate ˙OH. Wheat straw and hardwood biochars were tested for the decomposition of 1-methyl-1-cyclohexane carboxylic acid based on $H_2O_2$ (Devi et al., 2020). Wheat, maize straw, and pine needle biochars removed diethyl phthalate completely. $O_2$ was converted first into $O_2^{˙-}$ by electron transfer from the radicals present on the biochar followed by conversion into $H_2O_2$. Later, $H_2O_2$ reacted with PFRs by single electron transfer on the biochar surface to generate ˙OH (Fang et al., 2015). Sewage sludge biochar also activated $H_2O_2$ to decompose norfloxacin (98.8% degradation efficiency) (Liu et al., 2018).

Biochar has also played an important role in photo-Fenton reactions during heterogeneous catalysis. Particle aggregation is a serious issue in heterogeneous catalysis which mitigates the effectiveness of the light. The role of biochar as support improves the effect of light and the heterogeneous catalysis performance. For example, Xin et al. (2021) degraded tetracycline using $CuFeO_2$/biochar as the photo-Fenton catalyst. The spread of $CuFeO_2$ onto the biochar surface was confirmed by a band structure study. The particle spread narrows the bandgap and promotes electron transport. The high catalytic efficiency of $CuFeO_2$/biochar is due to its longer photoelectron lifetime and greater carrier density. The observed rate constant for $CuFeO_2$/biochar-mediated tetracycline removal was twice that observed for $CuFeO_2$. Mechanistically, $\cdot OH$ was the major active species, while $O_2^{\cdot -}$ and photoelectrons were secondary species (Xin et al., 2021). Similarly, visible-light-driven tetracycline removal was studied using $MnFe_2O_3$ nanoparticles and a $MnFe_2O_3$/biochar composite. The catalyst activated $H_2O_2$ under visible light and $\cdot OH$ degraded tetracycline (95%) (Lai et al., 2019). Metronidazole was also degraded (96.3%) using $H_2O_2$ activated by $CuFe_2O_4$-coated biochar ($CuFe_2O_4$@PBC) under visible irradiation (Cai et al., 2021).

## 3.2 Persulfate-activated AOPs

Persulfate-activated AOPs have been explored recently for water treatment. Generally, persulfate activation is performed by carbonaceous materials, heat, microwave, transition metal catalysis, UV-light, and ultrasound. Biochar can be employed in persulfate activation to stimulate it via oxygen-containing functional groups and persistent free radicals. In sludge-derived biochar, ammonia, organic nitrogen, and metal ions promote persulfate activation. The activation routes depend upon the type and amount of catalyst, oxidant, pH of the medium, and the nature of the pollutant (Zhao et al., 2021; Zhou et al., 2021) (Fig. 23.3B). The production of ROSs in persulfate-activated AOPs may systematically be presented by the following Eqs. (23.9)–(23.13) (Zhou et al., 2021).

$$Biochar_{surface}\text{-}OOH + S_2O_8^{2-} \rightarrow Biochar_{surface}\text{-}OO^{\cdot} + SO_4^{\cdot -} + HSO_4^- \quad (23.9)$$

$$Biochar_{surface}\text{-}OH + S_2O_8^{2-} \rightarrow Biochar_{surface}\text{-}O^{\cdot} + SO_4^{\cdot -} + HSO_4^- \quad (23.10)$$

$$SO_4^{\cdot -} + H_2O_2/HO^- \rightarrow HSO_4^-/SO_4^{2-} + \cdot OH \quad (23.11)$$

$$M^{n+} + S_2O_8^{2-} \rightarrow SO_4^{\cdot -} + SO_4^{2-} + M^{(n+1)+} \quad (23.12)$$

$$M^{(n+1)+} + SO_4^{\cdot -} \rightarrow SO_4^{2-} + M^{n+} \quad (23.13)$$

Here, we summarize a few representative studies on biochar-assisted persulfate activation in AOPs. For instance, Huang et al. (2021) removed malachite green (MG) using the $CuFe_2O_4$@biochar composite ($CuFe_2O_4$@BC) and $S_2O_8^{2-}$. The catalytic efficiency of the composite was 98.9% and the dominant oxidizing species were $O_2^{\cdot -}$ and $SO_4^{\cdot -}$ (Huang et al., 2021). Acid orange 7 (AO7) dye was removed by sawdust and rice husk biochar catalysts through $S_2O_8^{2-}$ activation. Sawdust biochar was better at remediating AO7 than rice husk biochar due to the availability of OCFGs. $\cdot OH$ and $SO_4^{\cdot -}$ participated in this reaction and $\cdot OH$ played a major role (Li et al., 2020).

Phenols (phenol, bisphenol A, and 2,4-dichlorophenol) were removed using a series of Fe/Mn co-doped biochar catalysts which activated $HSO_5^-$ (Kou et al., 2021). The radical and nonradical oxidation mechanisms worked simultaneously for organic pollutant removal. Among the ROSs ($^1O_2$, $O_2^{\cdot -}$, $SO_4^{\cdot -}$, $\cdot OH$), the nonradical species ($^1O_2$) was dominant in catalytic oxidation. Besides, the electron transfer mechanism may also have promoted pollutant removal (Kou et al., 2021). Moreover, phenol degradation was achieved by a Ca- and KOH-modified mesoporous biochar (Ca/BS-800-KOH) and $S_2O_8^{2-}$. The nonradical ($^1O_2$) species and radical pathways ($SO_4^{\cdot -}$ and $\cdot OH$) were both involved in oxidation (Zhang et al., 2020).

Antibiotics (tetracycline, chlortetracycline, and doxycycline) were degraded by poplar and pine sawdust biochars which activated $HSO_5^-$ and generated $SO_4^{\cdot -}$ and $\cdot OH$ (Zhang et al., 2021b). The PFRs and defect-rich biochar structure activated $HSO_5^-$ at 300–500°C and 800–900°C, respectively. The PFRs were carbon-centered radicals or carbon-centered radicals containing oxygen atoms. Singlet oxygen (nonradical pathway) had little effect on antibiotic degradation (Zhang et al., 2021a, b). Li and coworkers removed tetracycline using the KOH-activated biochar which stimulated $S_2O_8^{2-}$ via singlet oxygen (Li et al., 2021a, b).

### 3.3 Photocatalytic AOPs

Photocatalysis is a sustainable practice to manage emerging environmental problems. Recently, photocatalytic systems have been investigated. The light sensitivity of semiconductors is exploited in photocatalytic processes. The holes ($h^+$), $\cdot OH$, $O_2^{\cdot}$, and $^1O_2$ are the reactive species formed in photocatalysis, Eqs. (23.14)–(23.21) (Zhou et al., 2021).

$$\text{Semiconductor} + h\nu \to \text{semiconductor} + (h^+ + e^-) \quad (23.14)$$

$$h^+ + H_2O \to {}^\bullet OH + H^+ \quad (23.15)$$

$$h^+ + OH^- \to {}^\bullet OH \quad (23.16)$$

$$e^- + O_2 \to O_2^{\bullet -} \quad (23.17)$$

$$h^+ + O_2^{\bullet -} \to {}^1O_2 \quad (23.18)$$

$$O_2^{\bullet -} + H_2O \to HO_2^\bullet + HO^- \quad (23.19)$$

$$HO_2^\bullet + H_2O \to H_2O_2 + {}^\bullet OH \quad (23.20)$$

$$H_2O_2 \to 2\,{}^\bullet OH \quad (23.21)$$

The primary role of biochar in photocatalysis is:
(1) to provide support to photosensitizers and avoid aggregation.
(2) to provide a hydrophobic environment (to improve catalyst adsorption capacity).
(3) to act as an electron shuttle (to facilitate electron and hole separation).
(4) to reduce the bandgap through doping into the photosensitizer's structure.

The biochar retains pollutants on its surface and increases the efficiency of photocatalysts (Ahmaruzzaman, 2021). A biochar/ZnO composite degraded methylene blue under UV–visible light (98% degradation within 100 min) (He et al., 2021). The Zn-O-C (a bond formed at the interface of ZnO nanoparticles and biochar) facilitates a lower recombination rate of $e^-/h^+$ pairs. Biochar and ZnO composite provide a synergistic effect versus pure ZnO. The contributions of $h^+$, $^\bullet OH$, and $O_2^{\bullet -}$ in methylene blue degradation were confirmed by trapping experiments. Yu et al. (2021) activated $S_2O_8^{2-}$ via biochar composite with silver nanoparticles and removed rhodamine B almost completely under visible light within 10 min. Active species involved in degradation were $SO_4^{\bullet -}$, $^\bullet OH$, $^1O_2$, and $O_2^{\bullet -}$ (Yu et al., 2021).

Similarly, co-doped ($Ag_3PO_4/Fe_3O_4$) bamboo-activated biochar (Ag-Fe@BAB) photo-activated $S_2O_8^{2-}$ and degraded bisphenol A (95.6% within 60 min) under visible light. The species $O_2^{\bullet -}$ dominated, as confirmed by a benzoquinone scavenging effect. Further, methanol and t-butyl alcohol (radical scavengers) indicated the presence of $SO_4^{\bullet -}$ and $^\bullet OH$ species contributing toward bisphenol A degradation (Talukdar et al., 2020) (Fig. 23.3C). Bismuth-containing photocatalytic biochar degraded estrone by ˙OH under UV–Vis exposure (highest removal efficiency: 94.9%); the efficiency of this catalyst was enhanced by biochar due to adsorption and charge separation (Zhu et al., 2020c).

Iodine-doped biochar adsorbed and photocatalytically degraded phenol and tetracycline under visible light (Wang et al., 2021b). The iodine doping improved adsorption, created extra pores, enhanced pollutant oxidation, and reduced charge carrier transfer resistance via strong photo-induced excitation. Iodine-doped biochar more efficiently catalyzed phenol photodegradation. The species $O_2^{˙-}$, $H_2O_2$, $^1O_2$, ˙OH, electrons, and holes played active roles in photocatalysis and $O_2^{˙-}$ was the major contributor (Wang et al., 2021b). Tetracycline was degraded by carbon nitride-modified Enteromorpha biochar-activated $HSO_5^-$ in the presence of visible light (~90% of tetracycline removal within 1 h was achieved). The modified biochar provided electron-withdrawing groups to adjust carbon nitride (g-$C_3N_4$), and induced more π-π interaction to mitigate photocarrier recombination. The species $O_2^{˙-}$ and $^1O_2$ played a significant role in tetracycline photocatalytic degradation as compared to $SO_4^{˙-}$ and ˙OH, and $h^+$ (Tang et al., 2022).

The photocatalytic AOP has also been used to remediate carbamazepine, ibuprofen, ciprofloxacin, ofloxacin, and iohexol (Table 23.1).

## 3.4 Other AOPs

In addition to the processes discussed above, sonocatalytic (Gholami et al., 2020), electrochemical (Cao et al., 2020), and microwave-assisted (Sun et al., 2020) AOPs are also reported. Biochar application in these systems is in its infancy and needs in-depth research. The application of ultrasound in organic pollutant degradation is the most promising due to the ease of operation, time savings, and eco-friendly nature. The chemical effect of sonolysis produces acoustic cavitation involving the generation, growth, and collapse of microbubbles that are created as a result of ultrasound crossing through the aquatic medium. The final outcome of this process is the formation of localized hot spots having high pressure and temperature that exist for very short time periods. These hot spots provide an appropriate condition for the homolytic dissociation of water molecules forming strong oxidizing agents (˙OH and $H_2O_2$).

$$H_2O \longrightarrow ˙OH + ˙H \text{ (sonocatalysis)} \quad (23.22)$$

$$˙OH + ˙OH \rightarrow H_2O_2 \quad (23.23)$$

The application of sonolysis alone requires a long time to degrade an organic pollutant; hence, solid catalysts need to be employed to facilitate the catalysis process. Solid surfaces act as nuclei that nucleate the cavitation bubble formation. A few reports have been published on biochar-mediated sonocatalytic degradation of pollutants. For example, a sonocatalyst nanocomposite (Fe-Cu-layered double hydroxide/biochar) was used to degrade sodium cefazolin (Gholami et al., 2020). The composite showed a superior sonocatalytic performance (97.6% of degradation) versus bare biochar and neat Fe-Cu-layered double hydroxide. ·OH played a dominant role in sodium cefazolin degradation (Gholami et al., 2020). Bisphenol A was degraded in water through an ultrasound-supported sludge biochar catalyst and $S_2O_8^{2-}$ system (98% of BPA degraded within 80 min) (Diao et al., 2020). The reactive $SO_4^{·-}$ and ·OH species contributed to this degradation with $SO_4^{·-}$ dominance (Fig. 23.3D) (Diao et al., 2020). Bisphenol A was also removed by an ultrasound-assisted heterogeneous Fenton process with good degradation efficiencies (Chu et al., 2021). The magnetically modified biochar was used to produce ·OH radicals—the dominant oxidizing species (Chu et al., 2021). Jung et al. (2019) employed a $MnO_2$/biochar composite for bisphenol A where ·OH and $O_2^{·-}$ were the major active moieties (Jung et al., 2019). Zinc oxide nanorods supported on biochar removed gemifloxacin better in the presence of ultrasound (96.1%) as compared to pure ZnO nanorods and bare biochar (Gholami et al., 2019). The degradation mechanism involved ·OH as the major reactive species.

The electrochemical AOPs (eAOPs) are also an effective water treatment technology. The highly reactive ·OH can decompose recalcitrant pollutants. The eAOPs usually proceed via direct oxidation and oxidization by a redox mediator. Few reports are available on this method. Zn/Fe-modified biochar removed nitrobenzene better during electrolysis versus bare biochar under 2 V electrolysis and became dominant under 8 and 11 V. Both adsorption and electrochemical decomposition mechanisms worked for nitrobenzene degradation. ·OH participated in degradation using Zn-modified and Fe-modified biochar. The role of ·OH in Zn-modified biochar-mediated degradation, without Fe present, was insignificant due to its greater porosity and surface area. A porous material may adsorb ·OH obvious preventing contact between nitrobenzene and ·OH (Liu et al., 2021). Biochar-mediated electrolysis also removed nitrobenzene via electron transfer from the electrode. The mechanism involved a reduction of nitrobenzene to aniline followed by its ·OH-mediated degradation on the biochar surface or subsequently mineralizing aniline (Liu et al., 2020).

2,4-Dichlorophenol was electrochemically degraded using hydrochar as an anode and cathode material. Hydrochar without a proton exchange membrane separation achieved better degradation (6, 9V) versus adsorption or electrolysis alone. The anode-generated ·OH and Cl· or $H_2O_2$ also participated in this degradation. Hydrochar also contained OCFGs and PFRs, which induced ROSs. These reacted with 2,4-dichlorophenol (Cao et al., 2020). Similarly, 4-chlorophenol was electrochemically degraded at biochar-loaded particle electrodes with excellent efficiency (99.93%) (Xie et al., 2020).

Recently, microwave water treatments have been used as it is a selective, volumetric, and uniform heating method (Sun et al., 2020). This technology uses frequency in the range of 0.3 to 300 GHz and wavelengths ranging from 1 to 0.001 m. It provides non-contact heating where microwave absorbing materials can be rapidly heated throughout via nongradient temperature ramping. This irradiation produces hot spots or microplasmas that ionize the surrounding environment and accelerate the generation of ·OH. In comparison to other heating technologies, microwave irradiation is easy to operate, has short reaction times and better processing efficiency for small volumes, consumes less energy and chemicals, and mineralizes organic contaminants completely. For example, 2,4-dichlorophenol was removed using oak and apple tree biomass biochar prepared under $N_2$ and $CO_2$ environments (at 900°C) (Sun et al., 2020). The $CO_2$ environment creates a hierarchical porous structure, improves surface hydrophilicity, polarity, and acidity, and provides higher densities of biochar functionalities. Okra biochar's (OBC-900) graphitic structure and medium-developed microporous structure were more efficient at removing 2,4-D under microwave irradiation (81.6%, 1 min, 90°C) in comparison to room temperature (33.9%, 1 min) conditions. Highly reactive ·OH, $^1O_2$, and $O_2^{·-}$ radicals contributed to this degradative removal (Sun et al., 2020). Sludge-derived magnetic biochar catalytically oxidized bisphenol S under microwaves in the presence of $H_2O_2$. The species ·OH and $h^+$ played major roles in the reaction. Hot spots activated $H_2O_2$ and caused a doping effect due to charge transfer (Lv et al., 2019).

## 4 Summary and future perspectives

Biochar is a carbonaceous catalyst that also has high electrical conductivity, surface area, chemistry, and porous structure. Low cost and easy modifications of biochar into many composites with metal oxides, heteroatom, and co-doping make biochar an attractive catalyst for organic pollutant degradation through advanced oxidation processes (AOPs). Biochar can be used as a support for other catalysts used in water remediation. Biochar and biochar-supported catalysts have a huge scope of use in AOPs. This includes Fenton-like oxidation, persulfate activation, photocatalysis, sonocatalysis, and electrocatalysis. Biochar contains persistent free radicals, activates $H_2O_2$, $S_2O_8^{2-}$, and $HSO_5^-$, and generates reactive oxygen species which is the key contributor in AOPs. AOP-based wastewater treatments successfully remove dyes, phenols, and pharmaceuticals effectively among other organic pollutants. Metal and nonmetal modified biochar synergize reactive oxygen species generation, electron transfer, adsorption, and oxidation of pollutants.

Metal-free biochar-based AOPs are in their infancy, and the application of biochar in electrocatalysis and sonocatalysis must be critically explored. Sonocatalytic, electrocatalytic, and microwave-assisted degradation of contaminants via biochar/composites is an emerging area. Biochar-based catalysts may be designed and prepared by established methodologies exploiting biochar's tunability, where the improvement of specific surface area, porosity, and the enrichment of surface functional groups is possible. Future research must focus on reusability, stability, selectivity, scaling up, process development, economic aspects, and long-term usage of biochar. The degree of graphitization, pore structure, and specific surface area of biochar affect all AOPs because these variables influence the adsorption capacity and electron transfer properties, hence the catalytic performance.

Complex surface chemistry and heterogeneity compromise biochar's catalytic selectivity and pollutant removal efficiency. The performance of biochar-based catalysts may be improved by enhancing their catalytic selectivity and stability. The future research, thus, may focus on the following topics:

(1) Cost-effective biochar-based catalysts with potential scalable production for AOP application.
(2) Reusable biochar-assisted catalysts must be demonstrated and used.
(3) Biomass properties, modification of surface, catalytic activity, and biochar catalytic mechanisms.
(4) Development and application of sophisticated chemical methods for biochar synthesis and effects of biomass pretreatment on the structure.
(5) Theoretical studies of biochar-based catalysts in environmental remediation.
(6) Catalytic application of bare biochar in AOPs making the process cost-effective.

(7) Ecotoxicity aspects in AOPs considering the presence of PFRs.
(8) Influence of pollutant types and biochars in AOPs.
(9) Biochar-based catalysis in real wastewater treatments.

## Questions and problems

1. What is the expended form of AOP?
2. Which of the following is used to activate $H_2O_2$ in a biochar-based Fenton process to achieve best wastewater remediation results?
   (a) Only biochar
   (b) Only Fenton catalyst
   (c) Biochar + Fenton system
   (d) UV light only
3. What is meant by "heterogeneous Fenton process?"
4. How does a porous biochar make an efficient heterogeneous Fenton like processes for the removal of organic pollutants from wastewater?
5. Why are the heterogeneous Fenton processes more efficient than the homogeneous Fenton processes?
6. Why sulfate radical anion is a better oxidizing agent than hydroxyl radical for the degradation of organic pollutants?
7. What is the Fenton reagent?
8. Give examples of AOPs?
9. Is the rate of contaminant degradation affected by temperature? If yes, then how?
10. How do persistent free radicals (PFRs) present on a biochar-based heterogeneous catalyst help in the degradation of contaminants?
11. Name some commonly used oxidants to generate sulfate free radical ($SO_4^{\bullet}$) in advanced oxidation processes (AOPs).
12. A biochar-based catalyst was used to degrade an antibiotic sulfamethoxazole (SMX). The initial concentration of sulfamethoxazole in wastewater was 20 mg/L and after retention time t, the remaining concentration of sulfamethoxazole in wastewater was 2 mg/L. Calculate the percent sulfamethoxazole removal.
13. The degradation of tetracycline in wastewater was performed by peroxymonosulfate activated by metal-loaded biochar catalyst. The initial Total Organic Carbon (TOC) in the solution was 15 mg/L. After degradation of tetracycline over time t, the TOC was 3 mg/L. Calculate the percent mineralization of tetracycline.

# References

Ahmaruzzaman, M., 2021. Biochar based nanocomposites for photocatalytic degradation of emerging organic pollutants from water and wastewater. Mater. Res. Bull. 140, 111262.

Cai, H., Ma, Z., Zhao, T., 2021. Fabrication of magnetic CuFe2O4@ PBC composite and efficient removal of metronidazole by the photo-Fenton process in a wide pH range. J. Environ. Manag. 300, 113677.

Cao, W., Zeng, C., Guo, X., Liu, Q., Zhang, X., Mameda, N., 2020. Enhanced electrochemical degradation of 2, 4-dichlorophenol with the assist of hydrochar. Chemosphere 260, 127643.

Cha, J.S., Park, S.H., Jung, S.C., Ryu, C., Jeon, J.K., Shin, M.C., Park, Y.K., 2016. Production and utilization of biochar: a review. J. Ind. Eng. Chem. 40, 1–15.

Chen, X.L., Li, F., Chen, H., Wang, H., Li, G., 2020. Fe2O3/TiO2 functionalized biochar as a heterogeneous catalyst for dyes degradation in water under Fenton processes. J. Environ. Chem. Eng. 8 (4), 103905.

Chu, J.H., Kang, J.K., Park, S.J., Lee, C.G., 2021. Enhanced sonocatalytic degradation of bisphenol A with a magnetically recoverable biochar composite using rice husk and rice bran as substrate. J. Environ. Chem. Eng. 9 (4), 105284.

Deng, J., Dong, H., Zhang, C., Jiang, Z., Cheng, Y., Hou, K., Zhang, L., Fan, C., 2018. Nanoscale zero-valent iron/biochar composite as an activator for Fenton-like removal of sulfamethazine. Sep. Purif. Technol. 202, 130–137.

Deng, R., Luo, H., Huang, D., Zhang, C., 2020. Biochar-mediated Fenton-like reaction for the degradation of sulfamethazine: role of environmentally persistent free radicals. Chemosphere 255, 126975.

Devi, P., Dalai, A.K., Chaurasia, S.P., 2020. Activity and stability of biochar in hydrogen peroxide based oxidation system for degradation of naphthenic acid. Chemosphere 241, 125007.

Diao, Z.H., Dong, F.X., Yan, L., Chen, Z.L., Qian, W., Kong, L.J., Zhang, Z.W., Zhang, T., Tao, X.Q., Du, J.J., Jiang, D., 2020. Synergistic oxidation of bisphenol A in a heterogeneous ultrasound-enhanced sludge biochar catalyst/persulfate process: reactivity and mechanism. J. Hazard. Mater. 384, 121385.

Ding, D., Yang, S., Qian, X., Chen, L., Cai, T., 2020. Nitrogen-doping positively whilst sulfur-doping negatively affect the catalytic activity of biochar for the degradation of organic contaminant. Appl. Catal. B Environ. 263, 118348.

Ding, J., Xu, W., Liu, S., Liu, Y., Tan, X., Li, X., Li, Z., Zhang, P., Du, L., Li, M., 2021. Activation of persulfate by nanoscale zero-valent iron loaded porous graphitized biochar for the removal of 17β-estradiol: synthesis, performance and mechanism. J. Colloid Interface Sci. 588, 776–786.

Du, L., Xu, W., Liu, S., Li, X., Huang, D., Tan, X., Liu, Y., 2020. Activation of persulfate by graphitized biochar for sulfamethoxazole removal: the roles of graphitic carbon structure and carbonyl group. J. Colloid Interface Sci. 577, 419–430.

Fan, Z., Zhang, Q., Li, M., Sang, W., Qiu, Y., Xie, C., 2019. Activation of persulfate by manganese oxide-modified sludge-derived biochar to degrade Orange G in aqueous solution. Environ. Pollut. Bioavail. 31 (1), 70–79.

Fan, X., Cao, Q., Meng, F., Song, B., Bai, Z., Zhao, Y., Chen, D., Zhou, Y., Song, M., 2021. A Fenton-like system of biochar loading Fe–Al layered double hydroxides (FeAl-LDH@ BC)/H2O2 for phenol removal. Chemosphere 266, 128992.

Fang, G., Gao, J., Liu, C., Dionysiou, D.D., Wang, Y., Zhou, D., 2014. Key role of persistent free radicals in hydrogen peroxide activation by biochar: implications to organic contaminant degradation. Environ. Sci. Technol. 48 (3), 1902–1910.

Fang, G., Zhu, C., Dionysiou, D.D., Gao, J., Zhou, D., 2015. Mechanism of hydroxyl radical generation from biochar suspensions: implications to diethyl phthalate degradation. Bioresour. Technol. 176, 210–217.

Fang, G., Liu, C., Wang, Y., Dionysiou, D.D., Zhou, D., 2017. Photogeneration of reactive oxygen species from biochar suspension for diethyl phthalate degradation. Appl. Catal. B Environ. 214, 34–45.

Feng, D., Lü, J., Guo, S., Li, J., 2021. Biochar enhanced the degradation of organic pollutants through a Fenton process using trace aqueous iron. J. Environ. Chem. Eng. 9 (1), 104677.

Fu, H., Zhao, P., Xu, S., Cheng, G., Li, Z., Li, Y., Li, K., Ma, S., 2019. Fabrication of Fe3O4 and graphitized porous biochar composites for activating peroxymonosulfate to degrade p-hydroxybenzoic acid: insights on the mechanism. Chem. Eng. J. 375, 121980.

Gan, Q., Hou, H., Liang, S., Qiu, J., Tao, S., Yang, L., Yu, W., Xiao, K., Liu, B., Hu, J., Wang, Y., 2020. Sludge-derived biochar with multivalent iron as an efficient Fenton catalyst for degradation of 4-Chlorophenol. Sci. Total Environ. 725, 138299.

Gholami, P., Dinpazhoh, L., Khataee, A., Orooji, Y., 2019. Sonocatalytic activity of biochar-supported ZnO nanorods in degradation of gemifloxacin: synergy study, effect of parameters and phytotoxicity evaluation. Ultrason. Sonochem. 55, 44–56.

Gholami, P., Dinpazhoh, L., Khataee, A., Hassani, A., Bhatnagar, A., 2020. Facile hydrothermal synthesis of novel Fe-Cu layered double hydroxide/biochar nanocomposite with enhanced sonocatalytic activity for degradation of cefazolin sodium. J. Hazard. Mater. 381, 120742.

Hashemian, S., 2013. Fenton-like oxidation of malachite green solutions: kinetic and thermodynamic study. J. Chem. 2013, 1–7. 809318.

He, J., Xiao, Y., Tang, J., Chen, H., Sun, H., 2019. Persulfate activation with sawdust biochar in aqueous solution by enhanced electron donor-transfer effect. Sci. Total Environ. 690, 768–777.

He, Y., Wang, Y., Hu, J., Wang, K., Zhai, Y., Chen, Y., Duan, Y., Wang, Y., Zhang, W., 2021. Photocatalytic property correlated with microstructural evolution of the biochar/ZnO composites. J. Mater. Res. Technol. 11, 1308–1321.

Ho, S.H., Li, R., Zhang, C., Ge, Y., Cao, G., Ma, M., Duan, X., Wang, S., Ren, N.Q., 2019. N-doped graphitic biochars from C-phycocyanin extracted Spirulina residue for catalytic persulfate activation toward nonradical disinfection and organic oxidation. Water Res. 159, 77–86.

Huang, D., Wang, Y., Zhang, C., Zeng, G., Lai, C., Wan, J., Qin, L., Zeng, Y., 2016. Influence of morphological and chemical features of biochar on hydrogen peroxide activation: implications on sulfamethazine degradation. RSC Adv. 6 (77), 73186–73196.

Huang, D., Luo, H., Zhang, C., Zeng, G., Lai, C., Cheng, M., Wang, R., Deng, R., Xue, W., Gong, X., Guo, X., 2019. Nonnegligible role of biomass types and its compositions on the formation of persistent free radicals in biochar: insight into the influences on Fenton-like process. Chem. Eng. J. 361, 353–363.

Huang, Y.F., Huang, Y.Y., Chiueh, P.T., Lo, S.L., 2020. Heterogeneous Fenton oxidation of trichloroethylene catalyzed by sewage sludge biochar: experimental study and life cycle assessment. Chemosphere 249, 126139.

Huang, Q., Chen, C., Zhao, X., Bu, X., Liao, X., Fan, H., Gao, W., Hu, H., Zhang, Y., Huang, Z., 2021. Malachite green degradation by persulfate activation with CuFe2O4@ biochar composite: efficiency, stability and mechanism. J. Environ. Chem. Eng. 9, 105800.

Im, J.K., Boateng, L.K., Flora, J.R., Her, N., Zoh, K.D., Son, A., Yoon, Y., 2014. Enhanced ultrasonic degradation of acetaminophen and naproxen in the presence of powdered activated carbon and biochar adsorbents. Sep. Purif. Technol. 123, 96–105.

Ji, Q., Cheng, X., Sun, D., Wu, Y., Kong, X., He, H., Xu, Z., Xu, C., Qi, C., Liu, Y., Li, S., 2021. Persulfate enhanced visible light photocatalytic degradation of iohexol by surface-loaded perylene diimide/acidified biochar. Chem. Eng. J. 414, 128793.

Jian, H., Yang, F., Gao, Y., Zhen, K., Tang, X., Zhang, P., Wang, Y., Wang, C., Sun, H., 2021. Efficient removal of pyrene by biochar supported iron oxide in heterogeneous Fenton-like reaction via radicals and high-valent iron-oxo species. Sep. Purif. Technol. 265, 118518.

Jung, K.W., Lee, S.Y., Lee, Y.J., Choi, J.W., 2019. Ultrasound-assisted heterogeneous Fenton-like process for bisphenol A removal at neutral pH using hierarchically structured manganese dioxide/biochar nanocomposites as catalysts. Ultrason. Sonochem. 57, 22–28.

Kou, L., Wang, J., Zhao, L., Jiang, K., Xu, X., 2021. Coupling of $KMnO_4$-assisted sludge dewatering and pyrolysis to prepare Mn, Fe-codoped biochar catalysts for peroxymonosulfate-induced elimination of phenolic pollutants. Chem. Eng. J. 411, 128459.

Kumar, A., Saini, K., Bhaskar, T., 2020. Advances in design strategies for preparation of biochar based catalytic system for production of high value chemicals. Bioresour. Technol. 299, 122564.

Kurian, M., 2021. Advanced oxidation processes and nanomaterials—a review. Clean. Eng. Technol. 2, 100090.

Lai, C., Huang, F., Zeng, G., Huang, D., Qin, L., Cheng, M., Zhang, C., Li, B., Yi, H., Liu, S., Li, L., 2019. Fabrication of novel magnetic $MnFe_2O_4$/bio-char composite and heterogeneous photo-Fenton degradation of tetracycline in near neutral pH. Chemosphere 224, 910–921.

Li, L., Lai, C., Huang, F., Cheng, M., Zeng, G., Huang, D., Li, B., Liu, S., Zhang, M., Qin, L., Li, M., 2019a. Degradation of naphthalene with magnetic bio-char activate hydrogen peroxide: synergism of bio-char and Fe–Mn binary oxides. Water Res. 160, 238–248.

Li, J., Pan, L., Yu, G., Xie, S., Li, C., Lai, D., Li, Z., You, F., Wang, Y., 2019b. The synthesis of heterogeneous Fenton-like catalyst using sewage sludge biochar and its application for ciprofloxacin degradation. Sci. Total Environ. 654, 1284–1292.

Li, F., Duan, F., Ji, W., Gui, X., 2020. Biochar-activated persulfate for organic contaminants removal: efficiency, mechanisms and influencing factors. Ecotoxicol. Environ. Saf. 198, 110653.

Li, L., Liu, S., Cheng, M., Lai, C., Zeng, G., Qin, L., Liu, X., Li, B., Zhang, W., Yi, Y., Zhang, M., 2021a. Improving the Fenton-like catalytic performance of $MnO_x$-$Fe_3O_4$/biochar using reducing agents: a comparative study. J. Hazard. Mater. 406, 124333.

Li, J., Liu, Y., Ren, X., Dong, W., Chen, H., Cai, T., Zeng, W., Li, W., Tang, L., 2021b. Soybean residue based biochar prepared by ball milling assisted alkali activation to activate peroxydisulfate for the degradation of tetracycline. J. Colloid Interface Sci. 599, 631–641.

Liang, J., Xu, X., Zaman, W.Q., Hu, X., Zhao, L., Qiu, H., Cao, X., 2019. Different mechanisms between biochar and activated carbon for the persulfate catalytic degradation of sulfamethoxazole: roles of radicals in solution or solid phase. Chem. Eng. J. 375, 121908.

Liu, W.J., Jiang, H., Yu, H.Q., 2015. Development of biochar-based functional materials: toward a sustainable platform carbon material. Chem. Rev. 115 (22), 12251–12285.

Liu, J.J., Diao, Z.H., Liu, C.M., Jiang, D., Kong, L.J., Xu, X.R., 2018. Synergistic reduction of copper (II) and oxidation of norfloxacin over a novel sewage sludge-derived char-based catalyst: performance, fate and mechanism. J. Clean. Prod. 182, 794–804.

Liu, C., Chen, L., Ding, D., Cai, T., 2019. From rice straw to magnetically recoverable nitrogen doped biochar: efficient activation of peroxymonosulfate for the degradation of metolachlor. Appl. Catal. B Environ. 254, 312–320.

Liu, Q., Bai, X., Su, X., Huang, B., Wang, B., Zhang, X., Ruan, X., Cao, W., Xu, Y., Qian, G., 2020. The promotion effect of biochar on electrochemical degradation of nitrobenzene. J. Clean. Prod. 244, 118890.

Liu, Q., Jiang, S., Su, X., Zhang, X., Cao, W., Xu, Y., 2021. Role of the biochar modified with $ZnCl_2$ and $FeCl_3$ on the electrochemical degradation of nitrobenzene. Chemosphere 275, 129966.

Luo, H., Lin, Q., Zhang, X., Huang, Z., Liu, S., Jiang, J., Xiao, R., Liao, X., 2019. New insights into the formation and transformation of active species in nZVI/BC activated persulfate in alkaline solutions. Chem. Eng. J. 359, 1215–1223.

Lv, Y., Zhang, J., Asgodom, M.E., Liu, D., Xie, H., Qu, H., 2019. Study on the degradation of accumulated bisphenol S and regeneration of magnetic sludge-derived biochar upon microwave irritation in the presence of hydrogen peroxide for application in integrated process. Bioresour. Technol. 293, 122072.

Ma, D., Yang, Y., Liu, B., Xie, G., Chen, C., Ren, N., Xing, D., 2021. Zero-valent iron and biochar composite with high specific surface area via $K_2FeO_4$ fabrication enhances sulfadiazine removal by persulfate activation. Chem. Eng. J. 408, 127992.

Mian, M.M., Liu, G., Fu, B., Song, Y., 2019. Facile synthesis of sludge-derived $MnO_x$-N-biochar as an efficient catalyst for peroxymonosulfate activation. Appl. Catal. B Environ. 255, 117765.

Miklos, D.B., Remy, C., Jekel, M., Linden, K.G., Drewes, J.E., Hübner, U., 2018. Evaluation of advanced oxidation processes for water and wastewater treatment—a critical review. Water Res. 139, 118–131.

Ouyang, D., Chen, Y., Yan, J., Qian, L., Han, L., Chen, M., 2019. Activation mechanism of peroxymonosulfate by biochar for catalytic degradation of 1, 4-dioxane: important role of biochar defect structures. Chem. Eng. J. 370, 614–624.

Pan, X., Gu, Z., Chen, W., Li, Q., 2021. Preparation of biochar and biochar composites and their application in a Fenton-like process for wastewater decontamination: a review. Sci. Total Environ. 754, 142104.

Qi, K., Song, M., Xie, X., Wen, Y., Wang, Z., Wei, B., Wang, Z., 2022. CQDs/biochar from reed straw modified Z-scheme $MgIn_2S_4$/BiOCl with enhanced visible-light photocatalytic performance for carbamazepine degradation in water. Chemosphere 287, 132192.

Qin, F., Peng, Y., Song, G., Fang, Q., Wang, R., Zhang, C., Zeng, G., Huang, D., Lai, C., Zhou, Y., Tan, X., 2020. Degradation of sulfamethazine by biochar-supported bimetallic oxide/persulfate system in natural water: performance and reaction mechanism. J. Hazard. Mater. 398, 122816.

Shan, R., Han, J., Gu, J., Yuan, H., Luo, B., Chen, Y., 2020. A review of recent developments in catalytic applications of biochar-based materials. Resour. Conserv. Recycl. 162, 105036.

Siara, S., Elvis, C., Harishkumar, R., Chellam, P.V., 2022. $ZnAl_2O_4$ supported on lychee-biochar applied to ibuprofen photodegradation. Mater. Res. Bull. 145, 111530.

Sun, Y., Iris, K.M., Tsang, D.C., Fan, J., Clark, J.H., Luo, G., Zhang, S., Khan, E., Graham, N.J., 2020. Tailored design of graphitic biochar for high-efficiency and chemical-free microwave-assisted removal of refractory organic contaminants. Chem. Eng. J. 398, 125505.

Talukdar, K., Jun, B.M., Yoon, Y., Kim, Y., Fayyaz, A., Park, C.M., 2020. Novel Z-scheme $Ag_3PO_4$/$Fe_3O_4$-activated biochar photocatalyst with enhanced visible-light catalytic performance toward degradation of bisphenol A. J. Hazard. Mater. 398, 123025.

Tang, R., Gong, D., Deng, Y., Xiong, S., Zheng, J., Li, L., Zhou, Z., Su, L., Zhao, J., 2022. π-π stacking derived from graphene-like biochar/g-C3N4 with tunable band structure for photocatalytic antibiotics degradation via peroxymonosulfate activation. J. Hazard. Mater. 423, 126944.

Wan, Z., Sun, Y., Tsang, D.C., Iris, K.M., Fan, J., Clark, J.H., Zhou, Y., Cao, X., Gao, B., Ok, Y.S., 2019. A sustainable biochar catalyst synergized with copper heteroatoms and CO2 for singlet oxygenation and electron transfer routes. Green Chem. 21 (17), 4800–4814.

Wang, S., Wang, J., 2020. Peroxymonosulfate activation by Co9S8@ S and N co-doped biochar for sulfamethoxazole degradation. Chem. Eng. J. 385, 123933.

Wang, H., Guo, W., Liu, B., Wu, Q., Luo, H., Zhao, Q., Si, Q., Sseguya, F., Ren, N., 2019b. Edge-nitrogenated biochar for efficient peroxydisulfate activation: an electron transfer mechanism. Water Res. 160, 405–414.

Wang, R.Z., Huang, D.L., Liu, Y.G., Zhang, C., Lai, C., Wang, X., Zeng, G.M., Gong, X.-M., Duan, A., Zhang, Q., Xu, P., 2019a. Recent advances in biochar-based catalysts: properties, applications and mechanisms for pollution remediation. Chem. Eng. J. 371, 380–403.

Wang, Y., Zhu, X., Feng, D., Hodge, A.K., Hu, L., Lü, J., Li, J., 2019c. Biochar-supported FeS/Fe3O4 composite for catalyzed Fenton-type degradation of ciprofloxacin. Catalysts 9 (12), 1062.

Wang, T., Dissanayake, P.D., Sun, M., Tao, Z., Han, W., An, N., Gu, Q., Xia, D., Tian, B., Ok, Y.S., Shang, J., 2021b. Adsorption and visible-light photocatalytic degradation of organic pollutants by functionalized biochar: role of iodine doping and reactive species. Environ. Res. 197, 111026.

Wang, Z., Cai, X., Xie, X., Li, S., Zhang, X., Wang, Z., 2021c. Visible-LED-light-driven photocatalytic degradation of ofloxacin and ciprofloxacin by magnetic biochar modified flower-like Bi2WO6: the synergistic effects, mechanism insights and degradation pathways. Sci. Total Environ. 764, 142879.

Wang, H., Guo, W., Si, Q., Liu, B., Zhao, Q., Luo, H., Ren, N., 2021a. Multipath elimination of bisphenol A over bifunctional polymeric carbon nitride/biochar hybrids in the presence of persulfate and visible light. J. Hazard. Mater. 417, 126008.

Xie, Y., Wang, X., Tong, W., Hu, W., Li, P., Dai, L., Wang, Y., Zhang, Y., 2020. FexP/biochar composites induced oxygen-driven Fenton-like reaction for sulfamethoxazole removal: performance and reaction mechanism. Chem. Eng. J. 396, 125321.

Xin, S., Ma, B., Liu, G., Ma, X., Zhang, C., Ma, X., Gao, M., Xin, Y., 2021. Enhanced heterogeneous photo-Fenton-like degradation of tetracycline over CuFeO2/biochar catalyst through accelerating electron transfer under visible light. J. Environ. Manag. 285, 112093.

Xu, H., Zhang, Y., Li, J., Hao, Q., Li, X., Liu, F., 2020a. Heterogeneous activation of peroxymonosulfate by a biochar-supported Co3O4 composite for efficient degradation of chloramphenicols. Environ. Pollut. 257, 113610.

Xu, L., Fu, B., Sun, Y., Jin, P., Bai, X., Jin, X., Shi, X., Wang, Y., Nie, S., 2020b. Degradation of organic pollutants by Fe/N co-doped biochar via peroxymonosulfate activation: synthesis, performance, mechanism and its potential for practical application. Chem. Eng. J. 400, 125870.

Yang, Z., Zhu, P., Yan, C., Wang, D., Fang, D., Zhou, L., 2021. Biosynthesized Schwertmannite@ biochar composite as a heterogeneous Fenton-like catalyst for the degradation of sulfanilamide antibiotics. Chemosphere 266, 129175.

Yi, Y., Tu, G., Tsang, P.E., Fang, Z., 2020. Insight into the influence of pyrolysis temperature on Fenton-like catalytic performance of magnetic biochar. Chem. Eng. J. 380, 122518.

Yu, J., Tang, L., Pang, Y., Zeng, G., Wang, J., Deng, Y., Liu, Y., Feng, H., Chen, S., Ren, X., 2019. Magnetic nitrogen-doped sludge-derived biochar catalysts for persulfate activation: internal electron transfer mechanism. Chem. Eng. J. 364, 146–159.

Yu, C., Tang, J., Li, D., Chen, Y., 2021. Green synthesized nanosilver-biochar photocatalyst for persulfate activation under visible-light illumination. Chemosphere 284, 131237.

Zhang, H., Tang, L., Wang, J., Yu, J., Feng, H., Lu, Y., Chen, Y., Liu, Y., Wang, J., Xie, Q., 2020. Enhanced surface activation process of persulfate by modified bagasse biochar for degradation of phenol in water and soil: active sites and electron transfer mechanism. Colloids Surf. A Physicochem. Eng. Asp. 599, 124904.

Zhang, Y., Jiang, Q., Jiang, S., Li, H., Zhang, R., Qu, J., Zhang, S., Han, W., 2021a. One-step synthesis of biochar supported nZVI composites for highly efficient activating persulfate to oxidatively degrade atrazine. Chem. Eng. J. 420, 129868.

Zhang, Y., Xu, M., Liang, S., Feng, Z., Zhao, J., 2021b. Mechanism of persulfate activation by biochar for the catalytic degradation of antibiotics: synergistic effects of environmentally persistent free radicals and the defective structure of biochar. Sci. Total Environ. 794, 148707.

Zhang, S., Wei, Y., Metz, J., He, S., Alvarez, P.J., Long, M., 2022. Persistent free radicals in biochar enhance superoxide-mediated Fe (III)/Fe (II) cycling and the efficacy of CaO2 Fenton-like treatment. J. Hazard. Mater. 421, 126805.

Zhao, Y., Yuan, X., Li, X., Jiang, L., Wang, H., 2021. Burgeoning prospects of biochar and its composite in persulfate-advanced oxidation process. J. Hazard. Mater. 409, 124893.

Zhong, Q., Lin, Q., Huang, R., Fu, H., Zhang, X., Luo, H., Xiao, R., 2020. Oxidative degradation of tetracycline using persulfate activated by N and Cu codoped biochar. Chem. Eng. J. 380, 122608.

Zhou, X., Zhu, Y., Niu, Q., Zeng, G., Lai, C., Liu, S., Huang, D., Qin, L., Liu, X., Li, B., Yi, H., 2021. New notion of biochar: a review on the mechanism of biochar applications in advannced oxidation processes. Chem. Eng. J. 416, 129027.

Zhu, N., Li, C., Bu, L., Tang, C., Wang, S., Duan, P., Yao, L., Tang, J., Dionysiou, D.D., Wu, Y., 2020c. Bismuth impregnated biochar for efficient estrone degradation: the synergistic effect between biochar and bi/Bi2O3 for a high photocatalytic performance. J. Hazard. Mater. 384, 121258.

Zhu, F., Wu, Y., Liang, Y., Li, H., Liang, W., 2020b. Degradation mechanism of norfloxacin in water using persulfate activated by BC@ nZVI/Ni. Chem. Eng. J. 389, 124276.

Zhu, X., Li, J., Xie, B., Feng, D., Li, Y., 2020a. Accelerating effects of biochar for pyrite-catalyzed Fenton-like oxidation of herbicide 2, 4-D. Chem. Eng. J. 391, 123605.

Zhu, J., Song, Y., Wang, L., Zhang, Z., Gao, J., Tsang, D.C., Ok, Y.S., Hou, D., 2022. Green remediation of benzene contaminated groundwater using persulfate activated by biochar composite loaded with iron sulfide minerals. Chem. Eng. J. 429, 132292.

## Further reading

Wang, S., Wang, J., 2019. Activation of peroxymonosulfate by sludge-derived biochar for the degradation of triclosan in water and wastewater. Chem. Eng. J. 356, 350–358.

Xie, S., Li, M., Liao, Y., Qin, Q., Sun, S., Tan, Y., 2021. In-situ preparation of biochar-loaded particle electrode and its application in the electrochemical degradation of 4-chlorophenol in wastewater. Chemosphere 273, 128506.

Zhang, P., Tan, X., Liu, S., Liu, Y., Zeng, G., Ye, S., Yin, Z., Hu, X., Liu, N., 2019. Catalytic degradation of estrogen by persulfate activated with iron-doped graphitic biochar: process variables effects and matrix effects. Chem. Eng. J. 378, 122141.

#  Index

Note: Page numbers followed by *f* indicate figures, *t* indicate tables, and *b* indicate boxes.

## A

Acid activation, 262
Acid Blue 25, 434–435, 435*f*
Acid blue 158, 437–438, 438*f*
Acid dyes, 434–435, 435*f*
  adsorption
    on biochars, 448–449*t*, 455
    on modified biochars, 452–453*t*
Acid effect, 47
Acidic functionalization, 779
Acid orange 7 (AO7), 448–449*t*, 455
Actinides
  applications, 322
  aqueous media, behavior in, 322
  characteristics, 321
  definition, 321
  elements, classification of, 321
  and lanthanides, removal and recovery from aqueous media (*see* Inner transition elements (ITEs))
  paramagnetic and pyrophoric, 321
Activated carbon (AC), 210, 212, 225, 395–396, 398–406, 420–421, 443–444, 531–532, 737–739, 767
  microplastic retention, 374–386
  per-/poly-fluoroalkyl substances (PFAS) remediation, 561–562
  from spruce bark, 382–383
Adam retort, 63, 63*f*
Adams-Bohart model, 190–191
Adsorbates, functional moieties in, 646–648
Adsorbents, 328–332*t*, 412–414*t*, 442–443, 749–750
  arsenic, 280
  biochar, 398–406, 577–581*t*
  dosage, 335–336
  parameters, 450–451*t*
  pharmaceutical adsorption performance, 399–405*t*
  properties, 442
  recycling, 304–306
  regeneration, 304–306
  safe disposal, 304–306
  in single-element aqueous systems, 293–297
Adsorption, 280, 439, 442–443, 528–530, 529*t*
  in batch reactors, 745–758
    dissolved organic matter (DOM) removal, 750
    organic micropollutants (OMPs) removal, 750–753
    ultraviolet absorbance at 254nm ($UVA_{254}$) removal, 750
  in flow-through column adsorbers, 748–749*f*, 753–755
    dissolved organic matter (DOM) removal, 754
    organic micropollutants (OMPs) removal, 754–755
    ultraviolet absorbance at 254nm ($UVA_{254}$) removal, 754
  mechanism, 442
  performance, 745–758
Adsorption kinetics
  inner transition element adsorption, 336–338, 339*f*
  models, 173–174
    Bangham equation, 178
    Elovich equation, 176–177
    intraparticle diffusion (IPD) equation, 177, 177*f*
    linear film diffusion equation, 178
    mixed-order (MO) model, 176
    pseudo-first order (PFO) model, 174
    pseudo-second-order (PSO) model, 175
    revised pseudo-second-order (rPSO) model, 175–176
  per-/poly-fluoroalkyl substances (PFAS) sorption, 570–572, 573–576*t*
Advanced oxidation process (AOP), 440–441, 776
  biochar application, 781–784*t*, 786–798
  biochar-based catalysts, application of, 781–784*t*
  fenton-like AOP, 786–793
  functionalities, 787*t*
  mechanism/active species in, 787*t*
  other, 796–798
  overview, 776, 776*f*
  ozone-based, 776
  persulfate-activated, 793–794
  photocatalytic, 794–796
  reaction mechanisms, 788–789*f*

Aerobic digestion process, 212
Aerodynamic equivalent diameter (AED), 100–101
Agrichar, 549
Alkali- and alkali-earth metal (AAEM) pretreatment, 765–766
Alkali index, 25–27, 115–116, 117–118$b$
AlOOH nanoflake biochar composite, 257
Aluminum electrodes, 439–440
Aluminum-enriched tetrapak-derived biochar, 303
Aluminum-modified peanut shell biochar, 514–515
American Society for Testing Materials (ASTM) standards, 103–104, 104$t$, 108$t$
Ammonium/ammonia nitrogen ($NH_4^+$-N) removal, 500–501, 509, 510$f$, 511, 518
  batch sorption kinetics tests, 493–495, 496$f$
  influent concentration (IC) effect, 511, 512$f$
Anaerobic digestion, 212, 441–442
Anaerobic toxicity assay (ATA), 16
Angle of friction, 30–31, 102
Angle of repose, 30, 101–102
Aniline dyes, 437, 437$f$
Animal waste, 487–488
Anion exchange capacity (AEC), 140, 238–239, 241–242, 495
Anionic dye, 435, 448–449$t$, 455, 458–459
ANOVA test, 501–519
Antibiotics, 397, 794
AOP. *See* Advanced oxidation process (AOP)
Aqueous contaminant removal, nanobiochar for
  inorganic contaminants removal, 691
    anions, 691
    heavy metals, 692–694, 695–696$t$
  organic contaminant removal
    conventional organic contaminants, removal of, 684–687
    emerging contaminants, adsorption of, 687–691, 688–689$t$
Aqueous firefighting foam (AFFF), 556, 561–562, 570
Argentine kiln, 54–55, 55$f$
Aromaticity, 408–409
Arsenate, 234–235
Arsenic, 277–280, 622–623
  overview, 277–280
  removal
    adsorption, 280
    biochar-based composite sorbents, 284–285
    biochar production, 281–282
    engineering strategies for enhancement, 282–291
    mechanisms, 290$f$
    scaled-up drinking water arsenic treatment (*see* Scaled-up drinking water arsenic treatment)
    traditional techniques, 280
Artificial neural network analysis, 326
Ash content, 19–21, 45–46, 106–107
Ash-free matter (AFM) biochar, 600–602, 601$f$
Ash pretreatment, 762–764
$AsO_4^{3-}$ removal, 251, 252–254$t$, 257
$AsO_4^{3-}$ sorption, 245–248
Attenuated total reflectance (ATR), 370$t$, 372
Attritor ball mills, 670–671
Auger pyrolysis reactor, 67–70
Azo dyes, 435–436

# B
Backbone screening, 291–297
Bacteria, 475–476
Ball-milled nanobiochar, 678, 684–685, 686$f$, 690–691, 693–694
  Ni(II) adsorption, 692
  nitrogen doping, 674–675, 685–686
  thiol-modified ball-milled biochar, 675, 686–687
  volatile organic compounds (VOCs) adsorption, 687
Ball milling, 250, 668–672, 690, 694–697
Ball-to-powder ratio, 671
Bamboo biochars, 246–247, 344, 349–351, 456
Banana pseudostem biochar, 415
Bangham model, 178
Bare biochar, 777–778
Bark-based activated carbon, 379
Base-to-acid ratio, 25–27, 116, 117–118$b$
Basic dyes
  adsorption
    on biochars, 446–447$t$
    on modified biochars, 450–451$t$
Batch sorption, 153–154
Bed agglomeration index (BAI), 26–27, 116–118, 117–118$b$
Bed-depth-service-time (BDST) model, 187–189, 189$f$
Bio-based sorbents, 530
Biochar (BC), 39, 643
  advantages, 739
  analysis, 118–121
  application, 717
  chemical formula, 132
  classification, 668
  constructed treatment wetland (CTW), for CAFO wastewater treatment (*see* Constructed treatment wetland (CTW), for wastewater treatment)

definition, 323–324
distinctive characteristic, 643
enhancement methods,
 758–767
 alkali- and alkali-earth
  metal (AAEM)
  pretreatment, 765–766
 ash pretreatment, 762–764
 co-pyrolysis thermal air
  activation (CPTA),
  759–760
 double heating, 760–762
 regeneration, 760–762
for environmental application,
 135–136
functional moieties in,
 646–648
loading, 285
metal loadings on, 780,
 781–784t
nanobiochar
 (see Nanobiochar)
for oil sorption (see Oil
 sorption)
in photocatalysis, 795
physical properties, 98
 angle of friction, 102
 angle of repose, 101–102
 bulk/apparent density,
  98–99
 heat capacity, 102–103
 particle density, 98
 particle size and particle size
  distribution (PSD), 99–101
 thermal conductivity,
  102–103
production, 479, 480f
 biomass, structural changes,
  644–646
 subsequent biochar use,
  132–133
 thermal conversion
  processes, 95–98
 via gasification, 120–121b,
  132–133
properties, 460, 479, 480f
as sorption and plant growth
 media, 488–490
sorption properties, 643

surface characteristics, 662
thermal related properties,
 103
 eutectic point analysis,
  114–118
 heating value analysis,
  109–113
 heat of combustion, 111
 proximate analysis
  (see Proximate analysis)
 stoichiometry of reactions,
  111–113
 ultimate analysis, 107–109
 Van Krevelen plot, 107–109
transportation, 717
Biochar adsorption system
 designs, 193–194
 batch sorption, 153–154
 continuous flow studies,
  153–154
 continuous stirred tank, types
  of
  multistage downflow,
   153–154, 154–155f
  multistage upflow, 153–154,
   154–155f
  multistage upflow stirred
   columns, 153–154,
   154–155f
  single-stage downflow,
   153–154, 154–155f
  single-stage upflow,
   153–154, 154–155f
 dynamic sorption, 153–154
 sorption column design
  fixed-bed adsorbers
   (see Fixed-bed/expanded
   bed adsorber)
  moving/fluidized-bed
   adsorber, 181–182, 182f
 sorption isotherm models
  (see Sorption isotherm
  models)
 sorption kinetic models,
  173–174
  Bangham equation, 178
  Elovich equation, 176–177
  intraparticle diffusion (IPD)
   equation, 177, 177f

  linear film diffusion
   equation, 178
  mixed-order (MO) model,
   176
  pseudo-first order (PFO)
   model, 174
  pseudo-second-order (PSO)
   model, 175
  revised pseudo-second-
   order (rPSO) model,
   175–176
Biochar-assisted filters, 480
Biochar-augmented sand filter,
 478–479, 479f
Biochar-based catalysts
 bare biochar, 777–778
 modified biochar, 778–786
  co-doped biochar, 785–786
  doped with heteroatoms,
   785
  loaded with metals, 780–785
Biochar-based sorptive
 interactions, 648–654,
 650f
 catalytic degradation by
  radicals, 654
 cation bridging, 653
 cation exchange, 652–653
 cation-π bonding, 649–651
 electron donor-acceptor
  interactions, 649
 electrostatic interactions,
  652–653
 factors effecting, 655
  feedstock, 655–656
  modification, 657–659
  pH effect, 659–660
  pyrolysis method, 656–657
  pyrolysis temperature,
   656–657
 hydrogen bonding, 651–652
 hydrophobic interaction, 654
 pore filling, 653
 precipitation interaction,
  652–653
 surface complexation, 653
Biochar characterization
 ash concentration, 139–140
 chemical properties, 137

Biochar characterization (*Continued*)
electrical conductivity, 137
H/C$_{org}$ ratio, 139
inorganic carbon, 139
ion exchange capacity, 140–141
pH, 137
physical properties, 141, 142$t$
pore size distribution, 141–142
sampling and storage, 137
specific surface area (SSA), 143
spectroscopic analysis, 143
fourier-transform infrared (FTIR) spectroscopy, 143–144, 145$t$
solid-state $^{13}$C NMR spectroscopy, 145–146, 146$t$
X-ray photoelectron spectroscopy, 144
total carbon, 137–138
total porosity, 141–142
Biochar conversion
activated carbon, 121–129
adsorption isotherms for, 123–129
chemical activation process, 122–123
physical activation processes, 122
Biochar-supported nZVI (nZVI@BC), 611–612
advantages, 611–612, 618–619
characterization
scanning electron microscopy (SEM), 619$f$, 620
transmission electron microscopy (TEM), 619$f$, 620–621, 621$f$
X-ray diffraction (XRD) analysis, 621–622, 622$f$
X-ray photoelectron spectroscopy (XPS) analysis, 622, 622$f$
properties of, 614

synthesis
borohydride (BH$_4^-$) reduction, 613$f$, 614, 615$t$
bottom-up approach, 612–613
carbothermal reduction, 614–618
green synthesis strategy, 612–613
morphology of, 612–613, 613$f$
top-down synthesis, 612–613
for water remediation
heavy metal(loid)s, removal of, 622–626, 623$f$, 625$t$
organic contaminant removal, 627–631, 628–630$f$, 632$f$
Biochemical methane potential (BMP), 14–15
efficiency, 15–16, 15–16$b$
Biodiesel production, 13–14
Bioethanol production, 8–13
Biofilm formation, 483–484
Biogas production, 14–16
Biological conversion, preparation for
biomass properties, 6–16
biodiesel production, 13–14
bioethanol production, 8–13
biogas production, 14–16
Biological oxygen demand (BOD), 492–493
Biomass
ash, fusibility of, 118–119, 120$t$
conversion, 3–6
biological conversion processes, 5
into energy, 2
into liquid fuels, 2–3
thermal conversion processes, 3–5
feedstock, 255–257
materials, 1
properties (*see* Biomass properties)
role of, 1–2, 2$f$

Biomass iodine number (BIN), 124–126
Biomass properties
angle of friction, 30–31
angle of repose, 30
biological conversion, preparation for, 6–16
biodiesel production, 13–14
bioethanol production, 8–13
biogas production, 14–16
bulk density, 27–28
heat capacity, 31–32
particle density, 28–29
particle size distribution (PSD), 29–30
thermal conductivity, 31–32
thermochemical conversion, preparation for, 16–32
eutectic point analysis, 23–27
heating value analysis, 17–18
proximate analysis, 18–20
ultimate analysis, 21–23
Bioremediation, 528–530, 529$t$
Biosolids, 213–214, 583
Biosorption, 439
Bisphenol A, 366–367
Borohydride (BH$_4^-$) reduction, 613$f$, 614, 615$t$
Boyd diffusion model, 543–544
Brazilian beehive kiln, 54–55, 55$f$
Breakthrough curves, 182–184, 183$f$, 349–351
Bricarbras Furnaces, 64
Brick kilns, 53–59
Brunauer-Emmett-Teller (BET) isotherm model, 167
theory, 122, 599, 600$f$
Bulk density, 27–28, 98–99
Butler and Ockrent model, 168–169
By-product gasification chars, 74–75

C
CaAl-LDH/IPB composites, 343–346

Caberra-Mott model simulation, 620
Cadmium (Cd), 490
CAHB. *See* Charge-assisted H-bonding (CAHB)
*Calotropis gigantea* fibers (CGFs), 565
Canola/rapeseed oil, 13
Carbamazepine adsorption, 687–690
Carbohydrates, 10–11
Carbon-based methods
    for microplastic purification, 374–378
Carbonization, 39
    equipment, 50–79
    hydrothermal, 77–79
    reactions, 49–50
    wood
        brick kilns, 53–59
        concrete kilns, 53–59
        earth kilns, 52–53
        metal kilns, 59–61
Carbon to nitrogen ratio, 14–15
Carbonyl groups, 647
Carbothermal reduction, 614–618
Carboxylic acids, 647
Catalytic degradation by radicals, 654
Category indicator, 711
Cation bridging, 650$f$, 653
Cation exchange, 652–653
Cation exchange capacity (CEC), 140–141, 241–242
Cationic dye, 445–454, 446–447$t$, 457
Cation-$\pi$ bonding, 649–651, 650$f$
Cellulose, 49–50, 644
    biochar, 379
    decomposition, 562–563
Centrifugation, 668–669, 672–674
Characterization factor, 711
Charcoal
    as by-product, 67
        auger pyrolysis reactors, 67–70
        fast pyrolysis reactors, 71
        flash/high-pressure pyrolysis reactors, 70–71
        rotary drum reactors, 67–70
        solar pyrolysis reactors, 71, 72–73$t$
    by-product gasification chars, 74–75
    for environmental remediation, 72–79
    historical milestone, 42$t$
    from hydrothermal carbonization, 77–79
    intermediate gasification chars, 75–77
    for other uses, 72–79
    physicochemical properties, 43–44
    production, 52–66
        in Brazil, 41–42, 46$t$, 55–56
        equipment, 51
        retorts, 61–66, 65–66$t$, 67–68$f$
        wood carbonization kilns, 52–61
    yield and quality
        acid effect on, 47
        ash content effect on, 45–46
        heating rate effect on, 47–48
        pressure effect on, 48
        temperature effect on, 44–45
Charge-assisted H-bonding (CAHB), 650$f$, 651–652, 661
Chemical activation, 778–779
Chemical oxygen demand (COD), 492–493, 500–501, 517–518
    influent and effluent concentrations, 507–508, 508$f$
    removal, 507–509, 508$f$
    of wastewater, 124, 125$t$, 130
Chemisorption, 442, 694
    interactions, 661
    oxyanion, 263–264
Chimney kiln, 59–60
Chitosan, 261
Chlorinated organic compounds (COCs), 627
4-Chlorophenol (4-CP) removal, 630, 630$f$
Ciprofloxacin, 406–407, 409, 410$f$, 411, 412–414$t$
Clamps, 40
Clark model, 192–193
Coagulation, 280, 439
Coal
    fouling factor, 116$t$, 132
    slagging factor, 116$t$, 132
Coconut shells, 95
Co-doped biochar, 785–786
Colloidal biochar, 672–673
Colloid filtration theory (CFT), 481
Combined heat and power (CHP), 74
Completely mixed batch reactors (CMBR), 745, 749–750
Completely mixed reactors (CMRs), 309–310
Concentrated animal feeding operations (CAFOs)
    biochar-enhanced CTW, for wastewater treatment, 492–493
    ammonium, nitrate, and phosphate sorption, batch sorption tests for, 493–495, 496$f$
    grain size distribution, 493, 494$f$
    greenhouse study (*see* Greenhouse CTW tests, for CAFO wastewater treatment)
    definition, 487–488
    nutrient contamination, source of, 487–488
    nutrient management plans (NMPs), 488
    treatment technologies, need for, 488
    waste management, 488
Concrete kilns, 53–59
Congo red dye, 433–434, 434$f$, 455–456

Constructed treatment wetland (CTW), for wastewater treatment, 519–521
  aerated VFCWs, for low C/N wastewater treatment, 492
  biochar-amended soils, effectiveness of, 491
  CAFO wastewater treatment, biochar-enhanced CTW for, 492–493
    ammonium, nitrate, and phosphate sorption, batch sorption tests for, 493–495, 496$f$
    grain size distribution, 493, 494$f$
    greenhouse study (see Greenhouse CTW tests, for CAFO wastewater treatment)
  disadvantages, 490–491
  floating treatment wetlands (FTWs), 492
  natural/engineered geomedia, augmented with, 491–492
  nitrate removal rates, in FWS and SSF wetlands, 492
  phosphate removal, 490–491
  reactive media, use of, 491–492
  sorption and plant growth media, biochar application as, 488–491
Constructed wetland (CW), 491
Contaminant adsorption, 415
Contaminant-biochar interactions, prediction of, 660–662
Contamination methods, 475–477
Convection drying ovens, 6–7
Conventional activated sludge (CAS) process, 374–375
Conventional anaerobic pyrolysis (CAP) biochars, 759
Conventional biofilters, 479
Conversion efficiency, 17
Cooking oil adsorption, 545–546
Coprecipitation, 239–240
Co-pyrolysis thermal air activation (CPTA), 759–760
Corncob-derived biochar, 324–325, 327–333
Corn stalk-based biochar, 626
Corn stover biochar production, 719–727
  goal, 720–721
  interpretation, 727
  life cycle impact assessment (LCIA), 723–727, 733
  life cycle inventory (LCI), 721–723
    application, 722
    avoided fertilizer production, 722–723, 724$t$
    collection, 721
    handling, 722
    transportation, 722
  scope, 720–721
Corn straw biochar, 379
Corn-straw-derived biochar, PFAS sorption, 562–563, 562$f$
  pH dependence, 569
Cotton gin trash (CGT) ash, 115, 116$t$
  agglomeration index, 117–118$b$
  alkali index, 117–118$b$
  base-to-acid ratio, 117–118$b$
  fouling factor, 117$b$
  slagging factor, 117$b$
Cotton stalk agricultural waste, 458–459
Cow-manure-derived biochar, 338
Crab shell activated biochar (CSAB), 538, 539–541$t$, 543
  petroleum-based diesel oil uptake, 545–546
Crab shell biochar (CSB), 538, 539–541$t$, 543
  petroleum-based diesel oil uptake, 545–546
Cradle-to-gate system boundary, 707–708
Cross-polarization (CP), 145–146
Crude oil, 544–546, 544$f$
Cyclohexanol, 602–605, 603–604$f$

D
Dairy manure ash, 115, 116$t$
  fouling factor, 116$t$, 132
  slagging factor, 116$t$, 132
*Daphnia magna*, 306
Dastgheib and Rockstraw model, 171–172
Date palm frond biochar, 446–447$t$, 457–458
Date stone-derived activated carbon, 455–456
Deashing treatments, 216
Decentralized wastewater treatment systems, 476–477
Deionized (DI) water, 599, 600$f$, 603–604
Demineralization, 216
Density functional theory, 599
Derjaguin-Landau-Verwey-Overbeek (DLVO) theory, 482
Desorption test, 495
2,4-Dichlorophenol, 798
*Dictyophora indusiate*-derived biochar supported sulfide NZVI (DI-SNZVI), 345, 347–348
Diesel oil, 543
Differential thermogravimetric (DTG) analysis, 562–563
Dipole–dipole interactions, 648–649
Direct dyes, 433–434
Direct polarization (DP), 145–146
Disperse dyes, 432
Dissolved organic carbon (DOC), 584–585
Dissolved organic matter (DOM), 598, 602–606, 603$f$
  removal, 750, 754
Dolomite, 259
Double heating, 760–762

Douglas fir biochar, 534, 538, 539–541*t*, 545–546, 547*f*, 548–549, 568–571
2,4-D pesticide, 599, 600*f*, 602
Drinking water treatment plants (DWTPs), 374–375
Dry ball milling, 672
Dry solid pyrolysis, 778–779
Dual coadsorption, 299
Dubinin-Radushkevich model, 164, 411
Dyes
  acid dye, 434–435, 435*f*
  adsorption technology, 442–443
  aniline, 437, 437*f*
  azo dyes, 435–436
  basic, 434
  classification, 431–439, 431*f*
  direct, 433–434
  discharge
    from dyehouse, 430, 430*f*
    effect on river quality, 430, 430*f*
  disperse, 432, 433*f*
  effluent treatment technologies, 430, 439
  metal complex, 437–438
  mordant, 438–439
  production, 429–430
  properties, 430
  reactive, 430*f*, 431–432
  removal technologies, 439–442
    biological processes, 441–442
    chemical processes, 440–441
    physical processes, 439–440
    using adsorption onto biochars, 443–459, 446–453*t*
  sulfur, 436, 437*f*
  vat, 433, 433*f*
Dynamic adsorption equilibrium, 410–411

## E
Earth kilns, 52–53
*E. coli*, 478–480, 482–483
Effluent flow rate, 500
Electrical conductivity, 137
Electrochemical advanced oxidation process (eAOPs), 797
Electrocoagulation, 280
Electrokinetic coagulation, 439–440
Electron donor-acceptor (EDA) interactions, 649, 650*f*
Electron rich carbonyl oxygen, 647
Electrostatic attractions (EAs), 646, 652, 662
Electrostatic interactions, 650*f*, 652–653
Elemental analysis, 21–23
Elovich model, 176–177, 571, 692
  Uranium (VI) adsorption, 337–338, 339*f*
Emission inventory, 710
Empty bed residence time (EBRT) model, 186–187
Endothermic petroleum oil adsorptions, 545–546
Energy-dispersive Xray spectroscopy (EDS), 619*f*, 620, 621*f*
Engine oil uptake, 539–541*t*, 542*f*
Enthalpy change ($\Delta H^0$) value, 339–340, 341–342*t*, 343
Entropy change ($\Delta S^0$) value, 339–340, 341–342*t*, 343
Equivalent background compound (EBC), 603–605, 604*f*
Ester hydrolysis, 538–543
17β-Estradiol adsorption, 690–691
Ethanol Red, 11
Ethers, 646–647
*Eucheuma spinosum*-derived biochar, 456
Eutectic point analysis, 23–27, 114–118
Evapotranspiration rates, 500–501
Exfoliated nanosheet biochar, 673
Exogenous functional groups, 251–255
Exothermic adsorption, of cooking oil, 545–546
EXY model, 309

## F
Fast pyrolysis
  biochar, 534–537, 536*f*
  reactors, 71
Fatty acids, 533, 550
Fe-based biochar composites, 285
Feedstock, 246–248
  collection, 715–716
  effect, 655–656
  handling, 716
  price, impacts of, 219–220
  processing, 716
  production, 715–716
Feedstock-inherent effects, 744–745
Fe-impregnated corn stover biochar, 257
Fe-modified biochar, 285–286, 379
Fenton-like advanced oxidation process, 786–793
Fenton method, 370, 371*f*
$Fe_3O_4$-modified biochars ($Fe_3O_4$/MB), 337, 384–385
  bamboo, 301, 302*f*
$Fe_3O_4$/nanobiochar composites, 690–691
Fe oxide-modified biochar, 286–288, 286–288*t*
Fe salts, 260–261
Filtralite, 491–492
Fisher activated carbon, 129
Fixed-bed/expanded bed adsorber, 179
  downflow and upflow fixed-bed adsorbers, 179, 179*f*
  downflow and upflow fixed-bed multiple columns
    in parallel, 179, 181*f*
    in series, 179, 180*f*
  fixed-bed downflow column set-up, 179–180, 182*f*

Fixed-bed/expanded bed adsorber (*Continued*)
  operation and design, 184
    Adams-Bohart model, 190–191
    bed-depth-service-time (BDST) model, 187–189, 189$f$
    breakpoint, 184
    breakthrough curve, 182–184, 183$f$
    Clark model, 192–193
    empty bed residence time (EBRT) model, 186–187
    mass transfer model, 184–186, 185$f$
    primary adsorption zone (PAZ), 182–183
    Thomas model, 190
    Wolborska model, 191
    Yoon-Nelson model, 191–192
  time- and distance-dependent process, 179–180
Fixed carbon (FC), 19–21, 106–107
Flash carbonization, 217, 221
Flash/high-pressure pyrolysis reactors, 70–71
Floating treatment wetlands (FTWs), 492, 511
Flocculation, 383, 439
Flow-through fixed-bed biochar columns, 598
Fluorotelomer sulfonic acids (FTSAs), 563–565, 564$f$
Fouling factor, 23–24
  for coal, 116$t$, 117$b$, 132
  cotton gin trash (CGT) ash, 117$b$
Fourier-transform infrared (FTIR) spectroscopy, 143–144, 145$t$, 324–325, 327–333, 347–348, 370$t$, 372
Free water surface (FWS) flow wetlands, 492
Freeze-drying system, 8

Frenkel-Halsey-Hill isotherm (FHH) model, 167
Freundlich equation, 124
Freundlich isotherm model, 338–339, 410–411, 545
  breakthrough time, 161
  heterogeneous adsorption systems, applied to, 161
  linear form, 161
  nonlinear form, 161
  nonuniform adsorption, applied to, 161
  organic contaminant (OC) concentration, 602
  per-/poly-fluoroalkyl substances (PFAS) sorption, 572
Fritz and Schlunder multicomponent model, 171
Functional unit (FU)
  corn stover biochar production, 720

## G

Gasification
  biochar production via, 120–121$b$, 132–133
  charcoal, 74–75
Gas/steam treatment, 136
Gaussian energy distribution, 164
Geographic location, 731
Gibbs free energy ($\Delta G^0$) value, 339–343, 341–342$t$
Glyphosate, 687–691
Granular activated carbon (GAC), 754–755
  adsorber, 746–748, 749$f$
  filtration, 375–377, 376$f$
Granular biochar, 223–224
Greenhouse CTW tests, for CAFO wastewater treatment
  ANOVA and Tukey'sHSD test results
    of initial treatment test, 502–503, 503$f$
    of second treatment test, 503–505, 504$f$, 506$t$,

506–508$f$, 507–509, 510$f$, 511–512, 512–518$f$, 514–518
  of third treatment test, 518–519
layout, of experimental reactors, 496–498, 497$f$
planted reactors, relative plant growth in
  during initial test, 499–500
  during second test, 500
  during third test, 500
reactors after initial planting, 498–499
sampling and data analysis, 500–502
Greenhouse gas (GHG) emissions, 223–224, 711
Green waste (GW), 655

## H

Haber-Bosch process, 237–238
Hach spectrophotometer method, 494, 500–501
Half-cylinder metallic kiln, 61
Halsey model, 164
Hardwood biochar, 379
Hardwood sawdust biochar (HWC), 585–587
$H_3AsO_3$, 278
Heat capacity, 31–32, 102–103
Heating rate effect, 47–48
Heating value analysis, 17–18, 109–113, 130
Heating value calculations, 132
Heat of combustion, 111
Hemicellulose, 49, 644–646, 655
  decomposition, 562–563
Henry constant, 160–161
Henry's law, 160–161
Hg(II) sorption, 693–694
Hierarchically microporous biochars (HMBs), 566, 570
High biochar loading (HBL), 502
High heating value (HHV), 17–18, 109–110
High-performance liquid chromatography (HPLC), 6–7

High-speed ball milling method, 612–613
Hill isotherm model, 163–164
Hills equation, 687–690
Hot-tail kiln, 54–55
HTC. *See* Hydrothermal carbonization (HTC)
Humic acids, 420
Hydraulic loading rate (HLR), 492, 511
Hydrochars, 39, 221–222, 250
Hydrogen bonding, 650$f$, 651–652
Hydrophobic interactions, 654
Hydrothermal carbonization (HTC), 39, 77–79, 219, 221–222
Hydroxy group (–OH), 646

## I
Ibuprofen, 407–415, 410$f$
Ideal adsorbed solution theory (IAST), 172–173
Ideal adsorbed solution theory-equivalent background compound (IAST-EBC) model, 603–605, 604$f$
Inductively coupled plasma (ICP) spectrometry, 43–44
Industrial rectangular kilns, 57–59, 58$f$
Influent concentration (IC), 505–507, 511
Inner transition elements (ITEs), 321
  applications, 322, 351
  environment, primary anthropogenic sources in, 322
  origin of, 322
  removal and recovery, from aqueous solutions, 351–352
    adsorbent dosage, 333, 335–336
    adsorption isotherms and thermodynamics, 338–343, 341–342$t$
    batch tests, 349–351
    biochar and engineered biochar's sorption capacity, 324–333, 328–332$t$
    causes for, 323
    columns/fixed-bed reactors, 349–351
    contact time and kinetic studies, 333, 336–338, 339$f$
    conventional methods, 323
    future research, 352
    multicomponent systems, 343–346
    pH dependency, 333–335
    regeneration, recycling, and reuse, 346–347
    sorption mechanisms, 347–349, 350$f$
Inorganic aqueous contaminants removal, nanobiochar for, 667–668
  anions, 691
  heavy metals, 692–694, 695–696$t$
Inorganic carbon, 139
Inorganic mineral sorbents, 530
Intermediate gasification chars, 75–77
Internal combustion engine (ICE), 97
Interpretation, 713–714, 727
Intraparticle diffusion (IPD) model, 177, 177$f$, 571
  Uranium (VI) adsorption, 337–338, 339$f$
Ion exchange, 280–282, 289, 648–649, 650$f$, 652–653
  capacity, 140
Ionic composition, 298–299
Ionic strength, 343–346, 652
Iron oxide permeated mesoporous rice husk nanobiochar (IPMN), 693
Iron oxide/persimmon tannin/graphene oxide ($Fe_3O_4$/PT/GO), 326–327, 347–348
Irradiation, 439–440
Isopropanol, 546, 547$f$
Isothermal titration calorimetry (ITC), 348–349
Isotherm model, 602–604

## J
Jain and Snoeyink model, 169–170
*Juniperus osteosperma*, 566

## K
Koble-Corrigan isotherm model, 166–167
KOH/$H_2O_2$ method, 370, 371$f$

## L
Laboratory analytical procedures (LAPs), 6–7, 7$t$
*Lagerstroemia* sp., 296–297
Landfill leachate (LL), 752–753, 757
Langmuir coefficient, 495
Langmuir cooking oil uptake capacity, 537–538
Langmuir isotherm model, 123, 338–339, 410–411, 495
  ammonium/ammonia nitrogen ($NH_4^+$-N) adsorption, 495, 496$f$
  gas-solid-phase adsorption, 162
  liquid-solid-phase systems, 162
  nonlinear and liner form, 162
  oil sorption, 545
  per-/poly-fluoroalkyl substances (PFAS) sorption, 572
  separation factor, 162
  Thomas model, 190
  types of equilibrium isotherms, 162, 163$f$
Lanthanides
  and actinides, removal and recovery from aqueous media (*see* Inner transition elements (ITEs))
  applications, 322

Lanthanides (*Continued*)
  aqueous media, behavior in, 322
  elements, classification of, 322
  properties, 322
Lauric acid, 13–14
Lauric acid biochar (LBC), 533–534, 534*f*, 538, 539–541*t*, 542*f*, 545–546, 547*f*, 548–549
Lauric-magnetic biochar (LMBC), 534, 538, 539–541*t*, 542*f*, 545–546, 548–549
Layered double hydroxide composites (LDH-biochar), 326–327
Lead (Pb), 490
Lead-loaded ball-milled straw biochar, 693
Life cycle analysis (LCA)
  of biochar production, 714
    from corn stover, 719–727
    goal, 714–715
    impact assessment, 718–719
    scope, 714–715
    stages, 715–718
  biochar use in wastewater treatment, 728–730
  definition, 707–708
  environmental impact assessment technique, 706
  fundamentals, 706–714
  goal, 707–708
  interpretation, 713–714
  life cycle impact assessment (LCIA), 710–713
    methods, 713
  life cycle inventory (LCI), 708–710
  midpoint *vs.* endpoint analysis, 712*f*
  overview, 705–706
  structure, 706, 707*f*
Life cycle impact assessment (LCIA), 710–713
  corn stover biochar production, 723–727, 733
  methods, 713

Life cycle inventory (LCI), 708–710, 721–723
Lignin, 8–11, 49
Linear film diffusion model, 178
Liquid chromatography-mass spectrometry, 598
Low-impact development (LID), 488–489
Low temperature-circulating fluidized-bed gasifier (LT-CFB), 75–77
Low-temperature pyrolysis, biochar, 562

# M
Machine oil uptake, 539–541*t*, 542*f*
MacMillan-Teller isotherm (MET) model, 167–168
Magnesium (Mg) modified biochar, 514–515
Magnetic activated carbon, 285–286
Magnetic biochar (MBC), 257–258, 285–286, 340–343, 341–342*t*, 534, 538, 539–541*t*, 542*f*, 545–546, 548–549
  one-step microwave synthesis, 285–286
Magnetic biochar-assisted system, 791
Magnetic food waste-derived biochar, 458
Magnetic iron oxide banana pseudostem biochar, 415
Magnetic-lauric biochar (MLBC), 534, 538, 539–541*t*, 542*f*, 545–546, 547*f*, 548–549
Magnetic materials, 565
Magnetic nanobiochar, 675–676
Magnetic nanoparticles, 325–326
Magnetite iron oxide nanoparticle, 533–534
Magnetite/magnetic carbonized *Calotropis gigantea* fiber (MC-CGF), 565, 570–572, 582–584

Magnetized nanobiochar/$Fe_3O_4$ nanoparticles, 675–676
Maize straw biochar (MBC), 407–408
Maple wood biochar, 539–541*t*, 549
Marcopollutants, 475
*Marine chlorella vulgaris*, 458
Masonry rectangular kilns, 57
Mass transfer zone (MTZ), 184–186, 185*f*
Mathews and Weber model, 170
µATR spectroscopy, 370*t*, 372
Maximum contaminant levels (MCLs), 559
Mechanical removal, of oil, 528–530
Membrane biofiltering (MBR), 374–375
Membrane filtration, 440
Membrane/reverse osmosis, 280, 306–307
3-Mercaptopropyltrimethoxysilane (3-MPTS), 675, 686–687, 693–694
Metal-based biochar catalyst, 800
Metal-free biochar-based advanced oxidation process, 799
Metal (oxy)hydroxides, 255–257
Metalloids, pretreatment with, 258
Metal oxide nanoparticle biochar composite, 258
Metals
  biochar loaded with, 780–785, 781–784*t*
  complex dyes, 437–438
  kilns, 59–61
  pretreatment with, 258
Methylene blue, 454–455, 792
Methylisoborneol (MIB), 751
Mg-Al oxides nanoparticle-biochar composite, 258
MgO-doped hardwood and softwood biochars, 257

Microbeads, 362
Microbes, 475, 477
Microbial activity stimulation, 489–490
Microbial contamination, in water, 474–475
Microbial decontamination mechanism, 478–484
Microplastics (MPs)
  analysis methods, 367–374, 369$f$
    cleanup, 369–370
    concentration, 369–370
    for identification, 370–372
    instrumental methods, 372–374
    sampling condition, 368–369
    sampling volumes, 368–369, 370$t$
    for verification, 370–372
  classification, 362
  in clothing and textiles, 363–364
  definition, 362
  density solutions, 372$t$, 374
  effects
    in aquatic environment, 361–367
    in organisms, 365–367
  hydrophobic/water-repellent, 366
  immobilization mechanisms, 380$f$
  ingestion, 366–367
  polymers, properties of, 367, 368$t$
  purification, carbon-based methods for, 374–378
  removal, 379–386
    biochar chemistry, 383–385
    biochar physical characteristics, 379–383
    mixing biochar with soil/sand, 385–386
    solution chemistry, 383–385
  retention, 374–386
  sources, 361–367
  in traffic, 365
  transportation, 364

Microwave
  pyrolysis, 668–669, 673–674
  water treatments, 798
Miscanthus nanobiochar, 678–684
Missouri-type kiln, 56–57
Mixed-order (MO) model, 176
$MnO_2$/orange peel biochar composite, 340–343, 341–342$t$
Modification effect, 657–659
Modified biochars, 459–461, 778–786
  co-doped biochar, 785–786
  doped with heteroatoms, 785
  loaded with metals, 780–785
Modified carbons, 444
Moisture content (MC), 9, 104–105
Mordant brown 35, 438, 438$f$
Mordant dyes, 438–439
Mound kilns, 52–53, 52$f$
Moving/fluidized-bed adsorber, 181–182, 182$f$
Multicomponent sorption equilibrium models, 168
  Butler and Ockrent model/extended Langmuir model, 168–169
  Dastgheib and Rockstraw model, 171–172
  Fritz and Schlunder model, 171
  ideal adsorbed solution theory (IAST), 172–173
  Jain and Snoeyink model, 169–170
  Mathews and Weber model, 170
  Sheindorf et al. model, 172
Multicomponent testing, 298–301

# N
Nanobiochar
  applications, 667–668
  characteristics, 676–677, 677$f$
  future research, recommendations for, 694–697
  inorganic aqueous contaminants removal, 667–668
    anions, 691
    heavy metals, 692–694, 695–696$t$
  organic aqueous contaminants removal, 667–668
    conventional organic contaminants, removal of, 684–687
    emerging contaminants, adsorption of, 687–691, 688–689$t$
  preparation methods, 669$f$
    ballmilling, 668–672
    bottom-up approach, 668–669
    centrifugation, 668–669, 672–674
    microwave pyrolysis, 668–669, 673–674
    modified nanobiochar, preparation of, 674–676
    thermal heat flash treatment, 668–669, 673
    top-down approach, 668–669
  and pristine physicochemical properties, comparative evaluation of, 676–684, 677$f$, 679–681$t$
  properties, 672, 678, 682–683$t$
Nanoscale zero-valent iron (nZVI) particles
  biochar-supported (see Biochar-supported nZVI (nZVI@BC))
  high surface area, 611
  redox potential ($E^0$), 611
  water remediation
    advantages, 611
    limiting issues, 611–612
Natural organic sorbents, 531–532
N-containing functional groups, 658–659

Negative charge assisted hydrogen bonding, 651–652
Ni(II) adsorption, 692
Ni/Fe LDH-biochar composite, 289
Nitrate, 236$f$, 237–238
Nitrate nitrogen ($NO_3^-$-N) adsorption, 500–501
   batch sorption kinetics tests, 493–495
   influent and effluent concentrations, 511, 512$f$
Nitrobenzene (NB) removal, 752–753, 761$f$, 763$f$
Nitrogen, 785
   doping, 674–675, 685–686
   gas analysis, 599
Nitro group (–$NO_2$), 647
Nonadsorptive qualities, 739
   availability, 742
   cost, 742
   environmental impacts, 739–741
   environmental sustainability, 739–741
   production techniques, 743–745
Nonmetal-doped biochar, 779–780
$NO_3^-$ removal, 251, 252–254$t$
Norit-activated carbon uptake capacity, 537
$NO_3^-$ sorption, 243–245, 244$f$, 247
Nutrient contamination, 487–488
Nutrient management plans (NMPs), 488
nZVI-embedded biochar, 285–286

## O

*Oedogonium* sp., 299
Oil sorption
   bio-based sorbents, 530
   biochar and biochar composites, 530, 550
      hydrophobicity, 533–538, 551
      isotherms and sorption thermodynamics, 539–541$t$, 545–546
      kinetics and stirring effects, 539–541$t$, 542$f$, 543–545, 544$f$
   natural organic sorbents, 531–532
   pH dependence, 538–543, 539–541$t$, 542$f$
   pyrolysis temperature, 532–533
   regeneration and recycling, 546–549, 547$f$
   seawater, adsorption from, 544$f$, 549
   uptake mechanisms, 550
   inorganic mineral sorbents, 530
   oleophilic sorbents, use of, 530
   synthetic polymer sorbents, 530
Oil spills
   biochar (BC), 531–532
   British Petroleum (BP) Deepwater Horizon oil rig, explosion at, 527–528
   cleanup methods, 528–530, 529$t$
   environmental impact, 527–528
   Exxon Valdez oil spill, 527–528
   fish and other marine species, impact on, 527–528
   pathways, 528, 528$f$
One-pot synthesis, 285–286
One-step microwave synthesis, 285–286
On-site screening, 293, 306–310
On-site wastewater treatment systems (OWTSs), 589
Orange G, 456–457
Orange peel's biochar, 458–459
Organic aqueous contaminants removal, nanobiochar for, 667–668
conventional organic contaminants, removal of, 684–687
emerging contaminants, adsorption of, 687–691, 688–689$t$
Organic contaminant adsorption, 415
Organic contaminant (OC) adsorption, 597–599, 608
   full-scale biochar column adsorber performance estimation
      batch *vs.* column use rates, 605–606, 606$f$
      pilot test, 605
      rapid small-scale column test (RSSCT), 605
   IAST-EBC model, 603–605, 604$f$
   isotherm model, 602–604
   surface area and hydrogen to carbon (H:C) ratios
      AFM-Dose$_{75}$ values *vs.* H:C ratios, 600–602, 601$f$
      2,4-D adsorption capacity *vs.* BET surface area, 599–600, 600$f$, 602
      density functional theory, 599
      $N_2$-BET surface area characterization, 599
      nonequilibrium batch adsorption tests, 599
      sulfamethoxazole (SMX) removal *vs.* BET surface area, 599–600, 600$f$, 602
      UVA$_{254}$ and SMX breakthrough curves, in surface water DOM, 606–608, 607$f$
Organic contaminant sorption, 408–409
Organic matter, 300
Organic micropollutants (OMPs) removal, 750–755
Oxyanions, 233
   removal, 250–265, 252–254$t$
   chemical sorption, 263–264
   mechanisms, 262–265, 263$f$
   physical sorption, 264
   posttreatment technologies, 259–262

precipitation, 265
pretreatment technologies, 255–259, 256f
surface modification, 255–262
unmodified and modified biochar toward, 251–255
sorption
chemical properties, 238–240
feedstock type, 246–248
physical properties, 248–250
pyrolysis temperatures, 241–246
Oxytetracycline, 687–690
Ozonation, 440–441

# P

PAC. *See* Powdered activated carbon (PAC)
Palmitic acid, 13–14
Palm oil mill effluent (POME) uptake, 538–545
Palm tree-derived biochar, 338
Particle density, 28–29, 98
Particle equivalent diameter (AED), 29–30
Particle size distribution (PSD), 29–30, 99–101
Peanut shell biochar, 246–247, 673
Perfluoroalkyl carboxylates (PFCAs), 563–565
Perfluoroalkyl sulfonates (PFSAs), 563–565, 564f
Perfluoro carboxylic acid (PFCA), 558–559
Perfluorooctane sulfonamide (FOSA), 563–565, 564f
Perfluorooctane sulfonic acid (PFOS), 555–556, 559, 561–562, 561f, 569
Perfluorooctanoic acid (PFOA), 555–556, 559, 561–562, 561f, 566–568, 567f
Perfluorosulphonic acid (PFSA), 558–559
Peroxydisulfate (PDS), 629–631

Per-/poly-fluoroalkyl substances (PFAS)
applications, 555
classification, 555
cleanup methods, 559–561, 560f
definition, 555
drinking water contamination, 559
groundwater and soil matrices pollution, 556
physical and chemical properties, 557–558t, 558–559
presence of, 556
public health concern, 556
sorption, biochar and biochar composites, 591
aggregation, 589–590, 590f
breakthrough column studies and pilot-scale testing, 585–589, 588f
competitive ion effects, 584–585
electrostatic attractions, 589–590, 590f
hydrophobic interactions, 589–590, 590f
isotherms and thermodynamics, 572–582, 577–581t, 592
kinetic data, 570–572, 573–576t
pH dependence, 569–570
regeneration, 582–583, 592
remediation, 561–566
short-chain PFAS removal, 566–569, 567f, 591
in water bodies, 559
Persistent free radicals (PFRs), 629, 777–778, 791–794
Persulfate-activated advanced oxidation process, 793–794
Petroleum-based diesel oil uptake, 545
Pharmaceuticals adsorbents

adsorption performance of, 399–405t
biochar
as adsorbent, 398–406
physicochemical properties, 407–409
solution pH effect, 409–410
classification, 396–397
emerging contaminants, 396–398
overview, 395–396
pathways, 398f
receptors, 398f
removal, 395–396, 410–415, 412–414t
sorption onto biochar, 399–405t, 406–420
interactions, 415
matrix effects on, 419–420
mechanisms, 415–419
sources, 398f
pH effect, 339f, 659–660
Phenols, 794
Phosphate, 235–237, 236f, 300–301
Phosphate biochar (PBs), 345–346
Phosphate ($PO_4^{3-}$-P) removal, 500–501, 511–515, 514f, 519
batch sorption kinetics tests, 493–495
influent and effluent concentrations, 511–512, 513f
Phosphoric acid pretreatment, 47
Photocatalytic advanced oxidation process, 794–796
Photochemical degradation, 441
pH point of zero charge ($pH_{PZC}$), 334–335
*Phragmites australis*, 566–568
Physical modification, 249–250, 255, 266
Physical sorption, oxyanion, 264
Physisorptive interactions, 661

Phytoextraction, of heavy metal contaminated soils, 490
Phytoremediation enhancer, biochar as, 492–493
Pine juniper (PJ) biochar, 566, 570
Pine needle (PN)-derived biochars, 344, 561–562
Pinewood biochar (PWC), 286–288, 537, 539–541$t$, 585
Pinewood fast pyrolysis biochar, 407–408
Pine wood nanobiochar, 687–690
*Pinus monophylla*, 566
Pit kilns, 50–52, 52$f$
Planetary ball mill, 670–671
Plant-based biochars, 246–247
Polarity, 408–409
Polyethylene beads, 362–363
Polyvinyl chloride (PVC) column, 586
$PO_4^{3-}$ removal, 251, 252–254$t$
Pores
  biochar, 141
  diffusion, 415–416
  filling, 653
  size distribution, 141–142
Positive charge assisted hydrogen bonding, 651–652
$PO_4^{3-}$ sorption, 245
Posttreatment technologies, 259–262
Poultry litter biochar (PLC), 586–587
Powdered activated carbon (PAC), 726–730, 728–729$f$, 739–740, 753
Precipitation
  interaction, 652–653
  oxyanion, 265
Pressure effect, 48
Pretreatment technologies, 255–259, 256$f$
Primary adsorption zone (PAZ), 182–183, 185–186
Printed circuit board (PCB), 5, 5$f$

Pristine
  biochar, 676–684, 677$f$, 679–681$t$, 694–697
  As removal efficiency, 286–288, 286–288$t$
Protozoans, 475
Proximate analysis, 18–20, 103–107
  ash, 106–107
  calculations, 131
  fixed carbon (FC), 106–107
  moisture content (MC), 104–105
  volatile combustible matter (VCM), 105
Pseudo-first order (PFO) model, 174, 543–544, 570–571, 573–576$t$
  Uranium (VI) adsorption, 336–338, 339$f$
Pseudo-second-order (PSO) model, 175, 570–571, 573–576$t$
  Uranium (VI) adsorption, 336–338, 339$f$
Pseudo-second-order models, 543–544
Pyrolysis, 39
  biochar-based sorptive interactions, 656–657
  biochar formation process, 645$f$
  of cellulose, 49–50
  heating rate, 47–48
  of hemicellulose, 644–645
  of lignin, 49–50
  reactors, characteristics of, 51$t$
  temperature, 241–246, 532–533, 562, 656–657, 719

# R
Radke-Prausnitz isotherm model, 166
Raman spectroscopy, 370$t$, 372
Rapid sand filtering (RSF), 382
Rapid small-scale column test (RSSCT), 587, 588$f$, 603–604

Rare earth elements (REEs), 322, 343, 349–351
Reactive dyes, 431–432, 432$f$
Rectangular kilns, 56
Redlich-Peterson (RP) isotherm model, 165–166, 572
Red mud modified biochar (RMSDN600), 565, 569
Reed straw-derived biochar (RESCA), 566–568, 567$f$, 570–571, 584–586
Remazol dyes, 456
Removal efficiency, 286–288$t$, 304–306
Resonance-assisted hydrogen bonding (RAHB), 651–652
Retention, microplastics, 374–386
Retorts, 61–66, 65–66$t$, 67–68$f$
Revised pseudo-second-order (rPSO) model, 175–176
Rhodamine B, 454–455, 457–458
Rice husk biochars (RHBCs), 534–537, 536$f$, 539–541$t$, 544–545, 549
Rice-straw-based biochar, 326, 340, 341–342$t$
Rima container furnaces (FCR), 64
Rotary drum reactors, 67–70

# S
*Saccharomyces cerevisiae*, 11
Salicylic acid sorption, 411
Salt removal, 213
Sawdust biochars, 534–537, 544–545, 549
Scaled-up drinking water arsenic treatment
  adsorbent performance testing, 291–310, 294$f$
  multicomponent testing, 298–301
  with naturally occurring As concentrations, 301–303
  recycling, 304–306
  regeneration, 304–306
  safe disposal, 304–306

in single-element aqueous
    systems, 293–297
As sorption in natural water
    samples, 303–304
biochar utilization, 306–310
Scanning electron microscopy
    (SEM), 619f, 620
Sewage sludge, 365, 377
Sheindorf et al. model, 172
*Shorea robusta*, 458
Sieved biochar, 672–673
Single-element aqueous system
    screening, 293–297
Single-stack rectangular
    masonry-block kilns, 56
Sips model, 165
    oil sorption, 545
    per-/poly-fluoroalkyl
        substances (PFAS)
        sorption, 574–582
Size exclusion effect, 653, 658
Slagging factor, 23–24
    for coal, 116t, 117b, 132
        cotton gin trash (CGT) ash,
            117b
Slow-pyrolysis biochar, 534–537,
    536f
Slow-release fertilizer, 489
Sludge biochars, 458–459
Sludge handling and dewatering,
    213
Sodium borohydride ($NaBH_4$),
    613f, 614
Soil biochar application, 135
Soil fertilizer, 489
Solar pyrolysis reactors, 71,
    72–73t
Solid-state $^{13}C$ NMR
    spectroscopy, 145–146,
    146t
Solution pH, 409–410, 652
Sonolysis, 797
Sorption isotherm models,
    494
    adsorbate mobility/retention
        phenomenon, 156
    adsorption capacity, 156
    comparative evaluation of
        adsorbents, 158, 158f

discontinuous batch
    adsorption operation,
    156–157, 156f
generalized adsorption
    isotherm, 157–158, 157f
Henry's law, 160–161
laboratory batch tests, 156
mass balance of adsorbate,
    157
mass of sorbate sorbed/
    adsorbent mass, 156
mathematical models, 160
multicomponent sorption
    equilibrium models, 168
    Butler and Ockrent model/
        extended Langmuir
        model, 168–169
    Dastgheib and Rockstraw
        model, 171–172
    Fritz and Schlunder model,
        171
    ideal adsorbed solution
        theory (IAST), 172–173
    Jain and Snoeyink model,
        169–170
    Mathews and Weber model,
        170
    Sheindorf et al. model,
        172
multilayer physisorption
    isotherms
    Brunauer-Emmett-Teller
        (BET) isotherm model,
        167
    Frenkel-Halsey-Hill
        isotherm (FHH) model,
        167
    MacMillan-Teller
        isotherm (MET) model,
        167–168
    for oil sorption, 539–541t,
        545–546
    organic contaminant (OC)
        adsorption, 602–604
    per-/poly-fluoroalkyl
        substances (PFAS)
        sorption, 572–582,
        577–581t, 592
    pristine sorbent, 157

three-parameter isotherm
    models
    Koble-Corrigan isotherm
        model, 166–167
    Radke-Prausnitz isotherm
        model, 166
    Redlich-Peterson isotherm
        model, 165–166
    Sips/Langmuir-Freundlich
        isotherm model, 165
    Toth isotherm model, 165
two-parameter isotherms
    Dubinin-Radushkevich
        isotherm model, 164
    Freundlich isotherm model,
        161
    Halsey model, 164
    Hill isotherm model,
        163–164
    Langmuir isotherm model,
        162
    Temkin isotherm model,
        162–163
type of isotherms, 158–160,
    159f
Sorption kinetics, 493–494
Soybean oil, 13
Soybean stover biochar,
    246–247
Specific surface area (SSA), 141,
    143, 248–249, 624–625,
    678
Spent biochar, 489, 494
Spinoff screening, 292–293,
    298–306
Spruce pine wood biochar
    (SWC), 586–587
Standard sieve screen sizes, 100,
    101t
Steam activation, 284
Stoichiometric air-to-fuel ratio
    (AFR) calculations, 132
Stoichiometry of reactions,
    111–113
Stripping agents, 346
Subsurface flow (SSF) wetlands,
    492, 511
Sugarcane bagasse ash (SCBA),
    72–74

Sulfamethoxazole (SMX), 690–691, 751–752
  breakthrough curves, in surface water DOM, 606–608, 607$f$
  removal
    AFM-Dose$_{75}$ values vs. H:C ratios, 600–602, 601$f$
    vs. BET surface area, 599–600, 600$f$, 602
  sorption, 409–410, 419–420
Sulfapyridine, 690–691
Sulfonation, 779
Sulfur, 648
  dyes, 436, 437$f$
Surface area analysis, 599–600, 600$f$, 602
Surface charge, 408–409
Surface complexation, 653
Surface complexation modeling (SCM), 348–349
Surface modification, 251–262
Surface water dissolved organic matter (SW DOM), 756–757
Switchgrass-derived biochar, 324–325, 349–351
Synthetic fibers, 365
Synthetic polymer sorbents, 530
System boundary
  corn stover biochar production, 720–721, 720$f$
System expansion method, 710, 717, 720–721

## T

Tapioca peel-derived biochar, 446–447$t$, 457–458
Techno-economic analysis (TEA)
  biochar markets
    drivers, 224–225
    and economics, 207–211
  biochar production, 217–224
    coproducts, impacts of, 223
    environmental incentives, impacts of, 223–224
    feedstock price, impacts of, 219–220
    technology, impacts of, 220–223
  components, 205–206
  cost analysis framework, 206$f$
  overview, 205–207, 225–226
  wastewater facilities, 211, 213–214, 215$f$
    capital cost, 211–213
    cost, impacts on, 214–217
    maintenance cost, 213
    operation cost, 213
    process design costs, 211–213
    properties, 214–217
*Tectona* sp., 296–297
Temkin isotherm model, 162–163
Temperature effect, 44–45
Tetracycline, 411
  sorption, 409–411, 419–420
Tetracycline hydrochloride adsorption capacity, 690
Textile sludge biochar (TSB), 537–543, 539–541$t$
  cooking oil, exothermic adsorption of, 545–546
  regeneration and recycling, 546
*Thalia dealbata* biochar, 261
Thermal conductivity, 31–32, 102–103
Thermal heat flash treatment, 668–669, 673
Thermochemical conversion, 716–717
  biomass properties, 16–32
    eutectic point analysis, 23–27
    heating value analysis, 17–18
    proximate analysis, 18–20
    ultimate analysis, 21–23
Thermogravimetric (TG) analysis, 562–563
*Thespesia populnea*, 456–457
Thiol-modified nanobiochar, 675, 686–687, 693–694
Thomas model, 190, 351–352, 566

Three-dimensional (3D) hierarchically microporous biochars (HMBs), 566
Toluene, 537, 549
Top-Lit UpDraft (TLUD), 742, 743$f$
Torrefaction, 95, 219, 222–223
Total Kjeldahl nitrogen (TKN), 500–501, 509, 510$f$
Total nitrogen (TN), 491
Total suspended solids (TSSs), 491–493, 500–501, 505
  influent and effluent concentrations, 505, 506$f$
  removal, from wastewater, 505–507, 507$f$
Toth isotherm model, 165
Transition metals, 780
Transmission electron microscopy (TEM), 619$f$, 620–621, 621$f$
Transmission oil uptake, 539–541$t$, 542$f$
Transportation, 364
2,4,6-Tricholorophenol (TCP), 630–631, 632$f$
Trihalomethanes (THMs), 751
Triple coadsorption, 299
Tropical Products Institute (TPI) kiln, 59, 60$f$
Tukey's honestly significant difference (Tukey's HSD) test, 501–519

## U

Ultimate analysis, 21–23, 107–109, 131
Ultraviolet absorbance at 254nm (UVA$_{254}$), 606–608, 607$f$
  removal, 750, 754
Up-flow anaerobic sludge blanket (UASB) reactor, 213
Uranium (VI)
  adsorption capacity
    kinetic studies, 336–338, 339$f$
    pH dependency, 333–334

phosphate-functionalized biochar, 345
in PRB column configuration, 349–351
in simulated wastewater and seawater, 345
switchgrass-derived biochar, 324–325
for wheat straw-derived biochar, 327
desorption, from pine needle-derived biochar, 346–347
removal efficiency, 345
  adsorbent dosage, effect of, 336
  ionic strength, effect of, 344
  pH dependency, 333–334
  sewage sludge-derived biochar, 335$f$
  wood-derived biochar, 326
Urban stormwater, nutrients in, 487

## V
Vacuum filtration system, 6–7
Van der Waals forces, 589–590
Van Krevelen plot, 131
Vat dyes, 433, 433$f$
Vertical-flow-constructed wetlands (VFCWs), 492
Vibrator ball mills, 670–671
V&M kiln, 57–59
Volatile combustible matter (VCM), 19–21, 105
Volatile organic compounds (VOCs), 687

## W
Warfarin (WFN), 751
Wastewater decontamination, 211–212
Wastewater dissolved organic matter (WW DOM), 755–757
Wastewater sludge-based biochars, 597–602, 604–605
Wastewater treatment plants (WWTPs), 374–375
Water contamination, 473
  by livestock, 477
  microbial contamination, 474–475
  by pathogenic microorganisms, 473–474
  by poultrymanure, 477
Water purification
  methods, 478
  treatments, 374–375
Water solubility, 431
Wet air oxidation (WAO), 440–441
Wet ball milling, 672
Wet biomass, 250
Wet pyrolysis. *See* Hydrothermal carbonization
Wheat straw biochar (WBC), 407–408, 678–684

Wheat straw-derived biochar, 327
Wolborska model, 191
Wood biochar, 730
Wood carbonization, 40
  brick kilns, 53–59
  concrete kilns, 53–59
  earth kilns, 52–53
  metal kilns, 59–61
Wood-derived biochar, 246, 324–325
Wood distillation industry, 40, 61–62
Woody biochar, 672–673

## X
X-ray absorption spectroscopy (XAS), 348–349
X-ray diffraction (XRD), 347–349, 621–622, 622$f$
X-ray photoelectron spectrometry (XPS), 144, 324–325, 347–349, 622, 622$f$

## Y
Yoon-Nelson model, 191–192

## Z
Zero-valent iron biochar (ZVI-biochar) composites, 285–286, 326–327, 337
Zinc, 490

Printed in the United States
by Baker & Taylor Publisher Services